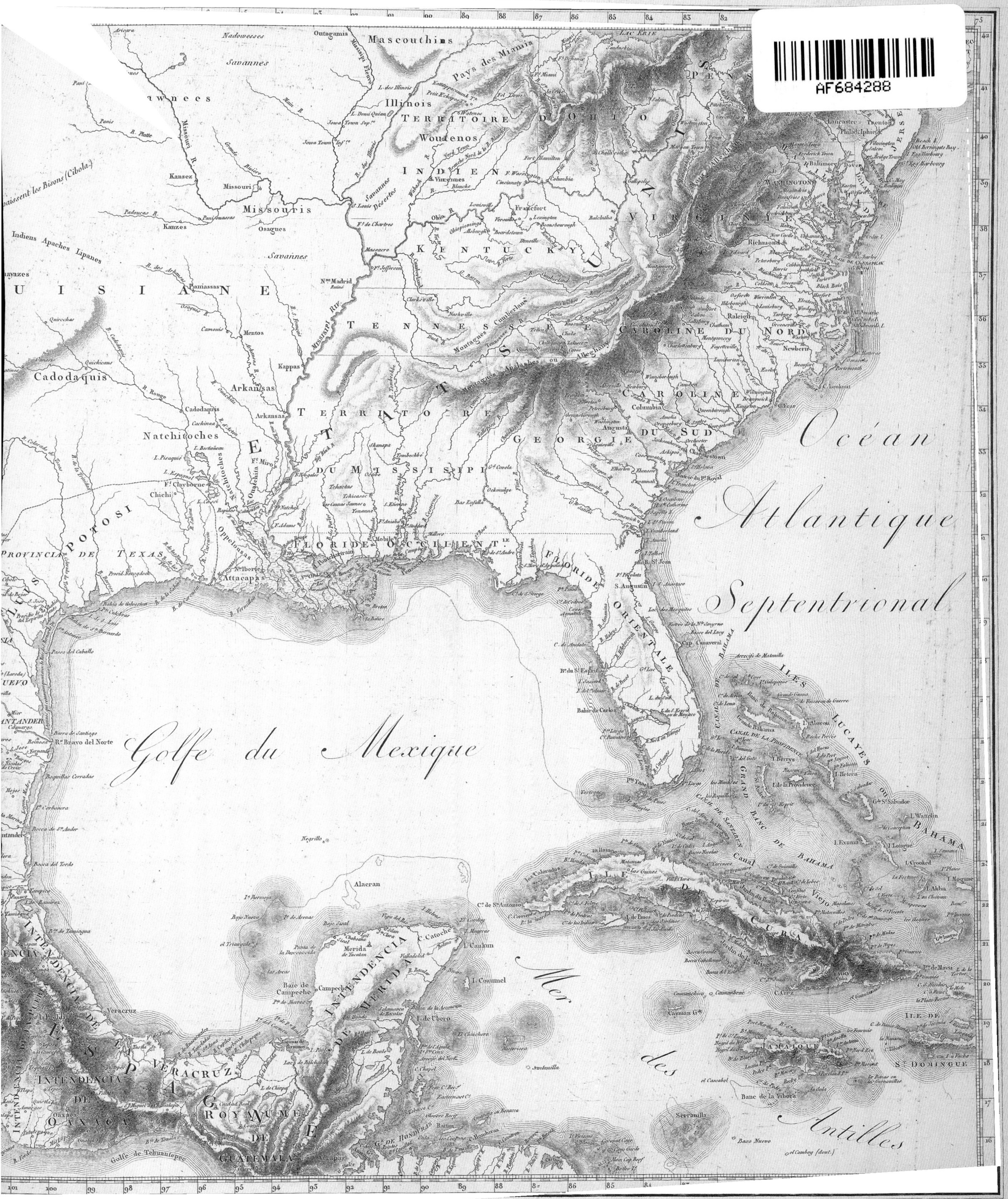
Golfe du Mexique
Océan Atlantique Septentrional
Mer des Antilles
Mascouthins
Pays des Miamis
Illinois
TERRITOIRE D'OHIO
INDIEN
Woutenos
Missouris
KENTUCKY
TENNESSEE
CAROLINE DU NORD
CAROLINE DU SUD
GEORGIE
VIRGINIE
TERRITOIRE DU MISSISIPI
ETATS UNIS
FLORIDE OCCIDENTALE
FLORIDE ORIENTALE
LOUISIANE
Cadodaquis
Natchitoches
Arkansas
PROVINCIA DE TEXAS
POTOSI
Attacapas
Oppelousas
Indiens Apaches Lipanes
Savannes
Kansez
Missouri
Richmond
Washington
Baltimore
Philadelphie
Trenton
Lancaster
Raleigh
Newbern
Columbia
Charlestown
Augusta
S. Augustin
Cap Canaveral
ILES LUCAYES OU BAHAMA
GRAND BANC DE BAHAMA
Canal Viejo de Bahama
ILE DE CUBA
JAMAIQUE
ILE DE ST. DOMINGUE
Cayman Gd.
Banc de la Vibora
Serranilla
Baxo Nuevo
INTENDENCIA DE MERIDA
Merida de Yucatan
Campeche
Baie de Campeche
Alacran
Negrillo
I. Cozumel
C. Catoche
Veracruz
VERACRUZ
INTENDENCIA DE OAXACA
ROYAUME DE GUATEMALA
Golfe de Tehuantepec
NOUVELLE ESPAGNE
Rº. Bravo del Norte
Barra de Santiago
Boca del Tordo
Santander
I. Exuma
I. Watteln
Gde. St. Salvador
I. Longue
I. Crooked
Mogane
Bahia de Galveston
Golfe de Honduras
Lac Erie
Pensilvanie
New Jersey
et l'écriture par L. Aubert.

ALEXANDER VON HUMBOLDT *and the* United States

ELEANOR JONES HARVEY

with a preface by
HANS-DIETER SUES

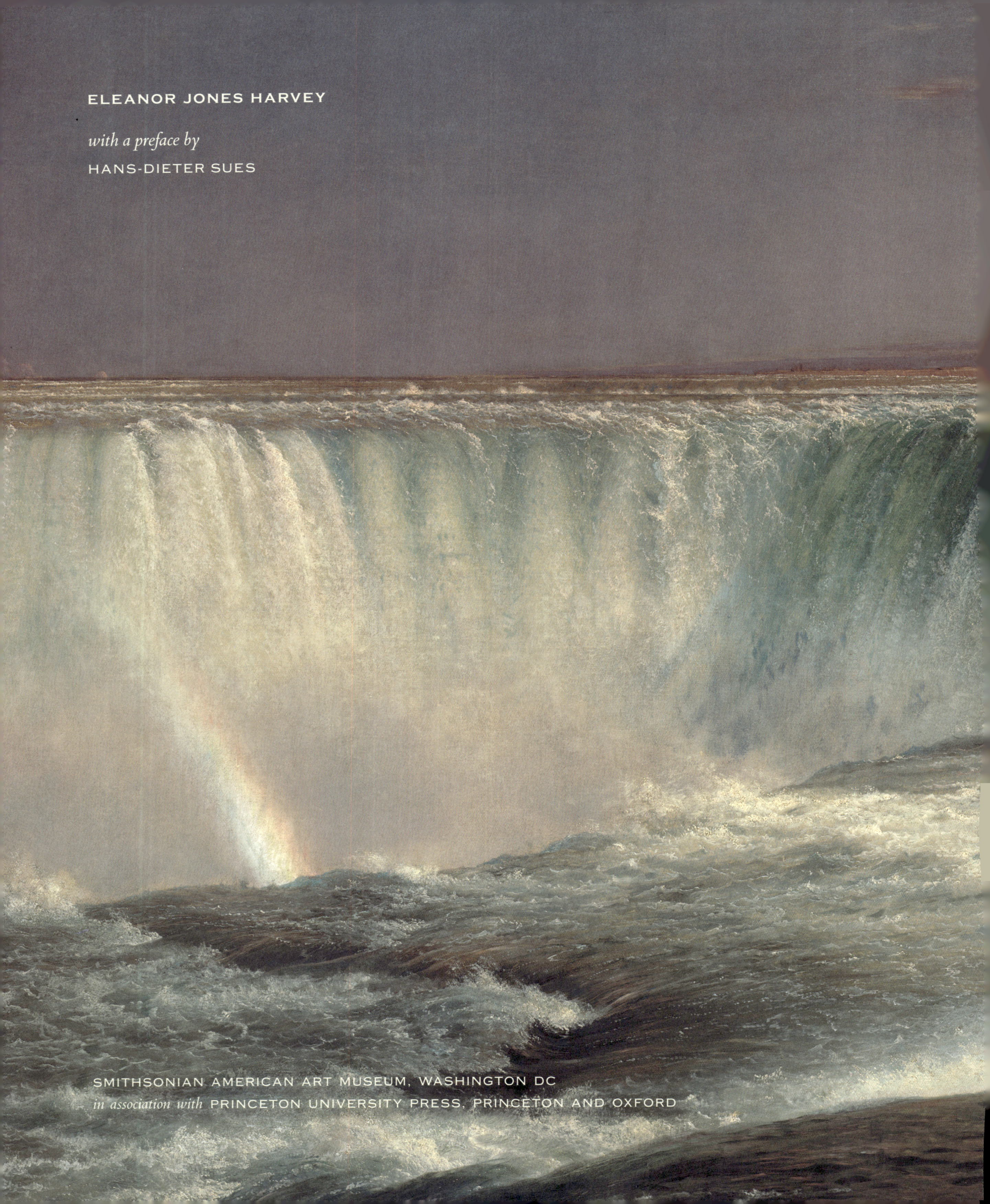

SMITHSONIAN AMERICAN ART MUSEUM, WASHINGTON DC
in association with PRINCETON UNIVERSITY PRESS, PRINCETON AND OXFORD

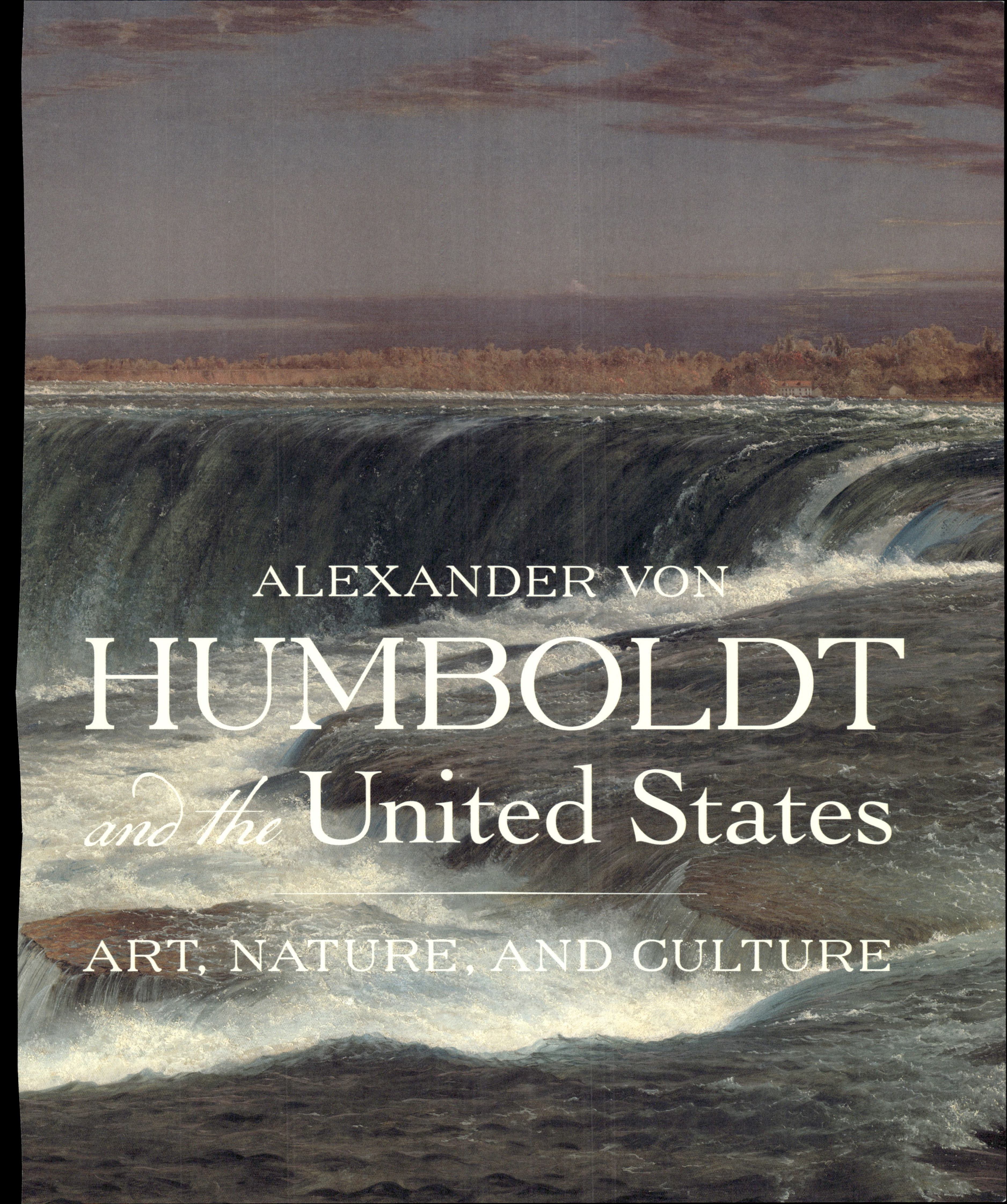

ALEXANDER VON

HUMBOLDT *and the* United States

ART, NATURE, AND CULTURE

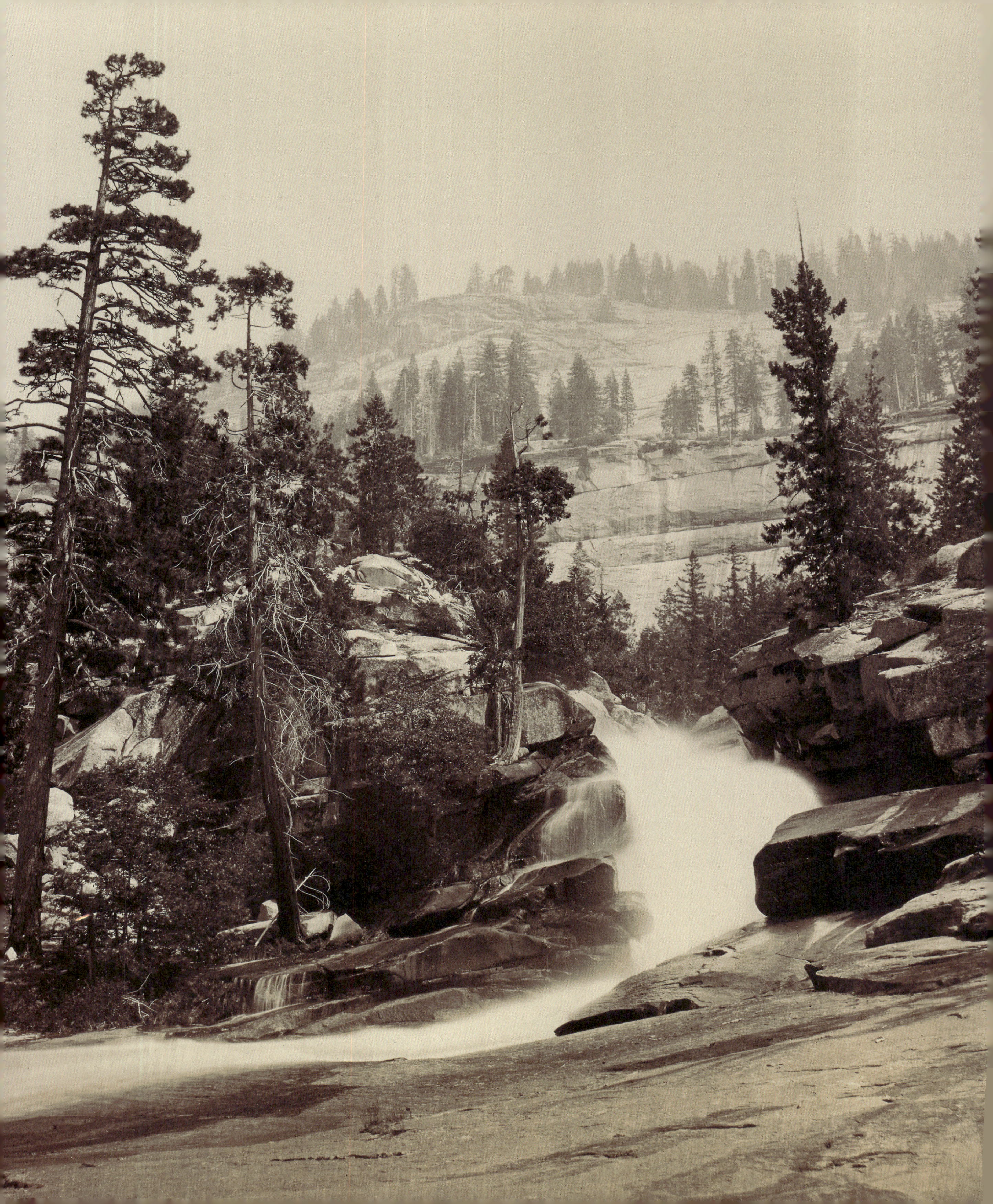

Alexander von Humboldt and the United States:
Art, Nature, and Culture
is organized by the Smithsonian American Art Museum
with generous support from

Joanne and Richard Brodie
Billings and John Cay
Fern and Hersh Cohen
Sheila Duignan and Mike Wilkins
Marie M. Halff
Liliane A. and Christian W. E. Haub
Raymond J. and Margaret Horowitz Endowment
Kandeo Asset Management
Maureen and Gene Kim
LATAM Trade Capital
Robert Lehman Foundation
Henry Luce Foundation
The Lunder Foundation—Peter and Paula Lunder Family
Provost of the Smithsonian
Lucy S. Rhame
Holly and Nick Ruffin
Jacquelyn and William Sheehan
Smithsonian Scholarly Studies Awards
Terra Foundation for American Art
Kelly Williams and Andrew Forsyth

This exhibition is supported by an indemnity from the
Federal Council on the Arts and the Humanities.

ROBERT LEHMAN FOUNDATION

HENRY LUCE FOUNDATION

TERRA FOUNDATION FOR AMERICAN ART

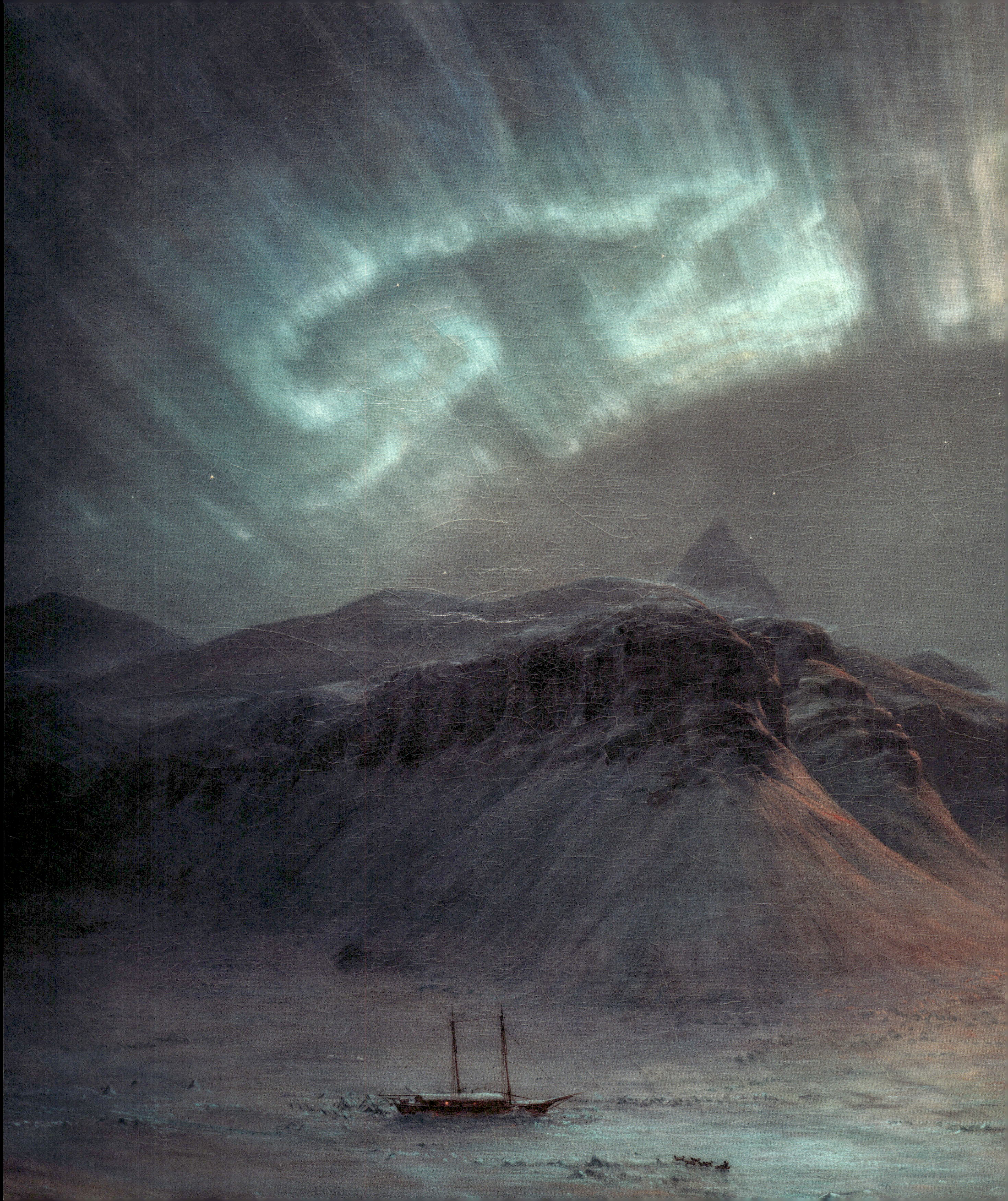

CONTENTS

Lenders to the Exhibition 9
Director's Foreword *Stephanie Stebich* 11
Preface: Alexander von Humboldt and American Science *Hans-Dieter Sues* 15
Introduction 25

CHAPTER 1 | HUMBOLDT IN THE UNITED STATES 39
CHAPTER 2 | MAPPING A NATIONAL IDENTITY 85
CHAPTER 3 | AMERICA'S WILDERNESS ICONS 135
CHAPTER 4 | HUMBOLDT IN EUROPE 175
CHAPTER 5 | HUMBOLDT AND AMERICAN SLAVERY 215
CHAPTER 6 | HUMBOLDT AND NATIVE AMERICAN ETHNOGRAPHY 255
CHAPTER 7 | EMBODYING COSMOS: FREDERIC EDWIN CHURCH 313
CHAPTER 8 | HUMBOLDT AND SMITHSON 373

Catalogue of the Exhibition 397
Selected Bibliography 403
Index 427
Acknowledgments 436
Image Credits 442

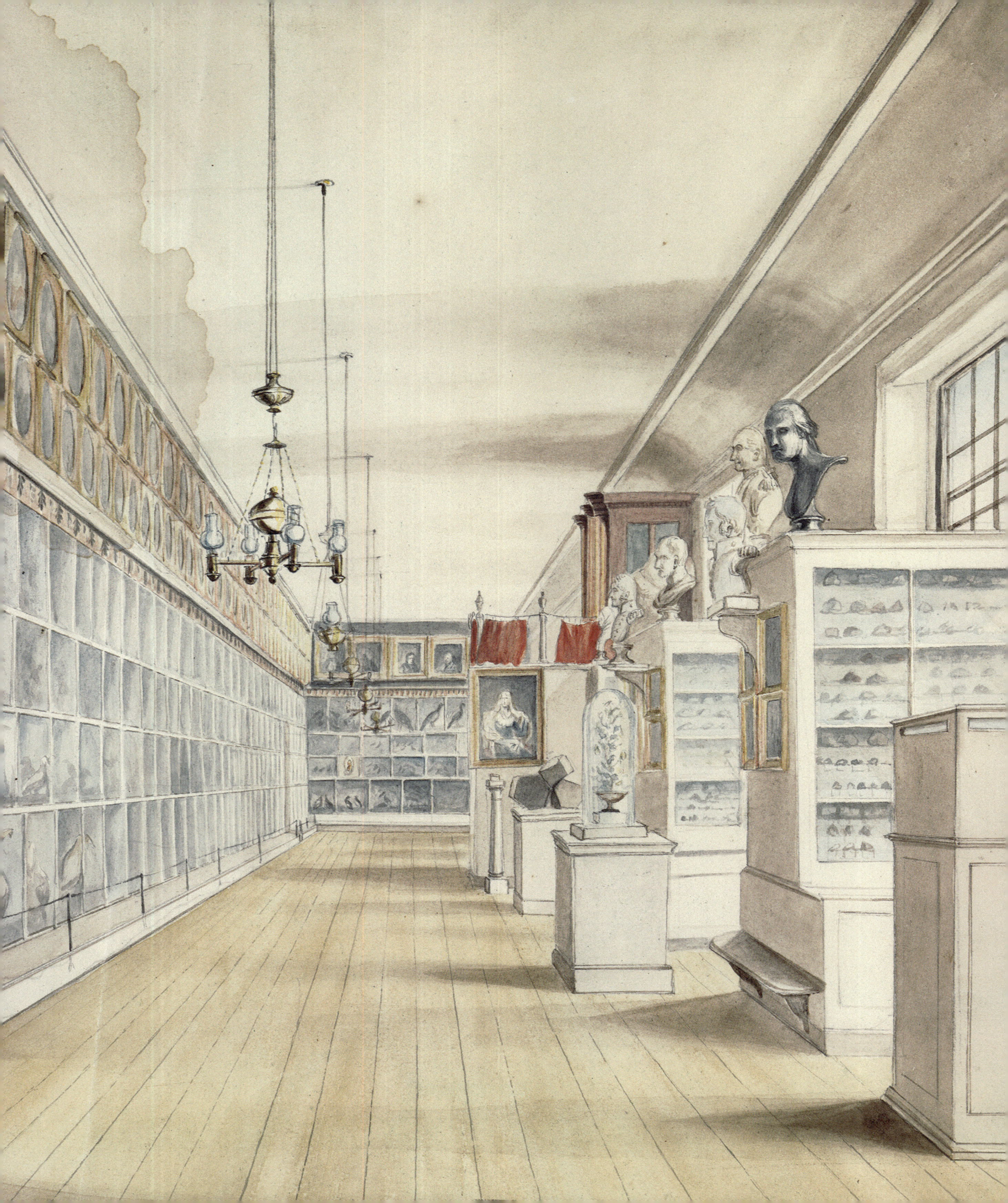

LENDERS TO THE EXHIBITION

Albany Institute of History and Art
American Philosophical Society
Autry Museum of the American West
Berlin-Brandenburgische Akademie der Wissenschaften, Archiv
Boston Athenæum
Cincinnati Art Museum
The College of Physicians of Philadelphia
Concord Free Public Library
Cooper Hewitt, Smithsonian Design Museum
Detroit Institute of Arts
Établissement public du château, du musée et du domaine national de Versailles
The Fralin Museum of Art at the University of Virginia
George Glazer Gallery
Hessisches Landesmuseum, Darmstadt
Robert W. Hoge and Immaculada Socias Hoge
Independence National Historic Park
Joslyn Art Museum
Library of Congress
Maryland Historical Society
Mount Vernon Ladies' Association
Museum of Fine Arts, Boston
National Gallery of Art
National Gallery of Art Library
National Museum of Natural History, Smithsonian Institution
National Portrait Gallery, Smithsonian Institution
The New-York Historical Society
New York State Museum
Mr. and Mrs. Robert F. Norfleet Jr.
Olana State Historic Site, New York State Office of Parks, Recreation and Historic Preservation
Pennsylvania Academy of Fine Arts
Royal Botanic Gardens, Kew
Smithsonian American Art Museum
Smithsonian Libraries
Smithsonian's National Museum of American History
Sons of the Revolution in the State of California
Staatliche Museen zu Berlin, Nationalgalerie
Terra Foundation for American Art
The Thomas Jefferson Foundation at Monticello
Timken Museum of Art
Alan V. Weinberg
and one Private Lender

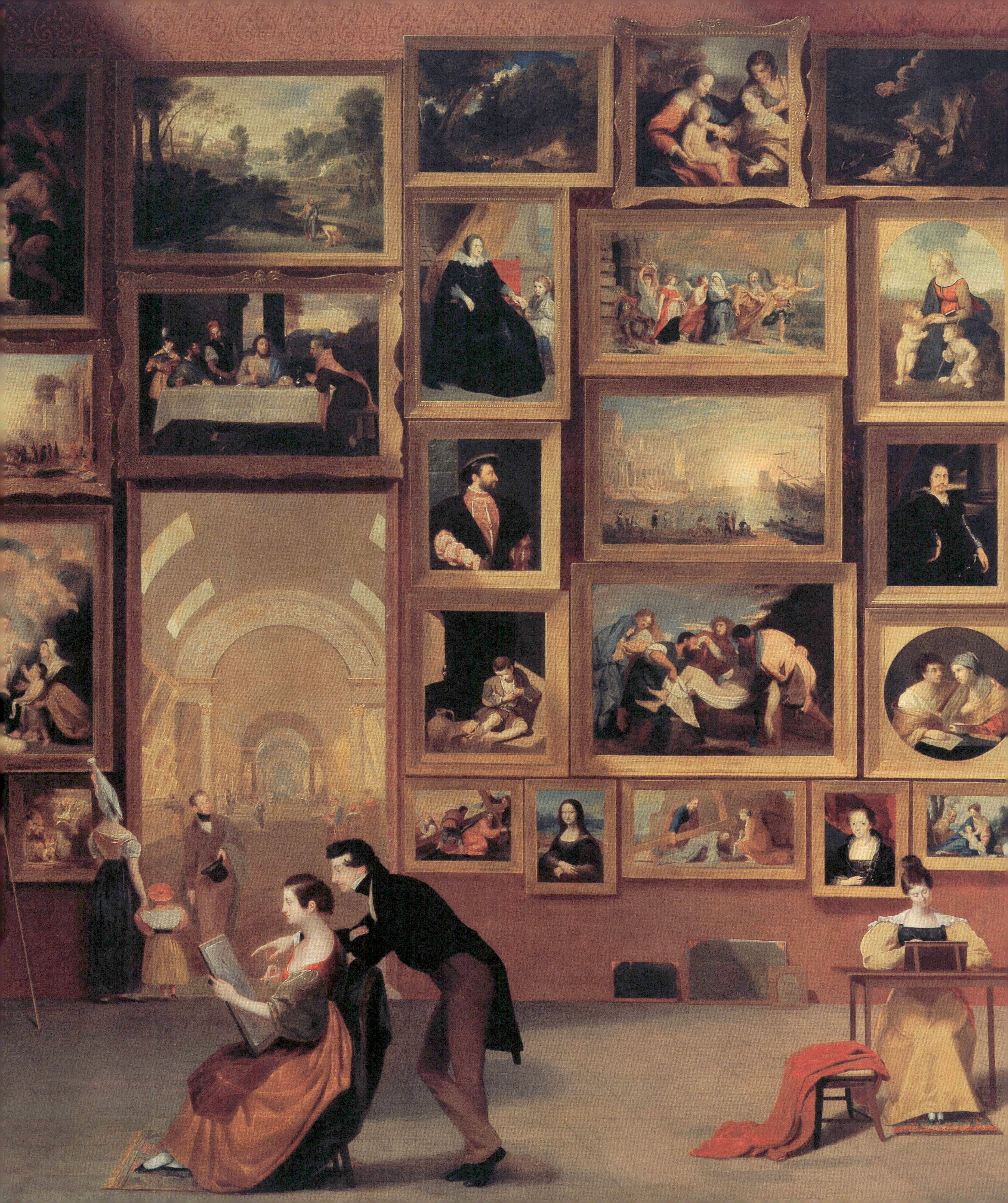

DIRECTOR'S FOREWORD

The Prussian naturalist Alexander von Humboldt was among the most widely recognized and admired public figures during the first half of the nineteenth century. He was internationally acclaimed for his groundbreaking concept of the unity of nature—that everything on the planet is interconnected. This was a radical idea at that time. In 1804 Humboldt made a six-week visit to the United States that established relationships that shaped American art and culture. America's nature-based sense of identity stems in large part from Humboldt's friendship with President Thomas Jefferson and their shared appreciation for natural landmarks as emblems of America's robust prospects in politics, and the arts and sciences. Humboldt fell in love with the country's democratic principles and considered himself "half an American." This exhibition and catalogue tell the mostly forgotten yet pivotal story of how this trip influenced both Humboldt and the United States.

Humboldt's magnum opus, *Cosmos*, published in five volumes between 1845 and 1859, was among the most widely read books in Europe and the United States. The scholarship on Humboldt is vast and multidisciplinary; yet relatively little has been written about his influence here in the United States. Eleanor Jones Harvey, the curator of the exhibition and author of this book, brilliantly analyzes iconic American artworks to reveal the immediate, sustained, and profound influence Humboldt had on the development of early American culture. In the process, she reveals how the American wilderness became emblematic of this country's distinctive character.

Harvey has dusted for Humboldt's fingerprints on American art and culture and found them nearly everywhere. Humboldt's friendship with

President Thomas Jefferson launched his involvement in American politics and his influence on our cultural identity. Humboldt's observations on nature inspired two generations of American artists and writers to herald landmarks in the American wilderness as national icons. His concern for the planet—the effects of deforestation and land use in particular—drove him to amass statistics that help us understand global weather patterns and see our planet as "one great whole, animated by the breath of life." Humboldt's encouragement of American democracy and exploration permeated the work of some of our finest artists, shaped American culture, and influenced the founding of the Smithsonian Institution.

The words from James Smithson's will that founded "an establishment for the increase and diffusion of knowledge" are a Humboldtian credo. Dr. Harvey demonstrates that many of the people involved in the Smithsonian's early direction had close ties to Humboldt. Among them was Smithson himself, the English chemist whose bequest established and named the institution. The Smithsonian has grown over almost 175 years to embrace the vast range of disciplines and multitude of inquiries that we find in Humboldt's books, across the arts, the sciences, and the study of human cultures. Humboldt was described as being an entire learned society all by himself. The collections and ongoing research in the nineteen museums, the National Zoo, and twenty-one research centers and programs that compose the Smithsonian encompass the breadth of Humboldt's interests. Considered in this light, the Smithsonian is the perfect setting for this expansive evaluation of Humboldt's influence on the art and culture of the United States. Using the fine arts as a lens through which to understand Humboldt's reach makes the Smithsonian American Art Museum the ideal museum to mount this show.

Humboldt was not solely concerned with nature. Importantly, he believed in the equality of all races. In his books and letters Humboldt expressed his concern for America's native communities and argued tirelessly for the abolition of slavery. The National Museum of the American Indian, in sight of the U.S. Capitol, relates the devastating impact of U.S. expansion across the continent and soundly affirms the vibrancy of native cultures. Humboldt might be gratified that the Smithsonian's recently opened National Museum of African American History and Culture, which tells the multivalent stories of African American contributions to the nation's history, stands on the National Mall in a direct line from the Thomas Jefferson Memorial.

The multidisciplinary character of this project was only possible thanks to collaborations with many of the Smithsonian's units. The staff, collections, and research resources at the Archives of American Art, Biodiversity Heritage Library, Cooper Hewitt, Smithsonian Design Museum, National Anthropological Archives, National Museum of African American History and Culture, National Museum of American History, National Museum of the American Indian, National Museum of Natural History, National Portrait Gallery, National Zoo, Smithsonian Archives, Smithsonian Channel, Smithsonian Libraries, Smithsonian Tropical Research Institute, and Smithsonian American Art Museum have played key roles in the development of the ideas central to this project. Many of our sister institutes are also lenders to the exhibition. For that we are exceedingly grateful. It is this collaborative spirit that Humboldt encouraged in his colleagues and serves as one of the Smithsonian's greatest strengths.

Projects such as this one require immense resources to be successful. The Smithsonian American Art Museum would like to recognize and thank the following individuals, foundations, and corporations for their generous support: Joanne and Richard Brodie, Billings and John Cay, Fern and Hersh Cohen, Sheila Duignan and Mike Wilkins, Marie M. Halff, Liliane A. and Christian W. E. Haub, Raymond J. and Margaret Horowitz Endowment, Kandeo Asset Management, Maureen and Gene Kim, LATAM Trade Capital, Robert Lehman Foundation, Henry Luce Foundation, The Lunder Foundation—Peter and Paula Lunder Family, Provost of the Smithsonian, Lucy S. Rhame, Holly and Nick Ruffin, Jacquelyn and William Sheehan, Smithsonian Scholarly Studies Awards, Terra Foundation for American Art, Kelly Williams and Andrew Forsyth, Federal Council on the Arts and the Humanities, and Furthermore: a program of the J. M. Kaplan Fund.

An exhibition of this magnitude requires similar generosity from the numerous museums and individuals willing to share their works. Many of these artworks are treasures in their home institutions, and we are especially grateful that they recognized the significance of this project. Foremost among them we would like to thank the Hessisches Landesmuseum in Darmstadt, Germany, which has displayed the original mastodon skeleton from Peale's Museum for over a century. Their enthusiasm for this loan provides us with the centerpiece for an exhibition devoted to understanding how nature became the source of America's primary cultural emblems: from the mastodon to Natural Bridge and Niagara Falls, to the crown jewels of the National Park Service.

Alexander von Humboldt believed in the benefits of collaboration. He created networks of colleagues, friends, and admirers. Befitting Humboldt, this project required Herculean effort on the part of the entire staff of the Smithsonian American Art Museum. All ten departments contributed to the success of this venture. This exhibition exemplifies the museum's mission of making American art a central part of our lives. The works of art in this exhibition provide insight into Humboldt's influence on our culture, and in so doing, highlight the importance of the Smithsonian as part of that legacy.

Stephanie Stebich

The Margaret and Terry Stent Director
Smithsonian American Art Museum

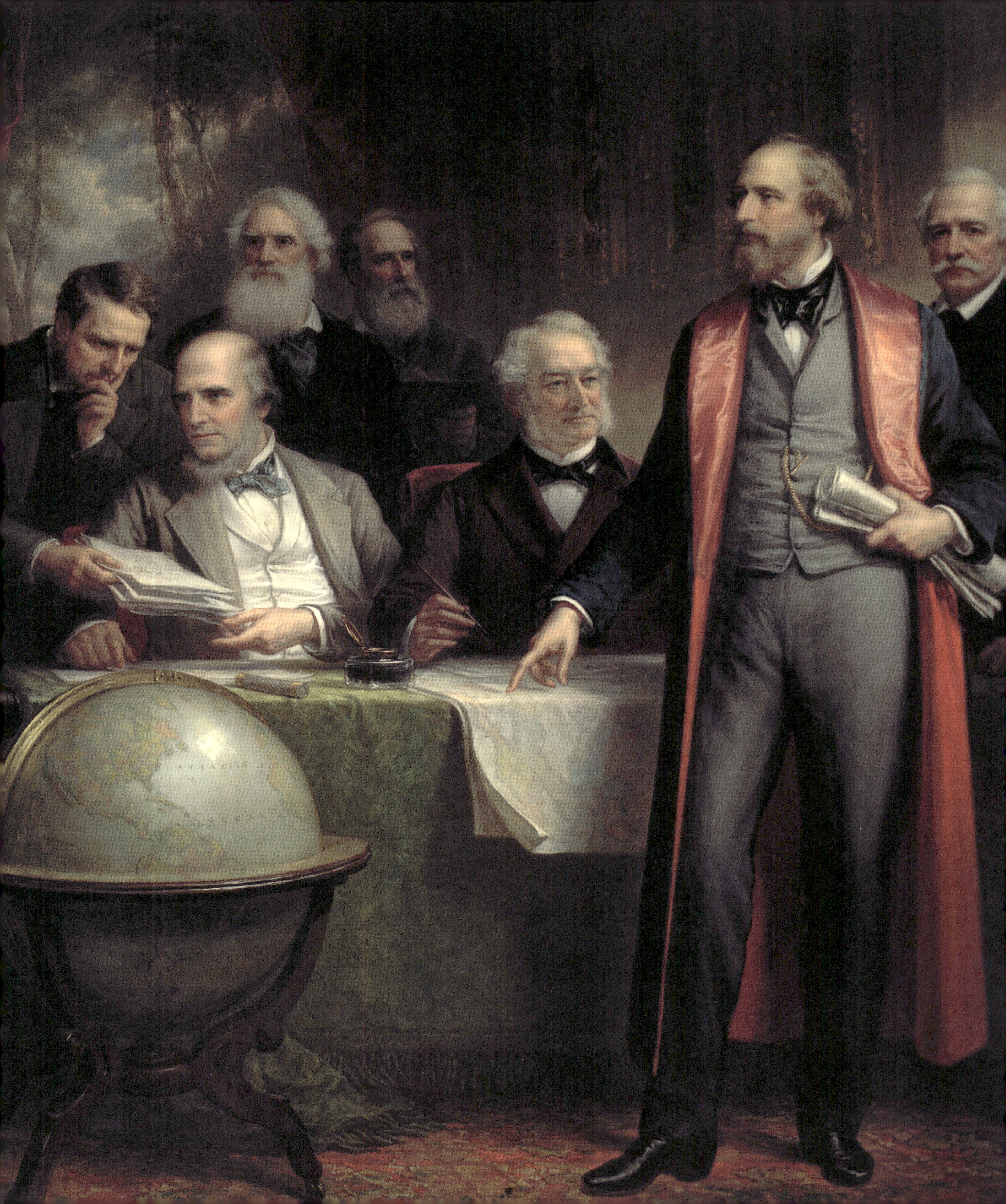
ATLANTIC
OCEAN

PREFACE | HUMBOLDT AND AMERICAN SCIENCE

Alexander von Humboldt (1769–1859) was one of the most celebrated explorers of all time.[1] Ralph Waldo Emerson proclaimed him "one of those wonders of the world."[2] Humboldt was also a protean scholar whose seemingly boundless interests included astronomy, botany, chemistry, economics, geography, geology, physics, politics, and zoology. He and his elder brother, Wilhelm (1767–1835), a prominent philologist, educational reformer, and the driving force behind the establishment of the University of Berlin (which was renamed in the brothers' honor in 1949), became leading intellectual figures in the nineteenth-century German states.

Although Alexander von Humboldt came from a Prussian aristocratic family, he had enlightened views on political and social issues and was a lifelong outspoken opponent of slavery. At an early age, he became deeply interested in nature and avidly read books about exploration. After studies at the University of Frankfurt an der Oder, the University of Göttingen, and the Mining Academy of Freiberg, Humboldt worked for five years as a Prussian mining official. This provided him with the opportunity to conduct a wide range of scientific experiments and undertake applied research, such as the development of a respirator and a safety lamp for miners.[3]

Humboldt was eager to embark on a major journey to an unexplored region of the world. Early plans for trips to Egypt and to the West Indies never materialized due to war in Europe. Finally, authorized by King Carlos IV of Spain, he and the French naturalist Aimé Bonpland (1773–1858) set off in 1799 on a self-financed five-year voyage of exploration to the "equinoctial regions" of the New World (Caribbean, northern South America, and Mexico), which established Humboldt's international reputation as an explorer.

Consistent with the precepts of the Enlightenment, Humboldt believed that the study of nature required observation and experimentation.[4] He advocated precise measurements of every physical parameter in a particular setting to elucidate the mutual interactions of physical forces. Humboldt argued that this should be done without any theoretical preconceptions. He obtained the best instruments available at the time for taking precise measurements. Humboldt believed that a wealth of such data would reveal nature as a dynamic equilibrium and the interconnectedness of physical forces. In the latter respect, he anticipated the Gaia hypothesis[5] proposed by James E. Lovelock and Lynn Margulis in the 1970s. At the same time, Humboldt also sought to connect this equilibrium with the aesthetic appreciation of cultured observers, which clearly linked him with the romantic tradition. To him, feeling and intellect represented complementary ways of understanding nature. Feeling provides the impetus for scientific exploration of natural phenomena and their connections. The knowledge gained from such studies, in turn, enhances appreciation of nature's beauty—allowing the observer to truly "see" the world.

HUMBOLDT'S IMPACT ON AMERICAN SCIENCE AND CULTURE

After his six-week visit to the United States in 1804, Humboldt stayed in contact with American politicians and scientists and maintained a keen interest in the political and intellectual development of the United States for the remainder of his life. On various occasions, he referred to himself as "half an American."[6] Humboldt's plan for a more extended stay in the United States was never realized. His *Personal Narrative of Travels to the Equinoctial Regions of the New Continent during the Years 1799–1804*[7] and especially his grand synthesis, *Cosmos*,[8] established him as a cultural icon in the United States during the first half of the nineteenth century. A fellow polymath and early American promoter of nature conservation, George Perkins Marsh (1801–1882) wrote admiringly that Humboldt's work

> *embraced the whole past history and present phase of every branch of physical research, and . . . was moreover graced with the elegances of all literature and dignified with the comprehensive wisdom of all philosophy.*[9]

Humboldt's view of nature as a harmonious, interdependent whole influenced a wide range of American painters, thinkers, and writers, including Frederic Edwin Church, Ralph Waldo Emerson, Edgar Allan Poe, Henry David Thoreau, and Walt Whitman.[10]

Humboldt was excited by the prospects for exploration as the American frontier rapidly moved westward. In one of his farewell letters written at the end of his American sojourn, he noted, "This country that stretches to the west of the mountains presents a vast area to conquer for science."[11] This pronouncement and Humboldt's writings, especially *Personal Narrative*, influenced John C. Frémont (1813–1890) and others in the United States who later set out to explore and survey the American West prior to the Civil War.

Inspired and guided by Humboldt's work, Frémont scientifically documented physical settings he encountered in the American West and their faunas and floras.[12] He was also an excellent cartographer whose maps would aid countless Americans migrating to the West and led to him becoming popularly known as "the Pathfinder." Frémont honored Humboldt by bestowing his name on various geographic features in the American West. Humboldt effusively praised Frémont's work and arranged for the award of a gold medal to him from the king of Prussia.

Humboldt was an indefatigable correspondent who wrote and received tens of thousands of letters during his lifetime. He built and actively maintained a vast international and interdisciplinary network of contacts who exchanged information and opinions on a wide range of topics.[13] This enabled Humboldt to connect and support many scientists. In turn, it provided him with a rich supply of observations for his own work from correspondents worldwide.

Humboldt's approach to exploration also influenced early research on the ocean. Scientists started collecting data on the physical and chemical parameters of the sea at various depths and in different regions of the globe during the first half of the nineteenth century. During his journey from Lima, Peru, to Ecuador, Humboldt himself measured the temperature and direction of the now-eponymous cold-water current in the Pacific along the western coast of South America.

As the officer in charge of the United States Navy office (the precursor of today's United States Naval Observatory) in Washington, D.C., Matthew Fontaine Maury (1806–1873)[14] established the systematic gathering of information on sea conditions and weather by American shipmasters. Influenced by Humboldt, Maury used a vast trove of data to generate global characterizations of geographic systems, developing what Humboldt termed a "physical geography of the sea." In 1847 Maury published his widely acclaimed *Wind and Current Chart of the North Atlantic*. He was also interested in global wind patterns and their impact on ocean currents. Rejecting the then-prevailing notion that winds were responsible for the major oceanic currents, Maury maintained that these currents were the result of temperature-related changes in the density of seawater, which generate systems of circulation between the cool and warm regions of the globe. Finally, he prepared the first bathymetric map of the Atlantic Ocean. While not very accurate, this map was important for the laying of the first transatlantic cable, a project that Humboldt had enthusiastically supported.

HUMBOLDT AS MENTOR TO AMERICAN SCIENTISTS

Humboldt showed a remarkable talent for spotting and advancing the careers of promising young scientists. He had realized that he could not possibly survey the entire world by himself and hoped his protégés would conduct such studies across the globe. In particular, Humboldt mentored two men who went on to become leading figures in nineteenth-century American science—Louis Agassiz and Arnold Guyot.

Louis Agassiz (1807–1873) was a Swiss-born naturalist who made many important contributions to the life and earth sciences.[15] Early in his career, based in Neuchâtel, Agassiz was particularly interested in the diversity and interrelationships of extant and extinct fishes and echinoderms. Following up on earlier work by the German Swiss naturalist Jean de Charpentier and especially the German naturalist Karl Friedrich Schimper, he also investigated the movement and structure of glaciers in the Swiss Alps. The three researchers tried to relate various geological phenomena in the Alps, such as gravel beds and massive erratic blocks, to moving glaciers. Based on this work and observations elsewhere in Europe, Agassiz proposed the theory that glaciers had once covered much of Europe due to a dramatic drop in temperatures.[16]

Agassiz first met Humboldt when he moved to Paris to study with the eminent paleontologist and zoologist Georges Cuvier (1769–1832). Impressed by the energetic young naturalist, Humboldt became his patron and even subsidized the costly production of Agassiz's magisterial *Recherches sur les poissons fossiles* (1833–44). With Humboldt's encouragement and with financial support from the king of Prussia (on Humboldt's recommendation), Agassiz traveled to the United States in 1846 to explore the animal and plant life of this country for two years. A course of public lectures at the Lowell Institute in Boston quickly generated new contacts and opportunities for Agassiz, and he soon decided to remain in America permanently. He married into one of Boston's leading families. His wife, Elizabeth Cabot Cary, later cofounded Radcliffe College. As a professor at Harvard's Lawrence Scientific School, Agassiz founded the Museum of Comparative Zoology and developed it into a leading center for the study of comparative anatomy and animal diversity.[17] An engaging lecturer, his teaching emphasized fieldwork and direct observation of specimens over traditional memorization of published facts. Although Agassiz organized various expeditions, including a voyage to Brazil in 1865 to find evidence contradicting Darwin's theory, he no longer produced groundbreaking scholarship later in his life. He did much, however, to advance the cause of natural history in the United States because he firmly believed that the study of nature could influence and inspire American politics. Agassiz was a major force behind the establishment of the National Academy of Sciences in 1863. In his capacity as a regent of the Smithsonian Institution, he helped guide the early development of the nation's foremost museum and research complex.

Although Agassiz always presented himself as a disciple and close friend of Humboldt, he was an outspoken proponent of "scientific" racism, which was anathema to his mentor and cast a deep shadow over his scholarly legacy. He also wanted to use science to advance a national agenda, unlike Humboldt's cosmopolitan concept of science.

During a visit to Philadelphia in 1846, Agassiz developed a visceral dislike of blacks, which, coupled with his obsession about the consequences of miscegenation, led him to argue that the different races of humans represented separate species. He was influenced by the polygenism of the physician Samuel George Morton (1799–1851) in Philadelphia. Morton had measured a large number of human skulls to determine their brain volume (without accounting for body size and sex) and, based on a highly biased analysis, asserted that whites had greater brain capacity than blacks and Native Americans.[18] Agassiz first expounded his view in the *Christian Examiner* in 1850. Because his claims had already been criticized for contradicting the biblical narrative of human creation, he argued that all humans shared a "unity of type" and that the scripture only referred to the origin of the white race from Adam and Eve. Agassiz believed that the human races had originated in the discrete geographic regions that they occupy in the present day (although migration had subsequently obscured some of the boundaries between these regions). In his view, the races all met the biological criteria to be considered separate species. Agassiz continued, "It seems to us to be mock-philanthropy and mock-philosophy to assume that all races have the same abilities, enjoy the same powers, and show the same natural dispositions, and that in consequence of this equality they are entitled to the same position in human society."[19]

Another proponent of "scientific" racism, Reverend John Bachman (1790–1874), a naturalist and minister in Charleston, South Carolina, took issue with Agassiz's polygenism. In Bachman's view, it not only contradicted the biblical account of human creation but was biologically unsound in its interpretation of human races as distinct species. He argued for the unity of the human species, citing Humboldt, although he did not accept this as an argument in support of racial equality.[20]

However, Agassiz kept Humboldt's intellectual legacy alive well after his mentor's passing. His 1850 book on Lake Superior and his ambitious but unfinished monograph series *Contributions to the Natural History of the United States of America* clearly reflected his mentor's comprehensive, empirical approach to the study of nature.[21]

Humboldt also mentored another Swiss-born naturalist, Arnold Guyot (1807–1884).[22] Guyot first met Agassiz as a young student in Karlsruhe, and the two men became lifelong friends. He became acquainted with Humboldt while studying in Berlin. Early in his career, Guyot examined glaciers in Switzerland and provided important information in support of Agassiz's glacial hypothesis. He later joined his friend on the faculty of the College of Neuchâtel, which had been established by the Prussian king in 1838. At Agassiz's urging, Guyot immigrated to the United States after the college's operations ceased following the revolutions of 1848. Like Agassiz, he delivered a course of invited lectures at the Lowell Institute in Boston. In 1854 Guyot received a faculty appointment at the College of New Jersey (later renamed Princeton University), where he taught for the remainder of his life. His research was concerned with geography, which was clearly influenced by Humboldt's thinking on this subject, and he was also interested in meteorology. His efforts to develop a national system for meteorological observations led to the establishment of the United States Weather Bureau (the precursor of today's National Weather Service) in 1870.

Like Agassiz, Guyot differed from Humboldt in his thinking about race. In his book *The Earth and Man* (first published in 1849), he twisted Humboldt's ideas concerning the relationships between plants and climate to argue that permanent racial differences were determined by geography. In Guyot's view, "[t]he people of the temperate continents will always be the men of intelligence, of activity, the brain of humanity, if I may venture to say so; the people of the tropical continents will always be the hands, the workmen, the sons of toil."[23]

HUMBOLDT'S CONTRIBUTIONS TO THE SCIENCES

Many hagiographic accounts have celebrated Alexander von Humboldt as a universal genius who laid the foundations for many sciences. However, as even his earliest scientific biographers pointed out, Humboldt never made a single groundbreaking scientific discovery nor did he formulate a major theory that posterity would come to firmly associate with his name. Working as a private scholar, he was not the founder of an academic school, but he served as mentor to many a young scientist. The sheer breadth of his interests led Humboldt to eschew in-depth pursuit of any particular line of scientific research. Instead he encouraged others to undertake the detailed examination of specific topics. A well-known example is Humboldt's discovery that the intensity of Earth's magnetic field decreases from the poles to the equator. He left it to his friend, the mathematician Carl Friedrich Gauss (1777–1855), to develop a method for determining the strength of Earth's magnetic field.[24] Thus, even his early scientific biographers considered Humboldt a heroic explorer, a tireless collector of data, and a gifted public communicator of science rather than an original creative mind.[25] Indeed, Humboldt is most widely known for his holistic thinking about nature and for introducing nature and world cultures to diverse audiences. However, he also could look back on numerous scientific accomplishments. The impact of much of Humboldt's research has yet to be critically reassessed by modern historians of science.

Humboldt and Bonpland's *Essai sur la géographie des plantes* (1805) is generally considered the foundation of plant biogeography.[26] Humboldt acknowledged earlier efforts, especially by his friends Georg Forster (1754–1794) and Karl Ludwig Willdenow (1765–1812), to document and interpret the geographic distribution of plants. Humboldt's work explored the spatial relationships of plants to one another and to geographically variable physical parameters such as altitude, humidity, and temperature. He contrasted his undertaking with traditional botanical research, which primarily focused on classifying plants based on their structural attributes and which he considered a critical first step in interpreting patterns of plant distribution. Humboldt's studies on the interactions of plants with one another and their relationships to their physical environment also established him as a pioneer in ecology,[27] although his contributions to this biological discipline (which was not formally designated until 1866) are often not fully recognized.

Together with Carl Ritter (1779–1859), Humboldt is regarded as the founder of modern geography.[28] Humboldt conducted regional studies with special consideration of conditions relevant to human geography as well as economic and political systems. Through this research, his innovative representations of geographic data (e.g., his famous illustration of the Andean volcano Chimborazo showing vegetation zones at different altitudes), and his exacting work as a cartographer, he contributed substantially to the development of geography as a science.

Carl Ritter became interested in the relationships between humanity and nature early in his career.[29] This interest deepened when he met with Humboldt in 1807. Later both scholars were active in Berlin. Ritter took a comparative approach to geography, aimed at establishing the interdependence of all phenomena on Earth's surface. He tried to demonstrate the individuality of continents, believing that their distinctive shapes and locations were based on a divine plan to facilitate the development of human civilization.[30] Like Humboldt, Ritter wanted to accumulate, without preconceived notions, as many observations as possible on particular regions of the globe in order to discover laws underlying the relationships between humans and the natural world. Unlike Humboldt, however, he was unabashedly Eurocentric in his views, regarding Europe as home to humanity's greatest achievements.[31] Ritter's methods for and writings about geographic research, especially social geography, influenced the historical development of this academic discipline.[32]

Humboldt made many empirical contributions across a wide range of sciences, including physics and geology. The extensive, well-documented collections of plants, many of them new to science, made by Bonpland and Humboldt in the Caribbean and South America laid the foundation for our understanding of neotropical plant diversity.[33]

Consistent with his egalitarian outlook, Humboldt tried to make his observations and thinking widely accessible.[34] His public lectures in Berlin in 1827 and 1828 attracted many women, who at that time were still not admitted to universities yet constituted a major community of readers, as well as people from a wide range of social backgrounds. The popular appeal of Humboldt's presentations and the great success of his books in Europe and the United States testified to the efficacy of his communication efforts.

Following Humboldt's passing on May 6, 1859, his influence in the United States rapidly waned. Cities across America still celebrated the centennial of his birth on September 14, 1869, with dedications of memorials and many speeches paying homage to his life and accomplishments.[35] However, a major 1890 review of the beginnings of American science[36] did not even mention Humboldt. By the end of the nineteenth century, only academic geographers still held Humboldt in high regard.[37] During much of the twentieth century, few Americans had ever heard of Humboldt. In recent years, however, there has been a major resurgence of interest in his life and work in the United States. Books by Aaron Sachs, Laura Dassow Walls, and Andrea Wulf[38] have reintroduced Humboldt and his influence on American cultural development to a wider readership.

Various reasons have been suggested for Humboldt's precipitous descent into obscurity in the United States during the second half of the nineteenth century.[39] His reputation remained undiminished in Germany although his work and thoughts were successively reinterpreted in light of the prevailing political ideologies during the turbulent modern history of that country.[40] Humboldt continued to be regarded as a major historical figure in Latin America, in part due to his friendship with Simón Bolívar, the liberator of the Spanish colonies in northern South America.[41] The two men had first met in Paris in 1804. Bolívar noted that "[Humboldt's] learning has done America more good than all of the conquistadores," and he considered Humboldt "the discoverer of the New World."[42]

An important factor behind Humboldt's descent into oblivion in the United States was increasing public hostility against the influx of immigrants from Germany, culminating in the anti-German hysteria during World War I. These campaigns helped erase Humboldt and other German cultural figures from the public memory for many years.

Starting in the mid-nineteenth century, Humboldt's ideas and methods increasingly were at odds with emerging scientific practice in the United States and Europe. The rapid expansion of knowledge and the professionalization of science led to profound changes. Individual scientific disciplines became the domains of full-time, academically trained specialists who focused on empirical studies of specific natural phenomena and processes and eschewed grand syntheses in the Humboldtian mode. Amateur naturalists continued to contribute to some disciplines (e.g., botany) by gathering data in the field, but other disciplines (e.g., physics) became closed to them. The life sciences witnessed a major shift in emphasis, in part due to their growing academic affiliation with medicine.[43] Laboratory-based experimental research in subjects such as cytology, embryology, and physiology soon eclipsed fieldwork exploring organic diversity in its natural settings. Concurrently, scientists adopted a technical style of writing that was aimed at fellow researchers rather than the reading public.

Through his writings, however, Humboldt continued to inspire many younger scientists to explore the interactions in the natural world. Charles Darwin (1809–1882) took Humboldt's *Personal Narrative* along on his voyage on the HMS *Beagle*.[44] He was strongly influenced by this book and modeled his own *Journal of Researches* on it. Darwin considered Humboldt an inspiration, who "like another Sun illumines everything I behold."[45] In 1881, one year before his death, Darwin wrote to his friend Joseph Dalton Hooker:

> *I believe that you are fully right in calling Humboldt the greatest scientific traveller who ever lived. You might truly call him the parent of a grand progeny of scientific travellers, who taken together have done much for science.*[46]

He had once met Humboldt in person but "remember[ed] nothing distinctly about [their] interview, except that Humboldt was very cheerful and talked much."[47] Like the elder scholar, Darwin perceived and appreciated nature aesthetically and intellectually. Both felt they were part of nature through their experiences and keenly appreciated the sublime beauty of natural phenomena. A recent study has documented how Darwin, during the earlier phases of his career, was profoundly influenced by not only Humboldt but also poets like John Milton and William Wordsworth.[48] Darwin's prose in his earlier works, such as his *Journal of Researches* and his magnum opus, *On the Origin of Species*, reflects that influence. Later, as a pillar of the scientific establishment, Darwin became more cautious in his expressions and adopted much less florid language.[49]

Humboldt and Darwin shared a holistic view of nature. However, deeply influenced by Charles Lyell's *Principles of Geology* (1830–33), Darwin went further than Humboldt in relating the development of the living world to geological processes and changes. He posited that all species evolved from a limited number of common ancestors and transformed over time—descent with modification.[50] According to Darwin, natural selection was the external force behind the changes that conferred advantages on organisms and species in their struggle for existence.

Darwin's theory of evolution by natural selection presented a view of life that stood in stark contrast to Humboldt's worldview.[51] Darwin's ever-changing world is one of unceasing struggle for survival, with individual organisms locked in competition with each other. By contrast, Humboldt's vision of nature emphasizes balance and harmony, with an integrated whole emerging from the complex interactions of countless animate and inanimate elements. In Darwin's world, evolutionary change does not foster harmony in nature, nor is it driven by some internal force. His vision won out, in part because it aligned well with the prevailing political and economic views in Europe and the United States during the late nineteenth and early twentieth centuries. There is no unambiguous record in Humboldt's writings that he ever entertained notions of evolutionary transformation. Indeed, he explicitly rejected efforts by other researchers to link present-day species to extinct ones.[52]

HUMBOLDT'S RENAISSANCE

In his writings, Humboldt repeatedly stressed the threat humans posed to nature. He first recognized the connection between deforestation and regional climate change and witnessed the adverse environmental impact of agricultural practices while traveling through Venezuela in 1800.[53] Humboldt saw humans as part of nature, not outside it, and warned that they were capable of destroying it. He viewed the atmosphere, land and sea, and all life—including humans—as parts of a harmonious, interconnected, and interdependent whole. Thus, Humboldt has been considered a proto-environmentalist.[54] Unfortunately, increasing scientific specialization during the nineteenth and first half of the twentieth century left no room for the development of an integrative research program based on a Humboldtian vision. Humboldt's thinking was also in conflict with the traditional view (informed by religious teaching) that nature had been created solely for humanity's benefit, enjoyment, and exploitation. However, his vision found enthusiastic reception by nineteenth-century American naturalists, including George Perkins Marsh and the "patron saint" of American environmentalism, John Muir (1838–1914). In his book *Man and Nature, or, Physical Geography as Modified by Human Action* (first published in 1864), Marsh discussed environmental degradation due to human agency in geographic and historical terms. He urged people to restore past damage inflicted on and prevent further destruction of nature.[55] Marsh had earlier stated, "[W]hereas Ritter and Guyot think that the earth made man, man in fact made the earth."[56]

Humboldt's vision of nature has become relevant again now that humanity faces a global environmental crisis. Most present-day scientists lack Humboldt's holistic perspective of nature. Instead they tend to focus on specific ecological changes without considering the complex web of interactions between humans and the environment. Furthermore, the current environmental movement often focuses on the preservation of animals, plants, or places rather than taking a more integrated approach that also considers humans.[57] Not enough is done to stimulate the public's appreciation of the natural world and to help people comprehend its vital importance for humanity. There is an urgent need for a Humboldtian perspective if we are to understand and address the unparalleled crisis now facing our species.

Hans-Dieter Sues

Chair of Paleobiology, Senior Research Geologist
and Curator of Vertebrate Paleontology
National Museum of Natural History, Smithsonian

NOTES

1 The most comprehensive treatment of Alexander von Humboldt's life and work is Hanno Beck's two-volume monograph: *Alexander von Humboldt, Band 1: Von der Bildungsreise zur Forschungsreise 1769–1804*; and *Alexander von Humboldt, Band 2: Vom Reisewerk zum "Kosmos" 1804–1859* (Wiesbaden, Germany: Franz Steiner Verlag, 1959).

2 "An abstract of Emerson's remarks made at the celebration of the centennial anniversary of the birth of Alexander von Humboldt, September 14, 1869," in Ralph Waldo Emerson, *The Complete Works,* vol. 11, *Miscellanea* (New York and Boston: Houghton Mifflin, 1904), 457. See https://www.bartleby.com/90/1124.html.

3 Ursula Klein, "The Prussian Mining Official Alexander von Humboldt," *Annals of Science* 69, no. 1 (2012): 27–68.

4 The classic study on "Humboldtian science" is Susan F. Cannon, *Science in Culture: The Early Victorian Period* (New York: Science History Publications, 1978), 73–110.

5 James E. Lovelock and Lynn Margulis, "Atmospheric Homeostasis by and for the Biosphere: The Gaia Hypothesis," *Tellus* 26, nos. 1–2 (1974): 2–10. See also James E. Lovelock, *The Ages of Gaia: A Biography of Our Living Earth* (New York: W. W. Norton, 1995).

6 "Sitzung vom 7. Mai 1859," *Zeitschrift für allgemeine Erdkunde,* Neue Folge 6 (1859): 414–16. A statement by the ambassador of the United States to Prussia, Joseph A. Wright, is cited on p. 415.

7 Alexander von Humboldt and Aimé Bonpland, *Personal Narrative of Travels to the Equinoctial Regions of the New Continent, during the Years 1799–1804*, 7 vols., trans. Helen Maria Williams (London, England: Longman, Rees, Orme, Brown, and Green, 1814–29).

8 Alexander von Humboldt, *Cosmos: A Sketch of the Physical Description of the Universe*, 5 vols., trans. E. C. Otté (London, England: Henry G. Bohn, 1849–70).

9 George Perkins Marsh, "The Study of Nature," *Christian Examiner* 68 (January 1860): 33–62. Quotation appears on p. 44.

10 Stephen Jay Gould, "Church, Humboldt, and Darwin: The Tension and Harmony of Art and Science," in *The Paintings of Frederic Edwin Church*, ed. Franklin Kelly (Washington, DC: National Gallery of Art, 1989), 94–107. See also Frank Baron, "From Alexander von Humboldt to Frederic Edwin Church: Voyages of Scientific Exploration and Artistic Creativity," *Internationale Zeitschrift für Humboldt-Studien* 6, no. 10 (2005): 2–15.

11 Humboldt to William Thornton, 20 June 1804. See Ulrike Moheit, ed., *Alexander von Humboldt: Briefe aus Amerika, 1799–1804* (Berlin, Germany: Akademie Verlag, 1995), 299–300.

12 For biographies of Frémont, see Ferol Egan, *Frémont: Explorer for a Restless Nation* (Garden City, NY: Doubleday, 1977); and Tom Chaffin, *Pathfinder: John Charles Frémont and the Course of American Empire* (New York: Hill and Wang, 2002).

13 Ottmar Ette, "Alexander von Humboldt, die Humboldtsche Wissenschaft und ihre Relevanz im Netzzeitalter," *Internationale Zeitschrift für Humboldt-Studien* 7, no. 12 (2006): 32–39.

14 For a recent review of Maury's contributions, see Jason W. Smith, *To Master the Boundless Sea: The U.S. Navy, the Marine Environment, and the Cartography of Empire* (Chapel Hill: University of North Carolina Press, 2018).

15 Edward Lurie, *Louis Agassiz: A Life in Science* (Chicago: University of Chicago Press, 1960).

16 See Louis Agassiz, *Études sur les glaciers* (Neuchâtel, Switzerland: Jean et Gassmann, 1840); and Louis Agassiz, *Théorie des glaces et ses progres les plus récents* (Genève, Switzerland: Bibliothèque Universelle de Genève, 1842).

17 For a history of the Museum of Comparative Zoology, see Mary P. Winsor, *Reading the Shape of Nature: Comparative Zoology at the Agassiz Museum* (Chicago: University of Chicago Press, 1991).

18 Stephen Jay Gould, *The Mismeasure of Man* (New York: W. W. Norton, 1981). Concerning Samuel George Morton, see pp. 50–69.

19 Louis Agassiz, "The Diversity of Origin of the Human Races," *Christian Examiner* 49 (July 1850): 110–45. Quotation appears on p. 143.

20 John Bachman, *The Doctrine of the Unity of the Human Race Examined on the Principles of Science* (Charleston, SC: C. Canning, 1850).

21 Louis Agassiz, *Lake Superior: Its Physical Character, Vegetation, and Animals, Compared with Those of Other and Similar Regions* (Boston: Gould, Kendall and Lincoln, 1850); and Louis Agassiz, *Contributions to the Natural History of the United States of America* (Boston: Little, Brown, 1857). Only four of a projected ten volumes were ever published.

22 James D. Dana, "Memoir of Arnold Guyot, 1807–1884," *Biographical Memoirs of the National Academy of Sciences* 2 (1886): 309–47.

23 Arnold Guyot, *The Earth and Man: Lectures on Comparative Physical Geography, in Its Relation to the History of Mankind*, trans. C. C. Felton (Boston: Gould and Lincoln, 1855). Quotation appears on p. 331.

24 Carl Friedrich Gauss, *Intensitas vis magneticae terrestris ad mensuram absolutam revocata* (Göttingen, Germany: Dietrich, 1833).

25 Karl Bruhns, ed., *Alexander von Humboldt: Eine wissenschaftliche Biographie*, 3 vols. (Leipzig, Germany: F. A. Brockhaus, 1872). The chapters in volume 3 review Humboldt's contributions to a range of scientific disciplines.

26 Frank N. Egerton, "A History of the Ecological Sciences. Part 32: Humboldt, Nature's Geographer," *Bulletin of the Ecological Society of America* 90 (2009): 253–82.

27 Donald Worster, *Nature's Economy: A History of Ecological Ideas* (Cambridge, UK: Cambridge University Press, 1985), 135.

28 Kent Mathewson, "Alexander von Humboldt's Image and Influence in North American Geography, 1804–2004," *Geographical Review* 96, no. 3 (2006): 416–38.

29 Uta Lindgren, "Ritter, Carl Georg," *Neue Deutsche Biographie* 21 (2003): 655–56. Online version at https://www.deutsche-biographie.de/pnd11860130X.html#ndbcontent.

30 Carl Ritter, *Über räumliche Anordnungen auf der Außenseite des Erdballs, und ihre Functionen im Entwicklungsgange der Geschichten* (Berlin, Germany: Ferdinand Dümmler's Buchhandlung, 1850).

31 Carl Ritter, *Europa, ein geographisch-historisch-statistisches Gemälde für Freunde und Lehrer der Geographie, für Jünglinge, die ihren Cursus vollendeten, bey jedem Lehrbuche zu gebrauchen. Nach den besten Quellen bearbeitet von C. Ritter*, 2 vols. (Frankfurt am Main, Germany: J. C. Hermannsche Buchhandlung, 1804 and 1807).

32 Karl Lenz, ed., *Carl Ritter—Geltung und Deutung. Beiträge des Symposiums anlässlich der Wiederkehr des 200. Geburtstages von Carl Ritter November 1979 in Berlin (West)* (Berlin, Germany: Dietrich Reimer Verlag, 1981).

33 William T. Stearn, ed., *Humboldt, Bonpland, Knuth and Tropical American Botany: A Miscellany on the "Nova genera et species plantarum"* (Lehre, Germany: J. Cramer, 1968).

34 Andreas Daum, *Wissenschaftspopularisierung im 19. Jahrhundert. Bürgerliche Kultur, naturwissenschaftliche Bildung und die deutsche Öffentlichkeit, 1848–1914* (Munich, Germany: R. Oldenbourg Verlag, 1998).

35 A frequently cited example is Louis Agassiz, *Address Delivered on the Centennial Anniversary of the Birth of Alexander von Humboldt* (Boston: Boston Society of Natural History, 1869).

36 George B. Goode, "The Origin of the National Scientific and Educational Institutions of the United States," *Papers of the American Historical Association* 4 (1890): 53–161.

37 Mathewson, "Alexander von Humboldt's Image and Influence in North American Geography, 1804–2004."

38 Aaron Sachs, *The Humboldt Current: Nineteenth-Century Exploration and the Roots of American Environmentalism* (New York: Viking, 2006); Laura Dassow Walls, *The Passage to Cosmos: Alexander von Humboldt and the Shaping of America* (Chicago: University of Chicago Press, 2009); Andrea Wulf, *The Invention of Nature: Alexander von Humboldt's New World* (New York: Knopf, 2015).

39 Sandra Nichols, "Why Was Humboldt Forgotten in the United States?," *Geographical Review* 96, no. 3 (2006): 399–415.

40 Nicolaas A. Rupke, *Alexander von Humboldt: A Metabiography* (Chicago: University of Chicago Press, 2008).

41 Gerd Kohlhepp, "Scientific Findings of Alexander von Humboldt's Expedition into the Spanish-American Tropics (1799–1804) from a Geographical Point of View," *Anais de Academia Brasileira de Ciências* 77, no. 2 (2005): 325–42.

42 J. Fred Rippy and E. R. Brann, "Alexander von Humboldt and Simón Bolívar," *American Historical Review* 52, no. 4 (1947): 697–703. Quotations appear on p. 699 and p. 701, respectively.

43 Philip J. Pauly, "The Appearance of Academic Biology in Late Nineteenth-Century America," *Journal of the History of Biology* 17, no. 3 (1984): 369–97.

44 "Books on the Beagle," Darwin Correspondence Project, University of Cambridge, https://www.darwinproject.ac.uk/people/about-darwin/what-darwin-read/books-beagle.

45 Darwin's diary entry for February 28, 1832. See Charles Darwin's *Beagle* diary 1831–1836, p. 42—Bahia, February 1832, at darwin-online.org.uk.

46 Darwin to Hooker, 6 August 1881. See classmark MS DAR95:518-23, letter number 13277, Cambridge Digital Library, https://cudl.lib.cam.ac.uk.

47 Nora Barlow, ed., *The Autobiography of Charles Darwin 1809–1882* (New York: W. W. Norton, 1953), 107.

48 Charles Morris Lansley, *Charles Darwin's Debt to the Romantics: How Alexander von Humboldt, Goethe and Wordsworth Helped Shape Darwin's View of Nature* (Oxford, UK: Peter Lang, 2018).

49 An example of Darwin's later style of writing is *The Descent of Man, and Selection in Relation to Sex* (London, England: John Murray, 1871).

50 Charles Darwin, *On the Origin of Species, or the Preservation of Favoured Races in the Struggle for Life* (London, England: John Murray, 1859).

51 Gould, "Church, Humboldt, and Darwin," 104–105.

52 Alexander von Humboldt, *Kosmos. Entwurf einer physischen Weltbeschreibung, Band 1* (Stuttgart, Germany: J. G. Cotta, 1845), 285–86.

53 Sachs, *Humboldt Current*, 77.

54 Ibid., 5.

55 Brett Clark and John Bellamy Foster, "George Perkins Marsh and the Transformation of Earth: An Introduction to Marsh's *Man and Nature*," *Organization & Environment* 15, no. 2 (2002): 164–69.

56 Marsh to Spencer Baird, 21 May 1860, University of Vermont Libraries. See http://cdi.uvm.edu/manuscript/uvmcdi-85191.

57 For a thoughtful assessment of issues facing environmentalism in the United States, see Sachs, *Humboldt Current*, 345–53.

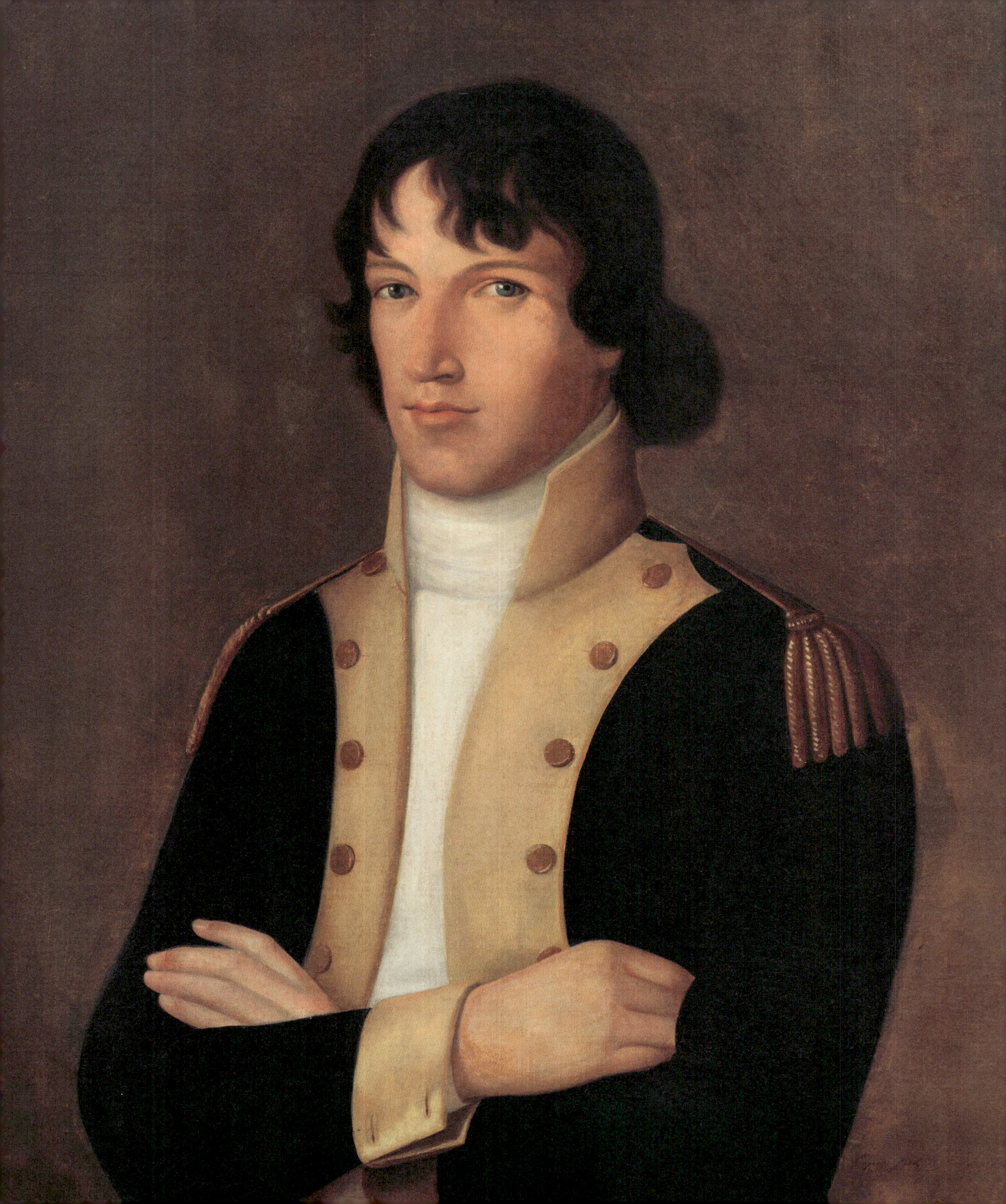

INTRODUCTION

Alexander von Humboldt is best remembered as an extraordinarily driven and omnivorously curious Prussian naturalist, whose radical ideas and prolific writings made him arguably the most important natural philosopher of the nineteenth century. Today, more than one hundred animals and more than three hundred plant species are named for him. Open a terrestrial atlas and you will find features named for Humboldt on every continent; open a celestial atlas and you will find his name attached to two asteroids and to a lunar mare (the Mare Humboldtianum) on Earth's moon. At first glance, Humboldt would not seem to have much relevance for American art and culture. But Humboldt understood the deep connections between his passion for science and the aesthetic gratification he derived from nature. He traveled with more than forty scientific instruments; among them was a cyanometer (see fig. 15), an invention designed to determine which of fifty-two shades of blue accurately matched the color of the sky as seen at different altitudes. He believed that artists needed to know enough natural history to render their sketches and paintings with great accuracy and that scientists needed to embrace an aesthetic appreciation for the natural world.[1] In the second volume of his magnum opus, *Cosmos*, he spent a third of the narrative giving advice to landscape painters, drawing on historical examples of the genre while encouraging artists to seek subjects beyond the familiar Mediterranean region. He believed that the awe-inspiring features of the hemispheric American landscape were worthy corollaries to their European counterparts and to the man-made wonders of the Old World.

This book takes as its starting point Alexander von Humboldt's six-week visit to the United States between late May and early July of 1804 and follows its downstream consequences for the next fifty years, examining American art as a way of understanding the roots of this country's deep cultural identification with nature. Humboldt's primary goal during his trip was to meet Thomas Jefferson, then president of the United States, to share ideas with a man he suspected was his intellectual equal. His interest in the United States reflected his fervent desire to see its democratic style of government flourish, to extend his South American explorations into the newly acquired Louisiana Territory, and to harness North American data in his emerging picture of the world's ecosystems.

Humboldt's brief stay in Philadelphia and Washington came at exactly the right moment to articulate the scientific and cultural promise of the United States. "This country that stretches to the west of the mountains presents a vast area to conquer for science!"[2] he wrote in a letter to Jefferson's friend, William Thornton, exulting in the potential for new information about natural history, cartography, and anthropology. Humboldt saw in nature the kind of geological and geographical monuments that rivaled anything made by man. In Europe, many scientists believed that the New World was lacking in civilization and that every life-form—plants, animals, and people—was smaller and weaker than its European counterpart. Jefferson had marshaled considerable data of his own to refute this theory in his book *Notes on the State of Virginia*, published first in French in 1785 and an American edition three years later. Humboldt had read Jefferson's book and, after seeing for himself the impressive topography of New Spain, along with its rich biodiversity and sophisticated pre-Columbian artifacts, he joined Jefferson in articulating a very different assessment of New World cultures. By seeing virtue in the distinctive features of the country's landscape and its inhabitants, he sped the United States along the path of deploying nature—and by extension landscape painting—as a core component of its cultural identity.

Humboldt's influence on the United States was immediate, sustained, and profound. For more than fifty years, when an American invoked Humboldt, it was to call on an unimpeachable authority on virtually any subject. Although his primary reputation rested on the ideas he developed from his scientific travels and experiments, Humboldt embraced the United States for more than its potential to enhance scientific knowledge. Humboldt envisioned the United States as an instruction manual for democracy, modeling its founding ideals for the rest of the world. His encouragement extended to the broader liberal arts and into American politics. Proclaiming himself "half an American," Humboldt became a magnet for American travelers in Europe.[3] The complete roster of influential men and women who became part of his international orbit is vast and exceeds the scope of this project, but each chapter examines a particular area of Humboldt's interest in the United States and the interconnected cast of characters who were inspired through their contact with him to give shape to America's cultural identity. This book explores the mutual influences between the United States and Alexander von Humboldt in conversation with each other, during a period when the country wanted his imprimatur and sought his approbation.

WHY HUMBOLDT?

Who was Alexander von Humboldt? Why and how did his life affect the United States to such a great extent? Despite the enormity of his influence during the nineteenth century, Humboldt's name and reputation faded in the United States after his death in 1859. Many of his new ideas simply became an accepted part of what we know about this planet; others were superseded by his colleagues and successors. However, between the 1820s and 1850s he had become one of the best known and widely admired public figures in the world. He lived into his ninetieth year, traveled on four continents, wrote more than thirty-six books and twenty-five thousand letters to a network of correspondents around the globe. He had an infectious personality and boundless curiosity, surrounded himself with some of the leading minds of his era, and never stopped talking. Charismatic, annoying, exuberant, caustic, but undeniably relevant, he straddled the Enlightenment penchant for wanting to know everything about everything and the establishment of modern scientific methods designed to query that accrued knowledge. He claimed to sleep only four hours a night and called coffee "concentrated sunbeams."[4] Among his many scientific achievements, Humboldt theorized the spreading of the continental landmasses through plate tectonics, mapped the distribution of plants on three continents, and charted the way air and water move to create bands of climate at different latitudes and altitudes. He tracked what became known as the Humboldt Current in the Pacific Ocean and created what he called isotherms to chart mean temperatures around the globe. He observed the relationship between deforestation and changes in local climate, located the magnetic equator, and found in the geological strata fossil remains of both plants and animals that he understood to be precursors to modern life forms, acknowledging extinction before many others.[5] Over the course of his adult life he developed a revolutionary theory that all aspects of the planet, from the outer atmosphere to the bottom of the oceans, were interconnected—a theory he called the "unity of nature."

It is hard to overstate how radical an idea this was in its day. After spending more than thirty years amassing data and testing ideas, Humboldt delivered a series of lectures in Berlin in 1827, describing theories that electrified his audience. From these lectures, he began drafting the book that would cement his lasting significance, as he described to his close friend, Varnhagen von Ense, in 1834:

> *I am going to press with my work,—the work of my life. The mad fancy has seized me of representing, in a single work, the whole material world,—all that is known to us of the phenomena of heavenly space and terrestrial life, from the nebulae of stars to the geographical distribution of mosses on granite rocks; and this in a work in which a lively style shall at once interest and charm. Each great and important principle, wherever it appears to lurk, is to be mentioned in connection with facts. . . . My title at present is* Kosmos; Outlines of a description of the physical World. . . . *I knew that Kosmos is very grand, and not without a certain tinge of affectation; but the title contains a striking word, meaning both heaven and earth.*[6]

Humboldt's singular text grew to fill five volumes, which were written in the last decade of his life to summarize all that he had learned in his scientific research based on his idea of the unity of nature. From the inaugural publication of the first volume in 1845, *Kosmos*—translated in English as *Cosmos: A Sketch of the Physical Description of the Universe*—was an international best-seller, with publishers vying for shipments of the book in at least twenty-six countries. *Cosmos* was translated almost as fast as it was published, was serialized in popular magazines, and inspired a generation of naturalists, explorers, artists, and authors. Some of the brightest minds and prominent scientific thinkers of the era embraced Humboldt's expansive thinking: inspired by Humboldt's early publications, Charles Lyell drew confidence in outlining his *Principles of Geology*; Charles Darwin idolized Humboldt, whose encouragement contributed to Darwin's developing theories regarding the evolution of species. Humboldt's friend Goethe proclaimed that he learned more from an hour in Humboldt's company than he did spending eight days reading other books.[7]

For Humboldt's U.S. audience, it was his travel narratives rather than his scientific monographs that ignited the imagination. Before *Cosmos*, Humboldt had published thirty-four other volumes, all sharing an evolving articulation of his underlying premise of the unity of nature. His *Essai sur la géographie des plantes (Essay on the Geography of Plants)* was published in 1805, followed by *Ansichten der Natur (Aspects of Nature)* in 1808. *Relation historique du voyage aux regions équinoxiales du nouveau continent* is an eleven-volume compendium of his travels through the Americas. Its English translation became known as the seven-volume *Personal Narrative of Travels to the Equinoctial Regions of the New Continent, during the Years 1799–1804* and four-volume *Essai politique sur le royaume de la Nouvelle-Espagne (Political Essay on the Kingdom of New Spain)*. In 1826 he published the *Essai politique sur l'île de Cuba (Political Essay on the Island of Cuba)* and, after that, several books stemming from his 1829 trip across Russia. On top of that, interspersed with these travel volumes he produced separate monographs devoted to astronomy, botany, geology, mineralogy, and zoology.[8]

Humboldt's innate curiosity and restless intellect were shaped by his early training with some of the best minds in Europe, including his mentors, Johann Friedrich Blumenbach and Christoph Daniel Ebeling, and explorer Georg Forster. When Humboldt traveled to England with Forster in 1790, he met a young chemist named James Smithson, who became another part of Humboldt's expanding network and, later in his life, was the founding benefactor of what became the Smithsonian Institution. It was Forster, who had sailed with Captain Cook on his second voyage, who inspired Humboldt to travel widely and pursue evidence to support his unity of nature idea.

Humboldt spent five years traveling across South America, Mexico, and Cuba between 1799 and 1804. Along the way Humboldt did more than gather plant specimens and artifacts; he witnessed the Transit of Mercury and discovered the location of the magnetic equator. That signature measurement allowed him to recalibrate his equipment and take the most accurate readings to that point of longitude and latitude in the Americas. Humboldt's trip corrected the location of numerous cities across South America and Mexico, literally recalibrating American cartography.[9] On top of that, he constructed the most detailed map of central North America, extending north from Mexico to the Canadian border. Sharing that map with Jefferson may have been the single most significant contribution Humboldt made to American geopolitics.

One measure of Humboldt's deep impact in the United States is the outpouring of grief when news of the eminent naturalist's death spread across the globe in 1859. In the United States, the *New York Times* and *Harper's Weekly* (fig. 1) devoted their entire front pages to his eulogy, enumerating Humboldt's achievements, extolling his significance, and amplifying the emotional response to news of his death. Ten years later, in 1869—the centennial of Humboldt's birth—the world again gave itself over to celebrating Humboldt's name and reputation and remarking on the progress others had made standing on his broad shoulders. Once again Humboldt dominated the front page of American newspapers. The *New York Times* devoted extensive coverage to what was being called the "Humboldt celebration."[10] In Boston, Harvard professor Louis Agassiz, perhaps the leading scientist of his generation and a Humboldt protégé, delivered a heartfelt address and choreographed a program of eulogies and inspirational speeches by the leading authors and scientists of the day. It was clear, both in 1859 and in 1869, that this country owed much to Humboldt's curiosity, writings, support, and international networks of influential people.

Ralph Waldo Emerson, who spent more than thirty years reading Humboldt, was among the speakers at the 1869 Boston celebration. In his description, we get a sense of how even Emerson struggled to express the magnitude of Humboldt's accomplishments:

> *HUMBOLDT was one of those wonders of the world, like Aristotle, like Julius Caesar, like the Admirable Crichton, who appear from time to time, as if to show us the possibilities of the human mind, the force and the range of the faculties,—a universal man, not only possessed of great particular talents, but they were symmetrical, his parts were well put together. As we know, a man's natural powers are often a sort of committee that slowly, one at a time, give their attention and action; but Humboldt's were all united, one electric chain, so that a university, a whole French Academy, travelled in his shoes. With great propriety, he named his sketch of the results of science* Cosmos. *There is no other such survey or surveyor.*[11]

Emerson set the tone for the continued appreciation of Humboldt, eliding *Cosmos* with its author as he spoke of the overwhelming importance of the man and his signature accomplishment.

CHANNELING HUMBOLDT IN THE UNITED STATES

The structure of this book follows the path of discovery I traveled. In my work on American landscape painting, I knew that Humboldt had inspired American artist Frederic Edwin Church to travel to South America and that his influential advice to landscape painters formed a substantial component of his second volume of *Cosmos*, published in English in 1849. While researching the American Civil War, I discovered that Humboldt was a lifelong abolitionist and had supported John C. Frémont's 1856 presidential campaign. When I learned Humboldt had visited the United

States and met with Thomas Jefferson, I knew I needed to follow his connections with this country wherever they led. Those connections were myriad and complex, much like everything else in Humboldt's long life. In excavating Humboldt's influence in the United States, my purpose is to dive deep into Humboldt's world and then step back to understand how his brief time in the United States emboldened this country to continue adopting nature—the concept of wilderness, specifically—as an emblem of the scale and scope of our cultural ambitions. The central story in this volume addresses how Humboldt's six-week visit shaped the cultural identity of the United States in the image of its landscape, from majestic formations like Natural Bridge and Niagara to an appreciation of the Great Plains as a "shoreless ocean" rather than a "great desert."[12]

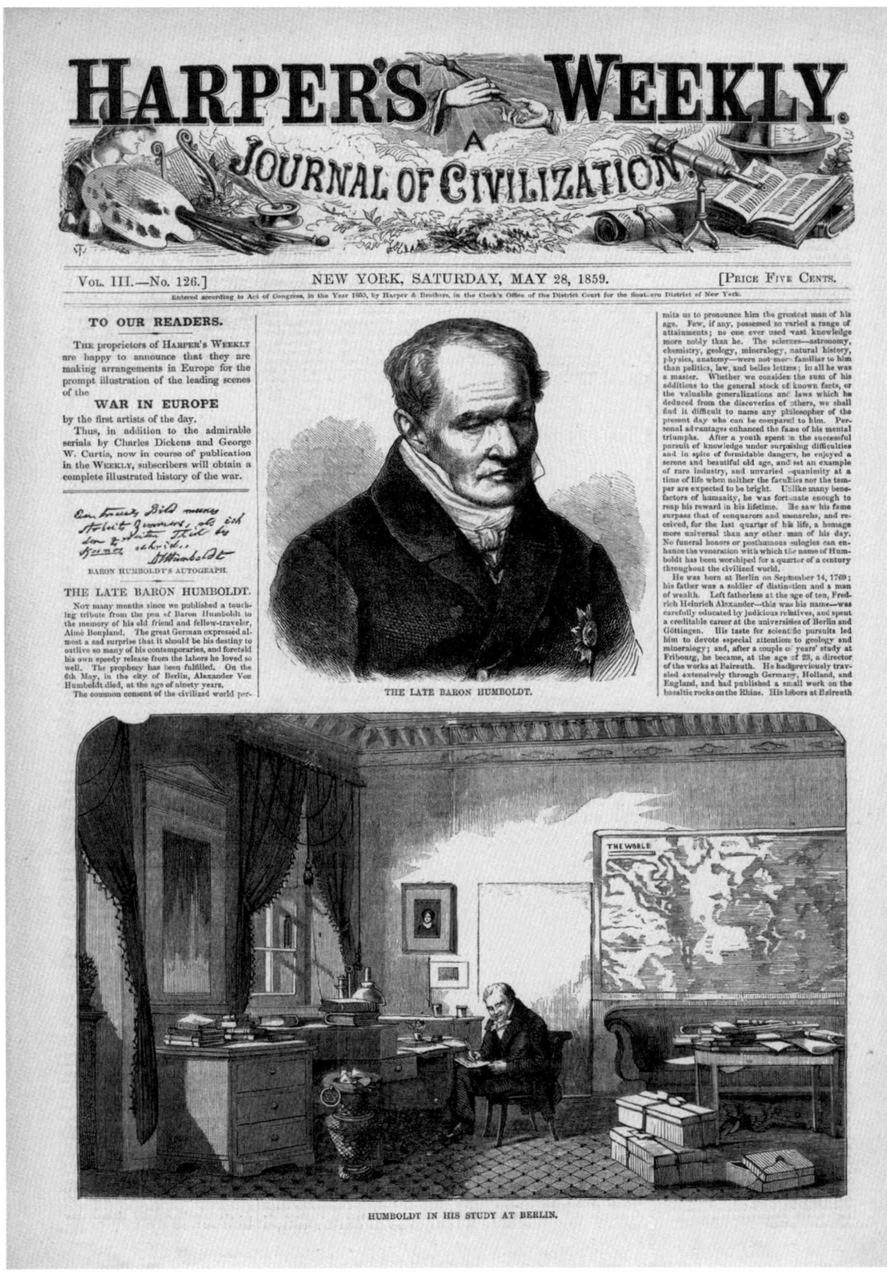
HARPER'S WEEKLY.
A JOURNAL OF CIVILIZATION.

Vol. III.—No. 126.] NEW YORK, SATURDAY, MAY 28, 1859. [Price Five Cents.

Entered according to Act of Congress, in the Year 1859, by Harper & Brothers, in the Clerk's Office of the District Court for the Southern District of New York.

TO OUR READERS.

The proprietors of Harper's Weekly are happy to announce that they are making arrangements in Europe for the prompt illustration of the leading scenes of the

WAR IN EUROPE

by the first artists of the day.

Thus, in addition to the admirable serials by Charles Dickens and George W. Curtis, now in course of publication in the Weekly, subscribers will obtain a complete illustrated history of the war.

BARON HUMBOLDT'S AUTOGRAPH.

THE LATE BARON HUMBOLDT.

Not many months since we published a touching tribute from the pen of Baron Humboldt to the memory of his old friend and fellow-traveler, Aimé Bonpland. The great German expressed almost a sad surprise that it should be his destiny to outlive so many of his contemporaries, and foretold his own speedy release from the labors he loved so well. The prophecy has been fulfilled. On the 6th May, in the city of Berlin, Alexander Von Humboldt died, at the age of ninety years.

The common consent of the civilized world permits us to pronounce him the greatest man of his age. Few, if any, possessed so varied a range of attainments; no one ever used vast knowledge more nobly than he. The sciences—astronomy, chemistry, geology, mineralogy, natural history, physics, anatomy—were not more familiar to him than politics, law, and belles lettres; in all he was a master. Whether we consider the sum of his additions to the general stock of known facts, or the valuable generalizations and laws which he deduced from the discoveries of others, we shall find it difficult to name any philosopher of the present day who can be compared to him. Personal advantages enhanced the fame of his mental triumphs. After a youth spent in the successful pursuit of knowledge under surprising difficulties and in spite of formidable dangers, he enjoyed a serene and beautiful old age, and set an example of rare industry, and unvaried equanimity at a time of life when neither the faculties nor the temper are expected to be bright. Unlike many benefactors of humanity, he was fortunate enough to reap his reward in his lifetime. He saw his fame surpass that of conquerors and monarchs, and received, for the last quarter of his life, a homage more universal than any other man of his day. No funeral honors or posthumous eulogies can enhance the veneration with which the name of Humboldt has been worshiped for a quarter of a century throughout the civilized world.

He was born at Berlin on September 14, 1769; his father was a soldier of distinction and a man of wealth. Left fatherless at the age of ten, Fredrich Heinrich Alexander—this was his name—was carefully educated by judicious relatives, and spent a creditable career at the universities of Berlin and Göttingen. His taste for scientific pursuits led him to devote especial attention to geology and mineralogy; and, after a couple of years' study at Fribourg, he became, at the age of 23, a director of the works at Baireuth. He had previously traveled extensively through Germany, Holland, and England, and had published a small work on the basaltic rocks on the Rhine. His labors at Baireuth

THE LATE BARON HUMBOLDT.

HUMBOLDT IN HIS STUDY AT BERLIN.

Fig. 1
"The Late Baron Humboldt," *Harper's Weekly*, vol. 3, no. 126, May 28, 1859, printer's ink on paper, approx. 16 × 11 in., University of Michigan Library

Chapter one maps the contours of Humboldt's brief stay in the United States in 1804. This trip established the foundation of his extensive network of friends and admirers there. Meriwether Lewis and William Clark had just embarked on their exploration of the newly acquired Louisiana Purchase, and Jefferson was in the midst of contentious international negotiations with France and Spain over the new southern and western borders of the United States. Humboldt arrived with maps and statistics that helped Jefferson and his cabinet think strategically about those negotiations. The Prussian traveler's effusive personality and unbounded curiosity about American geography, culture, and politics sparked lifelong friendships with some of the key figures of American history.

Beyond politics, Humboldt inspired artist Charles Willson Peale to resume his dormant painting career to paint Humboldt's portrait for his museum. In Philadelphia he was feted by the scientific community. His early publications already graced the shelves of the library of the American Philosophical Society, which made him a member. However, it was after this visit that Humboldt would become a force of nature himself. For the remaining fifty years of his life, people in the United States became part of Humboldt's global network of friends, allies, and scientific partners. Those alliances helped define the nation; America's presence on the international stage shone brighter with Humboldt's approbation, an imprimatur many in the United States assiduously cultivated.

Chapter two examines the influence of Humboldt's early publications that appeared in the years immediately following his South American travels. Humboldt's lectures and books established his reputation as a leading mind in the natural sciences. His eagerness to absorb the new information from the United States added another dimension to American exploration. American explorers knew their maps, measurements, statistics, and expedition narratives would make their way into his hands. Updated maps and illustrated books were the lingua franca of expedition reports. Each American contribution to this international enterprise found its way into Humboldt's growing library, and details from them appeared in the Prussian baron's works. Further, Humboldt encouraged the addition of artists as members of those expeditions. In particular, Stephen Harriman Long and John C. Frémont conducted expeditions using Humboldt's ideas and books as their inspiration. The published report from the Long Expedition later served as a foundation for literary descriptions of the American interior that would in turn become an important aspect of the Hudson River school landscape aesthetic. Frémont's narratives helped create his persona as the Pathfinder and earned him the appellation among explorers of "the American Humboldt." During the nineteenth century, the scientific journey became an epistolary venture in which distance became a metaphor for reach.

Humboldt had always intended to return to the United States, but each successive venture he undertook and each new volume he published delayed and ultimately defeated that goal. Thus Humboldt cultivated proxies—explorers who traveled to the United States in his stead and with his support. The information assembled from these journeys flowed directly to Humboldt—population statistics, ethnographic information and artifacts, natural history specimens, and cartographic measurements. All of this was designed to fill the gaps in his increasingly comprehensive understanding of landforms; the global distribution of plants, animals, and people; and how climate operated as a force on everything. This path to the increase and diffusion of knowledge—something of a buzzword during the Enlightenment—was navigated through lavishly illustrated publications. The market for these books crested with the wave of popularity experienced by Humboldt as he wrote, illustrated, and published volume after volume based on his five years in the Americas—an enterprise that ruined him financially but contributed to his global fame.

The third chapter addresses Humboldt's belief that the New World should not be measured using the standard of architectural wonders found in the Old World. Europeans looked to the built environment—like cathedrals and universities—as evidence of cultural significance. As such, they saw the Americas as continents devoid of history. Instead Humboldt argued, "Nature herself is sublimely eloquent," applying aesthetic theory and vocabulary to descriptions of the natural monuments boasted by the New World.[13] His embrace of nature as an impressive attribute symbolic of cultural prowess encouraged the development of a wilderness aesthetic in the United States. As early as the 1780s the nascent United States of America had tentatively adopted a nature-inflected sense of identity thanks to Thomas Jefferson's *Notes on the State of Virginia*, which was read widely in French and English among the literary and scientific elite on two continents. In this slim volume, Jefferson enumerated the myriad ways American geography, agriculture, commerce, and people were in no way inferior to their European counterparts. His narrative and statistics refuted the statements made by the influential European naturalist George-Louis Leclerc, the comte de Buffon, that all aspects of the New World were smaller, weaker, and more degenerate than their European counterparts. The discovery of the bones of mastodons—at that time recognized as the largest terrestrial creature known on the planet—in present-day Kentucky and upstate New York seemed further proof that Buffon's theories were at best false and pernicious.

Jefferson's book staked out the position that America's cultural prospects were tied to the awe-inspiring scale and uniqueness of the very things found within its borders. He further argued that features like Virginia's Natural Bridge and New York's Niagara Falls were evidence of American geographic superiority. In doing so, he laid a foundation for erecting a cultural identity grounded in that strength. Humboldt amplified and encouraged these values in the first books he published after visiting the Americas, in which he began to articulate his concept of nature. His imprimatur contributed to the development of America's wilderness aesthetic, which in turn gave rise to the Hudson River school. This generation of artists elevated the genre of landscape painting by presenting landmark features as emblems of America's cultural ambitions. Humboldt's keen eye, observant curiosity, and scientific rigor provided the secular framework for exploration and admiration of nature; America's belief in the ability to find God in the wilderness joined hands with data gained through exploration to blend scientific relevance with divine inspiration.

If Humboldt's books were guides to the New World, he was one of the principal destinations for travelers to the Old World. Following the War of 1812, the vogue for visiting Humboldt in Europe grew. Chapter four explores how he became the center of an interconnected web of correspondents, colleagues, and admirers, many of whom were Americans. From his perch in Paris, Humboldt played a central role in French scientific societies. With each publication, the world took greater notice of Humboldt's ideas. By the 1820s Humboldt's words and images became an integral part of the American school curriculum, and lengthy excerpts from his books appeared frequently in the leading literary and scientific journals. The litany of American luminaries who beat a path to his door is an astounding array of politicians, statesmen, authors, intellectuals, artists, and scientists.

Here we encounter Humboldt, "half an American" by his own reckoning, as a man who admired and espoused American ideals. In Paris, Humboldt and the Marquis de Lafayette stood at the center of a group of liberal thinkers who supported the United States and welcomed American travelers. Both men saw in American democracy a template for saving Europe from monarchical and dictatorial ruin. Humboldt's liberal politics and outspoken support of America endeared him to this country while placing him at odds with the French emperor. The mutual dislike between Humboldt and Napoleon serves as a framework for understanding how and why Humboldt sought faster and more reliable networks of communication across Paris, the continent, and ultimately the Atlantic Ocean. He befriended Americans who were able to enhance the establishment of those relays. Washington Irving, James Fenimore Cooper, and Samuel F. B. Morse formed part of that network. Humboldt's eagerness to advocate for Morse's telegraph and subsequently, the laying of the transatlantic cable, spoke to his desire to be in contact with his allies and advocates instantaneously and without Napoleonic interference. For Humboldt, knowledge was intended to be shared—disseminated as widely as possible without artificial restrictions. Humboldt's connections stretched like telegraph wires into American politics as well as the fine arts and the sciences.

Humboldt's advocacy for the United States was not uncritical. Chapter five considers Humboldt's unequivocal stance on the one issue on which he and the United States parted ways: American slavery. An adamant believer in racial equality, Humboldt railed against colonial rule and enslavement. He associated nature with an inherent right to individual freedom for all humankind, and he believed societies and governments must protect that right. Although he sidestepped engaging directly with Jefferson on the issue, he spared little anger in his correspondence with those in his close circle. As early as 1825 he feared that the perpetuation of slavery in the United States would be the country's undoing, prescient thoughts that he shared with many in his American network.

Humboldt's fervent desire to see America as an exemplar of a true democracy kept him close to this country's leading figures but simultaneously left him frustrated at his inability to gain traction on this most important issue. That engagement with American politics peaked with Humboldt's vocal support for John C. Frémont's 1856 presidential campaign as the first Republican candidate, running on an abolitionist platform inspired by Humboldt. Frémont had conducted five of his own expeditions into the American West, displaying his admiration for Humboldt by naming as many landscape features for the explorer as he could. Frémont also played a role in California politics during the final push toward statehood. California unexpectedly entered the Union in 1850 as a free state, and the California landscape—notably that of Yosemite—became the emblem for the promise of liberty in a nation soon plunged into civil war.

Chapter six takes up another of Humboldt's humanitarian concerns. Before he had left the United States, he expressed concern for the cultural well-being of America's native populations, quizzing Jefferson on his relationship with the various nations. Humboldt's travels in South America had convinced him that the indigenous people he encountered were descendants of advanced civilizations destroyed by generations of Spanish colonial rule. Democracy, in Humboldt's mind, should extend to all inhabitants of a nation, regardless of race or standing. Humboldt's increased concern following President Andrew Jackson's Indian Removal Act of 1830 fueled his interest in fellow Prussian naturalist Prince Maximilian zu Wied's trip to the United States from 1832 to 1834 with artist Karl Bodmer. Prince Max, as he was often called, had trained with Humboldt's professors at Göttingen. Inspired by Humboldt, he spent two years in Brazil writing a highly acclaimed ethnography of the Botocudo (Aimoré) tribe. Prince Max and Bodmer's journey through the United States provided the research for a corollary study of the native communities they encountered in North America and helped satisfy Humboldt's unfulfilled desire to return to the United States.

Humboldt's admiration for Prince Max's ethnographic skills makes his friendship with American artist George Catlin all the more perplexing. Prince Max and artist Karl Bodmer viewed Catlin as a showman, poorly trained as both an artist and an ethnographer. When Catlin took his massive Indian Gallery to Europe, he engaged a group of traveling Iowa (Báxoǰe) Indians to perform for the audience. Humboldt met the American artist-impresario and the thirteen Iowa who accompanied him to Paris in 1845. For the first time, Humboldt had direct conversations with native peoples from the United States. Catlin and Humboldt established a friendship that remained strong the rest of Humboldt's life. Their friendship sheds light on one of Humboldt's core principles—that he could discern the value in a relationship without dismissing someone with whom he disagreed. In a similar manner, Humboldt was able to admire Louis Agassiz for his work on glaciation and admire his data on comparative anatomy and yet vehemently disagree with his protégé's conclusions on the superiority of the Caucasian race.

The publication of *Cosmos* made Alexander von Humboldt perhaps the best known public intellectual figure anywhere on the globe. In the United States *Cosmos* inspired Frederic Church's enthusiastic embrace of science and art, Emerson's seminal essay *Nature*, Henry David Thoreau's *Walden*, and Walt Whitman's poetic self-portrait in *Leaves of Grass*. During the 1850s there was a conscious effort on the part of these men to frame Humboldt as a distant mentor. Chapter seven explores in depth how Humboldt's views shaped the career of landscape painter Frederic Church. Church's debt to Humboldt is the best understood American artistic response to the naturalist's vision of the unity of nature. Reading Humboldt's advice to landscape painters calibrated Church's dual interests in science and art. His two trips to South America were directly inspired by Humboldt.

Humboldt's ideas inflected every aspect of Church's artistic production, including subjects far removed from the artist's South American subject matter. Church gladly embraced the opportunity to adopt a Humboldtian mantle for his artistic persona. In so doing he reaffirmed the significance of landscape painting as the genre most capable of conveying America's cultural ambitions. In a similarly immersive manner, transcendentalist writers Ralph Waldo Emerson and Henry David Thoreau embraced Humboldt, providing a literary counterpart to Church's expansive vision and attention to details in nature. Walt Whitman, ever his own idiom, deployed "Kosmos" as a metaphor describing his own persona. The idea that one could embody *Cosmos*, whether in a painterly or literary oeuvre, was a hallmark of the American fascination with Humboldt.

The final chapter considers Alexander von Humboldt's legacy in the actual and intellectual founding of the Smithsonian Institution. James Smithson's decision to leave his fortune to create an institution in Washington, D.C., "for the increase and diffusion of knowledge among men" was unexpected, but the roster of men charged with determining how best to live up to that lofty yet vague dictum were almost all Americans whom Humboldt had met or inspired. Humboldt's name was invoked repeatedly as the proponents of the institution hammered out its parameters. With the founding of the Smithsonian in 1846, America was able to deliver back to Humboldt the benefit of its own far flung scientific investigations. By this time Humboldt had his own emissaries to the Smithsonian, and every publication bearing its name found its way to him. Humboldt's imprimatur still mattered, as much as it had to Jefferson and Gallatin a half a century earlier. The desire for a national institute, a place that could absorb the collections of specimens and the accrual of knowledge from expeditions and scientific inquiries had been a tantalizing prospect as far back as the founding of Charles Willson Peale's museum in Philadelphia in 1786. Now the Smithsonian was poised to adopt a Humboldtian mantle. Over the years it would become the American repository for every branch of knowledge the Prussian traveler and natural philosopher had pursued.

TWO DEGREES OF HUMBOLDT

Humboldt's transition from an immediate, vital, and direct source of inspiration to a more cerebral ethos is echoed in the structure of this book. The first four chapters address Humboldt as a powerful intellect capable of inspiring the shape of American international relevance and domestic cultural identity. They center on the people who met Humboldt or corresponded with him, and over whom he had a direct and specific influence. This establishes the foundation for Humboldt's immediate impact on American art and ultimately the country's cultural identification with nature. The last four chapters build on that foundation, moving outward to explore Humboldt's wider sphere of influence over American art and culture. These chapters underscore his enduring impact but also point to the limits of that adulation. For all the Americans who revered Humboldt for his belief in the ideals of American democracy, the Prussian's advocacy for the essential human rights of all people proved to be a bridge too far for some of his friends and correspondents, as well as the U.S. government. To Humboldt's credit, he never abandoned his support for what he believed the United States could be. That idealism eventually branded Humboldt as a romantic rather than as the Enlightenment-inspired empirical thinker he most certainly was. Even in his last meetings with traveling Americans in the months before his passing, Humboldt was conveying messages back to the United States on the issues that mattered to him. When Humboldt died in 1859, it was as though part of the American conscience died with him.

Intriguingly, two of the figures central to this book never met the man they so admired, yet both of them were hailed as the "American Humboldt," one in exploration and the other in painting. John C. Frémont, who fashioned himself as an explorer in Humboldt's mold, traveled to Paris and met Humboldt's close friend François Arago, consulting with him about the instruments he should take on his next expedition; however, he did not travel to Berlin to meet his idol in person. Frederic Church had painted *Heart of the Andes* as an homage to his distant mentor but delayed traveling to Europe until the late 1860s, finding inspiration aplenty in the Americas. He may have been planning to accompany his painting to show to Humboldt in Berlin in 1859, but news of Humboldt's death changed his plans. The painting toured elsewhere, and Church spent that summer in Labrador, chasing icebergs for his next Great Picture. In each case, these men who crafted part of their lives and identities, incorporating Humboldtian values along the way, did not require a face-to-face encounter. In this sense, Humboldt transitioned from being a direct and immediate inspiration to providing a model for an expansive, restless, curious approach to living.

Throughout this book we are never more than two degrees of separation from Humboldt himself. Beyond that, Humboldt's influence became so diffuse that establishing a concrete, cognizant link back to the Prussian naturalist dissolves into the ether of widely accepted wisdom. John Ruskin's *Modern Painters* is a case in point. There is no doubt that Ruskin was aware of Humboldt's fusing of science and art in his aesthetic response to the wilderness landscape. Yet Humboldt's name does not appear in any of Ruskin's writings until the appendix to the third volume of *Modern Painters*, published in 1856, in which he vigorously protested charges that he had plagiarized Humboldt's ideas.[14] Although he admitted to having read Humboldt's *Cosmos*, Ruskin was certain that he had arrived at his ideas on science and art independently. By the time Ruskin published his own thoughts, Humboldt's were so widely accepted that they had been absorbed into the aesthetic and scientific appreciation of landscape without attribution. Nevertheless, in encouraging artists to blend science and art, Ruskin amplified and reinforced Humboldt's ideas.

This project recognizes that most Americans encountered Humboldt's ideas in translation. Humboldt paid close attention to the authorized translations of his works, but the complexity of his French and German prose did not guarantee that the resulting English text was accurate or easy to read. However, these were the texts most Americans used. As such, I have quoted from the translations made by Helen Maria Williams, Elizabeth Sabine, Elise Otté, John Black, and Thomasina Ross where my purpose was to indicate the specific text read by one of Humboldt's admirers. Where I have focused on Humboldt's ideas apart from his nineteenth-century audience, I have used the more elegant and accurate modern translations by Vera Kutzinski, Mark W. Person, and Sylvie Romanowski. However, there was no language barrier when it came to understanding Humboldt's illustrations. From the *Naturgemälde* in *Essay on the Geography of Plants* (see cats. 15, 16) to the black-and-white engravings in *Views of the Cordilleras* (see cat. 32, fig. 87) and the lavish, large-scale color prints found in his related *Picturesque Atlas* (see cats. 92, 101), Humboldt's illustrations allowed Americans to absorb directly the features he described.

THE ABSORPTION OF HUMBOLDT

American explorers across Humboldt's lifetime repaid his generosity in sharing his map and information with Jefferson—through the dissemination of maps, journals, published reports, and visual images. The movement of ideas and data from Humboldt out through his network of correspondents—and the return of new information to Humboldt, feeding his hunger for knowledge and the development of new theories—was a collaborative and reciprocal system. In his 1869 eulogy, Emerson remarked that where Humboldt walked, an entire scientific academy marched in his shoes. It was an apt metaphor, for Humboldt traveled across four continents and an ocean in search of the experiences and measurements that would anchor his ideas. He seemed happiest when he was working in the company of talented colleagues, remarking to fellow scientist Joseph Gay-Lussac that he did his best work when surrounded by those who were doing better work than he.[15] Humboldt's willingness to distribute his massive collection of field specimens to a network of trusted colleagues demonstrated his genuine passion for sharing knowledge and his pragmatic acceptance that alone he would not have the time to do all of the research himself. Humboldt spent his lifetime increasing knowledge, and as that knowledge diffused, so too did the connection to his name. During the nineteenth century, towns, counties, and streets across the United States bore his name; in the decade following his death, statues were erected in parks across the country. When the Nevada Territory petitioned for statehood in 1864, Humboldt was one of the options for the new state's name. The celebrations in his honor continued in the United States until the early twentieth century.

Louis Agassiz noted in 1869 that Humboldt's name was invoked less and less as the years passed, though his ideas continued to circulate widely. In his centennial address he remarked that every school child in America had been taught by Humboldt without ever knowing their teacher's name.[16] Ticking off a litany of areas in which Humboldt had revolutionized our understanding of the planet, Agassiz noted how others had been inspired by his writings to further investigate his observations. As the once unified field of "natural philosophy" branched into the full canopy of the individual physical and biological sciences, Humboldt was not claimed by any one discipline. Instead his work formed a foundation layer on which his friends, rivals, colleagues, and successors built the modern edifices of scientific inquiry and learning.

Although Humboldt's name and reputation was eclipsed in the United States, he was not forgotten elsewhere. He remains to this day a revered figure in South America, remembered for his friendship with Simón Bolívar, the Liberator. He is credited with inspiring the young revolutionary to return from Europe to free his homeland from Spanish rule. In Mexico, Humboldt was granted honorary citizenship for his contributions to the archaeological rediscovery and appreciation of pre-Columbian civilizations. In Germany he and his equally famous brother, Wilhelm, were honored by conferring their name upon what is now Humboldt University in Berlin. Marble statues of the eminent naturalist and linguist flank the gates, the two brothers together representing the full spectrum of the arts and the sciences. As this book goes to press, the Humboldt Forum on Museum Island in Berlin prepares to open its doors as a multidisciplinary, Humboldtian enterprise, timed to recognize the 250th anniversary of Alexander von Humboldt's birth.

In the United States, although Humboldt's name had vanished, his ideas did not. When Rachel Carson wrote *Silent Spring* in 1962, her argument for saving the American bald eagle by banning the use of DDT drew on the same logic of interrelated downstream consequences Humboldt had postulated regarding local human-induced climate change at Lake Valencia in Venezuela in 1800.[17] With the rise of the environmental and conservation movements of the twentieth century, Humboldt's ideas have gained renewed traction and gradually his name has become re-associated with those once radical ideas of planetary interconnectedness and the emergence of climate science in this era that some have designated as the Anthropocene. Alexander von Humboldt is experiencing a renaissance with this rise in eco-awareness, visible in contemporary fine arts practice as well as across the sciences, as befits his own broad reach.

DUSTING FOR HUMBOLDT'S FINGERPRINTS

Discerning Humboldt's influence in the United States might be described as an exercise in dusting for his fingerprints, and we find them on nearly every aspect of American culture. In exploring his American network and pondering those relationships, what became clear was Humboldt's far-reaching influence on the development of what has been called America's "wilderness aesthetic." Finding positive value in uncultivated territory—to merge the European aesthetics of the sublime with the beautiful as found in a wild landscape—is grounded in Humboldt's earliest writings on the Americas. Embraced by Jefferson, expanded by a generation of explorers who took his books and his ideas westward, and reimagined in early travel literature, Humboldt's appreciation for wilderness as a scientific wonderland helped change the perception of the interior of the North American continent and, by extension, the nation. George Perkins Marsh, himself deeply influenced by Humboldt, expanded on this essential quality. In his influential book *Man and Nature; or Physical Geography as Modified by Human Action*, published in 1864, he eloquently teased out the distinction between wilderness as a state of nature and the concept of wilderness as a desired state of mind.[18] In 1947 American author Wallace Stegner reiterated Humboldt's admiration for natural monuments as avatars of American identity. In *The Sound of Mountain Water*, he compiled his essays on the American West, concluding that the ultimate goal of the United States was "to create a society to match its scenery."[19] That is a sentiment of which Humboldt certainly would have approved.

This project sets out to recapture the immediate, sustained, and profound impact of Humboldt's ideas and understand how they shaped the works made by some of the most influential American artists of the nineteenth century. The artworks selected here include some of the strongest aesthetic statements about the nation's emerging cultural significance, created by artists who saw themselves as arbiters of that national conversation. Charles Willson Peale, Samuel F. B. Morse, and Frederic Church—among many others—deliberately wove into their major paintings messages about American cultural identity. In doing so, they advocated for the power of the fine arts to set the course for an aesthetic identity. Collectively their efforts created a nature-based aesthetic so well recognized today that its roots are obscured, much in the way Humboldt's concept of "Cosmos" has been accepted as received wisdom. That collaborative model lies at the heart of Humboldt's enterprise and at that of the Smithsonian as well. It is a model that encourages each of us to be an integral part of the never-ending search for knowledge. Humboldt's "unity of nature" urges us to use those data to learn about the consequences of our actions and take an active, informed role in the stewardship of our planet.

NOTES

1 Humboldt advocated for "the animated description of organic forms (plants and animals), in their local and picturesque relations to the varied surface of the earth (as a small fragment of the whole terrestrial life)," and believed that this kind of study was advanced "by the stimulus or impulse imparted to the mind when artistic treatment is applied to phenomena of nature on a great scale." Alexander von Humboldt, *Aspects of Nature, in Different Lands and Different Climates, with Scientific Elucidations*, trans. Elizabeth J. L. Sabine (London: Longman, Brown, Green, and Longmans, 1849), 1:261.

2 "Ce vaste pays qui s'étend à l'ouest des montagnes présente un vaste champs à conquérir pour les sciences!" Humboldt to Thornton, Philadelphia, 20 June 1804, in Ingo Schwarz, ed., *Alexander von Humboldt und die Vereinigten Staaten von Amerika: Briefwechsel* (Berlin: Akademie Verlag, 2004), 96–97, letter 9.

3 "Joseph Albert Wright, "Commemorative Remarks at the Geographical Society, Berlin, May 7, 1859," *Zeitschrift für allgemeine Erdkunde*, n.s., 6 (1859): 415; quoted in Ingo Schwarz, "'Shelter for a Reasonable Freedom,' or Cartesian Vortex: Aspects of Alexander von Humboldt's Relation to the United States of America," in "Alejandro de Humboldt y el mundo hispánico," special issue, *Debate y Perspectivas*, no. 1 (2000): 182.

4 Humboldt to Heinrich Christian Schumacher, Berlin, 2 August 1836, quoted in Karl Bruhns, ed., *Life of Alexander von Humboldt. Compiled in Commemoration of the Centenary of His Birth, by J. Löwenberg, Robert Avé-Lallemant, and Alfred Dove*, trans. Jane Lassell and Caroline Lassell (Boston: Lee and Shepard, 1873), 2:56.

5 Alexander von Humboldt and Aimé Bonpland, *Essai sur la géographie des plantes* (Paris: Chez Fr. Schoell, 1807), 22. Translated in Humboldt and Bonpland, *Essay on the Geography of Plants*, ed. Stephen T. Jackson, trans. Sylvie Romanowski (Chicago: University of Illinois Press, 2013), 69.

6 Humboldt to Varnhagen von Ense, Berlin, 27 October 1834, in Alexander von Humboldt, *Letters of Alexander von Humboldt, Written between the Years 1827 and 1858, to Varnhagen von Ense: Together with Extracts from Varnhagen's Diaries, and Letters from Varnhagen and Others to Humboldt* (London: Trübner, 1860), 15–19; quoted without ellipses in "Baron Humboldt's Letters," *Eclectic Magazine of Foreign Literature* 50, no. 2 (October 1860): 235.

7 Johann Wolfgang von Goethe to Karl August, Duke of Saxe-Weimar, March 1797, in *Goethe Begegnungen und Gespräche*, ed. Ernst Grumach and Rente Grumach (Berlin: De Gruyter, 2013), 4:288. Translated in Andrea Wulf, *The Invention of Nature: Alexander von Humboldt's New World* (New York: Knopf, 2015), 26.

8 Horst Fiedler and Ulrike Leitner have assembled the impressive compendium of Humboldt's published works, which covers each edition in each language. *Alexander von Humboldts Schriften: Bibliographie der selbständig erschienenen Werke* (Berlin: Akademie Verlag, 2000). During Humboldt's twenty-two years in Paris, he published sixteen volumes on botany, two volumes on zoology and anatomy, four volumes on astronomy and geophysics, two pictorial atlases, three volumes on his travels through New Spain, four volumes on the Kingdom of New Spain, and seven of the proposed eleven volumes of the *Personal Narrative*, in addition to *Aspects of Nature* and the *Geography of Plants*. This assessment is discussed in Stephen T. Jackson's introduction to *Essay on the Geography of Plants*, 3.

9 Alexander von Humboldt, *Political Essay on the Kingdom of New Spain*, trans. John Black (New York: Riley & Co., 1811), 1:xviii.

10 "HUMBOLDT. The One Hundredth Birthday of the Philosopher," *New York Times*, 15 September 1869, 1.

11 Ralph Waldo Emerson, "Humboldt: An Abstract of Mr. Emerson's Remarks Made at the Celebration of the Centennial Anniversary of the Birth of Alexander von Humboldt, September 14, 1869," *The Complete Works of Ralph Waldo Emerson: Miscellanies* (Boston: Houghton, Mifflin, 1904), 11:457. "The Admirable Crichton" was Scottish polymath James Crichton (1560–1582), widely admired for his extraordinary accomplishments in the arts, sciences, and languages. See Patrick Fraser Tytler, *Life of James Crichton of Cluny, Commonly Called the Admirable Crichton* (Edinburgh: James Ballantyne, 1819).

12 Alexander von Humboldt described the landscape of the steppes: "We seem to see before us a shoreless ocean." Humboldt, *Aspects of Nature*, trans. Sabine (London, 1849), 1:2.

13 *Views of the Cordilleras and Monuments of the Indigenous Peoples of the Americas, A Critical Edition*, ed. and trans. Vera M. Kutzinski and Ottmar Ette (Chicago: University of Chicago Press, 2012), 34.

14 "Appendix III: Plagiarism," in John Ruskin, *Modern Painters* (New York: E. P. Dutton, 1856), 3:371. The most astute study of this issue is Alicia Lubowski-Jahn, "A Comparative Analysis of the Landscape Aesthetics of Alexander von Humboldt and John Ruskin," *British Journal of Aesthetics* 51, no. 3 (July 2011): 321–33.

15 "I feel it more every day that I do good work only in an environment where others around me do even better." "Je sens tous les jours que l'on ne travaille bien que là où d'autres travaillent mieux autour de vous." Humboldt to François Gérard, Berlin, 12 February 1807, in Alexander von Humboldt, *Lettres américaines d'Alexandre de Humboldt*, with a preface and notes by Dr. E. T. Hamy (Paris: E. Guilmoto, 1905), 217. English translation by the author.

16 "Alexander von Humboldt: Celebration of the Centennial Anniversary of His Birth; Address by Professor Agassiz," *Littell's Living Age* 103, no. 1322 (2 October 1869): 48; reprinted from the *Boston Daily Advertiser*, 15 September 1859.

17 Alexander von Humboldt and Aimé Bonpland, *Personal Narrative of Travels to the Equinoctial Regions of the New Continent, during the Years 1799–1804*, trans. Helen Maria Williams (London: Longman, Hurst, Rees, Orme, and Brown, 1819), 4:142–50.

18 David Lowenthal does a masterful job of putting Marsh in context with Humboldt, Henry David Thoreau, John Muir, Frederick Law Olmsted, and other nineteenth-century ecologically minded Americans in *George Perkins Marsh: Prophet of Conservation* (Seattle: University of Washington Press, 2000), 416–19.

19 Wallace Stegner, *The Sound of Mountain Water: The Changing American West* (1947; repr., New York: Vintage, 2017), 32.

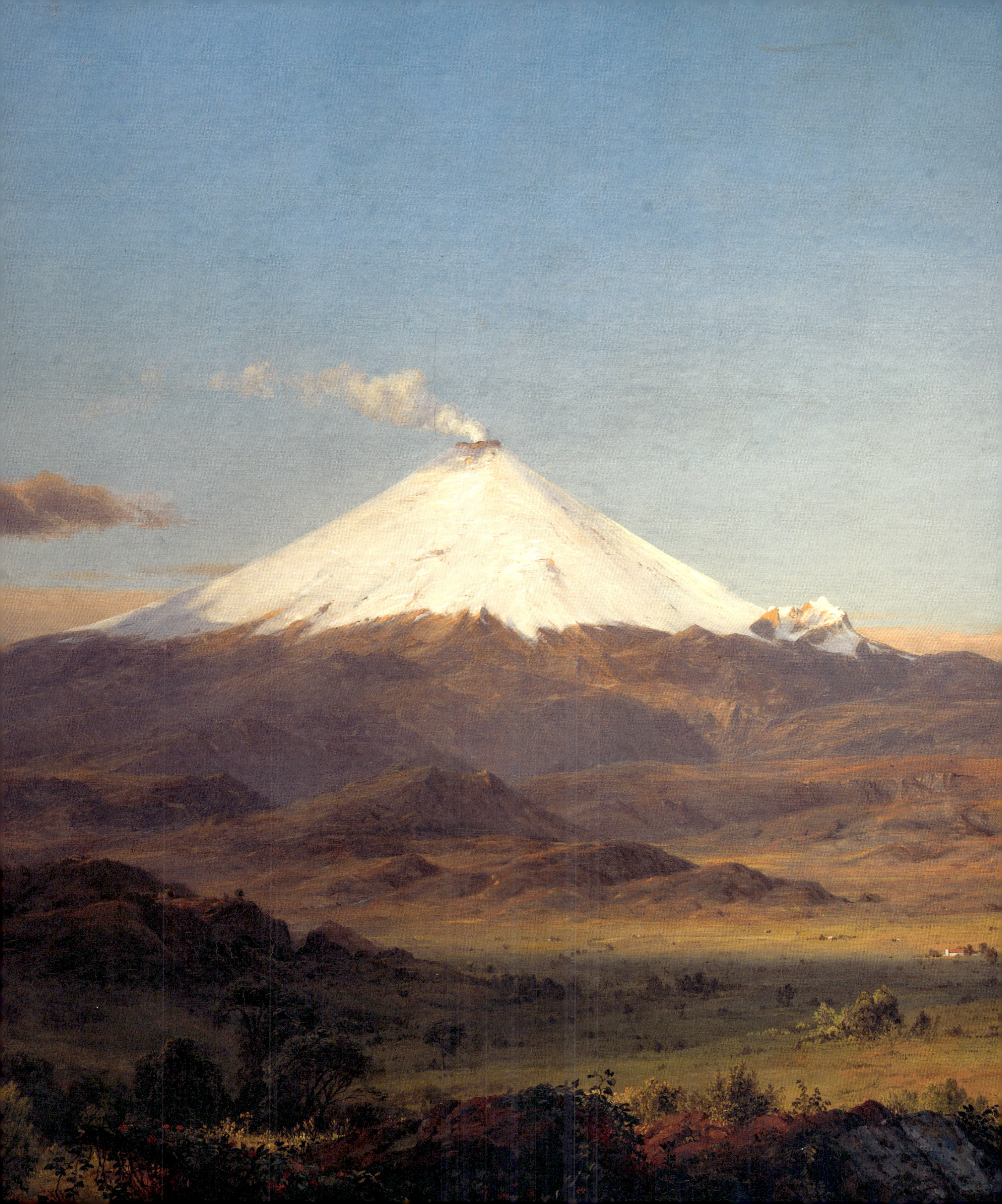

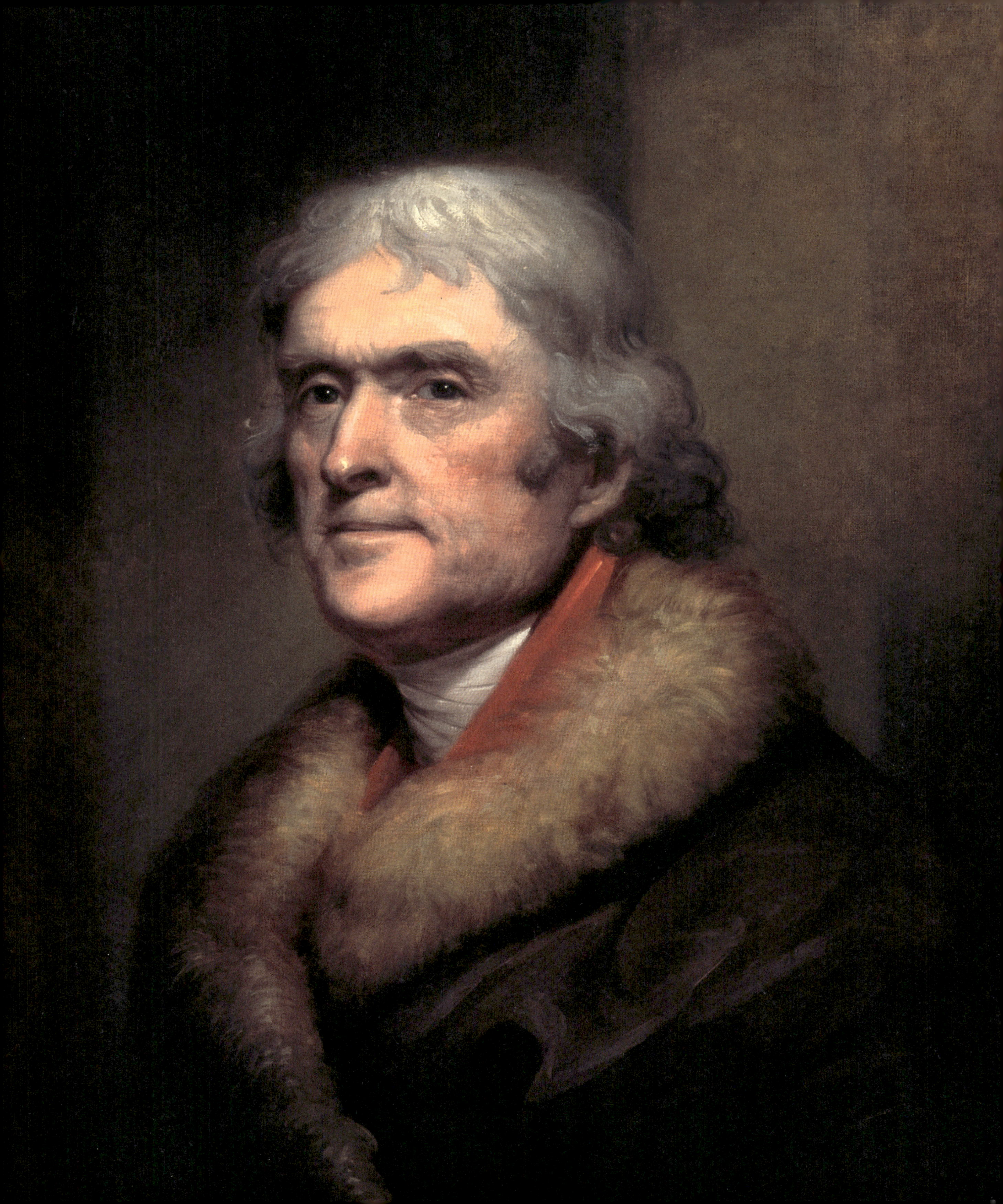

"I intend to enjoy the spectacle of a free people"

On May 19, 1804, the Spanish cargo ship *Concepción* cleared the capes of Delaware and sailed into the harbor at Philadelphia. On board was a thirty-four-year-old Prussian naturalist, Alexander von Humboldt, fresh from nearly five years spent traveling across South America and Mexico. Humboldt was eager to return to Paris, center of the European scientific world, but he had decided instead to extend his stay in the Americas and make a detour to the United States. Humboldt had several reasons for his decision. He was deeply interested in American democracy as an antidote to European monarchies. He was also curious to meet his North American scientific colleagues. But after learning of the sale of the Louisiana Territory to the United States, he was most interested in traveling to Washington to meet Thomas Jefferson. The American president had just doubled the size of the country and opened a new frontier for exploration. Humboldt believed that Americans would benefit from a deeper understanding of its new holdings, both in the name of science and in the cultural appreciation of the American landscape.

Jefferson had begun blazing that trail with his book *Notes on the State of Virginia*, published in 1785, arguing that the United States' natural wonders were equal to Europe's man-made monuments. Over time, icons of natural grandeur came to represent the scale and scope of America's cultural ambitions. Those landmarks became benchmarks of American society's potential for growth and prosperity. Appreciation of those natural monuments also served as a metaphor for the nation's aspirations. Humboldt recognized in the new country a political system in which the

Rembrandt Peale, *Thomas Jefferson* (detail), 1805. See p. 54.

CAT. 1 Friedrich Georg Weitsch, *Portrait of Alexander von Humboldt (1769–1859)*, 1806, oil on canvas, 49 ⅝ × 36 ⅜ in., Staatliche Museen zu Berlin, Nationalgalerie

freedom of its people could live up to its awe-inspiring landscape. He recognized an opportunity to explore that potential by meeting the American president and sharing with him his views.

The six short weeks that Humboldt spent in Philadelphia and Washington were of disproportionate significance to the United States. The timing of Humboldt's visit gave him an opportunity to shape U.S. negotiations with Spain over their shared border and provide concrete information regarding the natural and human resources of the newly added territory. That alone would have made his trip valuable to Jefferson. But Humboldt also saw America's future relevance vested in the quantity and caliber of its contributions to science, and more particularly to the explorer's theory of the unity of nature. Humboldt's enduring influence on Jefferson was to embolden his belief that America's cultural identity would be based in nature. Humboldt's interest in American scientific achievements and its political prospects would enhance the nation's standing in European circles as well. His stay in Philadelphia on both ends of this brief journey established his sustained presence in the American scientific community. His reception in Washington, however, cemented political friendships that would last for decades and influence American policies, ultimately helping to shape the nation's cultural identity.

For Humboldt, the trip had earned him the respect of the American scientific community and the promise of the statistics and measurements he did not have time to collect himself. For the United States, it offered validation of the new country by a man poised to become the most widely regarded public intellectual of his age. As part of Humboldt's far-flung network of colleagues and peers, American statesmen, explorers, artists, and writers would cultivate his friendship and actively seek his approbation. By the time Humboldt set sail for Le Havre in early July, the United States had adopted him as one of their own. The downstream consequences of that visit would resonate at home and abroad for the next fifty years.

HUMBOLDT'S DETOUR

First, Alexander von Humboldt (**CAT. 1**) had to travel there. He was in Cuba in April 1804, having just concluded five years of travel and research across South America and Mexico. He had dispatched trunks of materials along the way, and he kept voluminous diaries in which he recorded his observations, measurements, and ideas about how the entire planet functioned as one interconnected ecosystem—a radical concept at the time, and one that would define his significance as a natural philosopher for the rest of his long life. Humboldt was making arrangements to sail back to Europe when he met Vincent Gray, the American consul stationed in Havana. Gray discovered that the gregarious Prussian traveler was steeped in the available knowledge of the United States and had read Thomas Jefferson's *Notes on the State of Virginia* as part of his education in Hamburg. Humboldt expressed his admiration for Jefferson and for the promise of American democracy. The voluble scientist-explorer also revealed that he had just spent a year poring over documents in Mexican archives and had created a map of continental North America that showed—in greater detail than was known in the United States—exactly what the country had just acquired in the Louisiana Purchase.

Humboldt had undoubtedly heard the news of the acquisition of Louisiana while he was in Mexico. The Spanish king's fury at Napoleon's decision to sell his largest holdings in the New World to the Americans sent shock waves through Mexico. Humboldt had already been toying with the idea of making a side trip to Charleston, South Carolina, and Philadelphia, home to several of the country's most significant botanical gardens.[1] Gray recognized the geopolitical significance of Humboldt's map and his statistical data on the region. Obtaining a passport for Humboldt's entry into the United States had proved no problem.[2] However, he offered to send letters of introduction to President Thomas Jefferson and Secretary of State James Madison, thereby encouraging Humboldt to include Washington in his plans.

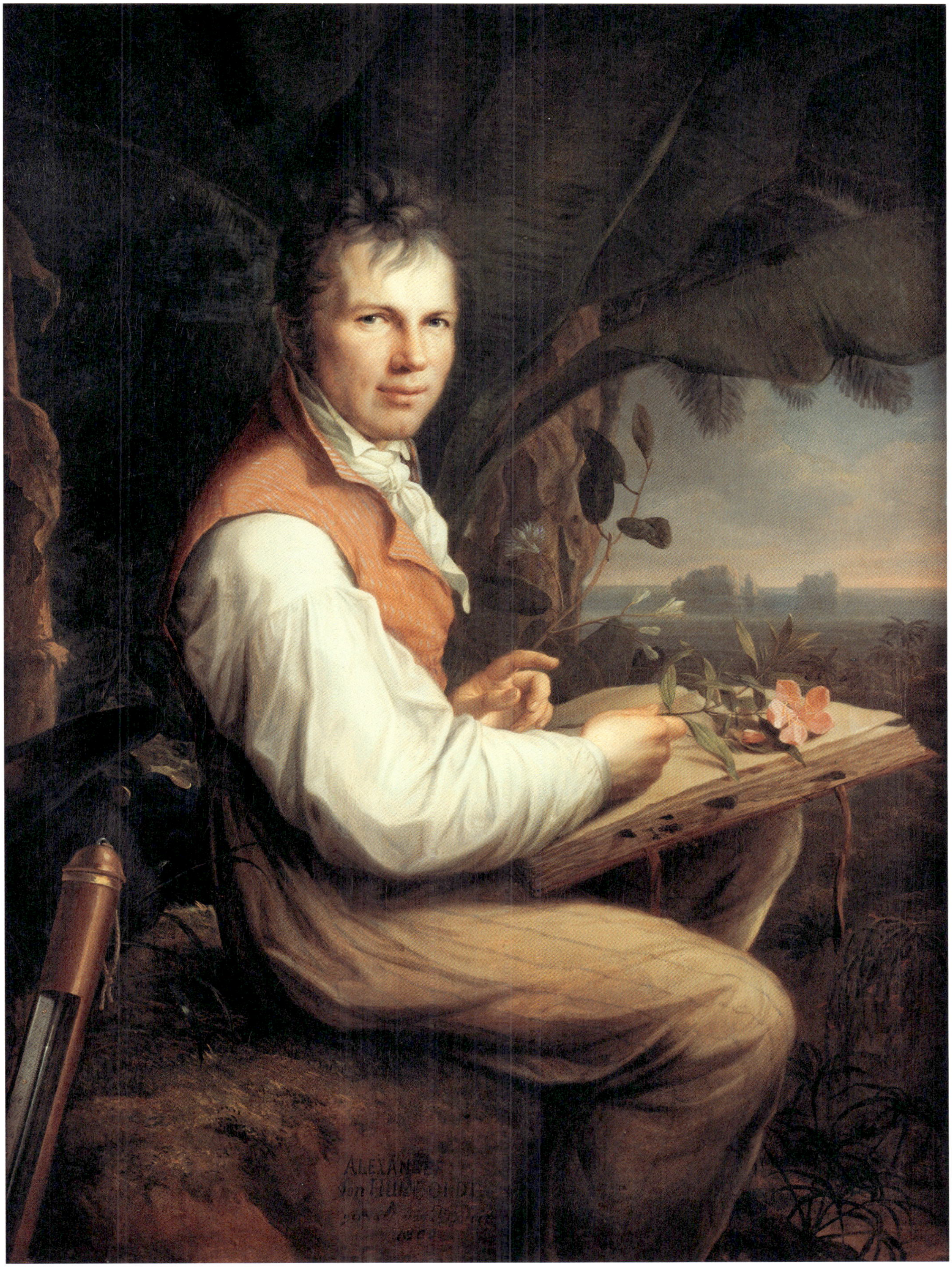

His astute thinking set in motion perhaps the most consequential visit by a European traveler in early U.S. history. In the larger scheme of things, the detour would cost Humboldt very little additional time and would certainly be mutually beneficial for the naturalist and the American president. How could Humboldt refuse?

Alexander von Humboldt had just completed an unprecedented study of New Spain, a part of the world usually closed to European travelers. Napoleon's incursions across Europe and his foray into Egypt during the 1790s had cut off most of the young Prussian's obvious travel routes. In 1799 Humboldt headed for Madrid and presented himself to the Spanish court, requesting permission to travel to the colonies in New Spain. Humboldt made a persuasive case, enlisting influential allies who had the king's ear, and offering to cover all of the costs and absorb all of the risks attendant on his ambitious venture. King Carlos IV of Spain relented and provided the Prussian explorer and his companion, French botanist Aimé Jacques Alexandre Bonpland, passports securing their entry into the Spanish colonies in South America. Five years and six thousand miles later, the two travelers, now accompanied by the young Ecuadorian nobleman Carlos Montúfar, completed their journey and arrived in Havana.[3] Humboldt and Bonpland had cached materials from an earlier part of their trip on the island, which they now retrieved.[4] Thanks to Gray's persuasiveness, all of this—the trunks, the journals, and the travelers—were bound for Philadelphia on a Spanish frigate.

Humboldt knew from Gray that the passage would not be an easy one; a British naval blockade extended along the East Coast of the United States, fallout from the abandonment of the Treaty of Amiens.[5] On top of that, Humboldt also feared everything he was traveling with might be seized by Spanish authorities before he reached Europe. But those hazards presented a less immediate danger than did the weather, as an early Atlantic storm threatened to sink the *Concepción* off the coast of Georgia, causing Humboldt to second-guess the wisdom of his impulsive decision to extend his voyage. In his journal Humboldt fretted:

I felt very much stirred up. To see myself perish on the eve of so many joys, to watch all the fruits of my labors going to pieces, to cause the death of my two companions, to perish during a voyage to Philadelphia which seemed by no means necessary (though undertaken in order to save our manuscripts and collections from the perfidious Spanish policies).[6]

The explorer's mood, however, lifted with the weather, making him impatient for the first sight of land on May 19, 1804, as the ship made its way past the capes of Delaware and into the harbor at Philadelphia. The entire passage had lasted almost a month, about half of the time Humboldt would spend on the mainland. While the trio waited to clear quarantine,[7] Humboldt wrote his letters to Thomas Jefferson and James Madison requesting permission to visit them in Washington.

Jefferson and Madison were waiting for those letters. While Humboldt and his companions were weathering the storm, Vincent Gray had alerted his boss, Secretary of State James Madison, of their imminent arrival. Mindful of the possibility of mail intercepted by British ships, Gray kept his comments fairly neutral, simply observing that he was confident Madison would find Humboldt a charming visitor and worthy of a warm welcome. His first dispatch described him as "Baron Humboldt, Supreme counsellor of Mines in the service of His Prussian Majesty and Member of the Royal Academy of Science at Berlin,"[8] which was not entirely accurate. Humboldt would eventually earn both the title and the full membership; for now, this description of his credentials was seemingly intended to bolster his chances of an audience with the American president.

On May 8, Gray wrote a second note to Madison, alerting him to more politically sensitive information on the way, cryptically noting,

In No. 220. you will find some Remarks of Baron Humboldt, who I lately recommended to your attention and protection; these can not be otherwise than interesting at this moment; particularly as he is now on his route to the city of Washington, and will have it in his power to give you much useful information relative to the country adjoining Louisiana.[9]

In that consular dispatch, Gray expanded on the underlying significance of Humboldt's visit: he was carrying statistics on New Spain that he had gathered during a year of research in Mexican archives. More important, Humboldt had drawn the single most detailed map of the North American continent's Spanish holdings, from Mexico all the way north to the British Canadian border. It contained far more detail than any map available to Jefferson and Madison when they had sent Meriwether Lewis and William Clark out West earlier that spring. But more important, Gray was convinced that Humboldt's map would provide information critical to the ongoing negotiations between the United States and the king of Spain to determine the new boundary between their two countries.

A career diplomat who had investments and land on the Spanish colony of Cuba, Gray understood the stakes involved. He recognized the potential significance of Humboldt's extended observations across New Spain, a part of the world normally walled off to outsiders. He also could not have failed to observe that Humboldt liked to talk—a lot—and he now had an audience appreciative of his stories and his plans for publishing his findings. Gray likely heard Humboldt expound on stories of where he'd been, what kind of governance he'd experienced, all of the interesting statistics he had amassed, what he had learned about pre-Columbian civilizations, and how he had been able to correct the location of landmarks and cities in South America and Mexico on existing maps, based on his having located the magnetic equator. Acapulco, he discovered, had been plotted a full four degrees off its actual location, and Humboldt's new measurements would prompt cartographers to completely update and redraw maps and charts for the New World continents.

Humboldt in turn would have been eager to catch up on current events—among them the successful overthrow of the French colony on the island of Saint-Domingue just four months earlier. Toussaint Louverture and General Jean-Jacques Dessalines had fought to achieve self-rule; after Louverture's capture and deportation to France, Dessalines had led the black army that established the independent republic of Haiti, the first country in the Americas governed under free black leadership. The repercussions were just beginning to be felt across the Caribbean islands and along the East Coast of the United States.

It is worth considering just how significant the acquisition of the Louisiana Territory was for the United States. Just three years earlier, in 1801, Spain had sold this tract to France, comfortable in believing it would remain in friendly European hands. But the racial unrest on Saint-Domingue erupted into outright revolution. The expulsion of the French troops, coming right on the heels of the French Revolution, so unnerved Napoleon that he sought to jettison his American holdings and pocket a sum capable of offsetting his growing war debts. In 1803 Napoleon shocked King Carlos IV of Spain by abandoning Saint-Domingue and selling a significant swath of the North American interior to the United States. The political landscape of the pan-Caribbean world, including the southern half of the United States, was in flux. The Spanish king was in no mood to provide terms favorable to Jefferson when determining the exact location of the new border.

Jefferson, too, had his eye on the politics in the Caribbean. Vincent Gray kept a close watch on developments in Cuba and Saint-Domingue and was likely aware in advance of Jefferson's plans to acquire Louisiana. Following a series of clandestine negotiations with France, the United States doubled the size of its holdings. But Jefferson and Madison knew precious little about the region composed of present-day Louisiana and Texas. As a result, Jefferson's purchase came in for serious criticism from his political enemies, who rightfully noted that he had little idea what he had just acquired, and even less leverage in negotiating the details with Spain. Humboldt's map would level that playing field, providing the Americans with just the advantage they needed to know what was at stake in their negotiations. Credible information regarding the topography, resources, and peoples in this territory would strengthen Jefferson's hand politically and provide needed intelligence for the somewhat hostile negotiations with the Spanish emissaries. As such, the timing of Humboldt's visit could not have been better. Jefferson and Madison had every incentive to welcome the Prussian traveler to Washington. Very rapidly they realized that there were plenty of other reasons to welcome him as well.

25

Printed by JOHN BIOREN, *No* 88, *Chesnut Street, For* SAMUEL EMERY, *Ship Broker, Next door to the Custom House, Philadelphia.*

BAGGAGE ENTRY.

Entry *made by* Baron Humbott *of* Baggage *being the Property of* himself, Mr Bonpland & Montufar *imported in the* Ship Concepcion *whereof is Master* M I de Acosta *from* Havana

MARKS.	NUMBERS.	PACKAGES AND CONTENTS.
		Two Trunks wearing apparell
		three Beds
		~~Twenty seven Boxes Plants & Collections for Natural History~~

District of Philadelphia.

I Baron Humbott do solemnly, sincerely and truly swear that this Entry contains to the best of my knowledge and belief, a just and true account of the contents of the Trunks mentioned herein, imported in the Ship Concepcion whereof is master M I Da Acosta from Havana that the contain no goods, Wares or Merchandise whatever other than wearing apparel and other personal Baggage & Collections as aforesaid being the Property of myself Mr Bonpland & Montufar who have in said vessel arrived in the United States, and are not, directly or indirectly, imported for any other Person or Persons, or intended for Sale.

SO HELP ME GOD.

Sworn this 24 Day of May 1804

before me COLLECTOR.

Baron of Humbolt

CAT. 2 Alexander von Humboldt, Baggage entry statement to the U.S. Customs, May 24, 1804, ink on paper, 9 ⅜ × 7 ⅞ in., American Philosophical Society

AMERICA'S SCIENTIFIC CAPITAL

Philadelphia was the scientific and cultural capital of the United States, despite having lost the designation of the nation's capital to Washington in 1800. Benjamin Franklin had founded the American Philosophical Society (APS) in 1743, the first professional scientific society in the British colonies. Even then the APS had been a destination for European scientific travelers and extended honorary memberships to a number of them.[10] This was an early form of intellectual networking and a show of respect designed to forge close ties among scientists and facilitate the sharing of papers and books between institutions. Humboldt was already familiar with the organization, whose members had been part of an international exchange of ideas with the Royal Society in London and the Institut National in Paris. The members of the APS were Humboldt's peers, and they were eager to meet him. One of them served as the customs inspector on the Philadelphia docks. Zacchaeus Collins was out of town when Humboldt penned him a note, impatient to clear quarantine. When Collins returned three days later and read Humboldt's letter, he moved with alacrity to clear the *Concepción*, appending a note to their permit that read

> *it is with regret they find that men to whom Science & learning are so much indebted as H[umboldt], M[ontúfar] & B[onpland] should have been subjected to the slighter inconvenience.*[11]

When they finally cleared customs on May 24 with their personal belongings, APS member Charles Willson Peale greeted them on the docks. Humboldt's baggage entry statement (**CAT. 2**) lists him and his companions and twenty-seven trunks of natural history specimens that he arranged to keep stored there for retrieval upon their eventual departure for France.[12]

Humboldt posted his letters to Jefferson and Madison, and the recently arrived trio began their tour of Philadelphia, often accompanied by members of the APS. Over the next five days they visited each of the city's scientific and cultural centers. When the capital had moved to Washington, the rooms in Independence Hall that had been used by Congress had become part of Charles Willson Peale's museum. Next door was Philosophical Hall, home to the APS. Facing the APS was the Library Company, and nearby was the College of Physicians. In each place Humboldt was introduced to the most current American scientific research.[13] News traveled quickly; within a few days the local newspapers started printing excerpts from letters Humboldt had sent to colleagues in Europe during his lengthy absence. In one scientific journal Humboldt read with relief of the "Arrival of M. de Humboldt's manuscripts at his brother's house in Paris, by way of Spain."[14] Now Humboldt knew that at least one advance shipment of materials had survived its transatlantic voyage.

Outdoors, Humboldt, Bonpland, and Montúfar toured several internationally famous horticultural enterprises. One of their first stops was Bartram's Garden, "the usual resort of botanists."[15] John Bartram had been the most widely recognized American participant in the robust international plant and seed exchange. Now his sons William and John maintained the business. Botany had been one of Humboldt and Bonpland's primary motivators for travel, as they collected and compared species of plants found at a similar altitude in Europe and the Americas. Humboldt had set out for New Spain convinced he could use plant specimens to draw connections between disparate places around the globe, and in his trunks was the proof of that concept. After his return to Paris, Humboldt's expanding idea of the unity of nature would take its earliest visual form in 1805, in his *Naturgemälde*, or Picture of Nature (see cats. 15, 16), where he began to correlate climate and altitude with characteristic plants in every place he had visited. On this brief visit to the United States, Humboldt was unable to spend time exploring. However, he could learn a great deal from Bartram's Garden about how North American plants would fit into his emergent global paradigm.

CAT. 3 Charles Willson Peale, *Self-Portrait with Mastodon Bone*, 1824, oil on canvas, 26 ¼ × 22 in., New-York Historical Society, Purchase, James B. Wilbur Fund

There was another kind of garden in Philadelphia, one that carried a deliberate patriotic message. While Congress was meeting in Independence Hall, gardener Samuel Vaughan had envisioned a space that brought together distinctive plants and trees from all thirteen of the original states as an American garden through which to teach North American botany and horticulture.[16] The identification, cultivation, and propagation of plants native to the United States would educate Americans about the worth of these features. Particularly at this moment, identifying unique plants, animals, and landscape features drove exploration and scientific inquiry, part of the process of laying claim to a specific American identity. Humboldt had focused on how native plants occurred naturally in a landscape. Cultivated gardens, on the other hand, could manipulate those plantings to visualize a political message. Humboldt would have understood the desire to remove any European species of plants, both in support of this pedagogy and to subtly affirm the solidarity of the former colonies after declaring their independence.

Among the eminent foreign scientists to whom the APS had extended honorary membership was Johann Friedrich Blumenbach, Humboldt's mentor at Göttingen. The same members who had voted Blumenbach in back in 1798 were now showing Humboldt around Philadelphia.[17] Several of his hosts had strong ties linking them to Humboldt's network of European colleagues. American botanist Benjamin Smith Barton had earned a medical degree from Göttingen in 1789, the same year Humboldt began his studies there. James Woodhouse maintained close ties with members of the Royal Society in London who had met Humboldt during his visit in 1790. Physician and budding paleontologist Caspar Wistar Jr. corresponded with Humboldt's colleague Georges Cuvier at the Jardin des Plantes in Paris.

Humboldt dove eagerly into the intellectual community in Philadelphia. He had for five years immersed himself in fieldwork, travel, and research. Finally, on the cusp of returning home, he had an audience that would appreciate his stories of his trek through the Andes and his ascent of Chimborazo. As fellow men of science, they appreciated Humboldt's emphasis on accurate measurements and his attention to the care of his delicate instruments. He explained how his measurements of longitude and latitude would correct the placement of landmarks and major cities on maps and how his elaborate drawings of pre-Columbian imagery— including an Aztec calendar—would overturn long-held prejudices that such civilizations lacked cultural sophistication. He enjoyed telling them how he nearly poisoned himself with curare and how he and Bonpland had shocked themselves handling electric eels. During this first week that Humboldt, Bonpland, and Montúfar spent in Philadelphia, Humboldt breathed new life into the city. John Vaughan shared Humboldt's interest in indigenous peoples' languages and cultures, and Vaughan had supported an early effort to send French botanist André Michaux west to bring back information on new plant species.[18] Several of the members had been instrumental in teaching Meriwether Lewis some of the skills he needed to lead the Corps of Discovery, among them Benjamin Rush, who had coached Lewis on medical techniques and essential supplies, and Robert Patterson, who had taught Lewis celestial navigation and had helped choose the instruments Lewis took with him.[19]

By the time he and his companions embarked for Washington, Humboldt had endeared himself to these men and their colleagues, pouring out his stories of exploration and adventure, proudly announcing his discoveries and the significance of his measurements. Language was only an intermittent and temporary barrier; Humboldt spoke German and French fluently and had a decent command of English and Spanish. By the end of his life he would be capable of understanding no fewer than eight languages. Rather than stop to search for a word, he simply switched back and forth between languages. Although some in this highly educated orbit understood French as well as English, APS librarian Nicholas Collin ruefully noted the limited linguistic ability of so many

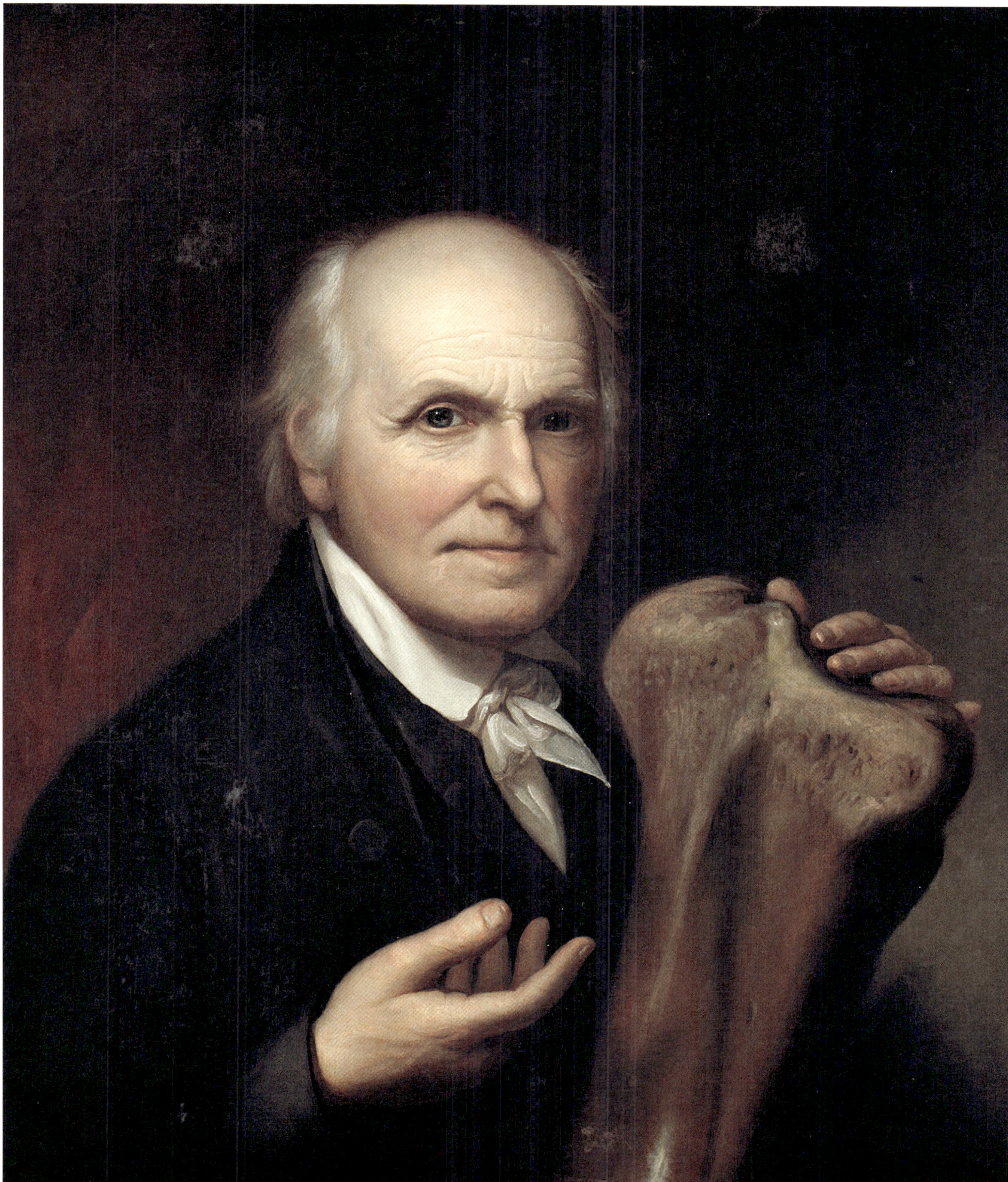

Americans, especially when it came to German, and how that limitation influenced which publications ended up on the APS library shelves.[20]

Charles Willson Peale (CAT. 3) seemed impressed by Humboldt's multilingual fluidity, noting in his journal:

> *The Baron spoke English very well, in the German dialect.... he possessed a surprising fluency of Speach, & it was amusing to hear him Speak English, French and the Spanish Languages, mixing them together in rapid Speach. He is very communicative and possesses a surprising fund of knowledge, in botany mineralogy astronamy Philosophy and Natural History: with a liberal Education, he has been collecting information from learned men of allmost all quarters of the world; for he has been travelling ever since he was 11 years of age and never lived in any one place more than 6 months together, as he informed us.*[21]

Although Humboldt continued to combine languages in conversation the rest of his life, his choice of languages here was not arbitrary: Bonpland was fluent in French and Spanish, Montúfar in Spanish alone; however, Peale observed that neither man could speak or understand much English.[22] Humboldt's constant switching between languages might have been as much a courtesy to them as evidence of his ebullient polylingual style.

"THE PHILOSOPHIC PARTY" GOES TO WASHINGTON

The members of the APS were reluctant to say goodbye to Humboldt. Several of them decided to accompany him to Washington, each having discovered shared interests with their guest. John Vaughan dubbed them "The Philosophic Party." Charles Willson Peale made the travel arrangements, eagerly taking responsibility for getting Humboldt and his companions to the President's House. The other three members of the APS were James Woodhouse, Anthony Fothergill, and Nicholas Collin. Woodhouse had been hired as a professor of chemistry in 1795 at the University of Pennsylvania, where he ran a laboratory and thrived on conducting sometimes dangerous experiments, much as Humboldt was wont to do with his own galvanic studies.[23] Woodhouse had experimented with Peruvian bark, or China bark, a colloquial term for the bark of a cinchona tree, which yielded quinine, which was used to treat yellow fever. Yellow fever swept through the city, as well as the Southern states, nearly every fall beginning in 1793.[24] The cinchona was a topic of particular interest to Humboldt as well as to these Philadelphia scientists. All across South America, Humboldt and Bonpland had compared different varieties of cinchona trees, proudly concluding "nobody has inspected so many different types of China bark as we have."[25] Woodhouse also shared Humboldt's interest in the origins of the basalt columns on the isle of Staffa, now part of Scotland, and was conversant about purported volcanic formations in North Carolina.[26]

In 1802 Woodhouse had traveled to London, conferring with members of the Royal Society who had met Humboldt in 1790.[27] There he had met Anthony Fothergill, a medical doctor from Bath, England, who immigrated to Philadelphia in 1803. Fothergill had been a member of the Royal Society since 1778, and in 1792 had been made a foreign member of the APS. Like many of his colleagues, he had kept a weather diary both in England and then in the United States, amassing data Humboldt would fold into his emerging vision of global ecosystems.[28] A relative newcomer to the United States, Fothergill was taking advantage of the opportunity to meet Jefferson as well, and he was carrying a letter of introduction from his close friend, Dr. Benjamin Rush.[29]

Nicholas Collin, a native of Sweden, had been close friends with APS founder Benjamin Franklin and was rector of the Swedish Church in Philadelphia. His interest in what was called "natural philosophy" inspired him to deliver an address at the APS on April 3, 1789, titled "An Essay on Natural Philosophy and Its Relationship to the Development of the New World," in which he argued passionately for the preservation of mature forests because he considered them emblematic of the American character: "Our stately forests are a national treasure, deserving the solicitous care of the patriotic philosopher and politician." He then urged his audience to "regard them as principal ornaments of their country; and while they clear a part for the purpose of agriculture, leave those hills crowned with towering pines, and stately oaks." He observed that the present generation had been so consumed by the revolution and the need to establish the nation that there was little time devoted to admiring America's natural wonders. He felt that future generations would likely see this issue more clearly, being "of better taste," and assumed they would be able to step back and admire the country's natural beauty.[30] Collin identified the key component in America's appreciation of its wilderness: the time spent in aesthetic reflection, fostering the identification of nature as a core component of civic and national pride.

For Collin, the vast swaths of dense forests were a landmark feature of the New World. Old-growth forests in Europe had fallen to the ax or represented a heathen danger to the cultivated social order. Jefferson had similarly argued that natural landmarks of all kinds deserved national recognition and admiration. In Venezuela, Humboldt had witnessed changes in the local climate due to extensive deforestation, and he was alarmed at the speed with which the United States was converting old-growth forests to farmland.[31] Collin shared Humboldt's concern for the deleterious effects of deforestation, demonstrating an early awareness of the consequences of poor land management practices. Collin was a frequent speaker at the APS, and many of his topics aligned with Humboldt's interests, notably on climate, natural philosophy, botany, meteorology, archaeology, ethnography, and the cause of yellow fever.[32] He shared Humboldt's enthusiasm for the potential for new discoveries in natural history in the United States, exclaiming, "What treasures may we not then expect in this new and vast division of the globe!"[33]

Collin believed the United States needed museums and gardens to teach the populace. Charles Willson Peale had founded just such a museum in Philadelphia in 1786 and had opened a second public exhibition space in Baltimore in 1796. The sign over the museum's front door in Philadelphia encouraged visitors to enter "the great school of nature," and "with the book of Nature open, explore the wondrous world, an institute of laws eternal."[34] Peale's stated goal was to provide Americans with an education that was both practical and uplifting, bringing together natural history specimens, American Indian artifacts, and a gallery of portraits of eminent American military heroes, scientists, and intellectuals. He envisioned it as "an institute of laws eternal" and a shrine to post-revolutionary America, an exemplar of what a democratic citizenry should know. Visitors learned civics surrounded by American flora and fauna and were encouraged to contribute specimens that would enhance the collections and "in such manner influence public opinions that republicanism will be highly promoted."[35] Integral to the mission of Peale's Museum was cultivating an awareness of American national identity. In that sense, he deemed its educational mission worthy of federal support. He had made this case to Jefferson on numerous occasions in the past, arguing for the need for a national institute based on natural history. Now he hoped that Humboldt might champion his cause as well.

Jefferson had received his letter from Humboldt on May 28, 1804, and although he penned an immediate reply, his letter welcoming the baron had not arrived by the time that the Philosophic Party had left for Washington.[36] Partly on account of this delay, and partly to express their delight in having made Humboldt's acquaintance, several APS members wrote their own letters of introduction, to be hand-delivered to Jefferson when the party arrived, to ensure that audience. John Vaughan detailed the group's travel plans, alerting Jefferson,

> *Mr. Peale, who with Dr. Collin, Dr. Fothergill, The famous Mr. Humboldt of the Berlin Acady. who has been so long travelling in South Ama. & his two Companions of his travels…leave this place tomorrow for Baltimore & Washington—The latter three are on their way to Europe in order to give publicity to their observations & the results of their travels.*[37]

Caspar Wistar Jr. sketched the topics of conversation that Humboldt had shared with the group, focusing on exploration and discoveries that he found exciting. He urged the president to welcome

> *[t]he Baron Humboldt [who] has just returned from an expedition to South America & to Mexico where he has spent five or six years passing from the height of Chimborasso & Phinchinka to the bottom of the Mines, navigating the Oronoko and the River of the Amazon traversing forests of the Chincona* [sic] *or Peruvian Bark Tree & investigating the history & antiquities of the ancient Mexicans—These Gentlemen will recommend themselves much more than it is in my power to do.*[38]

Monsieur le President,

Arrivé depuis le Mexique sur le sol heureux de cette Republique dont le Pouvoir executif à été confié a Vos lumieres, c'est un doux devoir pour moi de Vous présenter mes respects et l'hommage de la haute admiration que Vos écrits, Vos actions et la liberalité de Vos idées m'ont inspiré dès ma plus tendre jeunesse. Je me flattais de pouvoir Vous les

CAT. 4 Letter from Alexander von Humboldt to Thomas Jefferson, May 24, 1804, ink on paper, 9 ⅛ × 7 ¾ in., Library of Congress, Manuscript Division, Thomas Jefferson Papers

Humboldt's ascent to 19,413 feet on Chimborazo, then considered the tallest mountain in the world, was a truly impressive feat, accomplished in period dress, carrying bulky instruments, and with no supplemental oxygen. It was a mountaineering record that stood for thirty years. Benjamin Smith Barton noted:

I am persuaded that I need not offer any apology for introducing to your knowledge and attentions, the explorer of South America, and one of the most intelligent and active philosophers of our time. We all regret that his stay among us is to be so short.[39]

Barton's wistful remark expressed what each man experienced: he felt energized and relevant in Humboldt's company.

On the morning of May 29, this group boarded the first leg of the mail stage toward Baltimore.[40] Peale took command of the travel plans, which gave him the chance to be at the center of events. Peale carried with him a polygraph machine he had invented, allowing Jefferson to make duplicate copies of his correspondence. Now he could ensure that the delicate instrument would make the trip safely.[41] Together the six of them traveled by stage and ferry overland from Philadelphia to Baltimore, where they stopped and stayed two days, meeting with two of Peale's sons, Rembrandt and Rubens, and his daughter, Angelica, who were living there. Together with Angelica and her husband, Andrew Robinson, the group toured the city, went to dinner, and attended the theater.[42] They also indulged their curiosity by boarding a packet boat, a new form of public transportation that avoided Maryland's notoriously poor roads.[43] As they bounced along the rutted stage route, Peale worried whether his polygraph machine would survive the trip. Humboldt had similarly fretted over his delicate instruments packed on mules as he and his companions crossed the Andes. Finding better, smoother, and faster ways to transport people and instruments and communicate ideas became one of Humboldt's lifelong interests.

A stickler for details, Peale maintained a daily diary in which he kept track of the cost of each stage and ferry, made sure his charges stayed together en route, and worried about everything from the food to the state of the roads. Peale's diary is simultaneously the most complete account of Humboldt's trip to the United States and the most vexing. Focused on transportation logistics, worries about his dentures, and his discomfiture over others' drinking habits, Peale often leaves his reader wondering about the substance of the conversations he witnessed among Humboldt and the scientists and statesmen that he met. Although Peale does not mention it in his diary, he took Humboldt and his companions to the nascent Peale museum in Baltimore, an extension of the family enterprise that was run by Rembrandt and Rubens.[44]

"MONSIEUR LE PRÉSIDENT…"

Humboldt had carefully crafted his letters to Jefferson and Madison, both to ensure his invitation to meet them and to press on the issues that concerned him most about the fledgling nation. In both cases Humboldt introduced himself by outlining the highlights of his travels, stressing his accomplishments. Then he touched on every topic he felt would pique their interest and ensure a favorable response.[45] To Madison, he stressed the benefits of American democracy as a champion of human liberty:

I happen to pass through this country on my way to France because of my intense interest in the human welfare of the United States which resulted from your wise legislation and the civic virtues of your magistrates. After having witnessed the great spectacle of the majestic Andes and the grandeur of the physical world I intended to enjoy the spectacle of a free people worthy of a great destiny. You, Sir, who as statesman exerted such a benevolent influence on this destiny will, I pray, receive the expression of my respectful affection.[46]

Humboldt maintained a lifelong commitment to social justice and his belief in the equality among all races. His travels had deepened his appreciation for the potential of America's democracy as an alternative to oppressive European monarchies and the damage inflicted by colonial rule.

His letter to Jefferson (CAT. 4) was much longer and far more detailed, both in its enumeration of his accomplishments and his reasons for visiting. He opened with a statement that expressed his admiration for Jefferson and alluded to the similarities he already perceived between the two of them:

Mr. President, Arrived from Mexico on the blessed ground of this republic, whose executive powers were placed in your hands, I feel it my pleasant duty to present my respects and express my high admiration for your writings, your actions, and the liberalism of your ideas, which have inspired me from my earliest youth.[47]

In a single sentence Humboldt had described their shared traits, notably a willingness to actively pursue their democratic ideals in word and deed. He continued with a variant on the language he had used in writing to Madison:

I could not resist the moral obligation to see the United States and enjoy the consoling aspects of a people who understand the precious gift of Liberty.[48]

Now, Humboldt continued, he had delayed his journey home at some risk to his life and his findings in order to pay his respects in person to Jefferson—as an author, a statesman, and most important, as a fellow natural philosopher. He invoked the topic on which he hoped to make it impossible for Jefferson to pass up the opportunity for the two men to meet:

As a friend of science, you will excuse the indulgence of my admiration. I would love to talk to you about a subject that you have treated so ingeniously in your work on Virginia, the teeth of mammoth which we discovered in the Andes of the southern hemisphere at 1,700 toises *above the level of the Pacific Ocean. My friend Citoyen Cuvier may be expected to furnish me with an anatomical description. I hesitate to take more of your time and will confine myself to sending you the assurance of my deepest respect with which I have the honor to be, Mr. President, your very humble and very obedient servant, Baron Humboldt, Member of the Berlin Academy of Science.*[49]

Humboldt offered to fold Jefferson into his scientific network in Paris by mentioning Cuvier as the man who could make sense out of the mammoth. Having just seen Peale's nearly complete skeleton of the behemoth in Philadelphia, Humboldt knew this was a topic of much significance to the president. Being at that time neither a baron nor a full member of the Academy, Humboldt may have doubted that he had arrived at a stature that would automatically confer an invitation to meet the American president. Yes, he had been publishing his early research since the 1790s, and the APS held some of those essays and papers in its library, but he had no guarantee that Jefferson was aware of them. However, he had taken Jefferson's measure well, emphasizing the kindred nature of their personalities and the extent of their shared interests.

Humboldt had traveled to Washington specifically to meet Thomas Jefferson, but even that goal had multiple aims. Jefferson's *Notes on the State of Virginia* was a case study for American independence couched in the physical geography of his home state. The author had written a polemic specifically to address the derisive assertions of leading French natural philosopher George-Louis Leclerc, comte de Buffon, regarding the Americas; it simultaneously served as a patriotic manifesto for his home audience. Humboldt shared Jefferson's opinion of Buffon, who at that time was among the most highly regarded European naturalists, but who had never set foot on American soil. Humboldt had amassed his own data while traveling in the Americas that naturally expanded on Jefferson's narrower defense. Jefferson had deployed his data to support his argument for the robust prospects for the new nation's politics as well as its agricultural potential. Using the American landscape as a synecdoche for the nation, Jefferson had taken the first steps toward defining its goals in metaphorical rather than cultural terms. He would find in Humboldt an influential ally.

Humboldt would not have the chance to see Jefferson's response until the end of his trip in the United States. In it, the president delicately touched on the information about the Louisiana Territory with which Humboldt traveled. He expressed his hope that Humboldt would find American democracy to his liking and ruefully noted the comparatively unfinished state of Washington as the new capital, compared with Philadelphia. In welcoming him to Washington, Jefferson wrote:

[T]he countries you have visited are of those least known, and most interesting, and a lively desire will be felt generally to receive the information you will be able to give. No one will feel it more strongly than myself because no one perhaps views this new world with more partial hopes of it's [sic] exhibiting an ameliorating state of the human condition. In the new position in which the seat of our government is fixed we have nothing curious to attract the observation of a traveller and can only substitute in it's place the welcome with which we should receive your visit, should you find it convenient to add so much to your journey.[50]

Humboldt's primary concerns—to meet a fellow man of natural science, to experience American democracy, and to extend his growing network of correspondents who could assist him in building his concept of the unity of nature—made him a welcome presence in Washington.

HUMBOLDT IN WASHINGTON, THE "VAST SERBONIAN BOG"

On the evening of June 1, the Philosophic Party arrived in Washington, D.C., along the Bladensburg stage route. Their destination was Stelle's Hotel and City Tavern, an enormous complex that served as a hotel, Congressional boardinghouse, stables, and site of the celebratory party Jefferson had hosted upon acquiring Louisiana.[51] Peale got the group settled in their lodgings and alerted Jefferson's adviser, the doctor and architect William Thornton, as well as newspaper editor Samuel Harrison Smith of their arrival.[52] The following morning Peale made the two-mile-long trek down to the White House to arrange Jefferson's first visit with Humboldt, returning to the hotel to retrieve the group and escort them to meet the president.[53]

Fig. 2
Benjamin Latrobe, *North view of the President's House in the city of Washington*, ca. 1810, drawing, Library of Congress, Prints and Photographs Division, Frances Benjamin Johnston Collection

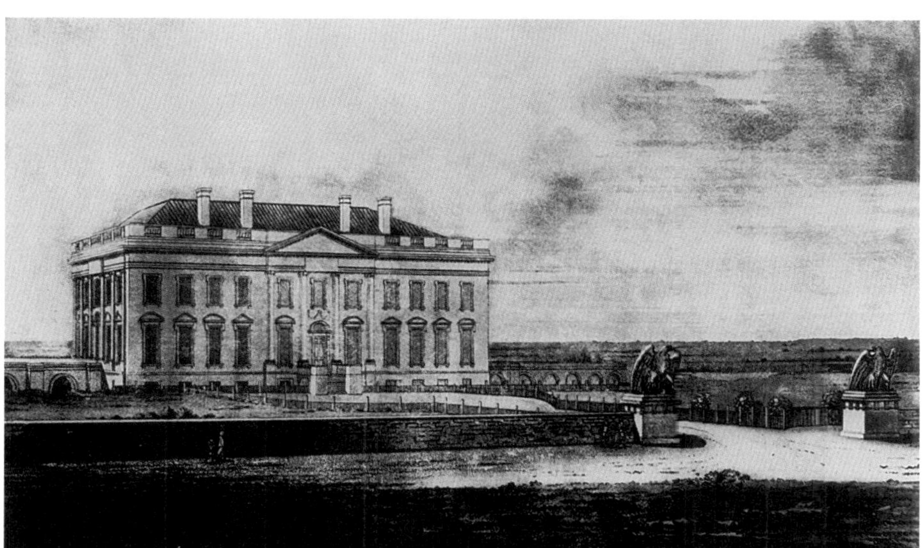

Washington was more an aspiration than a realized city; surveyors Andrew Ellicott and his skilled African American assistant, Benjamin Banneker, had drawn up a plan for its controlled growth and systematic layout of streets, but at present those streets were unpaved, undulating, and prone to flooding.[54] George Washington had taken a personal role in situating the newly designated Federal City and engaged French-American military engineer Pierre L'Enfant to design a city emblematic of American democracy. Congress occupied the highest ground, not the president, intimating that power resided with the people and not with a single head of state. The Mall established the primary axis, a vast landscaped area open to the public. Washington envisioned the wide boulevards and monumental buildings as symbols of American democratic ideals, suggesting the durability of the new nation. Peale did note with approval some of Jefferson's early improvements, among them two rows of black poplars planted on either side of the carriage road designated as Pennsylvania Avenue. As Jefferson had noted, the trees would take longer to grow and perforce must precede the establishment of paved roads. The choice of poplars was no accident; Jefferson knew that in Italy the Lombardy poplar was affectionately referred to as the "tree of liberty." The trees were acquired from George Washington's estate, Mount Vernon, lending an additional symbolic flourish to the new capital.[55]

Pierre L'Enfant had started to raise the newly widened roadbed using stone chips and laid a gravel path on one side and stone pavers on the other, suggesting the beginnings of passable sidewalks between the Capitol and the President's House.[56] However, Tiber Creek still cut across the landscape, making transportation dodgy, and both the Capitol and the President's House (fig. 2) were still under construction. It was a far cry from the more polished Philadelphia, and much smaller, with a population estimated at four thousand compared to Philadelphia's seventy-five thousand individuals.[57] Humboldt was familiar with the

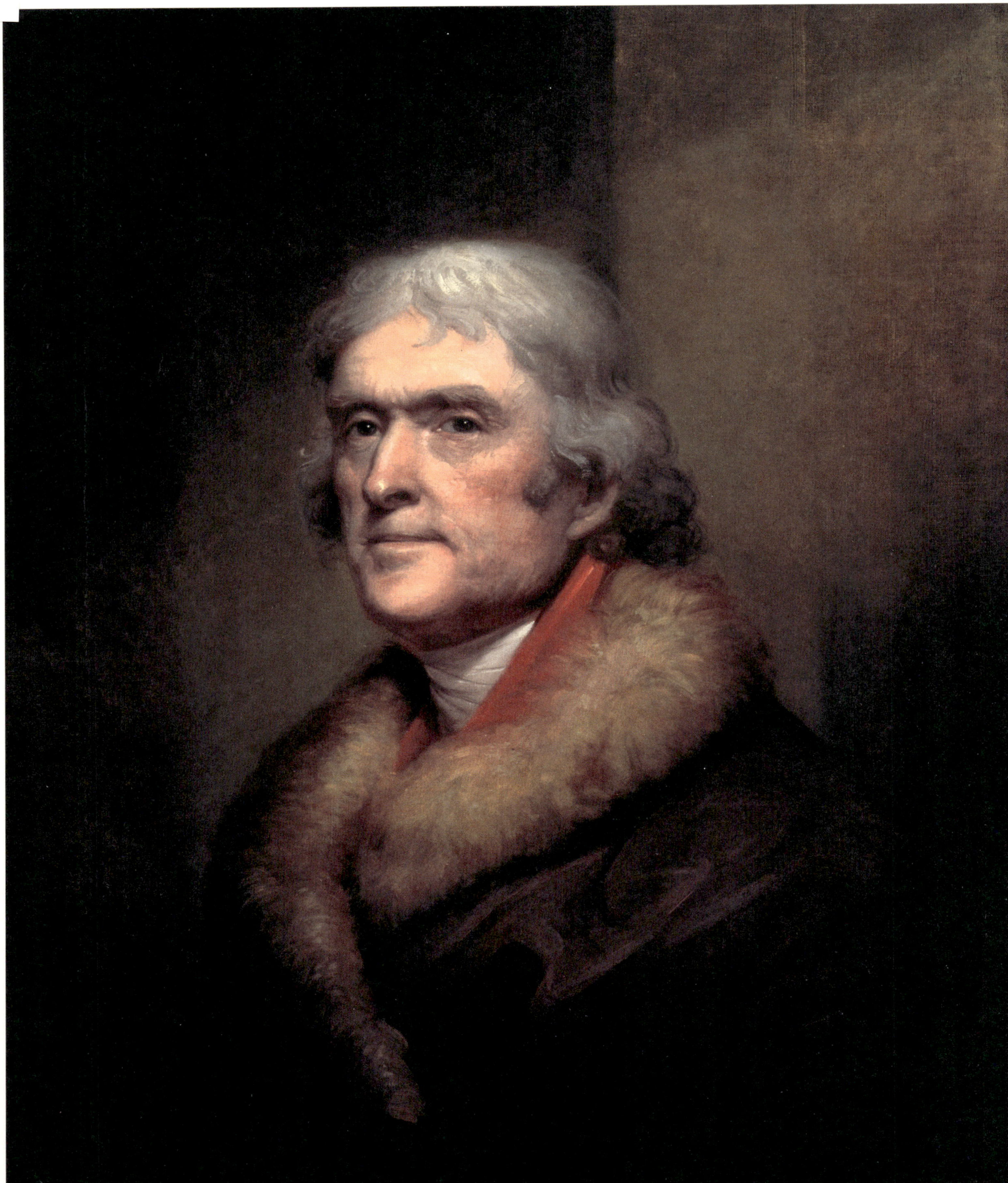

CAT. 5 Rembrandt Peale, *Thomas Jefferson* (1743–1826), 1805, oil on linen, 28 × 23 ½ in., New-York Historical Society, Gift of Thomas Jefferson Bryan

great European capitals of Berlin, Paris, and London; more recently he had just spent a year in Mexico City. Washington, D.C., retained a refreshing sense of wilderness in its rude contours and untamed meadows dotted with unfinished buildings. It, like the country, was still very much a work in progress.

Throughout the next week, Peale ricocheted across the city, making sure everything was in order, suggesting his somewhat nervous and compulsive need to provide structure and meaning to the trip. The first meeting between Jefferson and Humboldt was a study in contrasts. Jefferson, the tall, lanky, sixty-year-old president, his red hair turning gray, maintained a pleasantly informal demeanor. Rembrandt Peale's portrait of him (CAT. 5), made the following year, stressed that relaxed persona.[58] His inattention to fancy clothes and strict protocol vexed his enemies and even some of his friends, but the man who became known as the Sage of Monticello did not divert much energy to appearances.[59] William Burwell, Jefferson's private secretary now that Meriwether Lewis was in the field, described Humboldt as "about thirty of small figure, well made, agreeable Physiognomy, simple unaffected manners, remarkably sprightly, vehement in conversation, & sometimes eloquent; he spoke a variety of languages, the English among the rest."[60] The two men took a liking to each other immediately. Peale noted, "The President had known before the high reputation of the Baron as a Philosopher and ingenious observer of nature, that had traveled a great deal."[61] Humboldt's delight must have been palpable; having not received Jefferson's letter welcoming him to Washington, he could now lay that fear to rest. During the next two weeks, the two men established a lasting friendship, their personalities and interests uncannily similar. Jefferson's comfortable informality was both appealing and indicative of the kindred nature of their friendship.

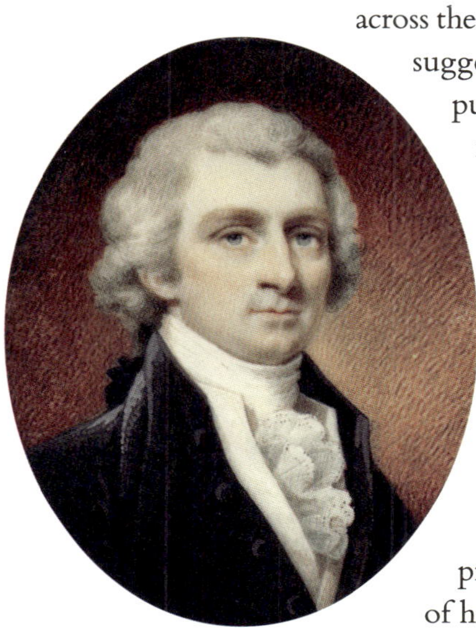

Fig. 3 Robert Field, *William Thornton* (1759–1828), ca. 1800, watercolor on ivory, 2 ⅞ × 2 ⅜ in., Smithsonian American Art Museum, Museum purchase through the Catherine Walden Myer Fund

Aside from two diplomatic encounters with French consul Louis-André, Baron Pichon and British ambassador Anthony Merry, Humboldt spent his days visiting the most impressive buildings in Washington, which housed the political, military, and intellectual prospects for the new nation, often in the company of some of Washington's most important residents.[62] The Navy Yard, where Humboldt inspected the armaments and boarded one of the frigates being outfitted for service in the Mediterranean, a reminder of the problems of piracy near Tripoli, represented the military arm of the city.[63] One morning Peale toured the Patent Office Building with Humboldt, where William Thornton (fig. 3) served as the commissioner of patents in addition to his duties as architect and adviser to Jefferson.[64] In this building, models of American inventions were on display to encourage American ingenuity and creativity. The fine arts were being encouraged, to a degree; Gilbert Stuart had taken a temporary studio in the city, and on two separate occasions Humboldt tagged along while Peale and the other members of the APS dropped in to do what they could to encourage the prickly artist to complete his posthumous portrait of Joseph Priestley. Vaughan had written to Thornton to see if he could prod Stuart into completing the painting, now months overdue; but as Peale noted of his rival, "Mr. Steuart is seldom in a hurry to finish his pictures."[65]

During George Washington's lifetime, the plan for the city was ambitious and designed to convey a sense of command, control, and dignity for a nation born in revolution. The decision to move the capital of the United States to the site of Washington, D.C.—beyond the deal struck by Jefferson, Madison, and Hamilton to enact that decision—was to create a purpose-built city intentionally designed to be a monument to the potential for the new nation. The President's House and the Capitol building represented the seat of political power. Thomas Law, a real estate developer and one of Jefferson's influential supporters, gave the group a tour of the unfinished Capitol building.[66] There they walked through the library and the

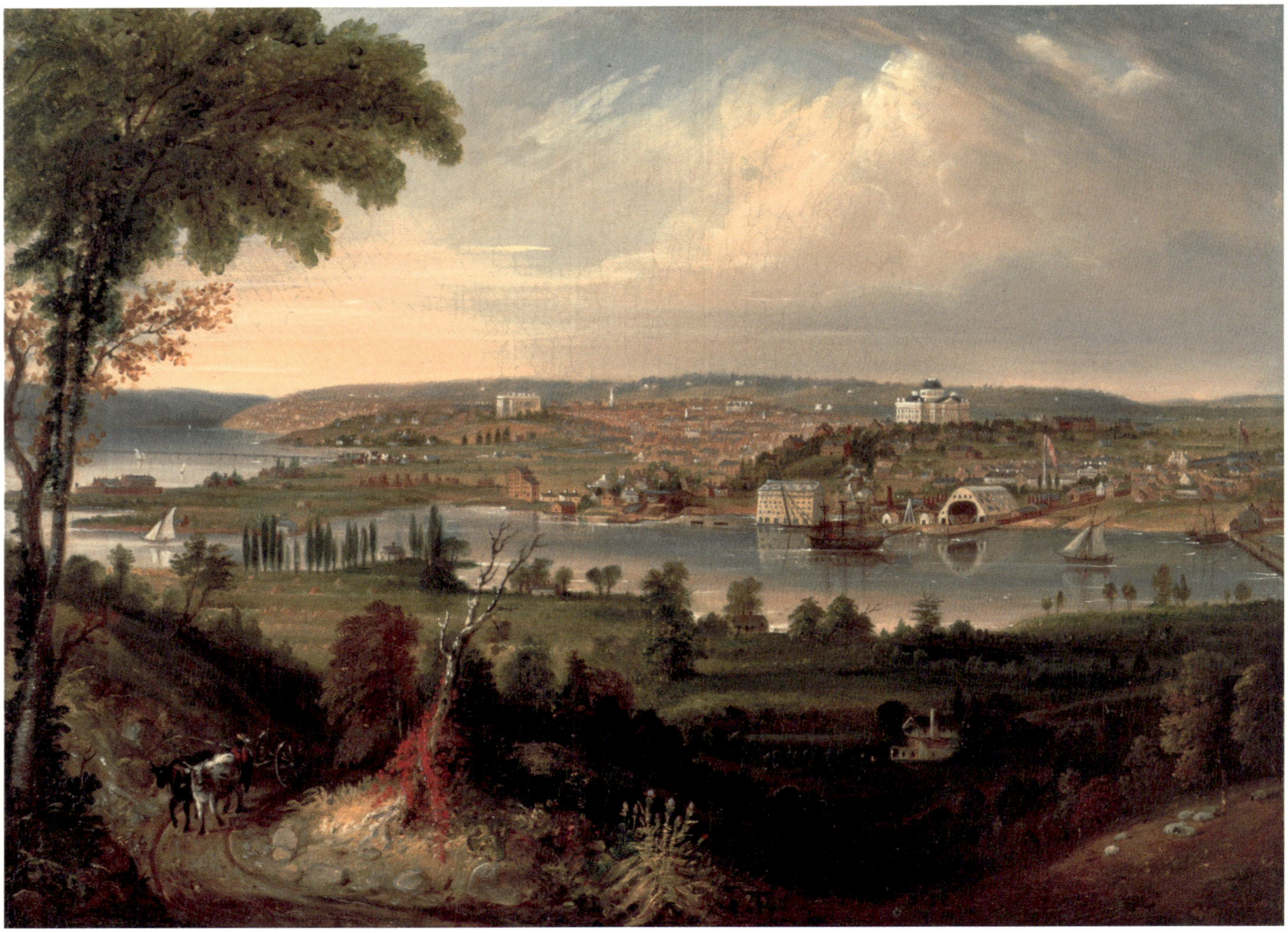

Fig. 4
George Cooke, *The City of Washington from Beyond the Navy Yard*, 1833, oil on canvas, 18 × 25 in., White House Collection/ White House Historical Association, Gift of the Morris and Gwendolyn Cafritz Foundation

separate chambers for the House and the Senate, the structure of the building reinforcing the allocation of political power in its elected representatives rather than in a single leader.

From the top of the Capitol building, Humboldt and his companions enjoyed an elevated view of the city.[67] Humboldt had a 360-degree prospect of the Eastern (now Anacostia) River's wooded hills, the tidal river flats stretching beyond the President's House, and the Potomac palisades extending from Georgetown to Great Falls (fig. 4). The raw qualities of the site remained readily apparent, but Humboldt did not seem fazed by the work yet to be done. The citizens of Washington, D.C., disparagingly called the end of Pennsylvania Avenue closest to the President's House a "Serbonian bog" until it was filled in beginning in 1816, a reflection of the embarrassment many residents felt in the unfinished city.[68] The metaphor enjoyed wide use in Washington, implying that the solidity of the ground—literally and perhaps politically—was but an illusion, masking a quagmire that could swallow the whole enterprise.

One city historian, reflecting on Humboldt's visit, wrote, "Only Baron von Humboldt saw a glorious city in the 'vast Serbonian bog' which the capital then presented." That self-deprecating remark spoke to the insecurities of a new nation that was attempting to build a culture out of raw nature. For Humboldt, wild nature was not a source of fear or embarrassment, but an invitation to explore. "He is on record as remarking when Mr. Law took him to the highest point on Capitol Hill, and pointed out the shining river rolling between its picturesque, verdant banks, the lovely isles dotting its surface and the splendid woodlands stretching along the valley of Tiber Creek, that never had he beheld a more beautiful panorama."[69] Here Humboldt was not just being diplomatic. Attuned to the ways in which nature had shaped South American civilizations, he was inclined to see the future of the

United States defined by its natural resources, rather than its built environment. His enthusiasm and optimism fueled his hosts' growing appreciation for his perspective. The United States already had begun to see its identity bound up in patriotic horticulture and in its natural history, but the nation still felt the pressure of European expectations for the growth of a civilization. The implications were clear: Humboldt was asked to see the nation's prospects embodied in its buildings; instead, he saw in the features of the landscape itself the potential for the nation's identity.

HUMBOLDT'S MAP

Humboldt had not mentioned the map he had made of North America in his letters to Jefferson or Madison, but it soon became clear that these men were eager to know as much as they could about the interior of the continent. On June 4 they hosted Humboldt, Peale, and the entire group of travelers for their first afternoon dinner at the President's House, an affair Peale enjoyed:

> *Not a single toast was given or called for, or Politicks touched on, but subjects of Natural History, and improvements of the conveniences of Life. Manners of the different nations described, or other agreable conversation animated the whole company.*[70]

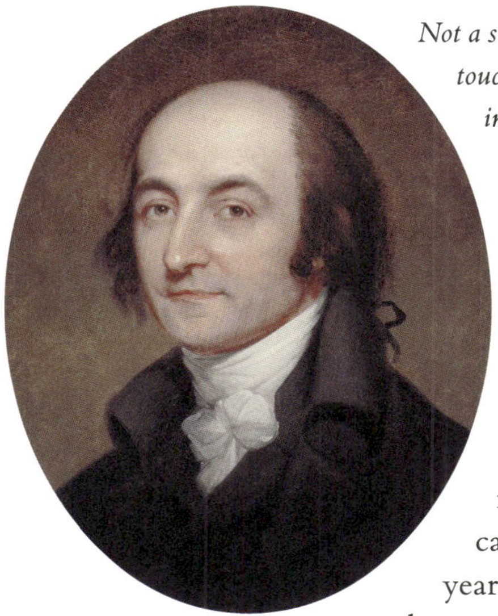

Fig. 5
Rembrandt Peale, *Portrait of Albert Gallatin* (1761–1849), 1805, oil on canvas, 25 × 20 in., Courtesy of Independence National Historic Park Collection, Philadelphia, PA

The courting of Humboldt had begun. On June 5, Humboldt spent the better part of the day at the President's House with Jefferson, Albert Gallatin, and Thornton, regaling them with stories and poring over the maps and statistical information he had compiled during his year in Mexico.[71] What Humboldt provided them with was nothing short of a deeply enhanced understanding of what the Louisiana Territory contained. Humboldt's hand-drawn map, compiled from all of the resources at his disposal in Mexico, showed the usual topographical features: waterways and mountains, along with cities and towns. He also included the locations of individual silver mines, churches, and chapels and marked the territorial boundaries within Mexico (CAT. 6). Readily apparent was why the boundary dispute between Mexico and the United States centered on the Sabine River versus the North River (Rio Grande); the Sabine placed the boundary along the western edge of modern-day Louisiana, while the Rio Grande defined present-day Texas's southern border.[72]

Humboldt's arrival provided an unexpected opportunity to rethink the implications of the acquisition of Louisiana. As the assembled group quizzed Humboldt about specific features and the overall topography, Albert Gallatin (fig. 5) became caught up in Humboldt's willingness to share what he knew. He had been initially unimpressed by the naturalist, but in a letter to his wife, Gallatin described the encounter as "an exquisite intellectual treat," explaining that despite Humboldt's propensity for monopolizing the conversation (and in a rapid mixing of four languages), "I was really delighted & swallowed more information of various kinds in less than two hours than I had for two years past in all that I had read or heard." Gallatin warmed to Humboldt as he comprehended the significance of his information, observing,

> *The extent of his reading and scientific knowledge is astonishing. I must acknowledge, in order to account for my enthusiasm, that he was surrounded with maps, statements, &c, all new to me, and several of which he has literally permitted us to transcribe.*

Humboldt's intelligence had culminated in, as Gallatin described it, "a map of natural philosophical & productions & statistics of that country better known than those of most European countries."[73] Jefferson and Madison had been unable to find a map of the region that would assist them with the upcoming negotiations with Spain. Humboldt's map filled those blank areas with a network of mountains, rivers, and settlements.

Humboldt allowed Gallatin to make a copy of his map of North America with the promise that the American statesman would return the original when he was done. His trust was such that when Jefferson began asking him serious questions about the geopolitics of Louisiana, Humboldt was prepared to share everything he knew. In writing, Jefferson queried:

> *Thos. Jefferson asks leave to observe to Baron de Humboldt that the question of limits of Louisiana between Spain & the US is this. They claim to hold to the river Mexicana or*

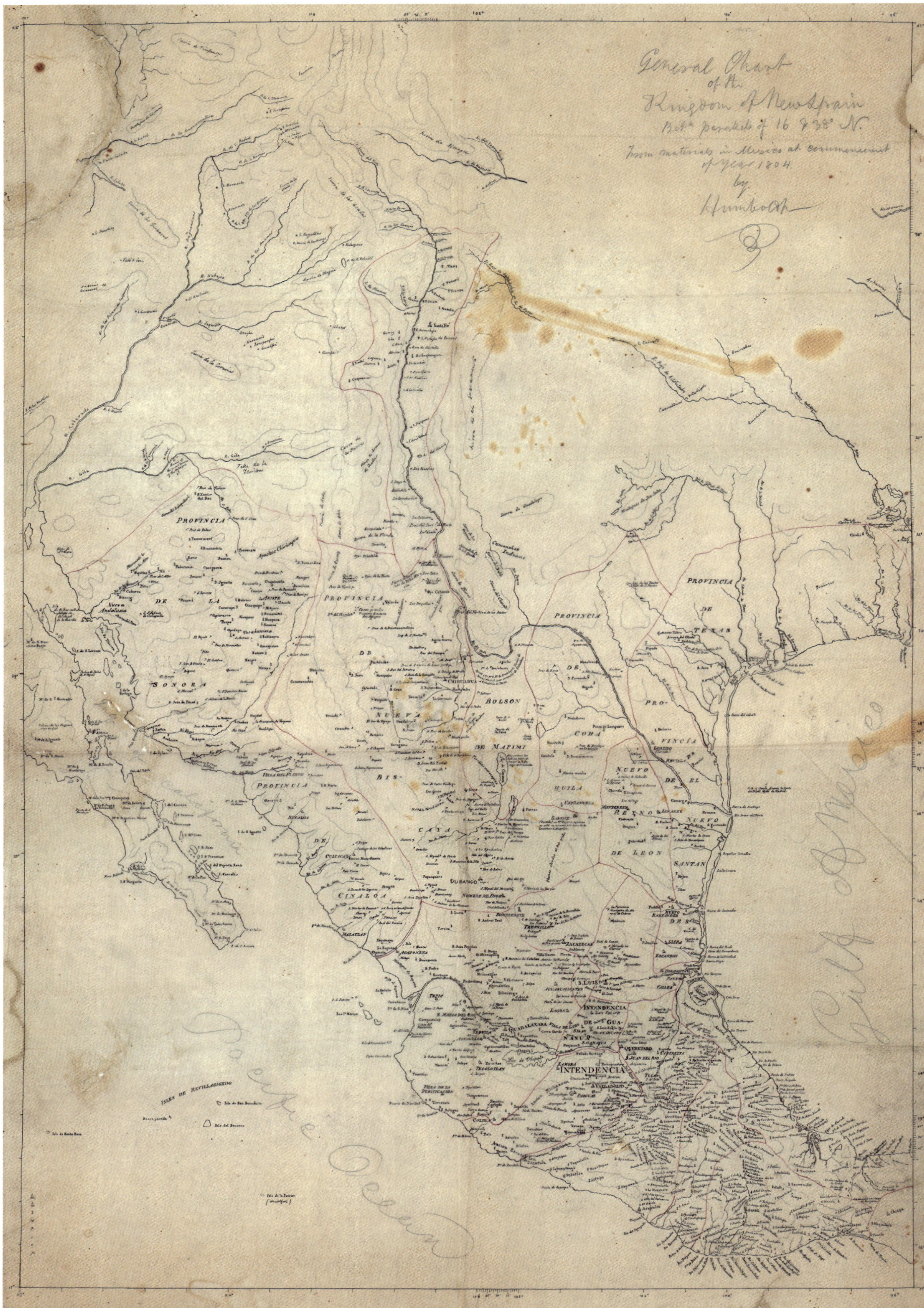

General Chart
of the
Kingdom of New Spain
Betn parallels of 16 & 38° N.
From materials in Mexico at commencement
of year 1804
by
Humboldt
PROVINCIA DE LA SONORA
PROVINCIA DE NUEVA BISCAYA
PROVINCIA DE TEXAS
BOLSON DE MAPIMI
COAHUILA
NUEVO REYNO DE LEON
DURANGO
CINALOA
INTENDENCIA
Pacific Ocean
Gulf of Mexico

CAT. 6 Copy after Alexander von Humboldt, *General Chart of the Kingdom of New Spain between Parallels of 16 & 38° N.*, 1804, pencil and ink on tracing paper, 37 ¾ × 26 in., Library of Congress, Geography and Map Division

Sabine & from the head of that Northwardly along the heads of the waters of the Missipi to the head of the Red river & so on. We claim to the North river from it's [sic] mouth to the source either of it's Eastern or Western branch, thence to the head of Red river & so on. Can the Baron inform me what population may be between those lines, of white, red, or black people? And whether any & what mines are within them?[74]

This Humboldt did. In addition, he supplied the president and his advisers with a copy of the *Tableau statistique du royaume de la Nouvelle Espagne*, a summary he had prepared for the Spanish viceroy that would be presented to him upon Humboldt's return to Europe.[75] In a brief letter to Caspar Wistar, Jefferson remarked:

I have omitted to state above, the extreme satisfaction I have received from Baron Humbodt's [sic] communications. The treasures of information which he possesses are inestimable, and fill us with impatience for their appearance in print.[76]

The three men came away with a renewed appreciation for the geopolitical significance of Humboldt's visit and his generosity in sharing what he knew.

Humboldt did not appear to be concerned that he was divulging confidential information, even though his entire voyage had been conducted with the permission of the Spanish king and under a Spanish passport. First, he had funded the entire voyage himself; but more important to Humboldt, he believed in the free exchange of all knowledge. Sensibly enough, he planned to publish his map when he arrived in Europe. Sharing it with Jefferson may have jumped the gun by a few years, but the information was intended to be public all along. Jefferson had been relieved to acquire Louisiana, thereby removing the French presence along his western border; however, he already harbored dreams of expanding the United States even further. As Jefferson made clear in their exchange of subsequent letters, he was already angling to extend the territory of the United States to minimize the number of international boundaries, thereby protecting the country from unwanted foreign entanglements. A decade before the articulation of the Monroe Doctrine, Jefferson was already directing the United States down a path that acknowledged hemispheric awareness and attempted to avoid international disputes with the continually warring European powers.

Jefferson made sure the local newspaper provided flattering coverage of Humboldt's visit. The June 8 issue of the *National Intelligencer* ran a lengthy article heralding Humboldt's arrival and visit to Washington. It touched on the baron's exploits and appears to have been based on the letter Humboldt had sent Jefferson by way of introduction. It enumerated the highlights of Humboldt's South American journey, stressed his accomplishments, and sent a signal that this was a visitor deserving of respect:

We are pleased to add that the citizens of Washington, and more especially the President, fully impressed with the great and disinterested services of the Baron, have, by their attentions, paid his talents the respect to which they are so eminently entitled.[77]

Those "great and disinterested services" to which editor Samuel Harrison Smith referred were the map and the written report that Humboldt graciously shared with Jefferson. Humboldt's generosity was in keeping with the collaborative behavior he cultivated within his scientific networks. By sharing his information and requesting American statistics in return, Humboldt had laid the groundwork for a half-century of reciprocal communication with the United States. Before Humboldt left for Europe, Gallatin sent him information on the population, exports, and navigation of the United States, including detailed statistics on the populations of American Indian tribes, free and enslaved blacks, and the tonnage of American-owned ships engaged in domestic and international trade, wryly remarking to Jefferson that the summary of exports he had prepared for Humboldt "is better arranged than the usual reports to Congress."[78]

The time Humboldt spent with Jefferson served both political and personal aims. Over the course of the week, Humboldt and Jefferson spent parts of each day together, swapping stories and poring over maps and documents. The vast accumulation of books, maps, fossils, and other natural history specimens in Jefferson's study foreshadowed the look of Humboldt's own study and library spaces in Paris and Berlin (see CAT. 44). Jefferson's notorious informality won him Humboldt's friendship and resulted in Jefferson extending to Humboldt the run of the President's House at any hour.[79] Never had a monarch or viceroy treated him this way. Jefferson's philosophy was, "When brought together in society, all are perfectly equal, whether foreign or domestic, titled or untitled, in or out of office."[80] For Humboldt, who had initially feared he was not significant enough to secure an audience with the American president, this behavior encouraged his belief that Jefferson might actually embody the democratic ideals of the nation he now led.

NATURAL HISTORY AS AN AMERICAN LANDMARK

Humboldt's friendship with Jefferson prompted Peale to engage the visiting naturalist as his advocate regarding his museum. Charles Willson Peale had founded his museum with the new nation's democratic ideals in mind. Foremost, Peale acknowledged that democracy depended on an educated electorate. In celebrating America's rich natural history, his displays were intended to help shape America's emergent national and cultural identity. He had moved his museum into a place already redolent with civic pride—the same chambers where Congress had debated the wording of the Constitution. He now hoped to have the federal government purchase the entire enterprise and move it to Washington as a national institute. The data Jefferson had deployed in *Notes on the State of Virginia* supported his narrative upholding the significance of the nation. The artifacts Peale continued to amass complemented and expanded that polemic. In 1801 he had overseen the excavation of two nearly complete mastodon skeletons in upstate New York, the first of which he installed in his museum in Philadelphia with great fanfare. In his introductory letter, Humboldt had recognized Jefferson's interest in the mammoth as a defining symbol of the nation; its discovery had been a defining moment for the Peale family museum enterprise. Humboldt's conversation with Rembrandt Peale concerning the mastodon was covered in an article that appeared the following week in the *National Intelligencer*.[81]

The impressive scale of this "largest terrestrial being" was a point of pride both for Peale and for Jefferson, who had featured the mammoth in his *Notes on the State of Virginia* as the strongest evidence refuting Buffon's argument regarding intrinsic American degeneracy. The year 1802 had been one for "mammoth" to enter the American vernacular; in 1804 perhaps it was time for that mammoth, and Peale's undertaking, to become a truly national landmark of its own. Humboldt's fascination with the mammoth became another means by which Peale might be able to make his point. In his letter to Jefferson, Humboldt had invoked the mammoth as a shared interest, even suggesting that his friend Georges Cuvier could assist with the proper identification of the bones found in the Americas. The mammoth, as it was universally called at this time, had established for Jefferson proof that a land mammal in North America had been larger than any related species found in Europe.[82] As such it became a concrete symbol of Jefferson's attempts to elevate the status of the new nation through natural history. Peale aspired to see his museum become another monument to that cause.

Before acquiring the mammoth skeleton, Peale had already organized his museum in accordance with the prevalent belief in a Great Chain of Being, in which all creatures on earth were arranged into a hierarchy at which man stood at the top (just below God). Peale hung his portraits of eminent American statesmen, military leaders, and scientists around the top of the gallery walls, surmounting four rows of cases, each of which contained stuffed specimens in natural settings. These cases were arrayed in keeping with this hierarchy as well. In his painted dioramas and taxidermy specimens Peale had striven to "bring into one view a world in Miniature."[83] When he first opened the museum in 1786, Peale intended it to serve as a visual textbook for Americans to understand the natural world. His goal was to provide Americans with access to a comprehensive education on their native flora and fauna and the lifeways of the indigenous peoples. As a multivalent civics and natural history lesson, a visit to Peale's Museum was designed to inculcate pride in America's natural bounty and its cultural achievements. The founder's goal was aligned with the political goals for the new nation, taking a cue from Jefferson's book in presenting worthy specimens for study and admiration.

When the capital was moved to Washington, space opened up in Philadelphia's State House complex. Peale moved his museum there, leasing three rooms, but he could not, despite repeated entreaties, get either full sponsorship or approval for the APS to buy the museum. Even after becoming the third U.S. president, Jefferson was still serving as president of the APS, albeit at a distance. In 1802 Peale began sending letters to him and Madison, suggesting that the government purchase his museum for the nation and move it to Washington as a national institute.[84] Jefferson deftly deflected the conversation, noting only that he was unsure as to the constitutionality of using federal funds in such a manner.[85]

Peale envisioned his museum as part of a city and a nation coming into focus.[86] He could see the traction Humboldt had with Jefferson, Madison, and Gallatin. Now he hoped that Humboldt might be able to accomplish what he so far had not. He had observed how well Jefferson and Humboldt got along, and how close their friendship was becoming. Peale must have broached this topic with Humboldt, for late one night when Humboldt returned from a visit to the President's House, Peale recalled in his diary:

> *[T]he Baron came to my room & told me that he had been conversing with the President about me & my Museum, that he wondered that the Government did not secure it by a purchase [of] it—for such opertunities of getting so complete collections of natural subjects seldom occurred. The President replyed that it was his ardent wish and he hoped the period was not far distant & he thought that each of the States would contribute means and thus it might be made a National Museum. The Baron seemed pleased with the subject of their conversation & altho he could not relate all that passed yet he assured me that the President was very much my friend.*[87]

Humboldt's diplomatic and noncommittal answer temporarily satisfied Peale. But the larger issue of the political and financial feasibility of a national institute would recur, remaining unresolved until the founding of the Smithsonian in 1846.[88]

MOUNT VERNON

A very different landmark to an equally monumental figure awaited the Prussian traveler and his companions south and west of the city, at Mount Vernon (CAT. 7). Home to the late General and President George Washington, Mount Vernon combined patriotism, politics, and the use of native plants to signify national identity. It was a shrine to the founding principles of the new nation—and to the memory of a man remembered as "first in war, first in peace, and first in the hearts of his countrymen."[89] Yet two distinct and ideologically inconsistent facets of the American experiment were deeply intertwined in the fabric of the Mount Vernon landscape: the seemingly impossible duality of a nation founded on individual liberty and yet unwilling to abolish slavery. In his letter to Jefferson, Humboldt had broached his belief in the importance of human equality as a core democratic principle, delicately referencing his opposition to slavery, couching it in the metaphor of enjoying the air in a country in which its people could breathe more freely. His education with Blumenbach at Göttingen, coupled with his experiences in the Americas, had shaped his generally liberal political tendencies. These were about to be tested.

In the morning the assembled group set off in two carriages: Humboldt, Bonpland, Montúfar, Collin, Fothergill, Woodhouse, and Peale. They crossed the Potomac River on Mason's Ferry at Georgetown, traversing Mason's Island[90] (now Roosevelt Island) and "traveld a hilly road that frequently gave us a view of the river Potomack."[91] Their carriage ride followed the bluffs of the river eleven miles south of Alexandria, approaching Mount Vernon through the gardens George Washington had worked on for the last decade of his life. As they navigated the approach to the west entrance to the mansion, Humboldt would have caught intermittent glimpses of the house between carefully orchestrated clusters of trees and shrubs lining the way. These gardens gave the appearance of being natural, but in truth each plant had been chosen to present collectively a message about national unity. The trees and shrubs came from all thirteen of the nation's original states, planted so as to use hardier varieties to protect the more delicate species, visualizing the nation's new motto, *E pluribus unum*. Through botany, Washington cultivated the effect he so desired in the body politic. Washington had borrowed this idea from Samuel Vaughan, who had designed the gardens outside the Philadelphia State House in 1785. Vaughan believed that the American Revolution had ensured "the permanent security of the rights of mankind,"[92] and inspired by that ideal, he announced, "I am now planting trees and shrubs in the State House Square; and as I wish to collect there a specimen of every sort in America that will grow in this state."[93]

Two years later, during the summer of 1787, the members of the Constitutional Convention had walked through that garden in Philadelphia each day to meet inside that State House. Washington and Vaughan exchanged ideas, and during the convention Vaughan traveled to Mount Vernon and sketched a plan for a similar "American" garden on a much grander scale (fig. 6). More impressive still was the primary source of plants for both of these patriotic gardens—John Bartram's garden in Philadelphia, referred to as "the Seminary of American vegetables."[94] Bartram's emphasis was on comprehensiveness for commercial purposes, supplying plants and seeds to an international market.[95] His sons had inherited their father's business, and it became the source of many of the plants that both Vaughan and Washington used in their own gardens.[96] William Hamilton's Woodlands, a spectacular garden full of internationally sourced plants, was another supplier for Washington; he also requested individual species from Governor DeWitt Clinton in New York and Steven Elliott in South Carolina.[97]

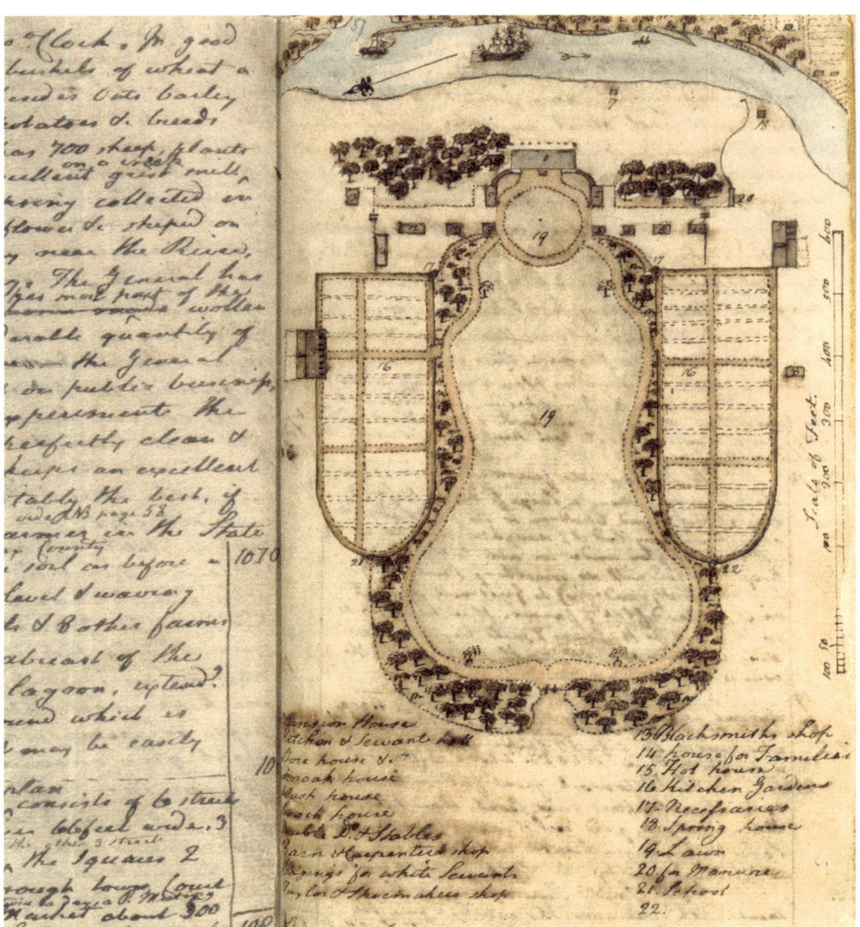

Fig. 6
Samuel Vaughan, *Journal with sketch of the Mount Vernon landscape* (detail), 1787, ink on paper in a bound journal, open: 7 9/16 x 10 9/16 in., Mount Vernon Ladies' Association, Gift of Mrs. E. A. Conrad, Mrs. Lawrence Howe, Mrs. William Loring Vaughan, Mr. Samuel Vaughan, Jr., and purchase from Mrs. George W. Gibson, 1989

Now, a scant five years after his death, Washington's gardens were still intact. Humboldt was keenly interested in botany and, from the early 1790s, had begun amassing the information that would allow him to postulate a world whose plants provide clues to local climate, soil, and altitude. He had already visited both Bartram's Garden and Samuel Vaughan's American garden. By combining plantings from thirteen very different states, across multiple climate zones, Washington had used his plantings to send a message Humboldt would have immediately understood: that the physical geography of the American landscape could convey the nation's patriotic aspirations.

The day spent at Mount Vernon affords a cloudy window into Humboldt's other agenda—that of urging the United States to live up to its ideals, expressed in Jefferson's words in the Declaration of Independence, that all men are created equal. As much as Humboldt admired Jefferson and the ideals he espoused, slavery was the issue that remained awkward between them. Humboldt had been horrified watching a slave auction in Cumaná's main square shortly after his arrival in Venezuela in July 1799.[98] His experiences in the Americas cemented his lifelong virulent opposition to slavery, a topic he would return to in his published works and private correspondence. He did not, however, raise this topic directly with Jefferson, reverting to courtly diplomacy when expressing his admiration for America as a land destined for liberty—and only hinting at measures needed for it to serve as an exemplar for a true democracy. In Philadelphia

CAT. 7 Eastman Johnson, *The Old Mount Vernon*, 1857, oil on board, 23 ⅜ × 34 ½ in., Mount Vernon Ladies' Association, Purchased with funds courtesy of an anonymous donor and the Mount Vernon Licensing Fund, 2009

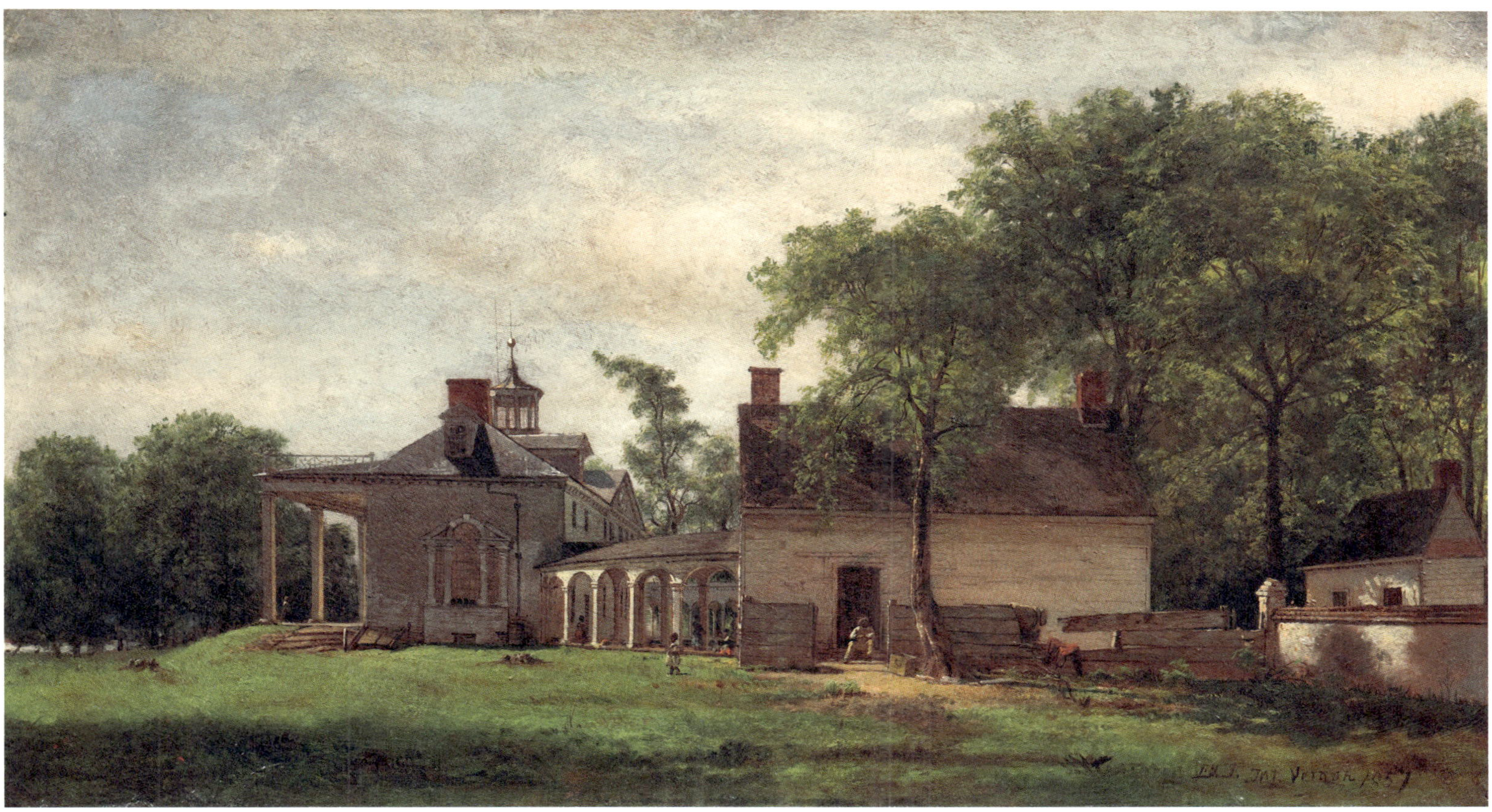

he had been surrounded by like-minded Quakers and abolitionists; now he was seeing enslaved people building the U.S. Capitol building, serving in the president's household, and working in the surrounding fields. Humboldt needed to understand more about this facet of the American experience.

Throughout South America, Humboldt had been sensitive to the treatment of indigenous peoples and had sought out indigenous guides, hoping to learn about nature from individuals he believed had a better understanding of it than their colonial overlords. Now he wanted to meet the formerly enslaved people who had served the Washington family. As Peale noted, the only man still alive who had served George Washington was the elderly William Lee, whom he described as "his faithfull attendant through the war....I found him now making shoes, he was now a cripple," the result of several accidents that displaced his kneecaps.[99] Although there is no known record of a conversation between Humboldt and Lee, it seems highly unlikely Humboldt would have missed this opportunity.

Peale certainly echoed Humboldt's revulsion with enslavement, writing,

> *It is surely a miserable situation to be surrounded with a number of Slaves, however kindly they may be used, yet the very Idea of Slavery is horrible—and I much fear the consequence should the Negroes in Virginia N & S Carolina and in Georgia revolt—and however diligent or vigilent their masters may be yet it will certainly take place sooner or later.*[100]

Peale's diary likely reflects a conversation among the group. Woodhouse, Fothergill, and Collin were also abolitionists, with Collin being described by Jacques Pierre Brissot, a leading French abolitionist and the founder of the Société des amis des Noirs (Society of the Friends of the Blacks), as "a fervid apostle of liberty" who disapproved of slavery and advocated for the freeing of all slaves in his books and religious tracts.[101] The divide in the United States over the issue of slavery would only grow deeper over time. Samuel Vaughan had planted his State House garden convinced the Revolutionary War had secured "the permanent security of the rights of mankind," but the jarring picture of enslaved labor planting that same garden at Mount Vernon was impossible to reconcile. This hypocrisy would be a recurring source of anger and frustration for Humboldt, framing his strongest criticism of a country he otherwise deeply admired.

The varying qualities of freedom were all embedded in the fabric of Mount Vernon and in the developing hagiography surrounding the person who had lived there. Peale was an avid gardener and fully aware of the symbolic weight associated with the plantings in Philadelphia and at Mount Vernon. He had been close to George Washington, serving with him at Valley Forge, and this trip brought back memories. Passing through the house, Peale observed, "our company seemed to enjoy themselves in the grand Portico from which they had an extensive view of the Potomack."[102] Outside there were sculpted gardens carrying a patriotic message, while those

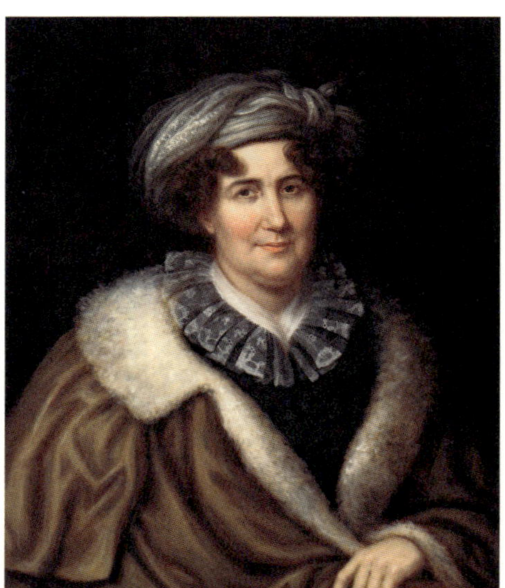

Fig. 7
Charles Bird King, *Margaret Bayard Smith* (1778–1844), ca. 1829, oil on canvas, 30 × 25 in., Redwood Library and Athenaeum, Newport, Rhode Island, Gift of Charles Bird King

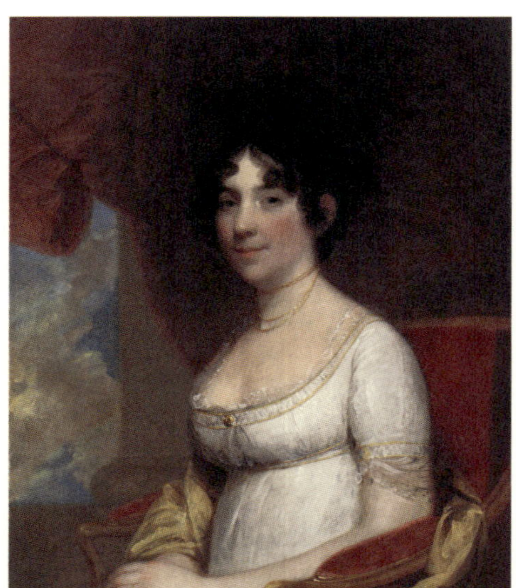

Fig. 8
Gilbert Stuart, *Dolley Madison* (1768–1849), 1804, oil on canvas, 29 3/16 × 24 1/8 in., White House Collection/White House Historical Association, Gift of the Walter H. & Phyllis J. Shorenstein Foundation in memory of Phyllis J. Shorenstein

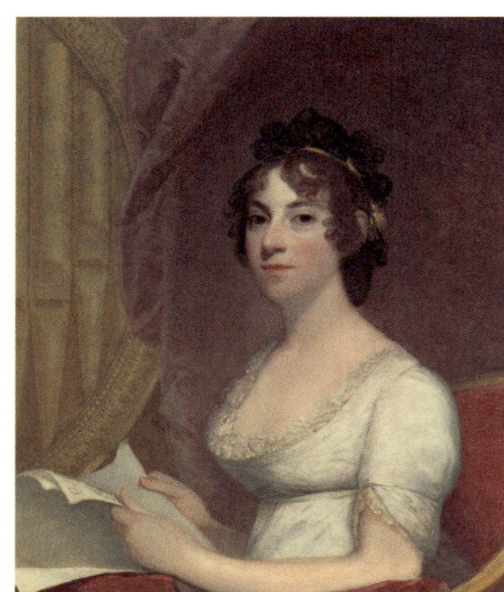

Fig. 9
Gilbert Stuart, *Anna Maria Brodeau Thornton (Mrs. William Thornton)* (1775?–1865), 1804, oil on canvas, 28 13/16 × 24 1/8 in., National Gallery of Art, Andrew W. Mellon Collection

plantings and the house itself were maintained by enslaved labor. Inside the house, the group would have seen the key to the Bastille, a gift from Lafayette after he oversaw its destruction during the French Revolution. These contradictions would not have been easy to absorb. One of Humboldt's great consistencies throughout his long life was his unshakable belief in equality among the races, and that skin color or geographic point of origin did not determine an individual's intelligence or potential. The trip back to the city took several hours, and the company arrived after nine at night. It had been a long and thought-provoking day.[103]

WASHINGTON'S SALON SOCIETY

In Philadelphia, Humboldt's hosts were the men of the APS, and the topics of conversation centered on the sciences. In Washington, the social gatherings around Humboldt were more commonly hosted by women, with a broader discussion that strayed into politics and the arts. Throughout his long life, Humboldt cultivated relationships with highly intelligent and ambitious women. In Paris and Berlin he would frequent the salons, which were largely orchestrated by women with intellectual ambitions who gained access to the leading political, scientific, and artistic circles by hosting them in their living rooms.[104] During Jefferson's tenure as ambassador to France, he made the rounds of Parisian salon society, moving in the liberal circles that supported American democracy and hoped for a similar outcome in France. As president, Jefferson requested that the wives of his cabinet secretaries and the city's civic leaders create a similar social environment in Washington.[105] Bright, witty, and popular, Margaret Bayard Smith, Dolley Madison, and Anna Thornton (figs. 7–9) formed the core of the Washington social scene. On the nights Humboldt was not dining with the president, they took turns hosting him and his companions in their homes.

When it was the Madisons' turn, on June 5, 1804, Peale recalled it as "a sumtious dinner, & our Company praised his Wines, of which he produced a great variety."[106] Humboldt had just spent the entire day with Jefferson at the President's House, along with Gallatin and William Thornton, going over the map of North America that they had all been so eager to see. The conversation that night must have been lively and engaging; both Dolley Madison and Albert Gallatin wrote admiring letters about Humboldt's company, and across town, John Quincy Adams's wife, Louisa, wrote,

> *there are a number of Savants here one of whom is a Dr. Fothergill from England and another a Baron Rumboul [Humboldt] from Prussia I am not sure that this is the mans name but Kitty who dined with them at Mr. Maddissons yesterday says it was so pronounced.*[107]

Anna Thornton was equally enamored of Humboldt. She hosted an elaborate dinner on June 7, noting in her daily record she purchased a gallon of Madeira, six bottles of claret, strawberries, vegetables, fish, honey, bacon, peppers, and lamb for the occasion.[108] The guests included Humboldt, Bonpland, Montúfar, Fothergill, Woodhouse, Collin, Peale, the Madisons, and the Smiths;[109] and Peale noted,

> *the repast was elligant and the close harmonious, for Mrs. Thornton played on her Fortepiano and sang pleasing songs in English, french & Italian, in a high stile of Execution—and before Dinner the Doctor amused with the sight of a numerous collection of Plants which he had drawn from Nature and finished as neatly as any that I have seen of the kind—also a handsome collection of Prints which the Doctr keept in portfolio's.*[110]

Humboldt became particularly excited with Thornton's drawing of a cinchona plant and asked if Thornton would make him a copy. Anna Thornton noted that, on June 25, her husband mailed the drawing to Humboldt in Philadelphia.[111]

Margaret Bayard Smith also developed a great appreciation for Humboldt during his stay. She and her husband Samuel Harrison Smith had established the *National Intelligencer* after moving to Washington upon Jefferson's election to the presidency.[112] She often wrote for her husband's paper, albeit anonymously, and her Common Books, which served as intermittent daily diaries, formed the basis of her articles and books that provided a window into early Washington society and its eminent residents and visitors, including anecdotes from Humboldt's visit. The Smiths hosted the travelers on numerous occasions, for breakfast, tea, or dinner, at times inviting Margaret's

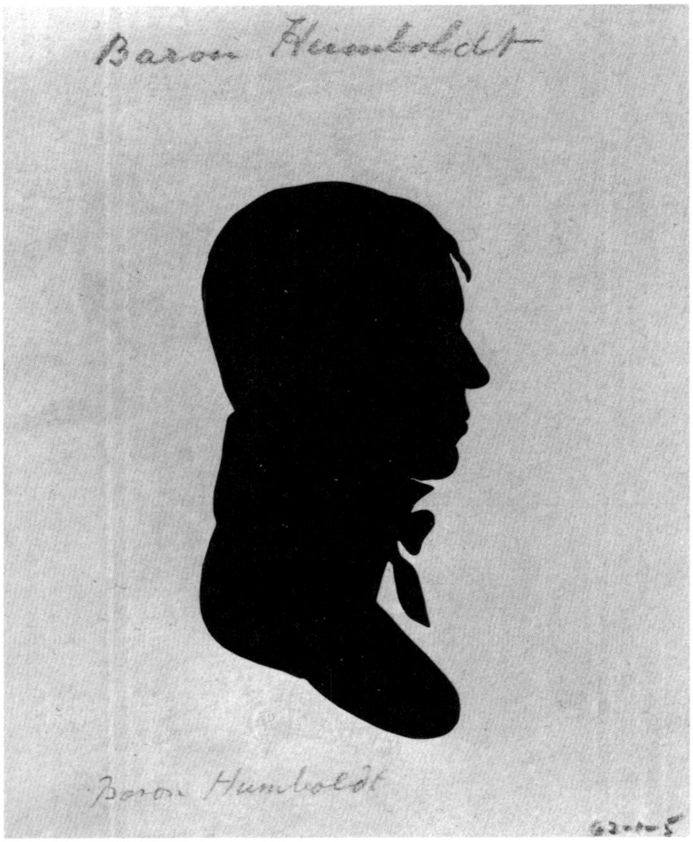
Baron Humboldt
Baron Humboldt

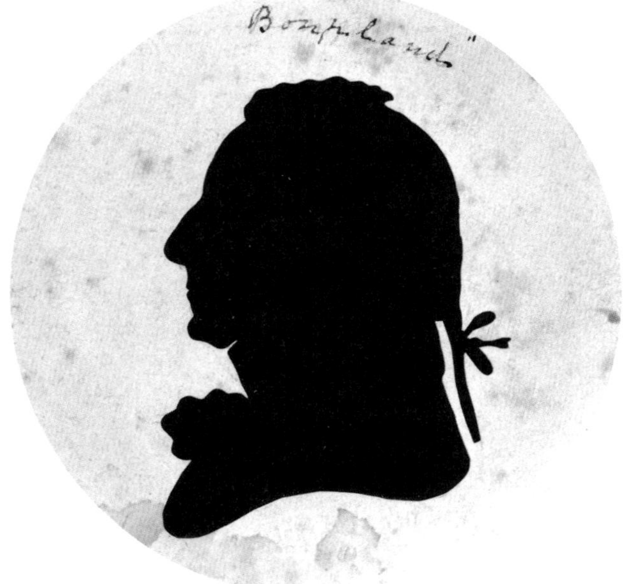
Bonpland"

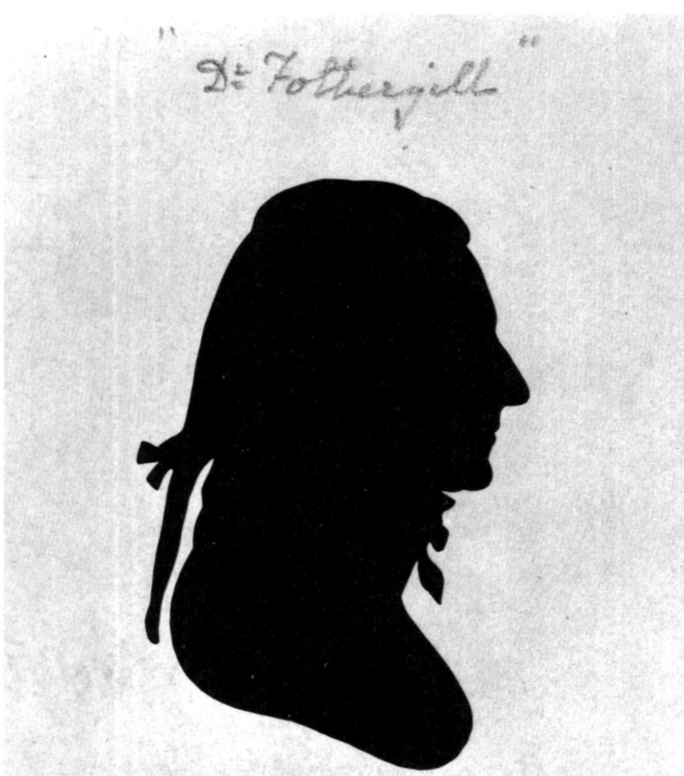
"Dr Fothergill"
Doctr Fothergill

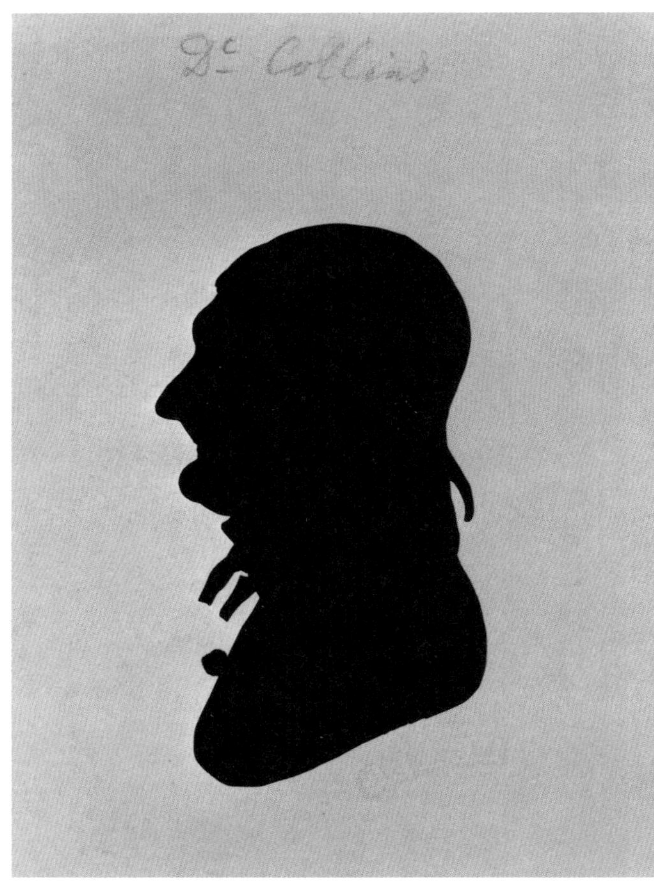
Dr Collins

CAT. 8 Charles Willson Peale, *Alexander von Humboldt*, 1804, paper silhouette, 4 ⅝ × 4 in., The Thomas Jefferson Foundation at Monticello

CAT. 9 Charles Willson Peale, *Aimé Jacques Alexandre Bonpland* (1773–1858), 1804, paper silhouette, 3 ¾ × 3 ¾ in., The Thomas Jefferson Foundation at Monticello

CAT. 10 Charles Willson Peale, *Anthony Fothergill* (1732–1813), 1804, paper silhouette, 4 ⅝ × 3 ⅞ in., The Thomas Jefferson Foundation at Monticello

CAT. 11 Charles Willson Peale, *Nicholas Collin* (1746–1831), 1804, paper silhouette, 4 ⅝ × 4 in., The Thomas Jefferson Foundation at Monticello

CAT. 12 Raphaelle Peale, *Thomas Jefferson*, 1804, paper silhouette, 4 ⅝ × 3 ⅞ in., The Thomas Jefferson Foundation at Monticello

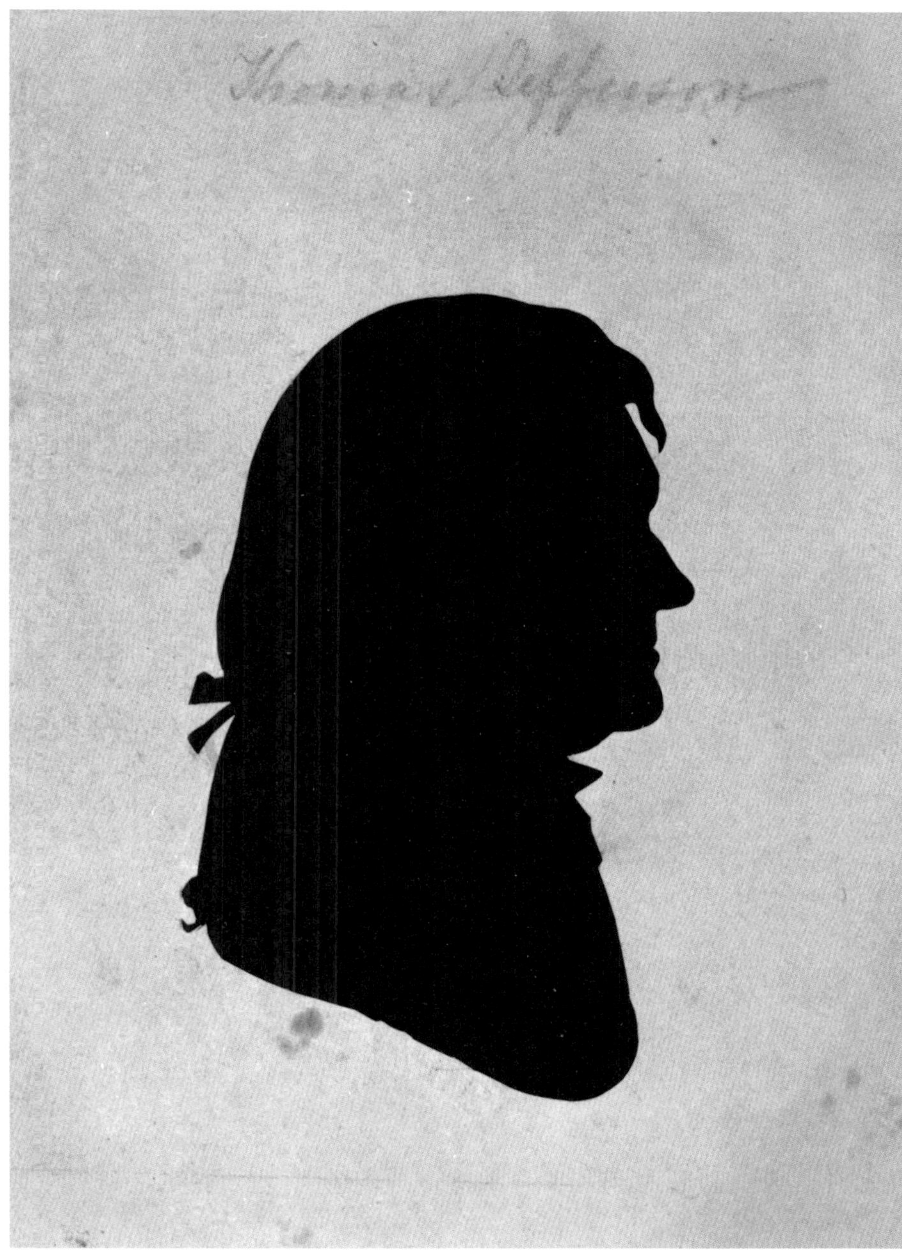

sisters to join the group. On all of these occasions, Jefferson was a quiet absentee. As much as he had enjoyed the salon life in Paris, he preferred to meet with Humboldt apart from these engagements.

"OUR BARON HUMBOLDT"

Before leaving Philadelphia, Peale had arranged for the Philosophic Party to sit for silhouettes at his museum. Silhouettes had become fashionable in Paris at the end of the eighteenth century, and the trend quickly flourished in the United States. Peale had installed a physionotrace in both of his museums, allowing visitors to sit for silhouettes to purchase as souvenirs.[113] At a breakfast hosted by the Smiths on the morning of June 8, Peale handed out sets of those silhouettes as souvenirs, promising to distribute more of them later in the day to those that wanted them.[114] Anna Thornton and Margaret Bayard Smith each claimed a set, preserving them with their diaries and personal papers. Jefferson also received a complete set. The profiles of Humboldt and Bonpland were joined by those of Collin and Fothergill (CATS. 8–11). Although Peale mentioned a silhouette of himself, none appear in the preserved sets; and even more curious, if there was a silhouette made of Carlos Montúfar (fig. 10), none have come to light.[115] Jefferson's silhouette (CAT. 12), preserved with his set at Monticello, appears to have been made at a later date in Washington, at Peale's suggestion, and sent to Peale so that it could be cut in multiples and handed out as a souvenir to museum visitors.[116]

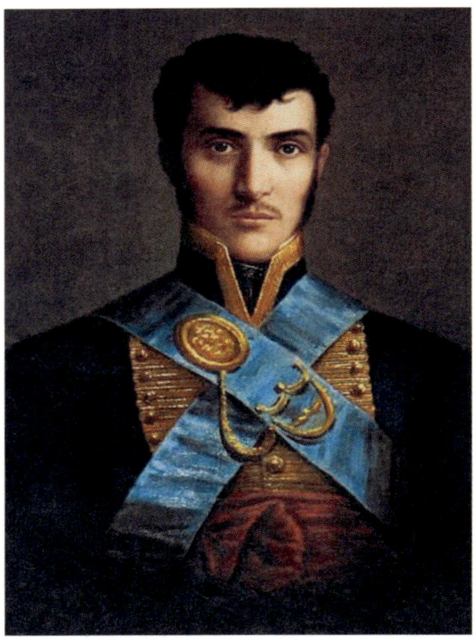

Fig. 10
Manuel Salas Alzamora, *Carlos de Montúfar y Larrea*, ca. 1920, oil on canvas, Palacio de Najas, Salón de los Próceres, Quito, Ecuador

Two days later on June 10, Humboldt paid a call to the Smiths in the evening. Margaret Bayard Smith then wrote a letter to her sister-in-law in Philadelphia, describing Humboldt as

a charming man, and we have the singular pleasure of enjoying a great deal of his company. His hours have been fully occupied by attentions which I have no doubt will make some impression on his heart. An enlightened mind has already made him an American, and we are not without hopes, that after having scratched his curiosity with travel he will spend the remainder of his days in the United States. This will be a great acquisition.[117]

Margaret Bayard Smith had identified "[a]n enlightened mind" as the characteristic that distinguished being an American. It was a phrase that just as easily described Jefferson as Humboldt. It celebrated the ethos of rational thought that characterized period descriptions of the Enlightenment, while capturing the sense of wonder and pride derived from their intellectual curiosity of the world around them. If Enlightenment sensibilities defined Americans, then Humboldt certainly qualified.

As he made plans to depart, Humboldt continued to make the rounds in the evenings, paying social calls and charming his hosts. The idea that he might be swayed to return to the United States—or even better, remain permanently—is a recurring thread through the letters of the people he met. Fancying acquiring Humboldt, like acquiring his silhouette, was becoming a pastime in Washington. As he prepared to leave Washington, Dolley Madison expressed her admiration for Humboldt in a letter to her sister, Anna Cutts:

We have lately had a great treat in the company of a charming Prussian Baron. All the ladies say they are in love with him.... He is the most polite, modest, well-informed, and interesting traveller we have ever met, and is much pleased with America. I hope one day you will become acquainted with our charming Baron Humboldt. He sails in a few days for France with his companions, and is going to publish an account of his travels in South America, where he lived five years; proposing to return here again. He had with him a train of philosophers, who, though clever and entertaining, did not compare to the Baron.[118]

Dolley Madison's desire for his swift return was genuine. Her fond expression, "our Baron Humboldt," summed up the response from the people who met the Prussian traveler during his brief stay.

William Thornton expressed a similar sentiment, writing to John Vaughan:

I have not been for many years so much gratified as by the visit of the wise men of the East whom you sent to us. But I am sorry the interesting Baron has pocketed all South

America. I wish he could have rested his limbs a while & published his works here.—The Treasures of Knowledge he has amassed are worth more than the richest gold mine.[119]

Thornton's poetic reference to Humboldt, Bonpland, and Montúfar as the "wise men of the East" conveys both the exotic nature of this trilingual group from two different corners of the globe and the sense of awe at their presence. Humboldt had struck a nerve and, in response, began to refer to himself as "half an American." Humboldt apparently encouraged this line of thought, whether in sincerity or to appeal to his hosts. He was proving himself a deft networker, and his diplomatic skills served him well. Humboldt's enthusiasm for the United States would encourage Americans to visit him in Europe for the next fifty years. Clearly in character, intellect, and curiosity, the traveler inspired a growing group of people to envision "our Baron Humboldt" as an intellectual monument for Washington and the new nation.

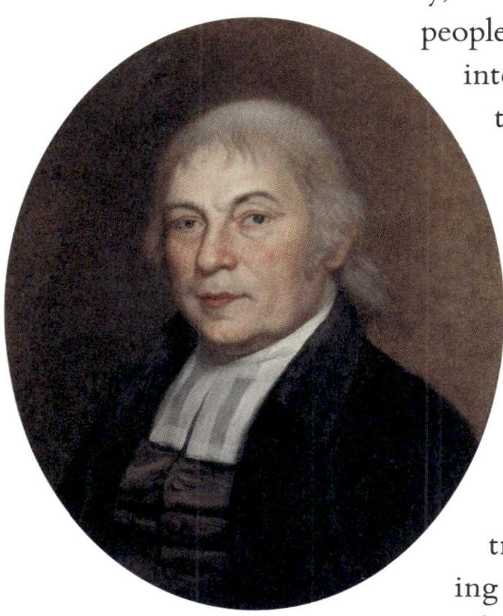

Fig. 11
Charles Willson Peale, *Portrait of Gotthilf Heinrich Ernst Mühlenberg* (1753–1815), 1810, oil on canvas, 24 ¾ × 20 ¼ in., Courtesy of Independence National Historic Park Collection, Philadelphia, PA

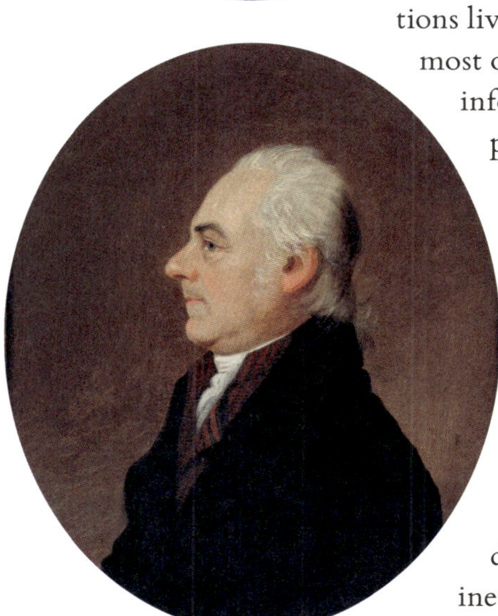

Fig. 12
Jacob Eichholtz, *Andrew Ellicott* (1754–1820), 1809, oil on wood panel, 9 × 7 in., New-York Historical Society, Gift of Mrs. Thomas Clapham

Peale and his colleagues from the APS left Washington to return to Philadelphia early in the morning of June 9, leaving Humboldt, Bonpland, and Montúfar with Jefferson and Gallatin for their last few days in the city.[120] For the next four days, Jefferson had Humboldt nearly to himself. He immediately began peppering the traveler for concrete information regarding the border with Mexico and the populations living there. On June 11, Humboldt spent most of the day again with Jefferson, and the information Humboldt provided to the president and his advisers served as the basis for renewed negotiations with Spain that would continue for the next two years.[121] Gallatin immediately engaged a local cartographer named Kraft to make a copy of Humboldt's map; the original remained with Gallatin for another year before being returned to the explorer.[122]

That night Jefferson hosted a farewell dinner for the travelers, inviting his cabinet to join them one last time. General James Wilkinson, a fellow APS member, declined, citing health reasons, but in his note ruefully stated,

I lament the sudden departure of Baron Humboldt as I feel a strong interest in having his answers to quere? which I take the Liberty to inclose you, because by such answers we shall be able to determine the accuracy of His information.[123]

Wilkinson's queries echoed Jefferson's, confirming the value attributed to Humboldt's information and their appreciation for his willingness to share what he knew.

A SIDE TRIP TO LANCASTER

Humboldt and his two companions did not head directly back to Philadelphia. Instead they made a detour to Lancaster, Pennsylvania, home to surveyor Andrew Ellicott and the internationally recognized botanist Henry Mühlenberg.[124] Given Humboldt's sustained interest in botany, to visit this part of the United States and not pay a call would have been unthinkable. Dr. Gotthilf Heinrich Ernst Mühlenberg (fig. 11)—known as Henry—was respectfully referred to as "the American Linnaeus," for his skill in plant classification.[125] He had been a member of the APS since 1785 and was an internationally recognized figure in the study and exchange of plants and seeds. His herbarium was the earliest professional one collected by a native-born American. By 1811 Mühlenberg wrote that he had collected more than 320 species of grasses and reeds alone.[126] In 1803 he had written to a colleague, "My desire to know every N[orth] American plant with certainty is insatiable." He and Humboldt shared more than the study of botany. Mühlenberg's rationale was decidedly patriotic, as he preferred to study plants that "can persist in our free air,"[127] echoing Humboldt's metaphor describing his preference for a democratic form of government.

Andrew Ellicott (fig. 12) described himself as an indifferent botanist, but he had a particular interest in hearing about Humboldt's map. Ellicott had mounted the most recent survey of the lower Mississippi River valley from the Ohio River to New Orleans between 1796 and 1800, conducting political reconnaissance at the same time he was making cartographic measurements and calculating location using his astronomical readings. The two men shared a passion for instrumentation. Humboldt had carried more than forty instruments across South America, each in its own case, and his location of the magnetic equator would recalibrate the location of specific sites on maps of the Americas. Trained as an astronomer and surveyor, Ellicott had built himself an observation tower in the Spanish colony of Florida in order to witness the transit of Mercury in 1799—the same astronomical event Humboldt had chronicled while he was in Callao, Peru.[128] Humboldt's sustained interest in cartography as an aspect of his botanical research would lead to his developing the field of plant geography, which intrinsically linked location and climate with characteristic plant species.

As Humboldt left Lancaster he became part of Mühlenberg's international botanical network, agreeing to carry the seeds for grasses that Mühlenberg had collected around Lancaster, promising to check them against French botanist André Michaux's archive, housed in Paris.[129] From Philadelphia, Humboldt wrote to thank him for his hospitality and promised to write back with Michaux's classifications. He closed by saying, "We saw the Hamilton garden with astonished delight. Yours, most gratefully, Humboldt."[130] Even though Humboldt was coming to the end of his sojourn, he could not resist the impulse to add side trips. There was simply so much to see. William Hamilton's Woodlands (fig. 13) was yet another well-known botanical repository.[131] Each experience left Humboldt wanting more from his time in the United States. He toyed with the idea of a quick trip to New York to see David Hosack's Elgin Gardens.[132] Humboldt also briefly considered a side trip to South Carolina, again tempted by the network of botanists there. Regretfully he concluded these would have to wait for a return trip to the United States.

That botanical network was uppermost in Humboldt's mind in Lancaster. In a fragment of a travel diary, Humboldt made a list of American botanists and where they lived, as though scripting the itinerary for yet another set of excursions. He also jotted down an incomplete tally of his expenses in Philadelphia, including meals, rented horses, and clothing, as well as the price of porter beer. This fragment raises a tantalizing prospect. If Humboldt kept a journal during his stay in the United States, similar to his bound travel journals from South America and Mexico, it has not survived intact. Single pages bearing information

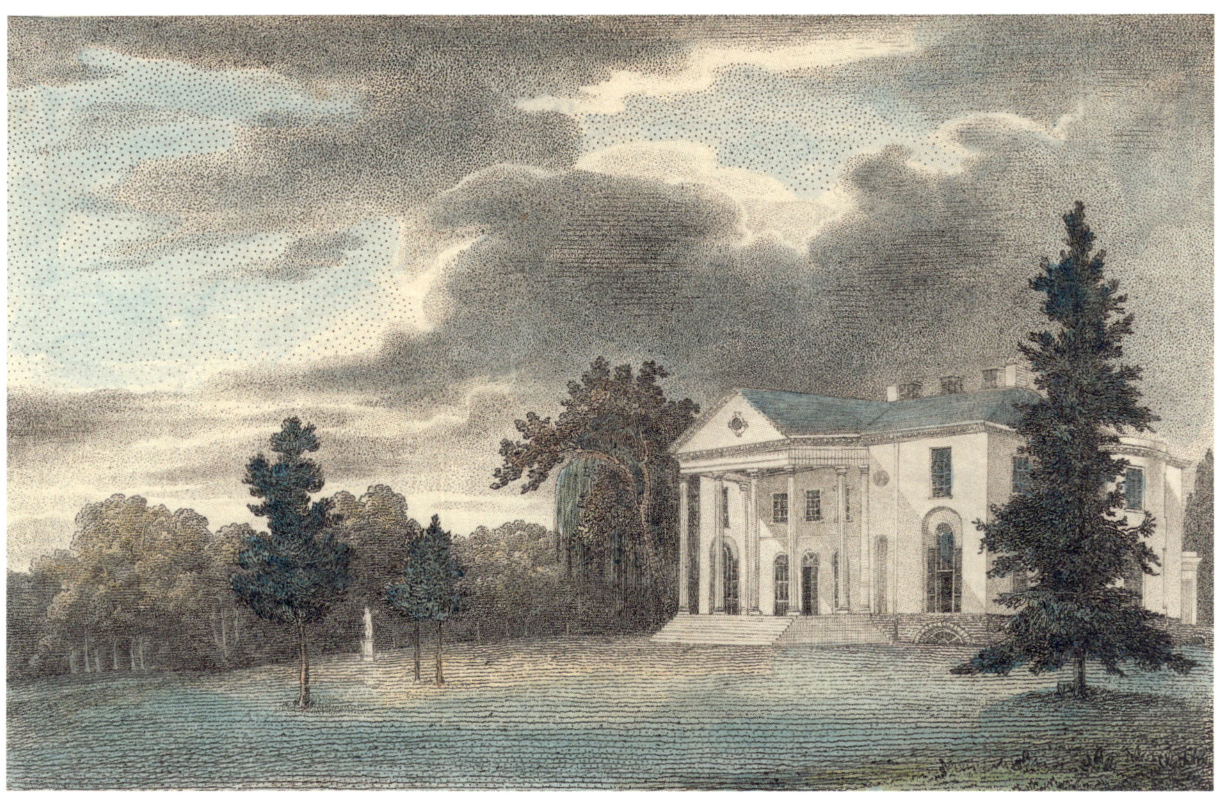

Fig. 13
William Russell Birch, *Woodlands, the Seat of Mr. Wm. Hamilton, Pennsylv.* from *Country Seats of the United States of North America*, 1809, hand-colored engraving, 6 ½ × 8 in., Library Company of Philadelphia, Gift of Marguerite Brenner

accrued while Humboldt was in the United States have surfaced amid the mass of papers left at his death, but in many cases these have been cut up and stored along with other fragments on a similar subject in envelopes designed as part of his personal filing system.[133] Humboldt's obsessive collecting of data—on every topic imaginable—became a hallmark of his practice, with every detail contributing to his understanding of an underlying natural law. For Humboldt, even this short stay in the United States had yielded a small trove of information relevant to the larger work he envisioned.

FAREWELLS IN PHILADELPHIA

While Humboldt had been in Lancaster, Caspar Wistar was making his own plans to keep Humboldt around. At the June 15 meeting of the APS, it was "Resolved, That Dr. Wistar be permitted to introduce A. Von Humboldt as a Visitor at any subsequent meeting."[134] When Humboldt returned to Philadelphia on June 18, Wistar was not about to let Humboldt leave without honoring him at one of his eponymous "Wistar parties," usually held on Saturday evenings. Concerned that Humboldt might depart before the upcoming weekend, Wistar issued invitations to a special Thursday evening party in Humboldt's honor.[135] Wistar parties were legendary in Europe and America as a highlight of a visit by scientifically minded individuals. A decade later, when Duke Bernhard of Saxe-Weimar visited the United States, he made a point of visiting Jefferson at Monticello and attending a Wistar party in Philadelphia. He described an event composed "mostly of scientific and literary gentlemen. This assembly is called a Wistar Party. It is a small, learned circle which owes its existence to a Quaker physician, Dr. Wistar, who assembled all the literati and public characters of Philadelphia at his house every Saturday evening, where all well recommended foreigners were introduced."[136]

Wistar's instincts were sound. There were few vessels headed to France, and when Humboldt learned that the *Favorite* was scheduled to sail to Le Havre on June 27, he booked passage for himself, his companions, and his trunks of specimens. Word of Humboldt's departure plans kicked off a week of receptions and parties. In short order, Dr. Benjamin Rush and John Francis Mifflin each hosted events.[137] Peale had remained in Baltimore after leaving Washington and made his return to Philadelphia on June 21. His top priority now was to arrange to have Humboldt sit for his portrait.[138] Peale scrambled to get these sittings in place. He was understandably excited and nervous; he had all but ceased painting as a profession years before. On June 23, 1804, Charles wrote to his sons, Rembrandt and Raphaelle, "Baron Humboldt will sett tomorrow; I am prepairing the Room below, your Uncle James has supplyed me with canvis & colours."[139]

Over the course of three days, Peale painted a charming, lively portrait of the Prussian (CAT. 13), wherein we see the departing Humboldt through Peale's eyes. The artist captured the attentive, alert, multi-focused aspect of Humboldt, in pronounced contrast to the more relaxed visages of most of his sitters on display in his museum. Humboldt's sidelong glance looks both at the viewer and beyond him, as though part of him had already moved on, capturing the evanescence of his visit. In a letter to his father-in-law, John DePeyster, Peale wrote:

> *I have been richly rewarded for the expence and trouble of a journey by the agreable conversation of Baron Humboldt who is without exception the most extraordinary traveller I ever met with; he is the fountain of knowledge which flows in copius streams—to drop this metaphor to take another, he is a great luminary defusing light on every branch of science—I say defusing, because he is <the most> so communicative of his knowledge which he has treasured up by his travels of upwards 19 Years. His company is courted by the learned where ever he goes.*[140]

Pronouncing himself relieved and pleased with the results, the artist hung Humboldt's portrait in his museum, part of his gallery of American leaders and eminent scientists.[141] If Humboldt insisted on returning to Europe, at least his portrait would remain here, among his scientific colleagues. In scale, format, and presentation, Peale's portrait of Humboldt identified him as an American luminary. He appeared in the company of George Washington and Thomas Jefferson, along with the leading members of the APS. Peale ruefully concluded, "[W]e enjoyed much in a short period, 3 weeks

CAT. 13 Charles Willson Peale, *Portrait of Baron von Humboldt*, 1804, oil on canvas, 21 × 17 in., The College of Physicians of Philadelphia

was like only so many days."[142] Peale echoed the sentiment expressed by Thornton, Vaughan, and others—that the baron's stay had been far too short; that the days and weeks had sped by. In the decades that followed, Humboldt's backward glance would become a full-force stare as the naturalist established his reputation and assumed a dominant role in international public life. For now, Peale had claimed him as America's Baron Humboldt.

A delay in their sailing date permitted the travelers another week in Philadelphia beyond June 27—time for one last celebratory event in their honor. This was a dinner held under the rib cage of the mammoth skeleton in Peale's Museum. Although Peale did not mention this event in his diary, a young German American university student, John Bachman, did:

I was a youth of sixteen, when Humboldt visited America. Wilson, the Ornithologist, procured for me an invitation to be present at a reception given in Philadelphia to this eminent philosopher and naturalist. During my stay there, I saw him every day. Before leaving, he inserted in his note-book the names of his acquaintances in Philadelphia, and, to my surprise, mine was included on the list.[143]

Bachman described the dinner more fully in a eulogy he delivered following Humboldt's death in 1859:

A dinner had been prepared for the occasion in Peal's [sic] *Museum. Among the few naturalists who attended were the two Bartrams, Wilson, the Ornithologist, Lawson, his engraver, George Ord, and a few others, whose names have now escaped my recollection. To this small group was added a considerable number of men who were eminent in the various departments of literature and science.*[144]

Peale had hosted a dinner beneath the mammoth in 1801 when it first went on view, celebrating it as an emblem of America's robust character. Now Peale celebrated the man whose stature, like that of the mammoth, lent undeniable weight to America's cultural relevance.

Between June 20 and 27, Humboldt wrote his letters of farewell to Jefferson, Madison, Gallatin, and Thornton, conveying to each his appreciation for the warmth of his reception and expressing a desire to return to the United States. Humboldt would remember fondly his time in Philadelphia and Washington and in future letters would ask to be remembered to the people who had made his stay so rewarding.[145] He arranged with Madison passports for the three travelers, asking that Bonpland's be carefully worded to protect the Frenchman from being arrested in the event their ship was boarded by the British Royal Navy. The naval blockades along the Atlantic coast also put Humboldt's safe return in jeopardy, and the explorer worried that his trunks of specimens might be seized. Madison was in a position to provide a measure of immunity to such circumstances.[146] He also reminded the statesman to make sure Gallatin returned his map and alluded to another topic that had been under discussion in Washington: the benefits of a canal cut across Panama to facilitate travel across the Americas. It was a topic with far-reaching, hemispheric ramifications, and one Humboldt would come back to repeatedly during his lifetime. For now, it was a placeholder in lieu of a longer, more substantive conversation Humboldt hoped to have on his return.[147]

Freedom was also very much on Humboldt's mind. In his letter to Gallatin, he observed: "I prefer your climate to any other. The more wholesome air is where it is breathed more freely."[148] Humboldt would often invoke this metaphor, knowing that it would be readily understood by his intended audience. His only direct mention of slavery was in his letter to Thornton, suggesting that Humboldt felt he might have more traction working through Jefferson's adviser than with the president directly.[149] As he had in his introductory letter to Jefferson, Humboldt now took extra effort with his parting letter. In it, he wrote:

I have had the good fortune to see the first Magistrate of this great republic living with the simplicity of a philosopher who received me with that profound kindness that makes for a lasting friendship. My circumstances oblige me to leave but I carry with me the consoling experience of having

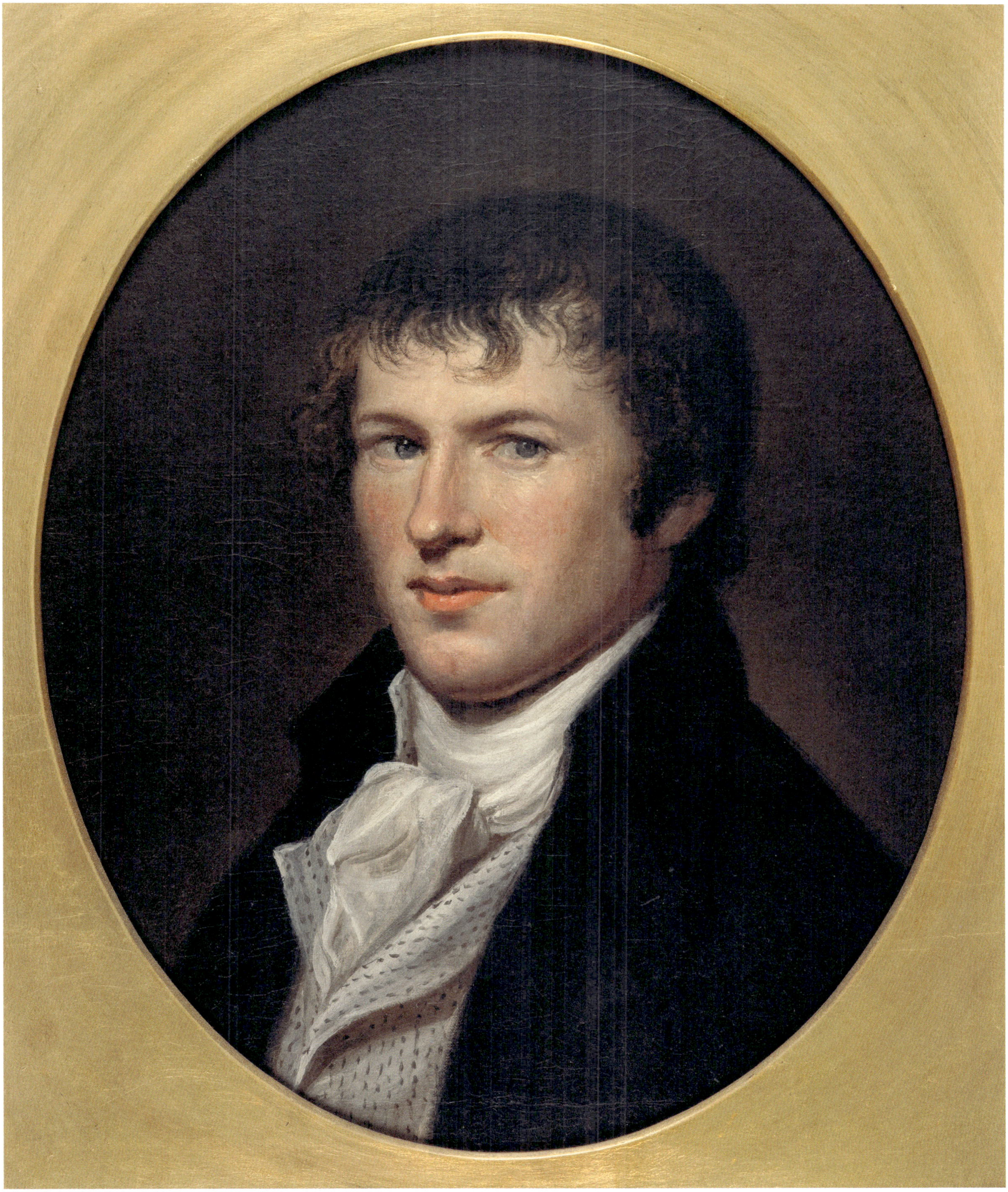

CAT. 14 American Philosophical Society, *Alexander von Humboldt's membership certificate*, 1804, ink on paper bound in an album, closed: 24 3/16 × 17 5/16 × 1 9/16 in., Berlin-Brandenburgischen Akademie der Wissenschaften, Archiv

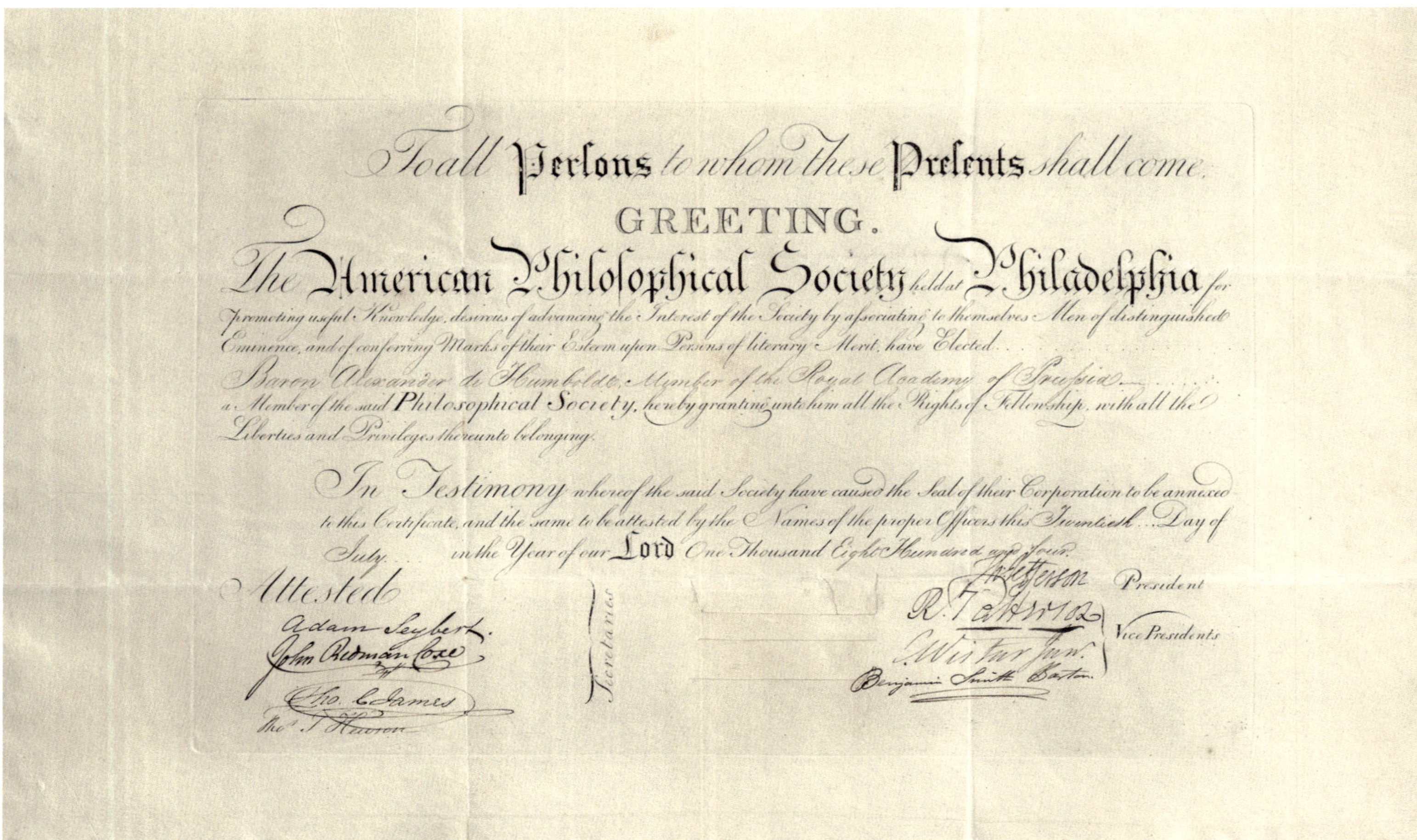
To all Persons to whom these Presents shall come.

GREETING.

The American Philosophical Society held at Philadelphia for promoting useful Knowledge, desirous of advancing the Interest of the Society by associating to themselves Men of distinguished Eminence, and of conferring Marks of their Esteem upon Persons of literary Merit, have Elected Baron Alexander de Humboldt, Member of the Royal Academy of Prussia a Member of the said Philosophical Society, hereby granting unto him all the Rights of Fellowship, with all the Liberties and Privileges thereunto belonging.

In Testimony whereof the said Society have caused the Seal of their Corporation to be annexed to this Certificate, and the same to be attested by the Names of the proper Officers this Twentieth Day of July in the Year of our Lord One Thousand Eight Hundred and four.

Attested

Adam Seybert

John Redman Coxe

Secretaries

President

Vice Presidents

Benjamin Smith Barton

seen the people of this continent advance with great strides towards the perfection of social conditions, whereas Europe presents an immoral and melancholy spectacle.[150]

Jefferson in turn remembered Humboldt as "the most scientific man of his age he had ever seen,"[151] and Thornton read in Humboldt's farewell letter his exuberant pronouncement, "This country that stretches to the west of the mountains presents a vast area to conquer for science!"[152] Humboldt's visit to the United States might have been the coda to five years of travel, but its influence—both on the traveler and on his American audience—would have lasting ramifications on American culture. American exploration would bear Humboldt's stamp going forward. The nation's cultural identity would become more deeply linked to the features in the landscape. Humboldt himself would be invoked as an authority on a wide range of matters, his name and his ideas used to strengthen America's position on the world stage.

Finally, at John Vaughan's request, Humboldt wrote a twenty-page summary, in French, of his travels through the Americas. Vaughan translated it into English and published it in *The Literary Magazine, and American Register* as a supplement in their July 1804 issue. As was often the case with articles of broad interest, other periodicals printed versions of them as well. The *Weekly Visitor, or Ladies' Miscellany* picked up Humboldt's narrative and reprinted it, in its entirety, across three installments in their September issues.[153] In this narrative, Humboldt described his journey with Aimé Bonpland as their "travels in the two hemispheres, by land and sea, probably the most extensive which any *individual* has ever undertaken."[154] Two weeks after his departure, Humboldt was voted a full member of the American Philosophical Society.[155] His certificate bore the signature of his friend, Thomas Jefferson (CAT. 14). It would be the first of eleven American scientific societies that would honor Humboldt with membership.

The French frigate *Favorite* sailed from the Philadelphia docks on July 7, clearing the capes of Delaware two days later. Humboldt was on his way home after five full years away from his European friends and colleagues. He carried with him two letters, one from Peale to Étienne Geoffroy Saint-Hilaire, a zoologist at the prestigious Institut National in Paris. In it, Peale offered to exchange specimens for their mutual advantage.[156] Peale had tried and failed to establish this relationship before, but he now hoped that Humboldt's involvement might change the outcome. Humboldt also carried a letter from Jefferson introducing him to his friend and fellow revolutionary, the Marquis de Lafayette. The friendship that would blossom between the two Europeans would bind Humboldt more closely to Jefferson and to the United States.

On August 4 the *Favorite* docked at Le Havre, and Humboldt once again cleared customs, this time retrieving everything he had brought with him. He, Bonpland, and Montúfar headed for Paris. There, Humboldt would quickly become a fixture in the European scientific community. But he did not forget his experiences in the United States, and over the next fifty years, a steady stream of Americans would become friends, visitors, and correspondents. Humboldt's sphere of influence in the United States would grow ever wider with each passing decade.

Humboldt's interest in American geography, culture, and politics sparked lifelong friendships with some of the key figures in American history. The brief duration of his trip precluded his being able to conduct any kind of research or exploration, but it did allow him to establish the relationships that would facilitate the acquisition of information about American ethnology, linguistics, and cultural anthropology. For the remaining fifty-five years of Humboldt's life, the United States became part of Humboldt's global network of friends, allies, and scientific partners. Humboldt's enthusiasm for the United States encouraged American confidence in designating natural landmarks as emblems of the scale and scope of its cultural ambitions, and Humboldt himself would incorporate references to these distinctive features in his own writings. America's presence on the international stage shone brighter with Humboldt's approbation, an imprimatur assiduously cultivated in the United States. Humboldt was becoming an American monument, part of the fabric of America's cultural identity.

1 Humboldt met the Scottish-born botanist John Fraser of Charleston, South Carolina, in Havana in 1801 and offered to help ship material to Sir Joseph Banks at the Royal Society in London. Afterwards, Humboldt wrote to his brother, Wilhelm von Humboldt, and his friend and colleague, Karl Willdenow, of his plans to sail from Havana to Charleston and travel overland from there to Philadelphia. For Fraser, see Marcus B. Simpson Jr., Stephen W. Moran, and Sallie W. Simpson, "Biographical Notes on John Fraser (1750–1811): Plant Nurseryman, Explorer, and Royal Botanical Collector to the Czar of Russia," *Archives of Natural History* 24, no. 1 (July 2010): 1–18. For Humboldt's letter describing his arrangements with Fraser, see Humboldt to Karl Willdenow, Havana, 21 February 1801, in Ulrike Moheit, ed., *Alexander von Humboldt: Briefe aus Amerika 1799–1804* (Berlin: Akademie Verlag, 1993), 122–30, letter 41; for Humboldt's proposed itinerary to the United States, see Humboldt to Christiane von Haeften, Havana, 27 March 1804, in Moheit, ed., *Alexander von Humboldt*, 283–84, letter 133.

2 Sandra Rebok has researched the evolution of Humboldt's interest in a side trip to the United States in "'In Whose Heads I Would Like My Ideas to Live': Humboldt's Interest in the United States," *Iberoamericana* 19, no. 70 (2019): 113–30.

3 Carlos Montúfar y Larrea (1780–1816) was the grandson of a *marqués* sent from Spain to be the president of Quito, Ecuador, in 1753. Montúfar joined Humboldt and Bonpland in January 1802 for the remainder of their journey through the Americas, climbing Pichincha, Cotopaxi, and Chimborazo together. After arriving in Paris in 1804, he returned to Quito in 1810, supported independence from Spain in 1811, and joined Simón Bolívar in December 1814 for the liberation of Bogotá. He did not survive the wars for independence. Captured after a battle, he was executed on July 31, 1816. See Reinhard Andress, "Alexander von Humboldt und Carlos Montúfar als Reisegefährten: Ein Vergleich ihrer Tagebücher zum Chimborazo-Aufstieg," *Internationale Zeitschrift für Humboldt-Studien* 12, no. 22 (May 2011): 10–22. Montúfar kept a diary of part of this trip: "Biaje de Carlos Montúfar, con el Baron de Humboldt y Don Alexandro Bonpland," Original manuscript in the Lilly Library, Indiana University, Latin American MSS—Peru (1802).

4 In 1801 Humboldt and Bonpland stopped in Havana with over two years' worth of packed specimens. At that time they stored some of the materials they had amassed and divided the rest into three collections, each packed with a duplicate copy of their precious travel journals. These they arranged to ship to Paris, London, and Berlin. "We have arranged the plants in three collections, as we have duplicates and triplicates of everything. We carry a small herbarium with us around the world to allow us to make comparisons. A second one (it belongs to Bonpland, with whom I share everything, of course) has already been sent to France, and the third ...I am sending today through Mr. John Fraser to London by way of Charleston. Through duplication we minimize the danger." Humboldt to Karl Willdenow, 21 February 1801, in which he describes the duplicate shipments as a hedge against piracy or shipwreck. Translated and quoted in H. Walter Lack, *Alexander von Humboldt and the Botanical Exploration of the Americas* (Munich: Prestel, 2009), 30–34; original in Alexander von Humboldt, *Lettres américaines d'Alexandre de Humboldt*, preface and notes by Dr. E. T. Hamy (Paris: E. Guilmoto, 1905), 108. "Nous avons distribun les plantes dans trois collections, car nous avons deux ou trois exemplaires de chaque plante. Nous emportons avec nous, autour du monde, un petit herbier, pour pouvoir faire des comparaisons. Un second, celui de Bonpland, avec qui je partage naturellement tout, est parti pour la France, et j'envoie aujourd'hui le troisième, avec M. John Fraser, par Charleston à Londres....Nous diminuons le danger par le dispersement des collections."

5 The Treaty of Amiens signaled a temporary cessation of the war between France and England during the French Revolutionary Wars. Signed on March 25, 1802, the treaty lasted until May of 1803, when England again declared war on France in an effort to restrain Napoleon. The Royal Navy renewed its blockade of the Atlantic, including the trade routes to the islands of Cuba and Saint-Domingue.

6 "Je me sentais très émû. Me voir périr à la veille de tant de jouissances, voir périr avec moi tous les fruits de mes travaux, être la cause de la mort des deux personnes qui m'accompagnaient, périr dans un voyage de Philadelphie qui ne paraissait pas de toute nécessité (quoique entrepris pour sauver nos m[anu]s[crit]s et collections contre la perfide politique espagnole)." Margot Faak, ed., *Alexander von Humboldt, Reise auf dem Rio Magdalena, durch die Anden und Mexico, Part 1: Texts* (Berlin: Akademie-Verlag, 1986), 397–98; translated from the French by Ingo Schwarz, "Alexander von Humboldt's Visit to Washington and Philadelphia, His Friendship with Jefferson, and His Fascination with the United States," in "Alexander von Humboldt's Natural History Legacy and Its Relevance for Today," special issue, *Northeastern Naturalist* 8, no. 1 (November 2001): 43.

7 Humboldt to Zaccheus Collins, On the Delaware River, 20 May 1804. Humboldt wrote to Collins, the customs inspector for Philadelphia, hoping to avoid quarantine and delay. Ingo Schwarz, ed., *Alexander von Humboldt und die Vereinigten Staaten von Amerika: Briefwechsel* (Berlin: Akademie Verlag, 2004), 87, letter 1.

8 Gray to Madison, Havana, 28 April 1804, James Madison Papers, Series 1, General Correspondence, 1723–1859, Microfilm Reel 8, MSS Division, Library of Congress. Noted in Herman Friis, "Baron Alexander von Humboldt's Visit to Washington, D.C., June 1 through June 13, 1804," *Records of the Columbia Historical Society* 60/62 (1960/62): 4.

9 Gray followed up with a second letter to Madison on May 8, 1804, from Havana. Consular Dispatches, Havana, vol. 1, December 11, 1788–October 2, 1807, The National Archives and Records Administration, 4 pps. ALS. James Madison, *The Papers of James Madison*, Secretary of State Series, vol. 7, *2 April–31 August 1804*, ed. David B. Mattern et al. (Charlottesville: University of Virginia Press, 2005), 190–92.

10 See Anne Wharton, "The American Philosophical Society," *Atlantic Monthly* 61, no. 367 (May 1888): 611–24.

11 Humboldt to Collins, On the Delaware River, 20 May 1804; and Collins to Humboldt, Philadelphia, 23 May 1804. Both letters in the Academy of Natural Sciences MS Collection, Zaccheus Collins Papers, letters 225 and 226; Schwarz, *Alexander von Humboldt und die Vereinigten Staaten von Amerika*, 87–88, letters 1, 2.

12 The baggage entry form Humboldt signed for the three travelers lists "twenty-seven boxes [illeg] collections for natural history." Baggage entry statement to the U.S. Customs, May 24, 1804, Alexander von Humboldt Papers, American Philosophical Society. Collins was also one of many Philadelphia-area plant specialists who desired to include Humboldt and Bonpland in their international botanical network.

13 Jean Théodoridès, "Les sejours aux Etats-Unis de deux savants européens du XIXe siècle: Alexander von Humboldt et Victor Jacquemond," *Archives internationales d'histoire des sciences* 16, no. 64 (1963): 290.

14 "I happened, while at the public library of Philadelphia, to cast my eyes on the table of contents of a scientific review, where I found these words: 'Arrival of M. de Humboldt's manuscripts at his brother's house at Paris, by way of Spain!' I could scarcely suppress an exclamation of joy; never did a table of contents appear to me better made." Alexander von Humboldt and Aimé Bonpland, *Personal Narrative of Travels to the Equinoctial Regions of the New Continent, during the Years 1799–1804*, trans. Helen Maria Williams (London: Longman, Rees, Orme, Brown, and Green, 1829), 7:287. This anecdote is also recounted in Richard Henry Stoddard, *The Life, Travels, and Books of Alexander von Humboldt* (New York: Rudd & Carleton, 1859), 303.

15 Quoted in Catherine L. Bachman, *John Bachman, D.D., LL.D., Ph.D., The Pastor of St. John's Lutheran Church, Charleston* (Charleston, SC: Walker, Evans & Cogswell, 1888), 391. See also Lester D. Stephens, *Science, Race, and Religion in the American South: John Bachman and the Charleston Circle of Naturalists, 1815–1895* (Chapel Hill: University of North Carolina Press, 2000).

16 Samuel Vaughan was good friends with Benjamin Franklin. A London businessman, he had investments in Jamaica and immigrated to Philadelphia, where he met George Washington in 1783. See David Beck Ryder, *West Indian Slavery and British Abolition, 1783–1807* (New York: Cambridge University Press, 2009), 64.

17 Archives of the American Philosophical Society, Minutes, 1793–1798, pp. 220 and 222; Frank William Peter Dougherty, *The Correspondence of Johann Friedrich Blumenbach*, ed. Norbert Klatt (Göttingen: Norbert Klatt, 2013), 5:242–44, letter 1139. Present at that meeting were Thomas Jefferson (APS president), Nicholas Collin, Caspar Wistar, James Woodhouse, Benjamin Smith Barton, John Vaughan, Charles Willson Peale, and Samuel Harrison Smith—all of whom Humboldt met on this trip.

18 François André Michaux (1770–1855) had been appointed by the French government to lead a scientific mission to catalogue North American trees, with an eye to finding hardy species to replenish French forests. He arrived in 1785, and in 1792 Jefferson enlisted the APS to sponsor Michaux to "explore the interior country of North America from the Missisipi [*sic*] along the Missouri, and Westwardly to the Pacific ocean." American Philosophical Society's Subscription Agreement for André Michaux's Western Expedition [ca. 22 January 1793]. Thomas Jefferson, *The Papers of Thomas Jefferson*, vol. 25, *1 January–10 May 1793*, ed. John Catanzariti (Princeton, NJ: Princeton University Press, 1992), 81–84. The Genet Affair scuttled those plans. Michaux spent ten years in the United States gathering information and specimens for publications that remain significant achievements in American botany.

19 Patterson was a mathematician who taught Lewis celestial navigation. Andrew Ellicott, who taught Lewis surveying techniques, had been one of Patterson's students in 1768. They remained close friends the rest of their lives.

20 Collin noted that most Americans could not read another language, and of the APS, "English books of all kinds are imported in considerable quantities and a large number of French, but not many Germans, for this language falls into disuse more and more and the majority of that people [the Germans] do not bother about learning it." Collin to Aurivillius, 30 November 1807, cited in Amandus Johnson, *The Journal and Biography of Nicholas Collin, 1764–1831* (Philadelphia: New Jersey Society of Pennsylvania, 1936), 95, 98. Collin also served as the APS librarian and curator. See *Early Proceedings of the American Philosophical Society 1744–1838*, ed. Henry Phillips Jr. (Philadelphia: Press of McCalla & Stavely, 1884), 142–612.

21 Charles Willson Peale diary, original spelling intact, in Lillian Miller et al., eds., *The Selected Papers of Charles Willson Peale and His Family*, vol. 2, *Charles Willson Peale: The Artist as Museum Keeper, 1791–1810*, part 2 (New Haven, CT: Yale University Press for the National Portrait Gallery, 1988), 683.

22 Of Bonpland, Humboldt attested: "I am extremely satisfied with my travelling companion [Aimé] Bonpland.... He arranges himself readily with foreign customs and people, speaks Spanish very well, is very plucky and courageous. He possesses splendid qualities for a travelling naturalist." Humboldt to Karl Willdenow, 21 February 1802, Havana. "Bonpland et moi nous avons toutes les raisons d'être extrémement contens....il est actif, travailleur, il s'adapte facilement aux moeurs et usages des hommes, parle très bien l'espagnol, et il est courageux et intrépide—en un mot, il a des qualités exquises pour un voyageur naturaliste." Humboldt, *Lettres américaines d'Alexandre de Humboldt*, 114; translated in H. Walter Lack, "Humboldt and Bonpland," in *Alexander von Humboldt and the Botanical Exploration of the Americas* (Munich: Prestel, 2009), 27.

23 In 1792 Woodhouse's thesis for his Doctor of Medicine degree was "On the Chemical and Medical Properties of the Persimmon Tree, and the Analysis of Astringent Vegetables," in which his experiments teased out the medicinal and useful properties of these plants, including how beneficial persimmon resin is for treating "fevers, for hemorrhoids, for dysentery, for diabetes, for 'spungy, swelled gums and loose teeth,' for 'stone in the urinary passages,' and for chronic ulcers" along with more mundane applications in the tanning of leather, for making ink and black dye, and to ferment it with yeast to make bread, or alcoholic spirits. Edgar F. Smith, *James Woodhouse: A Pioneer in Chemistry (1770–1809)* (Philadelphia: J. C. Winston, 1918; New York: Arno Press, 1980), 27–33.

24 Smith, *James Woodhouse*, 50–55

25 Alexander von Humboldt, *Reise auf dem Río Magdalena, durch die Anden und Mexico, Part Two: Übersetzung, Ammerkungen, Register*, ed. Margot Faak (Berlin: Akademie Verlag, 1990), 126. Translated in Maren Meinhardt, *A Longing for Wide and Unknown Things: The Life of Alexander von Humboldt* (London: Hurst, 2018), 202. Quinine from cinchona bark, native to South America, was used to suppress the symptoms of yellow fever. It remains the primary treatment for malaria. For the discovery of quinine, see James E. McClellan III, *Colonialism and Science: Saint Domingue in the Old Regime* (Chicago: University of Chicago Press, 1992), 161–62.

26 Smith, *James Woodhouse*, 94–95.

27 In England Humboldt had become acquainted with Sir Joseph Banks and Charles Blagden, the president and secretary, respectively, of the Royal Society, and met a young chemist, James Smithson. This network of scientists forms the basis of chapter 8 in this volume.

28 See A. Fothergill, *Letterbook*, one volume, 258 pp, MSS.B. F823, American Philosophical Society Archive.

29 Rush to Jefferson, Philadelphia, 28 May 1804, in Thomas Jefferson, *The Papers of Thomas Jefferson*, vol. 43, *11 March to 30 June 1804*, ed. James P. McClure (Princeton, NJ: Princeton University Press, 2017), 502–503.

30 Dr. Nicholas Collin, "An Essay on Those Inquiries in Natural Philosophy Which at Present Are Most Beneficial to the United States of America," delivered before the APS, April 3, 1789, in Gilbert Chinard, "The American Philosophical Society and the Early History of Forestry in America," *Proceedings of the American Philosophical Society* 89, no. 2, Symposium on Forestry and the Public Welfare (18 July 1945): 461; essay reprinted in Whitfield J. Bell and Nicholas Collin, "Nicholas Collin's Appeal to American Scientists," *William and Mary Quarterly* 13, no. 4 (October 1956): 531, 533.

31 In his diary for 7 February 1800, Humboldt described the deforestation at Lake Valencia as having triggered changes in the water evaporation cycle and in patterns of erosion. He concluded that cutting down the trees accelerated changes in the local climate. Alexander von Humboldt and Aimé Bonpland, *Personal Narrative of Travels to the Equinoctial Regions of the New Continent, during the Years 1799–1804*, trans. Helen Maria Williams (London: Longman, Hurst, Rees, Orme, and Brown, 1819), 4:143–44.

32 Johnson, *Journal and Biography of Nicholas Collin*, 52–53, 68, 111.

33 Essay reprinted in Bell and Collin, "Nicholas Collin's Appeal to American Scientists," 537.

34 Quoted in Hans Huth, *Nature and the American: Three Centuries of Changing Attitudes* (Lincoln: University of Nebraska Press, 1972), 17.

35 Peale to Timothy Matlack, Philadelphia, 9 March 1800, in Miller et al., *Selected Papers of Charles Willson Peale and His Family*, vol. 2, part 1, 283

36 William Thornton to John Vaughan, City of Washington, 6 July 1804, American Philosophical Society Archive.

37 Vaughan to Jefferson, Philadelphia, 29 May 1804, in Jefferson, *Papers of Thomas Jefferson*, 43:508–509. The letter alerts Jefferson to Humboldt's travel plans.

38 Wistar to Jefferson, Philadelphia, 29 May 1804, in Jefferson, *Papers of Thomas Jefferson*, 43:509.

39 Barton to Jefferson, Philadelphia, 28 May 1804, in Jefferson, *Papers of Thomas Jefferson*, 43:499.

40 Charles Willson Peale diary, 29 May 1804, in Miller et al., *Selected Papers of Charles Willson Peale and His Family*, vol. 2, part 2, 680–82. The route took them across the Schuylkill River in Philadelphia at Gray's Ferry into Delaware, along the stage route through Chester, Wilmington, Elktown, and Charlestown, to Havre de Grace, where they had supper before crossing the Susquehanna River by ferry, and resumed riding a stage through Joppa, to Baltimore.

41 Charles Willson Peale diary, in Miller et al., *Selected Papers of Charles Willson Peale and His Family*, vol. 2, part 2, 690–91.

42 Charles Coleman Sellers, *Charles Willson Peale*, vol. 2, *Later Life, 1790–1827* (Philadelphia: American Philosophical Society, 1947), 183.

43 Charles Willson Peale diary, 29 May–21 June 1804, in Miller et al., *Selected Papers of Charles Willson Peale and His Family*, vol. 2, part 2, 685, 688.

44 In three letters written to Jefferson, on June 15, 18, and 28, 1804, Peale makes reference to "[m]y son Rembrandt now at Baltimore exhibiting his Skeleton of the Mammoth." Jefferson, *Papers of Thomas Jefferson*, 43:601, 616, 663.

45 For a thoughtful and detailed assessment of this letter, see Ingo Schwarz, "Alexander von Humboldt's Visit to Washington and Philadelphia, His Friendship with Jefferson, and His Fascination with the United States," 43–56.

46 Pénétré de l'intérêt le plus vif pour la prospérité de l'espèce humaine sur le Sol des États unis, fruit de Votre sage législation et des vertus civiques de Vos Magistrats, j'ai pris la route de ces pays-ci pour me rendre en France. C'est une idée bien consolante pour moi qu'après avoir été témoin des grands phénomènes que présente la Nature magestueuse de la Cordillère des Andes, qu'après avoir vu ce qui est grand dans le monde physique, je puisse jouir du Spectacle moral que présente un peuple libre et digne de sa belle destinée. Vous, Monsieur, qui en homme d'état avez eu une influence si bienfaisante sur cette destinée,

recevez les hommages du respectueux attachement avec lequel j'ai l'honneur d'être Monsieur." Humboldt to Madison, Philadelphia, 24 May 1804, in Madison, *Papers of James Madison,* Secretary of State Series, 7:247–48; and Schwarz, ed., *Alexander von Humboldt und die Vereinigten Staaten von Amerika*, 90–91, letter 4. Translation in Helmut de Terra, "Alexander von Humboldt's Correspondence with Jefferson, Madison, and Gallatin," *Proceedings of the American Philosophical Society* 103, no. 6 (15 December 1959): 796.

47 "Monsieur le Président, Arrivé depuis le Mexique sur le sol heureux de cette Republique dont le Pouvoir exécutif a été confié à Vos lumières, c'est un doux devoir pour moi de Vous présenter mes respects et l'hommage de la haute admiration que Vos écrits, Vos actions et la libéralité de Vos idées m'ont inspiré dès ma plus tender jeunesse." Humboldt to Jefferson, Philadelphia, 24 May 1804, in Schwarz, *Alexander von Humboldt und die Vereinigten Staaten von Amerika*, 88–90, letter 3. Translated in de Terra, "Alexander von Humboldt's Correspondence with Jefferson, Madison, and Gallatin," 787–88. Humboldt had, indeed, read Jefferson's book while studying in Hamburg under Daniel Christoph Ebeling, who maintained an impressive library of materials sympathetic to the United States.

48 "je n'ai pas pu résister à l'intérêt moral de voir les États unis et de jouir de l'aspect consolant d'un peuple, qui sait apprécier le don précieux de la Liberté. Puisse-t-il m'être permis de Vous présenter personnellement mes respects et d'admirer en Vous de près un Magistrat philosophe, qui réunit les suffrages des deux Continens! Daignez excuser, Mr. le Président, la confiance et la longueur de cette lettre. J'ignore si mon nom Vous et connu par mon ouvrage sur le Galvanisme ou par les Mémoires de l'Institut National de Paris." Humboldt to Jefferson, Philadelphia, 24 May 1804, in Schwarz, *Alexander von Humboldt und die Vereinigten Staaten von Amerika,* 88–90, letter 3. Translated in de Terra, "Alexander von Humboldt's Correspondence with Jefferson, Madison, and Gallatin," 787–88.

49 "Ami des sciences Vous recevrez avec indulgence les hommages de mon dévouement. J'aimerais Vous parler encore d'un objet que Vous avez si ingénieusement traité dans Votre ouvrage sur la Virginie, des dents de Mammot que nous avons découverts dans les Andes de l'Hémisphère *austral* à 1700 toises de hauteur sur l'Océan Pacifique. Mon ami le C. Cuvier en donnera la description anatomique. Ce serait abuser de Vos bontés que de Vous entretenir plus longtems et je me borne à Vous répéter les assurances de la profonde vénération avec laquelle je serai toute me vie, Mr. le Président / Votre t[rès] h[umble] et t[rès] o[béissant] serviteur / Le Baron de Humboldt de l'Académie des Sciences de Berlin." Humboldt to Jefferson, Philadelphia, 24 May 1804, in Schwarz, ed., *Alexander von Humboldt und die Vereinigten Staaten von Amerika*, 88–90, letter 3. Translated in de Terra, "Alexander von Humboldt's Correspondence with Jefferson, Madison, and Gallatin," 787–88. A toise was a unit of measure in prerevolutionary France, corresponding to about six feet, or 72 inches.

50 Jefferson to Humboldt, Washington, 28 May 1804, in Jefferson, *Papers of Thomas Jefferson*, 43:501. Schwarz, *Alexander von Humboldt und die Vereinigten Staaten von Amerika*, 91–92, letter 5.

51 In 1804 tavern keeper Pontius D. Stelle bought the Washington City Hotel at the corner of First Street and New Jersey Avenue, on the site of the current Supreme Court building on Capitol Hill, and renamed it Stelle's Hotel. See Nina Tristani, "The Many Lives of the Supreme Court Site," *Capitol Hill Restoration Society News* (September 2015): 1. See also Charles O. Paullin, "History of the Site of Congressional and Folger Libraries, 1730–1950," *Records of the Columbia Historical Society* 37/38 (1937): 179.

52 Peale was carrying letters to both men. Thornton's was from John Vaughan, in which he wrote, "Mr Peale D Fothergill & D Collins, with the famous traveller Humbold & two of his 'Companions de Voyage' visit your city." 29 May 1804, William Thornton Papers, vol. 3, pp. 438–39, MSS division, Library of Congress.

53 Peale wrote to Jefferson: "The Baron requests me to present his compliments that he will do himself the pleasure to wait on you with Messrs. Bonpland & Montufar. Doctr Woodhouse also desires me to include his respects that he will also wait on you. Doctr Fothergill is not at present within, but I shall see him this afternoon—& I believe he will isteem [*sic*] your invitation, an honour not to be neglected—& therefore I answer for his attendance—. Ammediately [*sic*] after I have dined I will remove to the Tavern in your vicinity & there [*sic*] wait on you—." Note from Peale to Jefferson, Sunday [June] 3rd, 1804, in Jefferson, *Papers of Thomas Jefferson*, 43:532. William Thornton was also present for this first meeting. Anna Thornton Diary, 2 June 1804, in Anna Maria Brodeau Thornton Papers, 1793–1861, MMC-3412, Library of Congress.

54 Banneker and Ellicott knew each other growing up; Ellicott chose Banneker as his surveying assistant. *The George Town Weekly Ledger* noted: "Sometime last month arrived in this town Mr. Andrew Ellicot, a gentleman of superior astronomical abilities. He was employed by the President of the United States of America, to lay off a tract of land, 10 miles square, on the Potowmack, for the use of Congress;—is now engaged in this business, and hopes soon to accomplish the object of his mission. He is attended by Benjamin Banniker, an Ethiopian, whose abilities as a surveyor, and as an astronomer, clearly prove that Mr. Jefferson's concluding that that race of men was void of mental endowments was without foundation." Silvio A. Bedini, *The Life of Benjamin Banneker: The First African-American Man of Science* (New York: Scribner, 1971), 122; cited in George W. Paulson, *William Thornton, M.D.: Gentleman of the Enlightenment* (Columbus, OH: George W. Paulson, 2007): 200–201.

55 Letter from Thomas Munroe to Jefferson, 14 March 1803; Jefferson to Munroe, 21 March 1803. Cited in Saul Padover, ed., *Thomas Jefferson and the National Capital, Containing Notes and Correspondence Exchanged between Jefferson [et al.] Relating to the Founding, Surveying, Planning, Designing, Constructing, and Administering of the City of Washington, 1783–1818* (Washington, DC: Government Printing Office, 1946): 297–300. For a more extensive discussion of the metaphorical meaning of these plantings, see Andrea Wulf, *Founding Gardeners: The Revolutionary Generation, Nature, and the Shaping of the American Nation* (New York: Vintage Books, 2011), 146–49.

56 Charles Willson Peale diary, 29 May–21 June 1804, in Miller et al., *Selected Papers of Charles Willson Peale and His Family*, vol. 2, part 2, 692. For a history of the city, see Wilhelmus Bogart Bryan, *A History of the National Capital from Its Foundation through the Period of the Adoption of the Organic Act*, 2 vols. (New York: Macmillan, 1914–16).

57 See Friis, "Baron Alexander von Humboldt's Visit to Washington, D.C., June 1 through June 13, 1804," 4–8, for a comparison of the two cities at this time. Andrew Ellicott viewed dimly the choice of Washington, D.C., for the capital, confiding to his wife that "this country, intended for the permanent residence of Congress, bears no more proportion to the country about Philadelphia and Germantown, for either wealth or fertility, than a crane does to a stall-fed ox." Quoted in Anne Hollingsworth Wharton, *Social Life in the Early Republic* (Philadelphia: J. B. Lippincott, 1902), 26–27.

58 Jefferson sat for Rembrandt Peale on January 23, 24, and 31 in 1805, just prior to his inauguration for his second term as president. On the eve of the inauguration, March 3, 1805, the painting was on display at Peale's Museum in Philadelphia, under artificial light for the extended viewing. The painting remained there until the museum's dispersal. See Alfred L. Bush, *The Life Portraits of Thomas Jefferson* (Charlottesville, VA: Thomas Jefferson Memorial Foundation, 1987).

59 Margaret Bayard Smith acerbically noted: "Baron Humboldt, formed not his estimate of men and manners, by their habiliments and conventionalisms, and refined as were his tastes, and polished as were his manners, he was neither shocked nor disgusted, as was the case with the British Minister (Mr. Foster) by the old fashioned form, ill-chosen colours, or simple material of the President's dress. Neither did he remark the deficiency of elegance in his person, or of polish in his manners, but indifferent to those external and extrinsic circumstances, he easily discerned, and most highly appreciated the intrinsic qualities of the Philosophic Statesman through even the homely costume, which had concealed them from the ken of the fastidious diplomat." Margaret Bayard Smith MSS "The President's house forty years ago" in her *Commonplace Book,* pp. 15–16, Papers of Mrs. Samuel Harrison Smith, MSS Division, Library of Congress. See also Margaret Bayard Smith, *The First Forty Years of Washington Society, Portrayed by the Family Letters of Mrs. Samuel Harrison Smith (Margaret Bayard) from the Collection of her Grandson, J. Henley Smith*, ed. Gaillard Hunt (New York: C. Scribner's Sons, 1906), 395–96.

60 Transcribed in Gerard W. Gawalt, "'Strict Truth': The Narrative of William Armistead Burwell," *Virginia Magazine of History and Biography* 101, no. 1 (January 1993): 114–15.

61 Charles Willson Peale diary, 29 May–21 June, 1804, in Miller et al., *Selected Papers of Charles Willson Peale and His Family*, vol. 2, part 2, 690.

62 The day Humboldt met Pichon, Peale noted in his diary that he couldn't find a ride and ended up walking the three miles from Georgetown back to their hotel "in broiling Sun in a suit of Black cloaths," ahead of the group. Charles Willson Peale diary, in Miller et al., *Selected Papers of Charles Willson Peale and His Family*, vol. 2, part 2, 691. For Humboldt's visit with Merry, see Anna Thornton Diary, 9 June 1804, Anna Maria Brodeau Thornton Papers, 1793–1861, MMC 3412, Library of Congress.

63 Charles Willson Peale diary, in Miller et al., *Selected Papers of Charles Willson Peale and His Family*, vol. 2, part 2, 691.

64 William Thornton (1759–1828) was born in the Virgin Islands and educated in Great Britain. He met James Smithson on a trip to Staffa in 1784 (see chapter 8 in this volume). He immigrated to Philadelphia in 1786 and moved to Washington in the 1790s. Thornton was the architect for the Octagon and Tudor Place, and he battled with Benjamin Latrobe over the design for the U.S. Capitol. Charles Willson Peale diary, in Miller et al., *Selected Papers of Charles Willson Peale and His Family*, vol. 2, part 2, 697. In 1802 Madison established the Patent Office within the State Department and appointed Thornton as its superintendent, a position he held the rest of his life. Paulson, *William Thornton, M.D.*, 244.

65 Charles Willson Peale diary, in Miller et al., *Selected Papers of Charles Willson Peale and His Family*, vol. 2, part 2, 693, 698. It should be noted that when several members of the Peale family had gathered to paint George Washington from life in 1795, Stuart had commented to Martha Washington that her husband was being "Peale'd all around." See William Dunlap, *History of the Rise and Progress of the Arts and Design in the United States* (New York: George P. Scott, 1832), 1:206. Dolley Madison expressed her relief that her portraits and that of her husband James were finished; she noted that friends worried "Stuart is all the rage, he is almost worked to death, and every one afraid that they will be the last to be finished." Dolly Madison, *Memoirs and Letters of Dolly Madison: Wife of James Madison, President of the United States*, ed. Lucia B. Cutts (Boston: Houghton Mifflin, 1886), 35.

66 Thomas Law (1756–1834) was the son of the bishop of Carlisle in Great Britain. He made a considerable fortune during his youth in India, and arrived in Washington, D.C., during the early 1790s. In 1796 he married Eliza Parke Custis, Martha Washington's granddaughter.

67 "We first went to the Library, where Mr. Beckley received us with politeness, and also accompanied us through the other apartments. The Library is a spacious and handsome Room, and although lately organized, already contained a number of valuable books in the best taste of binding—we next visited the Chamber where the senate satt [*sic*]. it is an Elegant room with a very lofty Galleries adorned with large and Massy Pillars—This room is warmed by flues from fires made beneath it. We see the Room where the House of Deligates [Representatives] used to set—a Tempory building within the walls of the foundation of south wing—This the workmen were beginning to pull down. We also went to the Top of the Capitol & had a delightful view of all the surrounding Country." Charles Willson Peale diary, in Miller et al., *Selected Papers of Charles Willson Peale and His Family*, vol. 2, part 2, 692.

68 A "Serbonian bog" refers to the Serbonis Lake in Egypt, as described by Herodotus. Because sand blew across the lake's surface, it appeared to be solid ground but was, in fact, a marshy bog. It was a favorite metaphor during this period; Edmund Burke used it to describe the failed aspirations of the French Revolution: "Here end all the deceitful dreams and visions of the equality and rights of men. In 'the Serbonian bog' of this base oligarchy they are all absorbed, sunk, and lost for ever." Edmund Burke, *Reflections on the Revolution in France* (1790) (New York: Arlington House, 1965), 242. For the description of Pennsylvania Avenue as a Serbonian bog, see Ihna Thayer Frary, *They Built the Capitol* (Freeport, NY: Books for Libraries Press, 1969), 58.

69 Margaret Brent Downing, "Literary Landmarks," *Records of the Columbia Historical Society* 19 (1916): 28; quoted in Friis, "Baron Alexander von Humboldt's Visit to Washington, D.C., June 1 through June 13, 1804," 11.

70 Charles Willson Peale diary, in Miller et al., *Selected Papers of Charles Willson Peale and His Family*, vol. 2, part 2, 693.

71 Anna Thornton's June 5, 1804, diary entry confirms Thornton's presence. Anna Maria Brodeau Thornton Papers, 1793–1861, MMC3412, Library of Congress.

72 As late as March 1803, Jefferson and Gallatin were exchanging letters, trying to determine the course of the Rio Norte. Gallatin had arranged to have cartographer Nicholas King draw a map compiling the information from as many extant maps as possible. On March 20, Jefferson replied with some frustration: "I do not find in my library any thing which can throw light on the geography of the Rio Norte. I do not believe that in modern times any thing has been added to the information given as to that river in early times." The arrangements for the purchase of the Louisiana Territory were signed on April 30, 1803. Jefferson's instructions to Lewis included picking up whatever information he could about the course of the river. See Joel Kovarsky, *The True Geography of Our Country: Thomas Jefferson's Cartographic Vision* (Charlottesville: University of Virginia Press, 2014), esp. 60–61, 120.

73 Gallatin ended his letter by stating, "From mind I fall to body, and you will be better pleased in hearing that I enjoy perfect health then with the seven thousand new plants of Humboldt." Albert Gallatin to Hannah Gallatin, 6 June 1804, in Albert Gallatin Papers, 1794–1952, The New-York Historical Society.

74 Jefferson to Humboldt, 9 June 1804, in Jefferson, *Papers of Thomas Jefferson*, 43:554. Schwarz, *Alexander von Humboldt und die Vereinigten Staaten von Amerika*, 92–93, letter 6.

75 "Memorandum from Alexander von Humboldt," [on or after June 9, 1804], in Jefferson, *Papers of Thomas Jefferson*, 43:555–57.

76 Jefferson to Wistar, Washington, 7 June 1908, in Jefferson, *Papers of Thomas Jefferson*, 43:548–49.

77 "On Saturday last arrived in this city the BARON HUMBOLDT, accompanied by two of his fellow travellers in South America, and by Dr. Fothergill, of Bath, (England,) D. Woodhouse, Dr. Collin, and Mr. Peale of Philadelphia. The Baron Humboldt is a distinguished philosopher of Prussia, of which country he is a native. Animated by the wide field for research presented by the New World … he has been actively engaged in exploring the interior of South America; in making himself acquainted with the manners of the inhabitants, their numbers and their political economy; in making astronomical observations, for which he was provided with the best instruments; in examining the animal and vegetable kingdoms; and in becoming acquainted, from personal observation, with the mountains, volcanoes and mines, so numerously scattered over the country. He has had the singular good fortune of having endured, for a considerable length of time, a degree of heat, in which the mercury stood at 105 degrees, without experiencing one day's sickness; and of passing through a greater variety of climates than are to be found in any other quarter of the globe.—Among his proudest trophies is the having gained a higher eminence of the Chimborasso, (itself by far the highest mountain in the world) than had ever before been reached by natives or foreigners. He is about returning to Europe to publish the result of his researches, which we have no doubt will not only give new views of a large portion of South America, but likewise rescue the accounts of that region from the fiction and incorrectness with which they are at present enveloped." *National Intelligencer and Washington Advertiser*, 8 June 1804.

78 Albert Gallatin to Jefferson, Washington, 27 June 1804, in Jefferson, *Papers of Thomas Jefferson*, 43:659–60.

79 Samuel Harrison Smith's wife, Margaret Bayard Smith, recalled: "His visits to the President's-House, were unshackled by mere ceremony and not limited to any particular hour. One evening he called about twilight and being shown into the drawing room without being announced, he found Mr. Jefferson seated on the floor, surrounded by half a dozen of his little grandchildren so eagerly and noisily engaged in a game of romps that for some moments his entrance was not perceived. When his presence was discovered Mr. Jefferson rose up and shaking hands with him, said, 'you have found me playing the fool Baron, but I am sure to you I need make no apology.'" Smith, *The First Forty Years of Washington Society*, 396; and Margaret Bayard Smith, *A Winter in Washington; or, Memoirs of the Seymour Family* (New York: E. Bliss & E. White, 1824), 2:34.

80 Thomas Jefferson, "Rules of Etiquette," November 1803. Quoted in Paul L. Ford, ed., *Autobiography of Thomas Jefferson* (Philadelphia: University of Pennsylvania Press, 1892–1899), 8:276; cited in Andrea Wulf, *Founding Gardeners*, 146.

81 "It is with some interest that I have learned from the Baron Humboldt, who has been five years travelling through South America, that among other curious animal remains, he has (and he is the first that has) discovered several specimens of the Mammoth, perfectly distinguished by the great carnivorous teeth. He has found them as far as latitude 33 south, but always on the highest mountains, which the Baron take to be satisfactory proof that this Great Unknown must have been, as has been generally supposed, the inhabitant of a cold climate. In North America none of these bones have ever been found but in comparatively low situations; this is to be expected of an animal, which in a cold climate would inhabit the vallies, and in a warm one would seek the cold retreats of the mountains." Rembrandt Peale, "Communication," *National Intelligencer and Washington Advertiser*, 6 June 1804.

82 See chapter 3 in this volume for an extended discussion of mastodon bones found in the United States.

83 Charles Willson Peale Autobiography, typescript page 272, American Philosophical Society Archive. Peale had this published in a column titled "People of America and Citizens of Philadelphia," *Aurora* (27 January 1800); quoted in Miller et al., *Selected Papers of Charles Willson Peale and His Family*, vol. 2, part 1, 274–75.

84 Peale to Jefferson, from the Museum, 12 January 1802, expresses the hope that "in the end these labours would be crowned in a National Establishment of my Museum...I wish to know your sentiments on this subject, whither the United States would give an encouragement, and make provision for the establishment of this Museum in the City of Washington." Thomas Jefferson, *The Papers of Thomas Jefferson*, vol. 36, *1 December 1801–3 March 1802*, ed. Barbara B. Oberg (Princeton, NJ: Princeton University Press, 2009), 365–66. Peale enclosed a draft of a statement he wanted to submit to the Pennsylvania State Legislature, giving them first right of refusal to absorb the museum or subsidize it. Jefferson, *Papers of Thomas Jefferson*, 36:367. Peale had already floated this idea to Madison on 21 May 1802: "I am anxious to discover from the most respectable sources whether any and what expectations may be indulged of the Museum ever being taken under Public patronage. This, as it has been my grand Ultimatim, has occupied much of my thoughts; and I am convinced that a vast deal may be done in a permanent way with but a trifling expence—no needlessly expensive building and a small fund." C. W. Peale to James Madison, Museum, 21 May 1801, in James Madison, *The Papers of James Madison*, Secretary of State Series, vol. 1, *4 March–31 July 1801*, ed. Robert J. Brugger et al. (Charlottesville: University Press of Virginia, 1986), 218–19.

85 Jefferson to Charles Willson Peale, Washington, 16 January 1802, in which he commends Peale for his unwavering devotion to the museum but cautions, "you know that one of the great questions which has divided political opinion in this country is Whether Congress are authorised by the constitution to apply the public money to any but the purposes specially enumerated in the Constitution?" Jefferson does raise the idea that if he were to found a university, he would want Peale's Museum to be a part of it, "but the moment is not arrived for proposing this with a hope of success." Jefferson, *Papers of Thomas Jefferson*, 36:385–86.

86 After touring Humboldt through the city and commenting on how well construction was progressing, Peale noted in his diary: "All these Circumstances made me at this moment think it very probable that some provision will be made to induce me to remoove my Museum to this City, and I think the time not far distant that some proposials will be made to that purpose." Charles Willson Peale diary, in Miller et al., *Selected Papers of Charles Willson Peale and His Family*, vol. 2, part 2, 693.

87 Charles Willson Peale diary, in Miller et al., *Selected Papers of Charles Willson Peale and His Family*, vol. 2, part 2, 694.

88 The network of people and ideas connecting Humboldt and the founding of the Smithsonian is the subject of chapter 8 in this volume.

89 Virginia Governor Major-General Henry "Light-Horse Harry" Lee, *Funeral Oration on the Death of George Washington* (Boston: Joseph Nancrede and Manning & Lorining, 1799), 14.

90 Friis, "Baron Alexander von Humboldt's Visit to Washington, D.C., June 1 through June 13, 1804," 24.

91 Charles Willson Peale diary, in Miller et al., *Selected Papers of Charles Willson Peale and His Family*, vol. 2, part 2, 694.

92 Jenny Graham, *The Nation, The Law, and The King: Reform Politics in England, 1789–1799* (New York: University Press of America, 2000), 1:51.

93 Samuel Vaughan to Humphry Marshall, 14 May 1785, Philadelphia, Series X, Manuscripts, Box 10/4, Humphry Marshall Papers, USDA History Collection, Special Collections, National Agricultural Library, Washington, D.C. Published in William Darlington, *A Memorial of John Bartram and Humphry Marshall* (Philadelphia: Lindsay and Blakiston, 1849): 556–57.

94 Quoted in Matthias Schönhofer, *Letters from an American Botanist: The Correspondence of Gotthilf Heinrich Ernst Mühlenberg (1753–1815)* (Stuttgart: Franz Steiner Verlag, 2014): 179n104.

95 In *Founding Gardeners*, Andrea Wulf describes an excursion to Bartram's Garden that summer by delegates to the convention. Frustrated by an impasse in the negotiations over how each state would be represented in the new union, a group of these men made an early morning excursion to John Bartram's garden to clear their heads. The display of "horticultural unity" would have had powerful resonance in a group made up primarily of agrarian-minded delegates. See Wulf, *Founding Gardeners*, 73–76.

96 See the plant lists in Susan Schoelwer, ed., *The General and the Garden: George Washington's Landscape at Mount Vernon* (Mount Vernon, VA: Mount Vernon Ladies' Association, 2015), 143–61. Original lists in George Washington Papers, Series 4, General Correspondence, no. 1697–1799, Library of Congress.

97 George Washington visited William Hamilton at a different property, "Bush Hill," many times while he was working on Mount Vernon. Washington to Samuel Vaughan, 14 January 1784; Washington to William Hamilton, 15 January and 6 April 1784. John C. Fitzpatrick, ed., *The Writings of George Washington from the Original Manuscript Sources 1745–1799* (Washington, DC: United States Government Printing Office, 1938), 27:297–99, 303, 388. Cited in Sarah P. Stetson, "William Hamilton and His 'Woodlands,'" *Pennsylvania Magazine of History and Biography* 73, no. 1 (January 1949): 29.

98 See Alexander von Humboldt with Aimé Bonpland, *Personal Narrative of Travels to the Equinoctocal Regions of America, during the Years 1799–1804*, trans. Helen Maria Williams (London: Longman, Hurst, Rees, Orme, & Brown, 1822), 2:239–40.

99 Charles Willson Peale diary, in Miller et al., *Selected Papers of Charles Willson Peale and His Family*, vol. 2, part 2, 696.

100 Charles Willson Peale diary, in Miller et al., *Selected Papers of Charles Willson Peale and His Family*, vol. 2, part 2, 696.

101 Brissot wrote, "Le docteur Collins...est un savant d'un rare mérite....Le docteur est un fervent apôtre de la liberté." Jacques Pierre, Brissot de Warville, *Nouveau voyage dans les États-Unis de l'Amérique Septentrionale, etc. (fait en 1788)* 2:52; quoted in Johnson, *Journal and Biography of Nicholas Collin, 1764–1831*, 193. For Collin on the abolition of slavery, see Collin to Professor Thunberg of Uppsala, 12 May 1801, in Johnson, *Journal and Biography of Nicholas Collin*, 196. The Marquis de Lafayette was also a vocal member of Brissot's abolitionist society in Paris.

102 Charles Willson Peale diary, in Miller et al., *Selected Papers of Charles Willson Peale and His Family*, vol. 2, part 2, 696, 695.

103 Peale goes into great detail regarding the trip back into Washington, which involved tipsy drivers engaged in a race to see who could block the other's attempts to pass, with Fothergill urging on his driver to take more risks, and Peale and Collin pleading for restraint. Charles Willson Peale diary, in Miller et al., *Selected Papers of Charles Willson Peale and His Family*, vol. 2, part 2, 696–97.

104 The importance of the Paris salons for Humboldt and American travelers is a topic in chapter 4 in this volume.

105 G. S. Wilson, *Jefferson on Display: Attire, Etiquette, and the Art of Presentation* (Charlottesville: University of Virginia Press, 2018), 145.

106 Charles Willson Peale diary, in Miller et al., *Selected Papers of Charles Willson Peale and His Family*, vol. 2, part 2, 693.

107 Louisa Catherine Johnson Adams to John Quincy Adams, 6 June 1804, from Washington, D.C. Her husband responded from Quincy on June 17: "The name of the Prussian Gentleman, whom you mention as having been at Washington is *Humboldt*—I think I recollect hearing of him, and of his voyage to South America while we were at Berlin." John Quincy Adams to Louisa Catherine Johnson Adams, Quincy, MA, 17 June 1804, Adams Papers, Massachusetts Historical Society, Boston.

108 Anna Thornton daily record book (1793–1804), entries for 6 and 7 June 1804, in Anna Maria Brodeau Thornton Papers, 1793–1861, MMC3412, Library of Congress.

109 Anna Thornton daily record book (1793–1804), Anna Maria Brodeau Thornton Papers, 1793–1861, Library of Congress. Charles Willson Peale's diary mentions Gilbert Stuart but leaves out several others.

110 Charles Willson Peale diary, in Miller et al., *Selected Papers of Charles Willson Peale and His Family*, vol. 2, part 2, 698.

111 Anna Thornton daily record book (1793–1804), entries for 7 and 25 June 1804, in Anna Maria Brodeau Thornton Papers, 1793–1861, Library of Congress. Thornton's diary for Monday, 25 June: "The drawing of the Cinchona finished by Dr. T. & Sent to the Baron Humboldt." Cinchona was used to combat yellow fever, and as such was a plant of particular interest to Humboldt, as well as to Thornton and Benjamin Rush, among other members of the APS.

112 Samuel Harrison Smith had been secretary of the APS, and when Jefferson became president, he and Margaret Bayard married and then moved to Washington. Margaret's political views at first ran counter to Jefferson's, but over time she became one of his fiercest advocates. Together they became two of Jefferson's most reliable supporters. In return, Jefferson steered to the paper coverage of Congress, making the *National Intelligencer* the precursor to the *Congressional Record*. That gave Smith unparalleled access to the political powers in the city and made his stories among those most often quoted by newspapers in other American cities.

113 For a thoughtful discussion of how silhouettes functioned in Philadelphia at this time, see David R. Brigham, *Public Culture in the Early Republic: Peale's Museum and Its Audience* (Washington, DC: Smithsonian Institution Press, 1995), 68–82.

114 "I had brought with me sundry Profiles of the Baron, & of Doctr Fothergill & myself—Doctr Collin, these I distributed to such Persons as I thought they would be most acceptable with Mrs. Smith her Sister & sister in Law in conversation about our departure from Washington the next morning Mrs. Smith requested some of said profiles, her Sister having got those I had presented in the morning—." Charles Willson Peale diary, in Miller et al., *Selected Papers of Charles Willson Peale and His Family*, vol. 2, part 2, 699.

115 Peale mentioned in his diary that he had brought his own silhouette to distribute; however, his likeness does not appear in any of the known sets either. James Woodhouse is also not included in this set of silhouettes, suggesting he may have been a late addition to the group, possibly making his own way to Washington.

116 Peale to Jefferson, Baltimore, 15 June 1804, in Miller et al., *Selected Papers of Charles Willson Peale and His Family*, vol. 2, part 2, 711–12. Peale intended to have copies available in Baltimore "in the Room of my Sons Exhibition of the Skeleton of the Mammoth." Peale to Jefferson, Baltimore, 18 June 1804. Jefferson replied, sending two profiles made by William Burwell. Jefferson to Peale, Washington, 20 June 1804, in Jefferson, *Papers of Thomas Jefferson*, 43:601, 616, 625.

117 Margaret Bayard Smith to Mary Ann Smith, Washington, 10 June 1804, in the Papers of Mrs. Samuel Harrison Smith, vol. 8, folios 67147–67148, MSS. division, Library of Congress; quoted in Friis, "Baron Alexander von Humboldt's Visit to Washington, D.C., June 1 through June 13, 1804," 30.

118 Dolley Payne Todd Madison to Anna Payne Cutts, June 1804, in Madison, *Memoirs and Letters of Dolly Madison*, 45–46.

119 Thornton to Vaughan, City of Washington, 6 July 1804 in the American Philosophical Society Archive.

120 Charles Willson Peale diary, in Miller et al., *Selected Papers of Charles Willson Peale and His Family*, vol. 2, part 2, 700–701. Peale noted that he, Collin, and Woodhouse returned to Philadelphia by way of Annapolis; Anna Thornton noted that Dr. Fothergill was also part of the group. Anna Thornton daily record book, entry for 9 June 1804, in Anna Maria Brodeau Thornton Papers, 1793–1861, Library of Congress. Fothergill sent a note to Jefferson on the eve of their departure reading, "Dr. Fothergill cannot leave the City of Washington without offering his grateful acknowledgments to the President of the U.S. for his repeated civilities, & kind hospitality. Wishing him uninterrupted health & prosperity, begs leave to request his acceptance of these little tracts as a slender token of remembrance." Letter sent through "Mr. Maddison," 8 June 1804, in Thomas Jefferson Papers, Library of Congress; and Jefferson, *Papers of Thomas Jefferson*, 43:550.

121 "I propose to-day to take an ultimate view of our instructions to our negociators in Spain, and chiefly to decide whether any later views, and particularly those from Baron Humboldt should occasion alterations of opinion in any point." Thomas Jefferson to Albert Gallatin, 3 July 1804, Gallatin Papers, New-York Historical Society.

122 There appear to have been at least two copies of the map, one commissioned by Gallatin from Mr. Kraft, the other made surreptitiously by General James Wilkinson for use by Zebulon Pike. On February 23, 1805, Gallatin wrote to Jefferson indicating he still had Humboldt's original map, and with Kraft's death, could not get another copy made. Albert Gallatin to Thomas Jefferson, 23 February 1805, Papers of Thomas Jefferson, Library of Congress.

123 Wilkinson's questions generally mirror Jefferson's from June 9 on population statistics and distances between cities and towns. Wilkinson to Jefferson, Washington, 11 June 1804, in Thomas Jefferson Papers, Library of Congress; and Jefferson, *Papers of Thomas Jefferson*, 43:573–74.

124 Mühlenberg was one of Jefferson's strong supporters. In a letter dated 14 February 1806, to George Logan, he wrote, "With infinite Satisfaction I daily hear, that the Gentleman I so long respected, Mr. Jefferson, the Friend of the People and Sciences is indefatigable in preserving national Prosperity and Dignity, and enlarging the Boundaries of Science, especially natural History." Logan Family Papers, 1664–1871, Collection 379, Historical Society of Pennsylvania, Philadelphia; cited in Matthias Schönhofer, "'Runs, Creeks, and Rivers Join': The Correspondence Network of Gotthilf Henry Ernst Mühlenberg," in *A Peculiar Mixture: German-Language Cultures and Identities in Eighteenth-Century America*, ed. Jan Stievermann and Oliver Scheiding (University Park: Pennsylvania State University Press, 2013), 178n88.

125 Michaux called Mühlenberg "one of the most learned botanists that America has produced, who deserves a distinguished place among the most eminent in that pleasing science in Europe." François André Michaux, *The North American Sylva* (Philadelphia: J. Dobson, 1841), xviii.

126 Henry Ernest Mühlenberg papers, Collection 1442, 1781–1816, Historical Society of Pennsylvania, Philadelphia, http://www2.hsp.org/collections/manuscripts/m/Muhlenberg0443.html.

127 Mühlenberg to Dawson Turner, 21 February 1803, and Mühlenberg to Schreber, 4 October 1802, UAE Briefnachlass Schreber. Both quoted in Matthias Schönhofer, "'Runs, Creeks, and Rivers Join,'" 162, 162nn83–84.

128 Andrew Ellicott, *The Journal of Andrew Ellicott: Late Commissioner on Behalf of the United States during Part of the Year 1796, the Years 1797, 1798, 1799, and Part of the Year 1800 for Determining the Boundary between the United States and the Possessions of His Catholic Majesty in America* ... (Philadelphia: Printed by William Fry, 1814), 206–207. In Lancaster, Humboldt also met William Barton. He was Benjamin Smith Barton's brother, and in 1802 William had published *A Dissertation on the Freedom of Navigation and Maritime Commerce, and Such Rights of States Relative Thereto, as Are Founded on the Law of Nations* (Philadelphia: J. Conrad, 1802), which he dedicated to Thomas Jefferson, for his roles as president of both the United States and the APS.

129 There is no evidence in Mühlenberg's papers that Humboldt ever replied to him concerning his queries; however, there is an undated letter from Humboldt to Michaux regarding American trees at the APS, suggesting the two men were in contact after Humboldt's return from the United States. André Michaux Letters and Papers, 1783–1890, Mss.B.M58, http://amphilsoc.org/collections/view?docId=ead/Mss.B.M58-ead.xml#d133853e1563368095744.

130 "My dear and honored friend, I use these last moments before my departure tomorrow, to express once more my heartfelt thanks to you for the great kindness which you showed me and my friends in Lancaster. Your grasses and your kind letter have come to hand and I promise to let you know Michaux's names for them. Bonpland commends himself to you with gratitude for your kind remembrance. Yesterday we had a delightful noon with your brother the worthy General. Remember me to the good Ellicott and Mr. William Barton." "Mein theurer Verehrungswerther Freund, Ich benuze diesen letzen Augenblick vor unserer morgenden Abreise um Ihnen nochmals meinen herzlichsten Dank für die unbeschreibliche Güte zu sagen mit welcher. Sie mich und meine Freunde in Lancaster behandelt haben." Humboldt to Mühlenberg, 27 June 1804, APS Film 1097. Translated in Paul A. W. Wallace, "Henry Ernest Muhlenberg," *Proceedings of the American Philosophical Society* 92, no. 2 (5 May 1948): 109. "Your brother the general" was Peter Mühlenberg, to whom Jefferson wrote a letter of introduction for the Baron. If Humboldt sent Mühlenberg the information he sought, the botanist never received it: "Ich hatte Hrn Baron v Humboldt 160 Specimina der hiesig Graminum und plantar calam mitgegeb um sie in Paris mit Michauxs herbario zu vergleichen, habe aber nie irgend eine Antwort erhalten," translated as "I had given Baron von Humboldt 160 specimens of the local Graminum and plantar calam to compare them with Michaux's herbarium in Paris, but never received any answer." Mühlenberg to Schreber, 18 March 1806, UAE Briefnachlass Schreber, quoted in Schönhofer, *Letters from an American Botanist*, 302n197. English translation by the author.

131 Jefferson also sent seeds from the Lewis and Clark Expedition to Hamilton for propagation, believing Hamilton had the requisite skill to cultivate unfamiliar seeds. Jefferson to Hamilton, 6 November 1805 and 22 March 1807; and Jefferson to Bernard M'Mahon, 22 March 1807, in Thomas Jefferson Papers, Library of Congress; quoted in Edward Morris Betts, *Thomas Jefferson's Garden Book, 1766–1824, with Relevant Extracts from His Other Writings* (Charlottesville, VA: Thomas Jefferson Memorial Foundation, 1999), 307, 344. Also cited in Stetson, "William Hamilton and His 'Woodlands,'" 31.

132 After visiting Hamilton's garden, Humboldt expressed a desire to make a side trip to New York and meet David Hosack, an internationally admired gardener who was among the first Americans elected as members of the Royal Horticultural Society of England. Hosack's Elgin Gardens (which would become the New York Botanical Garden) had opened in 1801, and Hosack said, "[W]e have every reason to expect that the establishment of a botanic garden…will…in a short time meet with the general design of protecting useful knowledge, and those arts which are most essentially connected with human happiness." David Hosack, *Introduction to Medical Education* (New York, 1801); cited in Therese O'Malley, "Landscape Gardening in the Early National Period," in *Views and Visions: American Landscape before 1830*, ed. Edward J. Nygren with Bruce Robertson (Washington, DC: Corcoran Gallery of Art, 1986), 155.

133 Ulrich Päßler, "Plantae des États-Unis: Rediscovered Pages from Alexander von Humboldt's United States Diary," *Internationale Zeitschrift für Humboldt-Studien* 18, no. 35 (2017): 10. Humboldt routinely scissored his correspondence and his notes in order to collect information on specific topics in designated envelopes, using sealing wax to glue together bits of paper. This would become his filing system in preparation for writing *Cosmos*. For an image of Humboldt's collaged notes, see figure 14 in this volume.

134 *Proceedings of the American Philosophical Society* 22, no. 119, part III. Early Proceedings of the American Philosophical Society for the Promotion of Useful Knowledge, Compiled by One of the Secretaries, from the Manuscript Minutes of Its Meetings from 1744–1838 (July 1885), 358.

135 The invitation read, "Dear Sir: Baron Humbold [*sic*] has returned from Washington and engaged to spend the evening with us. If you are disengaged I will be very happy to see you, and am with great esteem / Your friend and servant / C. Wistar, Junr. / Thursday / June 21st, 1804." Letter to B. R. Morgan, Esq., member of the Philadelphia Bar; quoted in Hampton L. Carson, Esq., *The Centenary of the Wistar Party: An Historical Address* (Philadelphia: Wistar Association, 1918), 8.

136 Carson, *Centenary of the Wistar Party*, 17–18. Also see the Duke of Saxe-Weimar-Eisenach in his *Travels through North America during the Years 1825–26* (New York: G. & C. Carvill, 1828). See chapter 4 in this volume for discussion of the duke's travels to North America.

137 John F. Mifflin (1759–1813) was Charles Willson Peale's lawyer and commission merchant in Philadelphia. He was made a member of the APS in 1796.

138 Charles Willson Peale to Raphaelle and Rembrandt Peale, from the Museum, 21 June 1804, in Miller et al., *Selected Papers of Charles Willson Peale and His Family*, vol. 2, part 2, 716–17.

139 Charles Willson Peale to Raphaelle and Rembrandt Peale, from the Museum, 23 June 1804, in Miller et al., *Selected Papers of Charles Willson Peale and His Family*, vol. 2, part 2, 720–21.

140 Peale to John DePeyster, from the Museum, 27 June 1804, in Miller et al., *Selected Papers of Charles Willson Peale and His Family*, vol. 2, part 2, 725.

141 "The Portrait however meets with the approbation of every one that has yet seen it. I intended it as a proof of my being yet able to do any thing however difficult it might be that I could have done in an earlier time of life." *Poulson's American Daily Advertiser* illustrated the portrait in the issue for August 13, 1804. Charles Willson Peale diary, in Miller et al., *Selected Papers of Charles Willson Peale and His Family*, vol. 2, part 2, 728–29.

142 Peale to John DePeyster, from the Museum, 27 June 1804, in Miller et al., *Selected Papers of Charles Willson Peale and His Family*, vol. 2, part 2, 725.

143 Bachman, *John Bachman*, 173.

144 Quoted in Bachman, *John Bachman*, 391. Bachman's eulogy was first delivered to the German societies in Charleston, South Carolina, in 1869, and printed in *XIX Century* 1, no. 5 (October 1869): 402–405.

145 In 1805 Humboldt wrote to John Vaughan, asking him to convey regards to "l'aimable & savant M Barton, le bon très bon Mr Wistar, Peale, Hare, Pierce Butler, Woodhouse, Patterson Ellicott, Colin [*sic*] Mifflin. Mille respects si vous leur écrivez au Président & aux grands hommes d'état Gallatin & Madison. N'oubliez pas Mr. Smith & l'Ingénieux D Thornton à Washington, D Fothergill & le Quaker Samuel Rhoads," translated as "the amiable & learned M Barton, the very good Mr Wistar, Peale, Hare, Pierce Butler, Woodhouse, Patterson Ellicott, Colin [*sic*] Mifflin. A thousand respects if you write to the President & great statesmen Gallatin & Madison. Do not Forget Mr. Smith & Clever D Thornton in Washington, D Fothergill & Quaker Samuel Rhoads." Schwarz, *Alexander von Humboldt und die Vereinigten Staaten von Amerika,* 105–106, letter 19. Rhoads was a Philadelphia merchant Humboldt met on June 28. For Humboldt's visit, see Elizabeth Sandwith Drinker, *Diary of Elizabeth Sandwith Drinker*, ed. Elaine Forman Crane (Boston: Northeastern University Press, 1991), 3:1747. Entry for June 28, 1804, reads, "Clear, wind westerly.—S Swett spent the day—Henry [husband] and Jos Drinker this evening in the Office—WD and self took a walk all the way to S Rhoads little Sally has mist the fit of Chill and fever, and others are well—they are getting ready to go into the country—Baron Humboldt and Sarah Rhoads were there.—we came home between 9 and 10 o'clock." Elizabeth Sandwith Drinker's daughter, Mary, married Samuel Rhoads in 1796. After his business at the corner of Water and Penn Street failed, he left for Uruguay. In 1810, the year of the May Revolution in Uruguay, he was sailing from Montevideo on the *Resolution*, under Captain Eliphalet Davies, when it was lost at sea en route to Havana. For Rhoads's fate, see Henry Drinker Biddle, ed., *Drinker Family in America: To and Including the 8th Generation* (Philadelphia: J. B. Lippincott, 1893), 17. I am grateful to Cassandra Good for steering me to sources providing women's perspectives on Humboldt's visit.

146 "To James Madison from Alexander von Humboldt, Philadelphia, [ca. 19] June 1804," in Madison, *Papers of James Madison*, Secretary of Sate Series, 7:334–36. Schwarz, *Alexander von Humboldt und die Vereinigten Staaten von Amerika*, 93–94, letter 7.

147 Humboldt to Madison, Philadelphia, 21 June 1804 and 27 June 1804, in Schwarz, ed., *Alexander von Humboldt und die Vereinigten Staaten von Amerika,* 97–98, 102, letters 10 and 15. Humboldt would take up this idea in two of his early publications: *Political Essay on the Kingdom of New Spain* (published in English in 1811) and *Views of the Cordilleras* (published in English in 1814), and he continued to press for a canal the rest of his life. See chapter 4 in this volume for a lengthier discussion of this matter.

148 "je préfère Votre Climat à tout autre. Car l'air le plus sain est celui où *l'on respire le plus librement."* Humboldt to Gallatin, 20 June 1804, in Gallatin Correspondence, 1804, no. 45, New-York Historical Society; and Schwarz, *Alexander von Humboldt und die Vereinigten Staaten von Amerika*, 95, letter 8.

149 Chapter 5 in this volume addresses this letter and the broader topic of Humboldt's views on American slavery.

150 "J'ai eu le Bonheur de voir le premier Magistrat de cette grande République vivre avec la simplicité d'un Siècle philosophique et me recevoir avec cette bienfaisante bonté, qui attaché pour jamais. Je parts parce que ma position l'exige, mais j'emporte avec moi la Consolation, que tandisque l'Europe présente un spectacle immoral et mélancholique, le people de ce Continent marche à grands pas vers la perfection de l'état social." Humboldt to Jefferson, Philadelphia, 27 June 1804, in Schwarz, *Alexander von Humboldt und die Vereinigten Staaten von Amerika*, 101, letter 14.

151 "[H]is presence at Wtn. [Washington] attracted from Pa. all the men of science & learning, Drs. Woodhouse, Fothergill, Peale, &c. & for 2 weeks we had an assemblage at the Pts. [President's] of Literati, to whom I listen'd with pleasure; from this specimen I can readily conceive the enjoyment of those scenes described by Marmontel, who passed so much of his life in such society, & who has recorded in fascinating language thier most interesting incidents. Mr. J. appeard delighted with H[umboldt] & said he was the most scientific man of his age he had ever seen." Jean François Marmontel (1723–1799) was a popular French author and acquaintance of Jefferson's. See Gawalt, "'Strict Truth': The Narrative of William Armistead Burwell," 115.

152 "Ce pays qui s'étend à l'ouest des montagnes présente un vaste champs à conquérir pour les sciences!" Humboldt to Thornton, Philadelphia, 20 June 1804, in Schwarz, *Alexander von Humboldt und die Vereinigten Staaten von Amerika*, 96–97, letter 9.

153 *Literary Magazine, and American Register* 2, no. 10 (July 1804) 321–27; "Baron Humboldt," *Weekly Visitor* 2, no. 101 (8 September 1804): 388; 2, no. 102 (15 September 1804): 395–96; and 2, no. 103 (22 September 1804): continued from page 396.

154 *Literary Magazine, and American Register* 2, no. 10 (July 1804): 321, 327.

155 "Election of Four New Members: Wm. Short, Esq., of Virginia; Le Baron Alexandre de Humboldt, of Prussia; Jos. Willard, D.D., President Harvard College, Massachusetts; Zaccheus Collins, of Philadelphia," *Proceedings of the American Philosophical Society* 22, no. 119, Part III. *Early Proceedings of the American Philosophical Society for the Promotion of Useful Knowledge, Compiled by One of the Secretaries, from the Manuscript Minutes of Its Meetings from 1744–1838* (July 1885): 359.

156 Peale to Étienne Geoffroy Saint-Hilaire, from the Museum., 28 June 1804, in Miller et al., *Selected Papers of Charles Willson Peale and His Family*, vol. 2, part 2, 727–28.

[illegible] als sie vom Missouri herab kommen.
[illegible]-Canoes sind lang u. schmal, [illegible] mit einem [illegible]
[illegible] Schaufel. [illegible]
[illegible] Canadier (Engagés oder Trapper) zum Theil [illegible]
[illegible]
[illegible]
[illegible] General Clark [illegible]
[illegible]
[illegible]
Missouri, [illegible] Red-Bud, [illegible]
[illegible] N. America, [illegible]
[illegible] Dogwood, Red-bud, [illegible]
[illegible] u. a. [illegible]

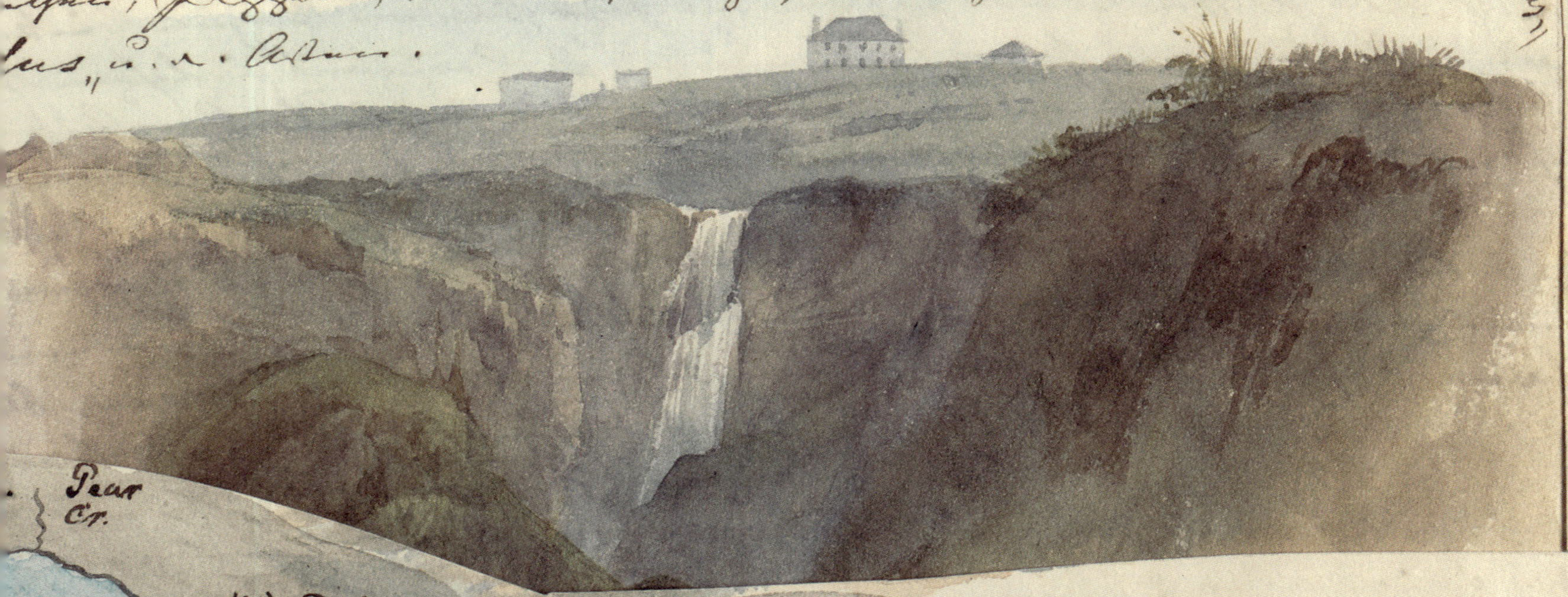

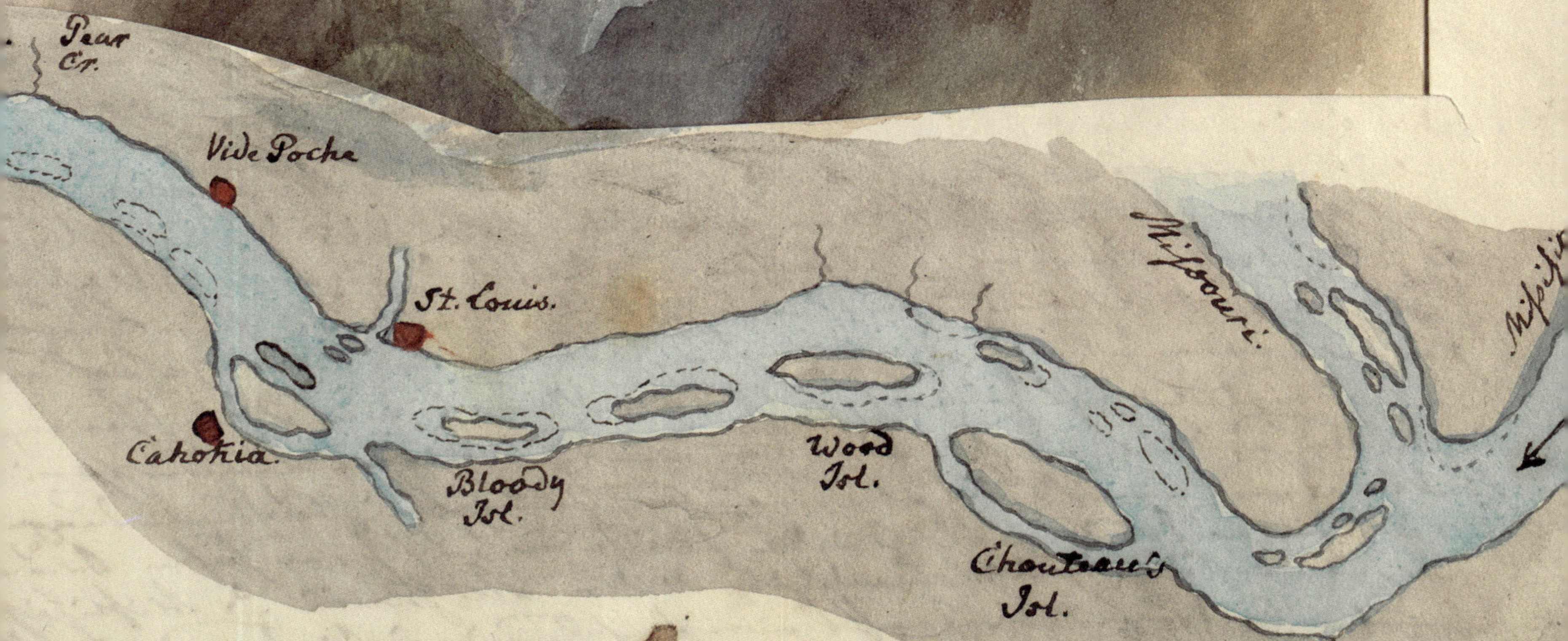

[illegible] Chouteau [illegible] Sanford. Pilcher
[illegible] Company, [illegible] Mackenzie u. [illegible]
[illegible] (2.) [illegible] Missouri [illegible] Fort Leav[illegible]
[illegible]
[illegible]

[illegible]

"When will I be able to venture into those immense areas of the west?"

Before Humboldt had left the United States, he already envisioned a return trip that would allow him to explore the new Louisiana Territory. In his farewell letter to William Thornton, he had exulted, "This country that stretches to the west of the mountains presents a vast area to conquer for science!"[1] His version of conquest involved scientific instruments, trunks of specimens, and keen observation. Inspired by the prospect of what Lewis and Clark would bring home from their expedition, Humboldt foresaw the relevance of the United States in new species, new maps, and new data to connect with what was known about the rest of the globe. As a result, the earliest evidence of Humboldt's direct influence on the United States appeared in the structure and goals of government-sponsored exploration of the American West.

Prince Maximilian zu Wied-Neuwied, *American Fur Co., Manuscript Journal of Prince Maximilian's Travels in North America* (detail), 1833. See p. 123.

The Lewis and Clark Expedition returned in 1806 with relatively little visual material to enhance their narrative; the principal value lay in their detailed route maps, artifacts, and descriptions of the Indian tribes they encountered. Maps had always been a central feature of exploration; increasingly, following Humboldt's scientific travel books, illustrations became important components of reports on these ventures. The addition of the visual record would enhance Americans' understanding of their own country and their identification with the newly added territory. The succession of lavish, illustrated volumes Humboldt published in the two decades immediately following his travels through the Americas spread his ideas and encouraged the publication of a wave of similarly impressive and equally expensive volumes in the United States. By the 1820s his ideas concerning the unity of nature and his data interwoven from a multitude of sources provided a new model for documenting exploratory travel—one that paired a compelling narrative with detailed charts and tables of data and that added black-and-white and hand-colored engravings to provide a layer of visual interest. Inspired by Humboldt's impressive illustrated atlas of his travels, the U.S. explorers leading these endeavors began taking along artists as well as topographers and naturalists.

The maps, sketches, watercolors, and daguerreotypes made in the field by the next generation of American explorers formed the foundation of a cultural identity in which the landscape itself played a significant role. The predilection for merging the cultural identity of the United States with the distinctive features found in its landscape had roots in Jefferson's *Notes on the State of Virginia*, amplified by the encouragement Humboldt provided for this exploration of an American identity. The published expedition reports fed more than government policy and scientific inquiry; in short order, the popularity of landscape painting would flower from both literary and scientific roots, inspired by artists willing to embrace study of the natural sciences as part of their profession. In the United States, Humboldt's ideas and his poetic prose became an influential factor in the confluence of natural history and cultural expression visualized in the new nation's physical geography.

The expansionist goals of the United States—to encompass the land between the Atlantic and Pacific Oceans—meant finding ways to assimilate unfamiliar topography and indigenous communities into the national identity and culture. Humboldt and Bonpland had together served as botanists, zoologists, ethnographers, and cartographers. The expanded model would distribute these tasks to a team of trained individuals. Three principal explorers consciously adopted a Humboldtian approach to their expedition equipment and goals. Stephen H. Long was the first to take along specialists who each could contribute a significant component of the final report. John C. Frémont bestowed Humboldt's name on a wide variety of landscape features and hired a photographer as his artist—the first explorer to do so. Humboldt encouraged the exchange of knowledge derived from these expeditions and praised each explorer's methods and data. On top of that, Humboldt encouraged a generation of European protégés who traveled to the United States to collect data in his stead, with the express aim of providing him with the results of their expeditions. Foremost among them was Prince Maximilian zu Wied-Neuwied, who, with artist Karl Bodmer, spent two years making a record of the American landscape and its peoples, eventually publishing elaborate volumes that rivaled Humboldt's in their wealth of information and accomplished illustrations. Those two advances, the team-oriented approach to exploration and the publication of illustrated books designed for a broad and general audience, would transform American exploration. Humboldt's earliest publications set the tone for those ambitious narratives.

HUMBOLDT'S *NATURGEMÄLDE*: THE GEOGRAPHY OF PLANTS

Humboldt was quick to publish his first two books in which he compiled and evaluated the material he and Bonpland returned with in 1804. These two volumes, *Essay on the Geography of Plants* (1805) and *Aspects of Nature* (1808) established Humboldt as a rising star in what was still known as "natural philosophy," a synthesis of art, science, and observation that served as the foundation for many of the scientific disciplines that exist today. Humboldt's third enterprise was far more ambitious: he envisioned an eleven-volume series he called *Relation historique du voyage aux régions équinoxiales de nouveau continent*. In its English translation, this became known as the *Personal Narrative of Travels to the Equinoctial Regions of the New Continent*, which, along with his four-volume *Essai politique sur le royaume de la Nouvelle-Espagne* (*Political Essay on the Kingdom of New Spain*), enjoyed wide distribution in the United States and shaped the goals of the Stephen H. Long Expedition of 1818 to 1820 and several of John C. Frémont's expeditions between 1842 and 1853.[2]

What set Humboldt's writings apart had more to do with how he synthesized information than any discovery he made. Even before he had embarked for South America, Humboldt had begun exploring the idea that would animate his entire career—that all of nature was interconnected. Humboldt believed that every living thing on the planet was interrelated and that he could prove, through empirical data and high-level reasoning, that there was a unity of nature that would allow for a complete reconceptualization of natural philosophy. In particular, he hypothesized that all of the world's plants, animals, and topographies could be correlated by altitude and climate to form a single, planetary ecosystem.[3] By 1790 Humboldt was convinced that his own empirical data, drawn from his early excursions across Europe and in his travels to come, would bear out this hypothesis. In a letter written shortly after his return from South America, he had explained his belief "that there is but One Spirit animating the whole of Nature from pole to pole—but One Life infused into stones, plants, and animals, and even into man himself."[4]

It is hard to recapture just how revolutionary this idea was at the time. He continued, "In all my wanderings I was impressed with the conviction… [that] my views of Nature had been elevated, and I had, as it were, become endowed with new perceptive faculties."[5] Those enhanced faculties would increasingly rely on the exchange of information across a vast network of friends and colleagues around the world. This concept of ecosystems was radical when Humboldt envisioned it; five years in the Americas with Bonpland provided him with the opportunity to test his thesis in the field and compare his new data with what was known already about Europe. Humboldt's earliest publications after his return to Paris staked out his concept that there were underlying reasons governing the distribution of plants and animals, largely tied to climate and altitude, and backed it with empirical evidence in his specimen trunks and voluminous journals.

Humboldt had returned to Europe in August 1804 to a hero's welcome. While his ship, the *Favorite*, cleared quarantine in the harbor near Bordeaux, Humboldt wrote letters to friends in Paris announcing his safe arrival. In each of these he included a variation on his description of his stay in the United States. Of his six weeks spent visiting Philadelphia, Baltimore, and Washington, he noted with pride that he "was loaded with marks of honour by Jefferson, the President of the States."[6] Americans had begun to cultivate an association with Humboldt, but just as important was Humboldt's standing with Jefferson. The uncertainty of his ambitious itinerary and the dangers of transatlantic travel had fostered rumors in the European papers that Humboldt had been lost at sea, had "perished among the savages," or had succumbed to yellow fever.[7] As news of his safe arrival spread, so too did the eagerness to hear his stories of his exploration, to see the massive number of specimens contained in his trunks, and to begin to assimilate this body of new information into various disciplines of the natural sciences. Humboldt shared their excitement. He had enjoyed regaling his American audience with his stories; he now had his second audience, his intellectual and scientific peers across the continent. To one colleague he mentioned his "thirty cases of treasures of all kinds, botanical, geological and astronomical," admitting "[i]t will take me years to bring out my great work."[8]

Humboldt wasted no time. As he unpacked his specimens, he began distributing them to his colleagues in the scientific community. One allotment of plant specimens went with Bonpland to the Jardin des Plantes, another to the Muséum national d'Histoire naturelle in Paris, and the third Humboldt passed along to his Prussian mentor Karl Willdenow and colleague Karl Sigismund Kunth.[9] Many of these plants were new to European botanists. In addition to plants, Humboldt had collected rocks, which he sorted, classified, and sent to colleagues in Prussia for further distribution.[10] In this manner Humboldt launched his reputation for crowdsourcing the vast quantities of his specimens for study and classification, taking advantage of the simultaneous and collective wisdom of the European scientific community.

From mid-January through early March of 1805 he delivered the first of a series of public lectures at the Institut National on the botany of South America. From these Humboldt and Bonpland developed the first book derived from their travels, *Essai sur la géographie des plantes* (*Essay on the Geography of Plants*).[11] In this relatively slim volume, Humboldt began to unpack some of his most revolutionary ideas, foremost among them his belief in what he called the "unity of nature," that everything on the planet, and by extension in the universe, was interconnected. Each of his plant specimens was tagged with unusual pieces of information: the altitude at which it was collected and the nature of the strata in which it was found. This information, Humboldt reasoned, would allow him to correlate location, altitude, climate, and substrate in order to extrapolate a better understanding of how these things were interrelated around the globe.

Humboldt's *Essay on the Geography of Plants* invokes a broad spectrum of modern disciplines—including geology, botany, and meteorology—to discern bands of climate that support characteristic plant forms. In this small volume, Humboldt attempted to map out his concept of the geographic distribution of plants as a way of understanding how the entire planet is interconnected. Botany ultimately provided the proof for Humboldt's thesis. His proto-ecological idea, while obvious today, was a distinct departure from the more hieratic Great Chain of Being that had anchored the study of the natural world for centuries. The Great Chain of Being stipulated that all life-forms on earth could be arranged in descending order, with man at the top. The order was predicated on the relative complexity of anatomy and traits associated with advanced brains.

Unity of nature based on botany was not Humboldt's only revolutionary idea to appear in this book. Plants are anchored in the earth, and the earth is composed of rock that, in Humboldt's mind, seemed far less stable and immutable after he experienced an earthquake in Cumaná (Venezuela) in 1799. If the very earth was unstable, fluid, and dynamic, how could anything be fixed? In his *Essay*, Humboldt postulated, if

> *[t]he most striking geological phenomena show that the whole crust of the earth was once in a liquid state; that the stratification and differentiation of rocks indicate that the formation of mountains and the crystallization of the great land masses around a common nucleus did not happen at the same time on the whole surface of the earth,*[12]

then subsequent cooling would vary over time and by location. Further, that process would be reflected in the local climate. By understanding the geological history of a specific area, Humboldt felt confident he could predict how the creation of land-forms would shape the local environment.

Humboldt was still not done. If plants anchored in the soil told him one story, the underlying rock strata told another. Piecing together the locations of similar geological formations suggested to him another way of understanding the distribution of plants and animals across the globe. He wrote, "Geology bases itself on the analogous structures in coastline, in the bottom of the Ocean and on the identity of animals inhabiting them." Believing that, he postulated,

> *In order to determine the ancient link between neighboring continents, geology bases itself on the analogy of the coastal structures.... The geography of plants...can show that the separation of Africa from South America occurred before the development of organized forms of beings.*[13]

This observation later inspired the embryonic thinking behind plate tectonics and provided the basis for British scientist Charles Lyell's *Principles of Geology*, published in 1830.

Humboldt's idea that plants and animals, originally on a single landmass but now occupying separate continents, had changed to adapt to their surroundings also inspired Charles Darwin as he began to formulate his theory of evolution. Humboldt read in the fossil record evidence of earlier life-forms, the potential for extinction, and suggestions of the antiquity of the earth.

> *The geography of plants delves into the interior of the earth: it looks at the ancient monuments that nature has left behind in petrifications, in wood fossils, and in coal strata that are the tomb of the initial plant life of our planet.*[14]

From this accrued body of knowledge he drew several prescient conclusions: first, the earth was ancient; second, individual continents appeared to have drifted from their original positions; and third, the earth's distribution of plant and animal life could be read as an index of the past and current state of climate and topography.

The resulting thirty-five-page essay could hardly do justice to Humboldt's revolutionary ideas, much less accommodate all the data required to support those statements. The second section of the book contained his essential data, presented in more traditional lists and tables. This material provided the evidence supporting his narrative. In this section Humboldt and Bonpland combined their field measurements with existing research that had been conducted by a raft of their colleagues and forerunners, bringing together information from multiple continents and numerous European expeditions. The third component was what Humboldt termed his "One Plate," an illustration intended to visualize both the ideas in the essay and the data in the tables. That tableau was an oversized, folded map that served as a concordance to the essay (CAT. 15), tucked in a sleeve in the back of the book. In the center of this large-scale, hand-colored engraving, Humboldt drew a profile rendering of Chimborazo and Cotopaxi against a blue sky ruffled with cirrus clouds, water lapping along the shoreline at the base of the mountains. Half of the mountains were engraved to convey the illusion of a landscape painting, the other half filled with Latin names of the plant species that populated the hillsides from the base of the snowcaps down to sea level. Half artwork and half raw data, Humboldt's tableau straddled art and science, incorporating more information than could be contained in a single set of tables or a suite of watercolors.

On either side of this image, Humboldt created tables of information, correlating the heights of prominent peaks in Europe, Asia, and South America; detailing atmospheric information; and creating an index of plants found at specific altitudes around the globe. Humboldt's plant geography map was the precursor to modern info-graphics: half art, half scientific observation, devised to bolster a daring hypothesis that the world was a singular ecosystem of interrelated parts. Humboldt credited William Playfair for the idea of combining visuals and data to express complex ideas simply. In Playfair's case, it was the Englishman's maps to explain the British National Debt;[15] in Humboldt's case, it was nothing less than a first attempt to explain the planet.

Humboldt created two versions of this map: one hand-colored to deepen the visual connection to landscape painting, the other in black and white (CAT. 16), a complex engraving using stippled areas and precise line work to foster the illusion of topographic space. The explorer called this map his *Naturgemälde*, which is literally translated as Picture of Nature. Humboldt also called it his Plant Geography Map. This image is a singular "Great Picture," to use the art historical term, that mapped the entire ecosystem around Chimborazo with comparative data drawn from multiple locations. In tandem with the written essay, it visualizes Humboldt's nascent ideas of interrelated ecosystems. For example, plant types found at the lower altitudes of Mont Blanc in the French Alps and Chimborazo in Ecuador are variations on the same species. Around the globe, broad-leaved plants in lower altitudes give way to mountain scrub, which in turn yields to mosses and lichens at the higher altitudes, and finally, bare rock and perpetual snow closest to the summits. Humboldt extrapolated from his data that, were he to climb mountains of similar height on any continent, he would find this same pattern of plant life. Consequently, he could determine the altitude at which those plants thrived and the climate that would sustain them. He further postulated that altitude and distance from the equator ran in parallel—that the biomes he encountered while climbing Chimborazo would be similar to those one would encounter as one moves closer to the poles. No one had thought to pursue such a systematic approach to global climate or flora before.

Humboldt stated his goal to find a way to include every aspect of his accumulated data in a single image:

> *namely, the vegetation, the animals, the geological facts, the cultivation of the soil, the temperature of the air, the limit of perpetual snows, the chemical constitution of the atmosphere, its electrical intensity, its barometric pressure, the decrement of gravitation, the intensity of the azure color of the sky, the diminution of the light during its passage through the successive strata of the air, the horizontal refractions, and the heat of boiling water at different heights.*

Humboldt hoped his approach would allow him to "show the simultaneous action and connecting links of the forces which pervade the universe."[16] Bringing together words and images allowed him to present his data in two forms, juxtaposed to provide a sense of the actual topography paired with the amplitude of information he had accrued. Humboldt inserted an aesthetic sensibility into his documentation, enlivening his tables of data by allowing some of it to take the form of artistic renderings. Humboldt's penchant for using multiple visual platforms for correlating information was

CAT. 15 Alexander von Humboldt and Aimé Bonpland, *Géographie des plantes Équinoxiales: Tableau physique des Andes et Pays voisins,* from *Essai sur la géographie des plantes*, 1805, hand-colored print, 24 × 36 in., Royal Botanic Gardens, Kew

CAT. 16 Alexander von Humboldt and Aimé Bonpland, *Geographie der pflanzen in den Tropen-Ländern: ein Naturgemälde der Anden,* 1811, black-and-white print, 24 × 36 in., Royal Botanic Gardens, Kew

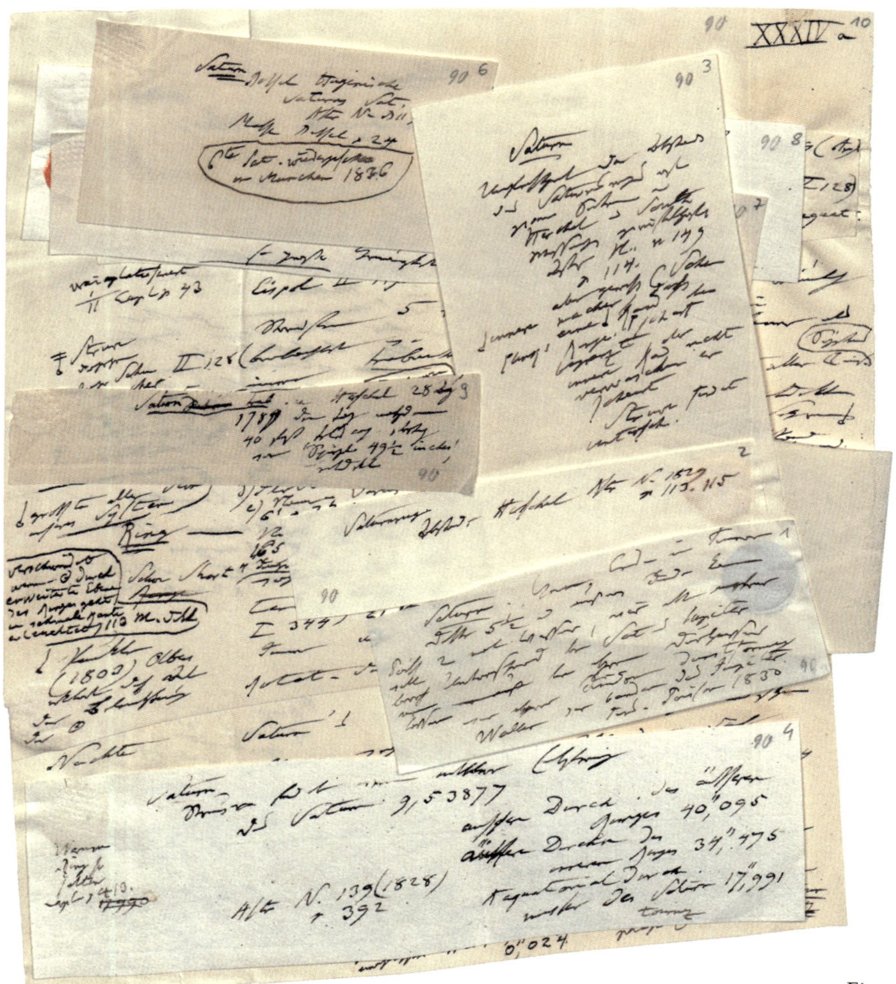

Fig. 14
Alexander von Humboldt, *Notes on Saturn,* ca. 1789–1830s, ink on paper and sealing wax, Staatsbibliothek zu Berlin, Stiftung Preussischer Kulturbesitz, Berlin, Germany

Following pages
Detail of Cat. 15

not just a clever way to incorporate as much data as possible on a single sheet—it was his working method for developing his larger concepts of global climate, plant distribution, and geographical interconnections.

Humboldt's *Naturgemälde* was a hybrid art form blending science and art in a manner consistent with Humboldt's dictum that an artist and scientist each needed to approach nature mindful of the other's discipline. It also played to his strengths: visualizing a complex group of variables operating within a system and laying them out to create a visually persuasive argument. In these cutaway maps, we see Humboldt grappling with his understanding of the planet, developing and refining his ideas. Humboldt was the first to bring together all of the individual measurements in the service of a larger idea; and in successive maps he refined and expanded his correlations. Over the next forty years he would continue to aggregate information—writing in his journals, transcribing excerpts from publications, using wax to attach related pieces of paper (fig. 14), and stuffing envelopes with fragments snipped from his voluminous correspondence with colleagues around the world. Together, word and image would form the fundament of his unity of nature concept, which underpinned every book he published, every map and infographic he drew.

It would make for a compelling presentation of his ideas and prove popular with his audiences. The concept of a unity of nature required amassing vast quantities of data, which in turn depended on developing collaborative relationships with a network of like-minded individuals around the globe. Humboldt's enthusiasm for expeditionary reports became an essential part of these relays among scientific data, travel narratives, and the visual record of discovery. The two principal products of Humboldt's travels—maps and illustrated books—served as the embodiment of the Enlightenment credo, the increase and diffusion of knowledge. The sharing of maps and data was an act of generosity Humboldt encouraged, one that would be reciprocated among Humboldt's friends and protégés, a sign of respect among members of a scientifically minded fraternity.

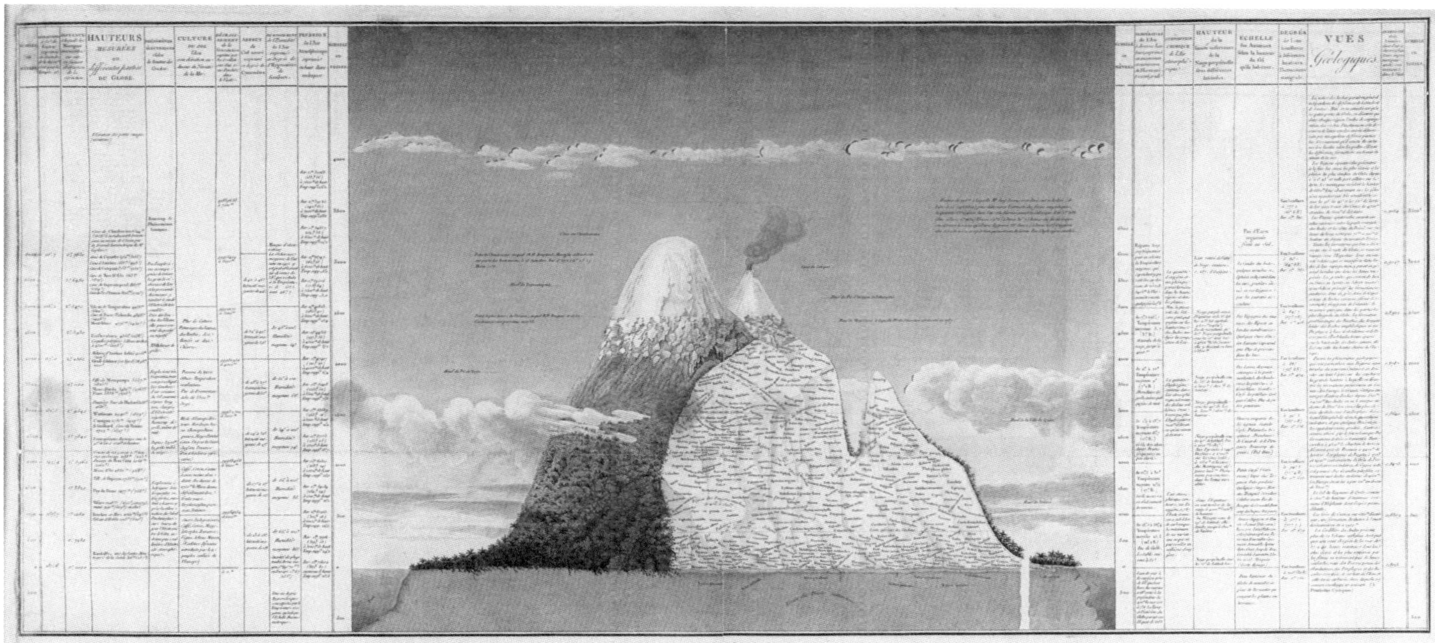

GÉOGRAPHIE DES PLANTES ÉQUINOXIALES.

Tableau physique des Andes et Pays voisins

Dressé d'après des Observations & des Mesures prises Sur les Lieux depuis le 10e degré de latitude boréale jusqu'au 10e de latitude australe en 1799, 1800, 1801, 1802 et 1803.

PAR

ALEXANDRE DE HUMBOLDT ET AIMÉ BONPLAND.

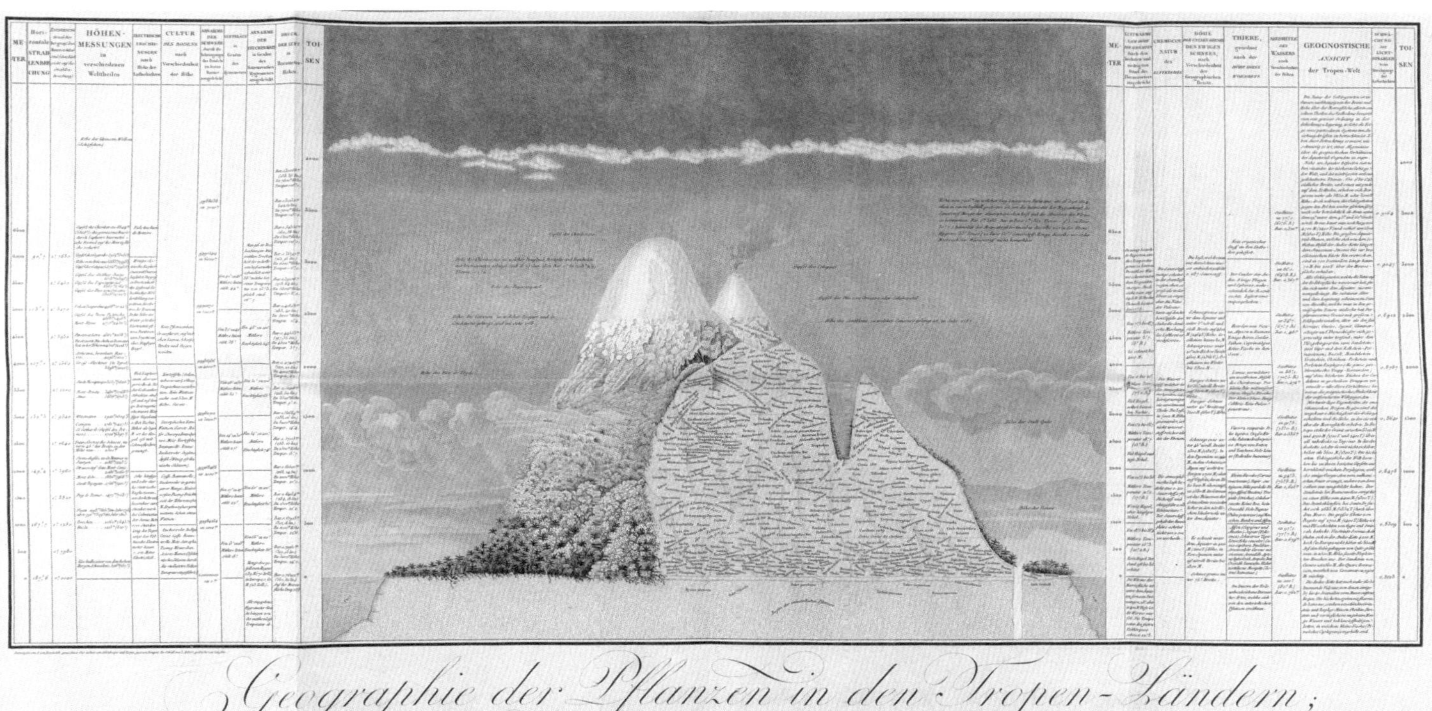

Geographie der Pflanzen in den Tropen-Ländern;

ein Naturgemälde der Anden,

gegründet auf Beobachtungen und Messungen, welche vom 10ten Grade nördlicher bis zum 10ten Grade südlicher Breite angestellt worden sind, in den Jahren 1799 bis 1803.

von ALEXANDER VON HUMBOLDT *und* A. G. BONPLAND.

…IS- …NT …ion …ar …tu- …mé …ule …	ASPECT du Ciel azuré exprimé en degrés du Cyanomètre.	DÉCROISSEMENT de l'Humidité de l'Air exprimée en Degrés de l'Hygromètre de Saußure.	PRESSION de l'Air Atmoſphérique exprimée en haut. Barométriques.	ÉCHELLE en TOISES.
				4000
			Bar. 0,m 30068. (133,li 36) à 7500m de haut. Temp. suppée –16°,0.	
…38			Bar. 0,m 32035. (142,li 61.) à 7000m de haut. Temp. suppée –13°,0	3500
			Bar. 0,m 34357. (152,li 38.) à 6500m de haut. Temp. suppée –10°,0.	
…04 m			Bar. 0,m 36747. (162,li 95.) à 6000m de haut. Temp. supp. –6°,0	3000
	de 40. à 46°. Intensité moyenne de 44°.	Manque d'observations. La Sécheresse moyenne de l'air sans nuages y est probablement au-dessous de 38°, qui réduits à la Température de 25°,3. sont 26°,7.	Bar. 0,m 39206. (173,li 84) à 5500m de haut. Temp. supp. –3,0	
…70 m			Bar. 0,m 41823. (185,li 40) à 5000m de haut. Temp. suppée 0°,4	2500
	de 32° à 42°. Intensité moyenne de 38°.	de 46° à 100°. Humidité moyenne 54°.	Bar. 0,m 44553. (197,li 55.) à 4500m de haut. Temp. suppée 3°,7	
…936 m			Bar. 0,m 47417 (210,li 20.) à 4000m de haut. Temp. suppée 6°,4	2000
	de 28° à 37°. Intensité moyenne de 32°.	de 51° à 100. Humidité moyenne 65°.	Bar. 0,m 50418. (223,li 50.) à 3500m de haut. Temp. suppée 9°,0.	
2702 m			Bar. 0,m 53689. (238,li 06.) à 3000li de haut. Temp. suppée 14°,4	1500
	de 24° à 30°. Intensité moyenne de 27°.	de 54° à 100° Humidité moyenne 74°.	Bar. 0,m 57073. (253,li 05.) à 2500m de haut Temp. suppée 18°,7.	
3468 m			Bar. 0,m 60501. (268,li 24) à 2000,m de haut Temp. suppée 20°,0.	1000
	de 17° à 27°. Intensité moyenne de 22°.	de 60°, à 100°. Humidité moyenne 80°.	Bar. 0,m 64134 (284,li 24) à 1500,m de haut Temp. suppée 21°,2	
4234 m			Bar. 0,m 67923. (301,li 18.) à 1000,m de haut. Temp. supp. 22°,6.	500
	de 13 à 23°. Intensité moyenne de 18°.	de 66°, à 100°. Humidité moyenne 86°.	Bar. 0,m 71961. (319,li 03.) à 500,m de haut. Temp. suppée 24°,0.	
…00 m		Quantité de pluye tombée, terme moyen 1m,89 (70 P.º) en Europe 0,m 67 (25 P.º)	Bar. 0,m 76202 (337,li 80.) au niveau de la mer Temp. suppée 25°,3	0
		Tous ces degrés Hygrométriques sont affectés par la Température mo-		

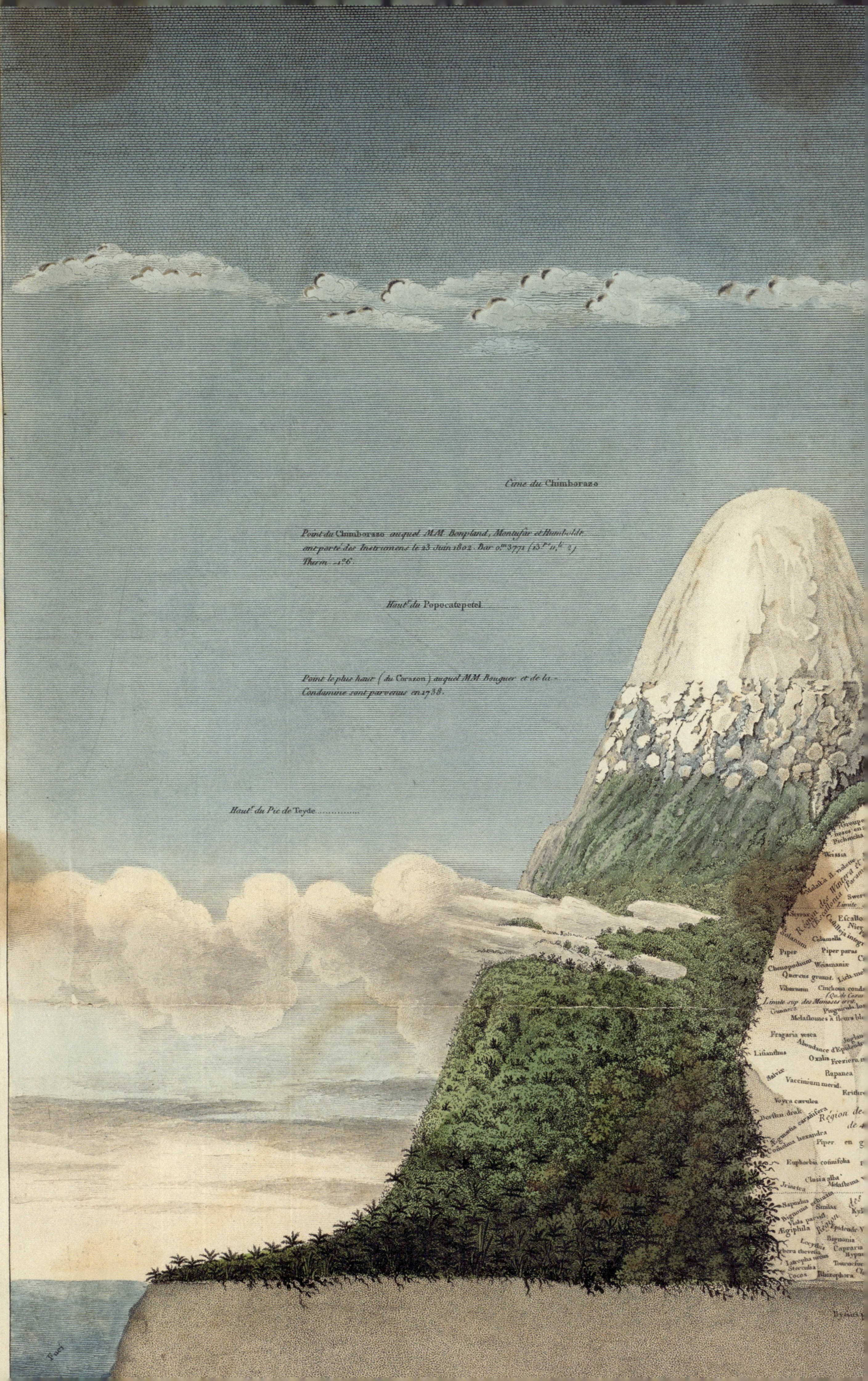

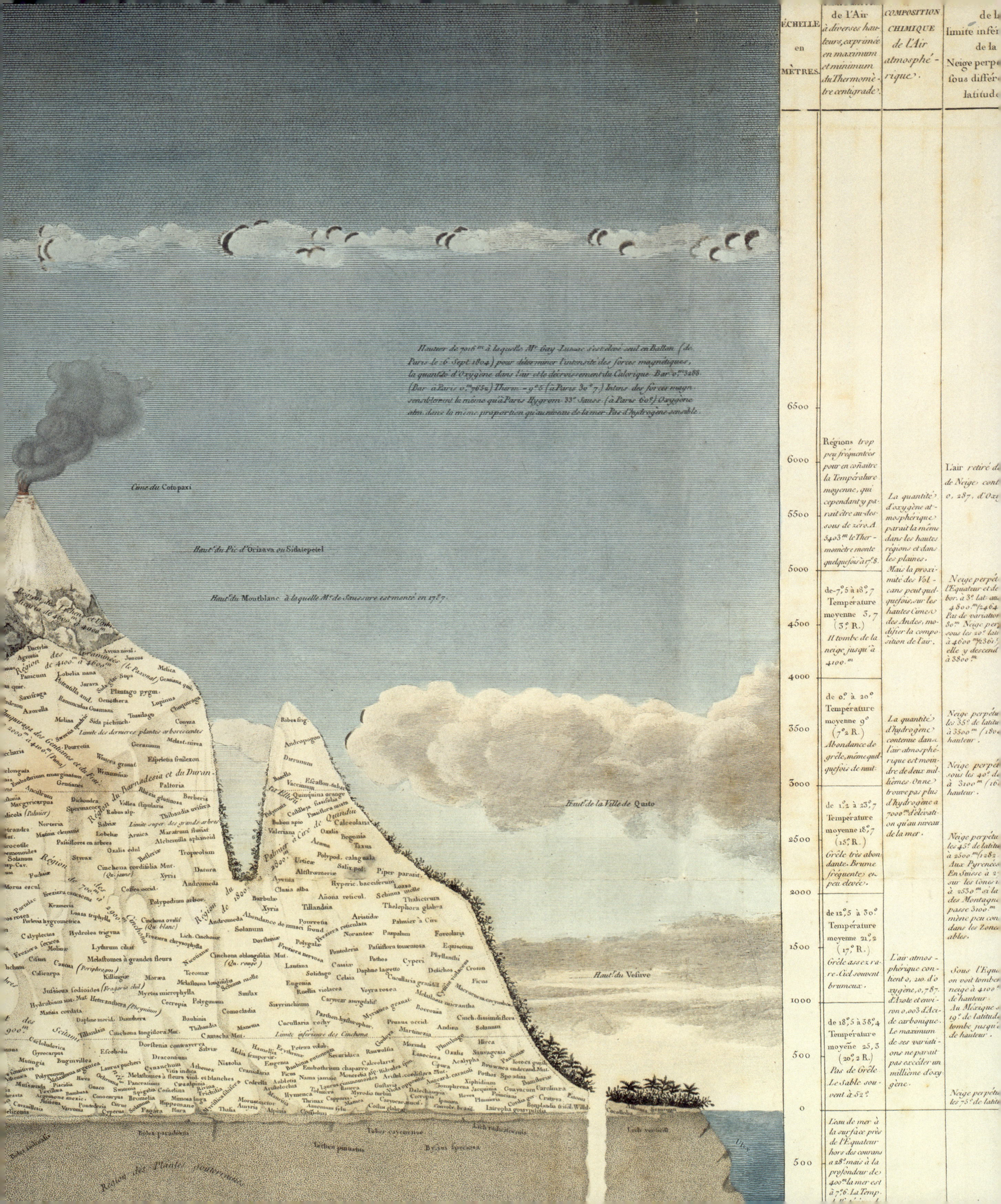

ÉCHELLE en MÈTRES.
de l'Air à diverses hauteurs, exprimée en maximum et minimum du Thermomètre centigrade.
COMPOSITION CHIMIQUE de l'Air atmosphérique.
de la limite inféri[eure] de la Neige perp[étuelle] sous différ[entes] latitude[s]
6500
6000
5500
5000
4500
4000
3500
3000
2500
2000
1500
1000
500
0
500
Régions trop peu fréquentées pour en coñaitre la Température moyenne, qui cependant y paraît être au-dessous de zéro. A 5403.m le Thermomètre monte quelquefois à 17°.8.
de -7°,5 à 18°,7 Température moyenne 3,7 (3° R.) Il tombe de la neige jusqu'à 4100.m
de 0° à 20° Température moyenne 9° (7°2 R.) Abondance de grêle, même quelquefois de nuit.
de 1°,2 à 23°,7 Température moyenne 18°,7 (15° R.) Grêle très abondante. Brume fréquente et peu élevée.
de 12°,5 à 30° Température moyenne 21°,2 (17° R.) Grêle assez rare. Ciel souvent brumeux.
de 18°,5 à 38°,4 Température moyeñe 25,3 (20°,2 R.) Pas de Grêle. Le Sable souvent à 52°.
L'eau de mer à la surface près de l'Equateur hors des courans a 28°. mais à la profondeur de 400.m la mer est à 7°,6. La Temp.
La quantité d'oxygène atmosphérique paraît la même dans les hautes régions et dans les plaines. Mais la proximité des Volcans peut quelquefois, sur les hautes Cimes des Andes, modifier la composition de l'air.
La quantité d'hydrogène contenue dans l'air atmosphérique est moindre de deux millièmes. On ne trouve pas plus d'hydrogène a 7000.m d'élévation qu'au niveau de la mer.
L'air atmosphérique contient 0, 210. d'oxygène, 0,787. d'Azote et environ 0,003 d'Acide carbonique. Le maximum de ses variations ne paraît pas excéder un millième d'oxygène.
Cime du Cotopaxi
Haut. du Pic d'Orizava ou Sitlaltepetl
Haut. du Montblanc à laquelle M.r de Saussure est monté en 1787.
Haut. de la Ville de Quito
Haut. du Vésuve

RECONCEIVING THE TRAVEL NARRATIVE: *ASPECTS OF NATURE*

Humboldt followed up his *Essay on the Geography of Plants* with the book he always regarded as his favorite: *Ansichten der Natur* (translated as *Aspects of Nature* or *Views of Nature*), which he wrote in 1807.[17] In this book Humboldt took his reader on a journey that expanded on the model of simultaneity around which he had organized his *Essay on the Geography of Plants*. For *Aspects of Nature*, Humboldt wrote eight essays recounting episodes drawn from South American travels, each of which he alternated with compilations of data points, statistics, and extrapolations that supported the ideas he advanced in his engaging prose. By presenting a travel narrative that outlined an idea and supporting his conclusion with the assembled pertinent data—the proof that showed his work—Humboldt could tell a compelling story whose narrative travelogue made the reader yearn for adventure, and he could provide the reader with the data, which took on enhanced meaning from the narrative context. No scientific book had been written as an adventure story, and the continued popularity of *Aspects of Nature* confirmed a broad public appetite for armchair travel with a dose of scientific rigor. *Aspects of Nature* would go through multiple printings in numerous languages. In each subsequent edition, Humboldt incorporated new information.[18] In particular, the third edition, translated into English in 1849, would have a lasting impact on American geopolitics at midcentury.

From his perch in Paris, Humboldt played a central role in the French scientific community. Demand for Humboldt's books supported translations from the original French into upward of twenty-five different languages, including English—making his work more widely accessible to a North American audience. In English, seven of these volumes appeared under the title *Relation historique du voyage aux régions équinoxiales du nouveau continent* (*Personal Narrative of Travels to the Equinoctial Regions of the New Continent, during the Years 1799–1804*), a compendium that read less like an ongoing travelogue and more like the beginnings of a comprehensive view of the world based on his early travels across Europe and his five years in the Americas.[19] Serving as his own writer, editor, illustrator, and publisher, Humboldt depleted his fortune making sure each of his volumes met his exacting standards. The large-format folios included lavishly illustrated maps, cutaway views of geological formations, illusionistic landscape vistas, and scientifically rendered botanical specimens. Curiosity about the South American continent and Mexico spurred much of the interest, carried along by Humboldt's engaging narrative. The oversize volumes with their hand-colored plates attracted an audience beyond scientists and explorers. Humboldt intended them to entice artists to travel and depict the Americas as well. Humboldt's *Personal Narrative* went on to enjoy a wide readership in the United States. He sent copies of the early volumes to Thomas Jefferson, and the series could be found in the libraries of scientific societies, universities, and private libraries. These books were essential reading for a generation of American explorers.

THE GEOPOLITICAL VALUE OF HUMBOLDT'S MAP OF NORTH AMERICA

From the very beginning, Humboldt incorporated information from the United States as part of his concept of the unity of nature. Jefferson, Madison, and Gallatin provided Humboldt with data and statistical information designed to amplify the North American continent's standing in his global view. Some of this data appeared in his *Essay on the Geography of Plants*.[20] In *Views of the Cordilleras*, one of the first volumes of his *Personal Narrative*, he made specific references to Jefferson's *Notes on the State of Virginia*. Here Humboldt compared South American landscape features to the equally impressive North American landmarks Natural Bridge and Niagara Falls. Humboldt pronounced that, "only when we have acquired the habit of viewing the Globe as a great whole," a whole that included the United States, could we see and appreciate the interconnections among all aspects of nature.[21] Through his writings, Humboldt made a case for the importance of including American landmarks and data as he developed his ideas. The exchange of information begun during his visit in 1804 would encourage future collaborations between Humboldt and American scientists, explorers, and artists.

Humboldt's boundless curiosity and abiding interest in new information about the North American continent influenced the character of the explorations conducted under the auspices of the United States government. Lewis and Clark had gone west seeking knowledge regarding the various branches of natural history, ethnography, linguistics, and economic resources, along with the identification of overland and riverine routes to the Pacific Ocean. Humboldt encouraged the accurate measurements of latitude, longitude, barometric pressure, and the observation of weather patterns. These became essential goals of each successive wave of expeditions, as America's borders expanded and its geopolitical stature changed. Humboldt also advocated for including artists as well as naturalists on these ventures, both for plein air fieldwork and to produce high-quality illustrations for the lavish tomes published afterward—books that often established the reputation of the explorer as well as underscored the significance of the expedition. The survey reports published by American explorers enhanced the nation's understanding of what lay west of the Mississippi and Missouri Rivers and provided new data for Humboldt's publications.

American exploration contributed to the construction of a cultural identity focused on an appreciation of nature, calling attention to the distinctive landmarks discovered along the way. American explorers set out looking for evidence of mammoths, cataloguing unique geographic features, identifying new species, and looking for trade routes that would connect the cities on the East Coast with the interior rivers and the Pacific Ocean cultures on the West Coast.[22] These explorers returned with better maps and a more thorough cataloguing of what existed on the American continent, all of which served the nation's geopolitical agenda. They also returned with specimens and artifacts that, in a museum setting, would give visual weight to the scientific and cultural riches of the United States. By including artists, these surveys provided a glimpse of the landscape more attuned to an aesthetic appreciation of nature. Curiosity about the American landscape translated into pride in its character. America was in the process of discovering itself, a process Humboldt encouraged that linked natural philosophy with national identity.

Humboldt had hoped to be one of those explorers, his quick mind sketching out an itinerary that he expected would fill many of the gaps in his knowledge of the continent. In his parting letter to Secretary of State James Madison, Humboldt thanked him for all of his kindnesses and announced:

I hope to return to this wonderful land within a few years. The route from Missouri to the Pacific coast will then already be open. The lakes, Canada and the immense plains that stretch from Pittsburgh to the mountains, which Fidler saw, offer a vast field for geologic studies. With some aid from your government one could do a great deal and as prolific a work and as important as that done by Pallas.

Humboldt invoked two of the leading geographic travelers of his day. Peter Fidler was a surveyor for the Hudson's Bay Company, who had surveyed part of the Missouri River and the Canadian Rockies; Peter Simon Pallas was a German naturalist who had traveled to Alaska and Siberia at the invitation of Catherine the Great in 1786. Humboldt continued, "I would venture to proceed north to Mt. Elias (as high as the mountains of Quito) and onward to the Russian possessions. All this could be carried out with courage and in good health."[23] His references suggest his ambitions, as he proposed to build on the research contributed by specific individuals, envisioning the accumulation of data that would provide him with accurate, concrete measurements to complement those he had taken in South America.

A year later, Humboldt was still mulling over his return trip to the United States. In his letter to John Vaughan at the American Philosophical Society (APS) in Philadelphia, he seemed frustrated at the delay and tantalized by the prospect: "When, when will I be able to venture into those immense areas of the west for which Mr. Jefferson (in his new place) [Monticello] would be better able to assist me? I have not abandoned my hopes." Of his itinerary, he pronounced, "I desire to visit Missouri, the Arctic, and Asia."[24] For Humboldt, and so many American explorers, "Missouri" was the de facto gateway to the entire western part of the North American continent. In 1804, Clark and Lewis (CATS. 17, 18) were exploring somewhere in this uncharted wilderness that would afterward be called the Missouri Territory. William Clark

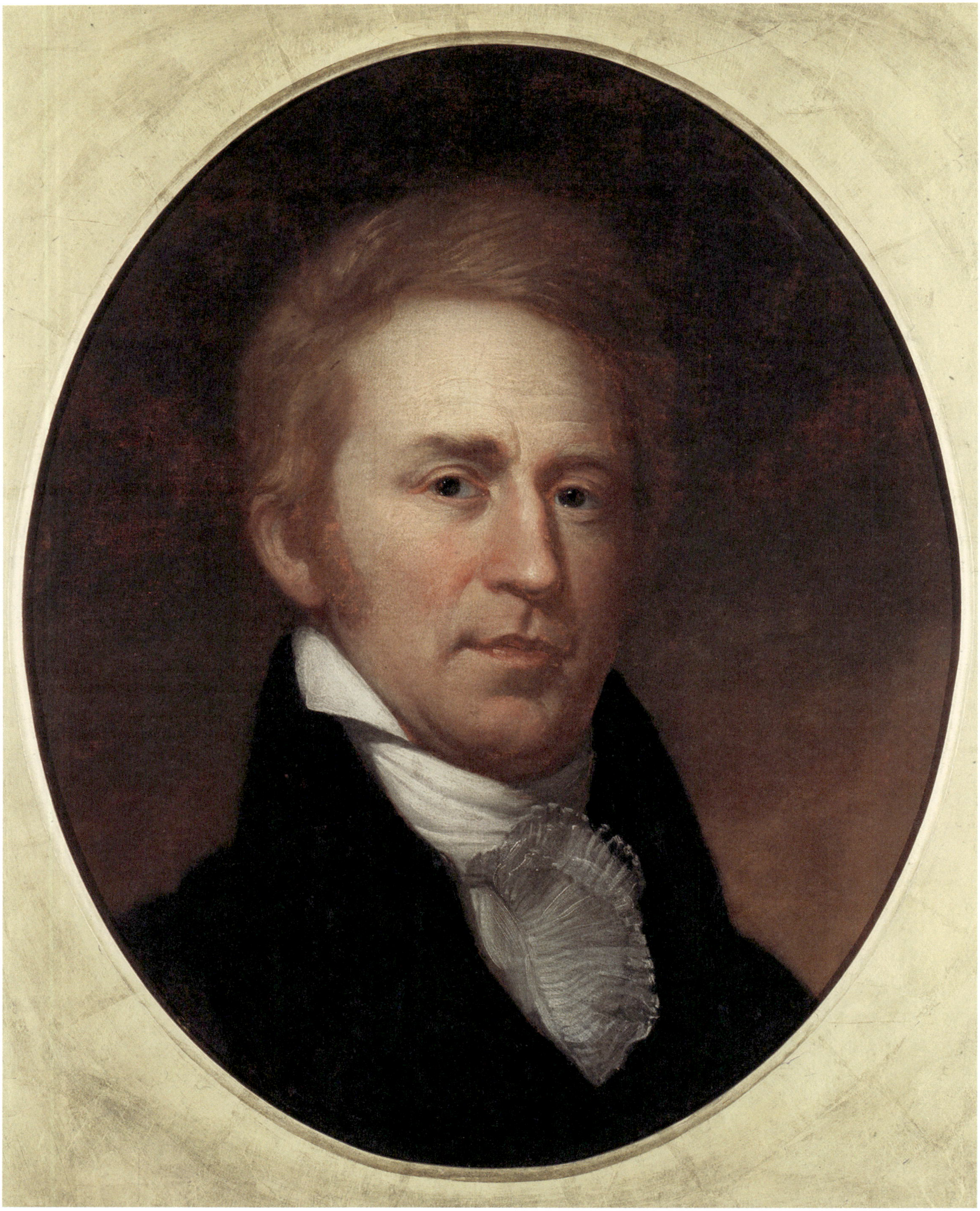

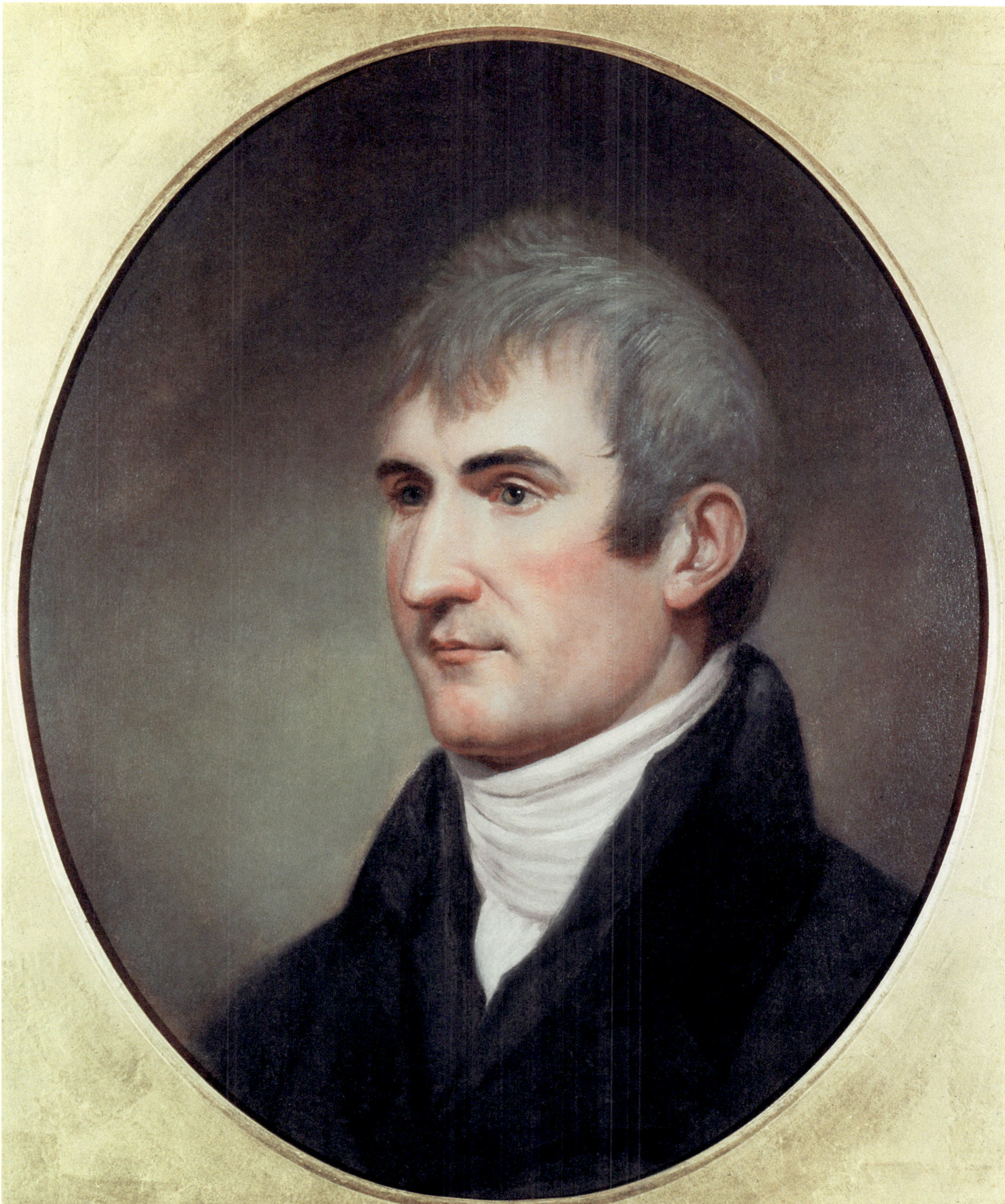

Preceding pages

CAT. 17 Charles Willson Peale, *Portrait of William Clark* (1770–1838), 1807, oil on paper mounted on canvas, 25 × 24 ½ in., Independence National Historic Park Collection, Philadelphia, PA

CAT. 18 Charles Willson Peale, *Portrait of Meriwether Lewis* (1774–1809), 1807, oil on wood panel, 25 × 20 in., Independence National Historic Park Collection, Philadelphia, PA

became its first territorial governor in 1813. His outpost in St. Louis later acted as an obligatory checkpoint for anyone serious about making the trek up the Missouri River and west toward the Rocky Mountains. "The Arctic" stretched from the Pacific Northwest coast up through Canada and into the Russian-held territory spanning the Bering Straits. And "Asia" covered an equally vast swath of Russia and China, from the Pacific coast back to Europe's borders. That trajectory would have greatly expanded Humboldt's global knowledge of physical geography and global climate. In 1829 Humboldt journeyed over 9,000 miles across Russia; for the remaining American territories, he had to settle for accumulating that information secondhand from the numerous subsequent government-sponsored expeditions across the United States and from his own protégés, whom he encouraged to explore the West Coast and the Southwest. Humboldt hoped the information they provided would correlate with his own data gathered in parts of Europe, South America, and Mexico.

In Washington, Humboldt had generously shared the map he had made in Mexico of North America, which included the most detailed information known for the central part of the continent (see cat. 6). In doing so, he opened a larger conversation about reciprocal generosity with information that would be a hallmark of his practice for the rest of his life. Humboldt clearly believed that all knowledge was best used when widely distributed, debated, and enhanced. Sharing his map and statistical information with the Americans was all part of this larger quest for the increase and diffusion of knowledge.

Humboldt recognized that Jefferson's ambitions in securing the Louisiana Territory were already in full expansionist mode. On June 11, one of Humboldt's last days in Washington, Jefferson's sympathetic press, the *National Intelligencer*, featured an extensive article on the importance of securing Louisiana as a way of protecting U.S. borders and interests, reducing or removing foreign governments from the continental landmass, and making it easier to prevent invasions.[25] It reflected Jefferson's twin geopolitical imperatives: to expand the territory to the Pacific Ocean, thereby reducing the extent and number of foreign borders; and to envision the American hemisphere equal to and apart from Europe's empires. Westward expansion was already in Jefferson's mind. Humboldt gave Jefferson the concrete information he needed to set that in motion quickly and effectively; but Jefferson's impulse was always there. In short, Humboldt's generosity did not spark expansionist ideas; it facilitated what was already underway before he arrived.

INSTRUMENTS OF DISCOVERY

Humboldt was particularly intrigued by the Lewis and Clark Expedition, which had just begun in early May of 1804, as Humboldt made his plans to arrive in the United States from Havana. With the acquisition of Louisiana, Jefferson had announced his intentions to expand the reach of the new republic while remaining close to home. Since home now stretched farther west, north, and south, it perforce needed to be explored and assimilated into the United States. Well in advance of this acquisition, Jefferson had been quietly arranging to send Meriwether Lewis and William Clark to investigate this region. While it was essentially a geopolitical expedition, Jefferson's letter to Lewis carried broader instructions, outlining the multitude of goals for the Corps of Discovery. In addition to finding a navigable river route to the Pacific Ocean, Jefferson tasked the expedition with making detailed maps and taking notes on the flora and fauna. Jefferson was particularly emphatic regarding the establishment of cordial relations with each indigenous community they encountered, in an effort to turn their loyalties away from the British or the French who dominated the international fur trade along the Missouri River.

In preparation for the venture, Lewis had studied with members of the APS. From Benjamin Rush, he learned field medicine. From Caspar Wistar, he learned the basics of collecting fossils. Robert Patterson provided a list of instruments for taking terrestrial measurements, and Patterson's pupil, Andrew Ellicott, taught Lewis celestial navigation. Lewis learned how to use instruments designed to measure altitude above sea level and how to pinpoint his location using astronomical calculations.[26] Ellicott's accomplishments in this arena far surpassed most Americans. He had conducted the most recent boundary survey between the United States and Spain from 1786 to 1800, in anticipation of the Louisiana Purchase. He and Benjamin Banneker had drawn up the most complete plan of the city of Washington, D.C.

Humboldt's approach to scientific exploration might best be understood in light of his obsession with his instruments. In South America, Humboldt had traveled with no fewer than forty-two separate measuring devices, each ensconced in its own velvet-lined mahogany box to protect it against the rigors of travel.[27] In his published works he described them as though they were alive, treating them as he did the human members of his entourage. Some of them were state-of-the-art, carefully calibrated instruments; others were of his own devising, notably a cyanometer (fig. 15) he and Horace-Bénédict Benjamin de Saussure created, which he used to measure the exact intensity and shade of blue in the sky.[28] The numbering system on the gradations of color assigned relative values to the sky, allowing Humboldt to ascertain the clarity of the air at different altitudes; however, it also signified the traveler's continual interest in the pictorial qualities of his surroundings. Humboldt's delight in the cyanometer aligned with his interest in capturing the aesthetic descriptions of the landscape. These would appear in many of Humboldt's volumes and culminate in an entire section of the second volume of *Cosmos* devoted to advice for landscape painters.

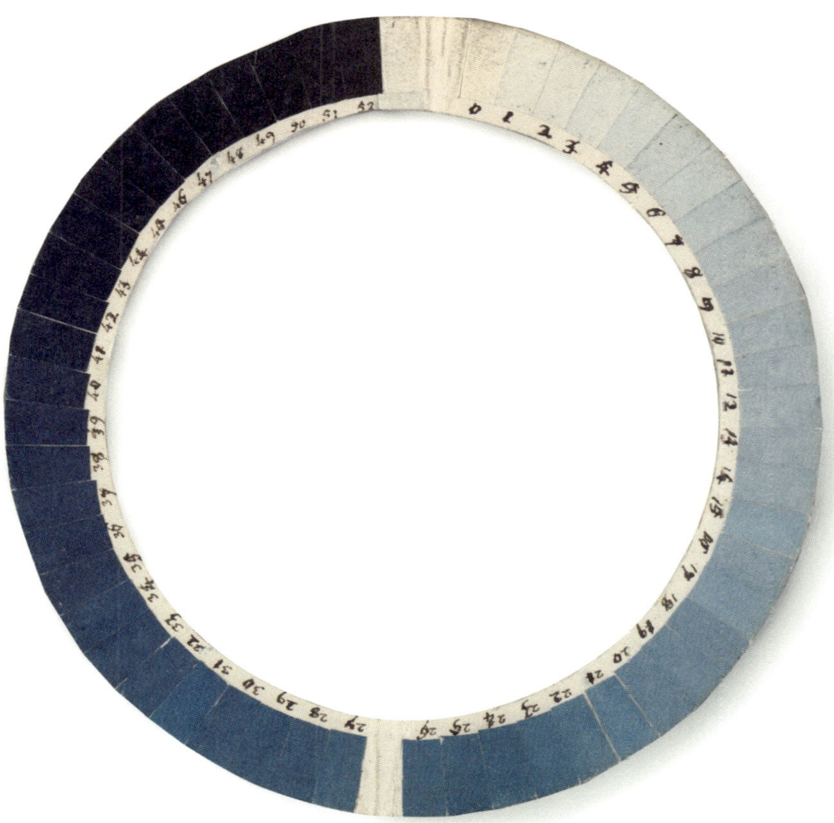

Fig. 15
Horace-Bénédict Benjamin de Saussure, *Cyanometer*, 1790, paper dyed with Prussian Blue, 7 3/16 in. in diameter, Bibliothèque de Genève

In Philadelphia and Lancaster, Humboldt had made a point to meet the scientists who had helped outfit and train the Corps of Discovery. He staked his reputation on the accuracy of his measurements, which he knew would determine his scientific colleagues' acceptance of his ideas and theories. And so Humboldt had peppered Jefferson with questions about the explorers' skills and their equipment, asking whether they had taken the right instruments and whether they could obtain measurements accurate enough for Humboldt to fold into his research. Jefferson apparently shared Humboldt's alarm, as he quietly wrote to Isaac Briggs and William Dunbar to reaffirm the accuracy and suitability of Lewis's equipment.[29] Humboldt's concern made sense: every time he had to rely on a proxy to be his eyes and ears, he had to be able to depend on the accuracy of that information, and he knew from recent experience that even the most seasoned explorers could make mistakes. Humboldt had taken satisfaction in correcting numerous errors of longitude and latitude in updating the cartography of Mexico, in part thanks to his own secure positioning of the magnetic equator. That was a point of calibration Lewis and Clark did not have, which meant every point they did measure would need to be recalibrated upon their return.

That recalibration resonated well beyond the matter of instrumentation. Humboldt believed in the limitless potential for America, so much so that he may have changed the stakes for Jefferson's expectations from their expedition. Humboldt quizzed Jefferson about their training, their astuteness, and their capacity for understanding, recording, collecting, and returning with the same kinds of specimens at the same level of scientific ability as Humboldt and Bonpland had amassed in their trunks and diaries. Humboldt wanted to know about their relationships with American Indians and Lewis and Clark's understanding of the geology, the topography, the characteristic plants and animals, the weather—in short, all of the categories of information Jefferson had outlined in his *Notes* and that Humboldt had amassed over the past five years. Perhaps most importantly, Humboldt effused about the value those discoveries would have for America's standing on the world stage and among scientific men: "Everything west of these mountains will be a vast area to conquer for science!" His interest conveyed his continued desire to be part of that process of discovery and dissemination of scientific knowledge, placing those in the United States as part of his expanding global network of correspondents.

Humboldt asked different questions than the ones Jefferson had posed in his instructions to Lewis, and by the time Humboldt met with Jefferson, the Corps of Discovery was beyond the president's reach. The explorers had no idea that Jefferson's vision of the world and of the country was shifting in their absence. Jefferson, Madison, Gallatin, and Humboldt had begun to reinvent America. They might still hope to find evidence of a living mammoth, but now this group sought to catalogue and understand America's natural resources and, by extension, the nation's cultural identity. Humboldt was already demonstrating what he could provide to the United States; before he left, he passed along a sixteen-page report to complement the map he had made, translating his data into French for Jefferson's use. He also wrote a lengthy narrative of his five-year sojourn, handing that to APS Secretary John Vaughan, so that it could be published in the American press. By the time Lewis and Clark returned in 1807, triumphant, Humboldt had published his *Essay on the Geography of Plants*, laying out the framework into which he would want copious American data to pour. Eager to maintain that transfer of knowledge, Jefferson pushed for Lewis to write up and publish the account of their expedition. Despite that pressure—or perhaps in response to it—Lewis was unable to accomplish this task before his untimely death. Jefferson then turned to Clark, the lesser wordsmith of the pair, to complete the task.

The content of the pair's journals confirms that their observations were not of the same scientific caliber as Humboldt's. In a letter to Humboldt, Jefferson alluded to these shortcomings, lamenting,

> *the discoveries of Govr. Lewis may perhaps furnish matter of value, if ever they can be brought forward. The mere journal of the voyage may be soon expected; but in what forwardness are the volumes of the botany, natural history, geography and meteorology of the journey, I am uninformed.*[30]

What had started out as a narrative of the journey had now grown in Jefferson's mind to encompass five or six volumes. His expanded ambitions for this trip far outstripped Lewis's ability to satisfy his mentor's desire for a publication of Humboldtian proportions. Lewis's writer's block may have owed its origins to Humboldt's very enthusiasm for what he hoped to learn from their trip.

ZEBULON PIKE AND THE PURLOINED MAP

Humboldt's map turned out to be a powerful piece of military intelligence for Jefferson as he negotiated with the Kingdom of Spain. But its use within Jefferson's administration threatened to derail the goodwill generated by Humboldt's gift. Humboldt had shared his information as the start of a reciprocal relationship. He expected to publish it himself after returning to Europe, knowing what an advance this map represented on the current knowledge of North America. But before that could happen, it became part of a more awkward and potentially damaging affair in American history. General James Wilkinson and Vice President Aaron Burr had concocted a plot to seize part of the American Southwest, including part of Mexico, in order to establish an independent republic of their own. For that, they needed intelligence about the terrain. Wilkinson suggested that while Lewis and Clark were exploring the northern reaches of the Louisiana Territory, Zebulon Pike be granted

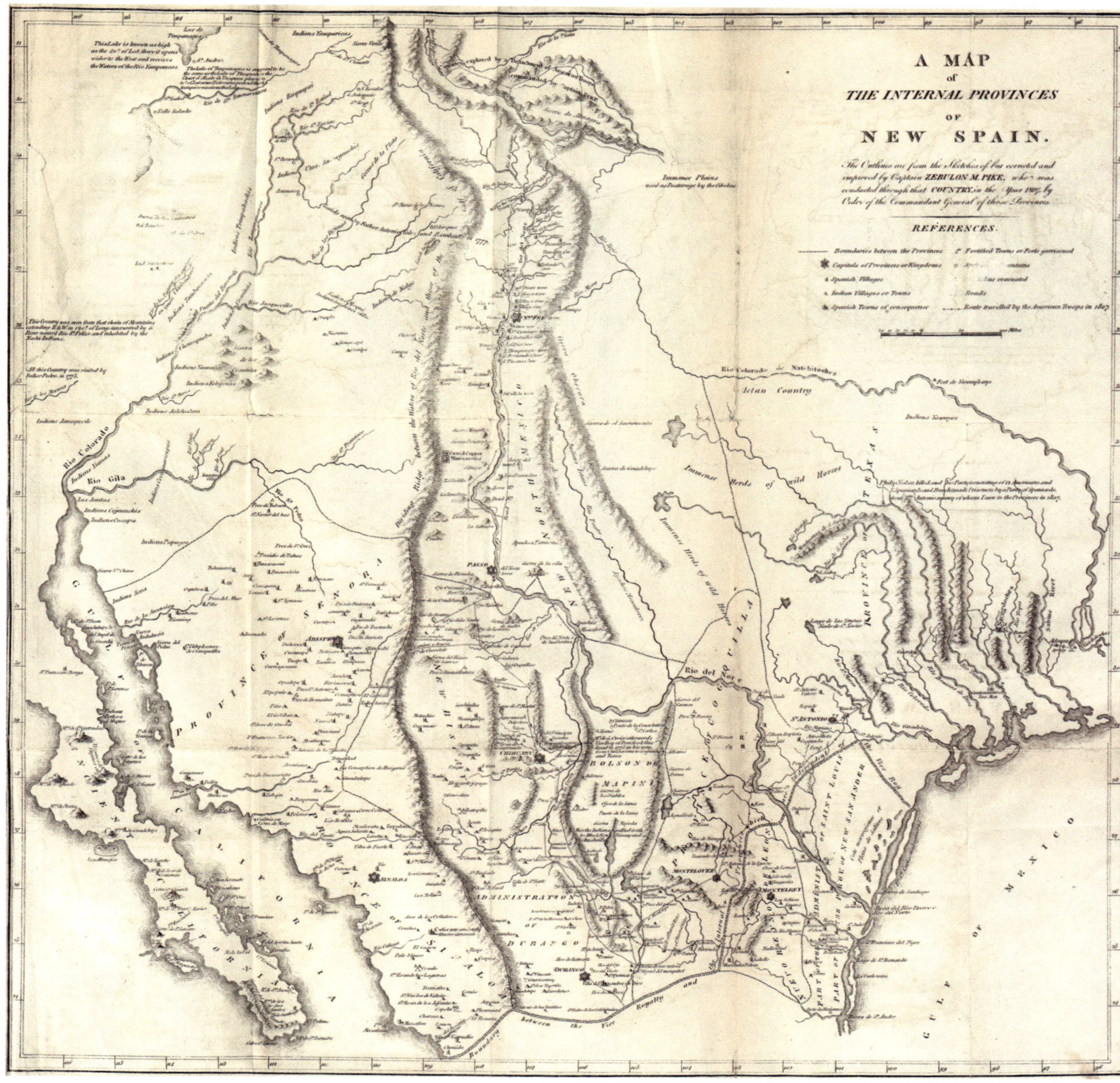

Fig. 16
Zebulon Montgomery Pike, *A map of the Internal Provinces of New Spain,* from *An Account of Expeditions to the Sources of the Mississippi,* 1810, engraving, 18 ⅛ × 18 ½ in., The David Rumsey Map Collection, David Rumsey Map Center, Stanford Libraries

permission to explore farther south. Jefferson had to be aware of Wilkinson's divided loyalties; on his own surveying trip to the lower Mississippi valley in 1798, Andrew Ellicott had sent a coded message to the State Department identifying four Americans who were taking annual stipends from Spain—one of whom was Wilkinson.[31] Still, Jefferson authorized Pike's expedition. In preparation, Wilkinson secretly secured a copy of Humboldt's map and allowed Pike to copy the salient features onto the map he had been issued for use on his survey.[32] Further, and unbeknownst to Jefferson as well, Pike had access to Humboldt's written report that he had entrusted to the president, containing statistical information on the Vice Royalty of Mexico.[33]

Pike's expedition of 1806 surveyed the territory from St. Louis in search of the headwaters of the Arkansas River, valuable information for the ongoing disputes over the border between the United States and Mexico. He added "Grand Peak," later named Pike's Peak, to the map, although he was unable to summit due to the deep November snow. When he strayed into Spanish territory in the Southwest, Pike was promptly arrested. After his release, the explorer published in 1810 an account of his travels that included Humboldt's map—without attribution (fig. 16). When Humboldt saw a copy of Pike's report, he was furious. The published volumes recounting explorations were the *lingua franca* of naturalists, establishing credit for accomplishments

CAT. 19 Alexander von Humboldt, *Map of Mexico and the Surrounding Territories*, plate 2, from *Atlas géographique et physique du royaume de la Nouvelle-Espagne*, 1811, engraving, open: 24 × 35 in., Smithsonian Libraries

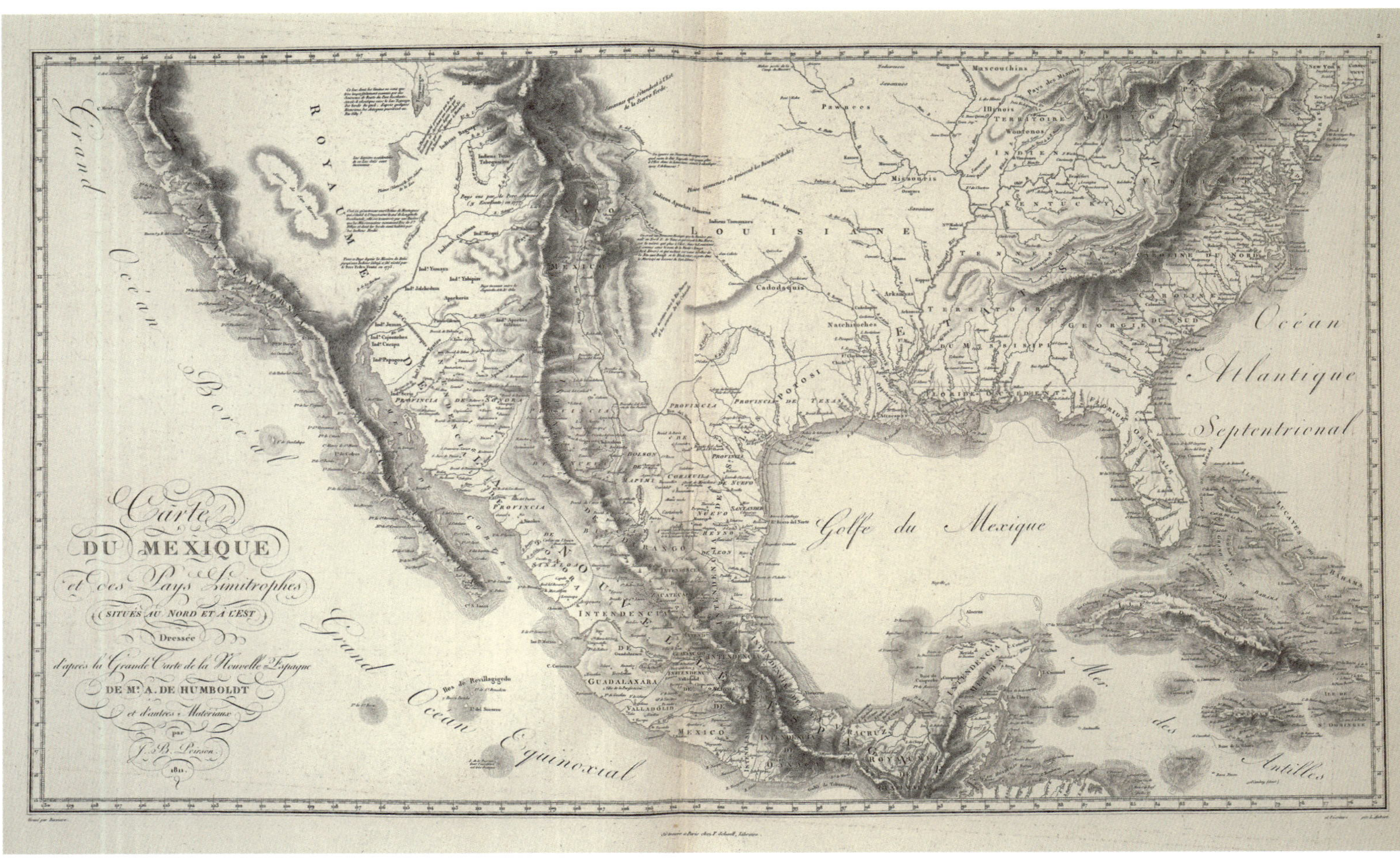

and discoveries and serving as a new benchmark for subsequent travelers. In a letter to Jefferson, Humboldt barely masked his outrage at the act of plagiarism:

Mr. Arrowsmith in London has stolen my large map of Mexico, and Mr. Pike has taken, rather ungraciously, my report which he undoubtedly obtained in Washington with the copy of this map, and besides, he also extracted from it all the names. I am sorry over my cause for complaint about a citizen of the United States who otherwise showed such fine courage. I do not find my name in his book and a quick glance at Mr. Pike's map may prove to you from where he got it.[34]

The timing of this transgression was particularly awkward for Jefferson. The onset of the War of 1812 and subsequent British naval blockades prevented him from being able to get a letter to Humboldt for almost two years. In the meantime, Humboldt finally published his version of the map with the title *Carte du Mexique et des pays limitrophes* (*Map of Mexico and the Surrounding Territories*) in his *Atlas géographique et physique du royaume de la Nouvelle-Espagne* (*Geographical and Physical Atlas of the Kingdom of New Spain*) of 1811 (CAT. 19). The atlas accompanied his *Essai politique sur le royaume de la Nouvelle-Espagne* (*Political Essay on the Kingdom of New Spain*), at the time the most comprehensive information on the American Southwest, based largely on Humboldt's research in the Mexican archives, where he pored over earlier Spanish explorers' reports. Humboldt went a step further, putting his charge of plagiarism in print in the introduction to the first volume of his *Personal Narrative*, published in French and translated into English in 1814:

The maps of Mexico, which are annexed to the narrative of his journey, are reduced from my great map of New Spain, of which I left a copy, in 1804, at the secretary of state's office at Washington.[35]

When Jefferson was finally able to respond, Pike was dead and Humboldt had made his point. Jefferson offered consoling words to his friend, blaming the British for the theft of his map:

That their Arrowsmith should have stolen your map of Mexico, was in the piratical spirit of his country. But I should be sincerely sorry if our Pike has made an ungenerous use of your candid communications here; and the more so as he died in the arms of victory gained over the enemies of his country.[36]

What the Prussian naturalist sought was reciprocity. In Washington he had shared his map and statistics and received from Albert Gallatin data on the United States in return. Humboldt published all of this material, along with his map, in 1811 in his *Political Essay on the Kingdom of New Spain* and repurposed it as part of his multivolume *Personal Narrative of Travels to the Equinoctial Regions of the New Continent*. In each case he dropped names and praised his collaborators. Pike's plagiarism was as much a breach of the standard protocol among generations of explorers as a personal awkwardness between two friends.

Humboldt had enclosed a copy of his 1811 book with his letter, somewhat softening the blow of his criticism by expressing his faith in what he would eventually learn when the report from the Lewis and Clark Expedition was finally published. In the text Humboldt mused at the prospect of the Missouri River providing transcontinental "communication" with the Pacific, expressing in cartographic terms one of Humboldt's fondest wishes, which was for stronger and more reliable networks for transit and communication of knowledge:

But we have not yet sufficient acquaintance with the nature of the ground through which the communication is proposed to be established, to pronounce upon the utility of these projects. The journal of Captain Lewis, at the expense of the Anglo-American government, on the Mississippi and Missoury, may throw considerable light on this interesting problem.[37]

Humboldt continued, paying Jefferson a compliment by name:

The wonderful journal of Captain Lewis was undertaken under the auspices of M. Jefferson, who by this important service rendered to science has added new claims on the gratitude of the savans [savants] of all nations.[38]

In response, Jefferson went on to apologize for something that pained him even more. Pike had been able to get his report published, but it would take until 1814 for Nicholas Biddle to salvage the project and see it through to publication. Jefferson was beside himself. Meriwether Lewis had been a close friend and confidant to the president. Jefferson lamented to Humboldt:

You will find it inconcievable [sic] that Lewis's journey to the Pacific should not yet have appeared; nor is it in my power to tell you the reason. The measures taken by his surviving companion Clarke[sic], for the publication, have not answered our wishes in point of dispatch. I think however, from what I have heard, that the mere journal will be out within a few weeks in 2. vols 8vo. These I will take care to send you with the tobacco seed you desired, if it be possible for them to escape the thousand ships of our enemies spread over the ocean. The botanical & zoological discoveries of Lewis will probably experience greater delay, and become known to the world thro' other channels before that volume will be ready. The Atlas, I believe, waits on the leisure of the engraver.[39]

When it was finally published, Humboldt absorbed the information collected by Lewis and Clark into his own ongoing research.

For Humboldt, gathering knowledge was but the first step in its increase and diffusion, with credit willingly shared among the contributors. That diffusion took the form of increasingly lavish books designed to encompass all of that data and expound on all of those ideas. Jefferson shared that sensibility, wanting to take justifiable pride in seeing Pike's volume—and Lewis and Clark's—become part of that international dialogue. In 1804 Dr. David Ramsay lauded the scientific potential for the addition of the Louisiana Territory, pronouncing,

When this country, hitherto untrod by the foot of either chymist or botanist, shall be carefully explored by our Waterhouses, Mitchells, and Bartons, such additions will be made to our present stock of knowledge, in natural history, as well both delight and benefit an admiring world.[40]

Contributing to the world's knowledge and, by extension, justifying America's prospects as a new power were goals Humboldt held in tandem.

Humboldt's anger over Pike's transgression was pointed but short-lived. Humboldt's contribution to the increase and diffusion of knowledge was in his profusely illustrated publications, setting a new standard for exploration narratives, an enterprise that made him the single best-known natural historian in the world. The meteoritic rise of Humboldt's reputation and the widespread impact of his multivolume *Personal Narrative* ensured that the subsequent waves of American explorations conducted by Stephen H. Long and John C. Frémont would be conceived and carried out with Humboldt's encouragement and by men steeped in his methods and his ideas. The American landscape became a primary subject of research, and appreciation for its distinctive features would inspire travel narratives and artistic representations of the expanding country. These surveys would have the benefit of Humboldt's emergent ideas about the unity of nature and the significance of what the U.S. researchers could add to the sum of world knowledge. His friendship with Thomas Jefferson and his continued correspondence with American cabinet officers and statesmen positioned Humboldt to reap the benefit of government-sponsored expeditions into the American interior. It was time to get started.

THE LONG EXPEDITION AND "HUMBOLDTIAN SCIENCE"

The War of 1812 had put western exploration on hold, and by the time hostilities ceased, the first volumes of Humboldt's *Personal Narrative* were in print. This became the new standard for scientific travel narratives designed to inform political leaders of the significance of their holdings. The paucity of visual material from the Corps of Discovery and the failure of several of their instruments meant that subsequent expeditions set off with the goal of correcting missing measurements and recording the visual character of the landscape. That mandate now had a broader range thanks to Humboldt as well: it required the expedition to collect information on plants; animals; geology; weather; and the customs, languages, and structures of indigenous communities; as well as cartographic measurements derived from sophisticated instruments and detailed, accurate drawings of everything.

By 1816 Stephen H. Long wanted to restart American exploration in a serious way. Long was college educated and trained as a military topographical engineer. He began by proposing itineraries for expeditions that would corroborate or correct the dead reckoning measurements during the Lewis and Clark Expedition. From 1810 to 1813, Long lived in Philadelphia and worked with members of the APS, many of whom had spent time with Humboldt in 1804. Robert Patterson, Andrew Ellicott's teacher, instructed Long on the arts of navigation. Librarians Nicholas Collin and John Vaughan gave Long access to the society's library, which included a complete set of Humboldt's publications. When Long talked Secretary of War John Calhoun into authorizing his expedition to the Rocky Mountains, he argued for taking along a team of specialists to add a robust natural history component to the venture. The Philadelphia scientific community supplied the rest of the expertise and training Long needed to accomplish his ambitious goals.

Calhoun finally authorized Long's expedition in 1818. His instructions to Long were lengthy and detailed. Central to the expedition was traditional cartographic exploration:

> *You will first explore the Missouri and its principle* [sic] *branches, and then, in succession, Red river, Arkansa and Mississippi, above the mouth of the Missouri. . . . permit nothing worthy of notice, to escape your attention.*[41]

Calhoun wanted Long to find and map the 49th parallel and correct or verify as many of Lewis's geographic measurements as possible. He also enumerated the scientific goals Long had suggested, among them making detailed notes on the soil, minerals, topography, flora, fauna, and each Indian tribe they encountered. Finally, he recommended that Long familiarize himself with Jefferson's instructions to Lewis and requested that members of the APS be consulted to weigh in with additional topics. Ultimately the expedition fulfilled two of Calhoun's imperatives: one was providing the most detailed and accurate account of American Indian customs as practiced by the Omaha, Pawnee, and Otoe tribes; the other was a geopolitical assessment of the viability of settling the Great Plains.

Long was the first American military officer to bring with him leading scientific experts drawn from the ranks of the APS and the newly founded Academy of Natural Sciences (ANS) in Philadelphia.[42] These men were all familiar with Humboldt's methods, his ideas, and his published writings. Long's expedition was also the first to designate a slot specifically for an artist, flagging two of the expedition's expanded goals: to create a visual record to complement the diaries each man was expected to keep, and to provide the basis for competent illustrations for the eventual publication of the expedition report. Humboldt's decade of publications to this point had set a high bar for expeditionary research. His emphasis on South America called for North American research to strengthen the place of the United States in this kind of cultural enterprise.

The Long Expedition to the Rocky Mountains from 1818 to 1820 was one of the first that made particular use of Humboldt's writings, his interests, and his influence within the scientific community in Philadelphia. As the explorer read up on expeditionary and natural history, he began thinking about transportation for his team and its equipment. Long had determined that a shallow draft steamboat might make navigation from Pittsburgh to St. Louis, then up the Mississippi and Missouri Rivers, easier than traveling entirely overland. With that in mind, he designed what would be built as the steamboat *Western Engineer*. In it he made room for a substantial library of reference books, chosen in consultation with the APS. Among them were Humboldt's *Personal Narrative* and his *Political Essay on the Kingdom of New Spain*, which contained his published map of the interior of North America.[43]

Out in the field, the expedition ran into significant difficulties. Long's plans for the *Western Engineer* resulted in a much larger vessel with a deeper draft than he had intended, making navigation in the strong currents of the Missouri and Mississippi Rivers extremely difficult. The neophyte participants often misjudged the amount of food, water, and supplies required to survive the overland portions of the trek. Their horses tended to run off or go lame. They were fortunate that most of the Indians they encountered tended to take pity on them and provided more help than hindrance. Through mishaps, they lost many of the notebooks of information. Their equipment failed, compromising many of their measurements. The lasting value of the enterprise resided more in the naturalists' and artists' contributions than in military intelligence or topographic measurements.[44] Zoologist

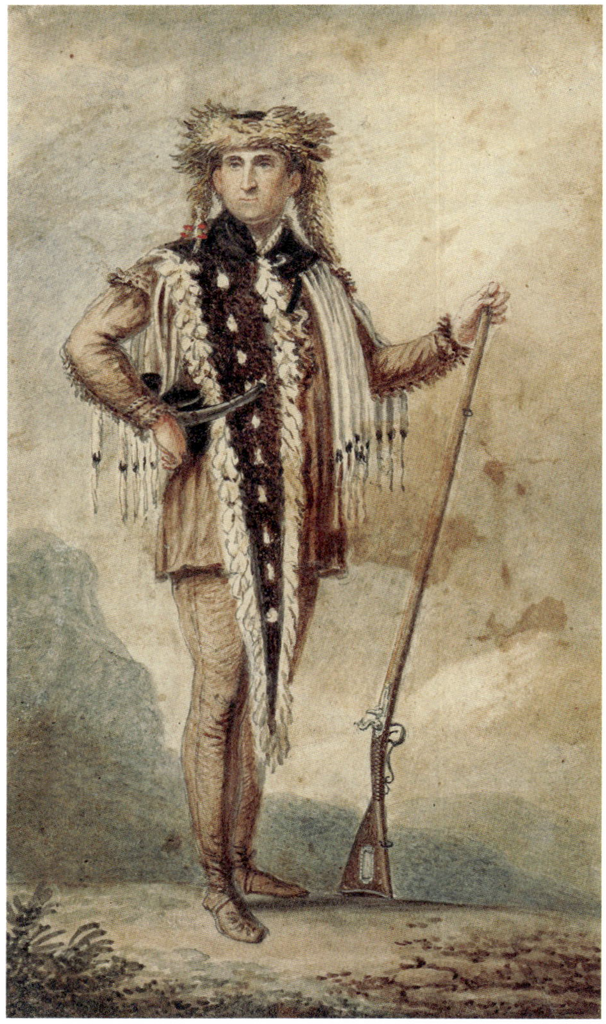

Fig. 17
Charles Balthazar Julien Fevret de Saint-Mémin, *Meriwether Lewis in Frontiersman's Regalia*, 1806–7, watercolor, ink, pencil, and gouache on paper, 6 ⅛ × 3 ¾ in., New-York Historical Society, Gift of the Heirs of Hall Park McCullough

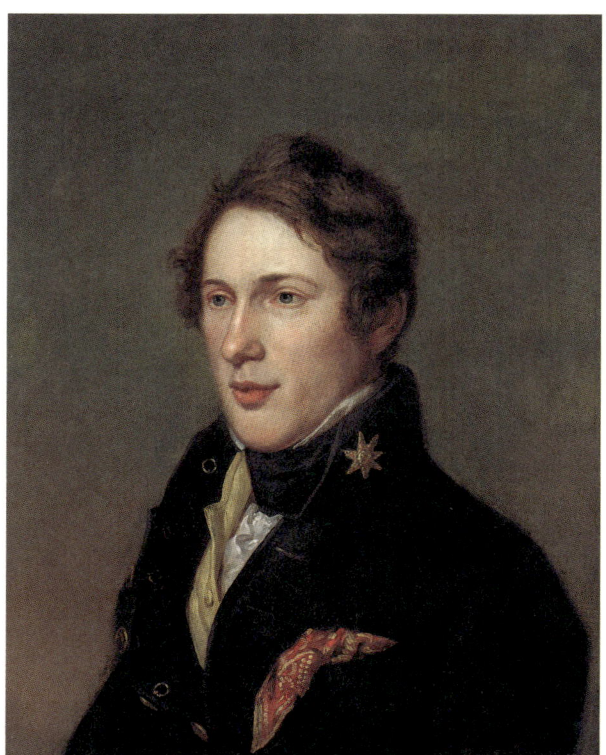

Fig. 18
Charles Willson Peale, *Titian Ramsay Peale II* (1799–1885), 1819, oil on canvas, 24 ¼ × 20 ⅛ in., Philadelphia Museum of Art, Promised gift of the McNeil Americana Collection

Thomas Say contributed much of the narrative, polished by Edwin James and published in 1823. All three volumes included illustrations of landscapes, plants, animals, and Native Americans by artists Samuel Seymour and Titian Ramsay Peale. The three-volume illustrated narrative expanded Americans' awareness of and appreciation for a wilderness landscape that was newly added to the country. This method defines what would come to be known in the modern era as "Humboldtian Science," a comprehensive approach to gathering information that contributed to the elevation and professionalization of separate disciplines and that relied on the joint expertise of teams of people whose areas of interest overlapped.[45]

The natural history specimens and Native American artifacts were earmarked for Peale's Museum, where they joined the artifacts from the Lewis and Clark Expedition. Titian Ramsay Peale played a central role in promoting the Long Expedition's scientific and aesthetic significance. The Peale's family involvement with the Long Expedition was the natural extension of their engagement with the Corps of Discovery. In 1809 the artifacts from the Lewis and Clark Expedition had arrived at the museum. Charles Willson Peale installed them in the galleries as part of its educational mission. Peale had gone so far as to have a full-size wax mannequin made of Lewis, and he dressed it in the explorer's buckskin shirt and leggings (fig. 17). His youngest son, Titian Ramsay Peale (fig. 18), successfully petitioned Calhoun to join the Long Expedition as the junior naturalist. Titian was raised to be an artist-naturalist. He was born in Philosophical Hall in 1799, literally under the roof of his father's museum.[46] Although he had been only five years old when Humboldt visited the United States, he grew up steeped in his father and older brothers' stories about his visit. He had watched his brother, Rembrandt, set off for Europe to publicize the American mammoth and return with stories of spending time with Humboldt in Paris.[47]

Titian spent much of his childhood mounting specimens in his family's museum, and by 1816 he was running Peale's Museum in Philadelphia. There he made close friends in the broader scientific community, notably Thomas Say and the French naturalist Charles Lesueur, who lived in Philadelphia from 1817 to 1825. Titian's artistic talent had already made him invaluable to Say, who asked Peale to provide the illustrations for his first proposed volume on entomology. In 1817 Titian was attending Caspar Wistar's lectures on anatomy at the APS, and in the fall, he and George Ord and Say accompanied a group of scientists led by William McClure to Florida and Georgia. There they retraced part of William Bartram's itinerary for the purpose of collecting natural history specimens. McClure was active in the newly established ANS, and in November 1817, Peale was elected a member.[48]

When Titian Peale returned from the Long Expedition to Philadelphia in 1821, he brought back at least twenty-two finished drawings and 120 sketches.[49] He proved adept at rendering a wide range of subjects well. His detailed botanical watercolors (fig. 19) demonstrate an acuity of observation resembling the botanical drawings Humboldt had published in his scientific volumes (fig. 20). His landscapes often extended across the two bound pages (fig. 21), conveying the expansiveness of the wide horizons he experienced for the first time on the Great Plains. A number of Peale's animal studies carried notations on proportions and scale, which would be helpful when he returned home to prepare animals and birds for display.

He exhibited a group of his finished watercolors at the Pennsylvania Academy of Fine Arts. Several of them evoke the illusionistic museum dioramas that he helped construct and now maintained. In his watercolor of magpies (fig. 22), Peale presents four birds, each demonstrating a different characteristic behavior (in flight, perched, picking at carrion), each rendered in a different scale, and each shown from a different perspective.[50] The goal was to provide an educational lesson in visual form. He gave equal attention to the landscape setting, working from the expedition materials to conjure a specific place or ecosystem as well. Charles Willson Peale had noted with satisfaction in his autobiography,

> *It is not customary in Europe, it is said, to paint skys and landscapes in their cases of birds and other animals, and it may have a neat and clean appearance to line them only with white paper, but on the other hand it is not only pleasing to see a sketch of a landscape, but by showing the nest, hollow, cave or a particular view of the country from which they came, some instances of the habits may be given.*[51]

What Peale envisioned with his museum was more than a *Wunderkammer*, or a collection of specimens and curiosities. He saw his museum as an educational platform for inculcating pride in American

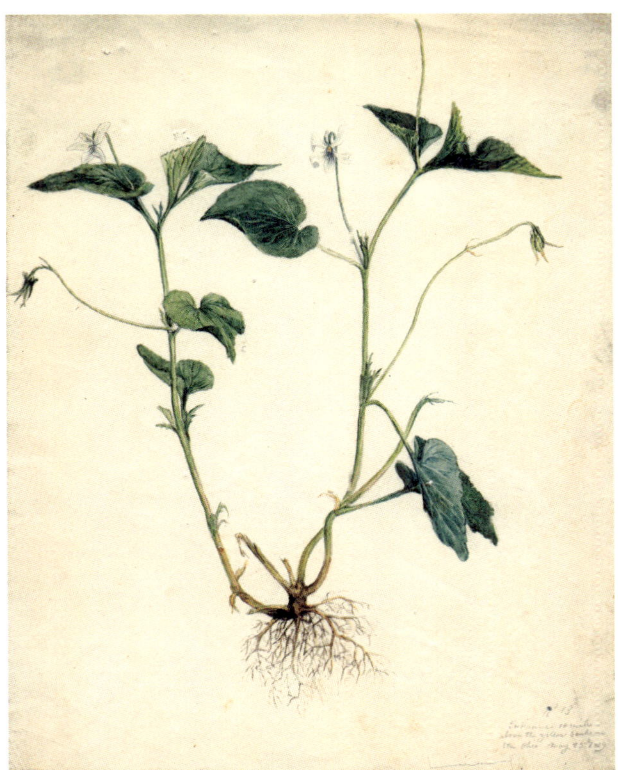

Fig. 19
Titian Ramsay Peale, *White Violet, Indiana 10 miles above the yellow banks on the Ohio,* 1819, watercolor and pencil on paper, 9 ⅞ × 7 ⅝ in., American Philosophical Society

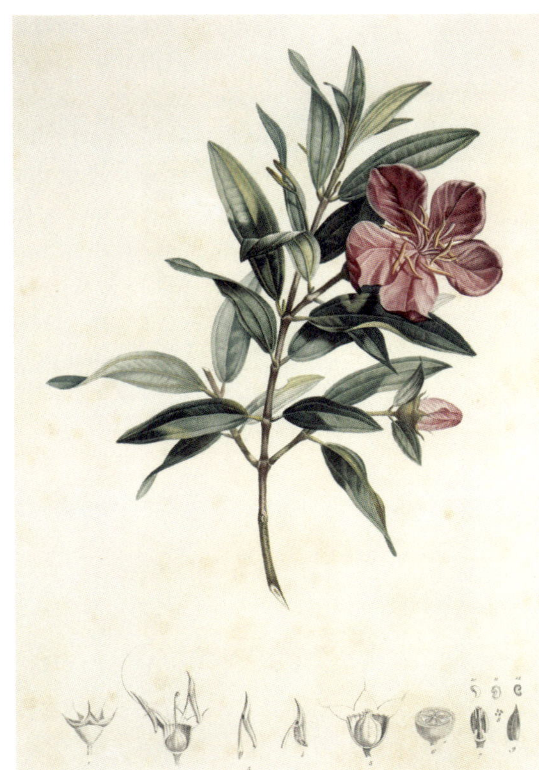

Fig. 20
Alexander von Humboldt and Aimé Bonpland, *Rhexia Speciosa*, from *Relation historique du voyage aux régions équinoxiales de nouveau continent,* vol. 1, plate 6, 1806, hand-colored engraving, The Natural History Museum, London

civic virtues and its natural history. By tying these two themes together in the same physical space, Peale set up a paradigm that would link the two in the development of an American cultural identity. Peale's elaborate habitats and dioramas and his lifelike taxidermy presented an integrated natural history lesson for his visitors, a visual corollary to Humboldt's writings on the interconnectedness of ecosystems. The portraits of distinguished Americans, with Humboldt included prominently among them, surmounted the stacked cases, a capstone to the experience.

Titian turned his artistic attention to providing detailed illustrations for two of the leading naturalists in his orbit. He drew many of the birds found in Charles Lucien Jules Laurent Bonaparte's 1823 successor volume to Alexander Wilson's *American Ornithology*. Thomas Say's first volume of his *Entomology* was published in 1824, and all eighteen of the color prints were based on Peale's meticulously crafted watercolors of insects. Peale made his own signature contribution to natural history through his study of butterflies (fig. 23). He pursued this passion, building a collection of butterflies mounted in shadowboxes and a portfolio of detailed watercolors intended for a never-realized publication on Lepidoptera.[52] Each of these publications added to the growing compendium of knowledge of American natural history, appealing both to scientists and the general public, impressive volumes that reinforced to the reader the riches to be found in American nature. Peale and his father may also have seen them as complementary appendices to the collections in their museum, extensions of the lessons built into the displays.

Charles Willson Peale had hoped that his museum would gather together representative samples of American flora and fauna as a means of teaching his audience about the natural sciences as well as about their country's history. Philadelphia was the cradle of this scientific and cultural outreach, supporting several scientific societies and art institutions. Growing familiarity with what at the outset was new and unknown fostered an appreciation of the unique qualities of American natural history and Indian cultures, making it easier to assimilate it all into an expanding definition of America's cultural identity. When Titian was appointed as Long's assistant naturalist, he and his father negotiated the arrangement that would direct the expedition's natural history collections to Peale's Museum. That agreement was crucial to the elder Peale. Staying current by guaranteeing that the latest discoveries would show up there first, Peale made the case for why the display of expeditionary discoveries was in fact the illustrated concordance to the published reports.

The importance of these published survey reports cannot be overstated. The illustrations proved to be of particular significance. Humboldt helped popularize illustrated accounts of topographic surveys. He had personally supervised the translation of his field sketches into professional, hand-colored engravings that attracted a broad, admiring audience. Early responses to his writings established that his readership extended beyond the scientific community, to those curious about discoveries made in previously uncharted parts of the world. There had been very little in the way of a visual record from the Lewis and Clark Expedition

Fig. 21
Titian Ramsay Peale, *Sunset on the Missouri River 370 miles above its mouth*, 1819, watercolor and pencil in a sketchbook, 5 ⅛ × 16 ½ in., American Philosophical Society

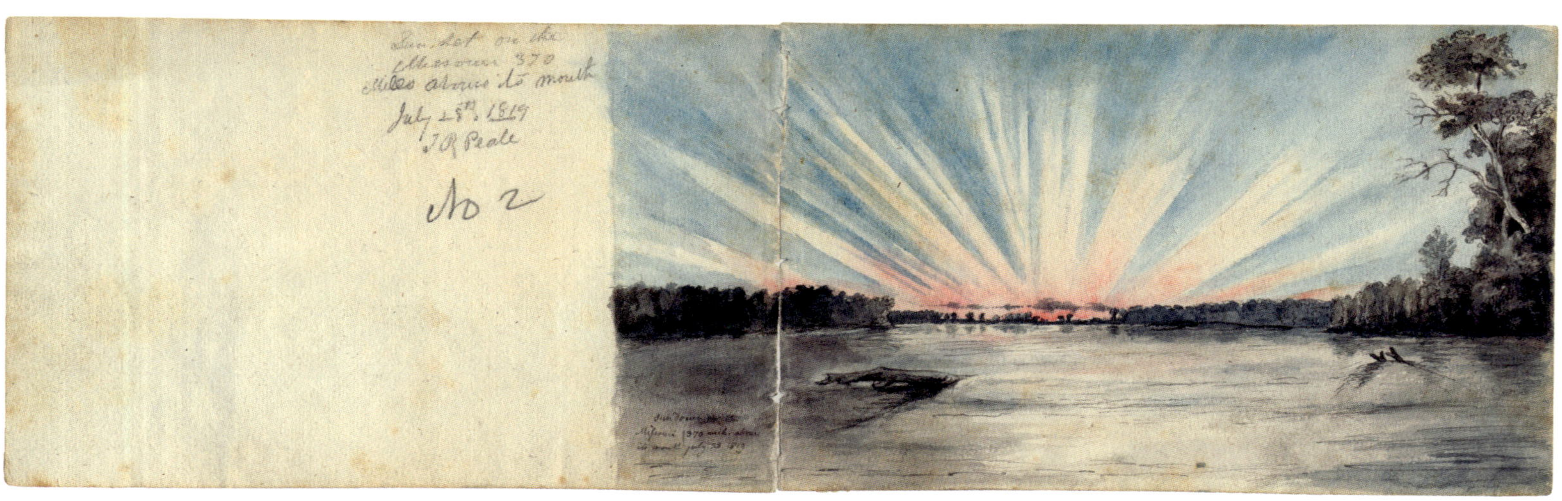

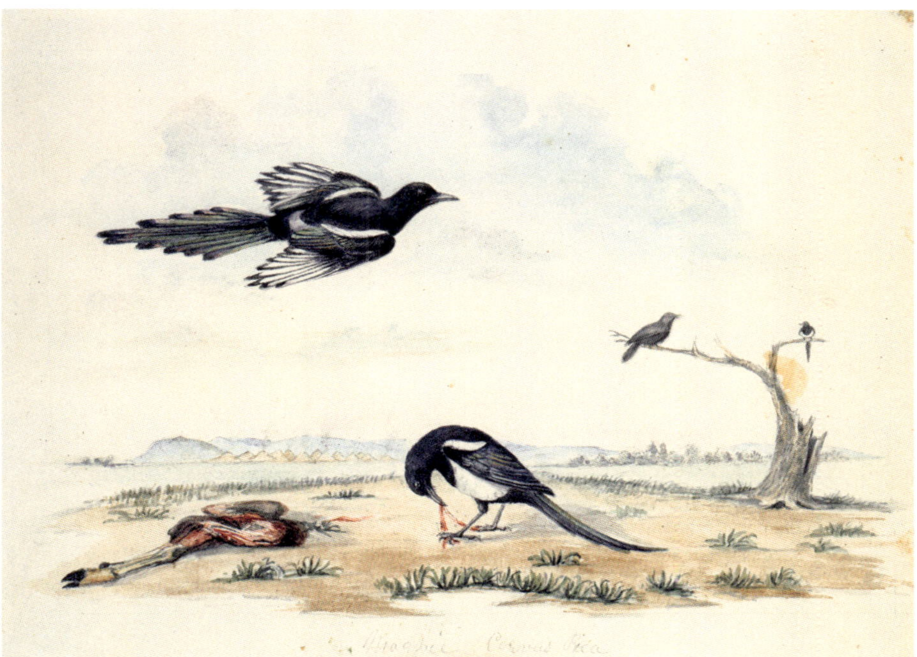

Fig. 22
Titian Ramsay Peale, *Corvus pica (magpie)*, 1822, watercolor and pencil on paper, 5 ⅜ × 7 ⅛ in., American Philosophical Society

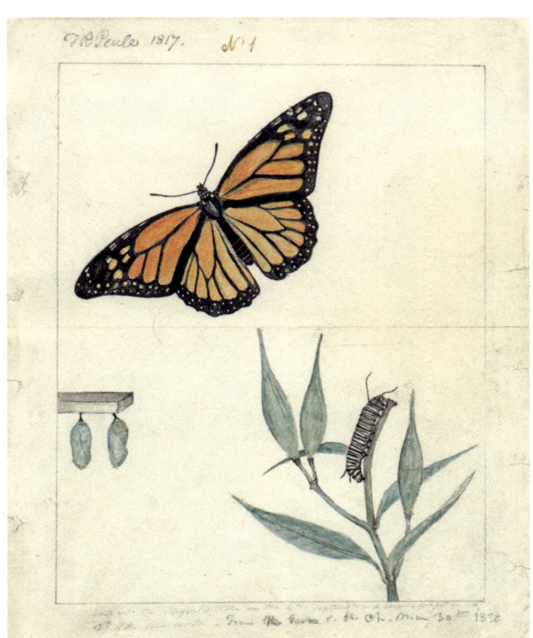

Fig. 23
Titian Ramsay Peale, *Monarch Butterfly*, 1817, watercolor and pencil on paper, 10 × 8 in., American Philosophical Society

beyond maps. Long was now in a position to address that lacuna. He and Calhoun hoped that Americans would be eager to learn what lay within their expanded borders. The published volumes chronicling the Long Expedition were illustrated with engravings based primarily on Samuel Seymour's drawings and paintings and several more by Peale. Thomas Say furnished much of the narrative material based on his diary, and the text enumerated all of the scientific and ethnographic information gathered by a team of specialists.[53] Their work represented a Humboldtian intersection between nature empirically observed and aesthetically appreciated and rendered for maximum intellectual and emotional effect.

The 1823 publication of the three-volume Long Expedition report had a lasting impact beyond the immediate connection to Humboldt. Although it had been plagued by unreliable instruments and lost specimens and notebooks, the expedition did expand on the information accrued by Lewis and Clark and by Pike. The illustrations complemented the narrative, and together they inspired other explorers and provided the basis for American literary descriptions of the country's interior. Visitors to Philadelphia could absorb the physical display of the country's natural history; readers of Long's report represented a much-expanded audience for the visualization of the landscape as a vicarious journey. In 1823 Edward Everett, editor of the influential *North American Review*, penned a highly complimentary review of these volumes, drawing them to the attention of Washington Irving and James Fenimore Cooper.[54] Everett admired Humboldt and drew a favorable comparison between his narratives and that of the Long Expedition. He lauded the ambition of the expedition team and decried the meager budget as unworthy of the aspirations of the nation's desire for information.

By tying Long's journey to the nation's priorities, Everett's praise helped expand attention to this publication beyond the small circle of explorers and naturalists who were its primary audience. For writers and artists, the descriptions of the change in the landscape—from heavily forested eastern mountains to the grasslands of the Midwest and the mountain orogeny west of the Mississippi and Missouri Rivers—opened up a narrative of distance and long-range vistas. The epistolary journey, like the expeditionary one, encompassed duration expressed by the endless, flat features of the prairie. As the country expanded, so too did knowledge of what it encompassed. Long's narrative defined the Great Plains for a generation, comparing its vastness to that of an ocean and its perceived desolation to a desert. Long's "Great Desert" proved to be more powerful as a literary metaphor than as an enticement to artistic treatment, much less to westward settlement. In the hands of Cooper and Irving, the Long Expedition report would provide the armature for a literary, and metaphorical, version of the American journey.

Long's team of naturalists represented one facet of Humboldt's imprint on American exploration. By expanding his roster to include an allied group of specialists, Long created the paradigm for American scientific, government-sponsored exploration. John C. Frémont (CAT. 20) would adopt a different tactic, which might best be described as a personality-driven expedition in which his own exploits set the tone for the discoveries and hardships endured along the way. His route to government sponsorship was through the competition to find viable transcontinental railroad routes, with the support of the powerful U.S. senator from Missouri, Thomas Hart Benton. The senator's support, while not without its limits, was in large part provided at the behest of his daughter, Jessie Benton Frémont (CAT. 21), who had eloped with the dashing explorer in 1841. She later shaped her husband's career as an explorer and politician.

With Jessie's encouragement, Frémont consciously aspired to be the "American Humboldt."[55] Born in Charleston, South Carolina, he had discovered Humboldt when Secretary of War Joel Poinsett introduced him to one of the Prussian explorer's friends and protégés, Joseph Nicholas Nicollet. Nicollet had studied with Humboldt in Paris and then immigrated to the United States in 1831. He and Humboldt were both aware of the lingering discrepancies between cartographic measurements taken during the Lewis and Clark Expedition and the Long Expedition. Nicollet wanted to take more sophisticated instruments out West and correct those measurements to create a more accurate geographic picture of the United States. Unable to obtain government sponsorship, the French émigré deployed Humboldt's model and embarked on his own self-funded, independent expedition up the Missouri River. When he presented his meticulous adjustments to Poinsett and asked for sponsorship for a second trip, Poinsett approved Nicollet's plan and suggested that he consider taking a young, ambitious assistant along with him.[56]

Frémont's association with Nicollet changed the course of his career and established his reputation as a daring explorer and writer in Humboldt's idiom. As Nicollet's assistant, Frémont learned Humboldt's techniques for the hypsometric measurement of altitude.[57] In 1838 the two men surveyed the region from Fort Snelling to the Pipestone Quarry, and in 1839 they traveled to Fort Pierre along the upper Missouri River. The resultant map of the hydrographic basin of the upper Missouri River was based on 90,000 instrument readings and 326 separate astronomical points, all derived from Humboldt's approach to charting mean temperatures.[58] Frémont understood the importance of his barometric readings. The lengths to which he went to preserve his instruments recall Humboldt's obsessive care in South America. When Frémont's barometer blew over in a windstorm, its thermometer broken, he figured out how to repair it in the field so he could record the altitude of the Wind River Peak, the highest point of elevation in the range, on which he then planted an American flag—a very Humboldtian splash of drama written into his otherwise scientific narrative.[59] Frémont began to emulate Humboldt, channeling the Prussian's drive and his willingness to take risks in the pursuit of discovery.

By the time he interviewed for his own command in 1842, Frémont described his ambitions in words reminiscent of Humboldt:

> *My mind had been quick to see a larger field and differing and greater results. It would be travel over a part of the world which still remained the New—the opening up of unknown lands; the making unknown countries known; and the study without books—the learning at first hand from nature herself; the drinking first at her unknown springs—became a source of never-ending delight to me.*[60]

That vaulting ambition and enthusiasm did not always translate into successful expeditions, but it did result in Frémont assigning Humboldt's name to numerous American landscape features, an act of admiration and approbation that did not go unnoticed by the Prussian baron. When he learned that Frémont had named a river in his honor, the Prussian quipped to a friend,

> *I wish you to know that I am a river about three hundred and fifty miles long; I have not many tributaries, nor much timber, but I am full of fish.*[61]

Having thus designated the Humboldt River and the Humboldt Mountains, Frémont wrote a bit more positively about these features:

Both the river and the mountain to which I gave his name are conspicuous objects; the river stretching across the Basin to the foot of the Sierra Nevada, and the mountain standing out in greater bulk and length than its neighbors, and being one of those which I have named fertile mountains, having on it abundant water and grass, and woods.[62]

Frémont's Prussian cartographer, Charles Preuss—himself a Humboldt devotee—convinced his commander that Humboldt's companion, French botanist Aimé Bonpland, also deserved recognition. In February 1844 Frémont and Preuss first saw what is today called Lake Tahoe, described as the largest lake in the Sierra Nevada. Although Frémont's 1845 report simply refers to this feature as "Mountain Lake," in a letter the explorer confirmed, "I gave to the basin river its name of Humboldt and to the mountain lake the name of his companion traveler, Bonpland, and so put it in the map of that expedition."[63] Preuss's map of 1848, widely distributed in Europe, identified it as Lake Bonpland (fig. 24).

Frémont's maps and travel reports cemented his reputation as "the Pathfinder" and earned him his favorite accolade—"the American Humboldt." His survey reports were ghostwritten by his talented and equally ambitious wife. Her childhood was steeped in state and national politics, to which she showed an early interest and aptitude for its machinations. Jessie Benton Frémont understood the power of a carefully crafted narrative and, unlike her husband, was a fine writer. "Every morning at nine I took my seat at the writing table and left it at one. Mr. Frémont had his notes all ready and dictated as he moved about the room.... And so, swiftly, the report of the first expedition was written."[64] Although she did not accompany her husband on any of his five expeditions, she fashioned accounts of his exploits that shaped his persona for American readers and voters. The accounts of Frémont's expeditions read less like scientific or military reports and more like dramatic travelogues, inspired, one suspects, by the popularity of Humboldt's works. The description of Frémont's heroic rescue efforts to save his barometer are most likely her words, designed to elevate the most mundane of circumstances to a heightened pitch. Although the reports were intended to satisfy the secretary of war, Senator Benton also saw in the raw material a story that would enhance commercial interest in the western half of the country, notably for determining a propitious route for a transcontinental railroad.

Humboldt took notice of Frémont's expeditions, undoubtedly flattered by the American's dedication to naming features in the landscape in honor of his distant mentor. In 1849, when Humboldt published his third edition of *Aspects of Nature*, the first edition of this work to be translated into English, he expanded his narrative to include several paragraphs complimenting Frémont for his measurements of the Rocky Mountains and the Wasatch Range. Humboldt then invoked his global comparisons, observing the similarities he saw between the plains of Tenerife and the American Great Basin.[65] Humboldt in particular admired Frémont's updated map and praised his use of the same altitudinal measuring technique he had pioneered. Ever obsessed with statistics, Humboldt noted approvingly the four hundred hypsomatic and astronomical measurements used to pinpoint longitude and latitude over a span of 3,600 miles from the Missouri River to the Pacific Ocean. Humboldt put this accomplishment in perspective, noting he was first to

represent, in geognostic profile, the form of entire countries—such as the Iberian Peninsula, the highlands of Mexico, and the Cordilleras of South America, . . . it has given me peculiar pleasure to see the graphical method of representing the form of the earth in a vertical direction, or the elevations of the solid portions of our planet above its watery covering, applied on so grand a scale as has been done in Frémont's map.[66]

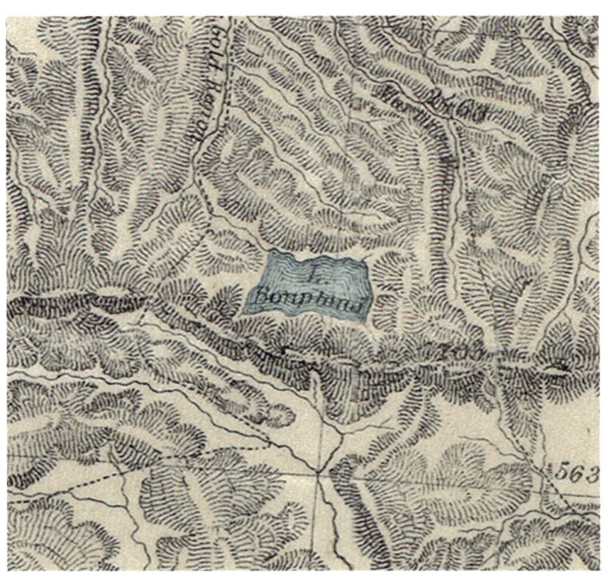

Fig. 24
Charles Preuss, *Map of Oregon and upper California from the surveys of John Charles Frémont and other authorities* (detail), 1848, engraving, full sheet: 33 × 26 ¾ in., Library of Congress, Geography and Map Division

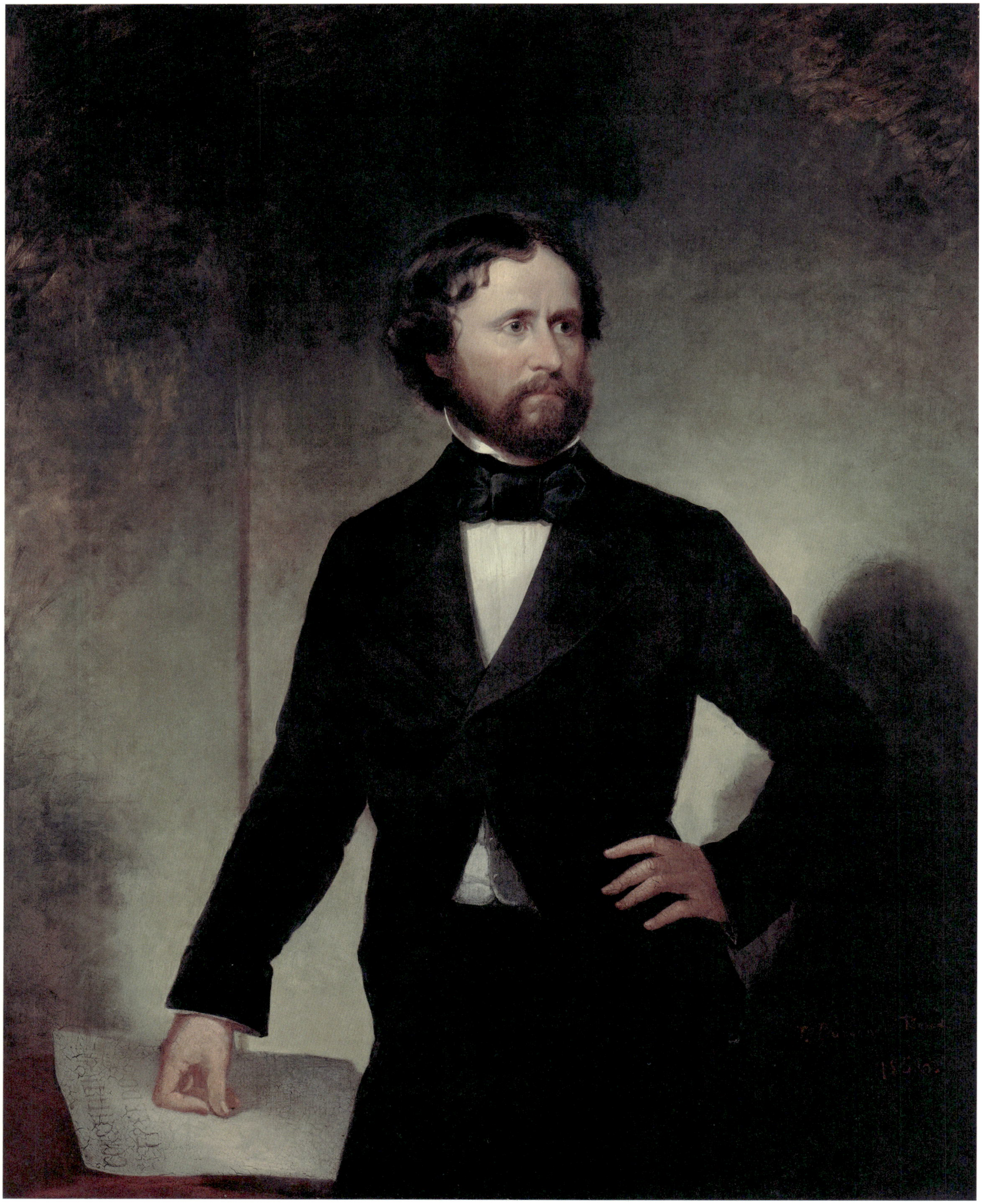

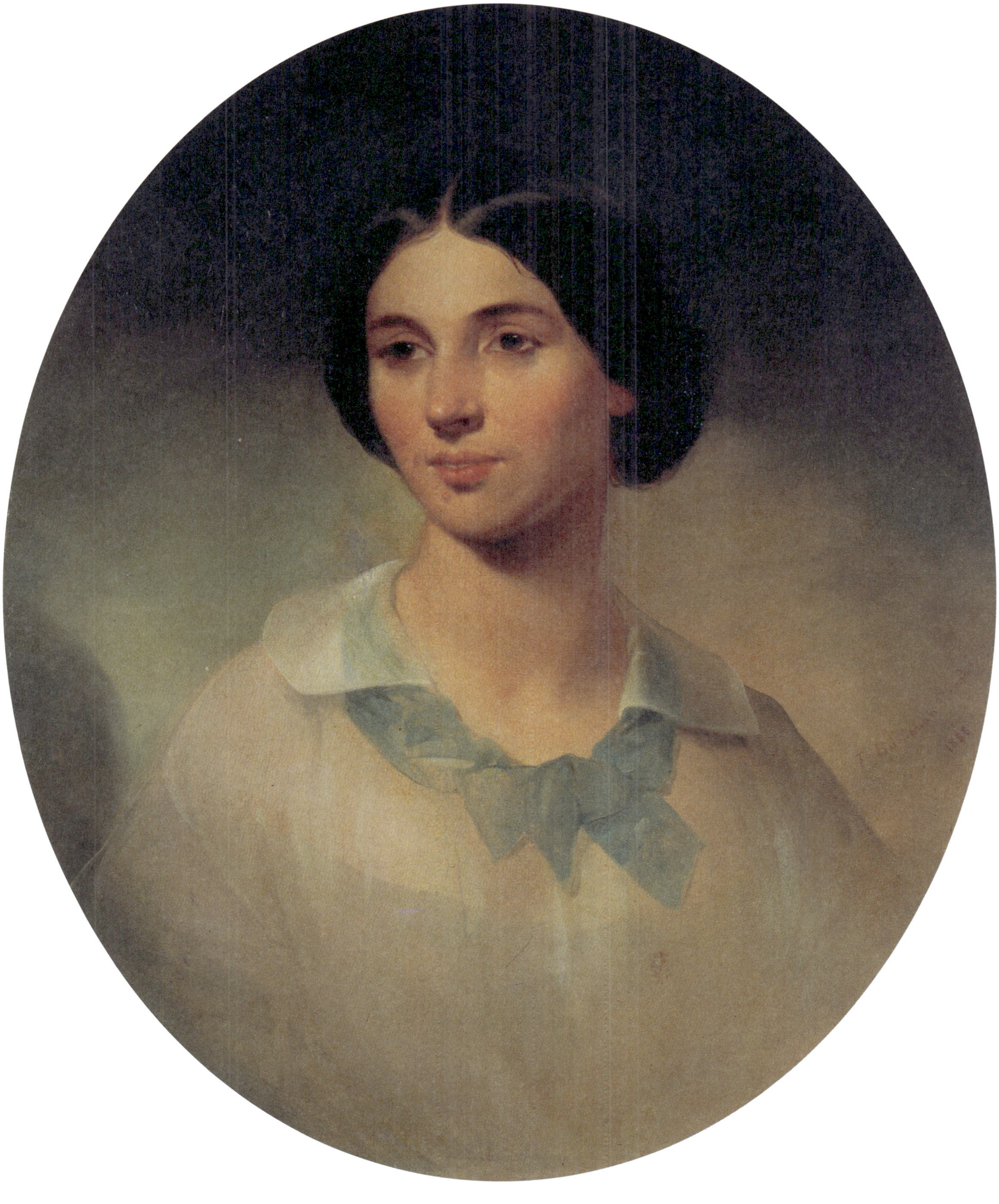

Preceding pages

CAT. 20 Thomas Buchanan Read, *Portrait of John C. Frémont* (1813–1890), 1856, oil on canvas, 40 × 30 in., Sons of the Revolution in the State of California

CAT. 21 Thomas Buchanan Read, *Portrait of Mrs. Jessie Benton Frémont* (1824–1902), *Wife of Gen. John Frémont*, 1856, oil on canvas, 30 × 24 ½ in., Braun Research Library Collection, Autry Museum of the American West, Gift of Elizabeth Benton Frémont

Frémont accepted Humboldt's praise as affirmation of his achievements, akin to Humboldt's praise of Jefferson in his earlier volumes. Both Meriwether Lewis and Stephen H. Long had encountered difficulties with faulty equipment. Frémont had quantified measurements that had until then been estimated by dead reckoning.

Humboldt was keeping a close watch on American expeditionary reports. In lieu of making a return trip to the United States, he had come to rely on American expedition narratives to supplement his research. He used *Aspects of Nature* to call out his preferred sources, writing,

> *The physical and geognostical [*the study of structural geology*] views entertained respecting the western part of North America have been rectified, in many respects, by the adventurous journey of Major Long, the excellent writings of his companion, Edward James, and more especially by the comprehensive observations of Captain Frémont.*[67]

Humboldt owned a substantial number of books focused on the United States, but made numerous references to Long and Frémont in his own letters and books. His personal library eventually encompassed more than ten thousand volumes and was a resource open to his colleagues and friends in Paris and Berlin.[68]

Frémont became a particular favorite of the Prussian baron, both for his expeditionary exploits and his shared belief in the need to abolish slavery in the United States.[69] With the entry of California as a free state in 1850, Frémont briefly served as its first U.S. senator. Upon hearing that news, Humboldt successfully petitioned the Prussian king to award to Frémont the Prussian Medal for Scientific Achievement, an accolade Frémont deeply cherished. Humboldt sent a letter congratulating the American explorer, which was widely reported in the U.S. press. That letter read in part,

> *After having given to you, in the new edition of my* Aspects of Nature*, the public testimony of the admiration which is due to your gigantic labors between St. Louis, of Missouri, and the coasts of the South Sea, I feel happy to offer you, in this living token, (*dans ce petit signe de vie*) the homage of my warm acknowledgment.*

Humboldt then went on to describe the attributes they held in common, noting,

> *You have displayed a noble courage in distant expeditions, braved all the dangers of colds and famine, enriched all the branches of the natural sciences, illustrated a vast country which was almost entirely unknown to us. . . . upon the proposition of the illustrious geographer, Chas. Ritter, the Geographical Society at Berlin has named you an honorary member.*

Finally, Humboldt graciously thanked his American acolyte "for the honour which you have done in attaching my name, and that of my fellow-laborer and intimate friend, Mr. Bonpland, to countries neighboring to those which have been the object of our labors."[70] Frémont had ensured that Humboldt's name would be indelibly stamped on the American landscape. In return, Humboldt and Ritter provided the American explorer and politician with their international scientific imprimatur. This kind of reciprocal approbation characterized the American admiration for Humboldt, recognizing in his accolades their own enhanced standing on the European stage.

Two years later, in 1852, John and Jessie Frémont visited Paris, meeting Humboldt's old friend, the mathematician and astronomer François Arago. According to Jessie, "Arago himself aided Mr. Frémont" on the selection of instruments he took with him on his next western expedition.[71] It was then the conversation turned to photography. In 1839 Arago had sponsored Louis-Jacques-Mandé Daguerre's invention of the eponymous photographic technique, securing for him a lifetime annuity for his contribution to science and technology.

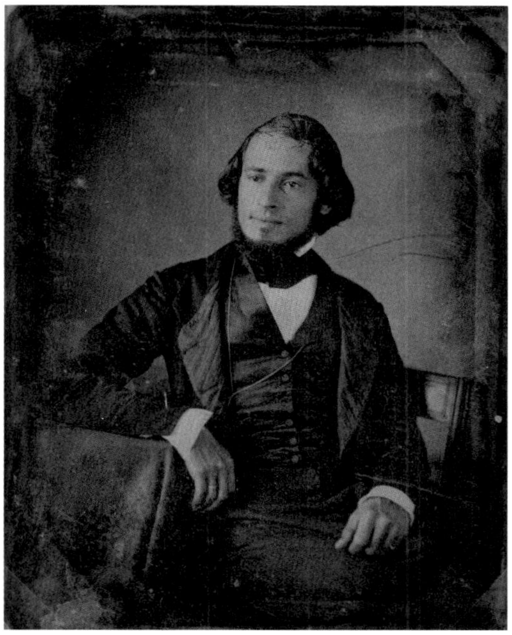

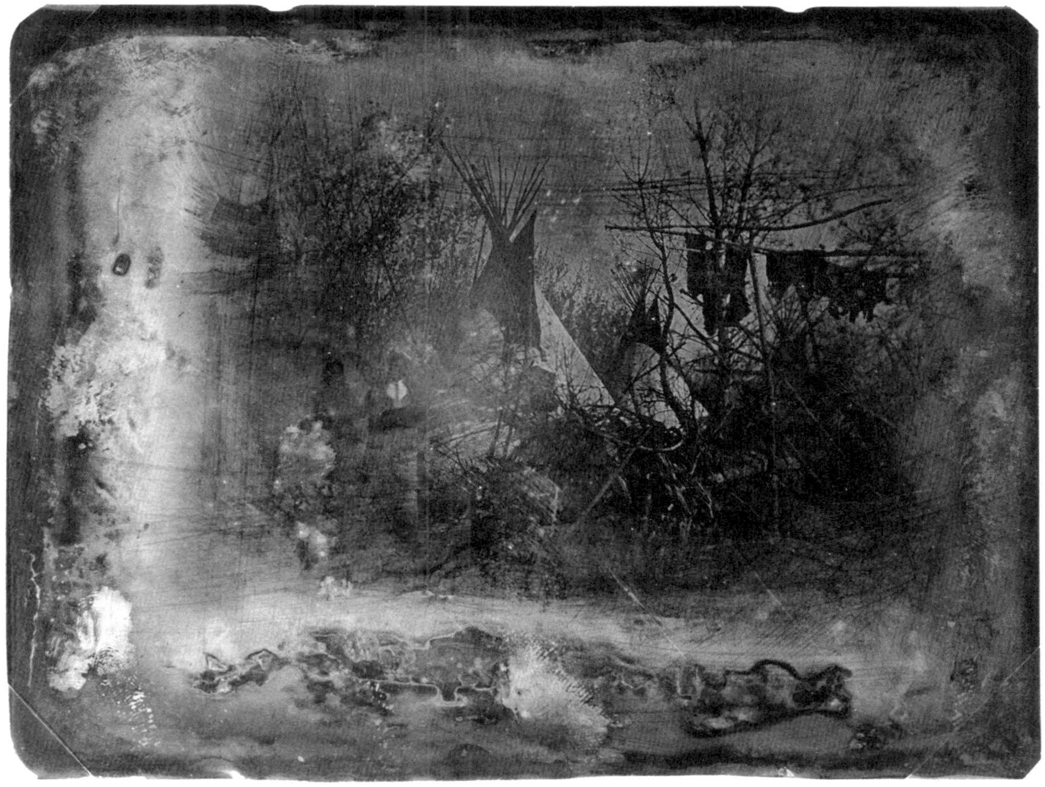

Fig. 25
Solomon Nunes Carvalho, *Self-portrait* (1815–1897), 1850, half-plate daguerreotype, 4 3⁄16 × 5 ½ in., Library of Congress, Prints and Photographs Division

Fig. 26
Solomon Nunes Carvalho, *Cheyenne Village*, 1853, half-plate daguerreotype, 4 3⁄16 × 5 ½ in., Library of Congress, Prints and Photographs Division

Shortly thereafter, Humboldt had observed as Daguerre made the first photographs of the surface of the moon, sparking a conversation as to whether it might be possible to calculate the height of its mountains using the length of the shadows cast by the sun.[72]

Perhaps further inspired by Humboldt's enthusiasm for the new technology, Frémont was the first to employ a photographer for a U.S. government expedition—his fifth and final expedition in 1853. Frémont hired Solomon Nunes Carvalho (fig. 25), who made close to three hundred daguerreotypes as part of Frémont's visual record. A number of his images served as the basis for engravings used in Frémont's survey report and in his memoirs.[73] Carvalho's *Cheyenne Village* (fig. 26) is the only extant image known, providing a sense of what he was able to accomplish in the field.[74] Although the plate is heavily scratched and silvered, it still conveys Carvalho's ability to compose a scene that includes information useful for the expedition's primary goals and a sensibility resonant of fine art landscapes. Two Cheyenne Indians occupy the left foreground, their forms slightly blurred. They stand in front of a tepee. The tangle of low trees and scrub resolve into a framework on the right used to dry hides, which are seen as dark abstract forms on the far right. Frémont was ecstatic with Carvalho's skills, writing to Jessie, "We are producing a line of pictures of exquisite beauty, which will admirably illustrate the country." Frémont wanted the veracity ascribed to the new medium of photography, while not wanting to sacrifice the allure of the scenery. At the conclusion of the expedition, Carvalho's daguerreotypes were converted to hand-drawn printing plates in Mathew Brady's photography studio. Sixty-six of these etchings derived from Carvalho's photographs survive, each one a sensitive adaptation of his subjects. Frémont had found a new way to bring together science and art in a way he was certain Humboldt would approve.

PRINCE MAX IN AMERICA

Humboldt had watched America's expansion and exploration from Europe, first in Paris and, as of 1827, mostly from Berlin. That move was hardly voluntary. Humboldt had used his sizable inheritance to pay for his travels and the enormous costs of overseeing the production of books that met his exacting standards. Although these volumes are what established and sustained his worldwide reputation as a leading scientific and intellectual force, Humboldt lost money on all of them. The Prussian king, who had unsuccessfully attempted to lure Humboldt back to Berlin, now offered him a stipend. Humboldt was in no position to refuse. If he wanted to continue his work at this level, he needed to come home. Reluctantly, he packed up his materials and relocated to Berlin.

In March 1833, Humboldt received a letter from James Madison, dusting off a relationship that had lain fallow since 1805. Madison expressed his hope that the explorer would make good on his desire to return to the United States.[75] By now, Humboldt was in his sixties, and although his exuberance and intellectual sharpness had not diminished, he was not enamored of the thought of another transatlantic crossing. Instead he widened his circle of influence yet again, inserting himself into the next wave of American exploration by encouraging and supporting younger German naturalists' exploration of the United States.

As Madison wrote his letter, Humboldt's protégé, Prince Maximilian zu Wied, had arrived in the United States. Prince Max (fig. 27), as he was called, had met Humboldt in Paris in 1804. Family members agreed that the younger man was smitten. In Paris, the young prince heard Humboldt and Bonpland's stories from South America and Mexico and their expectation that the specimens and diaries they brought back would fuel decades of research and writings. In the company of the Marquis de Lafayette and Humboldt, Prince Max would have heard stories of life in the United States and felt Humboldt's enthusiasm for a return trip that would permit him to travel west. In 1811 Max enrolled at University of Göttingen, where he studied with Humboldt's mentor, Johann Friedrich Blumenbach, and became fascinated by cultural anthropology. One of his classmates was William Backhouse Astor, son of John Jacob Astor, owner of the American Fur Company, the most powerful American presence in the international fur trade along the Missouri River. This fortuitous connection served Max in good stead as he planned his trip. Astor's company granted Prince Max permission to travel on their boats and work with their guides along the Missouri River.[76]

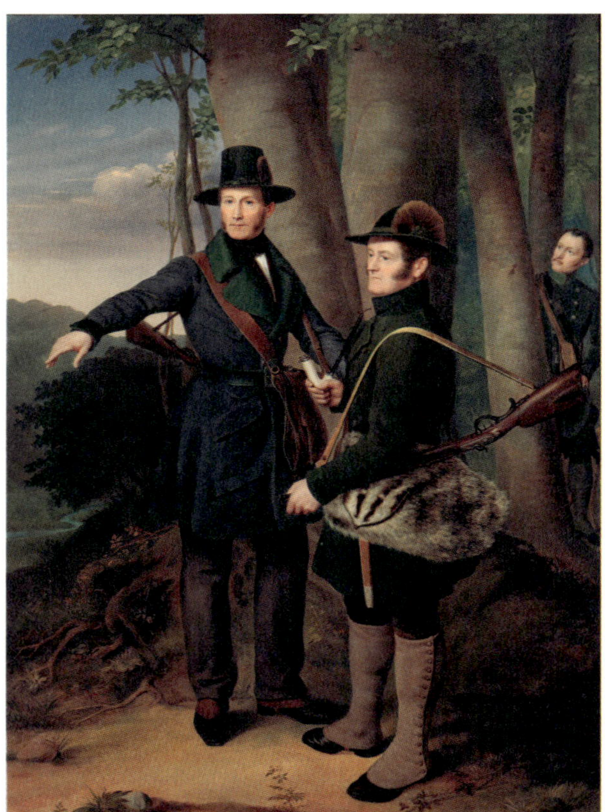

Fig. 27
Prince Charles zu Wied, *Prince Maximilian of Wied* (center, 1782–1867) *with His Brother Prince Charles* (left), ca. 1845, oil on canvas, framed: 51 ¼ × 38 ¼ × 2 ½ in., Joslyn Art Museum, Omaha, Nebraska, Gift of the Enron Art Foundation

Blumenbach created the field of cultural anthropology, focusing on determining whether all human races descended from a single ancestor or traced back to separate races. His conclusion, that there was a single point of origin for all human beings, was at the center of a larger debate regarding whether geographic factors might influence a person's physiognomy, skin tone, and intelligence.[77] He also taught a course on "Apodemics," or, the art of traveling.[78] Apodemics gave travel scientific purpose. His system inspired Humboldt's approach to assembling knowledge drawn from many different disciplines for the purpose of finding connections among things. Categories could be useful in classifying knowledge systematically, but Humboldt would focus as much on the underlying natural laws as individual discoveries. Humboldt was now one of Göttingen's most celebrated alumni. He was an exemplar of what education, inquisitiveness, and a restless desire to travel could accomplish. Humboldt's earliest publications demonstrated how that education could fuel a career in science. Blumenbach's interest in sending his students into the field to conduct research in anthropology and natural history inspired the explorer to chart a similar course.

From Göttingen, Max wrote to his friend, the Swiss zoologist Rudolf Schinz, "I am here in

Göttingen to improve my proficiency [in natural history] and then perhaps, if my fate will allow it, make a journey to North America"[79] Between Blumenbach and Humboldt, Max had two enthusiastic proponents for his goals, but they were delayed by Napoleon's invasions of Prussia. Prince Max served twice in the Prussian Army to combat the French, and in March of 1814 rode into Paris with the triumphant allied European forces.[80] Reunited with Humboldt and his colleagues, Max proudly boasted, "Humboldt and Bonpland now also belong to the number of my acquaintances."[81]

Napoleon's incursions into the Iberian Peninsula had forced the Portuguese royal family to flee Lisbon for their South American colony, Brazil. Exile spurred them to recognize the need to chart this vast colony and amass the same kind of information Humboldt had accrued during his travels through New Spain. Prince Max saw this as an opportunity to assist his mentor and fill another of the gaps in his world knowledge. In 1814 he successfully petitioned to conduct an expedition that spanned from 1815 to 1817.[82] Permission to enter Brazil must have been particularly satisfying for him. In 1799 the Portuguese had denied Humboldt access to Brazil, believing him to be a Spanish spy.[83] Prince Max's trip gave the younger man the chance to emulate his mentor. His two-year expedition through Brazil set the tone for his dual focus on natural history and a scholarly approach to studying indigenous peoples. Humboldt dove into Max's plans, recommending a Prussian botanist living in Rio de Janeiro as an essential traveling companion.[84] He undoubtedly shared much of his research to help frame Max's own objectives.

By the time the prince left for Brazil, he was prepared "to collect treasures that will certainly be useful also for my good friends," Humboldt among them, by studying aspects of the Brazilian ecosystem and indigenous cultures that would complement Humboldt's work on the northern part of the South American continent.[85] Prince Max spent considerable time among the Botocudo (Aimoré) Indians and made detailed notes on their culture and appearance that stand as the earliest and most complete study of this group. He returned with ample material for a two-volume illustrated narrative and atlas, published in 1820 and 1821. As Max wrote about his trip, he made constant reference to Humboldt's travels, observations, and publications, ever aware of his mentor's expectations. He clearly envisioned his work as a complement to Humboldt's, and together their expeditions mapped the larger part of the South American continent.[86]

Prince Max spent the rest of the 1820s planning his trip to the United States.[87] He was familiar with Humboldt's brief visit to Philadelphia and Washington and understood his desire to explore farther west. In 1824, the Marquis de Lafayette returned from his own reunion tour of the United States, during which he had been feted as a hero of the American Revolution. Prince Max's preparation followed Humboldt's template for international travel. He cast a wide net, soliciting advice from the Prussian scientific community, much as Stephen H. Long had done in Philadelphia. He began a yearlong correspondence with Duke Bernhardt of Saxe-Weimar, who had visited the United States from 1823 to 1824, had met President John Quincy Adams, stayed with Thomas Jefferson at Monticello, and noted with satisfaction seeing Humboldt's books on library shelves in Baltimore.[88] Humboldt and Jefferson had both been among the subscribers to the published chronicle of the duke's trip. Would he be amenable to sharing his advice on traveling to North America?[89] The duke sent a list of books on North America he thought would be helpful, and he recommended that the prince travel using a pseudonym, which he did, becoming the Baron Braunsberg for the duration of his travels.[90] Prince Reinhard of Solms also shared a list of recommended reading along with a list of the German émigrés he knew and where they lived in the United States—an informal network of friendly contacts he could call on as necessary.[91] A year later Prince Max requested and received from the duke

a letter of recommendation to General Clarke who travelled with Lewis to the Columbia River, whom Your Grace mentioned in the narrative of your travels. A few lines you might write to this man would make me very obliged to you.[92]

On one point, however, Humboldt was insistent: Prince Max needed to take a professionally trained artist. What that suggests is how significant both men understood the visual record to be from those travels.[93] In Brazil, Max, like Humboldt, had served as his own artist. Although his herpetological drawings were quite competent, his landscapes and drawings of the Botocudo tribe were not. Max's younger brother, Carl, and sister, Louise, ended up re-creating all of the illustrations for his publication. Humboldt wanted to ensure that he returned with highly accurate drawings of the American Indian tribes at the heart of his travel plans—drawings that would match in quality Max's written observations. By this time Humboldt had set a high bar for illustrated travel books with a scientific emphasis. The line between scientific illustration and art was blurring, requiring an artist who could combine both scientific rigor and aesthetic feeling in a single image.

To identify an artist capable of and interested in this assignment, Max solicited advice from his military friends, probably looking for someone with topographical training.[94] His search yielded him the services of Karl Bodmer, a Swiss landscape painter and engraver.[95] Bodmer had the requisite talent, plus a willingness to engage in rough travel while making field sketches and painting more detailed watercolors. During their two years in the United States, the pair accumulated trunks of natural history specimens and American Indian artifacts and shipped some of them home in advance. He also kept a detailed journal of the weather, terrain, flora, fauna, and people he encountered; however, he left the rendering of the landscape and the indigenous peoples to Bodmer. Bodmer's obsessive attention to detail when it came time to publish their venture matched Humboldt's own fastidiousness and resulted in a publication that, while published many years late and well over budget, was also lauded worldwide as the finest production of its type.

Prince Max served as Humboldt's proxy in the United States. His trip to North America conducted between May 1832 and July 1834 shared a similar structure to his Brazilian trip in that it was intended to combine natural history with a significant ethnographic focus. He knew from his research on Lewis and Clark and the Long Expedition that time spent with the American Indian tribes—understanding them in their environment, as he had the Botocudo—would complete the scope of his study. He was not trying to correct earlier maps or chart new terrain as much as he was interested in filling gaps in Humboldt's comprehension of nature's interrelated systems. Humboldt had encouraged him to think even bigger. Prince Max had additional help from another family in Humboldt's expansive network. One of Humboldt's closest friends, Friedrich von Gerolt, was then serving as the Prussian diplomatic consular-general to Mexico. He and his brother Carl, who lived in California, both received letters from a third family member in Linz asking that they take care of Prince Max should his travels bring him their way.[96] On July 4, 1832, Prince Max, Bodmer, and the Wied family hunter and taxidermist, David Dreidoppel, landed in Boston, traveling from there to Philadelphia before heading west through Pittsburgh to St. Louis.

Max had done his homework well. In Philadelphia he visited Titian Ramsay Peale at what was, after his father's death in 1827, his museum. Max's evaluation confirms the significance of the collection in shaping perceptions of the United States. Titian Peale struck him as a fellow traveler, steeped in the sciences and exploration. In a relatively new country, Prince Max commented in his journal unfavorably on the crudeness of American manners and customs and the paucity of scholarly libraries. Peale's Museum was the exception to that rule. He observed that the museum

> *contains the best collection of natural history in the United States. There is the fine large skeleton of the Ohio elephant (*Mastodon, Cuv.*), and likewise most of the animals of North America, pretty well stuffed. . . . Mr. Peale, the owner, accompanied the expedition under Major Long to the Rocky Mountains, where he procured part of these specimens himself. . . and the collection of Indian dresses, utensils, and arms, is, I think, the most important that I have yet seen. I was particularly interested by some oil paintings of Indian villages and scenery by Seymour. This artist also accompanied Major Long's expedition. Mr. Peale's collection deserves precedence above all the public museums in the United States, for its scientific arrangement, and because fewer trifling nicknacks have been admitted into it. Mr. Peale has also travelled in South America, and his health was still suffering from his visit to that country.*[97]

The range of Titian's travels must have intrigued his visitor, as did his commitment to a professional and educational emphasis with his displays. Titian's tenure with the Long Expedition impressed the German explorer, and by this time Long's account was widely considered in Europe and America as the most accomplished expedition report of its kind. Humboldt certainly thought so. Max's references to the mastodon would have brought Humboldt additionally to mind. And there in the Long Room, he would have also seen the portrait of his friend and mentor, Alexander von Humboldt, smiling down from Peale's gallery of eminent Americans.

With Peale's encouragement, Max spent time in Philadelphia with members of the APS and the ANS, which made him a corresponding member in 1834.[98] At the APS he met John Vaughan, who deeply admired Humboldt in 1804 when they met.[99] He discovered that Peale was good friends with another naturalist to whom Prince Max carried a letter of introduction, Charles Lucien Jules Laurent Bonaparte (1803–1857), the eldest son of Napoleon's older brother, Joseph.[100] Charles now lived at an estate in Bordentown, New Jersey, and welcomed the traveler. While the two men toured the property and discussed natural history, Bodmer painted the surrounding landscape. When Max left, he was carrying a letter from Charles, introducing him to entomologist Thomas Say, who was now living in New Harmony, Indiana.

New Harmony was located on a scenic bend of the Wabash River. It had been founded as a utopian community by Robert Owen, a Welsh textile manufacturer who immigrated to the United States. In 1825, ANS President William McClure had convinced several of his colleagues in the Philadelphia scientific community to leave the East Coast and follow him there to establish New Harmony as a scientific utopia. On December 8, their keelboat departed Pittsburgh. The scientists had christened it the *Philanthropist*; Owen quipped that it should be called the Boatload of Knowledge. Max had already heard of New Harmony—the Duke of Saxe-Weimar had visited there ten years earlier.[101] When he arrived, Max was overjoyed to find an "outstanding library for natural scientists… in which most of the European éditions de luxe and works containing copper-engravings may be found—this is, indeed, in this seclusion a rather unusual phenomenon!"[102] The travelers remained with Say and Lesueur for five months while Max recuperated from what might have been a mild case of cholera, a disease that was sweeping the country at that time. That gave Max plenty of time to read and establish friendships with Say and Lesueur.[103] Humboldt was one of their topics, as Max mentioned having seen his portrait hanging in Peale's Museum, along with those of his hosts.[104]

Wherever they traveled, Prince Maximilian took extensive notes and collected natural history specimens. Dreidoppel shot and skinned animals he later mounted for study and display. Bodmer used the opportunity to become familiar with the American landscape. In a series of lengthy letters to his siblings, Max gave his family insights into his American experience.[105] He was deeply moved by the stands of old-growth forests and shocked at how haphazardly land was cleared for lumber and settlement. Earlier travelers to America had admired the old-growth forests as emblematic of the untapped resources on the continent. Prince Max was now using his training as a naturalist and ethnographer to record these changes. He kept notes on everything, even treating his arrival in Boston as an ethnographic exercise. But he had not yet encountered any Indians, which made him all the more eager to reach St. Louis and present himself to William Clark.

Clark had become the most powerful man in St. Louis, serving as territorial governor and supervisor of all of the Indian agents assigned to work with specific tribes up the Missouri River watershed. He and Prince Max took a liking to each other. The evidence for this is nowhere more evident than in Clark's approval to have his nephew, Benjamin O'Fallon, copy the maps Clark had made for the Corps of Discovery for Max's use. These were route maps, or traverse maps, enumerating specific landmarks and other orientation points. O'Fallon gave him thirty-four hand-colored maps that detailed every curve of the river, marked every campsite Lewis and Clark had used, and included warnings regarding the inevitable sandbars and snags that made river travel so treacherous (CAT. 22A). Max understood the magnitude of the gesture. In his narrative he wrote, "Major Ofallon, whom we visited at his pleasant country seat, near St. Louis, had the kindness to furnish me with the map of the course of the Missouri, by Lewis and Clarke, on a large scale."[106] The details provided a bend-by-bend analysis of the

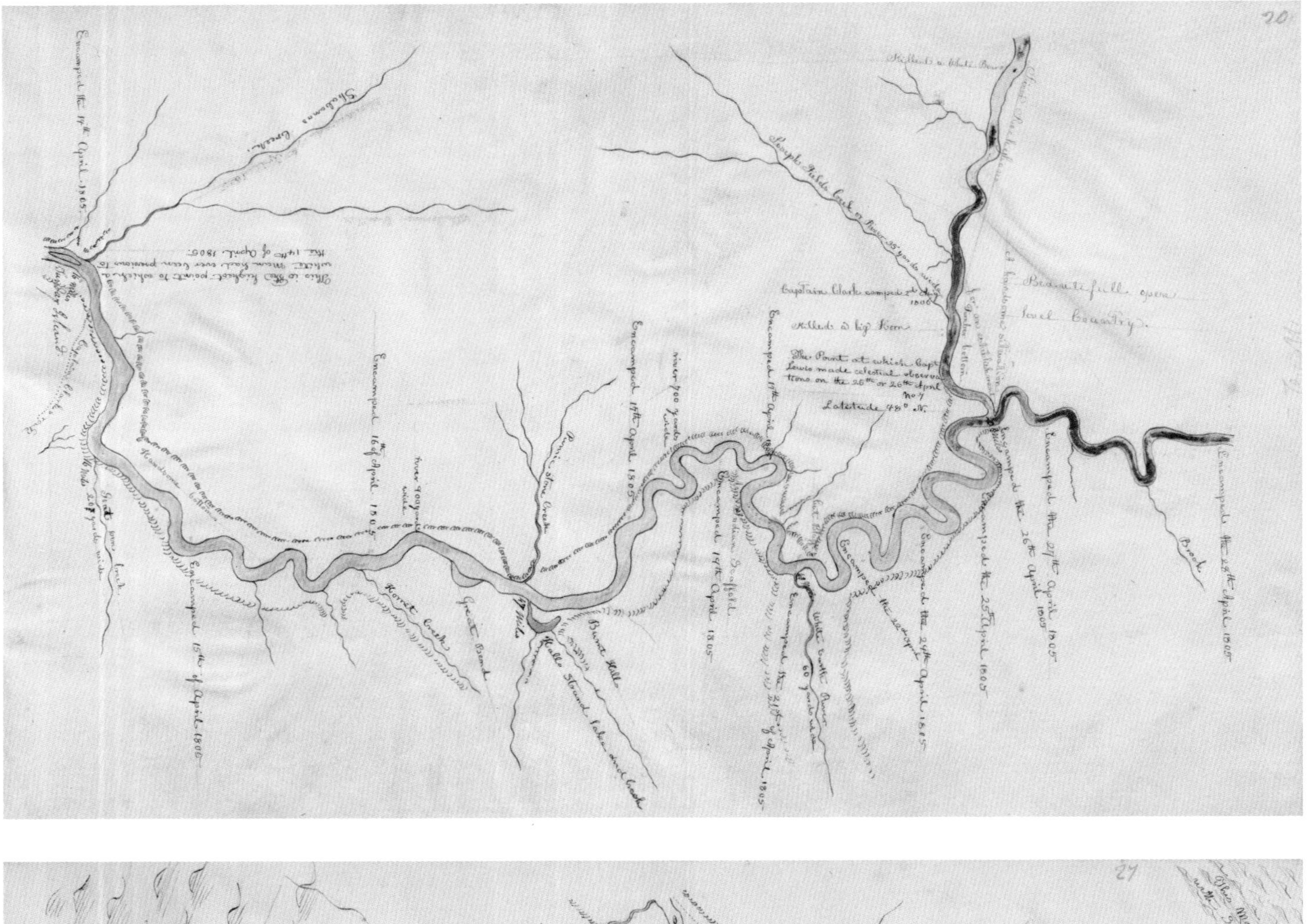
Encamped the 17th April 1805
Shabonos Creek
Captain Clark camped 23rd April 1805
Killed a big horn
The Point at which Capt. Lewis made celestial observations on the 25th & 26th April
Latitude 48° N.
Beautiful open level Country
Encamped 18th April 1805
Encamped 16th of April 1805
Encamped 15th of April 1805
Encamped 19th April 1805
Encamped 20th April 1805
Encamped the 22nd of April 1805
Encamped the 24th April 1805
Encamped the 25th April 1805
Encamped the 26th April 1805
Encamped the 27th April 1805
Encamped the 28th April 1805
White Earth River 60 yards wide
Goose Egg Lake
Burnt Hill
Grand Bend
Porcupine River
Brook

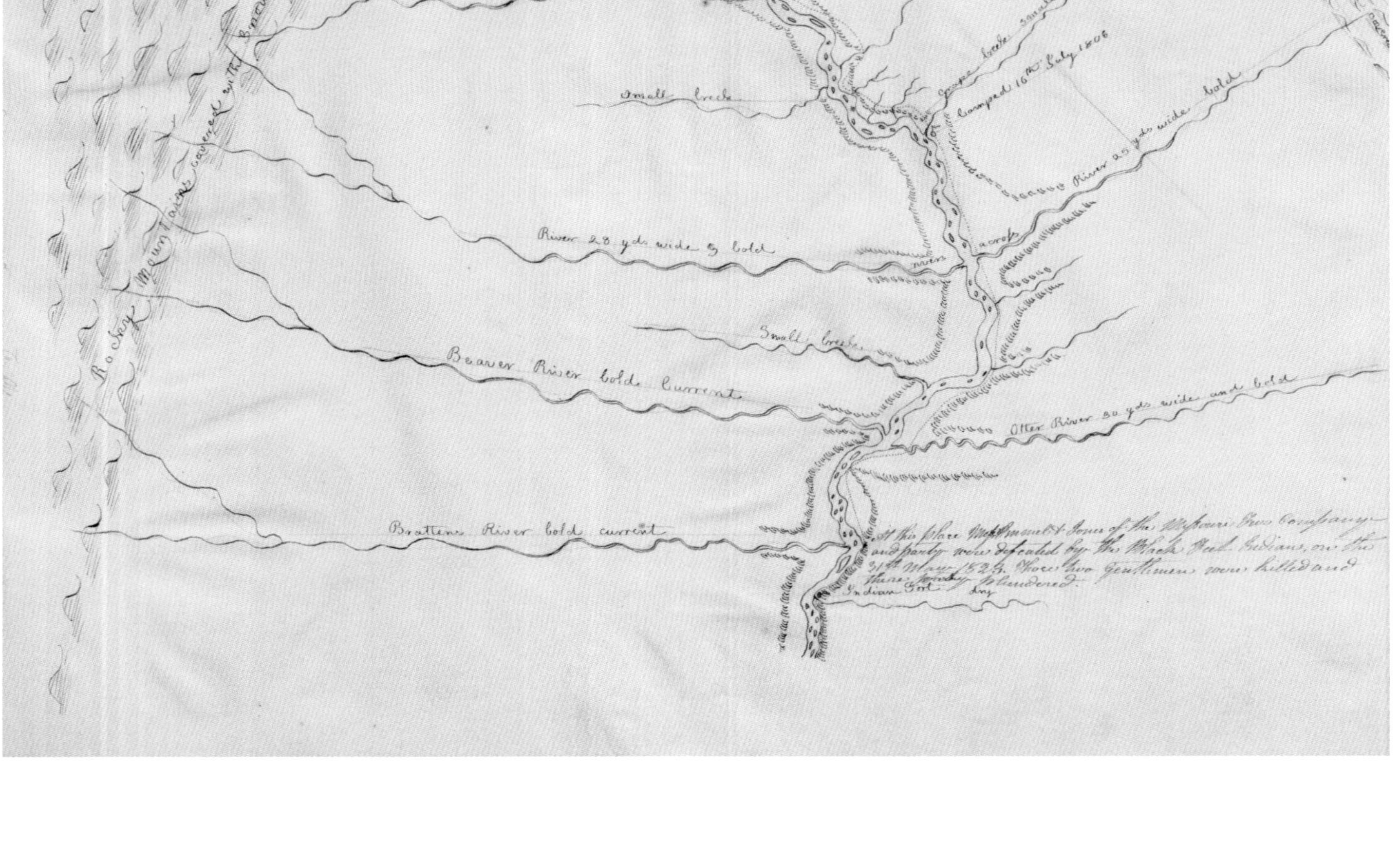
Small creek
Encamped 16th July 1806
River 25 yds wide bold
River 28 yds wide & bold
Small creek
Beaver River bold current
Otter River 30 yds wide and bold
Brattens River bold current
Indian Fort

CAT. 22A & B Benjamin O'Fallon, *Lewis and Clark's Special Map of the Missouri River*, pages 20 and 27, 1804–32, watercolor and ink on paper, single page: 10 ⅜ × 15 ¾ in., Joslyn Art Museum, Omaha, Nebraska, Gift of the Enron Art Foundation

river, indicating dangers and flagging resources all along the way. Increasing their value to the prince, O'Fallon had updated these maps. He had accompanied the Long Expedition as an Indian agent, tasked with traveling ahead of the group to ensure peaceful encounters with the Indian tribes. On one sheet he noted the location of an Indian massacre in 1823, evidence that he continued to add relevant information (CAT. 22B).

Acting as Humboldt's proxy, and relying on Clark's guidance, Prince Maximilian and his companions made their way up the Mississippi and Missouri Rivers, amassing collections of specimens, information, measurements, and cultural intelligence from Indians, trappers, scouts, and military men. Their return trip became an excursion up the Ohio River and to Niagara Falls (fig. 28) before concluding in New York City. Max was as impressed by Niagara Falls as he had been by Peale's Museum. The falls had long been recognized as one of North America's premier natural attractions, and in the United States it was common to make a pilgrimage to see them.

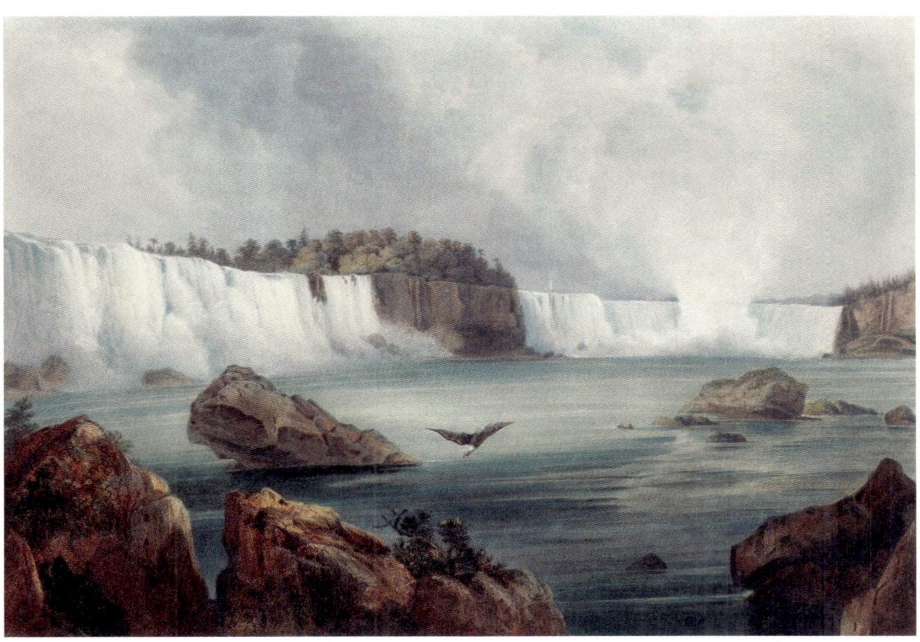

Fig 28
After Karl Bodmer, *Niagara Falls*, 1841, hand-colored aquatint, 16 ¼ × 21 ¼ in., Joslyn Art Museum, Omaha, Nebraska, Gift of the Enron Art Foundation

Jefferson had encouraged American artists to paint Niagara, helping to reinforce its primacy as an emblem of America's identity. In his finished aquatint, Bodmer included a bald eagle soaring over the cataract, the dual symbolism of the magnificent waterfall and the national bird not lost on his American nor his European audiences. The duo would repeat this symbolism in the cartouche accompanying the detailed map (CAT. 23A) tracing their route across the United States. The cartouche (CAT. 23B) acted as a distillation of their journey, with a herd of bison, an Indian village, a bear, and a beaver added to the crowded landscape. It served as an emblem of what they construed as quintessentially American about the landscape they had traversed.

The most important insights Max gleaned from two years in the United States were revealed in his detailed written and painted records of the individual Native Americans and tribal customs he and Bodmer encountered during their year along the Missouri River. That material is the subject of another chapter here.[107] Max and Bodmer's appreciation for the American landscape and their reliance on Clark's maps meant that they returned to Europe with a significant collection of natural history specimens and paintings of noteworthy geological features of equal interest to Humboldt. Max's journals are a compendium of densely written accounts with space left for footnotes and sidebar mentions (CAT. 24). Writing in an elegant Sütterlin script, he kept detailed notes on the weather, terrain, flora, and fauna he encountered along the way. His descriptions of the landscape remind us that, for travelers more accustomed to Europe—or even South America—the United States seemed strikingly different.

Prince Max often drew contrasts with Brazil, noting the dry air and sere grasslands of the upper Missouri region, which differed starkly from the tropical rainforest.[108] He was particularly unnerved by the hoodoos (fig. 29) in what the French-Canadian trappers called *les mauvaises terres*:

> *The mountains continued to increase in height; they were more and more naked and sterile; their colour was whitish-grey, grey-brown, often spotted with white, the upper part disposed in horizontal strata, or in narrow stripes.*

CAT. 23A & B *(full view and detail)* Karl Bodmer and Prince Maximilian zu Wied-Neuwied, *Map to Illustrate the Route of Prince Maximilian of Wied in the Interior of North America,* from *Reise in das innere Nord-America in den Jahren 1832 bis 1834,* 1839–41, hand-colored engraving on paper, 24 × 36 in., Smithsonian Libraries, Gift of Charlotte and Lloyd Wineland

CAT. 24 Prince Maximilian zu Wied-Neuwied, *American Fur Co., Manuscript Journal of Prince Maximilian's Travels in North America, Parts 7–14 (Vol. 2),* 1833, ink and watercolor on paper, single page: 12 ¾ × 7 ¾ in., Joslyn Art Museum, Omaha, Nebraska, Gift of the Enron Art Foundation

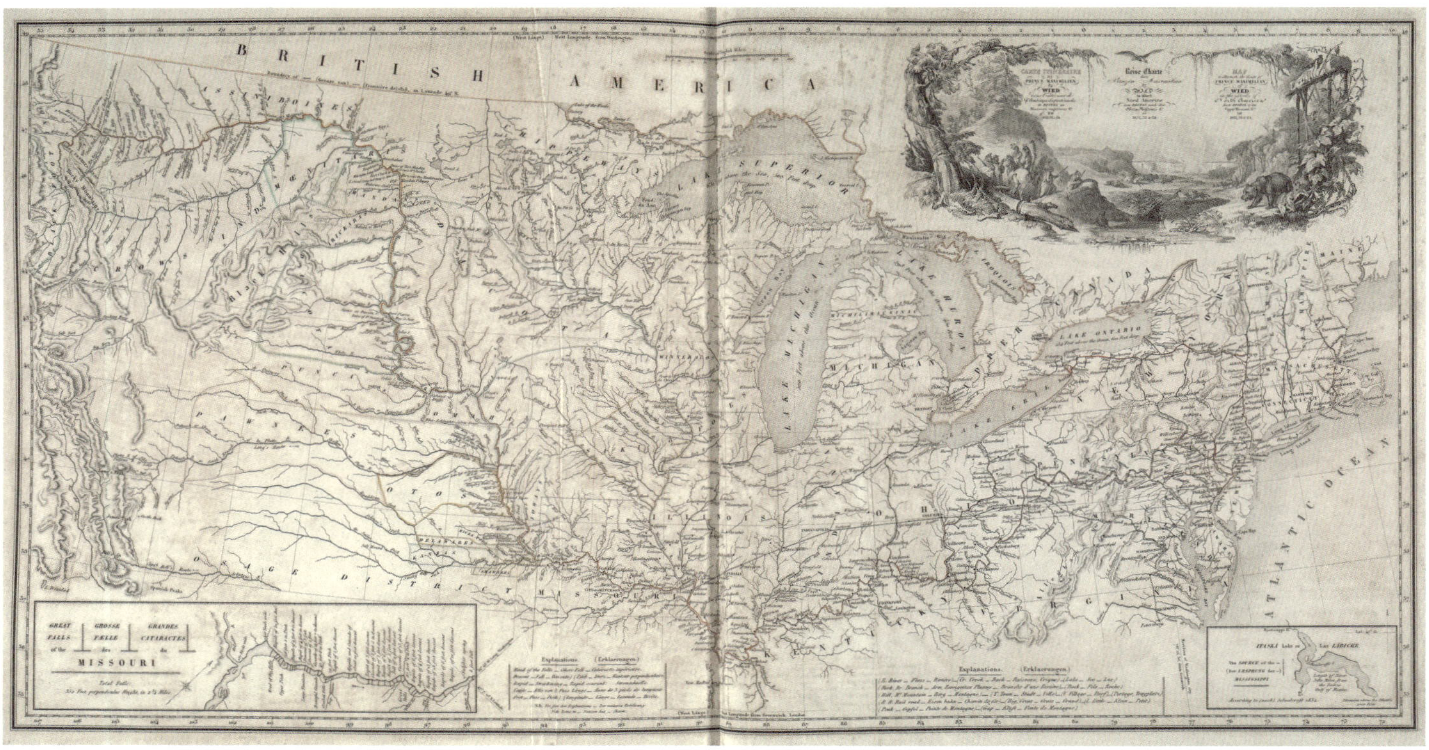

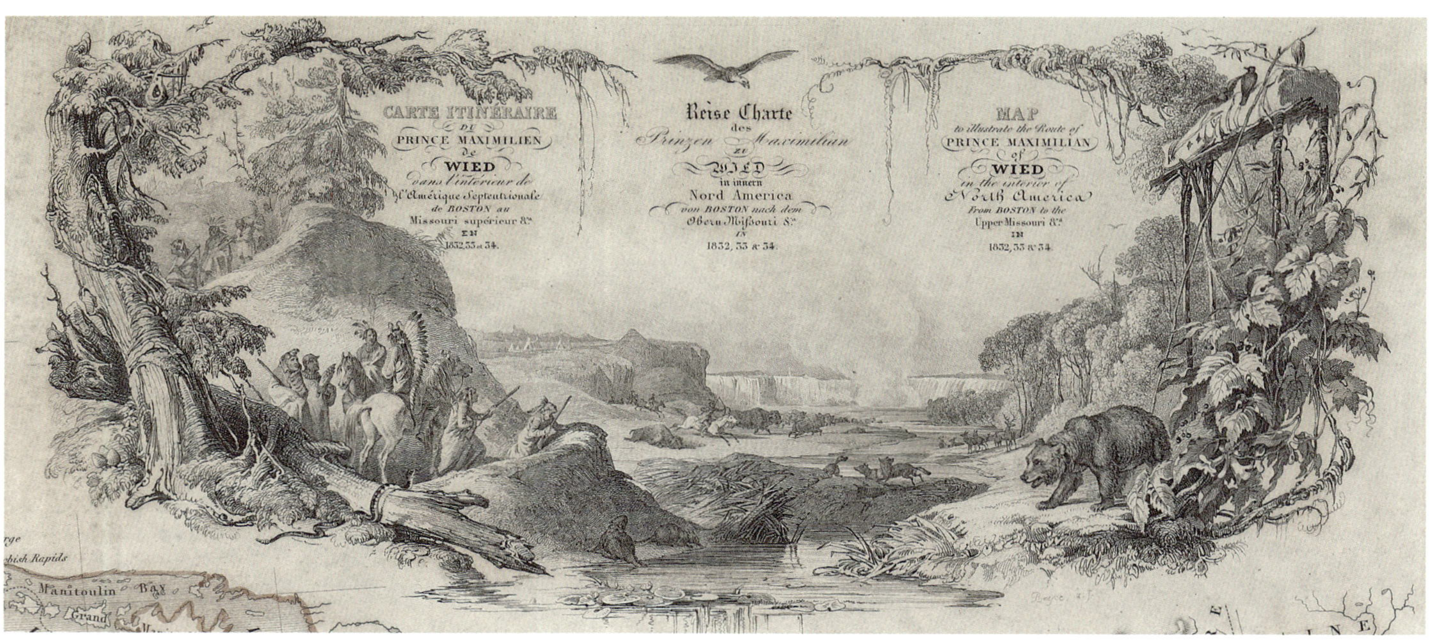

VII.

Reise von St. Louis bis zu den Grenzen der Ansiedlungen am Missouri, vom 10ten April bis zum 22ten.

Einfahrt in den Missouri – St. Charles – Gasconade River – Portland – Osage River – Jefferson City – Marion – Franklin – Boonville – La Mine River – Pierre à flèche – Grand River – Fox Prairie, wo die Missouris [illegible] – Gefährliche Schiffahrt – Nest des Bald Eagle – Ehemalige Wohnsitze der Osagen – Wir bleiben 3 Meilen von Fort Osage ~~[illegible]~~ am Fishing-creek auf einer Sandbank sitzen – Fort Osage – Die Osagen (Wasaschi) – Blue water River – Heftiges Gewitter – Liberty – Kanzas River – Die Konzas – Eine [illegible] Indianerfrau [illegible] – Gefährliche Lage – Little Platte River – Die Ayaways – Cantonment Leavenworth. –

Am 10ten April um ½11 Uhr setzte man die Dampfmaschine auf dem Yellow Stone in Bewegung, wo sich die Gesellschaft (1.) an Bord versammelt hatte. Herr Chouteau [illegible] Die Flagge des Yellow Stone [illegible] eine Flagge, wie hier abgebildet, mit den Initialen der Amer. Fur. Comp. [illegible] St. Louis in Masse versammelt. Die ~~Sakis~~ Saki ~~Ayaways~~ (2.) Kickapu Indianer in ihren originellen Aufzügen, die Männer in der Hand [illegible] standen am Ufer. Herr Bodmer skizzirte einige der Männer u. die Weiber [illegible], auch die 4 hier liegenden Canoes [illegible] Diese Canoes sind lang u. schmal [illegible] Oben auf dem [illegible] unseres Schiffes [illegible] 20 Canadier (Engagés oder Trapper) [illegible] Die Sakis in Procession [illegible] General Clark ein Council zu begehen [illegible] an den Missouri [illegible] der blühenden Red Bud [illegible] N. America, dass die meisten Bäume vor dem Laub blühen, [illegible] Kirschen, Pflaumen, Wallnüsse, Haseln, Dogwood, Red-bud, Prunus, Cerasus u. a. Arten.

Abfahrt von St. Louis den 10ten April.

Um ½ 8 Uhr [illegible]
$12\frac{3}{4}$° Mittags
$17\frac{1}{2}$° Abends.

Die Sakis ~~Ayaways~~ u. Kikapus.

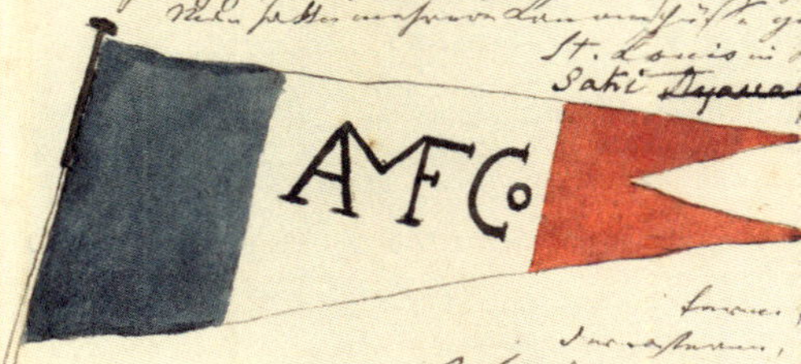

Die Männer sind Sakis, die Weiber Kickapus.

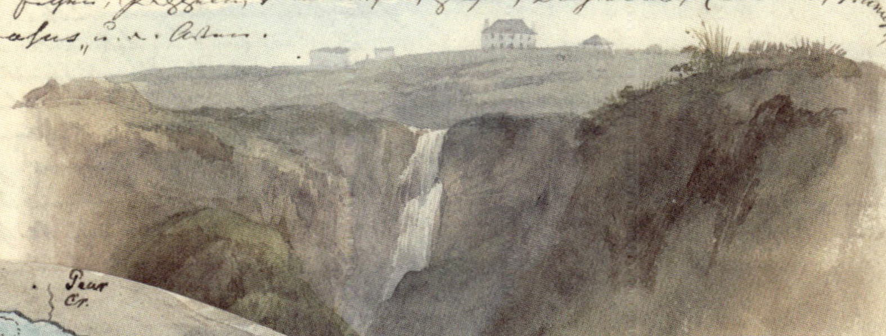

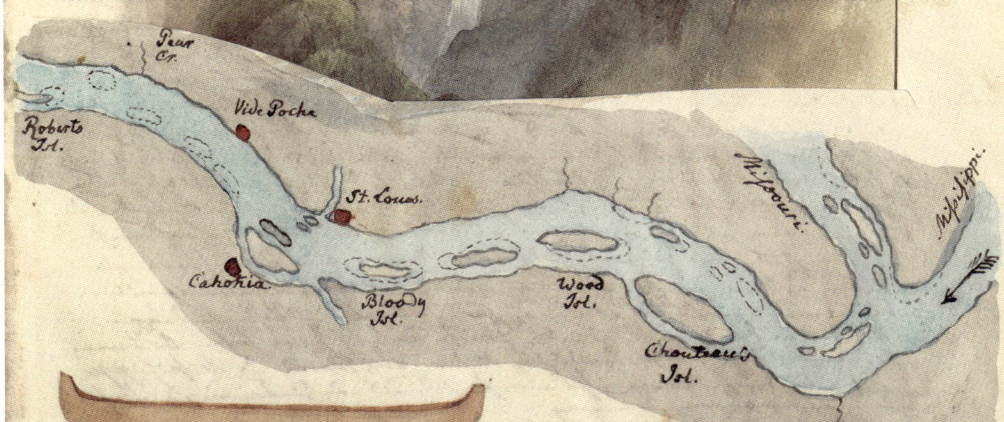

(1.) Die Gesellschaft bestand aus den Hr. Chouteau [illegible] einigen Angestellten der Company, [illegible] Sanford, Pilcher u. [illegible] (2.) Die ~~Ayaways~~ Sakis [illegible] Missouri [illegible] Ioways [illegible] Fort Leavenworth [illegible]

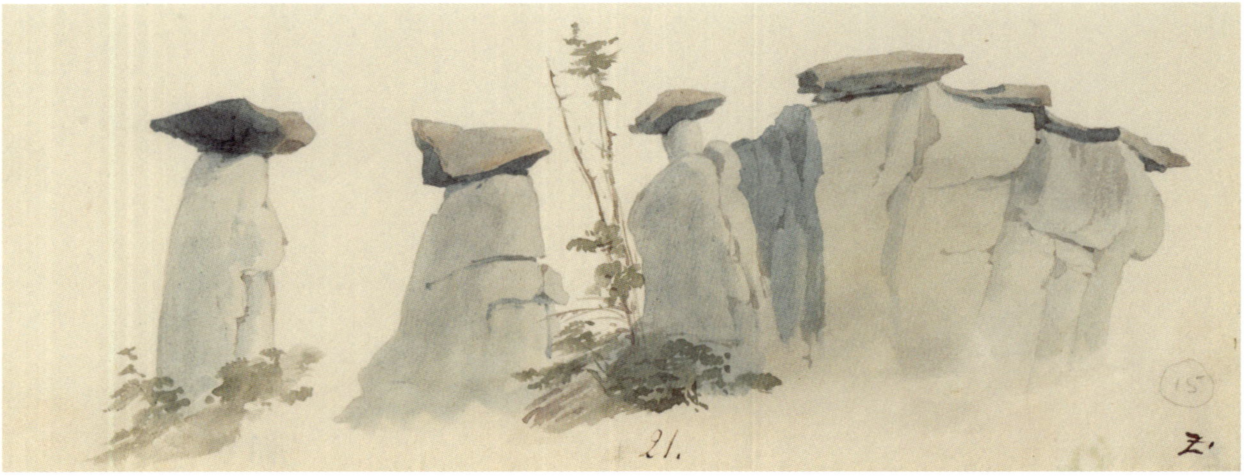

Fig. 29
Karl Bodmer, *Rock Formations on the Upper Missouri (detail)*, 1833, watercolor on paper, full sheet: 12 ½ × 7 ¾ in., Joslyn Art Museum, Omaha, Nebraska, Gift of the Enron Art Foundation

Farther up the river there were more “rude, naked mountains…” then,

> *two white mountain castles…[that] when seen from a distance, so perfectly resembled buildings raised by art, that we were deceived by them, till we were assured of our error.*[109]

Bodmer’s highly accomplished watercolors convey no sense of dread; rather, the artist seemed captivated by the formations, rendering them in delicate washes that present them as objects of wonder. The prince’s observation underscores a point Humboldt had made: that venerated monuments in the Old World tended to be man-made, whereas in the Americas, they were provided by nature. Hoodoos became castles, as Max recalibrated his frame of reference in the American West.

Back in Europe the two men collaborated on the finished prints. They compared notes to ensure the accuracy of the topography, wildlife, and weather. When he was composing *View of the Stone Walls on the Upper Missouri* (CAT. 25), Bodmer wrote to Max that he “vividly recall[ed] the nearly uninterrupted line of antelopes, we saw several times in spring in those areas.…I believed I had to use and include such an authentic motif in a picture of these otherwise so sad parts of the country.”[110] Max’s response to the Badlands was colored in part by having read the Long Expedition report as well as the published account of the Lewis and Clark Expedition. Accustomed to the more fertile agricultural regions of Europe and the lush Brazilian landscape, Prince Max reacted viscerally to Long’s “Great Desert” and the architectonic rock formations. Bodmer hired a group of the most accomplished printmakers in Paris and farmed out the work among them. He supervised the hand-coloring of the finished aquatints, using his watercolors as his frame of reference. These artists applied washes of color in gum arabic to each print to convey soft gradation of color on the friable rock formations. Max and Bodmer published their work in two volumes: one providing the narrative of their journey, the other a pictorial atlas that featured the color images. It was a monumental achievement that rivaled anything Humboldt had published.

Prince Maximilian’s travel journals and Karl Bodmer’s watercolors formed the basis of the single most significant illustrated record of the expanded American landscape of its day. *Reise in das innere Nord-America in den Jahren 1832 bis 1834* turned out to be just as expensive to produce as any of Humboldt’s travel narratives—and as fiscally ruinous. A notice in the New York *Commercial Advertiser* called it “the most elaborate and costly work on the geography and scenery of North America ever published.”[111] As a result, Max’s plans for a North American edition of his very expensive travel volumes foundered, and absent that edition, his images were not subject to any copyright restrictions in the United States. George R. Graham, founder of *Graham’s Magazine* in Philadelphia, leaped on this opportunity, pirating images and text to feed his audience excerpts from Max’s account. Between 1845 and 1850, *Graham’s Magazine* published no fewer than seventeen of Bodmer’s prints, some accompanied by snippets of Max’s descriptions, often unattributed. *Graham’s* boasted a circulation of forty thousand copies per issue by

CAT. 25 After Karl Bodmer, *View of the Stone Walls on the Upper Missouri*, 1840, hand-colored aquatint, plate size: 15 ⅞ × 20 ¾ in., Joslyn Art Museum, Omaha, Nebraska, Gift of the Enron Art Foundation

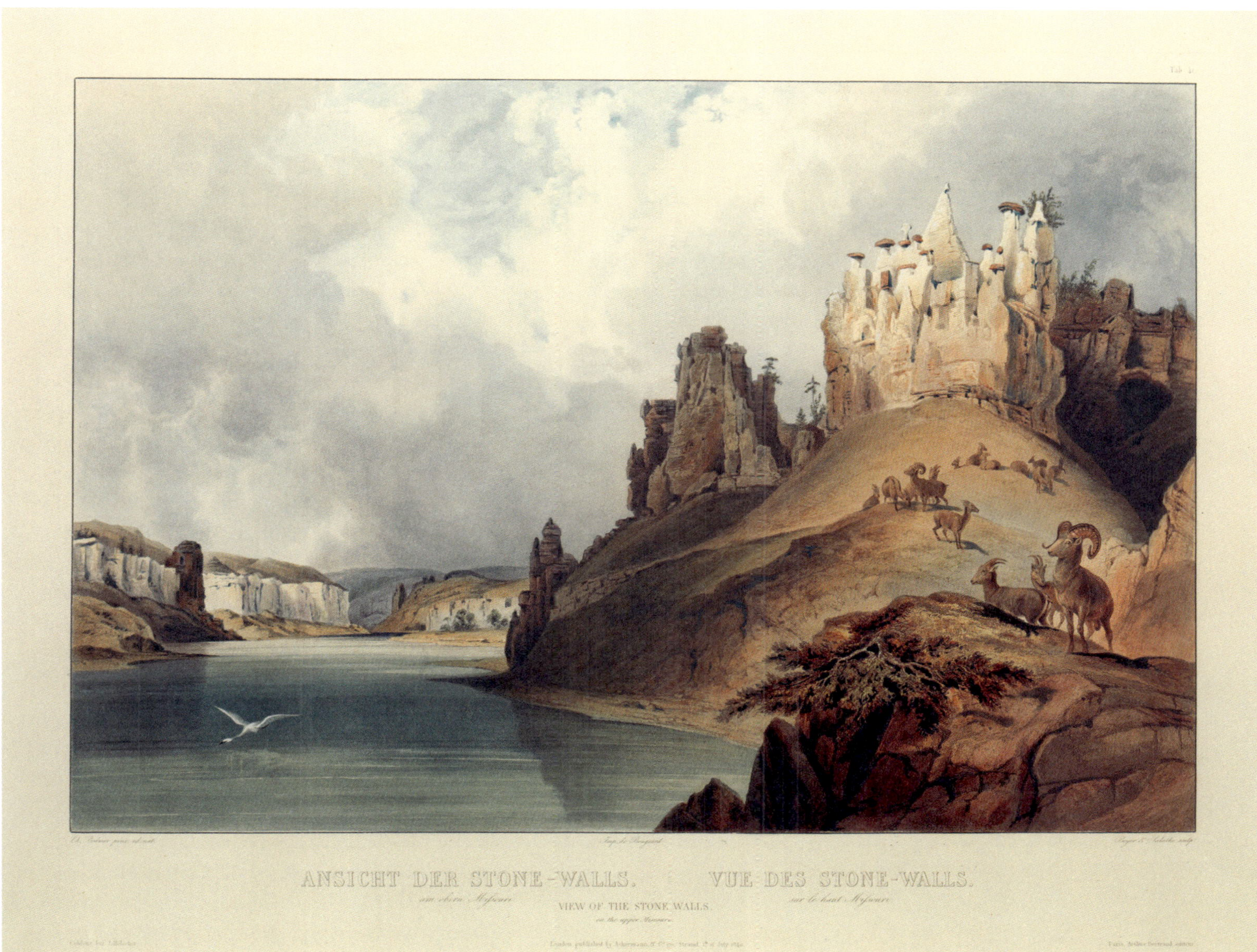

the time they began including Bodmer's works, which proved very popular.[112] Interest in exploration narratives fueled a market for illustrated articles in periodicals and individual gift books extolling the virtues of the American landscape.

HUMBOLDT'S LONG REACH

Humboldt had influenced other German travelers as well, notably Duke Paul of Württemberg. Duke Paul eagerly absorbed Humboldt's writings and was determined to devote his life to similar explorations. He, too, relied on the published account of the Long Expedition to plan his own inaugural trip up the Missouri River from 1823 to 1824. In 1829 his crowning achievement was his discovery of the source of the Missouri River.[113] Of Humboldt and Bonpland's example, the duke wrote "the extraordinary harvest to science which [their] journey brought consequently urged other natural scientists and geographers to follow in the paths blazed by these men."[114] One of the duke's protégés, Balduin Möllhausen (fig. 30), married the daughter of Humboldt's valet and became a surrogate son-in-law. Humboldt wrote letters to secure him a position on the Amiel Whipple Expedition from 1853 to 1854 and the Joseph Christmas Ives Expedition of 1857.[115] When Möllhausen chronicled his adventures, Humboldt wrote the introduction, lending his name to help publicize the work.[116]

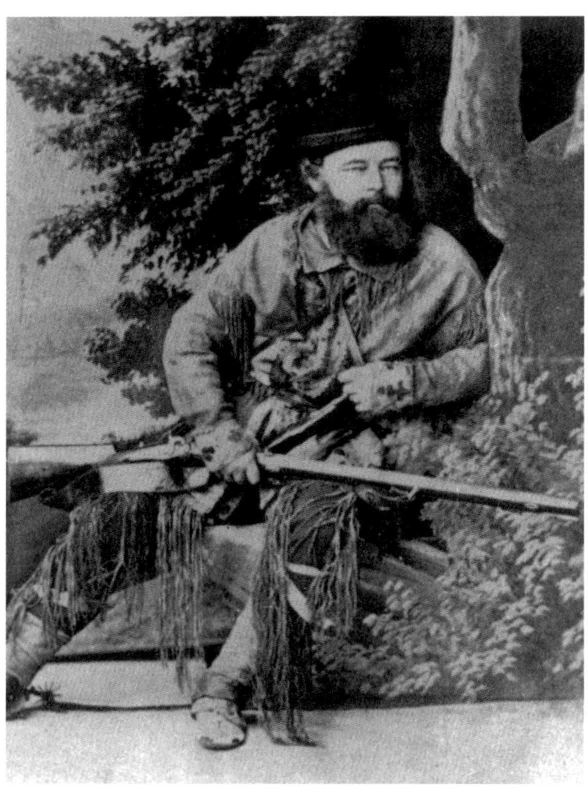

Fig. 30
Unidentified photographer, *Balduin Möllhausen* (1825–1905) *in frontier dress*, 1854, Kansas State Historical Society

Humboldt was selective about the testimonials he wrote, and he used this forum to expound on his stated goal of connecting all parts of the globe. In the preface for the English edition of Prince Adalbert of Prussia's *Travels in the South of Europe and in Brazil*, Humboldt wrote,

> *If human civilization, which is making such giant strides in North America, should ever penetrate into these fastnesses (a hope which St. Basil calls "day-dreams of man"), the structure of this great net of rivers, from north to south, may open channels of communication such as are unknown in any other part of the world.*[117]

For Humboldt, travel was a form of communication; it made possible the transmission of people, ideas, and information. Travel facilitated the increase and diffusion of knowledge, creating a network that connected all parts of the globe. As Prince Max and Bodmer had discovered, North American rivers represented their own network, spanning the interests of the international fur trade conducted along the Mississippi and Missouri Rivers. The use of Clark's route maps might have spared Stephen Long some of the difficulties he encountered with the *Western Engineer*. In Max's hands, they served as navigational tools and visual corroboration of Clark's endorsement of their journey, a kind of passport facilitating access as significant as the official papers required to traverse the territory.

Humboldt's unity of nature was more than a concept tied to plant geography or networks of rivers. Through his own travels and others' published volumes, Humboldt amassed the information he desired from Missouri to the Arctic and Asia. In doing so, he found a way to circle the globe, directly and by proxy. He folded all of this new information into each revised and expanded edition of his *Personal Narrative*. He also attempted to synthesize it during the last decade of his life in each of his five volumes of *Cosmos*. Humboldt's life works, like the generations of maps passed among explorers and statesmen, were a palimpsest of earlier versions. No two editions or translations of any of Humboldt's books would be identical. His reliance on published accounts in the United States—from Lewis and Clark, Stephen H. Long, John C. Frémont, and Prince Maximilian—fueled Humboldt's cumulative quest for knowledge. The information flowing from these journeys made its way straight to Humboldt—population statistics, ethnographic information and artifacts, natural history specimens, and, just as important, cartographic measurements, filling the gaps in Humboldt's increasingly comprehensive understanding of landforms and the global distribution of plants, animals, and people, and how climate operated as a global force on everything. Every one of those survey reports provided Humboldt with new information on everything from plant species to rock formations, weather diaries and astronomical measurements, cartographic data, and a wide variety of observations that contributed to his ever-expanding concept of the unity of nature.

Closer to home, these expeditions and their published narratives laid the groundwork for the establishment of an American cultural identity based on the landscape. The middle decades of the nineteenth century were defined by a half century of exploration tied to political and geographic expansion. Survey reports and illustrated travelogues fueled the imagination. Exploration narratives created a path for the literal journey to become a literary exploration; the visual component encouraged participation in the fine arts as well. American literature and art came into alignment, using the landscape as evidence and as metaphor. Landscape painters in particular will embrace Humboldt's desire to know what lay west of the Appalachian Mountains and invest their works with the cultural resonance Americans identified in the features unique to the country. European aesthetic conventions of the sublime, the beautiful, and the picturesque would be redefined in an American idiom. The wilderness would evolve from a feared protagonist to a path toward finding the hand of God in the wilderness. Singular landmarks like Niagara and Natural Bridge became national icons. As territorial expansion moved the nation ever westward, the post–Civil War surveys would popularize the astonishing features found in the landscapes of Yosemite, Yellowstone, and the Grand Canyon. Timothy O'Sullivan's photographs would make Clarence King's Fortieth Parallel Survey of 1867 memorable for his stunning views of previously little-known landmarks. The Hayden Survey of Yellowstone in 1871 would be remembered as much for Thomas Moran's paintings and William Henry Jackson's photographs as for the wealth of geological information obtained. In 1873 John Wesley Powell would ask Thomas Moran to accompany his expedition through the Grand Canyon, looking to draw attention that would help popularize it as a destination worth preserving. Art would play the pivotal role in the appreciation of and protections afforded to singular landscapes. The durability of this cultural identification with nature is a hallmark of America now as much as it was two centuries ago. The growth of the National Park Service, the establishment of national monuments, and the preservation of memorable features in the landscape have fostered an appreciation of what is thought of as the American wilderness.

1 "Ce pays qui s'étend à l'ouest des montagnes présente un vaste champs à conquérir pour les sciences!" Humboldt to Thornton, Philadelphia, 20 June 1804, in Ingo Schwarz, ed., *Alexander von Humboldt und die Vereinigten Staaten von Amerika: Briefwechsel* (Berlin: Akademie Verlag, 2004), 96–97, letter 9.

2 Horst Fiedler and Ulrike Leitner have assembled the impressive compendium of Humboldt's published works, which covers each edition in each language. *Alexander von Humboldts Schriften. Bibliographie der selbständig erschienenen Werke* (Berlin: Akademie Verlag, 2000). During Humboldt's twenty-two years in Paris, he published sixteen volumes on botany, two volumes on zoology and anatomy, four volumes on astronomy and geophysics, two pictorial atlases, three volumes on his travels through New Spain, four volumes on the Kingdom of New Spain, and seven of the proposed eleven volumes of the *Personal Narrative*, in addition to *Aspects of Nature* and the *Geography of Plants*. This assessment is discussed in Stephen T. Jackson's introduction to Alexander von Humboldt and Aimé Bonpland, *Essay on the Geography of Plants*, ed. Stephen T. Jackson, trans. Sylvie Romanowski (Chicago: University of Illinois Press, 2013), 3.

3 While in school at Göttingen he met and befriended Georg Forster, who had traveled around the world on Captain Cook's second voyage. Humboldt credited Forster with inspiring his own intense desire to travel, as well as his seeing the planet as a series of interconnected systems subject to underlying natural laws. Forster had noticed plants native to widely different locales that seemed to be from the same family, and he had begun to correlate those similarities. Humboldt picked up on this observation, and a decade before he left Europe in 1799 he had started sketching out his signature thesis: if similar plant species flourished around the globe at similar altitudes and under similar climatic conditions, then all of nature must, by extension, be "one great whole, animated by the breath of life"—what would come to be known as Humboldt's concept of the "Unity of Nature."

4 Humboldt to Frau Caroline von Wolzogen, Berlin, 14 May 1806, in Karl Bruhns, ed., *Life of Alexander von Humboldt. Compiled in Commemoration of the Centenary of His Birth,* trans. Jane Lassell and Caroline Lassell (Boston: Lee and Shepard, 1873), 1:359.

5 Ibid.

6 "Le président du Congrès, Jefferson, m'a comblé de témoignages d'honneur." Humboldt to Freiesleben, "Off Bordeaux, August 1, 1804, In haste," in Alexander von Humboldt, *Lettres américaines d'Alexandre de Humboldt*, with a preface and notes by Dr. E. T. Hamy (Paris: E. Guilmoto, 1905), 168–69; quoted in Bruhns, *Life of Alexander von Humboldt*, 1:338.

7 In the summer of 1803, a rumor circulated in Paris that Humboldt had "perished among the savages in North America." Adam Christian Gaspari, *Allgemeine geographische ephemeriden* 12 (1803): 239. The *Hamburger Correspondent* of 12 June 1804 announced, "We regret to learn that the celebrated traveller, Herr von Humboldt, has been attacked with yellow fever, and has died at Acapulco." *Allgemeine geographische ephemeriden* 14 (1804): 510; both quoted in Bruhns, *Life of Alexander von Humboldt*, 1:337n1.

8 "Je reviens avec trente-cinq caisses, et je suis chargé de trésors botaniques, astronomiques et géologiques; il me faudra des années pour publier mon grand ouvrage." Humboldt to Freiesleben, "Off Bordeaux, August 1, 1804, In haste," in Humboldt, *Lettres américaines d'Alexandre de Humboldt*, 168–69; cited in Bruhns, *Life of Alexander von Humboldt* 1:337–38. The actual number of trunks mentioned varies, as do the number of botanical and mineralogical specimens therein. The number seems far less relevant than the impressive range of new species to be classified.

9 Löwenberg cites a memorandum from Humboldt in 1858 summarizing the distribution of plants from this five-year trip. Humboldt and Bonpland brought back close to 6,000 plant specimens, which were catalogued with the altitude at which they were found, a new standard for plant classifications. Many of the plants eventually were purchased by the Berlin Royal Botanical Gardens; duplicates may be found in herbaria in Spain and England. Noted in Bruhns, *Life of Alexander von Humboldt* 1:398–99n2. Karl Ludwig Willdenow (1765–1812) was also in touch with Gotthilf Henry Ernst Mühlenberg in the United States during the 1790s, part of the international fraternity of botanists. In 1799 he sent word that Mühlenberg had been made a member of the Berlin Society of Friends of Natural History. See Matthias Schönhofer, "'Runs, Creeks, and Rivers Join': The Correspondence Network of Gotthilf Henry Ernst Mühlenberg," in *A Peculiar Mixture: German-Language Cultures and Identities in Eighteenth-Century America*, ed. Jan Stievermann and Oliver Scheiding (University Park: Pennsylvania State University Press, 2013), 160, 176–77nn69–70; and Matthias Schönhofer, *Letters from an American Botanist: The Correspondence of Gotthilf Heinrich Ernst Mühlenberg (1753–1815)* (Stuttgart: Franz Steiner Verlag, 2014), 246. Karl Sigismund Kunth (1788–1850) was Humboldt's peer and assistant in Paris. From Havana in 1801, Humboldt had sent earlier shipments of material to colleagues in Berlin, Madrid, London, and Paris. Humboldt to Willdenow, La Havane, 21 February 1801, original in Humboldt, *Lettres américaines d'Alexandre de Humboldt*, 110–14; translated and discussed in H. Walter Lack, *Alexander von Humboldt and the Botanical Exploration of the Americas* (Munich: Prestel, 2009), 30, 34, 93.

10 Plants were easier to carry than rocks, and a year earlier, in June 1803, Humboldt had sent a shipment of mineral specimens to the Institut National with a request that one of their geological experts begin classifying them: "Voilà les objets que nous avons l'honneur de vous présenter, Citoyens, et qui mériteront peut-être l'attention des citoyens Haüy, Vauquelin, Chaptal, Berthollet, Guyton et Fourcroy dont les travaux ont tant contribué au progrès de la Minéralogie et de la Chimie analytique." Humboldt and Bonpland to l'Institut National de France, a la capital du Mexique, 21 June 1803, in Humboldt, *Lettres américaines d'Alexandre de Humboldt*, 162. "Here are the objects we have the honor to present to you, Citizens, and which may deserve the attention of the citizens Haüy, Vauquelin, Chaptal, Berthollet, Guyton, and Fourcroy, whose work has contributed so much to the progress of Mineralogy and Analytical Chemistry." [English translation by the author.] Humboldt shipped some of the rocks to Hermann Karsten. Humboldt to Karsten, Paris, 10 March 1805; quoted in Bruhns, *Life of Alexander von Humboldt*, 1:349–50. As with the plants, Humboldt had noted the altitude and strata at which he collected each sample. He also sent his collection of pre-Columbian artifacts, including a feather painting and several Mexican statues. Bruhns, *Life of Alexander von Humboldt*, 1:349–50.

11 Alexander von Humboldt and Aimé Bonpland's *Essai sur la géographie des plantes* (*Essay on the Geography of Plants*) was inspired by a lecture delivered at the Institut National on 7 January 1805.

12 "Si les phénomènes les plus frappans de la géologie nous attestent que toute la croûte de notre planète fut jadis dans un état liquide; si la stratification et la différence des roches nous indiquent que la formation des montagnes et la cristallisation des grandes masses autour d'un noyau commun ne se sont point effectuées dans le même temps sur toute la surface du globe." Alexander von Humboldt and Aimé Bonpland, *Essai sur la géographie des plantes* (Paris: Chez Fr. Schoell, 1807), 23; translated in Humboldt and Bonpland, *Essay on the Geography of Plants*, 69.

13 "Pour prononcer sur l'ancienne liaison des contenens voisins, la géologie se fonde sur la structure analogue des côtes sur les bas-fonds de l'Océan.... La géographie des plantes...elle annonce que la séparation de l'Afrique et de l'Amérique méridionale s'est faite avant le développement des êtres organisés." Humboldt and Bonpland, *Essai sur la géographie des plantes*, 19; translated in Humboldt and Bonpland, *Essay on the Geography of Plants*, 67.

14 "La géographie des plantes descend dans l'intérieur du globe: elle y consulte les monumens antiques que la nature a laissés dans les pétrifactions, dans les bois fossiles et les couches de charbons de terre, qui sont le tombeau de la première végétation de notre planète." Humboldt and Bonpland, *Essai sur la géographie des plantes*, 22; translated in Humboldt and Bonpland, *Essay on the Geography of Plants*, 69.

15 Alexander von Humboldt, *Political Essay on the Kingdom of New Spain*, trans. John Black (London: Longman, Hurst, Rees, Orme, and Brown, 1811), 1:cxxxii, 7.

16 Quoted in Cara Giaimo, "The Exquisite 19th-Century Infographics That Explained the History of the Natural World," *Atlas Obscura*, 8 January 2016, http://atlas-obscura.com/articles/the-exquisite-19th-century-infographics-that-explained-the-history-of-the-natural-world.

17 Humboldt to Varnhagen von Ense, Potsdam, 15 October 1849, in Alexander von Humboldt, *Letters of Alexander von Humboldt to Varnhagen von Ense from 1827 to 1858*, trans. and ed. Friedrich Kapp (New York: Rudd & Carleton, 1860), 252.

18 Subsequent editions often praised the work of colleagues. For the third edition, translated into English in 1849, Humboldt incorporated descriptions of the United States drawn from the Long Expedition, Zebulon Pike's travels to the Rockies, and John C. Frémont's expeditions. Such updates not only kept the material fresh but also provided incentive for fellow scientists and explorers to continue feeding information to Humboldt as part of this evolutionary vision of the earth. In truth, no two editions of any of Humboldt's works were ever identical; he could not resist adding, tweaking, and adapting his own thinking over time.

19 Humboldt was inspired by Georg Forster and Sir Joseph Banks to travel widely and make connections between disparate places. Each man had sailed with Captain James Cook on his first two voyages. As Banks put it, "my Grand Tour shall be one round the whole globe." Edward Smith, *The Life of Sir Joseph Banks, President of the Royal Society* (London: J. Head, 1911), 16; quoted in Barbara Maria Stafford, *Voyage into Substance: Art, Science, Nature, and the Illustrated Travel Account, 1760–1840* (Cambridge, MA: MIT Press, 1984), 4.

20 Humboldt and Bonpland credited Benjamin Smith Barton with providing botanical information. They addressed Peale's mammoth skeleton and their own discovery of mammoth teeth in the Andes in the context of questioning how broadly this animal had once ranged.

21 Alexander von Humboldt and Aimé Bonpland, *Personal Narrative of Travels to the Equinoctial Regions of the New Continent, during the Years 1799–1804*, trans. Helen Maria Williams (London: Longman, Hurst, Rees, Orme, and Brown, 1818), 1:233.

22 The effects of exploration and exploitation on Indian tribes is the focus of chapter 6 in this volume.

23 "Il me paraît que je reverrai ce beau pays en peu d'années. Le chemin du Missouri aux côtés de l'Océan pacifique sera alors déjà ouvert. Les lacs, le Canada et l'imense bassin qui s'étend depuis Pittsburg aux montagnes vues par Fidler offrent un vaste champs aux recherches géologiques. Avec quelques secours de Votre Gouvernement on pourrait y faire un grand et beau travail, aussi important au moins que celui de Pallas. Je hazarderais même alors de pénétrer au Nord jusqu'à Mount Elias (aussi élevé que les montagnes de Quito) et aux posesions Russes! Avec de la santé et du courage tout cela pourrait s'éxécuter." Humboldt to Madison, Philadelphia, [19–20] June 1804, in James Madison, *The Papers of James Madison*, Secretary of State Series, vol. 7, *2 April–31 August 1804*, ed. David B. Mattern et al. (Charlottesville: University of Virginia Press, 2005), 334–36. Peter Fidler was a surveyor of the Hudson's Bay Company who had surveyed part of the Missouri River and the Canadian Rockies; Peter Simon Pallas was a German naturalist who had traveled to Alaska and Siberia at the invitation of Catherine the Great in 1786. Mt. Elias is in present day Alaska. Schwarz, *Alexander von Humboldt und die Vereinigten Staaten von Amerika*, 93–94, letter 7. Translated in Helmut de Terra, "Alexander von Humboldt's Correspondence with Jefferson, Madison, and Gallatin," *Proceedings of the American Philosophical Society* 103, no. 6 (15 December 1959): 797.

24 "Quand, quand serai-je de nouveau avec vous—quand pourrai-je m'enfoncer dans ces immenses régions de l'Ouest pour lesquels M Jefferson (dans Sa nouvelle place) seroit plus en état de procurer des Secours que jadis? Je n'ai bien perdu des espérances.... puis j'ai des projets du Missouri, le cercle Polaire & de l'Asie." Humboldt to John Vaughan, Rome, 10 June 1805, American Philosophical Society Archive, Philadelphia. Schwarz, *Alexander von Humboldt und die Vereinigten Staaten von Amerika*, 1056, letter 19.

25 David Ramsay, "An Oration, on the Cession of Louisiana to the United States, Delivered on the 12th of May, 1804, in St. Michael's Church, Charleston South Carolina, at the request of a number of the inhabitants, and published by their desire," *National Intelligencer and Daily Advertiser*, 11 June 1804, 1.

26 Ellicott to Jefferson, 6 March 1803, in Thomas Jefferson, *The Papers of Thomas Jefferson*, vol. 40, *4 March to 10 July 1803*, ed. Barbara B. Oberg (Princeton, NJ: Princeton University Press, 2013), 15–16. See chapter 1 for a discussion of Lewis's instruments.

27 According to Helferich, "Humboldt gathered perhaps the most sophisticated armamentarium of scientific instruments ever before assembled. Each of the forty-two instruments, nestled in its own velvet-lined box, was the most accurate and most portable of its kind yet devised. There were thermometers for measuring the temperature of air and water, barometers for fixing elevation above sea level, quadrants and sextants for determining geographic position (including a sextant small enough to fit in a pocket), telescopes, microscopes, a balance scale, chronometers, compasses, a rain gauge, substances for performing chemical assays, electric batteries, electrometers (for measuring electric current), a Leyden jar (a glass vessel capable of storing static electricity), theodolites (surveyors' instruments for measuring vertical and horizontal angles), hygrometers (for measuring atmospheric moisture), a dip needle (for measuring variations in the orientation of the earth's magnetic field), and eudiometers (for measuring the amount of oxygen in the air)." Gerard Helferich, *Humboldt's Cosmos: Alexander von Humboldt and the Latin American Journey That Changed the Way We See the World* (New York: Gotham Books, 2004), 25. Humboldt's own discussion of these instruments runs to seven pages in his and Bonpland's *Personal Narrative of Travels to the Equinoctial Regions*, trans. Williams (London, 1818), 1:34–40.

28 "Les instrumens sont encore en vie!" Humboldt to Count Ferdinand von Galen, 8 June 1829, in *Briefe aus Russland 1829*, ed. Eberhard Knobloch, Ingo Schwarz, and Christian Suckow (Berlin: Akademie Verlag, 2009), 134; translated by Vera M. Kutzinski in Ottmar Ette, "TransTropics: Alexander von Humboldt and Hemispheric Constructions," in *Alexander von Humboldt and the Americas*, ed. Vera M. Kutzinski, Ottmar Ette, and Laura Dassow Walls (Berlin: Verlag Walter Frey, 2012), 211.

H-B de Saussure, "Description d'un cyanomètre ou d'un appareil destiné à mesurer la transparence de l'air," *Mem. Acad. Roy. Turin* (1788–89): 4, 409. Humboldt recorded a measurement of 46 on a scale of 52 at an altitude of 19,413 feet (3,036 toises) the day he climbed Chimborazo. He published this in Humboldt and Bonpland, *Essai sur la géographie des plantes* (Paris: Levrault, Schoell, 1805), 103. Lord Byron mocked Humboldt's obsession with his scientific measurements, finding the intense observation of young women more to his liking. He penned the lines:

> "Humboldt, 'the first of travelers,' but not
> The last, if late accounts be accurate,
> Invented, by some name I have forgot,
> As well as the sublime discovery's date,
> An airy instrument, with which he sought
> To ascertain the atmospheric state,
> By measuring 'the *intensity of blue*:'
> Oh, Lady Daphne! let me measure you!"

George Gordon Byron (Lord Byron), "Don Juan, Canto IV., CXIII," in *The Works of Lord Byron: A New, Revised and Enlarged Edition with Illustrations*, ed. Ernest Hartley Coleridge (New York: Charles Scribner's Sons, 1903), 6:215–16.

29 Jefferson to Isaac Briggs, Washington, 11 June 1804; quoted in Herman R. Friis, "Baron Alexander von Humboldt's Visit to Washington, D.C., June 1 through June 13, 1804," *Records of the Columbia Historical Society* 60/62 (1960/62): 32. Lewis and Clark took measurements of the moon's distance from specific stars but never made the conversions to longitude; instead Lewis used dead reckoning calculations made by other explorers after he returned.

30 Jefferson to Humboldt, Monticello, 16 March 1814; quoted in Paul A. W. Wallace, "Henry Ernest Muhlenberg," *Proceedings of the American Philosophical Society* 92, no. 2 (5 May 1948): 110. Jefferson had previously expressed his frustration with the delay in a letter written to Humboldt on December 6, 1813. Schwarz, *Alexander von Humboldt und die Vereinigten Staaten von Amerika*, 132, letter 41.

31 Catherine Van Cortlandt Mathews, *Andrew Ellicott: His Life and Letters* (New York: Grafton Press, 1908), 161–63.

32 It is not clear how Pike came into possession of a copy of Humboldt's map, but as a member of Jefferson's cabinet, Wilkinson likely had access to it while Humboldt was in Washington. Wilkinson had been invited to the farewell dinner for Humboldt, hosted by Jefferson at the President's House but declined for health reasons. Wilkinson to Jefferson, Washington, 11 June 1804, in Thomas Jefferson, *The Papers of Thomas Jefferson*, vol. 43, *11 March to 30 June 1804*, ed. James P. McClure (Princeton, NJ: Princeton University Press, 2017), 573. It remains unclear whether Pike was complicit in Wilkinson's plans or was an unwitting dupe; Jefferson saw fit to give Pike the benefit of the doubt.

33 Humboldt appended his statistical survey of Mexico, translated into French, which he had compiled for the viceroy of New Spain, José de Iturrigaray: "Tableau statistique du Royaume de la Nouvelle Espagne," with the table of contents covering "Geological sketch of the country, Population, Political divisions, Agriculture, manufacture, commerce, State roads, Mines, Military Forces." He added two pages of information specifically about Louisiana in answer to Jefferson's questions. See Schwarz, *Alexander von Humboldt und die Vereinigten Staaten von Americka*, 484–95, dokument 2.

34 Humboldt to Jefferson, Paris, 20 December 1811; translated from the French in Helmut de Terra, "Alexander von Humboldt's Correspondence with Jefferson, Madison, and Gallatin," 792. "Mr. Arrowsmith à Londres m'a volé ma grande Carte du Mexique: Mr. Pike a profité d'une manière peu généreuse de la communication qui lui a été faite sans doute à Washington de la copie de ma Carte: d'ailleurs il a estropié tous les noms. Je suis affligé d'avoir à me plaindre d'un citoyen des États Unis qui d'ailleurs a déployé un si beau courage. Mon nom ne se trouve pas dans son livre et un léger coup d'oeil sur la Carte de Mr Pike Vous prouve d'où il a puisé. Ma fortune a souffert, moins par mes voyages que par des événemens politiques: c'est perdre bien peu que de perdre sa fortune. Je trouve des consolations dans le travail dans des souvenirs et dans l'estime des hommes qui reconnoissent la pureté de mes intentions." Humboldt to Jefferson, Paris, 20 December 1811, in Thomas Jefferson, *The Papers of Thomas Jefferson*, Retirement Series, vol. 4, *18 June 1811 to 30 April 1812*, ed. J. Jefferson Looney (Princeton, NJ: Princeton University Press, 2008), 352–54. Schwarz, *Alexander von Humboldt und die Vereinigten Staaten von Amerika*, 122–23, letter 33.

35 Humboldt had noted, "Mr. Pike displayed admirable courage in an important undertaking for the investigation of western Louisiana; but unprovided with instruments, and strictly watched on the road from Santa Fe to Natchitoches, he could do nothing towards the progress of the geography of the *provincias internas*." After chastising Pike, Humboldt turned his wrath to the London-based mapmaker, Arrowsmith, continuing, "My general map of the kingdom of New Spain, formed on astronomical observations, and on the whole of the materials which existed in Mexico in 1804, has been copied by Mr. Arrowsmith, who has appropriated it to himself, by publishing it on a larger scale, under the title of *New Map of Mexico, compiled from original Documents, by Arrowsmith*. It is very easy to recognize this map from the number of chalcographical errors with which it abounds." Humboldt included a table in his text pointing out all of the errors he found on Arrowsmith's map in his *Personal Narrative of Travels to the Equinoctial Regions*, trans. Williams (London, 1818), 1:xxvii–xxviii.

36 Pike had died eight months earlier. Jefferson to Humboldt, Montpelier, 6 December 1813, in Jefferson, *Papers of Thomas Jefferson*, Retirement Series, vol. 7, *28 November 1813 to 30 September 1814*, ed. J. Jefferson Looney (Princeton, NJ: Princeton University Press, 2011), 29–32. Schwarz, *Alexander von Humboldt und die Vereinigten Staaten von Amerika*, 130–31, letter 41.

37 Humboldt, *Political Essay on the Kingdom of New Spain*, trans. Black (London 1811), 1:21–22.

38 Ibid., 2:337.

39 Jefferson to Humboldt, Montpelier, 6 December 1813, in Jefferson, *Papers of Thomas Jefferson*, Retirement Series, 7:29–32. Schwarz, *Alexander von Humboldt und die Vereinigten Staaten von Amerika*, 130–31, letter 41.

40 Ramsay, "Oration."

41 Reuben G. Thwaites, ed., *Early Western Travels 1748–1846*, vol. 14, *Part 1 of James's Account of S. H. Long's Expedition* (Cleveland, OH: Arthur H. Clark, 1905), 37–38.

42 Members of the Long Expedition team included Augustus Edward Jessup (1797–1859), geologist; Major John Biddle (1792–1859), journalist; William Baldwin (1779–1819), physician and botanist; Thomas Say (1787–1834), zoologist; and Samuel Seymour (1775–1823), artist.

43 Kenneth Haltman, "Figures in a Western Landscape: Reading the Art of Titian Ramsay Peale from the Long Expedition to the Rocky Mountains, 1819–1820" (PhD diss., Yale University, 1992), 253. Appendix A lists the books and maps Long took with him on the expedition.

44 Roger L. Nichols and Patrick L. Halley, *Stephen Long and American Frontier Exploration* (Newark, NJ: University of Delaware Press, 1980), 119–35.

45 The term "Humboldtian Science" was coined by Susan Faye Cannon and described in detail in *Science in Culture: The Early Victorian Period* (New York: Science History Publications, 1978). Laura Dassow Walls succinctly described its tenets as "Explore, Collect, Measure, Connect," in *Seeing New Worlds: Henry David Thoreau and Nineteenth-Century Natural Science* (Madison: University of Wisconsin Press, 1995), 98–102, and she expands on it in *The Passage to Cosmos: Alexander von Humboldt and the Shaping of America* (Chicago: University of Chicago Press, 2009), 126–27.

46 Jessie Poesch, *Titian Ramsay Peale, 1799–1885, and His Journals of the Wilkes Expedition* (Philadelphia: American Philosophical Society, 1961), 6.

47 For consideration of Rembrandt Peale's travels in Europe, see chapters 3 and 4 in this volume.

48 Poesch, *Titian Ramsay Peale,* 17–20.

49 Ibid., 35. In 1822 Titian exhibited four of these watercolors at the Pennsylvania Academy of Fine Arts. See *Catalogue of the Eleventh Annual Exhibition of the Pennsylvania Academy of Fine Arts, May, 1822*, 7, 13; cited in Poesch, *Titian Ramsay Peale,* 44–45, 60.

50 Amy R. W. Meyers, "Imposing Order on the Wilderness: Natural History Illustration and Landscape Portrayal," in *Views and Visions: American Landscape Before 1830*, ed. Edward J. Nygren and Bruce Robertson (Washington, DC: Corcoran Gallery of Art, 1986), 126.

51 Charles Willson Peale Autobiography, typescript, page 318, Peale Papers, APS; quoted in Charles Coleman Sellers, *Mr. Peale's Museum: Charles Willson Peale and the First Popular Museum of Natural Science and Art* (New York: W. W. Norton, 1980), 28.

52 Titian's drawings served as the basis for engravings illustrating articles on the Long Expedition published in the *Journal of the Academy of Natural Science* in 1822. See also Poesch, *Titian Ramsay Peale,* 43–46. Peale's Lepidoptera collection went to the Academy of Natural Sciences and is housed today at Drexel University. Peale's research was recently published as *The Butterflies of North America: Titian Peale's Lost Manuscript* (New York: American Museum of Natural History, 2015).

53 Edwin James, who wrote up the Long Expedition, became Colorado's first botanist. He and Titian Ramsay Peale were the first white explorers to climb Pike's Peak. Clifford Merrill Drury, *Diary of Titian Ramsay Peale: Oregon to California, Overland Journey September and October, 1841* (Los Angeles: Glen Dawson, 1957).

54 Edward Everett, "Long's Expedition," *North American Review* 16, no. 39 (April 1823): 242–69. For the importance of the Long Expedition to the literary works of Irving and Cooper, see chapters 3 and 4 in this volume.

55 Report of Senator Breese of Illinois, U.S. Senate, 1 August 1848, regarding the merits of publication of Frémont's expedition to California and Oregon, to be funded by Congress, with no copyright: "An assistant of the celebrated Nicollet who was a distinguished member of the French National Institute, he has reached a most commanding position as a scientific explorer, and achieved for himself the designation of the American Humboldt." John Bigelow, *Memoir of the Life and Public Services of John Charles Fremont* (New York: Derby & Jackson, 1856), 468.

56 In 1838 Poinsett appointed Nicollet to lead an exploration of the triangle of land between the Mississippi and Missouri Rivers and the Canadian border as part of the Army Corps of Topographical Engineers. Nicollet was charged with creating an accurate map of the territory. See Edmund Bray and Martha Coleman Bray, trans. and eds., *Joseph N. Nicollet on the Plains and Prairies: The Expeditions of 1838–39 with Journals, Letters, and Notes on the Dakota Indians* (St. Paul: Minnesota Historical Society Press, 1976).

57 *Hypsometric* is a scientific term relating to the measurement of heights. The term originates from the Greek word ύφος (hypsos), meaning height, and the word *meter* is from the Greek μέτρον (métron), "a measure." A hypsometric curve is a histogram of elevations of a landscape.

58 J. N. Nicollet, "Report Intended to Illustrate a Map of the Hydrographic Basin of the Upper Missouri River," 26th Congress, 2nd Session, 16 February 1841, *Senate Executive Document 237* (1840–41) (Washington, DC: Blair and Rives, 1842). On page 95, Nicollet cites Humboldt in noting, "If only in every thousand years the mean temperature of the atmosphere and of the earth, in different latitudes, could have been determined…we should know in what ratio the heat of different climates has increased or diminished, and if any change has taken place in the height of the atmosphere."

59 "As soon as the camp was formed," he said, "I set about endeavoring to repair my barometer. As I have already said, this was a standard cistern barometer, of Troughton's construction. The glass cistern had been broken about midway; but as the instrument had been kept in a proper position, no air had found its way into the tube, the end of which had always remained covered. I had with me a number of vials of tolerably thick glass, some of which were of the same diameter as the cistern, and I spent the day in slowly working on these, endeavoring to cut them of the requisite length; but, as my instrument was a very rough file, I invariably broke them. A groove was cut in one of the trees, where the barometer was placed during the night, to be out of the way of any possible danger, and in the morning I commenced again. Among the powder horns in the camp, I found one which was very transparent, so that its contents could be almost as plainly seen as through glass. This I boiled and stretched on a piece of wood to the requisite diameter, and scraped it very thin, in order to increase to the utmost its transparency. I then secured it firmly in its place on the instrument, with strong glue made from a buffalo, and filled it with mercury, properly heated. A piece of skin, which had covered one of the vials, furnished a good pocket, which was well secured with strong thread and glue, and then the brass cover

was screwed to its place. The instrument was left some time to dry; and when I reversed it, a few hours after, I had the satisfaction to find it in perfect order; its indications being about the same as on the other side of the lake before it had been broken. Our success in this little incident diffused pleasure throughout the camp; and we immediately set about our preparations for ascending the mountains." Recounted in Bigelow, *Memoir of the Life and Public Services of John Charles Fremont*, 48–50.

60 John Charles Frémont, *Memoirs of My Life* (New York: Belford, Clarke & Company, 1887), 65.

61 Hon. Theodore S. Fay to Rev. R. C. Waterston, Berlin (Prussia), 26 August 1869; quoted in Louis J. Agassiz, *Address Delivered on the Centennial Anniversary of the Birth of Alexander von Humboldt* (Boston: Boston Society of Natural History, 1869), 91.

62 Frémont, *Memoirs of My Life*, 434.

63 Letter from Frémont to "My Dear Sirs," Prescott, Arizona Territory, 20 February 1881, in "Lake Tahoe's Name," in Myron Angel, *History of Placer County, California* (Oakland, CA: Thompson & West, 1882), 404.

64 Jessie Benton Frémont, *Memoirs*, Bancroft Library, University of California, Berkeley, 41–42; quoted in Tom Chaffin, *Pathfinder: John Charles Frémont and the Course of American Empire* (New York: Hill and Wang, 2002), 138–39.

65 Humboldt wrote, "In the middle latitudes of 37° to 43°, the Rocky Mountains present, besides the high snowy summits comparable with the Peak of Teneriffe in elevation, lofty plains of an extent hardly met with elsewhere on the surface of the earth, and almost twice as extensive, in an east and west direction, as that of the Mexican plateaux. From the group of mountains, which commences a little to the west of Fort Laramie, to beyond the Wahsatch mountains, there is an uninterrupted swelling of the ground from 5300 to 7400 English feet above the level of the sea. A similar elevation may even be said to occupy the whole space from 34° to 45° between the Rocky Mountains proper and the Californian snowy coast chain. This space, a kind of broad longitudinal valley like that of the Lake of Titicaca, has been called, by Joseph Walker, a traveller well acquainted with these western regions, and by Captain Frémont, 'The Great Basin.' It is a terra incognita of at least 128,000 square miles in extent, arid, almost entirely without human inhabitants, and full of salt lakes, the largest of which is 4200 English feet above the level of the sea, and is connected with the narrow lake of Utah. (Fremont, Report of the Exploring Expedition, pp. 154 and 273–276.) The last mentioned lake receives the abundant waters of the 'Rock River;' Timpan Ogo, in the Utah language." Alexander von Humboldt, *Aspects of Nature, in Different Lands and Different Climates; with Scientific Elucidations*, trans. Elizabeth J. L. Sabine, 3rd ed. (Philadelphia: Lea and Blanchard, 1849), 54–55.

66 Humboldt, *Aspects of Nature*, 54.

67 Bigelow, *Life and Public Services of John Charles Fremont*, 35–36; and Humboldt, *Aspects of Nature*, 50.

68 See Henry Stevens, *Catalogue of the First Portion of the Humboldt Library* (London: J. Davy & Sons, 1862).

69 Humboldt's support for Frémont's presidential bid in 1856 is covered in chapter 5 in this volume.

70 All quotes from the letter that appears in both the French original and this English translation in Bigelow, *Memoir of the Life and Public Services of John Charles Fremont*, 328–29. This letter in its English entirety and with no misspellings, is printed in Horace Greeley, ed., *Life of Col. Frémont* (New York: Greeley & M'Elrath, 1856), 50.

71 Jessie Benton Frémont, "Great Events in the Life of General John C. Frémont…and of Jessie Benton Frémont," an unpublished memoir by Jessie Benton Frémont, Bancroft Library, University of California, Berkeley, 179; cited in Sally Denton, *Passion and Principle: John and Jessie Frémont, the Couple Whose Power, Politics, and Love Shaped Nineteenth-Century America* (New York: Bloomsbury, 2007), 204. It remains unclear whether the Frémonts considered traveling to Berlin to meet Humboldt on this trip; however, despite their high regard for Humboldt, they never met him.

72 The short answer was no, not without better telescopes to measure accurately the distance to the sun. But as a terrestrial measuring technique, altitude could be determined based on the angle and length of the shadows cast in photographs of mountains. Another of Humboldt's protégés, Baron Egloffstein, wondered if Humboldt had in fact pioneered this technique for use in terrestrial calculations; Humboldt himself gave credit to others for developing this technique. Steven Rowan, *The Baron in the Grand Canyon: Friedrich Wilhelm von Egloffstein in the West* (Columbia: University of Missouri Press, 2012), 91.

73 "To make a daguerreotype view, generally occupied from one to two hours, the principal part of that time, however, was spent in packing, and reloading the animals. When we came up to the Butte, Mr. Fuller made barometrical observations, in order to ascertain its exact height." Solomon Nunes Carvalho, *Incidents of Travel and Adventure in the Far West; with Col. Fremont's Last Expedition across the Rocky Mountains*…(New York: Derby & Jackson, 1857), 76; quoted in Joan Sturhahn, *Carvalho: Artist, Photographer, Adventurer, Patriot: Portrait of a Forgotten American* (Merrick, NY: Richwood Publishing, 1976), 98, 89.

74 It is not clear whether the collection of daguerreotypes was destroyed in the fire that swept through the Frémont's home in October 1881 or survived and are now lost. See Sturhahn, *Carvalho*, 133–34.

75 Madison to Humboldt, Montpelier, 12 March 1833, in Schwarz, *Alexander von Humboldt und die Vereinigten Staaten von Amerika*, 186, letter 95.

76 See chapter 3 in this volume for a discussion of Washington Irving's connections to the Astor fur empire, and chapter 6 for a deeper discussion of the Astor family's connections to Humboldt.

77 Blumenbach's work on polygenesis is treated in greater depth in chapters 5 and 6 in this volume.

78 Ron Tyler, "Karl Bodmer and the American West," in *Karl Bodmer's North American Prints*, ed. Brandon K. Ruud (Lincoln: University of Nebraska Press, 2004), 6.

79 Prince Maximilian to Rudolf Schinz, 7 July 1811, Zentralbibliothek Zurich, Ms Car XV 174, Umschlag III. Translated by Michael G. Noll in "Prince Maximilian's America: The Narrated Landscapes of a German Explorer and Naturalist" (PhD diss., University of Kansas, 2000), 56n2.

80 After the defeat of Napoleon I, the Treaty of Paris of 30 May 1814 was signed between France on one side and Austria, Great Britain, Prussia, and Russia on the other.

81 Noll, "Prince Maximilian's America," 33n15. Prince Maximilian to Schinz, 28 May 1814. Max's descendants confirmed the closeness of this friendship: "The principal event of that time was his first meeting with Alexander von Humboldt who had returned in 1804 from his long voyage to Central and South America [but who had been denied entry into Brazil]. This encounter had a profound effect upon Maximilian. There can be no doubt that his predominant interest in the Americas was due to the influence of the celebrated older scientist, who remained his model, friend, and mentor." See Karl Viktor Prinz zu Wied, "Maximilian Prinz zu Wied: Sein Leben und seine Reisen," in *Maximilian Prinz zu Wied. Unveröffentliche Bilder und Handschriften zur Viilkerkunde Brasiliens*, ed. Josef Roder and Hermann Trimborn (Bonn-Hannover-Stuttgart: F. Dummler, 1954), 17. Translation in Paolo E. Vanzolini and Charles W. Myers, "The Herpetological Collections of Maximilian, Prince of Wied (1782–1867), with Special Reference to Brazilian Materials," *Bulletin of the American Museum of Natural History* 395 (26 June 2015): 9.

82 Noll, "Prince Maximilian's America." See also Bernhard Gondorf, "Die Expedition von Maximilian Prinz zu Wied und Karl Bodmer in das innere Nordamerika," in *Prärie- und Plainsindianer: Die Reise in das Nord-America von Maximilian zu Wied und Karl Bodmer* (Mainz: Verlag Hermann Schmidt, 1993); and Paul Schach, "Maximilan, Prince of Wied (1782–1867): Reconsidered," *Great Plains Quarterly* 14 (1994): 5–20.

83 A letter in Humboldt's papers from Baron von Eschwege, who had explored Brazil as a Portuguese agent, written from Lisbon on 27 March 1848, includes a copy of the warrant for Humboldt's arrest, dated 2 June 1800, on the grounds that he might "conceal plans for the spread of new ideas and dangerous principles among the faithful subjects of this realm." Eschwege informs Humboldt that this order was rescinded at the behest of Count de Barca, who sensed Humboldt's detainment could backfire as an international embarrassment. Eschwege goes on to note that Humboldt actually crossed into Brazil along the upper Orinoco, where he might have been imprisoned for more than a year had the warrant not been withdrawn. Humboldt scribbled on this warrant, "I desire that this order may in somewhere be published after my death." For the implications of this potential transgression, see Vera M. Kutzinski, "Facts and Fictions: Alexander von Humboldt in Brazil," in *Contributions to Theory and Comparative History of Historiography: German and Brazilian Perspectives*, ed. Liuz Estevam de Oliveira Fernandes, Luísa Rauter Pereira, and Sérgio de Mata (Frankfurt: Peter Lang, 2015), 59–86; quoted in Bruhns, *Life of Alexander von Humboldt* 1:395–96.

84 Botanist Friedrich Sellow (1789–1831) had worked with Humboldt's friend, Karl Willdenow, at the Botanic Garden of Berlin. In 1810 he moved to Paris, where he met Humboldt and worked at the Jardin des Plantes. In both places he had studied the specimens Humboldt and Bonpland had brought back from South America. Humboldt sponsored Sellow's study in London, and armed with a letter of recommendation from Humboldt, Sellow was recruited as a member of a Russian scientific exploration of Brazil in 1814. Humboldt provided Prince Maximilian with a letter of introduction to Sellow, who was living in Rio de Janeiro, asking if he would accompany the prince through Brazil.

85 Prince Maximilian to Schinz at the University of Zurich, 10 September 1814. Max wrote that he wanted to go to Brazil "to collect treasures that will certainly be useful also for my good friends," presumably Humboldt being among them. There are 425 letters from Max from 1810 to 1860 in the Zentralbibliothek Zurich. Maximilian-Bodmer Collection, Margre H. Durham Center for Western Studies, Joslyn Art Museum, Omaha, Nebraska.

86 "Publication of the books on his travels earned Maximilian recognition and many honors. The one that must have pleased him most came from Alexander von Humboldt who sent him his autographed portrait. Ever since, this portrait adorned the desk in the prince's study." Karl Viktor Prinz zu Wied, "Maximilian Prinz zu Wied," 21.

87 In his Brazilian notes, Max wrote, "It would be very interesting for me to see the North American Indian tribes for the purpose of comparing them with those of Brazil, and I intend therefore perhaps some day to undertake a journey there." See Josephine Huppertz, "Textkritische Analyse und Vergleich zwischen schriftlichem Nachlass und Reisewerk," in *Maximilian Prinz zu Wied, Unveröffentliche Bilder und Handschriften zur Völkerkunde Brasiliens,* ed. Josef Röder and Hermann Trimborn (Bonn-Hannover-Stuttgart: F. Dummler, 1954), 75.

88 Bernhard Herzog zu Sachsen-Weimar and Heinrich Luden, *Reise Sr. Hoheit des Herzogs Bernhard zu Sachsen-Weimar-Eisenach durch Nord-Amerika in den Jahren 1825 und 1826* (Weimar, 1828), 1:296–99.

89 Bernhard, Duke of Saxe-Weimar to Prince Maximilian, Luxembourg, 6 April 1831; and Prince Maximilian to Duke of Saxe-Weimar, Neuwied, 19 March 1832, Maximilian-Bodmer Collection, Margre H. Durham Center for Western Studies, Joslyn Art Museum, Omaha, Nebraska.

90 "For his travels Maximilian adopted the pseudonym Baron of Braunsberg, a name that referred to an old family line from the thirteenth century. His younger brother, Heinrich Victor, had earlier used this same name during the peninsular campaign against Napoleon in order to protect the family from possible repercussions. Maximilian continued to use this name during his travels to North America, either out of habit or [perhaps, as Noll suggests,] simply to honor his fallen brother, Victor." Noll, "Prince Maximilian's America," 36n22.

91 Prince Reinhard of Solms to Prince Maximilian, Laubach, 25 March 1832. In this letter he lists the books Max should read before going and all of the Germans he knows who are in America and where they settled. In particular he recommends Andre Michaux for North American botany [*Flora Boreali Americana* volumes 1 and 2, Paris, 1803]. Maximilian-Bodmer Collection, Margre H. Durham Center for Western Studies, Joslyn Art Museum, Omaha, Nebraska.

92 Prince Maximilian to Duke of Saxe-Weimar, Neuwied, 19 March 1832, Maximilian-Bodmer Collection, Margre H. Durham Center for Western Studies, Joslyn Art Museum, Omaha, Nebraska.

93 Georg Forster had noted with satisfaction that Captain Cook's voyages would be staffed by more than scientists. His father, John R. Forster, accompanied Cook on a voyage that also included artist William Hodges. Georg Forster, *A Voyage round the World in His Britannic Majesty's Sloop, Resolution, Commanded by Captain James Cook, during the Years 1772, 3, 4, and 5* (London: B. White, J. Robson, P. Elmsly, and G. Robinson, 1777), 1:iv.

94 Prince Maximilian to the eminent German botanist Karl Friedrich Philipp von Martius (1794–1868), who also explored Brazil, as follows: "Thoughts of another journey are now going through my head...and I am thinking strongly of northern America. Which region of this interesting land? I think the interior regions of the Missouri would be highly interesting because of its tribes?...Doesn't North America also have much of interest for botany? Then I would want to bring along a draftsman who would not be too much of a burden on my pocketbook but could pay at least part of his way—certainly a rarity which will not be easy to find! He must be a landscape painter but also able to depict figures correctly and accurately, especially the Indians." 26 February 1830, Bayerische Staatsbibliothek, Munich; quoted in William J. Orr, "Karl Bodmer, the Artist's Life," in *Karl Bodmer's America*, ed. David C. Hunt and Marsha V. Gallagher (Lincoln: Joslyn Art Museum and University of Nebraska Press, 1984), 352.

95 von Mühlbach to Prince Maximilian, Coblenz, 25 February 1832: "I forward to your Highness herewith the sketches I received today from the young master artist [Bodmer]. All of these oil sketches were done by him here. They show your Highness approximately his artistic talent and how much can be expected from him for our purpose. The master assures me that he sketches as fast in oil as others do with pencil....I believe—if Your Highness considers his talents satisfactory—that nobody is as apt as the little Mr. Bodmer." Maximilian-Bodmer Collection, Margre H. Durham Center for Western Studies, Joslyn Art Museum, Omaha, Nebraska.

96 R. von Gerolt wrote to his brother, Carl, in Santa Clara, upper California, c/o Mrs. Johanna Pachecho. Letter from Linz, 1 May 1832. He reminds Carl that he met the Prince in Neuwied, and hopes he'll see Maximilian in California. Von Gerolt wrote a second letter to his brother, Friedrich, the Royal Prussian General–Consular Secretary to Mexico, also dated 1 May 1832 from Linz: "Dear brother Fritz, The bearer of this letter, Mr. Maximilian von Braunsberg from Neuwied, well known as a natural scientist and through his voyage to Brazil, is about to make a journey to North America, Florida, and California accompanied by landscape-painter Mr. Bothmer [*sic*], where he will pay a visit to our brother Carl, then return to Europe across Mexico. In your official position and by reason of your long stay in Mexico, it will be easy for you to make Mr. Braunsberg's sojourn in Mexico and in the country, very pleasant and instructive." Maximilian-Bodmer Collection, Margre H. Durham Center for Western Studies, Joslyn Art Museum, Omaha, Nebraska.

97 Maximilian, Prince of Wied, *Travels in the Interior of North America, by Maximilian, Prince of Wied. With numerous engravings on wood, and a large map,* trans. H. Evans Lloyd (London: Ackermann and Co., 1843), 20.

98 Noll, "Prince Maximilian's America," 45.

99 "On 29 July at nine o'clock, Mr. Krumbhaar paid me a visit and afterward was kind enough to introduce me to Mr. Vaughan (pronounced "Wohn" [in German]), a secretary of the local learned society." Maximilian, Prinz von Wied, *The North American Journals of Prince Maximilian of Wied*, ed. Marsha V. Gallagher and Stephen S. Witte, trans. William J. Orr, Paul Schach, and Dieter Karch (Norman: University of Oklahoma Press in cooperation with Durham Center for Western Studies, Joslyn Art Museum, 2008), 1:86.

100 The Bonaparte family had an interesting history in the United States. Joseph, whom Napoleon had installed as the king of Spain following his invasion of the Iberian Peninsula, had immigrated to the United States in 1815 after Napoleon was exiled to Elba; Joseph's son, Charles, joined him in 1823. Joseph built an estate in Bordentown, New Jersey, known for its extensive gardens, and he lived in the United States for seventeen years before moving back to Europe. Michael J. Brodhead, "The Work of Charles Lucien Bonaparte in America," *Proceedings of the American Philosophical Society* 122, no. 4 (18 August 1978): 198–203; Charlemagne Tower, "Joseph Bonaparte in Philadelphia and Bordentown," *Pennsylvania Magazine of History and Biography* 42, no. 4 (1918): 289–309. See also letter from Prince Charles de Marignano to his father, the Count de Survilliers (Joseph Bonaparte), Rome, 15 April 1832, recommending Prince Max to him. Maximilian-Bodmer Collection, Margre H. Durham Center for Western Studies, Joslyn Art Museum, Omaha, Nebraska.

101 Prince Maximilian to his brother, August, New-Harmony, on the river Wabash, started 20 November 1832: "I went to see a natural scientist, who was living here quite separated from the world, Mr. Thomas Say, who was known through his work on insects and conchology of his home-country, as well as his journey with major Long to the West and the Rocky Mountains. Although Mr. Say is living here quietly and separated from the world, nonetheless, he is corresponding with all the natural scientists of Europe and America; also Mr. Maclure, proprietor of half New Harmony and its surroundings.... For Mr. Say I had a letter from C. Bonaparte and the acquaintance of this well informed, modest and honorable man was very agreeable to me." Maximilian-Bodmer Collection, Margre H. Durham Center for Western Studies, Joslyn Art Museum, Omaha, Nebraska.

102 Prince Maximilian to his brother, August, New-Harmony, on the river Wabash, started 20 November 1832, Maximilian-Bodmer Collection, Margre H. Durham Center for Western Studies, Joslyn Art Museum, Omaha, Nebraska.

103 Lesueur had an international reputation, having accompanied French explorer Nicolas Baudin on a voyage to Australia and the South Seas between 1800 and 1803. Humboldt had written to Baudin from South America, hoping to join the voyage in progress, but Baudin had already sailed from France before Humboldt's letter arrived. As a result, Humboldt instead spent the following year in Mexico, followed by his six weeks in the United States, before returning home to Europe. Humboldt to Baudin, Cartagena, 12 April 1801, letter 128, in Humboldt, *Letters of Alexander von Humboldt to Varnhagen von Ense*, trans. Kapp (New York, 1860), 235–38.

104 Prince Maximilian's diary, 2 March 1832: "After seven o'clock we went to Mr. Say's. We talked, among other things, about Peale's Museum in Philadelphia. The collection was first established by the late elder Peale, who was a saddler and eventually an enthusiastic portrait painter. He did a self-portrait with his saddler's hammer at his feet. This, as well as numerous pictures of distinguished men from the time of the American Revolution [and] distinguished scholars including von Humboldt of Messrs. Say, Lesueur, etc., are included in this collection, hung in a row." Maximilian, Prinz von Wied, *North American Journals of Prince Maximilian of Wied*, 1:317.

105 Prince Maximilian to his brother, August, New-Harmony, on the river Wabash, started 20 November 1832: "During my illness, Messrs Say and Lesueur were my faithful daily visitors, the latter one brought us always some of his countless copybooks with his drawings, which he had made in all continents, part of them ethnographic, others of natural historic character, there were also numberless landscapes. Could I show, however, to you dear August and to our good designer of landscapes, Carl, the drawings of Mr. Bodmer, you would exclaim sometimes, oh! how wonderful, beautiful, beautiful! He has now 70 pages of sketches, among these also 6–7 from New Harmony." Maximilian-Bodmer Collection, Margre H. Durham Center for Western Studies, Joslyn Art Museum, Omaha, Nebraska.

106 Maximilian, Prince of Wied, *Travels in the Interior of North America*, 110–11.

107 Prince Maximilian and Bodmer's study of American Indians is the subject of chapter 6 in this volume.

108 Prince Max's journal, 18 July 1833: "I could not cease thinking of my voyages on Brazilian rivers. There, where nature is so magnificent luxuriant, grand, and abundant…where here, as a rule, everything is still, dead, and solitary….If one meets buffalo herds in the prairie, then it is indeed animated, but more often one finds it desolate and dead….One cannot call them luxuriant grasslands or savannas; they are dry plains alternating with ranges of hills—often vast regions of gentle hills, often (at least near the upper Missouri) high, strange, [and] bare clay hills. Luxuriant grass is found in only a few isolated spots, [yet] interesting plants [occur] in most places." Maximilian, Prinz von Wied, *North American Journals of Prince Maximilian of Wied*, 2:280–81.

109 On August 6 1833, Prince Max notes in his journal that he had read the Lewis and Clark description of this area, now called the Stone Walls. He comments that the sight was so remarkable that no one left the deck for the day, so they could all admire the scenery: "Here, on both sides of the river, the most strange forms are seen, and you may fancy that you see colonnades, small round pillars with large globes or a flat slab at the top, little towers, pulpits, organs with their pipes, old ruins, fortresses, castles, churches, with pointed towers, &c., &c., almost every mountain bearing on its summit some similar structure….a painter who had leisure might fill whole volumes with these original landscapes…which Mr. Bodmer has very accurately represented….All these eminences are inhabited by numerous troops of the wild mountain sheep, of which we often saw thirty or fifty at a time climbing and springing over the sand-stone formations." Maximilian, Prince of Wied, *Travels in the Interior of North America*, 222, 236–37.

110 Bodmer to Prince Max, 22 January 1839; quoted in Ron Tyler, "Karl Bodmer and the American West," in Ruud, *Karl Bodmer's North American Prints*, 19.

111 New York *Commercial Advertiser*, 22 November 1843; quoted in Ron Tyler, "Karl Bodmer and the American West," *Studii și cercetări de istoria artei, Artă Plastică*, n.s., 5, no. 49 (2015): 25.

112 Ron Tyler goes into great detail regarding the dissemination of Bodmer's images and the costs associated with Max and Bodmer's published two-volume narrative and the pictorial atlas in "Karl Bodmer and the American West," *Studii și cercetări de istoria artei, Artă Plastică*, esp. 24–28.

113 Paul Wilhelm, Duke of Württemberg, *Travels in North America, 1822–1824*, ed. Savoie Lottinville, trans. W. Robert Nitske (Norman: University of Oklahoma Press, 1973), 128, 201, 203, 396.

114 Paul Wilhelm, Duke of Württemberg, *Travels in North America, 1822–1824*, 36–37; quoted in Walls, *Passage to Cosmos*, 131.

115 Humboldt to Secretary of War Jefferson Davis, Berlin, 24 March 1857, in Schwarz, *Alexander von Humboldt und die Vereinigten Staaten von Amerika*, 417–18, letter 273.

116 Alexander von Humboldt preface to *Diary of a Journey from the Mississippi to the Coasts of the Pacific, with a United States Government Expedition*, by Balduin Möllhausen, trans. Mrs. Percy Sinnett (London: Longman, Brown, Green, Longmans, and Roberts, 1858), 1:xi–xxv.

117 Adalbert, Prinz von Preussen, *Travels in the South of Europe and in Brazil: With a Voyage up the Amazon by His Royal Highness Prince Adalbert of Prussia*, trans. Sir Richard H. Schomburgh and J. E. Taylor (London: David Bogue, 1849), 1:vi; cited in Joseph Ewan, "Humboldt and American Botany," *Rhodora* 58, no. 691 (July 1956): 196.

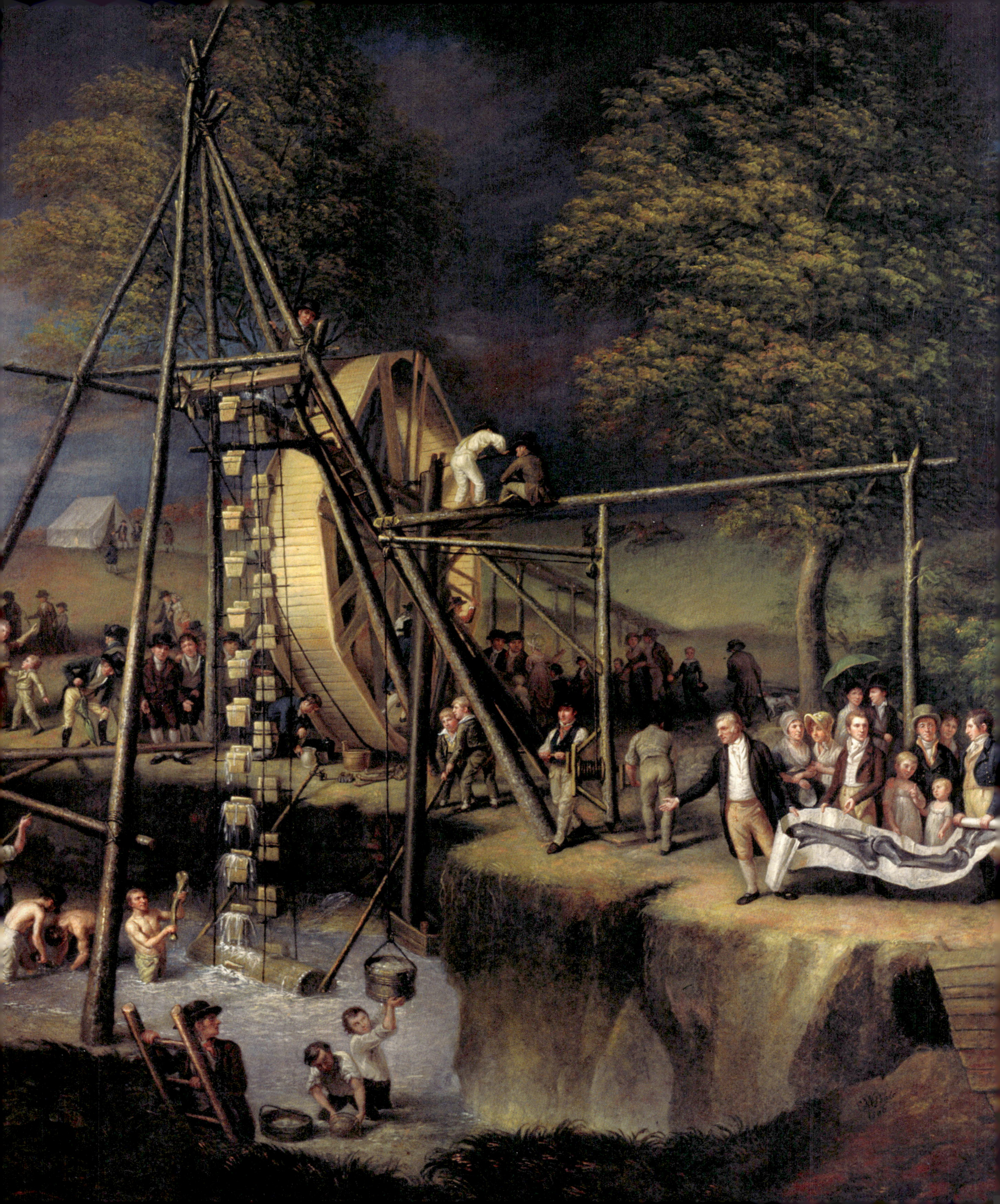

"Nature is the domain of liberty"

The American landscape has served as a vehicle for the cultural identity of the United States since its founding, finding its most powerful expression in the paintings of the Hudson River school and the writings of the transcendentalists during the middle decades of the nineteenth century. Admiration of natural landmarks anchored early travel within the original colonies, with places like Natural Bridge in Virginia and Niagara Falls in New York topping the list of destinations. In the first decades following the Revolutionary War, cultural leaders in the United States sought to define the characteristics of a new American nation. While Americans traveled to Europe to see castles and cathedrals, products of centuries of civilization, their ventures within the United States continued to focus on natural landmarks rather than the built environment. A series of creation myths, many of them derived from or encouraged by Humboldt's work, forged the country's affirmation of nature as the wellspring of its culture. What developed was a full-blown wilderness aesthetic that linked America's cultural ambitions to its inspirational landscape features and monumental natural history specimens. Pride in what was already here—rather than what man built upon it—formed the core of America's cultural identity.

According to the leading European natural philosophers, what was missing in America was centuries of accrued architecture, literature, and art, the standard by which they judged the sophistication of a civilization. Europe had identified the self-proclaimed Seven Wonders of the World, all products of advanced civilizations of the ancient Mediterranean, all of which were man-made features. These impressive and astonishing sights,

Charles Willson Peale, *Exhumation of the Mastodon* (detail), ca. 1806–1808. See p. 145.

which included the Great Pyramid of Giza, the Colossus of Rhodes, and the Statue of Zeus at Olympia, reinforced the idea that Western civilization had developed from this rich foundation. By these standards, it seemed the Americas lacked cultural significance. European disdain for what were widely perceived as relatively crude conditions on both American continents fostered cultural insecurity within the newly established United States, calling for a robust response. Thomas Jefferson had published his *Notes on the State of Virginia* in 1785 as a rebuttal to criticism of the paucity of impressive features and attributes of the former British colonies. It was an early salvo in the larger arena of establishing and defending America's place in global affairs.

The debate at the center of all of this was fomented by the French naturalist Georges-Louis Leclerc, comte de Buffon, and the Dutch naturalist Cornelius de Pauw, both of whom asserted the supremacy of European intellect and all forms of life on the European continent. Buffon, who was *intendant* of the Jardin des Plantes and keeper of the Royal Cabinet of Natural History in Paris, was the highly esteemed author of the *Histoire Naturelle, générale et particulière, avec la description du Cabinet du Roi*, an encyclopedic collection of thirty-six large volumes written between 1749 and 1788, which was possibly the most widely read series of books in eighteenth-century France.[1] However, for all of the impressive wealth of accurate knowledge Buffon amassed, he had published the erroneous assertion that every species found in the American hemisphere was a smaller, weaker, or more degenerate version of its European counterpart. His line of thinking was adapted from de Pauw's 1768 declaration:

> *The discovery of America is the most remarkable of all events witnessed by man. No matter how far we look back in human history, nothing compares to this event; & it is, without a doubt, a sublime & frightening scene to behold the fact that one half of our planet is so disgraced by nature that everything there is either degenerate or monstrous.*[2]

De Pauw believed that even modern Europeans who moved to the Americas withered in stature, intellect, and strength as a result of exposure to poor climate and soil.

Another prominent French cleric, Abbé Guillaume-Thomas Raynal, repeated the same arguments frequently and at great length.[3] Raynal dismissed more than the physical stature of Americans; he condemned their intellect as well, earning Jefferson's undying enmity with the words "one should not be surprised that America has yet to produce a good poet, a clever mathematician, a genius in even one art or science."[4] This trio held sway in the court of public opinion in Europe, despite none of them ever having set foot in the Americas. When de Pauw was asked to contribute a nineteen-page essay on America to the addendum to Denis Diderot's *Encyclopédie*, published in 1776, it cemented his stature as the leading authority on the New World.[5] The symbolism of that date was not lost on the former colonials. The success of the American Revolution threw into doubt the logic behind such pronouncements, couched as they were in the assurance that civilization could not flourish in the Americas, and the American experiment was, by extension, doomed to fail.[6]

This was the thinking that fueled Jefferson to compose his *Notes on the State of Virginia*. He wrote his only book in direct response to a letter sent to the Continental Congress in 1780 by François, Marquis de Barbé-Marbois, then secretary of the French delegation in Philadelphia.[7] His twenty-two queries addressed the points Buffon and his peers had made denigrating the flora, fauna, and human inhabitants of the United States. Marbois hoped to receive answers from all thirteen of the American states. The Virginia delegation handed off its copy of the queries to Jefferson. This was a wise move. Most of the Founders fumed at the trio of European savants' dismissal of the Americas, but only Jefferson could be described as possessing a scientific bent. "Nature," Jefferson later wrote to Pierre-Samuel Dupont in Paris, "intended me for the tranquil pursuits of science, rendering them my supreme delight."[8] As he amassed his data and built his story, Jefferson wrote a rational yet impassioned manifesto for the robust character of the American land, its governance, and its inhabitants.

As word circulated about the contents of Jefferson's manuscript, he was pressed to publish it. In 1785 the first edition of *Notes on the State of Virginia* appeared in French, while Jefferson was living in Paris; the authorized American edition appeared three years later. The published volume deftly presents Jefferson's home state as a paradigm for the entire country. Jefferson's narrative and the accompanying tables of data took on Buffon directly, referring to him by name no fewer than fifty times throughout the volume. It was a gauntlet thrown down that challenged European preconceptions with American empirical data.[9] This slim volume made a much stronger case for the superior size of American mammals, at the top of which was the mammoth; the superiority of the American political system, grounded in democracy; and the geological significance of impressive landscape features, like Natural Bridge.

By itself, Jefferson's text stands as a pivotal pronouncement. He described the impressive features in the landscape, calling out Natural Bridge in particular as the embodiment of the sublime. He pointed to the massive scale of the moose and the mammoth as proof of the robust indigenous life-forms. Jefferson enumerated the physical, geographical, and political structures that formed the solid foundation on which American democracy would be built, and in so doing he made the case for using these attributes as a measure of the scale and scope of America's cultural ambitions. Those features became a critical component of American literature and art, both of which presented the American landscape as the locus for the nation's cultural identity.

Twenty years after it was published, *Notes on the State of Virginia* enjoyed renewed significance when Alexander von Humboldt invoked it in his letter addressed to its author. Humboldt had read Jefferson's *Notes* while he was a student in Hamburg. His teacher, Daniel Christoph Ebeling, had helped instill in Humboldt an appreciation for American democracy and the American landscape. In a letter to American geographer Jedidiah Morse, Ebeling observed how little most Europeans knew of the United States and hoped in his own work to create "a faithfull [*sic*] picture of a truly free Republic founded upon the most solid foundations," conveying "such happiness as you enjoy."[10] Ebeling and Humboldt both looked to the New World as a potential exemplar for building a just society. Each read in the features of the American landscape a prospect for a bright future, contrasting its peaceful democracy with European monarchies' near-constant warring. Those false accounts of the flora, fauna, and physical geography of the Americas, promulgated by European naturalists, also required correcting. Humboldt's recently concluded travels through South America and Mexico had confirmed for him the extent of the ridiculous European misapprehensions about every aspect of life in the Americas. He now had his notes, his specimens, and his own ideas to publish in refutation of Buffon, Raynal, and de Pauw. Humboldt's introductory letter to Jefferson was an invitation for the two men to create a more united front in defense of the Americas.[11]

Alexander von Humboldt's letter of May 24, 1804, introducing himself to Thomas Jefferson signaled that he was prepared to engage in the European debate over the validity of an American civilization. Jefferson's focus in responding to Marbois's queries had been geopolitical. Humboldt brought new ammunition to the fight. As he traveled across South America, Humboldt had reflected on natural features as the New World equivalents to those architectural wonders elsewhere, writing with satisfaction, "Whilst contemplating these grand scenes, we feel little regret that the solitudes of the New World are not embellished with the monuments of antiquity."[12] In his assertion that not all monuments were man-made, he provided an intellectual armature for the appreciation of a different kind of monument, one that celebrated nature rather than culture.

Central to Humboldt's writings is an abiding love of nature. As he traveled, Humboldt developed an acute eye for noticing objects of scientific interest wherever he was; he also cultivated an aesthetic appreciation of his surroundings. His narrative style in his published writings conveyed his exhilaration in traveling far from civilization, as he described natural monuments sculpted by geological processes and the weather. For Humboldt, everything he encountered fit into an emergent understanding of planetary interconnectedness. The French writer François-René de Chateaubriand wrote admiringly of his friend that Humboldt

paints scenes of American nature with a striking authenticity. You believe you are surfing the waves with him, losing yourself with him in the depths of the woods which stretch out to the Ocean's seashores and reach the chain of the Cordillera;

he makes you see the great deserts in all shades of light and dark, and his descriptions insinuating ideas of a higher order, always reminisce on man or reflect on life.[13]

That connection between man and nature would prove to be one of Humboldt's enduring appeals in the United States and would contribute to the development of a wilderness aesthetic for a nation long on natural wonders and short on man-made monuments. That wilderness aesthetic in the United States would expand to form a kind of cultural nationalism.

THE AMERICAN INCOGNITUM: THE NINTH WONDER OF THE WORLD

In his letter Humboldt had emphasized two things that he knew would pique Jefferson's interest: first, that he had read the president's *Notes on the State of Virginia*, and second, that he shared his interest in mammoth bones.

I should like to talk to you more about a subject that you so astutely dealt with in your work on Virginia, some mammoth teeth that we discovered in the Andes of the Southern Hemisphere at 1,700 toises [10,523 feet] above the level of the Pacific Ocean.[14]

Humboldt did not include this comment solely to encourage a meeting between the two men; he used it to send a much more significant message—that he understood the political reasons behind the writing of his book and why Jefferson was so interested specifically in mammoths. In a single sentence Humboldt sketched an agenda that linked the two men in their mutual endeavors to defend American civilizations against European claims of its inherent degeneracy. Humboldt was offering to advocate on behalf of the New World as it was being disparaged by some of the leading scientific voices in Europe.

The mammoth was an excellent place to start. In *Notes*, Jefferson claimed with pride that the mammoth, also known as the "American incognitum," was "the largest of all terrestrial beings."[15] His description was targeted at Buffon's criticisms. Having described the geographic and temperature range in which the North American bones had been found—a climate that Buffon and de Pauw had deemed incapable of fostering an advanced civilization—Jefferson made the point that this creature far exceeded the size of any known species of elephant, its closest plausible relative. Building his argument, Jefferson started by stating, "The skeleton of the mammoth (for so the incognitum has been called) bespeaks an animal of five or six times the cubic volume of the elephant, as Mons. de Buffon has admitted."[16] Its teeth were similarly five times as large and configured differently than those of an elephant, suggesting it was possible this was a carnivorous beast, adding a hypothetical element of ferocity to its impressive physical stature. For this conclusion, Jefferson drew on the observations in an article published by the Royal Society in 1769, in which the British physician and anatomist William Hunter declared that the American incognitum was "some carnivorous animal, larger than an ordinary elephant."[17] That desire for something other than a placid herbivore would resurface several times during the ensuing decades.

The mammoth—universally recognized at this moment as the largest vertebrate land mammal ever found—was an American feature, its scale a direct rebuttal of Buffon's theory of American degeneracy. Individual bones had been excavated in North America as early as 1739, many of them from a site known as Big Bone Lick, along the Ohio River in modern-day Kentucky, which was then western Virginia.[18] Fragments of mammoth teeth had also been found in upstate New York, and during the American Revolution George Washington had been presented with one such fragment found by a soldier. During the months following the British surrender at Yorktown in 1781, when active fighting had ceased and the two armies remained encamped in the field, Washington was approached by Christian Friedrich Michaelis, a Hessian surgeon who had been serving in the British Army. Michaelis requested and received permission from the American general to dig for fossil bones, but bad weather thwarted his aims.

Charles Willson Peale, who was serving under Washington, knew that in 1766 Dr. John Morgan, a member of the American Philosophical Society (APS), had acquired a group of bones. Morgan was not willing to part with this collection, but Peale agreed to make a series of drawings of each bone for Michaelis, who took them home and gave them to his father, a professor at Göttingen.[19] Humboldt's mentor Johann Friedrich Blumenbach was aware of these drawings and made use of them as he developed his ideas concerning comparative anatomy, making it highly likely that Humboldt studied these drawings in 1789 as part of the curriculum, shortly after he had read Jefferson's *Notes*. Subsequently, in Paris Humboldt met Georges Cuvier, a young, brash, and ambitious naturalist also at work on comparative anatomy. He had access to the few mammoth bones that the Institut National had acquired from Siberia and North America. Cuvier was studying these fossils and contemporary elephant bones in an effort to identify exactly to what species they belonged. When Humboldt found mammoth teeth of his own in the Andes, he knew this discovery would add significantly to Cuvier's ability to understand this creature and that Jefferson would be equally interested in the results.

From the moment Peale first studied and drew Morgan's bones for Michaelis, he wanted a complete skeleton for display as an emblem of America. Peale had founded his museum, which he first opened to the public in 1786, on mammoth bones.[20] In 1799 Peale's friend and fellow APS member Nicholas Collin issued a directive to acquire and display a complete set of bones as proof of the mammoth's existence and its significance to science. Within a year, word spread that a farmer near Newburgh, New York, had come across large bones while draining a marl pit on his property. Peale immediately made arrangements to purchase the rights to excavate as many bones as possible, including securing a loan from the APS for equipment. Peale amassed a nearly complete skeleton from John Masten's pit and the better part of a second skeleton from the nearby Barber farm. Peale mounted the first skeleton (CAT. 26) and placed it on view in a designated "Mammoth Room" adjacent to his museum in Philadelphia.

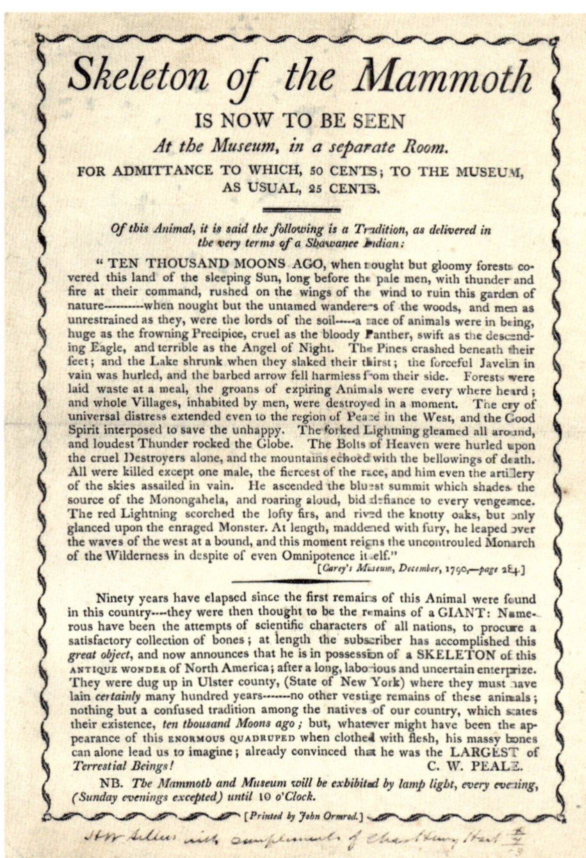

Skeleton of the Mammoth

IS NOW TO BE SEEN

At the Museum, in a separate Room.

FOR ADMITTANCE TO WHICH, 50 CENTS; TO THE MUSEUM, AS USUAL, 25 CENTS.

Of this Animal, it is said the following is a Tradition, as delivered in the very terms of a Shawanee Indian:

"TEN THOUSAND MOONS AGO, when nought but gloomy forests covered this land of the sleeping Sun, long before the pale men, with thunder and fire at their command, rushed on the wings of the wind to ruin this garden of nature---------when nought but the untamed wanderers of the woods, and men as unrestrained as they, were the lords of the soil-----a race of animals were in being, huge as the frowning Precipice, cruel as the bloody Panther, swift as the descending Eagle, and terrible as the Angel of Night. The Pines crashed beneath their feet; and the Lake shrunk when they slaked their thirst; the forceful Javelin in vain was hurled, and the barbed arrow fell harmless from their side. Forests were laid waste at a meal, the groans of expiring Animals were every where heard; and whole Villages, inhabited by men, were destroyed in a moment. The cry of universal distress extended even to the region of Peace in the West, and the Good Spirit interposed to save the unhappy. The forked Lightning gleamed all around, and loudest Thunder rocked the Globe. The Bolts of Heaven were hurled upon the cruel Destroyers alone, and the mountains echoed with the bellowings of death. All were killed except one male, the fiercest of the race, and him even the artillery of the skies assailed in vain. He ascended the bluest summit which shades the source of the Monongahela, and roaring aloud, bid defiance to every vengeance. The red Lightning scorched the lofty firs, and rived the knotty oaks, but only glanced upon the enraged Monster. At length, maddened with fury, he leaped over the waves of the west at a bound, and this moment reigns the uncontrouled Monarch of the Wilderness in despite of even Omnipotence itself."

[*Carey's Museum, December*, 1790,—*page* 284.]

Ninety years have elapsed since the first remains of this Animal were found in this country----they were then thought to be the remains of a GIANT: Numerous have been the attempts of scientific characters of all nations, to procure a satisfactory collection of bones; at length the subscriber has accomplished this *great object*, and now announces that he is in possession of a SKELETON of this ANTIQUE WONDER of North America; after a long, laborious and uncertain enterprize. They were dug up in Ulster county, (State of New York) where they must have lain *certainly* many hundred years-------no other vestige remains of these animals; nothing but a confused tradition among the natives of our country, which states their existence, *ten thousand Moons ago*; but, whatever might have been the appearance of this ENORMOUS QUADRUPED when clothed with flesh, his massy bones can alone lead us to imagine; already convinced that he was the LARGEST of *Terrestial Beings!*

C. W. PEALE.

NB. *The Mammoth and Museum will be exhibited by lamp light, every evening, (Sunday evenings excepted) until* 10 *o'Clock.*

[*Printed by John Ormrod.*]

Fig. 31
Charles Willson Peale, Broadside advertising the "Skeleton of the Mammoth," 1802, printer's ink on paper, 11 ¾ × 9 in., American Philosophical Society

The mammoth became the first emblem of North American identity. Its unprecedented scale set the tone for American ambitions of equivalent significance. As the American incognitum came into focus as the mammoth, that identifying noun also became a powerful adjective in the service of a national metaphor. On Christmas Eve 1801 Peale opened the Mammoth Room to the public (fig. 31), ushering in 1802 as the year of the mammoth in the United States. The Philadelphia *Gazette of the United States* heralded its debut as an "American Miracle."[21] Newspapers chronicled the creation of a mammoth cheese, followed by a sequence of mammoth-designated foodstuffs designed to celebrate the discovery.[22] Jefferson himself was deemed "the mammoth chief," although that sobriquet cut both ways.[23] The mammoth proved popular with museum visitors, attracted in part by Peale's promotional efforts. He advertised the mammoth skeleton using Jefferson's description of it as that of "the LARGEST of Terrestrial Beings," and "a SKELETON of this ANTIQUE WONDER of North America." During a parade through the streets of Philadelphia, the behemoth was described as "The 9th Wonder of the World."[24] He and Jefferson were eager to find New World equivalents to classical antiquities venerated in Europe. In this case, the antiquity to which he referred was the age of the fossil bones, the skeleton standing in for the closest thing to an architectural wonder the United States could provide.

The behemoth's excavation was a seminal moment in Peale's life and the nation's history. Peale had always considered his museum as part

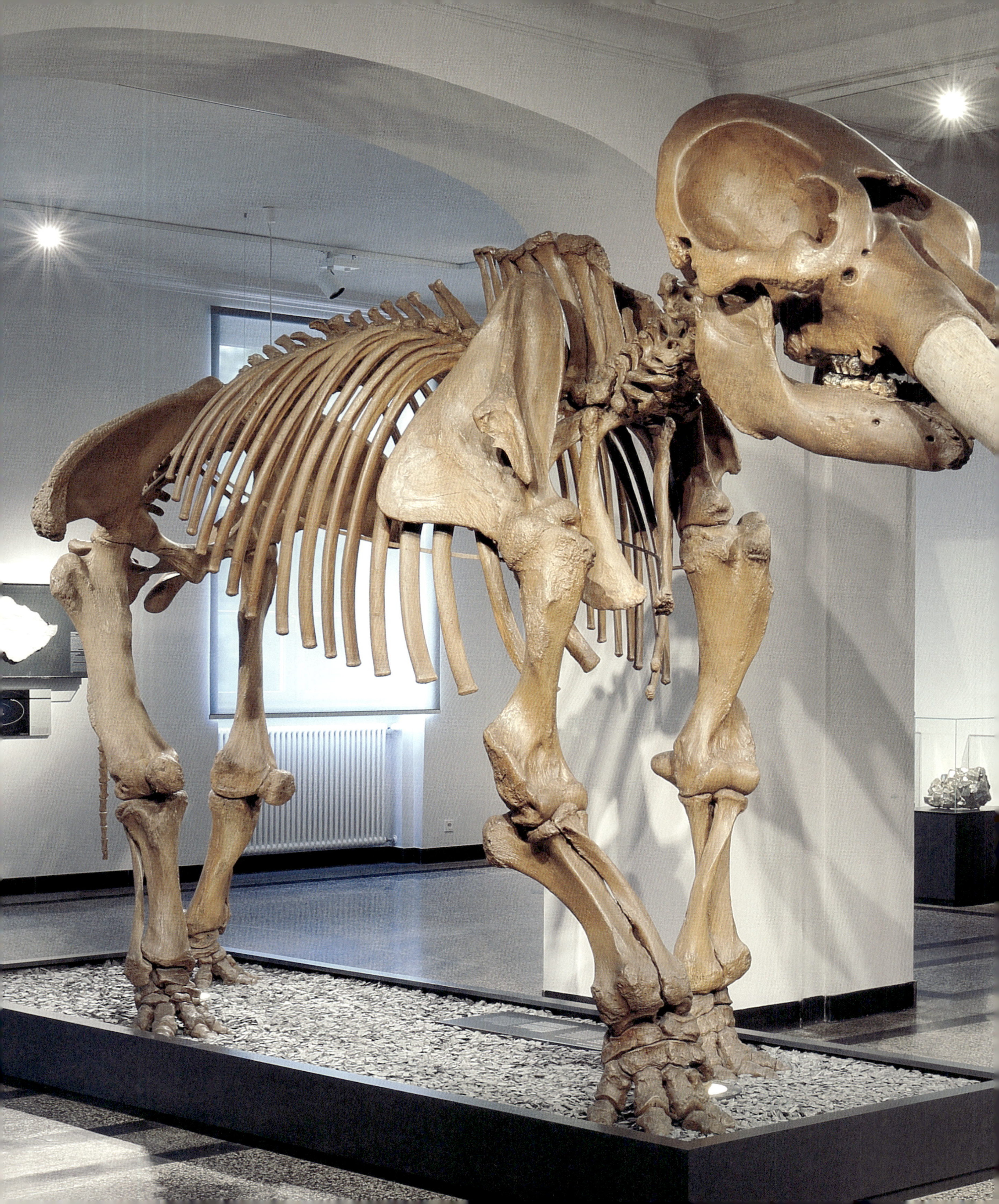

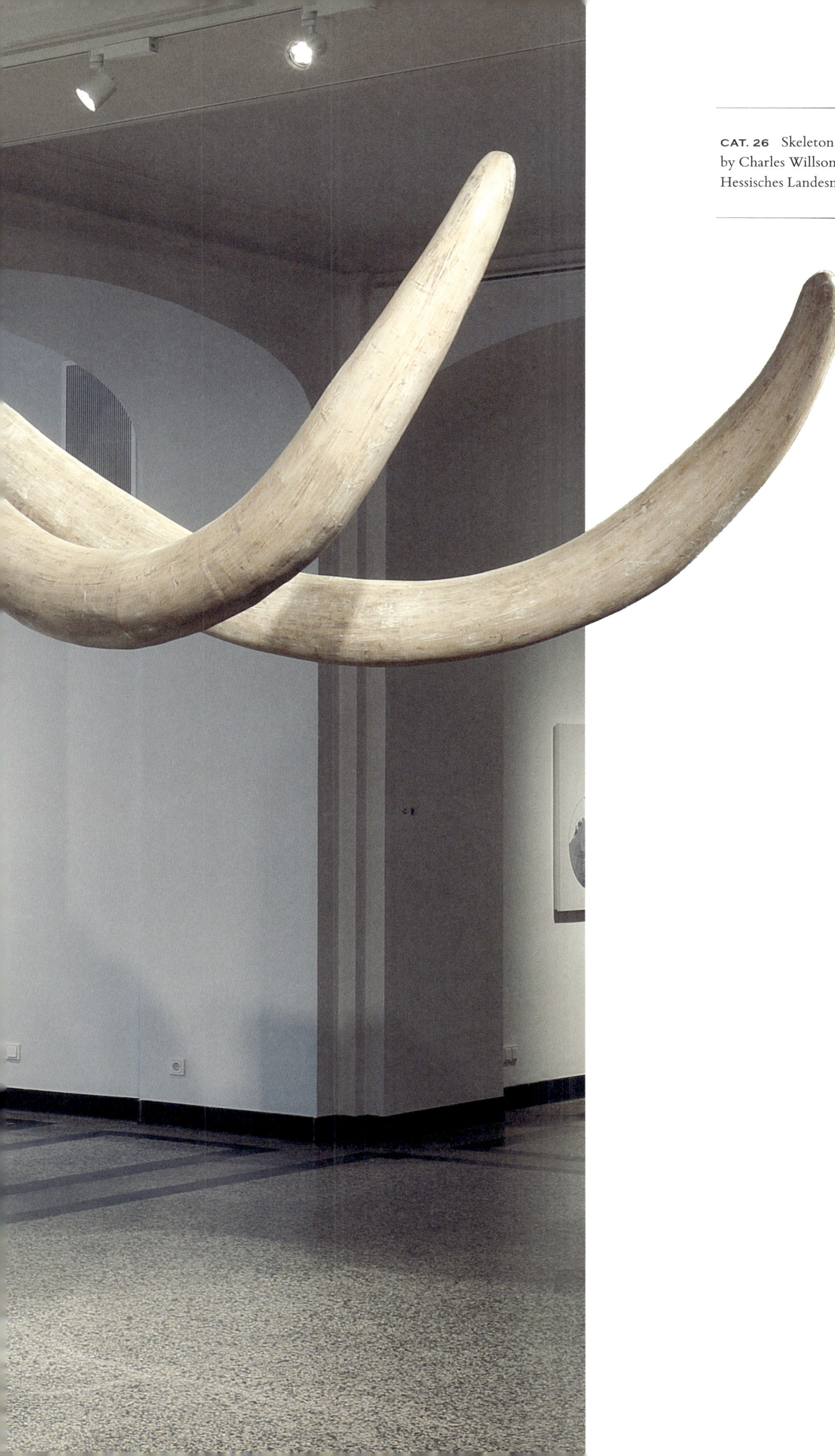

CAT. 26 Skeleton of the Mastodon, excavated 1801–2 by Charles Willson Peale, approx. 118 × 177 × 65 in., Hessisches Landesmuseum, Darmstadt, Germany

CAT. 27 Rembrandt Peale, *Sketch of the Elephant (Mastodon)*, ca. 1801, pencil on paper, 15 5/16 × 12 3/4 in., American Philosophical Society

of the defense against European assumptions about the lesser stature of the Americas. He envisioned displays that provided "a comparative view of each genus...of the old and new world,"[25] designed in part to be a visual affirmation of American significance, using nature as his proof. When Humboldt arrived in Philadelphia in 1804, he visited Peale's Museum and saw the first complete, articulated skeleton of Jefferson's incognitum. In Humboldt, Peale had an authority figure who could appreciate that comparative approach as part of Peale's underlying educational mission. With the recent addition of the mammoth display, Peale's Museum presented the scale as well as the span of American natural history, providing lasting visual form to this facet of America's developing cultural identity.

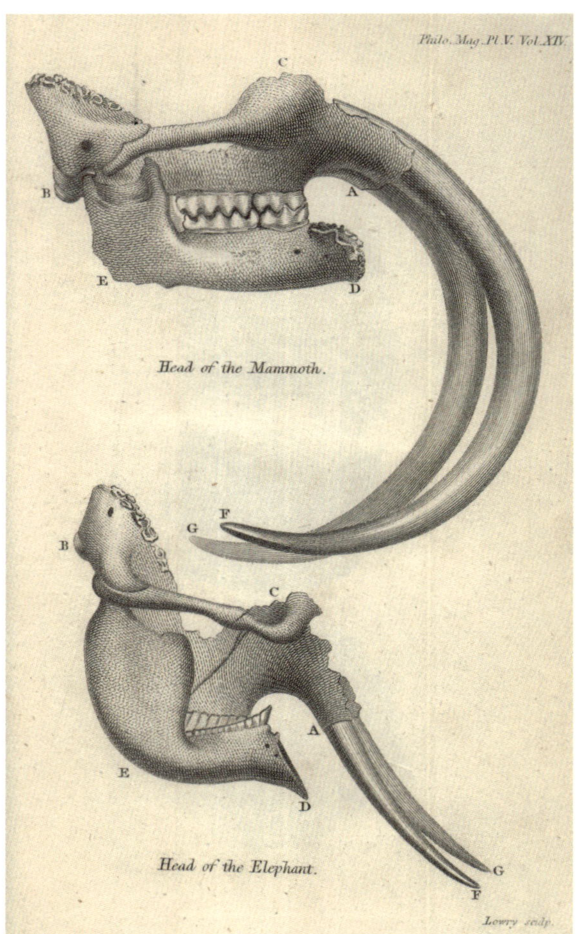

Fig. 32
Rembrandt Peale, *Mammoth*, from *Philosophical Magazine, London*, series 1, vol. 14, 1802, engraving, approx. 6 × 9 in., Penn State University Libraries

Shortly after Peale unveiled the mammoth in Philadelphia, his sons, Rembrandt and Rubens, took the second skeleton on tour through parts of Europe.[26] Their goal was twofold: first, to introduce the American incognitum to Europe directly, a concrete rebuttal to Buffon and his ilk; second, to generate revenue for the museum back home. Rembrandt wrote and printed a ninety-three-page *Historical Disquisition on the Mammoth*, an extended broadside he sold to visitors that served as a dramatic telling of an era when mammoths roamed the North American continent. This skeleton was not as complete as the one on view in Philadelphia, and for both skeletons Rembrandt took charge of carving the missing bones from wood, with help from Peale's free black assistant, Moses Williams, and Philadelphia sculptor William Rush.[27]

Bilateral symmetry allowed the men to carve mirror images of extant bones as guides for what was missing, but there was a lingering issue concerning the orientation of the tusks. Due to the weight of the tusks and the soft earth into which they originally sank, no one knew exactly how those massive tusks fit into the skull. Rembrandt Peale had sketched the skeleton of an elephant in 1801 (**CAT. 27**), perhaps to be used as a guide to the assembly of the mammoth bones after they emerged from Masten's pit. His drawing is a faithful reproduction of an engraving published in 1734, in which the tusks curve upward.[28] It made sense, given the close correlation of the rest of the bones to those of the elephant, that the orientation of the mammoth's tusks would be the same.

Charles Willson Peale had harbored a hope, along with Jefferson, that the mammoth was a carnivorous beast, based on the dentition patterns on the molars; but in Europe Rembrandt became enamored of the idea that, if the beast truly had been carnivorous, its tusks might have swept the ground like those of a saber-toothed cat (fig. 32). He published his theory while he was in London, comparing the skull and tusks of a modern elephant with those of the traveling Peale skeleton. His father expressed his concern and discomfiture upon hearing of this modification, writing to his sons, "I have seen your sketch of the comparission of the head of the Elephant and Mammoth in the Philosophical Magazine, and I am anxious to know your particular reasons for changing the tusks of

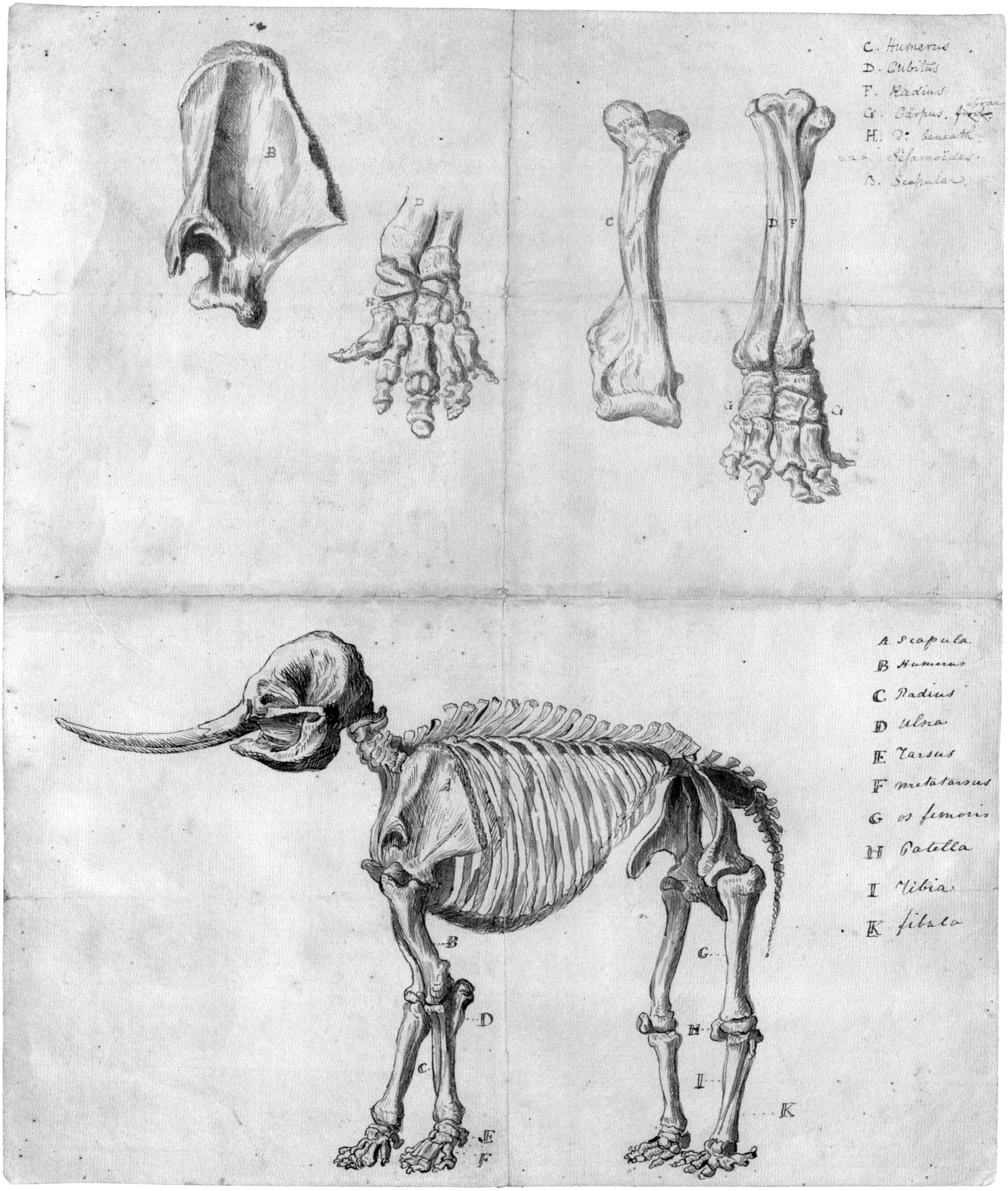

C. Humerus
D. Cubitus
F. Radius
G. Carpus, above
H. D: beneath
Sesamoides
B. Scapula
A Scapula
B Humerus
C Radius
D Ulna
E Tarsus
F metatarsus
G os femoris
H Patella
I Tibia
K fibula

the Mammoth."[29] The elder Peale was shocked seeing his son's bold theory in print. Their disagreement carries with it reverberations of the larger shifts taking place in the political climate of the United States. Despite having fought in a revolution to overthrow the British Empire, Charles Willson Peale adopted a more conservative approach to challenging authority in his museum. His sons proved far more daring in their presentation of untested ideas, very much in keeping with the ethos of Jacksonian America. When they took over the daily operations of the family museum, Peale's sons reoriented the tusks to sweep downward, where they remained as late as 1815.[30]

In 1806 Georges Cuvier announced the results of his research, after comparing the bones of the mammoth from three different continents with those of modern elephants. Using North American bones supplied by Jefferson, South American bones from Humboldt, and a group of Siberian bones already at the Institut, Cuvier now recognized two distinct species, the mammoth and the mastodon.[31] What Peale and Jefferson knew as the mammoth was now classified as a mastodon. Peale wrote to Jefferson:

In the report by Mr. Cuvier on the fossil bones which you presented to the National Institute I find the committee have given the name Mastodonte to the animal which we commonly call Mammoth. . . . I would willingly change the name of Mammoth if you think the Name which Mr. Cuvier has givin is appropriate. Therefore I would be much obleged by your opinion previous to a christening. And which (the thought of a moment) may be done in stile, by giving a dinner to 13 men seated within the Thorax.[32]

Before his sons had packed up the second skeleton to take on tour in Europe, Rembrandt had hosted a farewell dinner beneath it, which was covered in the Philadelphia press.[33] Now, faced with the prospect of renaming his beloved mammoth as a mastodon, Peale suggested to Jefferson that it might be appropriate to celebrate the rechristening with another dinner under the fossil's chest.

Peale's reluctance to adopt Cuvier's new scientific nomenclature without Jefferson's blessing provides insight into the metaphorical power of the term "mammoth." As much as Peale was a man of science, he was also enamored of the way his museum exhibits told a story of robust, monumental American ambition and potential. Peale had orchestrated his Mammoth Room as an exercise in metaphorical thinking. The Mammoth Room itself was dominated by the skeleton, which measured close to fifteen feet in height and sixteen feet in length. At its foot Peale had installed the diminutive skeleton of a mouse, while his human visitors provided the middle scale. His son Rembrandt's ninety-three-page essay was framed and hung, sheet by sheet, around three sides of the room, providing a narrative for visitors as they circled the behemoth. Now Peale had to contemplate letting go of that metaphor in favor of accurate scientific nomenclature. The power of that metaphor and the implications for its symbolic cultural weight were hard to relinquish.

Peale never did rename the Mammoth Room, although he did eventually embrace the designation "mastodon" elsewhere. In 1804 he was inspired by Humboldt's visit to take up painting again, years after he had effectively retired from the profession. After painting Humboldt's portrait in the days prior to his departure for Paris, he turned to a much more ambitious work that would be simultaneously a history painting, a landscape, and a creation myth for a powerful United States. Within a month of Humboldt's departure, Peale wrote to Jefferson, "I have in view the execution of one large historical picture, which perhaps may be my last work in that line."[34] For the next two years Peale played with the composition of what would simply be known as "The Mammoth Picture," for both its subject and its scale. Peale began with a small sketch that he sent via his son, Rembrandt, to Benjamin West in London. He later wrote to West, describing his "historical picture" that told the story of the exhumation.

Having the desire to represent the scene of getting up the Bones of the Mammoth, it being a very interesting article of the Museum, last summer I undertook a picture larger than one of which Rembrandt showed you the beginning when he was in London. This Picture contains a great number of figures, for while I was engaged in that labour, my exertions excited the admiration of all the people for a considerable distance round that country.[35]

Peale's painting, now titled *Exhumation of the Mastodon* (CAT. 28), encapsulates the process by which he was able to retrieve the skeleton from Masten's waterlogged pit.[36] His ingenious pulley

CAT. 28 Charles Willson Peale, *Exhumation of the Mastodon*, ca. 1806–8, oil on canvas, 49 × 61 ½ in., Maryland Historical Society, Baltimore City Life Museum Collection, Gift of Bertha White in memory of her husband, Harry White

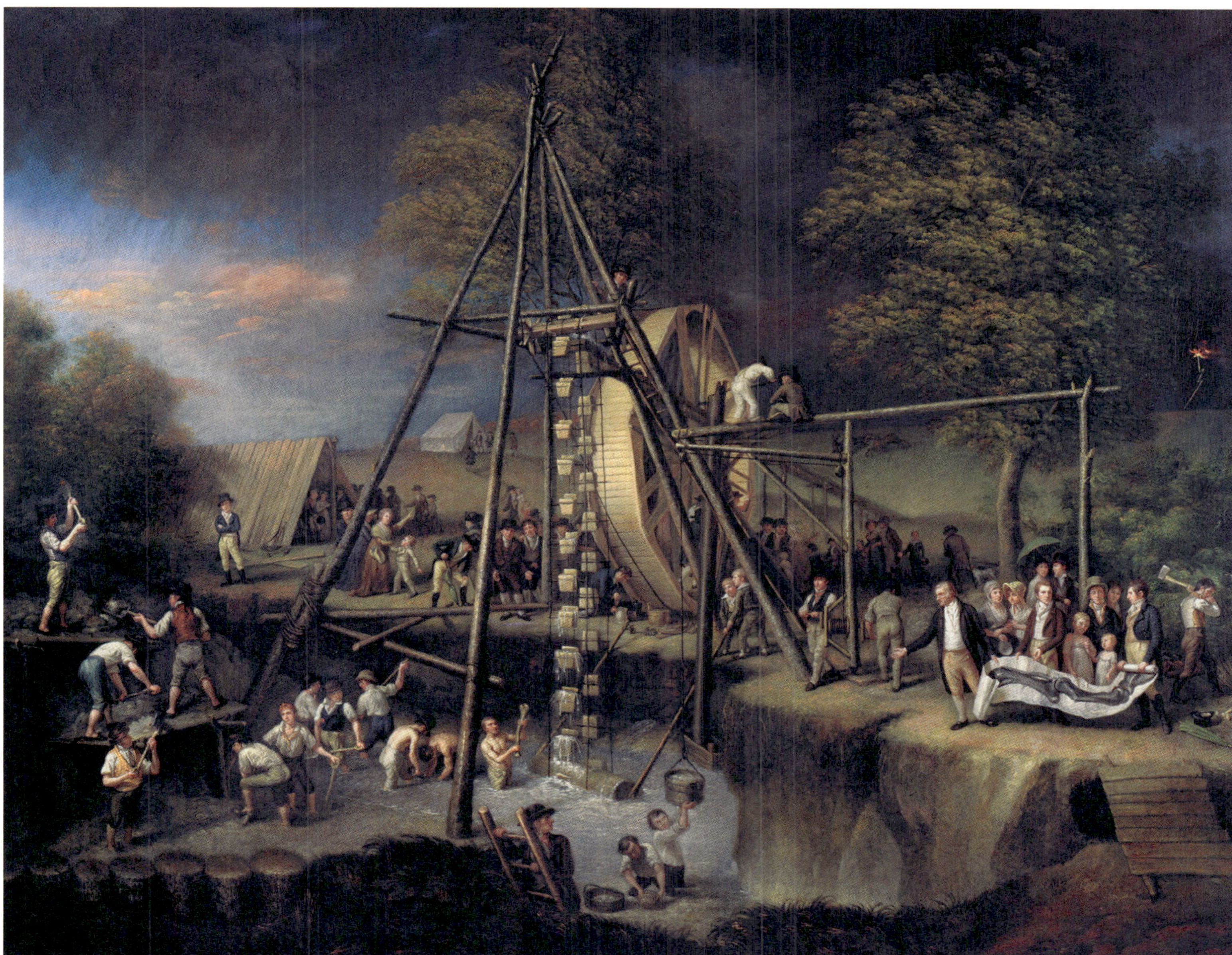

CAT. 29 Charles Willson Peale, *The Artist in His Museum*, 1822, oil on canvas, 103 ¾ × 79 ⅞ in., Pennsylvania Academy of Fine Arts, Philadelphia, Gift of Mrs. Sarah Harrison (The Joseph Harrison Jr. Collection)

system allowed his team of workers to lower the water level to unearth and retrieve the individual bones.[37] The crowd of spectators included fifteen members of the artist's family, making it clear Peale saw this event, and the entire museum apparatus, as a family affair.[38] Peale stands close to the edge of the pit. His sons, Rembrandt and Raphaelle, help him hold up the life-size drawing of the creature's leg bones, including the femur, to aid in the identification of each bone as it emerged from the pit. The massive scale of the waterwheel and pulley system on the far bank echoes that of the skeleton being lifted out of the primordial ooze. The painting's action alludes to a metaphoric Great Chain of Being, as the buckets rise in sequence on the right bank, and the workmen form a human chain while emerging from the pit or tending the waterwheel.

Despite the emphasis on the accuracy of the armature designed to facilitate the excavation, and Peale's fidelity to the incipient stormy weather he recorded in his diary, the artist took significant liberties in composing and staffing his painting. Although Dr. James Woodhouse had accompanied Peale to the site, it is Peale's close friend the ornithologist Alexander Wilson, arms folded, who is depicted watching the activity. Wilson's appearance here is that of a much-satisfied supporter, perhaps representing the Philadelphia scientific community's vested interest in the excavation.[39] Even more curious is the inclusion of Peale's deceased second wife, along with his current, very much alive third wife, lending the entire assembly the feel of a séance, to which all of the observers and participants have been drawn as they gather near the pit. In titling the work an "exhumation" rather than an "excavation," Peale extended that sense of yearning to reach beyond the grave. Within the painting, Peale himself leans in toward the pit, gesturing toward the workmen as though by sheer force of will he can conjure the bones to appear.

The recognizable thigh bone measured almost four feet in length, prompting the writer for the New York *Mercantile Advertiser* to proclaim,

> *[T]hese huge bones irresistibly force upon us by the power of associating ideas, a representation of an animal, a monster so vastly disproportionate to every creature or subject on which we have been accustomed to exercise our intellectual powers; as to induce a momentary suspension of every animal faculty but admiration and wonder.*[40]

In this painting Peale is literally excavating American identity out of the marl. The backdrop for this activity is a Hudson River valley landscape near the Catskills, complete with the threat of a coming storm. Beyond conveying the thrill of working under pressure and uncertain atmospheric conditions, the lightning storm makes a coy reference to Benjamin Franklin's scientific discovery of electricity—another founding moment in the American mythology linking culture and nature. The excavation of the bones near Newburgh carried its own historical resonance, as it had been the site of George Washington's headquarters during the Revolution. By 1801 Newburgh was a pilgrimage site for many veterans of the conflict who saw in it a landscape overlaid with symbolic meaning. Skillfully balancing human action and natural forces, the concrete and the symbolic, Peale provides the intellectual, emotional, and perceptual framework that invests nature and landscape painting with the gravitas and scale to hail American identity. Twenty years before Thomas Cole would inspire the Hudson River school of landscape painting, *Exhumation of the Mastodon* established the tone and the expectations for the setting, scope, and drama of a genre that would define the nation's cultural ambitions.

In 1809, Peale thought the *Exhumation* would be his last significant work of art. However, in 1822 he tackled his final large-scale painting, a self-portrait commissioned by the board of his museum. *The Artist in His Museum* (CAT. 29) presents the founder preparing to sweep away the curtain that prevents us from seeing the entire room. The mastodon looms large behind the artist, its massive bulk only partially visible. The tusks cannot be seen at all, suggesting they are once again in their upward configuration. Peale's concern for the scientific validity of his portrayal may have led him to hide them altogether,

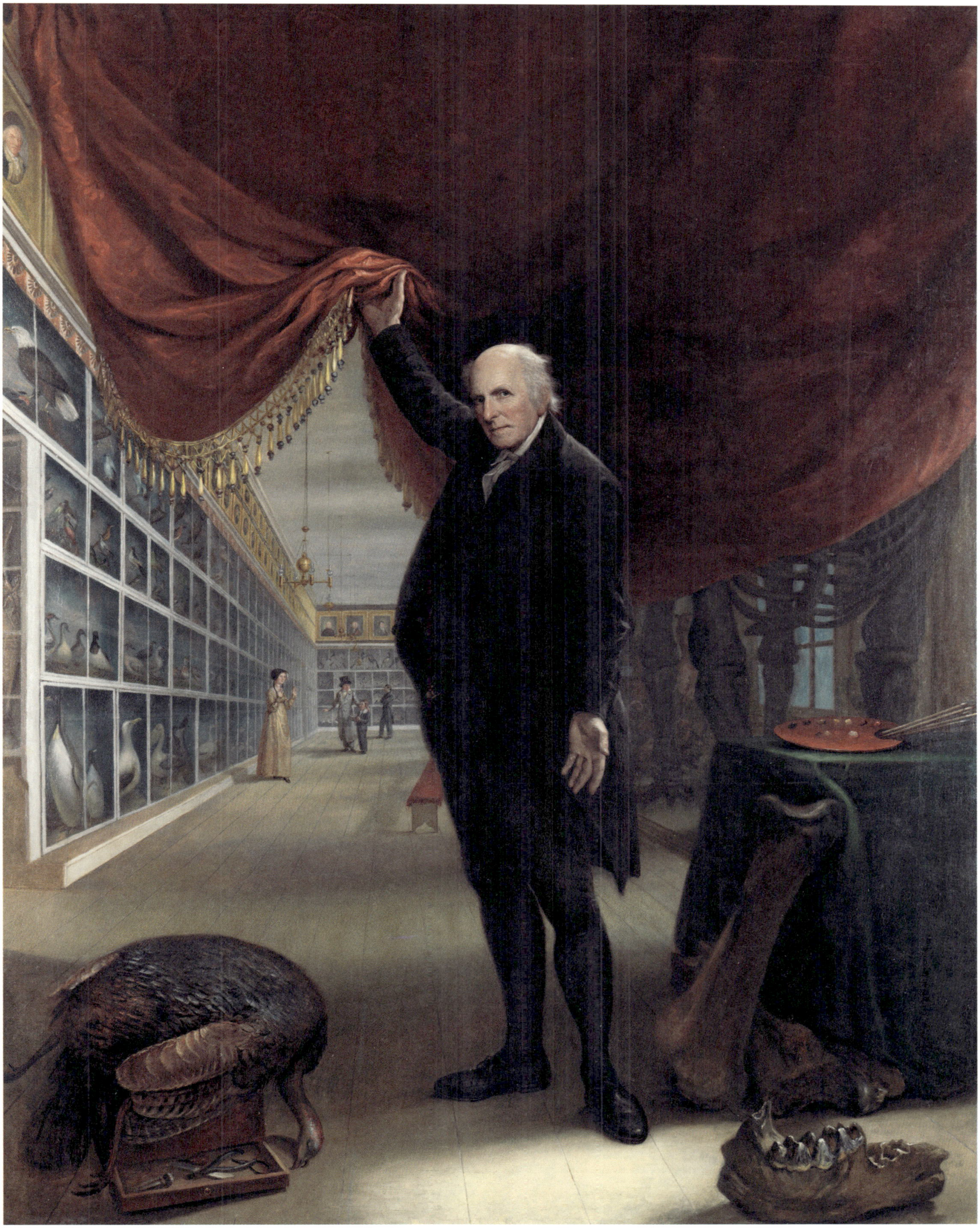

literally keeping them behind the curtain. His son, Titian Ramsay Peale, assisted his father with two critical aspects of this composition. In January 1821 he prepared a sensitive and detailed watercolor wash of the skeleton (CAT. 30) seen from the side. Its mass consumes the entire sheet, leaving no room for the tusks, perhaps suggesting that this was still an awkward topic of family conversation.

Titian also painted a watercolor of the Long Room (CAT. 31), meant to assist his father with the orthogonal perspective in this elongated space. The Long Room lived up to its name, being one hundred feet long and twelve feet high, installed floor to ceiling with four stacked levels of dioramas surmounted by a row of painted portraits wrapping around the top of the walls. Although both artists are credited with working on this piece, the singular touch of pencil and wash suggests the elder Peale's role was likely mentor and adviser. Here the immediate focus is on the registers of stuffed birds in their glass-fronted habitats, with the double row of portraits above, Humboldt's among them. Cases of minerals and shells line the opposite wall, illuminated by large windows. George Washington's bust, prominent in its darker coloration, turns toward the viewer. American Indian artifacts occupy a case recessed in the doorway leading to the adjacent room. The mastodon is nowhere to be seen; it was still on view in the Mammoth Room in the southeast chamber of Philosophical Hall. A topiary hummingbird tree stands under a bell jar on a pedestal, while paintings and sculptures round out the display. The carefully drawn orthogonals order the interior space; however, the delicate painting also contains a wealth of visual information reminiscent of the Humboldtian reach of the Long Expedition. The intersection of people—native, colonial, and foreign—and places are embodied in the specimens and artifacts that come together to create a cultural history lesson of America. Beyond its usefulness to the elder Peale, Titian's watercolor suggests an embrace of Humboldt's expansive vision for the United States within the family enterprise.

Charles Willson Peale's painting folds his son's perspective into a larger message. Peale had always intended his museum to teach Americans to see their democratic ideals and their natural history as an evocation of their cultural identity. He had occupied a significant role in all three arenas. This self-portrait is literally, symbolically, and metaphorically a meditation on all three. As a patriot, Peale had served with George Washington, whose portrait is the only one visible as the artist holds the curtain. Directly below is the case containing the bald eagles, the national symbol. Immediately below that, visible in the case connecting the two rooms, is the preserved paddlefish that Robert Patterson had given him as his first museum object. Peale's painting incorporates the first president, the first bird, and his first museum object. On the floor at Peale's feet, the lifeless carcass of a turkey, the bird Peale had championed as the nation's feathered emblem, waits to be stuffed and brought back to a lifelike state by the taxidermist's art. Peale's taxidermy tools and his loaded palette and brushes imply an artist still ready and capable. And next to him, propped against the table, is an additional femur of a mammoth—the index bone from which the height and mass of the entire creature could be extrapolated. As an artist, scientist, and American, Peale constructed a self-portrait that seamlessly merges his own identity with that of the museum. When it came to the mammoth's tusks, the elder Peale got the last word.

Jefferson had touted the mammoth's impressive size as the scale by which to measure the nation's cultural ambitions. However, his deep-seated unease with the concept of extinction complicated his enthusiasm for this creature as an emblem of the country's future. In his instructions to Meriwether Lewis, Jefferson had expressed a hope he would return with evidence that the mammoth yet roamed the American interior. Part of Buffon's degeneracy argument was predicated on the belief that for a creature to become extinct, it had to have been a weak species, not deserving of survival. That conclusion threatened to undermine the mammoth as an effective avatar of America's robust cultural ambitions, suggesting that the United States, too, would succumb to extinction.

Humboldt had an answer for that. Unlike Jefferson, he embraced the concept of extinction and saw in the fossil record evidence of deep time:

The whole globe appears to have undergone the same catastrophes. At a height superior to that of Mount Blanc, on the summit of the Andes, we find petrified sea-shells; fossile bones of elephants are spread over the equinoctial regions; and what is very remarkable, they are not discovered at the feet of the palm trees in the burning plains of the Orinoco, but on the coldest and most elevated regions of the Cordilleras. In the New World, as well as in the Old, generations of species long extinct have preceded those, which now people the earth, the waters, and the air.[41]

Humboldt's thinking reflected a new attitude in Europe, one that began to uncouple science from religion. His mentor Blumenbach had already dismissed the traditional concept of a Great Chain of Being, in which extinction was an impossibility and everything was thought to remain in a universal, static balance:

Nature will not go to pieces if one species of creature dies out, or another is newly created,—and it is more than merely probable, that both cases have happened before now,—and all this without the slightest danger to order, either in the physical or the moral world, or for religion in general.[42]

Closer to home, Jefferson's APS colleague Benjamin Smith Barton opined,

I speak of these animals as extinct. In doing this, I adopt the language of the first naturalists of the age. No naturalist, no philosopher; no one tolerably acquainted with the history of nature's works and operations, will subscribe to the puerile opinion, that Nature does not permit any of her species of animals, or of vegetables, to perish.

While certain of the reality of extinction, Barton also continued to champion a belief in the order of nature, which he referred to as "a harmony beautiful and divine." In addition, he accepted the idea of gradation of one species into the next. But he vehemently denied the essential connection between species that had been implied in the notions of plenitude and the economy of nature: "THERE IS NO SUCH THING AS A CHAIN OF NATURE: an absolutely necessary dependence (on this earth) of one species upon another."[43] On this, he and Humboldt were in hearty agreement.

The significance of *Exhumation of the Mastodon* lay in how Peale had articulated America's Herculean effort to create its own history and prehistory. For the new country to thrive, America needed to deploy its scientific and organizational resources to master its still largely unexplored territory, forge an alliance between science and industry to greater ends, and record it all for posterity. Landscape became the "ground" on which the American past and future were respectively unearthed and built. Exploration, invention, and scientific investigation permitted the country to know its past, and in so doing add incalculable knowledge to the understanding of the place of the United States in the world. Excavating mastodon bones signified American greatness emerging from the landscape itself—not what we built on it, but what we inherited from it. Here the character and content of the actual landscape became intrinsically linked to national and cultural identity. These bones were, as Humboldt had termed them, the ancient monuments of the Americas, worthy equals to their man-made European counterparts. As such, *Exhumation* is the genealogical founding ancestor for the new nation's cultural identification with the landscape.

Jefferson had sent Lewis and Clark out West with the hope they would return with evidence that the mastodon still roamed the plains. He was particularly proud of the mastodon femur and tusk that Clark had sent him for display at Monticello.[44] However, as the evidence mounted to the contrary, so did his interest in seeing America's future greatness associated with a different set of landmarks. Humboldt's admiration for the actual North American landscape—what one would see on it and excavate from it, more than what we would build on it—encouraged American appreciation of untamed nature. He was less interested in the architecture of the Capitol building or Mount Vernon than he was with the gardens and wooded hillsides. When Humboldt surveyed the prospect from Capitol Hill, his host, Thomas Law, despaired at the "vast Serbonian bog" that to him characterized the local landscape. Law's insecurity reflected a larger cultural blindness to the power of natural features to define the nation's ambitions. Humboldt saw it differently. Thomas Law's bog might as well have been John Masten's marl pit, out of which would emerge a cultural identity more solidly grounded in nature. That interdependence with nature, and the distinctiveness and variety among various civilizations across

CAT. 30 Titian Ramsay Peale, *The Gigantic Mastodon*, 1821, watercolor wash on paper, 14 ½ × 18 ⅞ in., American Philosophical Society

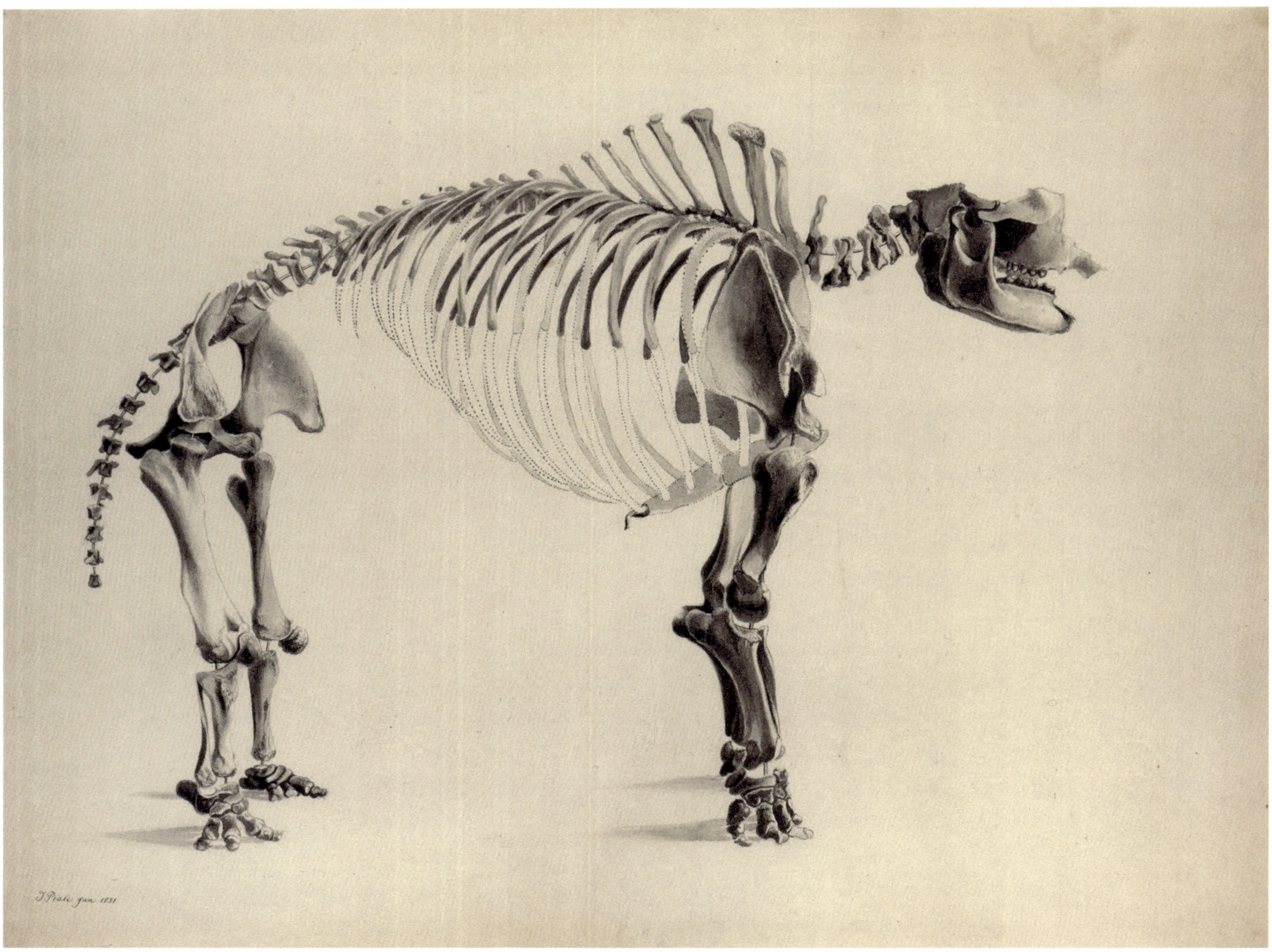

CAT. 31 Charles Willson Peale and Titian Ramsay Peale, *The Long Room, Interior of Front Room in Peale's Museum*, 1822, watercolor over pencil on paper, 14 × 20 ¾ in., Detroit Institute of Arts, Founders Society Purchase, Director's Discretionary Fund

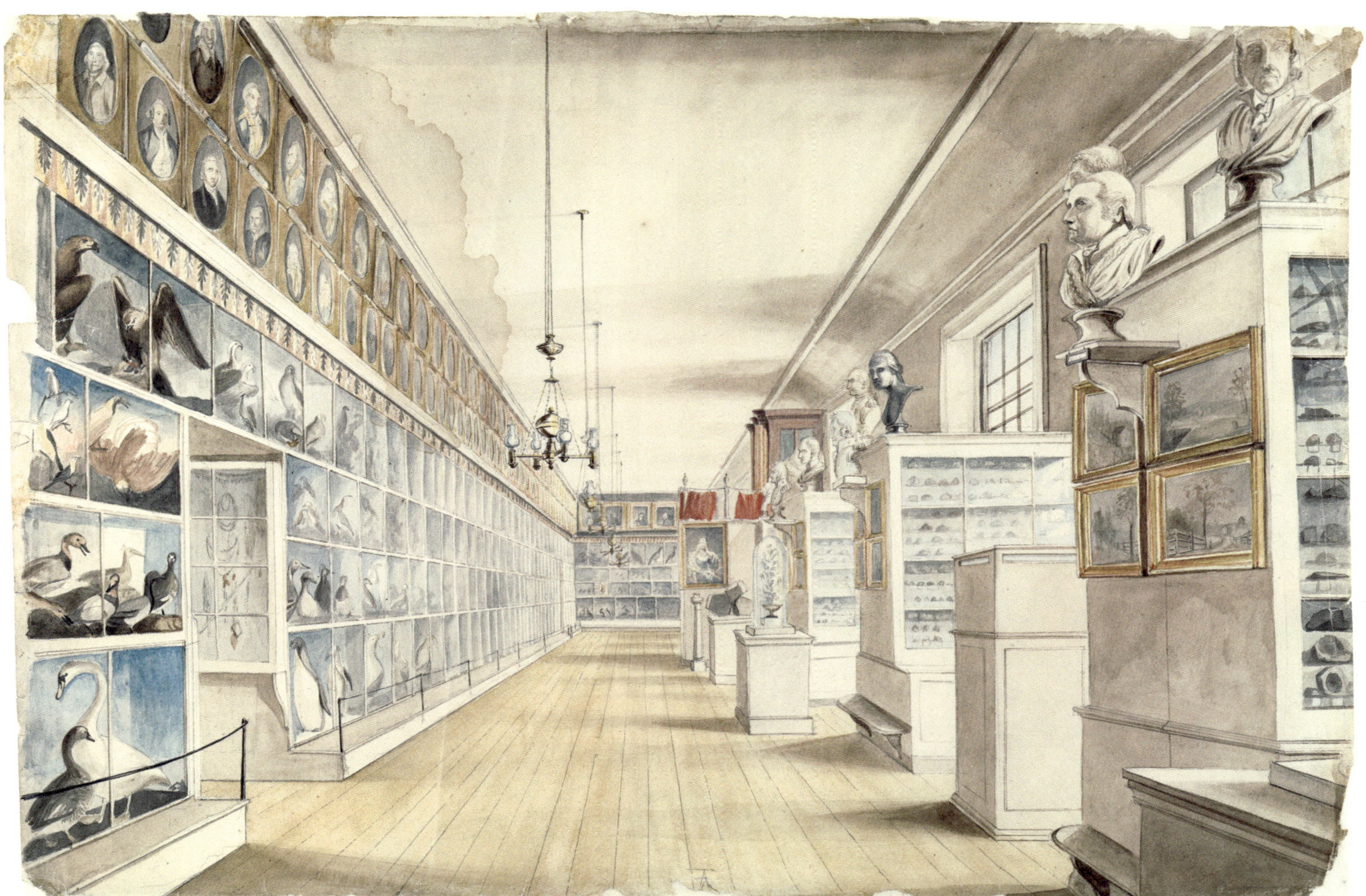

the globe, is what propelled the United States to formulate a cultural identity bound in an appreciation of uncultivated nature.[45]

Humboldt emboldened a natural tendency to identify unique features in the American landscape and claim them as emblematic of the country's identity. On June 11, 1804, Humboldt's last day in Washington, D.C., the *National Intelligencer*—Jefferson's sympathetic press vehicle—printed the text of a sermon that had been delivered in May in Charleston, South Carolina, extolling the benefits of acquiring Louisiana. In the opening paragraph, Dr. David Ramsay exulted,

> *Europe would be drained of half her waters, in furnishing supplies for such immense rivers as the Mississippi, St. Lawrence, La Plata, and Amazon, the last of which is the largest river in the known world. The most distinguished transatlantic cataracts, bear no comparison with the falls of Niagara. The forests of Europe, Asia, and Africa, are lost in the magnitude of those of America. The Elephant of the east is far inferior, in point of size, to the Mammoth of the west. We cannot suppose that this extraordinary physical grandeur of America, was created without design; and what design so probable, as to be subservient to some corresponding exalted state of man? May we not, therefore, indulge a hope, that the inhabitants of a country so eminently distinguished by the Author of Nature, are destined to form political affiliations of a larger size, and that these will enjoy an uncommon portion of happiness?*[46]

Ramsay made the point that everything about the expanded physical geography of the United States was bigger and more consequential than its European counterparts: the country, its rivers, waterfalls, and forests, its mammoths and its future prospects. The boundaries of the country and the ethos contained in the Declaration of Independence for life, liberty, and the pursuit of happiness was writ large in the addition of Louisiana.

Fig. 33
Mammut americanum (American Mastodon), Upper Jawbone excavated 1807, Big Bone Lick, Kentucky, 18 × 16 ½ × 11 ½ in., The University of Virginia, Department of Geology, on loan to The Thomas Jefferson Foundation at Monticello

Fig. 34
J. C. Stadler, after William Roberts, *Natural Bridge*, 1808, colored aquatint, 27 ¼ × 20 ½ in., The Thomas Jefferson Foundation at Monticello

Ramsay's questions were clearly rhetorical. With the acquisition of Louisiana, America's prospects were unbounded, ambitions for its future tied inextricably to the scale and magnificence of its natural landmarks. That confidence in the power of nature to embody the country's ethos would be echoed in Humboldt's own assertions regarding New World monuments. By the 1810s, American travel literature expressed and fostered an appreciation of nature outside the bounds of cultivated landscapes. The mammoth would give way to mammoth-scale features in the landscape. Descriptions and images of Natural Bridge, Niagara Falls, Mammoth Cave, Trenton Falls, and other notable landmarks were commonly found in issues of the *Columbian Magazine*, *Massachusetts Magazine*, *New York Magazine*, *Ladies' Magazine*, *Port Folio*, and the *American Museum*.[47] All of the natural wonders being extolled in popular magazines across the country shared the same distinctive characteristic: they were deemed "mammoth" in scale, out of proportion to the rest of the landscape, and worthy of admiration. They were exemplars of that "extraordinary physical grandeur" Ramsay had extolled in his sermon, made by the "Author of Nature." The robust physical, metaphorical, and spiritual elements in the American wilderness were its distinguished features, prognosticators of American happiness and the "exalted state of man."

NATURAL ICONS, NATIONAL ICONS: NATURAL BRIDGE AND NIAGARA FALLS

After departing the presidency, Jefferson used the rooms at Monticello to display evidence of his nascent confidence in finding cultural symbolism in natural features. His entry hall gradually became an American world in miniature, a microcosm of Peale's Museum, containing artifacts from Lewis and Clark, mastodon teeth (fig. 33), and maps of the expanding country. George Ticknor described the room, with its folding glass door entry:

> *On one side hang the head and horns of an elk, a deer, and a buffalo; another is covered with curiosities which Lewis and Clarke found in their wild and perilous expedition. On the third, among many other striking matters, was the head of a mammoth, or, as Cuvier calls it, a mastodon, containing the only* os frontis, *Mr. Jefferson tells me, that has yet been found. On the fourth side, in odd union with a fine painting of the Repentance of Saint Peter, is an Indian map on leather, of the southern waters of the Missouri, and an Indian representation of a bloody battle, handed down in their traditions.*[48]

Ticknor noted Jefferson's fondness for what he called "American antiquities," with particular emphasis on the mastodon bones.[49]

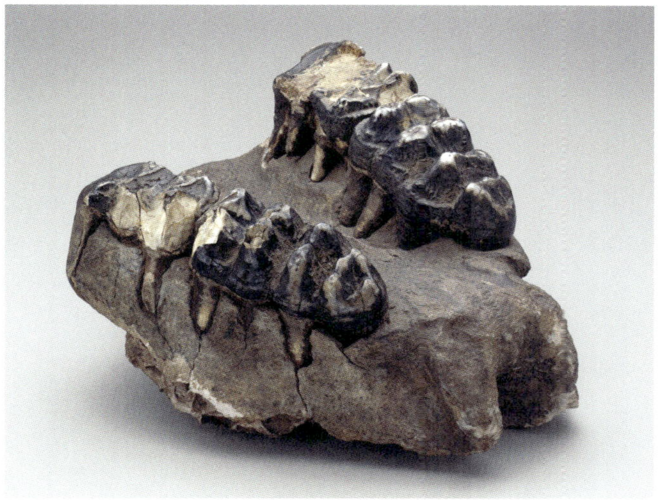

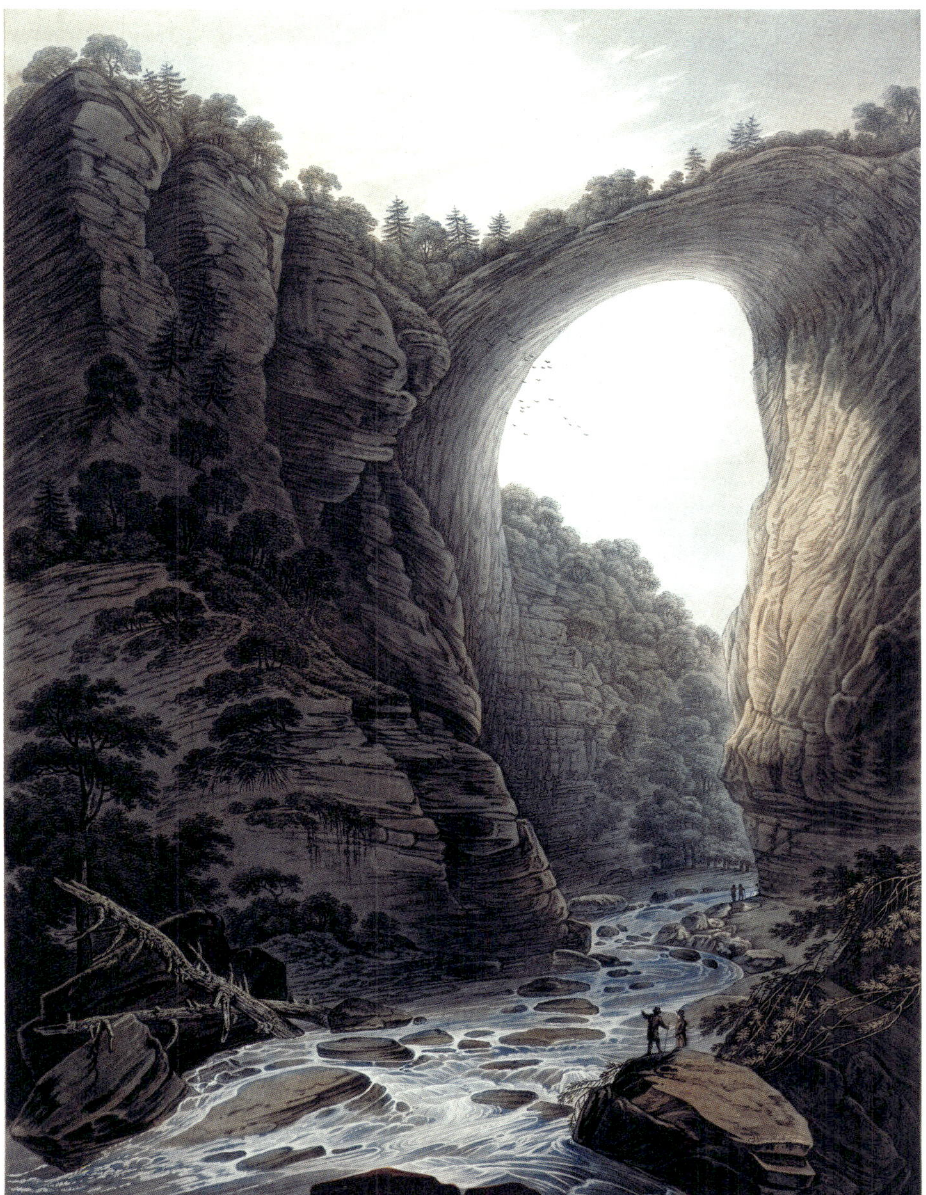

If Jefferson's entry hall at Monticello spoke to America's natural antiquities, his dining room extolled American landscape features to which the former president assigned emblematic value. William Roberts's now-lost painting of *Natural Bridge* graced the walls; in its place today hangs an engraved copy after it (fig. 34). Two different views of Niagara Falls, both by John Vanderlyn, hung together as a pair on an adjacent wall. Niagara had been deemed the "Eighth Wonder of the World," and Jefferson's interest in it was reminiscent of his urging American artists to lay claim to his beloved Natural Bridge, before some European artist botched it.[50] Thomas Birch's engraved view of Mount Vernon with a view of the Potomac River occupied the fulcrum in this visual essay on noteworthy and emblematic views.

Natural Bridge stands out as the earliest natural landmark to achieve iconic status in the United States. Jefferson had called attention to it in *Notes on the State of Virginia*, deeming it "the most sublime of Nature's works," and before long it became one of the most widely reproduced images extolling the American landscape.[51] Jefferson had owned the surrounding land, including the actual Natural Bridge, since 1773, but it had previously been closely associated with George Washington, who had surveyed the area for Lord Rockbridge in the 1750s. Local lore held that Washington had climbed the left-hand side of the arch to a height of twenty-five feet, where he had carved his initials; that became part of the arch's appeal to visitors.[52] Jefferson described the experience from the top of the arch as vertiginous and then observed,

> *descending then to the valley below, the sensation becomes delightful in the extreme. It is impossible for the emotions, arising from the sublime, to be felt beyond what they are here: so beautiful an arch, so elevated, so light, and springing, as it were, up to heaven, the rapture of the Spectator is really indescribable!*[53]

In 1805 it was one of five large-scale transparent scenery tableaux at Philadelphia's Chestnut Street Theatre, the premier venue for panoramic displays.[54]

Humboldt incorporated references to this American landmark in his own publications. In 1810 he published *Vues des Cordillères, et monuments des*

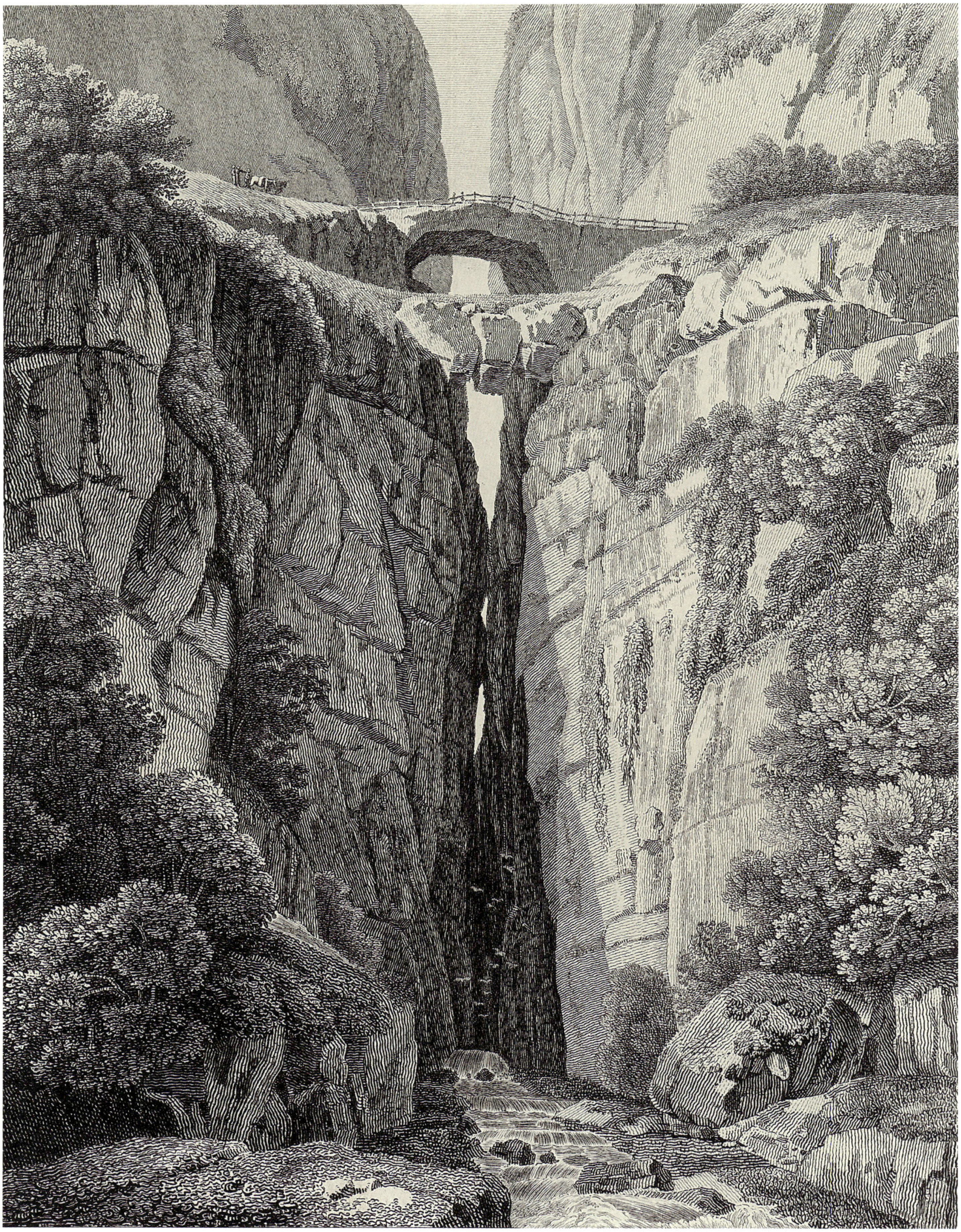

CAT. 32 Alexander von Humboldt, *Natural Bridge of the Icononzo*, Plate 4, from *Researches, Concerning the Institutions & Monuments of the Ancient Inhabitants of America*, 1814, engraving, closed: 9 × 6 × 1 in., Smithsonian Libraries, Gift of the Burndy Library

peuples indigènes de l'Amérique (*Views of the Cordilleras and Monuments of the Indigenous Peoples of America*), a wildly popular travelogue that folded his beloved measurements and observations into an engaging narrative. In 1814 it was translated into English and given a much more sensational title: *Researches, Concerning the Institutions and Monuments of the Ancient Inhabitants of America, with Descriptions and Views of Some of the Most Striking Scenes in the Cordilleras!* In the chapter on the natural bridges of Icononzo, Humboldt illustrated this impressive South American formation (**CAT. 32**) and compared it to Jefferson's extended commentary from his *Notes*. After providing a dramatic report of the abyss as experienced from the top of the arch, Humboldt reminded his reader that

> *A phenomenon, similar to the upper bridge, of which we have just given the description, exists in the mountains of Virginia, in the county of Rockbridge. This Mr. Jefferson has examined with an attention, that distinguishes all the observations of that excellent naturalist.*[55]

Humboldt validated a cherished American scenic landmark and the statesman who had written about it, folding the country and one of its presidents into his narrative. As the United States became part of Humboldt's worldview, he paired its singular features alongside unique formations elsewhere on the globe. His expressed admiration for the American landscape conveyed affirmation that the New World held treasures worthy of renown.

What Natural Bridge was to Virginia, Niagara Falls became for the nation. In *Views of the Cordilleras*, Humboldt described the massive volume of water delivered by these falls, this time in comparison to the taller vertical drop of the Tequendama Falls near Bogotá.[56] These two North American landmarks together came to stand for the United States. In 1822, cartographer Henry S. Tanner engraved the most up-to-date map of the country (fig. 35A). Maps like Tanner's contained more than geographic information. They were an expression of national identity and ambition—the fulcrum between what was known and what had yet to be defined, described, and owned. Tanner included Long's Peak, newly designated, and Astoria, John Jacob Astor's Columbia River fur trading post. He took the unprecedented step of identifying an Oregon Territory in the Pacific Northwest, laying cartographic claims to a contested part of the continent. Jared Sparks praised Tanner's map as a "troph[y] of American enterprise."[57]

The cartouche in the lower-right corner (fig. 35B) is particularly noteworthy as an emblem of cultural identity with natural landmarks. Natural Bridge and Niagara Falls appear together, elided into a single landscape. Natural Bridge appears in the foreground, only Cedar Creek has been replaced by the Niagara River, and to the right of the arch is the curve of Horseshoe Falls. Animals specifically native to the United States populate the vista and carry the metaphor further. Jefferson's moose and the nation's bald eagle share the landscape with a beaver, a bear, and a rattlesnake. The cartouche presented emblems associated with cultural identity—in this case, the bald eagle, Natural Bridge, and Niagara Falls—as the avatars of natural dominance. The *National Intelligencer* heralded Tanner's map, characterizing its aims as "patriotic as well as literary."[58] Together these two landmarks defined the scale and scope of America's cultural ambitions, inscribed onto the unfinished map of American expansion across the continent.[59]

AMERICA'S WILDERNESS AESTHETIC: "WILD AND COLOSSAL NATURE"

The heady confidence in assigning national significance to natural icons in Tanner's map was further evidence of the development of a wilderness aesthetic, an appreciation of the visual spectacle that set the United States apart from Europe in particular. Humboldt's statements about the appreciation of natural monuments in place of man-made monuments amplified his earlier comments to Jefferson, Madison, and Thornton about the inherent scientific value of "everything to the west of the mountains" and Humboldt's heartfelt desire to be a part of bringing those discoveries

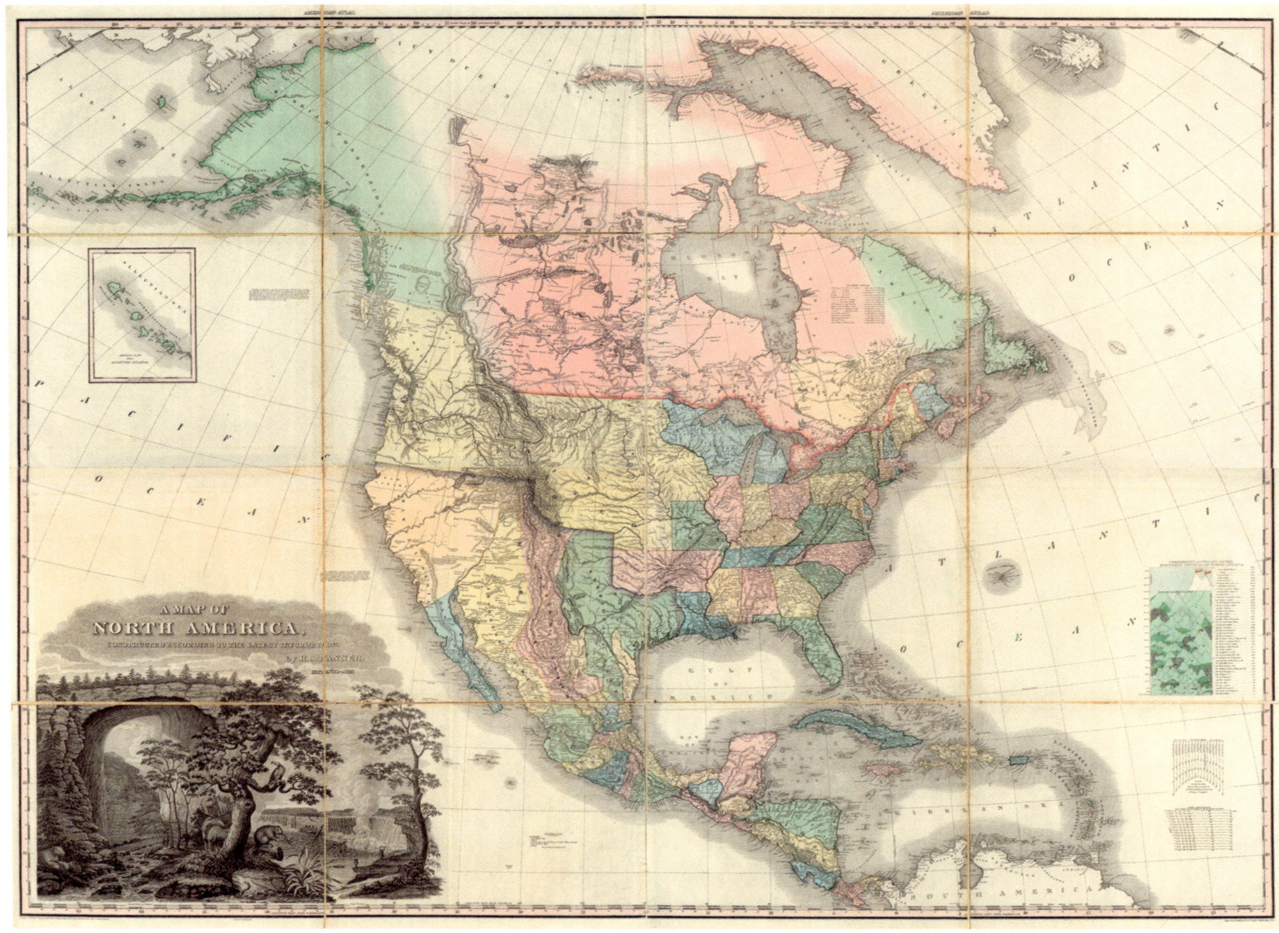

Fig. 35A
Henry S. Tanner,
A Map of North America,
from *A New American Atlas Containing Maps of the Several States of the North American Union*, 1825, hand-colored engraving, 42 15/16 × 57 7/8 in., The David Rumsey Map Collection, David Rumsey Map Center, Stanford Libraries

to light. His enthusiasm and curiosity had an additional salutary effect in the United States, providing a boost to a tentative, emergent appreciation for the qualities of raw nature. That would blossom as part of the acceptance of a wilderness aesthetic during the 1820s and flourish as the foundation for a landscape-oriented emphasis in literature, travel writing, and painting over the ensuing decades.

However, holding up individual features worthy of distinction was but a small step toward embracing all of wild nature as emblematic of American cultural identity. To do that required Americans to reconceptualize their relationship to nature. Instead of fear, or disinterest, they needed to seek out aesthetic, spiritual, and intellectual sustenance in the wilderness. The development of an American "wilderness aesthetic" required the encouragement and enthusiasm of someone like Humboldt to overcome deeply entrenched European attitudes that saw wild nature as evil, dangerous, or both. As Jefferson and Humboldt refuted European concepts of degeneracy, they replaced them with empirical evidence of robust life-forms and ecosystems. Curiosity began to supplant fear in what the wilderness promised.

The earliest accounts of encounters in the American wilderness required nature to be the adversary, following the puritanical path of seeing wilderness as a spiritual challenge, a life envisioned as adversity endured on the path to ultimate redemption. Centuries of European attitudes toward wilderness defined it in negative terms, signifying the struggle between the forces of chaos and man's spiritualizing effect on civilization. Uncultivated land was often perceived as useless, an attitude fed by this vision of dread. On top of that, the fear of Indians, who seemed at one with nature, able to blend into wild nature, was deeply unsettling. This multivalent defensiveness fed the ultimate fear: that wilderness would erode the fragile American social contract and bring out man's baser instincts.[60] The expansion of settlements gradually eased both the fear and the sense of being surrounded and outnumbered by hostile forces. Ultimately, the ability to find God in the wilderness became a critical element in the development of an American wilderness aesthetic. It tipped the balance from puritanical fear of the unknown to a pilgrimage into nature, in which fear would yield to Enlightenment curiosity, in the service of a spiritual epiphany.

If finding God in the wilderness represented one avenue toward a nature-based cultural ethos, science was the secularizing ingredient in the transformation of the wilderness from a heathen trap to a laboratory of wonder. Jefferson's *Notes on the State of Virginia* was his early affirmation of a kind of self-reliance that would prove to be a solid foundation for the development of positivist strains of wilderness appreciation based in the sciences, literature, and, eventually, the fine arts. In this regard he was not alone: noted botanists André Michaux and John Bartram added their own voices to a slowly growing chorus of naturalists who found in the wilderness a vast outdoor laboratory.[61] Their narratives incorporated data, measurements, and observations, reflecting the same training that Humboldt experienced in Prussia. Surveyors and natural historians led with their curiosity down this path. Each successive generation of naturalists found it easier to make peace with and find virtue in raw nature. Fear subsided with familiarity. Nature was an endlessly rewarding curriculum.

A childhood spent in the confines of settlements fostered a yearning for adventure and a curiosity, rather than fear, for what lay beyond the edges of cultivated lands. When John Bartram's son, William, traveled from 1773 to 1777, he exulted in the wilderness, standing atop a mountain in Georgia and describing how much he "enjoyed a view inexpressibly magnificent and comprehensive."[62] The younger Bartram encouraged his reader to revel in a landscape "as yet unmodified by the hand of man,... the unlimited, varied, and truly astonishing native wild scenes of landscape and perspective."[63] His peregrinations took him throughout the Southeast, and at Lake George, Florida, he felt "seduced by these sublime enchanting scenes of primitive nature."[64]

Positive attitudes toward nature evolved with safety, security, arability, and sustenance. As the easternmost landscape was settled and shaped by waves of European settlers, it became easier for the next generations to relax from the earlier point of vigilance. Being ensconced in a larger settled landscape environment encouraged a greater sense of curiosity about wilderness and the ability to perceive the aesthetic aspects of nature. As curiosity and wonder gradually replace fear and defiance, descriptions took on a more appreciative, admiring tone. Travelers wanting to make the trip to Niagara Falls considered the adventure as a kind of pilgrimage, one replete with discoveries and wonders. Nature provided a stimulus for travel; the reward for these excursions was described in intellectual, sensory, and spiritual terms. The British had been diffident concerning exploration and mapping of their American colonies; independence had fostered a curiosity only felt through ownership of the land. The American Revolution had vaulted the need for surveying higher in the emerging political consciousness, part of the need to understand and demarcate the boundaries of the new country. A similar process had taken place in the Netherlands, when an independent tradition of depicting the landscape in literature and art flourished after the end of the Eighty Years' War with Spain. The Golden Age of Dutch landscape painting developed after the cessation of hostilities, when the Dutch saw their surroundings with a renewed sense of patriotism and curiosity. In the United States, independence was fostering a kindred mind-set, and with the acquisition of Louisiana, a sense of urgency as well.

Fig. 35B
Detail of fig. 35A

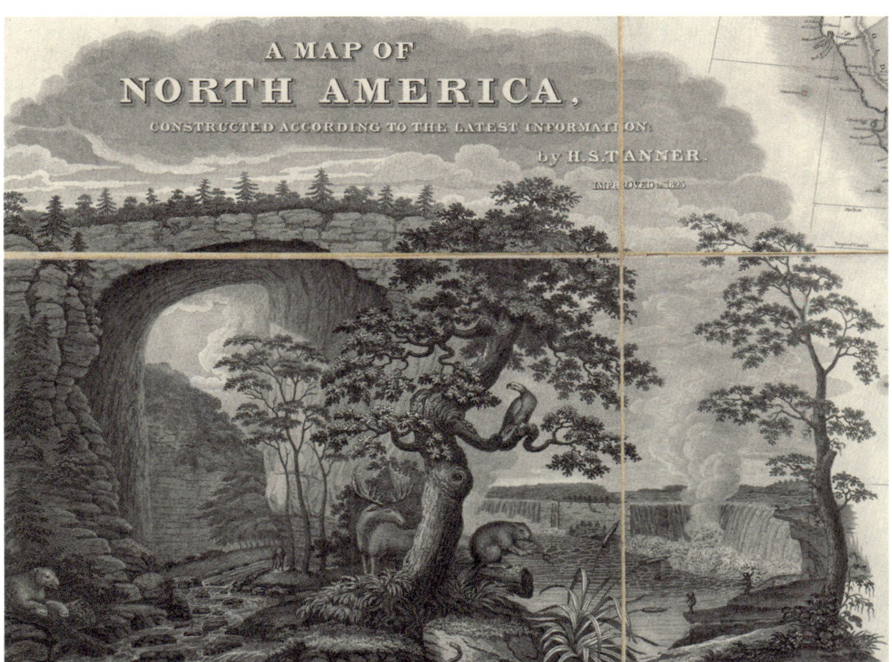

A handful of prominent European intellectuals came to America to experience the wilderness, among them Louis-Philippe, the future French king, who spent three years in exile in the United States between 1796 and 1800.[65] His travels took him into rural America, among the Seneca and Oneida Indian communities; his experiences planted a sympathetic seed for Americans that he retained once he ascended the throne.[66] The French poet and intellectual Chateaubriand incorporated elements from Bartram's travels in two of his works, *Atala* and *René*.[67] During his five-month stay from 1791 to 1792, he traveled through New York state, deeply moved by the landscape:

> *In vain does the imagination try to roam at large amidst our [Europe's] cultivated plains, . . . but in this deserted region the soul delights to bury and lose itself amidst boundless forests—it loves to wander, by the light of the stars, on the borders of immense lakes, to hover on the roaring gulph of terrific cataracts, to fall with the mighty mass of waters, to mix and confound itself, as it were, with the wild sublimities of Nature.*[68]

Decades later Chateaubriand, who knew Humboldt, shared the explorer's admiration of the monumental qualities of the American wilderness and linked them to American virtues, asserting, "Nothing is old in America except the trees, children of the soil, and liberty, the mother of all human society; these are worth as much as monuments and forebears."[69] That synergy among art, science, and nature became a hallmark of Humboldt's approach to understanding the physical world.

Humboldt's influence on America's emergent wilderness aesthetic had ramifications for the ways in which exploration contributed to America's developing cultural identity. Humboldt's concept of *Naturgemälde*—Picture of Nature—merged science, exploration, and the arts to encourage a developing appreciation for what was distinctively American about this continent and see those singularities as part of a larger, coeval whole. By embracing Humboldt's global vision, Americans embedded themselves in it, all the while retaining a sense of what made the landscape distinctive and, by implication, more significant to a nature-based American identity. The underlying concept of wilderness shifted from a rigid definition bound by scripture to a more scientific state of mind. As long as wilderness was perceived to be threatening, fear shaped Americans' attitudes toward the landscape. However, if wilderness appeared to be a state of nature that fostered uplift and promised an earthly sanctuary, the development of a strong, positive, spiritually resonant wilderness aesthetic could follow.

As part of this shift in appreciation, Americans had to redefine the traditional European aesthetic terms used to classify and describe landscapes. That meant a reappraisal of English aesthetician William Gilpin's love of the picturesque, centered in the cultivated landscape, of which there was still precious little in the United States; and Archibald Alison's associationism, which looked to features in the landscape that evoked the past in order to give meaning to the present—a past that was strikingly absent in America's short Eurocentric history. Gilpin summed up a general European dislike of raw nature when he wrote dismissively of forests: "The idea of a wild country, in a natural state, however picturesque, is to the generality of people but an unpleasing one."[70] Descriptions of the American wilderness would, on the whole, remain negative in Europe far longer than in the United States.

The rebuttals were swift in coming. In 1816 New York Governor De Witt Clinton gave the inaugural address at the new American Academy of Fine Arts, where he proclaimed that an American artist's inspiration

> *must derive its forms . . . from the country in which he was born. . . . [C]an there be a country in the world better calculated than ours to exercise and to exalt the imagination—? . . . Here Nature has conducted her operations on a magnificent scale: extensive and elevated mountains—lakes of oceanic size—rivers of prodigious magnitude—cataracts unequaled for volume of water—and boundless forests.*[71]

Clinton had a long-standing interest in natural philosophy, especially botany. He had sent George Washington seeds and saplings from New York state to be planted in the gardens flanking the carriage drive leading to the mansion at Mount Vernon, part of the general's plan for an American garden featuring specimens from all thirteen states.

In 1819 Benjamin Silliman (fig. 36), professor of natural history at Yale, wrote,

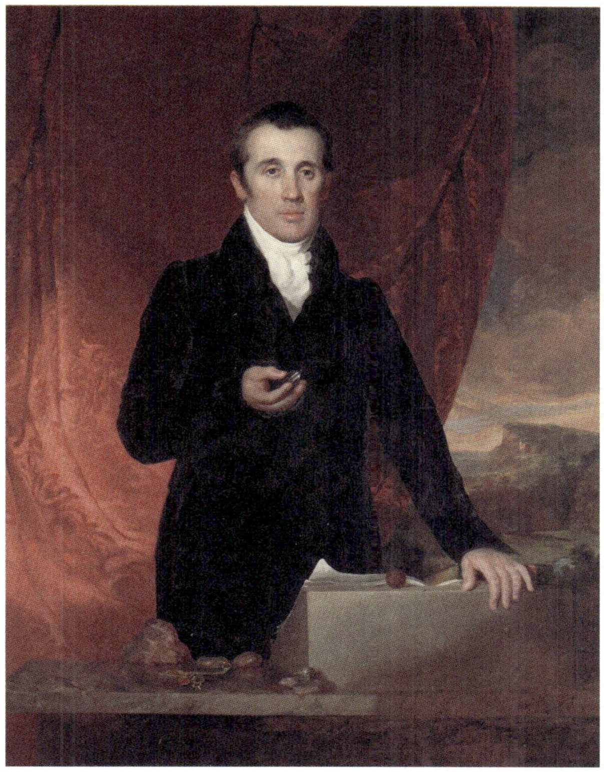

Fig. 36
Samuel F. B. Morse, *Benjamin Silliman Sr.* (1779–1864), 1825, oil on canvas, 55 ¼ × 44 ¼ in., Yale University

National character often receives its peculiar cast from natural scenery. . . . Natural scenery is therefore, always worthy of observation, and it will be a never-failing source of delight to those, who, perhaps not themselves painters or poets, participate in any degree in their faculties and perceptions; and find in mountains, plains, and valies [sic]*—in streams, lakes, and woods—in cataracts and caverns—in cultivated regions, and in untamed solitudes—in narrow defiles, and in the boundless horizon, every varying source of pleasure, and inexhaustible topics of admiration and praise. . . . Thus, natural scenery is intimately connected with taste, moral feeling, utility, and instruction.*[72]

By calling out "cataracts and caverns," Silliman made a pointed reference to Niagara Falls and Natural Bridge (which was believed to be the remnant of a collapsed cavern) as sources of "taste, moral feeling, utility and instruction. Nature's aesthetic gratification, spiritual uplift, and education became central tenets of the American wilderness aesthetic.

In 1821 selections from Humboldt's *Personal Narrative* made their way into numerous American popular journals. One such extract was introduced by the following paragraph:

There are some men whose names seem to irradiate the age in which they are born, whose every step in life forms an epoch in science, and who, as if Nature herself were sedulous to guard them as her historians, escape unhurt through perils that would alike appal the mind, and overwhelm the bodies, of less enthusiastic, less gifted individuals. Such is Humboldt, every addition to whose travels is an addition to our stock of knowledge. In him all the qualities that are requisite for a philosopher and a man of science are most happily combined, whilst the energies of his mind seem to transform themselves into physical powers of more than natural strength, to enable him to follow whither his ardour leads him.[73]

That physical robustness was compelling proof that the Americas did not sap a person's strength, but enhanced it. Humboldt was more than an adventurous explorer, he was a hero to all who valued the increase and diffusion of knowledge, wherever those journeys led. He became a staple feature in the American press, invoked by name as an authority worthy of respect and, by extension, as an ally and advocate for the value of natural sciences in the United States.

THE TRAVEL NARRATIVE AND AMERICAN LITERATURE

The popularity of Humboldt's early narratives, notably *Aspects of Nature* and the first four volumes of his *Personal Narrative*, encouraged this interpretive gloss on American nature. Travel literature was a genre Humboldt had reinvented with his second book, *Aspects of Nature*. By separating the engaging narrative from the supporting data, Humboldt had expanded his audience and satisfied both the amateur and the professional scientific traveler. Pairing a scientist's keen eye for observation and an artist's poetic touch for description, he painted word pictures that in his view answered the need to blend both. Starting in the 1810s, weekly newspapers and monthly journals began publishing excerpts from his work, often reprinting articles that appeared first in the London newspapers, helping to disseminate it beyond the more rarefied intellectual circles. *The Friends' Weekly*, *Littell's Living Age*, and *Godey's Lady's Book*, among others, aggregated information primarily from the New York and Boston newspapers much in the way that the Associated Press provides stories to a wide range of newspapers in the modern era.

Humboldt's name appeared in articles on a wide variety of subjects, his expertise invoked frequently. Often he was simply referred to by his last name, his identity assumed; occasionally this led to confusion with his older brother, the esteemed linguist Wilhelm von Humboldt, but for the most part, "Humboldt" in the United States was understood to be Alexander by default.

In Humboldt's view, the American trajectory toward enlightenment need not follow the same path as Europe's or Asia's for it to be relevant. In 1812 he expounded,

I am more than aware of how a traveler through America is at a disadvantage with respect to those who describe Greece, Egypt, the banks of the Euphrates or the South Sea islands. In the old world it is the civilized peoples and the gradations of their civilization that lend the picture its principal character; in the peoples of the new world, by contrast, the individual vanishes, as it were, together with his problems amidst a wild and colossal nature.[74]

That sentiment resonated in the United States. Humboldt spent decades convincing his audiences of the essential equality of the Old World and the New. American travel narratives soon borrowed Humboldt's idiom, crafting essays on the therapeutic value of a walk in the woods. *Port Folio* stated as its mission the task of introducing Americans to their own scenery:

It is our intention to devote a plate in each number of the Port Folio *to the description of American Scenery. To the pencil our country affords an inexhaustible abundance, which, for picturesque effect, cannot be surpassed in any part of the old world.*[75]

In the United States, travel literature widened the path for the American appreciation of wilderness as a national asset rather than a liability. In 1818 New Hampshire lawyer Estwick Evans embarked on a four-thousand-mile trip on foot across the United States, publishing his account the following year. In his narrative, he exulted,

How great are the advantages of solitude!—How sublime is the silence of nature's ever active energies! There is something in the very name of wilderness, which charms the ear, and soothes the spirit of man. There is religion in it.[76]

Evans's jubilation in the wilderness was an early augur of the move toward a transcendentalist experience of spiritual uplift in wild places, an effective counter to the puritanical preference for finding spirit in cultivated gardens. That concept, of finding God in the wilderness, rather than feeling abandoned by Him, may be found at the core of America's wilderness aesthetic in literature.

Curiosity regarding the wilderness and a burgeoning eagerness to take excursions into nature began in those settlements that provided a measure of protection from it. Early accounts of travel in search of scenery or spirit inflected American literature. In his early compilation of writings, *The Sketch-Book of Geoffrey Crayon, Gent.*, Washington Irving affirmed, "[N]ever need an American look beyond his own country for the sublime and beautiful of natural scenery."[77] He and James Fenimore Cooper both mined early American expedition narratives, notably the Long Expedition, for texture. Both Irving and Cooper spent considerable time overseas, but they used that distance to create a literary tradition for America that effectively provides a backstory, a missing past. Their published works, and the American scenery they invoked, did more than serve as a visual backdrop to the human condition. Following Peale's *Exhumation of the Mastodon*, their novels and short stories provided a literary creation myth for the nation, using tales from a timeless past designed to invest the American landscape with characters reminiscent of European fairy tales. Irving understood just how powerful this wilderness setting operated within his oeuvre. "[O]f all the scenery of the Hudson," Irving would write, "the Kaatskill Mountains had the most witching effect on my boyish imagination."[78]

HUMBOLDT'S ENCOURAGEMENT OF IRVING AND COOPER

Washington Irving (fig. 37) had met Alexander von Humboldt in London, and they were constant companions in Paris. Together they conducted research on the earliest explorations of the Americas and socialized in the same company at night. Irving lived seventeen years in Europe all told, during which time Humboldt's influence remained strong with the American author, inspiring Irving's interest in writing about the American western landscape and his proposed history of Mexico.[79] After spending the 1820s living in Germany, France, Spain, and London, Irving turned his attention west, making his one trip across the Great Plains. In 1832 he wrote *A Tour of the Prairies*, borrowing from the Long Expedition narrative to complement his own experiences. Emboldened by his success, Irving teamed up with John Jacob Astor, principal owner of the American Fur Company, to write *Astoria* (1836), a history of Astor's trading outpost, and on a larger scale, a celebration of expansionist enterprise as a founding principle of the United States. Of *Astoria*, Irving wrote, "I was sure…that a narrative…could not fail to be full of stirring interest, and to lay open regions and races of our country as yet but little known."[80] In 1836 he began a series of works known as the Bonneville adventures, chronicling Captain Benjamin Bonneville's exploits for a popular audience. The French-born Bonneville led three expeditions across the Rocky Mountains and is best remembered for establishing portions of the Oregon Trail. Irving's western novels drew on expedition reports to identify expansion as an economic and cultural goal for the United States. That narrative became enshrined in the policy and ethos known as Manifest Destiny.[81]

Fig. 37
John Wesley Jarvis, *Washington Irving* (1783–1859), 1809, oil on panel, 33 ½ × 26 ¾ in., Historic Hudson Valley Collection, Pocantico Hills, NY

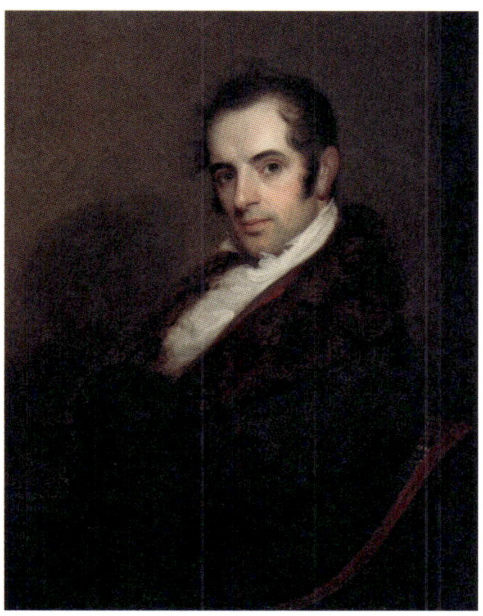

Fig. 38
John Wesley Jarvis, *James Fenimore Cooper* (1789–1851) (detail), 1822, oil on canvas, 30 × 25 in., Fenimore Art Museum, Cooperstown, NY, Gift of Dr. Henry S. F. Cooper

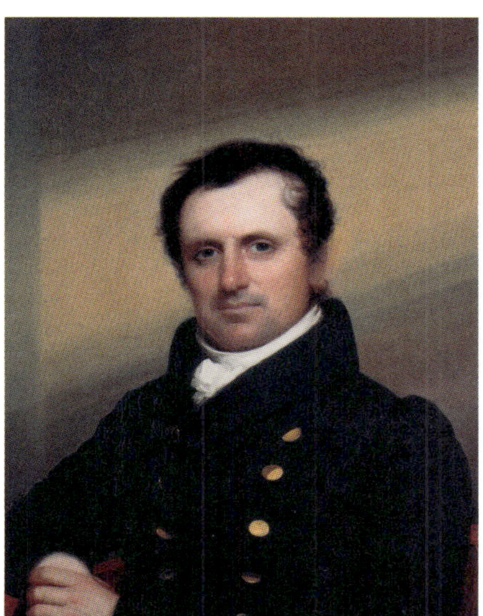

Cooper and Irving established literary production as a viable profession in the United States, and they did so with exploration narratives that featured the American landscape as a protagonist in shaping the American character. What they drew from reading illustrated expedition reports reflected the broader interest in the dramatic narrative potential of the American wilderness landscape, without the attendant scientific interest in the statistics. In doing so they changed the tone of the narrative, blending the suspense and drama of encounters on the edge of danger with a literary appreciation for the intrinsically moral qualities of the landscape. This extends to the delineation of several of their Indian characters, a reminder that many exploration narratives relied on white explorers developing a symbiotic relationship with Indian guides who were intimately familiar with the terrain and knew the influential men with whom each explorer must negotiate.[82]

James Fenimore Cooper (fig. 38) commented to a friend how deeply he had relied on passages from the published report of the Long Expedition for his own descriptions of the plains in his three Leatherstocking novels, beginning with *The Pioneers* (1823). *The Pioneers* had appeared in serial form in advance of the publication of the entire book, giving his intended audience a taste of what to expect. Those excerpts whetted the public appetite such that when *The Pioneers* was finally released, it sold 3,500 copies the first morning.[83] However, his extended stay in Paris from 1826 to 1833 placed him squarely in Humboldt's orbit.[84] He was in Paris when he began writing *The Last of the Mohicans* (1826) and *The Prairie* (1827), the final work in this series, in which the bleak landscape reinforces the tragic dimensions of the narrative. Cooper's portrayals of the Plains Indians and the melancholy setting in the Great Plains were drawn straight from the pages of Long's report, which described the vast middle of the country as a "great desert," in stark contrast to the fertile fields and thick, old-growth forests of Cooper's beloved Adirondacks. His novels' popularity attested to American readers' eagerness for stories set in the wilderness landscape.

Cooper and Irving extended and amplified the concept of the American wilderness as an essential component of the American character, an ethos that carried over from their published writings to their personal correspondence. From Paris in 1831, Cooper wrote,

> *Your Aunt Pomeroy is afraid we shall become too Europeanized for home. She knows little of our tastes or wishes. There are people who come here, who see us in the possession of advantages that certainly do not fall to all our countrymen, and as most Americans have an exaggerated idea of Europe, and especially of England, they fancy we cannot tear ourselves from a society they imagine so agreeable. Now my longing is for a Wilderness—Cooperstown is far too*

populous and artificial for me and it is my intention to plunge somewhere into the forest, for six months in the year, at my return. I will not quit my own state, but I shall seek some unsettled part of that.[85]

The idea that Cooperstown was too built up, that Cooper would need a refuge from his own refuge in order to restore his soul, speaks volumes about the therapeutic value associated with being in the wilderness. Wilderness, for Cooper, was a necessary corollary to, and antidote for, civilization. The idea was not to jettison one for the other, but to find balance, appreciating the value in both and understanding the limits of each. For Cooper and, by extension, his audience, wilderness was not evil in need of being conquered; it had moral standing, conveying character to those who lived in harmony with nature. When Natty Bumppo abandons civilization at the end of *The Pioneers*, he is convinced that settlers do not understand "how often the hand of God is seen in the wilderness."[86]

As Humboldt's travelogues arrived in print, Americans had a new paradigm for the genre: the appreciation of wilderness scenery on aesthetic grounds combined with marvel at the scientific details. Enthusiasm for wilderness was driven in part by invoking Humboldt's belief in the unity, harmony, and vastness of the universe. Humboldt spent his life examining and comparing world religions, looking for similarities, but rejecting specific dogma. In the United States, those attributes were placed in alignment with religion, in an effort to find God in the wilderness. American transcendentalism, especially in the works of poets and artists, explored that path. Transcendentalists took Humboldt's research and infused it with spirituality.

By the 1820s Americans were amenable to a wilderness aesthetic. The evocation of that appreciation appeared first in literature—both travelogues and fiction—before it became a dominating force in the fine arts. Texts with images praising American scenery occurred in magazines like *Port Folio* and George Ticknor's *North American Review*; every decade the language developed greater confidence, assertiveness, and eloquence in affirming the national character of the American wilderness. Ticknor, who knew Humboldt during his two years in Paris, published numerous articles extolling Humboldt's views on nature.[87] In 1820 *Picturesque Views of American Scenery* used Joshua Shaw's paintings engraved by John Hill to present "our lofty mountains...the unexampled magnitude of our cataracts, the wild grandeur of our western forests...unsurpassed by any of the boasted scenery of other countries."[88]

The laudatory descriptions in American prose and poetry suited local ears; however, Europeans visiting the United States at this time were shocked when they encountered the same landscape that had been lionized in print, perceiving it as horrifying to behold in person: "Instead of a garden, I found a wilderness."[89] It would be left to American artists and writers to invoke the hand of God in the wilderness as part of crafting a cultural identity extolling the beneficial aspects of wild nature.

THE DISCOVERY OF THOMAS COLE

In 1825 the leading figures in defining American artistic culture—William Dunlap, John Trumbull, and Samuel F. B. Morse—discovered Thomas Cole.[90] The story has become its own creation myth: that of a self-taught artist whose formative years were conducted in the wilderness, drawing raw scenery, traveling down rivers and across mountains to reach New York. Although he had studied briefly in Philadelphia and was familiar with the Peales, that prologue did not seem to augur the magnitude of what Cole (fig. 39) would accomplish in painting the American landscape. Cole's idiom of painting the American wilderness, replete with storms and dramatic lighting, has its precursor in Peale's *Exhumation*; but instead of the painting, here it is the discovery of the painter—Cole himself—that is the creation story. The response to three of Cole's paintings in a shop window in 1825 was like the appearance of the artist himself, literally emerging out of nowhere.

That creation myth for Cole—that he was reborn in his personal experience in the American wilderness, and now painted it as an extended emblem of the nation's inherent character—coincided with Humboldt's pervasive influence. Several of the men who were most impressed with Humboldt's ideas were Cole's close friends, including his patron Daniel Wadsworth, and Wadsworth's brother-in-law, Yale geologist Benjamin Silliman. Yet Cole himself seemed content to be carried along in this

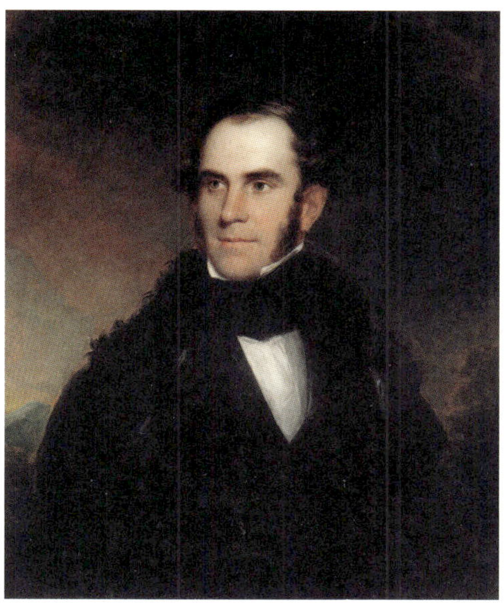

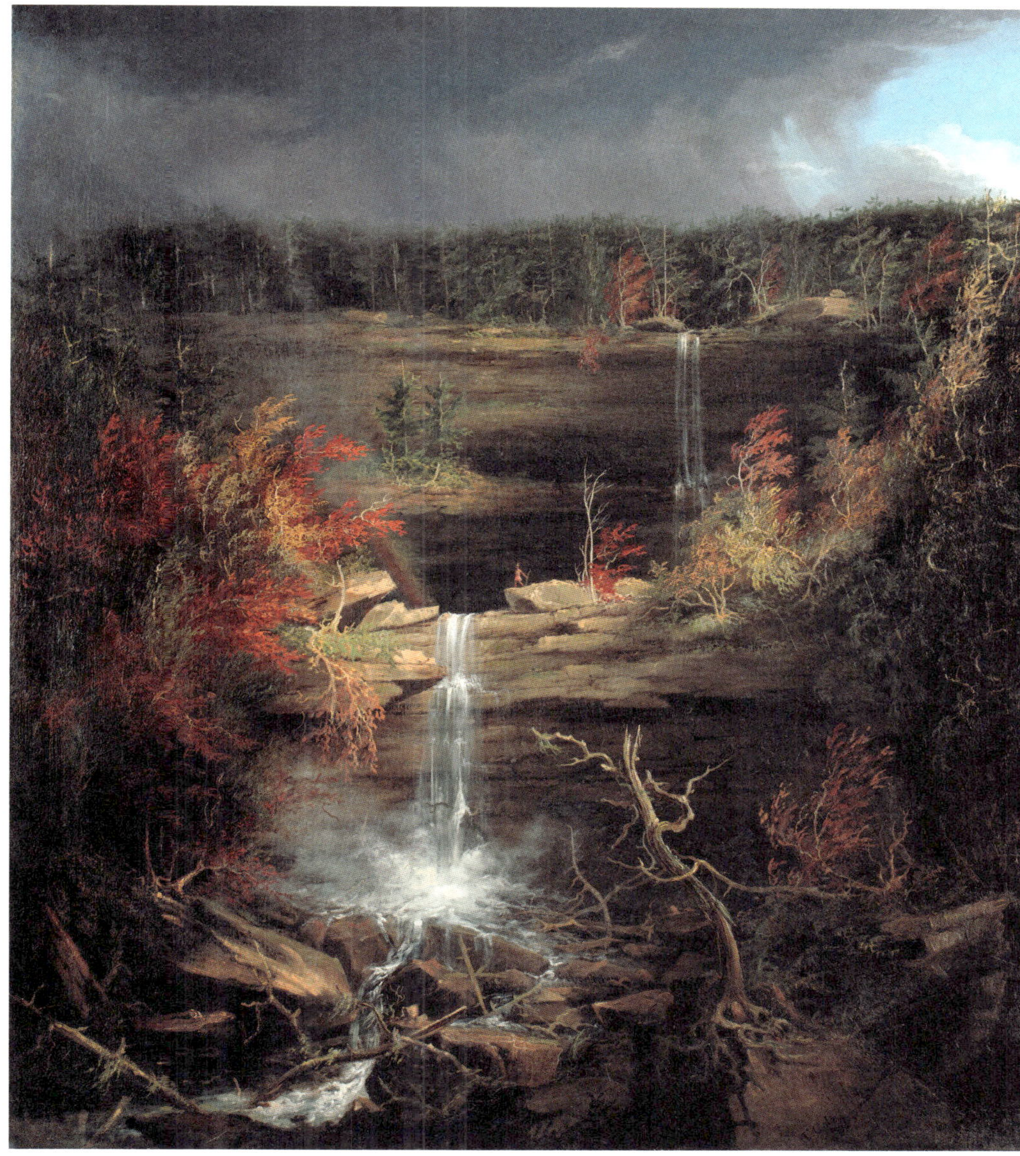

Fig. 39
Asher Brown Durand, *Portrait of Thomas Cole* (1801–1848), 1838, oil on canvas, 30 ½ × 25 in., Berkshire Museum, Pittsfield, MA

Fig. 40
Thomas Cole, *Kaaterskill Falls*, 1826, oil on canvas, 43 × 36 in., Private Collection

wake, without diving deeply into the Prussian naturalist's works. Lauded by men of culture, Cole was made a founding member of the National Academy of Design (NAD) in 1826, and titular head of what would become known as the Hudson River school. As Daniel Huntington said of Cole, he "studied to embody whatever was characteristic of the singular grandeur and wildness of mountain, lake and forest, in the American wilderness."[91] Cole had arrived in the United States from England with his family in 1818, after the War of 1812, at age seventeen. His family had settled in the Ohio River valley, where the predominantly self-taught artist drew extensively from excursions he took in the Alleghenies. As Cole remarked later, "The painter of American scenery has, indeed, privileges superior to any other. All nature here is new to art."[92]

In 1826, the year following his discovery, Cole painted *Kaaterskill Falls* (fig. 40). Combining the drama of steeply falling water with the natural arches of hollowed-out limestone caverns, pictorially this painting merges thoughts of Niagara and Natural Bridge under the same dramatic weather Peale used as his emotional backdrop for the *Exhumation of the Mastodon*. Cole became associated with the development of the wilderness landscape as an American idiom, although his wilderness was often adapted from Cooper's literary treatment. Cole incorporated scenes from Cooper's novels in his paintings, the diminutive figures playing a secondary role to the emotive power of the landscape. In one of the most interesting elisions among the actual, artistic, and literary American landscape, Cole painted scenes intended for the saloon of the steamboat *Albany*, which allowed passengers to move from the painted version to the windows that looked out onto the region that had inspired the works.[93] Cole's *Landscape with Figures: A Scene from "The Last of the Mohicans"* (1826, Terra Foundation, Chicago) (fig. 41) and *Landscape View near the Falls of the Kaaterskill* (1826, private collection), made famous in Cooper's *The Pioneers*, literally sailed up and down the Hudson River. In 1828 James Fenimore Cooper made the trip on the *Albany*, potentially seeing Cole's interpretations of the scenes from his books, which had been completed by December 1826.[94] The synergy that would forever link the Knickerbocker writers with the Hudson River school artists thus appeared to have been preordained.

The significance ascribed to Cole's nature-born, nature-inspired genius may be read in William Cullen Bryant's exhortation as Cole planned his first trip back to Europe since arriving with his family. In 1829 Bryant penned his sonnet "To Cole, The Painter, Departing for Europe," which was published in *The Talisman*. In his poem, Bryant recognized that Europe would offer opportunities for artistic training, but he feared that it might also blunt Cole's appreciation for American scenery. *The Talisman* was one of a growing number of specialty albums and books published between 1825 and 1865 to promote American scenery in literature and in engravings of paintings by Cole, Asher B. Durand, Thomas Doughty, and others. *The Atlantic Souvenir* and *The Token* were notable examples of this kind of illustrated book.[95] That Bryant feared Cole would stray from the path that had led him into the American wilderness reveals how closely American cultural figures identified wilderness landscapes as the defining idiom in American art and literature.

Bryant had assumed the mantle of editor of the *New York Evening Post* in 1826 and began taking his own trips through the Catskills and along the Susquehanna. He was an inveterate walker, taking twenty-five-mile-long excursions into nature as part of his inspiration. His need for the visual, for the experience in nature, was a fundamental part of the process behind his poems and his lectures. Bryant's nature poetry reveled in the features of the surrounding wilderness while Cole established many of the same locales pictorially, just as Cooper had made them relevant in literature.[96] Bryant published accounts of his trips in his newspaper, fostering an appreciation for nature that extended well beyond his own poetry and generating interest in Cooper's novels and Cole's paintings. All three men were founding members of the Bread and Cheese Club, and along with Samuel F. B. Morse created a tight-knit circle

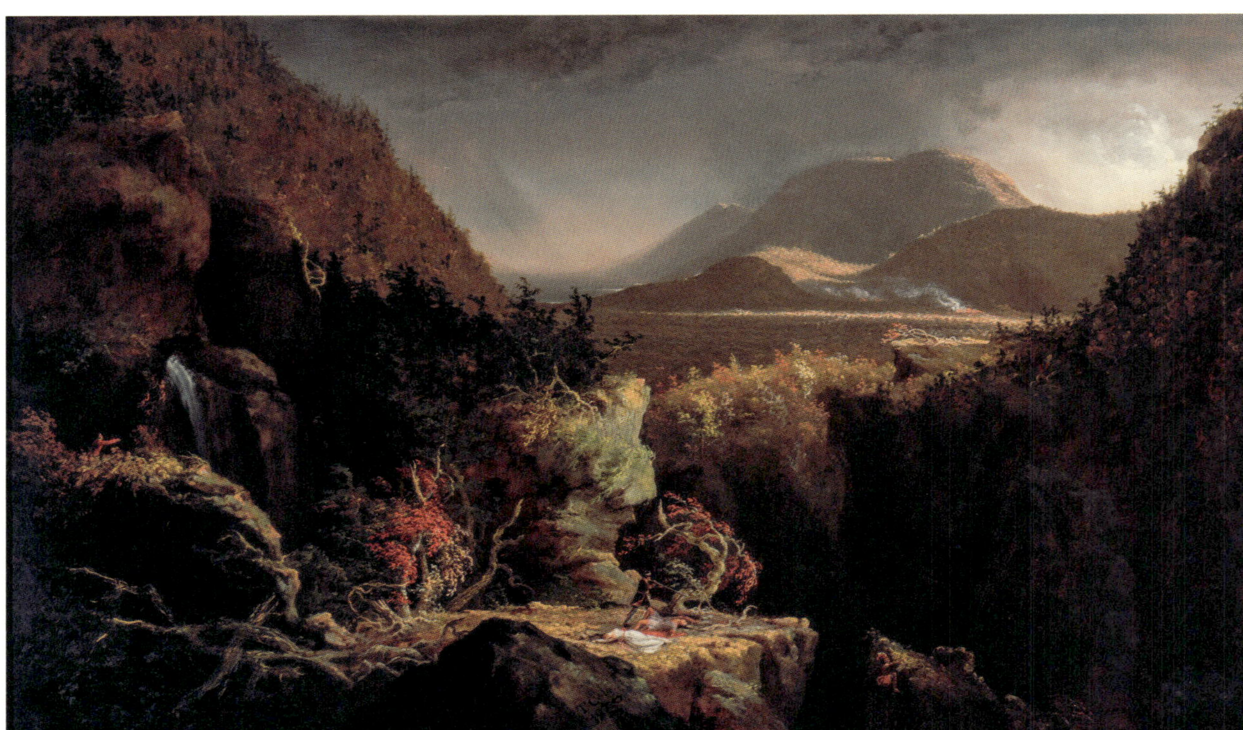

Fig. 41
Thomas Cole, *Landscape with Figures: A Scene from "The Last of the Mohicans,"* 1826, oil on panel, 26 1/8 × 43 1/16 in., Terra Foundation for American Art, Daniel J. Terra Collection, 1993.2

of advocates for American scenery as central to American cultural identity.

By 1830 Bryant and Durand collaborated on a book they titled *The American Landscape*. In the prospectus, Durand confidently asserted,

Nature is not less liberal of the characteristics of beauty and sublimity in the new world, than in the old. The perception of her charms is not less quick or vivid among our countrymen, nor will we believe that there is wanting, either taste to appreciate the truth and effect with which the features are copied, or willingness to reward those who execute the task with success.[97]

The volume contained sixteen plates "executed from Nature." Bryant's preface reads like a synopsis of Humboldt's *Essay on the Geography of Plants*, attuned to the specificities of rocks and climate native to a particular area, and the plants characteristic of each climate zone:

Every climate has its peculiar vegetation, its growth of trees and foliage, its own proper harvests, and its hues of the atmosphere colouring and transmuting all they rest upon. The living creatures also, the quadrupeds of the ground, and the birds of the air, by which the scenery is animated, are different in different regions: even man is not the same in his air and costume, his manner of constructing his habitations, and the tillage he gives the soil. You look on a series of landscapes, and say, this is English, this is Italian; the scene of this is laid in France, of that in Switzerland, of this in Holland, of that in Spain; here is an East India view, here a Persian, and here an African.[98]

After pointing out the geographic—and cultural—specificity of the landscapes around the world, Bryant and Durand then went on to describe the distinctive qualities of American scenery:

Our own scenery has its peculiarities not less strongly marked than those of the old continent. Among these, one of the most striking is the absence of those tamings and softenings of cultivation, continued for ages, which, while they change the general face of the landscape, at the same time break up the unity of its effect. In the masses of majestic forest also, which cover a large portion of our territory, are grouped trees, far exceeding in the number of their kinds those of Europe, and with a correspondent variety of general shape, branches and foliage, giving our scenery a magnificence of vegetable riches unknown to the eastern world.[99]

Bryant exhorted his reader to embrace the wilderness as providing a landscape that is more varied than more cultivated parts of the globe. By calling out as commendable the "absence of those tamings and softenings of cultivation," Bryant identified in untouched wilderness a character both unique and preferable to the dominant European paradigm.

Bryant continued, evoking the singular spiritual qualities associated with the American wilderness, noting,

Foreigners who have visited our country, particularly the mountainous parts, have spoken of a far-spread wildness, a look as if the new world was fresher from the hand of him who made it, the rocks and the very hillocks wearing the shape in which he fashioned them, the waters flowing where he marked their channels, the forests, enriched with a new creation of trees, standing where he planted them; in short, of something which, more than any scenery to which they had been accustomed, suggested the idea of unity and immensity, and abstracting the mind from the associations of human agency, carried it up to the idea of a mightier power, and to the great mystery of the origin of things.[100]

Bryant's invocation of "unity and immensity" tracks directly back to Humboldt. The poet argued for the premise that the American wilderness was bigger, better, more varied, and more interesting—and favored by the hand of its creator. Humboldt himself never invoked a creator or higher being, preferring to use "spirit" as an animating force stripped from religious dogma. But in the United States, transcendentalists followed Bryant's lead in adapting the Humboldtian enterprise and filtering it through a deistic or spiritualistic lens, finding in Humboldt's use of "spirit" a path to God. Bryant's goal was to extend those qualities in the landscape to the culture of its people, in so doing taking Humboldt to his most nationalistic conclusion, cementing nothing less than the cultural identification of America with the grandeur of its landscape features.

The continued synergy between artists and men of letters in extolling the cultural values and virtues of American wilderness took shape in the shared memberships in the Bread and Cheese Club, which evolved into the Sketch Club, and eventually was renamed and reconceived as the Century Association. In 1831 Cooper was named as an honorary academician in the NAD, acknowledging the influence of his novels on the appreciation for wilderness and the importance of his literary subjects to the development of American landscape painting. Icons in nature had thus become associated with national identity across the spectrum of the arts and sciences. Peale had established the parameters with the *Exhumation of the Mastodon*, invoking American history and cultural achievement, appreciated in the midst of a storm sweeping across the landscape. Early views of Natural Bridge and Niagara Falls impressed viewers with the magisterial proportions of American landscape features. Cole and Bryant expanded the canon, extolling the virtues of the Catskills and the Adirondacks as scenery exemplifying national imperatives. As the country expanded, so, too, did the repertoire of unique and notable landscape features deemed part of the nation's cultural identity.

EMERSON'S *NATURE* AND COLE'S "ESSAY ON AMERICAN SCENERY"

By the 1830s Americans had embraced Thomas Cole as the leading practitioner in the burgeoning arena of landscape painting, taking pride in his works while deeming them emblematic of U.S. cultural values. His career was buoyed by the coeval rise of transcendentalist literature concerned with the spiritual qualities associated with time spent in nature. Ralph Waldo Emerson (fig. 42) became the highly regarded exponent of the literary model of Humboldtian science allied with deep spirituality. As Edward Everett's student at Harvard, Emerson made his first notes in his journal about reading Humboldt in 1821.[101] In 1833 he visited Paris and was riveted by the Muséum national d'Histoire naturelle, the most impressive natural history museum of its kind, "without rival anywhere in Europe."[102] While he was there, Emerson was swept into the orbit of two of Humboldt's closest friends, François Arago and Joseph Gay-Lussac. He was deeply impressed by the Jardin des Plants, and like Humboldt, Emerson saw the grand order of nature embodied in the hierarchy of plant life. When he returned to Boston, he delivered an address titled "The Uses of Natural History," in which he pronounced, "The earth is a museum," and explained,

You are impressed with the inexhaustible gigantic riches of nature.... The universe is a more amazing puzzle than ever, as you look along this bewildering series of animated forms, —the hazy butterflies, the carved shells, the birds, beasts, insects, snakes, fish, and the upheaving principle of life everywhere so incipient, in the very rock aping organized forms.... [N]ot a form so grotesque, so savage, or so beautiful, but is an expression of something in man the observer.... [A]n occult relation between the very worm, the crawling scorpions, and the man. I am moved by strange sympathies. I say I will listen to this invitation. I will be a Naturalist.[103]

For Emerson, this pronouncement clarified the cosmic scope of this thinking.

Three years later, in 1836, he published the essay that stands as the most eloquent articulation of his nature-based philosophy. *Nature* introduced the most powerful metaphor for observation and reflection, in the belief that to truly appreciate nature, one must be immersed in it, literally, metaphorically, and spiritually:

We return to reason and faith. There I feel that nothing can befall me in life,—no disgrace, no calamity, (leaving me my eyes,) which nature cannot repair. Standing on the bare ground, —my head bathed by the blithe air, and uplifted into infinite spaces,—all mean egotism vanishes. I become a transparent eye-ball; I am nothing; I see all; the currents of the Universal Being circulate through me; I am part or parcel of God.[104]

Emerson's transparent eyeball would operate on a Humboldtian plane, and although in his writings about the unity of nature Humboldt never invoked a creator, he did embrace a similar uplift of spirit. Emerson's mature writings would remain deeply in tune with the way Humboldt envisioned the unity of nature. When Emerson wrote, "Nature is made to conspire with spirit to emancipate us,"[105] he echoed Humboldt's edict that "nature is the domain of liberty," a refrain that runs through Humboldt's writings, culminating in his magnum opus, *Cosmos*.

Emerson wrote *Nature* in 1835, just as Thomas Cole delivered an address titled *An Essay on American Scenery*, which argued that if the admiration of cultivated gardens distinguished the Old World, then an appreciation of the untamed wilderness is what defined the New. As Cole correctly asserted, a wilderness aesthetic indeed had come to be the defining characteristic of

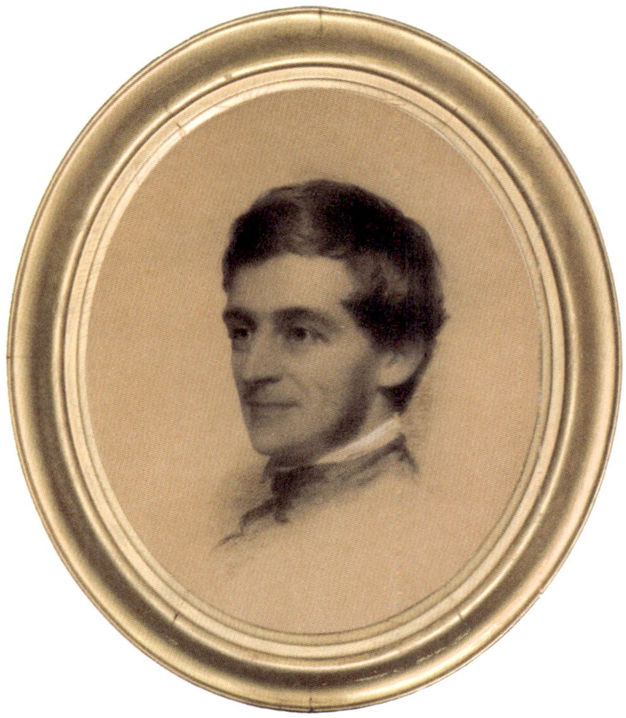

Fig. 42
Eastman Johnson, *Ralph Waldo Emerson* (1803–1882), 1846, charcoal on paper, 21 × 19 in. Courtesy National Park Service, Longfellow House, Washington's Headquarters National Historic Site

American landscape painting. Cole delivered this lecture at the American Lyceum on May 8, and at the NAD on May 16, 1835, and he then published it in the *American Monthly Magazine* in January 1836. In a synergy that cannot be pure coincidence, Cole and Emerson arrived at remarkably similar exhortations to their respective audiences at about the same time: that American nature, wilderness especially, was to be cherished as a critical part of the American character.[106] As Cole proclaimed, "The most distinctive, and perhaps the most impressive, characteristic of American scenery is its wildness."[107] Emerson's growing debt to Humboldt is well established, both in his appreciation of nature and in the essential belief that American society needed to abolish slavery before it could fulfill its promise as a true democracy.[108] Cole's religious faith and his appreciation of nature were essential to his character and his aesthetic philosophy. By this time, Humboldt's advocacy for an American identity based in nature had diffused through American society to such an extent that this was no longer a radical concept. Each man, in his own idiom, tied American cultural nationalism to an appreciation of the landscape.

In the immediate wake of his lecture on American scenery, Cole painted his *View from Mount Holyoke, Northampton, Massachusetts, after a Thunderstorm,* colloquially known as *The Oxbow* (fig. 43A). It was a vista that had been recommended to him by his patron Daniel Wadsworth. Wadsworth had illustrated Benjamin Silliman's published account of their excursion there in 1819.[109] On a monumental canvas measuring more than six feet in length, Cole composed a landscape that was at once a recognizable view and a meditation on Humboldt's view of man's relationship with the natural world. The division of the canvas into sweeping diagonals, with wilderness occupying the high ground and the cultivated landscape spread out below, puts the artist himself at the fulcrum between the new American aesthetic of wilderness appreciation and the European love for the picturesque view. The bend in the river acts as an ocular slingshot, sending the viewer's eye deep across the open expanse to the mountains in the distance, and back again.

The artist sits at the fulcrum between wilderness and cultivation. *The Oxbow* is not about choosing between those states of mind, but cultivating the ability to appreciate the merits of both and to understand that one without the other leaves American landscape aesthetics in a much diminished position. Inverting Archibald Alison, Cole asserts that American associations are about the future, not the past, where the creative tension between the two constructs the ideal environment. Emerson seemed profoundly moved by what Cole portrayed in *The Oxbow.* In his journal entry for October 13, 1837, he noted,

> *New Eyes. What is, appears. Go out to walk with a painter, and you shall see for the first time groups, colors, clouds and keepings, and shall have the pleasure of discovering resources in a hitherto barren ground, of finding as good as a new sense in such a skill to use an old one.*[110]

For Emerson, looking at Cole's landscape became a new lens through which to appreciate a familiar landscape. Cole's handling of color, composition, and weather were like a new sense, added to those Emerson was accustomed to deploying on his own walks across Boston Common.

Nestled in the foreground foliage, Cole smiles, equally comfortable painting both idioms (fig. 43B). In the underdrawing for this composition, Cole placed a red "X" to mark where he would insert himself into the landscape, using his furled umbrella as a marker to direct the viewer's eye toward his.[111]

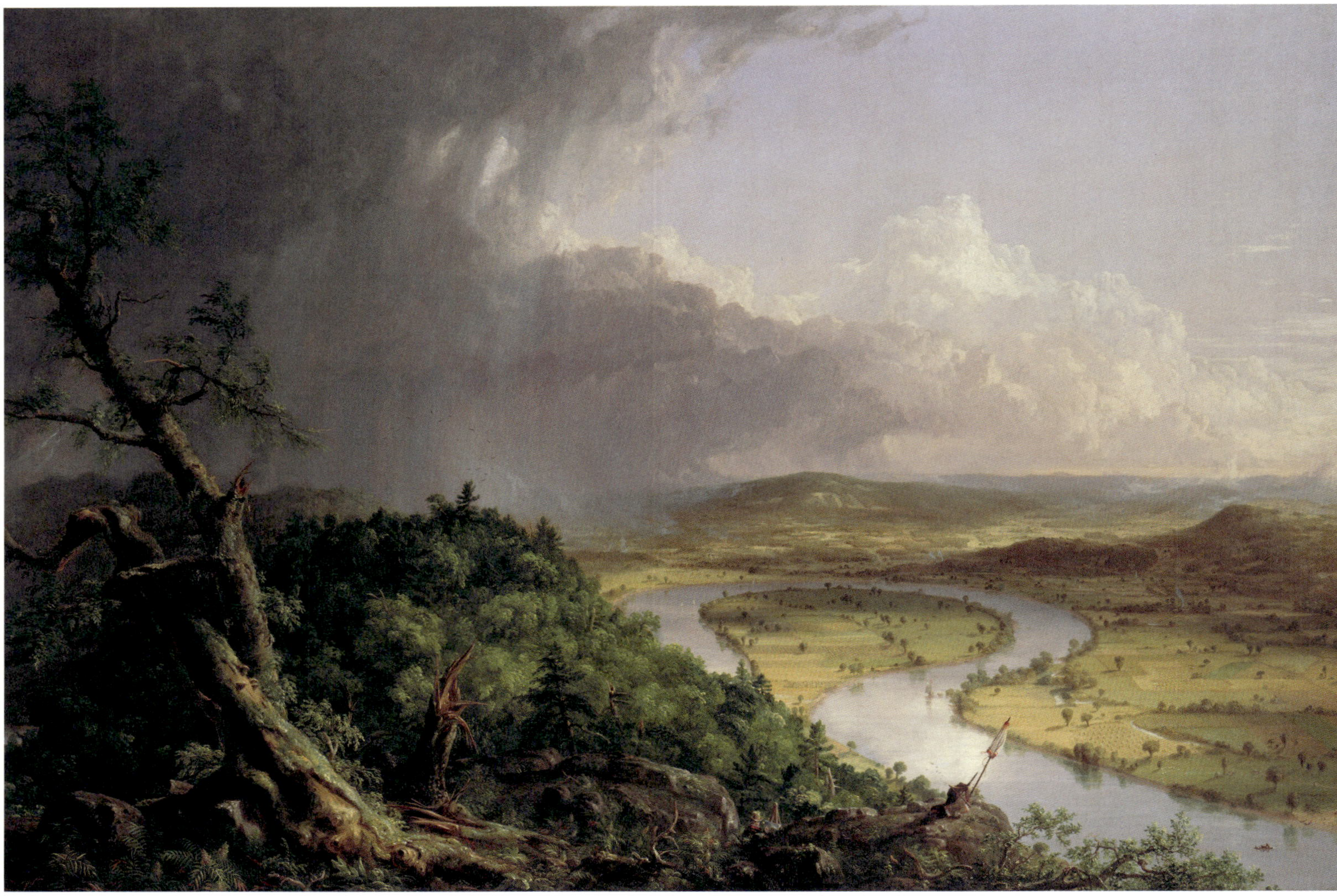

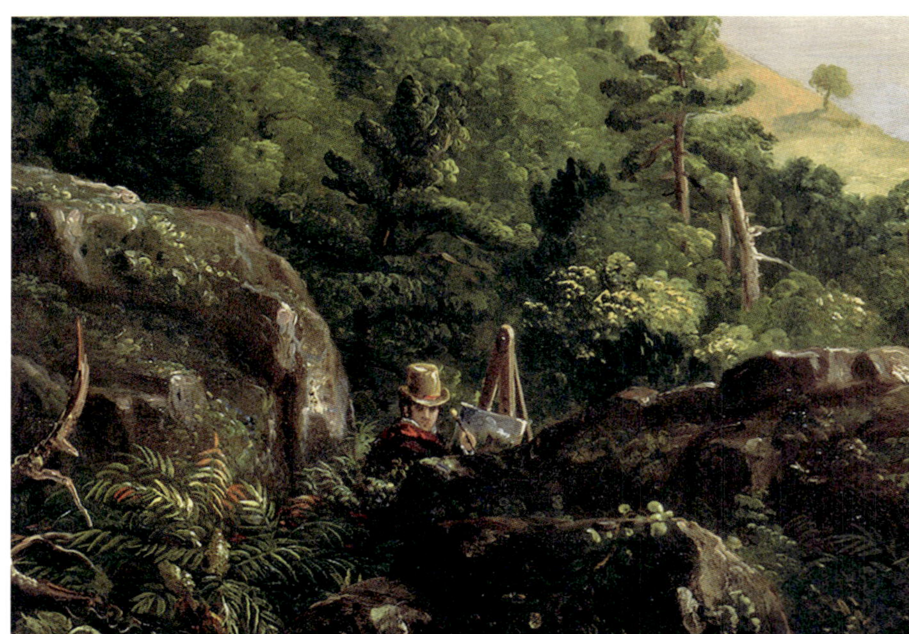

Fig. 43A
Thomas Cole, *View from Mount Holyoke, Northampton, Massachusetts, after a Thunderstorm—The Oxbow*, 1836, oil on canvas, 51 ½ × 76 in., The Metropolitan Museum of Art, NY, Gift of Mrs. Russell Sage, 1908

Fig. 43B
Detail of fig. 43A

Emerson, who had also hiked to the summit to experience this view from Mount Holyoke, wrote in *Nature* that in

> *Crossing a bare common, in snow puddles, at twilight, under a clouded sky, without having in my thoughts any occurrence of special good fortune, I have enjoyed a perfect exhilaration.*[112]

That act of crossing connotes an act of sanctification as well as transit. As Cole sits at the crossroads between nature and culture, so Emerson stands at the crossroads of urban and rural wilderness.[113] Emerson's transparent eyeball visualizes a Humboldtian unity of nature—where unity in this case is between man and his environment. Emerson expanded on his thinking the following year in his essay "The American Scholar," delivered initially as the Harvard Phi Beta Kappa address for 1837:

> *So much of nature as he is ignorant of, so much his own mind does he not yet possess. And, in fine, the ancient precept, "Know thyself," and the modern precept, "Study nature," become at last one maxim.*[114]

By 1840, the opening statement in *American Scenery* asserted, "Nature has wrought with a bolder hand in America.... a lavish and large-featured sublimity... quite dissimilar to the picturesque of all other countries."[115]

Jefferson's *Notes on the State of Virginia* initiated the United States' embrace of a wilderness aesthetic as a part of its cultural identity. The acquisition of Louisiana made curiosity a core constituent of the new nation's political purpose. American identity expanded to encompass what explorers found and politicians acquired. American ambitions took the shape and monumental scale of the landforms that mirrored the country's ideals.

The metaphorical mammoth may have yielded to the scientific nomenclature of the mastodon; however, "mammoth" would remain a popular and vivid term for the American landscape. In 1859 Horace Greeley traveled west to Yosemite, proclaiming the giant redwood trees to be "forest mastodons," evidence of American antiquity found in nature.[116] Two years later, in 1861 Carleton Watkins began using the largest camera ever made. He called his large-format photographs "mammoth-plates," in acknowledgment both of their impressive size and their outsize impact on viewers. His subject was Yosemite, itself composed of a series of mammoth-scale geological features. The magnitude of those individual geological features and the sweep of the continent would each lend themselves to such metaphorical pronouncements.

Humboldt had uncoupled the Great Chain of Being in natural history. Instead of accepting a deterministic hierarchy, he posited a more fluid set of interconnections. In a similar vein, American artists uncoupled the established hierarchy of genres, positioning landscape as a substitute for history painting, a metaphorical replacement for religious painting, and a form of self-portraiture. Landscape would blend the unheralded "view" with a "higher style of art" to form a genre that served as a corollary to the aftermath of the Revolution. Liberty would be couched in the structures of European civilization, but adapted to find virtue in what was here, rather than hewing to a set of value judgments that did not map onto the reality of the American experience. By mid-century the affiliation of landscape and identity was solidly fixed in the American psyche, deeply rooted enough to have shed Humboldt's specific imprimatur and become part of the fabric of American culture.

NOTES

1 Robert E. Schofield, "Charles Willson Peale and His Philadelphia Museum, 1784–1827: The Science Education of an Enlightened Entrepreneur," in "Technology, Medicine and Science in American Culture and Society," special issue, *American Studies* 30, no. 2 (Fall 1989): 23–24. Buffon is credited with the first thirty-six volumes before his death. The definitive edition included forty-four volumes.

2 Quoted in Ottmar Ette, "'Not just brought about by chance': Reflections on Globalization in Cornelius de Pauw and Alexander von Humboldt," *Studies in Travel Writing* 15, no. 1 (February 2011): 4–5. Translated from German by Mark Minnes, quoting de Pauw's *Discours preliminaire* in the first volume of his *Récherches philosophiques sur les Américains, ou Mémoires Intéressante pour servir à l'Histoire de l'Espèce Humaine* (Berlin, 1768).

3 Abbé Raynal, *Histoire philosophique et politique,* 6 vols. (Amsterdam, 1770; Paris, 1772). Over the next thirty years, subsequent editions were published in English, German, Dutch, Italian, and Spanish. Napoleon carried a set on his European campaigns, and numerous copies could be found in colonial American plantation homes. Cited and discussed in Lee Alan Dugatkin, *Mr. Jefferson and the Giant Moose: Natural History in Early America* (Chicago: University of Chicago Press, 2009), 39–40.

4 Raynal, *Histoire philosophique et politique,* 6:376. Translated by William Tucker (Louisville, KY) and quoted in Dugatkin, *Mr. Jefferson and the Giant Moose,* 44. Jefferson pushed back on Raynal, rebuking him for dismissing the contributions of Benjamin Franklin, America's foremost scientist, with whom Raynal was personally acquainted. Thomas Jefferson, "Query IV," in *Notes on the State of Virginia*, ed. William Peden (Chapel Hill: University of North Carolina Press for the Omohundro Institute of Early American History and Culture at Williamsburg, Virginia, 1954), 64–5.

5 Denis Diderot supervised publication of his *Encyclopédie, ou dictionnaire raisonné des sciences, des arts et des métiers* (*Encyclopedia, or a Systematic Dictionary of the Sciences, Arts, and Crafts*) in France between 1751 and 1772, with later supplements, revised editions, and translations. The early editions provided a cursory treatment of the Americas, which can be summed up as South America produced gold and silver, while North America furnished beaver pelts. The addendum published in 1776 expands this coverage to more than twenty pages written in part by de Pauw; here the language was less inflammatory but made many of the same points.

6 For a thorough and engaging discussion of these three naturalists and their milieu, see Dugatkin, *Mr. Jefferson and the Giant Moose.*

7 Jefferson knew Marbois. The two of them, along with George Washington, were all elected to membership in the APS in 1779. See Gilbert Chinard, "Jefferson and the American Philosophical Society," *Proceedings of the American Philosophical Society* 87 no. 3 (1943): 264. Marbois subsequently became treasurer to Napoleon and was involved in the financial negotiations for the Louisiana Purchase. He also became close friends with the Marquis de Lafayette and Humboldt in Paris. For more on these relationships, see chapter 4 in this volume.

8 Jefferson to Dupont, 2 March 1809, Thomas Jefferson Papers, Library of Congress; quoted in Dugatkin, *Mr. Jefferson and the Giant Moose*, 52.

9 The Abbé Raynal was loath to give up this argument. Jefferson liked to tell the story of a dinner Franklin hosted in about 1768 or 1769: "He [Franklin] had a party to dine with him one day at Passy, of whom one half were Americans, the other half French & among the last was the Abbé. During the dinner he got on his favorite theory of the degeneracy of animals, and even of man, in America, and urged it with his usual eloquence. The Doctor at length noting the accidental stature and position of his guests, at table, 'come, says he, M. l'Abbé, let us try this question by the fact before us. We are here one half Americans, & one half French, and it happens that the Americans have placed themselves on one side of the table, and our French friends are on the other. Let both parties rise, and we will see on which side nature has degenerated.' It happened that his American guests were Carmichael, Harmer, Humphreys and others of the finest stature and form, while those of the other side were remarkable diminutive, and the Abbé himself particularly, was a mere shrimp. He parried the appeal however, by a complimentary admission of exceptions, among which the Doctor himself was a conspicuous one." Jefferson included this anecdote in a letter to Robert Walsh, 4 December 1818. It was published in the Philadelphia *National Gazette and Literary Register* on 5 April 1820 and was widely reprinted. Thomas Jefferson, *The Papers of Thomas Jefferson*, Retirement Series, vol. 13, *22 April 1818 to 31 January 1819*, ed. J. Jefferson Looney (Princeton, NJ: Princeton University Press, 2016), 464–67.

10 Ebeling to "Rev. Sir" [Jedidiah Morse], Hamburg, 26 June 1794, Harvard College Library; quoted in Ralph H. Brown, "The American Geographies of Jedidiah Morse," *Annals of the Association of American Geographers* 31, no. 3 (September 1931): 155.

11 In a letter to his brother, Wilhelm, dated 21 September 1801, Alexander von Humboldt wrote that the Carib people were "le plus fort et le plus musclé que j'aie jamais vu; à eux seuls ils contredisent les reveries de Raynal et de Pauw sur la faiblesse et la dégénérescence de l'espèce humaine dans le Nouveau-Monde. Un Caraïbe adulte ressemble à un hercule fondu dans l'airain." Alexander von Humboldt, *Lettres américaines d'Alexandre de Humboldt*, with a preface and notes by Dr. E. T. Hamy (Paris: E. Guilmoto, 1905), 122. This passage translates as "a race of the largest and strongest people I have ever seen. This tribe alone disproves Raynal's and Pauw's wild theories of weakness and degeneration in the people of the New World. A fully grown Carib is like a cast-iron Hercules"; in Hanno Beck, "Alexander von Humboldt: Letters from His Travels," in *Alexander von Humboldt, 1769–1859* (Bonn: Inter Nationes, 1969), 138.

12 Alexander von Humboldt and Aimé Bonpland, *Personal Narrative of Travels to the Equinoctial Regions of America, during the Years 1799–1804*, trans. and ed. Thomasina Ross (London: Henry G. Bohn, 1852), 1:434.

13 "Il a peint avec une verité frappante les scènes de la nature américaine. On croit voguer avec lui sur les fleuves, se perdre avec lui dans la profondeur de ces bois qu'ont d'autres limites que les rivages de 'Océan et a chaîne des accidens de la lumière et de l'ombre, et toujours se descriptions, se rattachant à un ordre de choses plus élevé; manènènt quelque souvenir de l'homme, ou des reflexions sur la vie." François-René de Chateubriand, "On Monsieur de Humboldt's Voyage," *Le Conservateur: Le Roi, la Charte et les Honnêtes Gens* 6 (1820), 32; translated in *Cosmos and Colonialism: Alexander von Humboldt in Cultural Criticism*, ed. Rex Clark and Oliver Lubrich, trans. Naomi Lubrich (New York: Berghahn Books, 2012): 28–29.

14 "J'aimerais Vous parler encore d'un objet que Vous avez si ingénieusement traité dans Votre ouvrage sur la Virginie, des dents de Mammot que nous avons découverts dans les Andes de l'Hémisphère *austral* à 1700 toises de hauteur sur l'Océan Pacifique. Mon ami le C. Cuvier en donnera la description anatomique." Humboldt to Jefferson, 24 May 1804, in Ingo Schwarz, ed. *Alexander von Humboldt und die Vereinigten Staaten von Amerika: Briefwechsel* (Berlin: Akademie Verlag, 2004), 89–90, Letter 3. Translation by J. Rogers, F. Baron, and S. Rebok, Humboldt Digital Library.

15 Jefferson asserted, "But to whatever animal we ascribe these remains, it is certain such a one has existed in America, and that it has been the largest of all terrestrial beings. It should have sufficed to have rescued the earth it inhabited, and the atmosphere it breathed, from the imputation of impotence in the conception and nourishment of animal life on a large scale: to have stifled, in its birth, the opinion of a writer, the most learned too of all others in the science of animal history, that in the new world, 'La nature vivante est beaucoup moins agissante, beaucoup moins forte:' that nature is less active, less energetic on one side of the globe than she is on the other." Jefferson here quoted Buffon, and in doing so, destroyed his dismissive argument with empirical evidence. He concluded by further noting that Buffon erroneously believed "that the animals common both to the old and new world, are smaller in the latter." Jefferson, *Notes on the State of Virginia*, 47.

16 Ibid., 45.

17 William Hunter, "Observation on the Bones, Commonly Supposed to be Elephants Bones, Which Have Been Found near the Ohio River," *Philosophical Transactions of the Royal Society of London* 58 (1769): 37; quoted in Mark V. Barrow Jr., *Nature's Ghosts* (Chicago: University of Chicago Press, 2009), 29n60.

18 Charles Le Moyne had been part of a French exploratory party that uncovered mastodon bones. They packaged up several of them, including a femur and a tusk, and sent them to the Jardin du Roi in Paris.

C. Bertrand Schultz, Lloyd G. Tanner, Frank C. Whitmore Jr., Louis L. Ray, and Ellis C. Crawford, "Paleontological Investigations at Big Bone Lick State Park, Kentucky: A Preliminary Report," *Science*, n.s., 142, no. 3596 (29 November 1963): 1167.

19 Michaelis was a physician with the Hessian mercenaries during the American Revolution. Whitfield Bell, "A Box of Old Bones: A Note on the Identification of the Mastodon, 1766–1806, " *Proceedings of the American Philosophical Society* 93, no. 2 (16 May 1949): 171–74.

20 Peale's inaugural board had twenty-seven members. He included three members of President George Washington's cabinet and several of the wealthiest men in Philadelphia, two dozen of whom were APS members. Many of them had met Humboldt in Philadelphia or Washington: John Mifflin, Robert Patterson, David Rittenhouse, John Vaughan, Nicholas Collin, William Barton, Caspar Wistar Jr., James Madison, and Thomas Jefferson. For a description of each man, see Charles Coleman Sellers, *Mr. Peale's Museum: Charles Willson Peale and the First Popular Museum of Natural Science and Art* (New York: W. W. Norton, 1980), 56–58.

21 "American Miracle," *Gazette of the United States (Philadelphia)*, 23 February 1802, 2.

22 The residents of Cheshire, Massachusetts, distributed an "Ode to the Mammoth Cheese, Presented to Thomas Jefferson, President of the United States" on 1 January 1802. *The Western Star* in Lenox, Massachusetts, opined that a mammoth cheese deserved a "Mammoth Apple Pye" fifteen feet wide. A Washington resident reported that Jefferson had provided Congress with a large slice of the cheese to accompany a "Mammoth Loaf of Bread, eight or ten feet in length, and of the breadth and depth of a large Coffin" provided by a local baker. The previous October, a report surfaced in Boston that the Cheshire mammoth cheese was being made, and in the *Columbian Centinel* a letter from "An American" offered a "truly Republican Beet" weighing nine pounds, to be made into pickles to accompany the cheese. I am grateful to Dr. Nancy Siegel for all of the references to mammoth food products.

23 The mammoth and Natural Bridge had become so closely associated with Jefferson that an opposition newspaper in Philadelphia ran a story announcing, "A baker in this city offers *Mammoth bread* for sale. We suppose that his gigantic loaves were baked at a *Salt Lick*, and perhaps may form a great *rock bridge*, or *natural* arch, between the mouth and maw of a voracious republican." *Port Folio* 2, no. 4, 30 January 1802, 31. A few months later the newspaper referred to Jefferson as "the mammoth chief," in *Port Folio* 2, no. 18, 8 May 1802, 143.

24 The handbill is reproduced in Sellers, *Mr. Peale's Museum*, 147. Sellers describes the parade in Philadelphia to advertise the opening of the Mammoth Room during which the skeleton was proclaimed the "ninth wonder of the world." Sellers, *Mr. Peale's Museum*, 142.

25 Charles Willson Peale to Dr. Mitchell, 26 July 1803, *Charles Willson Peale Letterbook #4 (9 January 1803–25 December 1803)*, p. 72, American Philosophical Society, Peale-Sellers Collection; cited in Jessie Poesch, *Titian Ramsay Peale (1799–1885) and His Journals of the Wilkes Expedition* (Philadelphia: American Philosophical Society, 1961), 8. Peale's aspirations guided his organizational goals but were never realized in the collections.

26 Rembrandt Peale's proposed trip to Europe is discussed in the *Philadelphia Gazette*, 24 December 1802; cited in Paul Semonin, *American Monster: How the Nation's First Prehistoric Creature Became a Symbol of National Identity* (New York: New York University Press, 2000), 327–30; and in Sellers, *Mr. Peale's Museum*, 141–47. The tour was brief; on 22 November 1803, *Poulson's American Daily Advertiser* noted it had returned to the United States. See Sellers, *Mr. Peale's Museum*, 157.

27 "Mr Peale is making progress in his Skeleton of the Non Descript, he puts in no other Bones, than those found together, & without any Doubt belonging to the Same animal; —Corresponding parts Deficient, he imitates in Wood—his Son completing on the Same principle, another, from materials found in other Places, & proposes taking it to Europe." John Vaughan to Jefferson, Philadelphia, 19 November 1801, in Thomas Jefferson, *The Papers of Thomas Jefferson*, vol. 35, *1 August to 30 November 1801*, ed. Barbara B. Oberg (Princeton, NJ: Princeton University Press, 2008), 698–99.

28 *Mémoires pour servir à l'histoire naturelle des animaux, dressés par M. Perrault (Memoirs to serve the natural history of animals, prepared by Monsieur [Claude] Perrault)*, vol. 3, part 3 of the *Mémoires de l'Académie royale des sciences*, (Paris, 1734), 101–156, plates 23–24. An annotated copy of this belonged to John Vaughan and is in the APS Library collection. See Earle Spamer, "When Is a Mastodon Not?," APS Library Blog post, 23 October 2018, https://amphilsoc.org/blog/when-mastodon-not-1.

29 Charles Willson Peale to Rembrandt and Rubens Peale, Philadelphia 1 April 1803, Charles Willson Peale diary, (original spelling intact), in Lillian Miller et al., eds., *The Selected Papers of Charles Willson Peale and His Family*, vol. 2, *Charles Willson Peale: The Artist as Museum Keeper, 1791–1810*, part 1. (New Haven, CT: Yale University Press for the National Portrait Gallery, 1988), 517, 519. Peale had seen Rembrandt's article, "A Short Account of the Mammoth," *Philosophical Magazine* 14 (1803): 162–69, accompanied by the comparative engraving.

30 Visiting Frenchman Edouard Montule published an engraving of the Peale mastodon, tusks down, following his tour of the United States in 1815. Alexander Anderson's 1802 drawing erroneously placed the tusks emanating from the eye sockets, as though they were cow horns, above the creature's eyes (the eyes were located in the sockets for the tusks). This iteration was not widely accepted. See Semonin, *American Monster*, 333.

31 Howard C. Rice Jr., "Jefferson's Gift of Fossils to the Museum of Natural History in Paris," *Proceedings of the American Philosophical Society* 95, no. 6 (21 December 1951): 623. In 1796 Cuvier proposed that African and Indian elephants were separate species and that they were distinct from fossil mammoths and the American incognitum. In 1806 Cuvier named the American fossil a mastodon.

32 Peale to Jefferson, Philadelphia, 3 April 1809, in Lillian Miller et al., eds., *The Selected Papers of Charles Willson Peale and His Family*, vol. 2, *Charles Willson Peale: The Artist as Museum Keeper, 1791–1810*, part 2. (New Haven, CT: Yale University Press for the National Portrait Gallery, 1988), 1189–90. Peale had hosted a dinner under the skeleton in 1802, when it was first placed on display in his museum. As Rembrandt prepared to take the second skeleton to Europe in 1802, he assembled it sufficient to host thirteen people for a boozy dinner "WITHIN THE BREAST of the animal, all comfortably seated around a small table, and one of Mr. Hawkins's Patent Portable Piano's—after which the following toasts were drank, accompanied with music." "American Miracle," *Gazette of the United States (Philadelphia)*, 23 February 1802, 2; quoted in Miller et al., *Selected Papers of Charles Willson Peale and His Family*, vol. 2, part 1, 409.

33 The dinner, the mastodon, and by extension Jefferson and his supporters, the Peales, were mocked anonymously in the *Port Folio* 2, no. 9, 6 March 1802, 71–73. That parody was written by Samuel Ewing at the behest of Thomas Boylston Adams. Letter from Thomas Boylston Adams to William Smith Shaw, 4 May 1802, Adams Papers, Massachusetts Historical Society. The entire episode is reprinted and discussed in Miller et al., *Selected Papers of Charles Willson Peale*, vol. 2, part 1, 401–409.

34 Peale to Jefferson, The Museum, Philadelphia, 19 August 1804, in Miller et al., *Selected Papers of Charles Willson Peale*, vol. 2, part 2, 747–48.

35 Peale to Benjamin West, The Museum, Philadelphia, 16 December 1807; quoted in Charles Coleman Sellers, "Charles Willson Peale and 'The Mammoth Picture,'" in *Exhuming the First American Mastodon or the Mammoth Picture* (Baltimore: Peale Museum Historical Series, Publication number 7, February 1951), n.p.

36 David Ward notes that the original title was *Exhuming the Mammoth*; Peale changed the name in 1809 after accepting Georges Cuvier's reclassification of the American bones as those of a mastodon. David C. Ward, *Charles Willson Peale: Art and Selfhood in the Early Republic* (Berkeley: University of California Press, 2004), 148.

37 Peale provided Jefferson with a detailed description of the entire process, including the pump apparatus and the manpower required to remove the water and the bones. Peale to Jefferson, The Museum, Philadelphia, 11 October 1801, in Jefferson, *Papers of Thomas Jefferson*, 35:430–36.

38 For a thorough discussion of this painting as a blend of portraiture, genre, and history painting, see Lillian B. Miller, "Charles Willson Peale as History Painter: *The Exhumation of the Mastodon*," in *New Perspectives on Charles Willson Peale: A 250th Anniversary Celebration*, ed. Lillian B. Miller and David C. Ward (Pittsburgh, PA: University of Pittsburgh Press, 1991), 145–65.

39 The APS provided funds up front to allow Peale to purchase the rights to excavate the bones and keep what he found.

40 Quoted in Lorett Treese, "Monster Bones: Charles Willson Peale and the Mysterious Nondescript Animal," *Pennsylvania Heritage* 44, no. 2 (Spring 2018): 7.

41 Humboldt, *Researches, Concerning the Institutions and Monuments of the Ancient Inhabitants of America, with Descriptions and Views of Some of the Most Striking Scenes in the Cordilleras!* trans. Helen Maria Williams (London: Longman, Hurst, Ross, Orme & Brown, 1814), 1:12.

42 Quoted in Philip F. Rehbock, "Discovery of Species Extinction at the Beginning of the 19th Century and Its Acceptance in Britain" (paper presented at the History of Paleontology Conference, Cambridge, 1997), 8; cited in Barrow, *Nature's Ghosts*, 40.

43 Benjamin Smith Barton, "A Discourse on Some of the Principle Desiderata in Natural History, and on Best Means of Promoting the Study of This Science in the United States," (Philadelphia: Denham and Town, 1807), 20; quoted in Barrow, *Nature's Ghosts*, 42.

44 Jefferson to Caspar Wistar Jr., 19 December 1807, Thomas Jefferson Papers, Library of Congress.

45 Humboldt wrote of the power of "wild and colossal nature" to carry cultural meaning. Ette makes a nuanced case for reading Humboldt's comparisons and contrasts between Old World and New in Ottmar Ette, "Alexander von Humboldt: The American Hemisphere and Trans-Area Studies," *Iberoamericana*, n.s., 5, no. 20 (2005): 85–108.

46 "An Oration, on the Cession of Louisiana to the United States, Delivered on the 12th of May, 1804, in St. Michael's Church, Charleston, South Carolina, at the request of a number of the inhabitants, and published by their desire, by David Ramsay, M.D.," *National Intelligencer* (Washington, DC), 11 June 1804.

47 Hans Huth, *Nature and the American: Three Centuries of Changing Attitudes* (Lincoln: University of Nebraska Press, 1957), 26–27, 32.

48 Ticknor, journal entry, Charlottesville, 7 February 1815, in George Ticknor, *Life, Letters, and Journals of George Ticknor*, ed. George Hillard, Mrs. Anna Eliot Ticknor, and Miss Anna Eliot Ticknor, vol. 1 (Boston: Houghton Mifflin, 1909), 34.

49 Ibid., 1:35.

50 In a letter to John Trumbull, Jefferson hoped the artist would do "yourself and your country the honor of presenting to the world this singular landscape, which otherwise some bungling European will misrepresent." Jefferson to Trumbull, 20 February 1791, in Thomas Jefferson, *The Papers of Thomas Jefferson*, vol. 19, *24 January to 31 March, 1791*, ed. Julian P. Boyd (Princeton, NJ: Princeton University Press, 1974), 298–301; cited in Katie McKinney, "William Roberts's 'Excursion over the Mountains': Backcountry Landscapes 'from the pencil of a Virginian,'" *Journal of Early Southern Decorative Arts* 39 (January 2019): np., https://www.mesdajournal.org/ 2019/william-robertss-excursion-over-the-mountainsbackcountry-landscapes-from-the-pencil-of-a-virginian/.

51 Jefferson, *Notes on the State of Virginia*, 24.

52 "The Natural Bridge, in 1826 (an early account from the New York *Christian Herald*), in Edmund Tompkins and J. Lee Davis, *The Natural Bridge and Its Historical Surroundings* (Natural Bridge of Virginia, Inc., 1939), 34.

53 Jefferson, *Notes on the State of Virginia*, 25.

54 See Wendy Bellion, "City as Spectacle: William Birch's Views and the Chestnut Street Theatre," *Studies in the History of Gardens and Designed Landscapes* 32, no. 1 (2012): 15.

55 Alexander von Humboldt, *Researches, Concerning the Institutions and Monuments of the Ancient Inhabitants of America, with Descriptions and Views of Some of the Most Striking Scenes in the Cordilleras!* (London, 1814), 1:59.

56 Ibid., 1:76–77.

57 Jared Sparks, "A New American Atlas, Containing Maps of the Several States of the North American Union...," *North American Review* 18 (Boston, 1824): 390; quoted in James V. Walker, "Henry S. Tanner and Cartographic Expression of American Expansionism in the 1820s," Paulus Swaen Auction and Gallery, https://www. swaen.com/Henry-S-Tanner.php.

58 *National Intelligencer*, 3 January 1823; quoted in Walker, "Henry S. Tanner and Cartographic Expression of American Expansionism in the 1820s."

59 Tanner's map was published first in 1822 and subsequently in 1825; future editions kept the cartouche through 1845, when the map was used by the James K. Polk administration to justify the invasion of Mexico.

60 J. Hector St. John de Crèvecoeur, *Letters from an American Farmer* (Gloucester, MA: P. Smith, 1968), 272; quoted in Roderick Nash, *Wilderness and the American Mind*, rev. ed. (New Haven, CT: Yale University Press, 1973), 28.

61 Nash, *Wilderness and the American Mind*, 53. John Bartram, *Observations on the Inhabitants, Climate, Soil, Rivers, Productions, Animals, and other matters worthy of notice. Made by Mr. John Bartram, in his Travels from Pennsylvania to Onondago, Oswego and the Lake Ontario, in Canada* (London: J. Whiston and B. White, 1751), 16.

62 William Bartram, *The Travels of William Bartram: Naturalists' Edition*, ed. Francis Harper (New Haven, CT: Yale University Press, 1958), 212–13; quoted in Nash, *Wilderness and the American Mind*, 54.

63 William Bartram, *Travels through North and South Carolina, Georgia, East and West Florida* (London, 1792), 48; quoted in Huth, *Nature and the American*, 20–21.

64 Bartram, *Travels of William Bartram*, 69; quoted in Nash, *Wilderness and the American Mind*, 54.

65 See Louis-Philippe, King of France, *Diary of My Travels in America*, trans. Stephen Becker (New York: Delacorte Press, 1977).

66 Louis-Philippe and his time in North America is discussed in chapter 6 in this volume.

67 Gilbert Chinard, *L'exotisme américain dans l'oeuvre de Chateaubriand* (Paris: Hachette, 1918), 242–55; cited in Huth, *Nature and the American*, 21. Back in Europe, Chateaubriand later wrote two books based in Kentucky, including *Atala*. Although *Atala* is an American Indian narrative, it is heavily classicized, its wilderness setting a foil for the central Christian theme. Eugène Delacroix painted this subject while George Catlin and the Iowa Indians were in Paris; see chapter 6 in this volume.

68 François René Vicomte de Chateaubriand, *Recollections of Italy, England, and America* (London: Henry Colburn, 1815), 1:206; quoted in Nash, *Wilderness and the American Mind*, 49.

69 "Il n'y a de vieux en Amérique que les bois, enfants de la terre, et la liberté, mère de toute société humaine; cela vaut bien des monuments et des aieux." François René Vicomte de Chateaubriand, *Voyage en Amérique, suivi des Natchez; par m. le vicomte de Chateaubriand* (Paris, 1838), 43. Translation by author.

70 William Gilpin, *Remarks on Forest Scenery* (London: R. Blamire, 1794), 2:166; cited in Edward J. Nygren, "From View to Vision," in Edward J. Nygren with Bruce Robertson, *Views and Visions: American Landscape before 1830* (Washington, DC: Corcoran Gallery of Art, 1986), 18.

71 Thomas Seir Cummings, *Historic Annals of the National Academy of Design* (Philadelphia, 1865), 12.

72 Benjamin Silliman, *Remarks Made on a Short Tour between Hartford and Quebec in the Autumn of 1819* (New Haven, CT: S. Converse, 1820; 2nd ed., 1824), 18; cited in Bruce Robertson, "The Picturesque Traveler in America," in Nygren, *Views and Visions*, 206.

73 "Humboldt's Travels," *New Monthly Magazine and Literary Journal*, American ed. 2, no. 7 (1 January 1821), 314. Other excerpts appeared at this time in "ART. XII. —From Baron Humboldt's Personal Narrative. —Vol. 4," *The Port Folio* 12, no. 1 (September 1821), 136; "Humboldt's and Bonland's [*sic*] Travels," *Atheneum; or, Spirit of the English Magazines* 9, no. 12 (15 September 1821), 469; and "Travels of Alexander von Humboldt. Williams, Helen Maria," *Atheneum; or, Spirit of the English Magazines* 10, no. 4 (15 November 1821), 132.

74 Alexander von Humboldt, *Reise in die Äquinoktial-Gegenden des Neuen Kontinents*, ed. Ottmar Ette (Frankfurt and Leipzig: Insel Verlag, 1991), 1:35. Translated by Ottmar Ette in "Alexander von Humboldt: The American Hemisphere and TransArea Studies," 101.

75 *Port Folio* 1, February 1809, 101; cited in Nygren, "From View to Vision," 39.

76 Estwick Evans, *A Pedestrious Tour of Four Thousand Miles through the Western States and Territories during the Winter and Spring of 1818* (Concord, NH, 1819), 102; quoted in Nash, *Wilderness and the American Mind*, 56.

77 Washington Irving, "The Author's Account of Himself," in *The Sketch-Book of Geoffrey Crayon* (London: John Miller, 1820), 3.

78 Pierre M. Irving, *The Life and Letters of Washington Irving* (New York: G. P. Putnam, 1862), 1:42.

79 For a discussion of this aspect of Irving's career, see chapter 2 in this volume.

80 Washington Irving, "A Letter from Washington Irving to Henry R. Schoolcraft, Esq.," *Littell's Living Age*, January–March, 1852, 104.

81 Peter Antelyes does an outstanding job of chronicling Irving's western narratives, examining how they alternately celebrate and critique expansionist tendencies and the excesses of enterprise. *Tales of Adventurous Enterprise: Washington Irving and the Poetics of Western Expansion* (New York: Columbia University Press, 1990).

82 For an extended discussion of Humboldt and Americans' relationships with native communities, see chapter 6 in this volume.

83 Huth, *Nature and the American*, 33.

84 For a discussion of Cooper's relationship with Humboldt, see chapter 4 in this volume.

85 James Fenimore Cooper to Richard Cooper, Paris, 25 May 1831, letter 207, in James Fenimore Cooper, *Letters and Journals of James Fenimore Cooper*, ed. James Franklin Beard (Cambridge, MA: Belknap Press of Harvard University Press, 1960), 2:89.

86 James Fenimore Cooper, *The Pioneers* (New York, 1912), 302.

87 Ticknor's friendship with Humboldt is discussed in chapter 4 in this volume.

88 Joshua Shaw, quoted in Huth, *Nature and the American*, 47. This quotation is also mentioned in Andrea Wulf, *Founding Gardeners: The Revolutionary Generation, Nature, and the Shaping of the American Nation* (New York: Vintage Books, 2011), 171–72.

89 Adlard Welby, *A Visit to North America* (London: J. Drury, 1821), 65; quoted in Nash, *Wilderness and the American Mind*, 26.

90 The artists John Trumbull, Asher B. Durand, and William Dunlap each purchased one of the three landscapes by Cole placed on view by the picture dealer Mr. Colman, thereby establishing Cole as the leading practitioner of an emerging American school of landscape painting. William Dunlap, writing under the pseudonym "American," gave the first account of the discovery of Cole's talents: *New York Evening Post*, 22 November 1825, 2. Dunlap's article is quoted at length in Ellwood C. Parry III, *The Art of Thomas Cole: Ambition and Imagination* (Newark: University of Delaware Press, 1988), 25–26.

91 Quoted in Louis LeGrand Noble, *The Course of Empire: Voyage of Life, and Other Pictures of Thomas Cole, N. A., with Selections from His Letters and Miscellaneous Writings: Illustrative of His Life, Character, and Genius* (New York: Cornish, Lamport, 1853), 85.

92 Cited in ibid., 202.

93 Nichols noted that the paintings "decorated the sides [of the cabin] with intervening windows." Arthur St. Clair Nichols, "Notes and Remarks," (July 1827), typescript in Hudson River vertical file, Print Room Archives, Museum of the City of New York, pp. 8–9; quoted in Kenneth John Myers, "Art and Commerce in Jacksonian America: The Steamboat *Albany* Collection," *Art Bulletin* 82, no. 3 (September 2000): 510, 525n34.

94 James Fenimore Cooper, *Notions of the Americans: Picked Up by a Travelling Bachelor* (1828; reprint, New York: State University of New York Press, 1991), 174; cited in Myers, "Art and Commerce in Jacksonian America," 510, 525n34.

95 Huth, *Nature and the American*, 48.

96 James T. Callow's *Kindred Spirits: Knickerbocker Writers and American Artists, 1807–1835* (Chapel Hill: University of North Carolina Press, 1967) provides an excellent framework for understanding the synergy between the literary and visual arts at this moment, with particular regard for the cultural resonance of landscape imagery.

97 Asher B. Durand, "Prospectus," *The American Landscape* (New York: Elam Bliss, 1830), 3. Beinecke Rare Book and Manuscript Library, Yale University.

98 William Cullen Bryant, "Preface," *The American Landscape* (New York: Elam Bliss, 1830), 2–3. Beinecke Rare Book and Manuscript Library, Yale University.

99 Ibid.

100 Ibid.

101 Ralph Waldo Emerson, *Journals of Ralph Waldo Emerson, with Annotations, 1820–1824*, ed. Edward Waldo Emerson and Waldo Emerson Forbes (Boston: Houghton Mifflin, 1909), 83.

102 The museum was a major research center and began publishing in 1802. Claude Schnitter, "Le Dévellopement du Muséum national d'histoire naturelle de Paris au cours de la seconde moitié du XIXe siècle: 'Ses transformer ou périr,'" *Revue d'histoire des sciences* 49, no. 1 (1996): 54; quoted in Patrice L. R. Higonnet, *Paris: Capital of the World* (Cambridge, MA: Harvard University Press, 2002), 138.

103 "The Uses of Natural History," lecture delivered 5 November 1833 at the Boston Society of Natural History, in *The Selected Lectures of Ralph Waldo Emerson*, ed. Ronald A. Bosco and Joel Myerson (Athens: University of Georgia Press, 2005), 5; quoted in David Robinson, "Emerson's Natural Theology and the Paris Naturalists: Towards a Theory of Animated Nature," *Journal of the History of Ideas* 41 (1980): 69.

104 Ralph Waldo Emerson, "Nature," in *Essays and Lectures* (New York: Library of America, 1983), 10. Humboldt's influence on Emerson and Thoreau recurs in chapter 7 in this volume.

105 Ibid., 33.

106 See Tyler Green, *Emerson's Nature*, forthcoming. I am grateful to Tyler Green for sharing his observations on this synergy between Cole and Emerson.

107 Thomas Cole, "Proceedings of the American Lyceum: Essay on American Scenery, by Thomas Cole," *American Monthly Magazine*, n.s., 1 (January 1836), 5.

108 Laura Dassow Walls has written extensively and persuasively about Emerson's debt to Humboldt in several of her books. See especially *Emerson's Life in Science: The Culture of Truth* (Ithaca, NY: Cornell University Press, 2003); and *The Passage to Cosmos: Alexander von Humboldt and the Shaping of America* (Chicago: University of Chicago Press, 2009).

109 Silliman, *Remarks Made on a Short Tour between Hartford and Quebec*. Wadsworth was Silliman's brother-in-law. Cole's closeness to both men put him peripherally in Humboldt's orbit in the United States.

110 Ralph Waldo Emerson, *Journals of Ralph Waldo Emerson, with Annotations, 1836–1838*, ed. Edward Waldo Emerson and Waldo Emerson Forbes (Boston: Houghton Mifflin, 1910), 321.

111 Elizabeth Mankin Kornhauser, "Manifesto for an American Sublime," in *Thomas Cole's Journey: Transatlantic Crossings* (New Haven, CT: Yale University Press for the Metropolitan Museum of Art, 2018), 77–78.

112 Emerson, "Nature," 10.

113 I am grateful to Peter Betjemann for his observations on the synergy between Cole and Emerson as it relates to *The Oxbow* and *Nature*.

114 Ralph Waldo Emerson, "The American scholar"; an address, delivered by Ralph Waldo Emerson before the P[hi] B[eta] K[appa] Society, at Cambridge, August 1837 (New York: The Laurentian Press, 1901), 10–11.

115 Nathaniel P. Willis, ed., *American Scenery; or, Land, Lake, and River Illustrations of Transatlantic Nature* (London: George Virtue, 1840), 1:v.

116 Horace Greeley, *An Overland Journey From New York to San Francisco in the Summer of 1859* (New York: C. M. Saxton, Barker, 1860), 312.

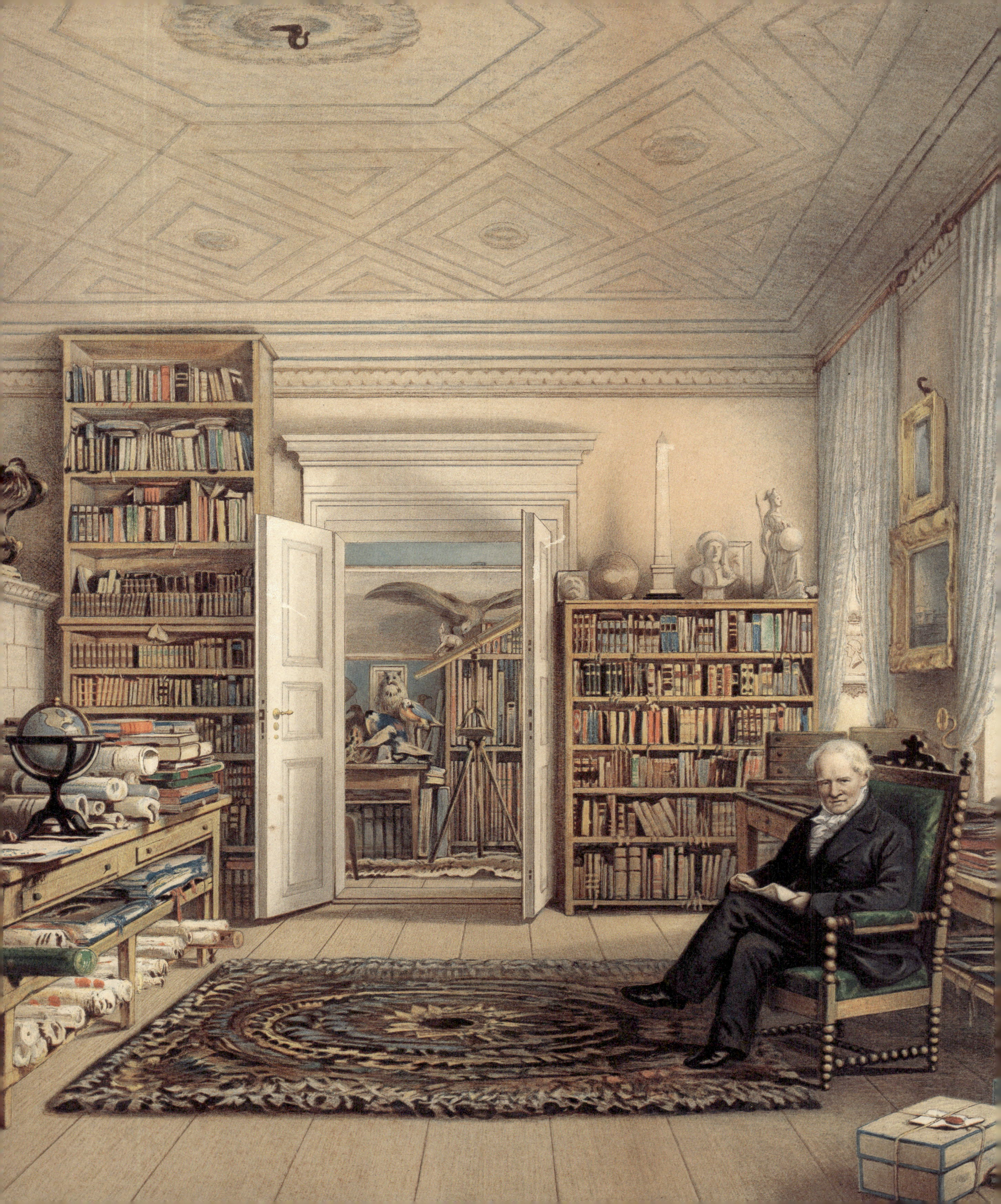

"I am half an American"

Humboldt saw the world as a global set of networks, connected intellectually by letters, books, and conversations; and connected geographically by maps and modes of transportation. Both systems were designed to facilitate the transmission of ideas—the increase and diffusion of knowledge—and Humboldt put immense effort into expanding them. Among people in the United States, there was a growing eagerness to be a part of Humboldt's networks. His reputation as the leading natural philosopher of his age, coupled with his charismatic personality, made him a favored destination for Americans traveling to Europe. In part this stemmed from an American desire for Humboldt's imprimatur on American cultural production. In a letter to John Vaughan in June 1805, Humboldt remarked, "This beautiful country that you inhabit and that presents such a beautiful moral picture you will see that in my writings I return to the United States. It is a passion in me to praise them."[1] To emphasize the connection he drew between his increasingly liberal political views and his support for the ideals of American democracy, Humboldt began referring to himself as "half an American," a tendency that those in the United States embraced as evidence of his high regard.

Both Humboldt and his U.S. admirers desired to be part of a worldwide network that fostered an international exchange of ideas and knowledge—without political interference. The age in which they lived complicated that goal. Naval blockades, piracy, and shipwrecks frequently disrupted transatlantic commerce. More damaging in Humboldt's mind was the direct interference of Napoleon. The French monarch-turned-emperor routinely intercepted mail intended for those who opposed him.

After Eduard Hildebrandt, *Humboldt in His Library* (detail), 1856. See p. 206.

Humboldt might not have posed an overt political threat to Napoleon's rule, but his vocal support for American democracy and his close friendships with like-minded men and women in Paris singled him out for particular suspicion. Increasingly Humboldt looked to the United States to develop conduits for faster and more reliable communications overland and across the oceans. His network of Americans would greatly facilitate that goal.

Humboldt certainly was aware that he had departed the United States a revered figure. William Thornton mourned to John Vaughan, "I wish he could have rested his limbs a while & published his works here.—The Treasures of Knowledge he has amassed are worth more than the richest gold mine."[2] Humboldt sent Jefferson and Gallatin copies of his publications, quizzing them about American political matters and eagerly awaiting their replies. As Humboldt's books began appearing in English, his ideas achieved broader distribution across American culture. As early as February 1805, Humboldt actively contemplated an English translation of what he estimated would be eleven volumes of his *Personal Narrative*, noting to his friend Marc-Auguste Pictet,

> *You might also drop a hint, that I have some idea of bringing out an edition in America, where I do not hesitate to say that among the anti-federal party the success of my expedition has produced quite an enthusiasm, apparent even in the newspapers that find their way here. The sale in the United States will be very large, and should it be requisite to find subscribers (a method that my friends Messrs. Jefferson, Madison, Galatin, Whister, Berton [Gallatin, Wistar, and Barton], and others, will be able to procure me a large number. An English edition, therefore, ought to at least consist of 4,000 copies.*[3]

That was an extraordinarily high number, and Humboldt's confidence in the size of his American audience suggests that his brief trip to the United States had buoyed his belief in his influence there.

Humboldt's estimate for his print run might have been too high, but his assessment of his relevance in the United States was not. Many of his American correspondents looked to him as an expert across the multiple fields of inquiry folded into his books. Humboldt thrived on connectivity, and from the beginning included Americans in his scholarly orbit. They in turn recognized that his approbation gave greater international relevance to their new country. That desire for Humboldt's attention and praise waxed during the decades following his brief visit to the United States. As he cultivated his American supporters, he often sought actively to influence politics as well as the direction of scientific research and exploration.

HUMBOLDT IN NAPOLEON'S SCIENTIFIC EMPIRE

During the five years he had been gone, Humboldt had not been completely out of touch while he was in South America. He had posted letters intermittently from cities in South America, Mexico, and Cuba, chatting up colleagues, recommending articles, and reaffirming his relevance despite the distance between them.[4] Before his ship had cleared quarantine in the harbor at Le Havre, he dispatched a flurry of letters to his colleagues in Paris announcing his arrival. There he threw himself into the French scientific community, delivering lectures at the Institut National on various aspects of his travels.[5] These lectures laid out the framework of Humboldt's belief that everything on earth existed in relationship to everything else, forming a web of life. Included in that web metaphor was Humboldt's expanding network of colleagues, allies, and friends on both sides of the Atlantic.

Humboldt's return to Europe launched his international reputation as one of the most significant public intellectuals of the first half of the nineteenth century. He made himself a central figure in the Parisian scientific community, which was, to Napoleon's credit, the envy of all Europe. Unfortunately for Humboldt, it also made Napoleon envious of the attention directed to this foreign scientist. Humboldt's return in August 1804 had coincided with Napoleon's rise to power. Thanks to his daring exploits during the French Revolution, the French monarch was accustomed to being the most important man alive; however, when news of Humboldt's ascent of the Andean mountain Chimborazo reached Paris, it temporarily eclipsed Napoleon's accomplishments, much to his annoyance.[6] The two men met in late October of 1804, when Humboldt was introduced to the monarch at a reception at the Tuileries. There Napoleon had

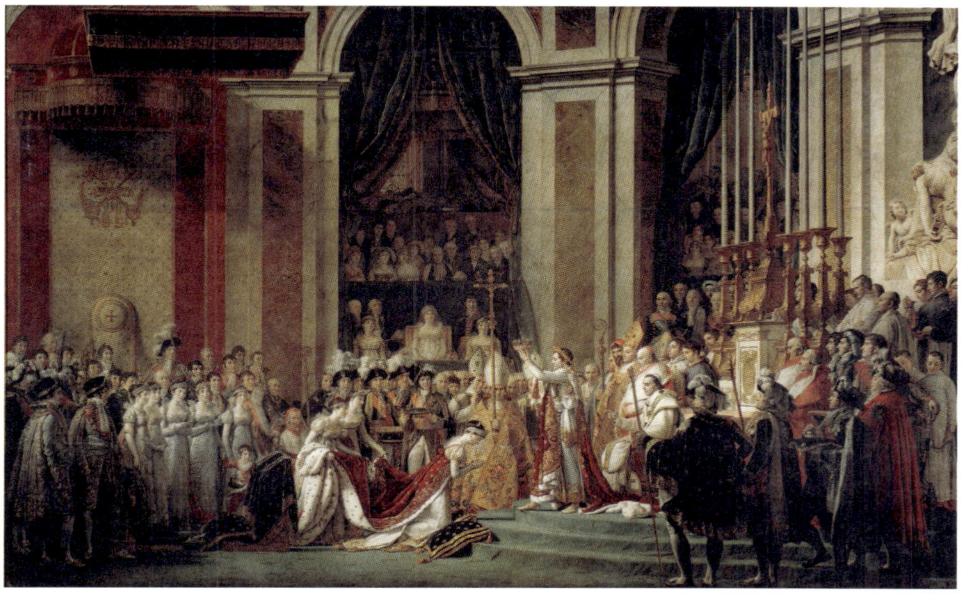

Fig. 44
Jacques-Louis David, *Consecration of the Emperor Napoleon I and Coronation of the Empress Joséphine in the Cathedral of Notre-Dame de Paris on December 2, 1804*, 1806–7, oil on canvas, 244 ½ × 385 ½ in., Musée du Louvre

sneered at Humboldt, comparing his international botanical research to the plant-collecting interests of his wife, Josephine.[7] Napoleon's insults suggest his displeasure in just how powerful a personality Humboldt was becoming so quickly after his return to France. Four months later, Napoleon crowned himself emperor (fig. 44), an audacious move that so shocked Ludwig van Beethoven that he removed Napoleon's name from the *Eroica* Symphony, which he had composed in his honor. Napoleon's brash exercise of power characterized his imperial reign and set the tone for his relationship with Humboldt.

The challenge for Humboldt was to find a way to coexist with the French monarch, which he achieved by embedding himself in the scientific community. This network afforded professional camaraderie and support, both of which Humboldt needed to thrive. As a foreign associate (*associé étranger*) of the French Institut, he became a fixture among the scientists and intellectuals who were an essential part of Paris's vibrant social life.[8] Perhaps more meaningful to Humboldt personally were the friendships he cultivated as a member of the Société d'Arcueil, founded in 1801 by his close friend, chemist Charles-Louis Berthollet, and funded by royal charter beginning in 1807. This was a small, handpicked group of about fifteen scientists, each at the top of his field, several of whom had their own sympathies for the United States. Humboldt's closest friends among the Société members included Georges Cuvier, the biologist and paleontologist who had taken the fossilized teeth sent to him by Jefferson and Humboldt and in 1806 teased out the distinction between mastodons and mammoths; chemist Jean-Antoine-Claude Chaptal, who in 1794 had entertained an offer from George Washington to emigrate to the United States;[9] and the two men closest to Humboldt, chemist and early balloonist Joseph Gay-Lussac, and astronomer François Arago. They met every other week year-round (except in December and January when the weather was bad) between 1801 and 1817 outside Paris at Berthollet's bucolic estate in the Bois de Boulogne.[10] This group encouraged and protected Humboldt, and of them Humboldt wrote, "I feel it more every day that I do good work only in an environment where others around me do even better."[11]

Socializing among the scientists resulted in important professional collaborations. Humboldt and Gay-Lussac coauthored a lengthy paper on magnetism in 1807, published in the Société's *Mémoires*, which Humboldt sent to Jefferson.[12] Magnetism would be one of Humboldt's lifelong interests, beginning with his early biological experiments with galvanic current and electric eels, and playing an increasingly important role in his involvement with electromagnetics regarding telegraphic communications. During 1805 Humboldt ricocheted around Europe, delivering lectures, meeting colleagues, and conducting experiments. In Rome in 1805 he was reunited with his older brother, Wilhelm, a celebrated linguist and diplomat, and during a side trip to Naples he witnessed the eruption of Vesuvius on August 12. After a nine-year absence, he returned to Berlin, where he was received by Prussian King Friedrich Wilhelm III. During the fall and winter of 1805 to 1806, he lectured frequently on his travels at the Berlin Academy of Science.[13] Trapped in Prussia following Napoleon's invasion of his homeland, Humboldt spent all of 1806 in Berlin, where he wrote his second book, *Ansichten der Natur* (*Aspects of Nature*). Humboldt dedicated his book to all the souls displaced by war and politics, an elliptical reference to Napoleon's campaigns.[14] When the peace treaties between Prussia and France were concluded in 1808, Humboldt was part of the Prussian delegation. The two rulers acceded to Humboldt's request to remain in Paris

CAT. 33 Rembrandt Peale, *Portrait of Alexander von Humboldt*, 1809–12, oil and encaustic on canvas, 22 ½ × 27 in., Collection of Robert W. Hoge and Immaculada Socias Hoge

with broad, undefined diplomatic duties. He would spend the next two decades refining his concept of the unity of nature and expanding his international networks.

Humboldt thrived in the Parisian intellectual community but chafed under Napoleon's increasingly autocratic rule. It was not simply that Humboldt was energetic, charismatic, and taking the intellectual community by storm. Many in Humboldt's network of correspondents were among Napoleon's most outspoken detractors. Tension between the two men intensified. For years, every time the they met, Napoleon would pretend he did not know who Humboldt was; deep down he harbored the belief that Humboldt was a Prussian spy.[15] In this regard, Humboldt was in good company. Among his acquaintances was the young South American activist Simón Bolívar (fig. 45), who was openly talking about revolution in his native South America[16] and casting aspersions on Napoleon's regime. Another was the most influential European advocate for the United States, the Marquis de Lafayette. Napoleon treated all three men as spies, in Lafayette's case, with good reason. Following the French Revolution, Lafayette had been freed from prison thanks to a young Napoleon's intercession, but he now refused to join the emperor's court, preferring, as he pointedly noted, to support a more democratic regime.[17] Lafayette made no secret of his support of representational government and was proud of his role in helping to secure America's independence from England. One of Napoleon's chancellors wryly noted that the marquis was the kind of man who could be found plotting publicly in a town square.[18]

Fig. 45
Rafael Salas, *Simón Bolívar* (1783–1830), 19th century, oil on canvas, Casa de la Cultura Ecuatoriana, Quito, Ecuador

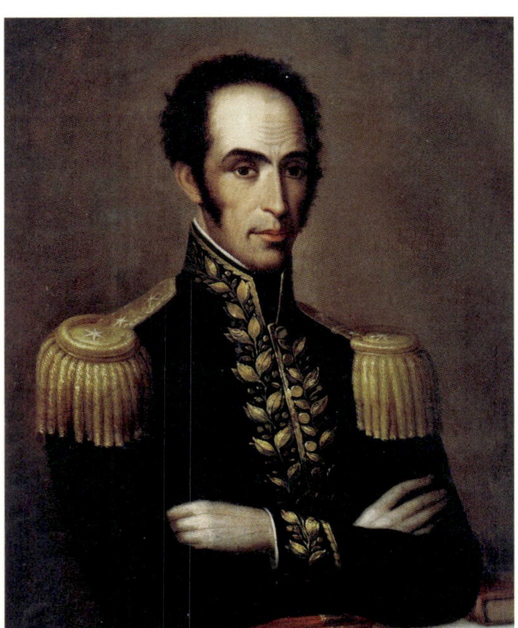

If Humboldt was half an American, the Marquis de Lafayette enjoyed the celebrity of being considered a full American. Hailed as a hero during the American Revolution, Lafayette had distinguished himself at George Washington's side. Washington considered him like a son and bestowed upon him honorary "full natural-born citizenship" of the United States. In Paris, Lafayette maintained an active correspondence with Jefferson—although his letters contain elliptical references to people and events, as both men knew Napoleon was intercepting his mail, as well as Humboldt's.[19] However, one letter Napoleon never saw was the one Jefferson gave to Humboldt on the eve of his departure from Philadelphia. In it Jefferson introduced Humboldt to Lafayette, fostering a close friendship that lasted the rest of the marquis's life.[20] In 1809 Lafayette wrote to Jefferson thanking him for that introduction, as the two Europeans had become fast friends.[21] Humboldt and Lafayette shared a great deal in common: a passion for the United States, their admiration for Jefferson, an abhorrence of slavery, their disdain for Napoleon, and a preference for democratic rule. Theirs was a friendship that would draw a generation of American visitors to Paris.

One of those early travelers was Rembrandt Peale, this time unencumbered by a mastodon skeleton. He toured Europe in 1808, making a point to spend as much time with Humboldt as he could. Humboldt introduced Peale around Paris,[22] and during the course of his stay Peale painted Humboldt's portrait (**CAT. 33**). For this work, Rembrandt used encaustic, mixing his oil paints with wax to provide depth and warmth to his colors, adding a more lifelike glow to his sitter's face. Viewers who knew Humboldt felt this to be among the finest likenesses of the naturalist, capturing his acute glance and quick smile. Humboldt's gaze is direct, appraising, and conveys a kind of compassionate gravitas.

Four years earlier in Philadelphia, Charles Willson Peale had painted a younger, more distant Humboldt, a man still in transit, coming into focus. Rembrandt had caught Humboldt as his reputation was rising rapidly, his popularity and relevance assured. In both portraits Humboldt exudes the same confidence, but here there is a warmth and engagement that suggests a man who sees a clear

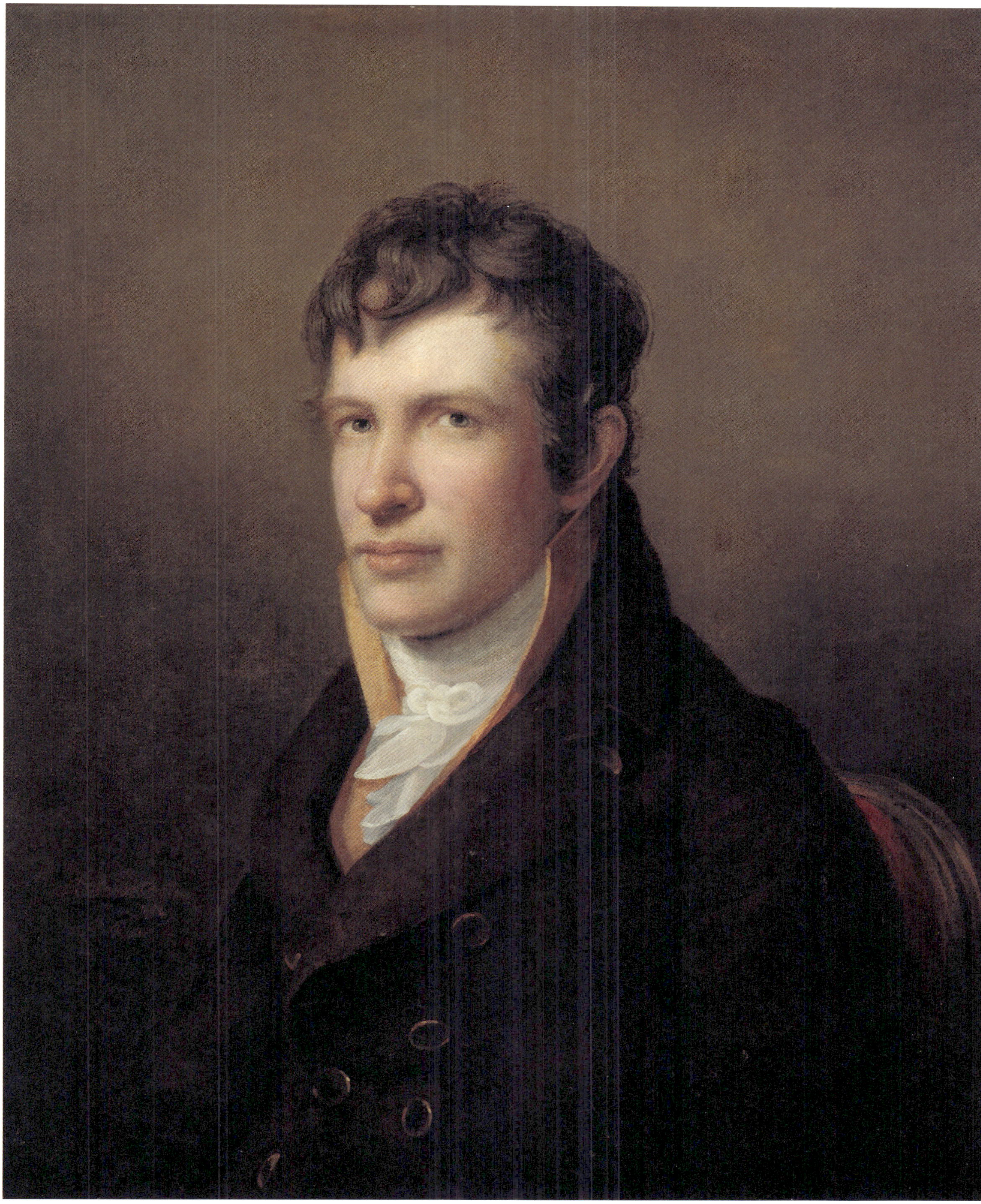

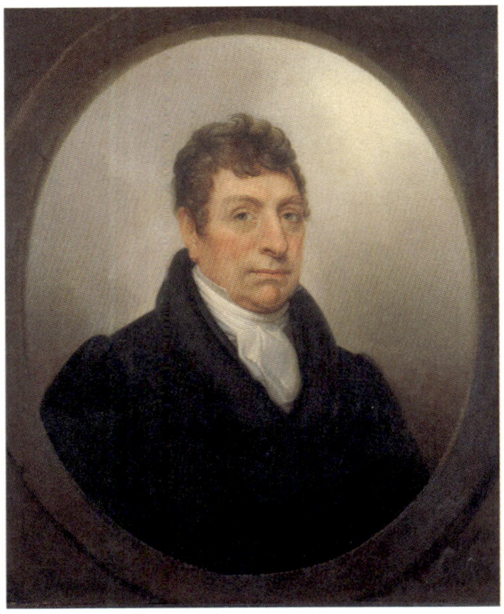

Fig. 46
Rembrandt Peale,
The Marquis de Lafayette (1757–1834), 1825,
oil on canvas, 34 ½ × 27 ⅜ in.,
The Metropolitan Museum of Art, NY, Rogers Fund, 1921

path forward. Humboldt likely introduced the young Peale to Lafayette, whose portrait he also painted in 1825 (fig. 46). Peale's likeness of Lafayette is flattering and sensitive, and the Frenchman must have been pleased to find his portrait framed in an illusionistic masonry tondo reminiscent of Rembrandt's "porthole" portraits of George Washington. Both of Peale's portraits later graced Peale's Museum gallery of notable men, joining the portrait of Jefferson that Rembrandt had completed in 1805 (see cat. 5).[23] All three of them conveyed what Charles Willson Peale hoped to inculcate in his museum visitors: a stalwart belief in democracy and the viability of the American experiment. After meeting Humboldt in Washington, Dolley Madison had hoped to convince "our Baron Humboldt" to stay in the United States; Jefferson had similarly tried to entice Lafayette to return. In both cases, the desire was to see each man become an American himself.

Peale was still in Paris in 1810, when Napoleon ordered his minister of police to expel the eminent naturalist and political gadfly. Humboldt appealed to his close friend, Chaptal, who was also the emperor's former minister of the interior, for help.[24] The French chemist scrambled to protect Humboldt and by extension the vibrancy of the scientific community in Paris. At a reception in the Tuileries, Chaptal convinced the angry monarch that to eject Humboldt would cripple the scientific community in Paris—of which Napoleon was justifiably proud—and besmirch the monarch personally. When Napoleon asked if the damage from expelling Humboldt would really be so great, Chaptal replied, "M. de Humboldt possesses all the sciences, and when he travels, the entire Academy of Sciences marches with him."[25] That fraternity of scientists now coalesced around Humboldt. Reluctantly, Napoleon withdrew his order, and Humboldt remained in Paris.[26] Shortly thereafter, on the eve of Rembrandt Peale's departure for the United States in September, Humboldt delivered a jab at Napoleon, promising the American artist, "I will come finish my life in your country, where I like the inhabitants, the land, and especially your wise constitution."[27] In the United States, he concluded, "the people really can breathe with more freedom."[28] For Humboldt, nature carried symbolic meaning, and the American landscape was imbued with democratic ideals. Paris, it seemed, might not be wide enough for both of them.

In 1814 two events shook the Atlantic world. In April, Napoleon abdicated the throne and went into exile; and over the summer, negotiations to end the War of 1812 between England and the United States culminated in the signing of the Treaty of Ghent on Christmas Eve. Humboldt had advocated for the United States in these negotiations, traveling to London on behalf of the Prussian government.[29] There he renewed his friendship with Albert Gallatin, a member of the United States negotiating team, along with John Quincy Adams and Henry Clay.[30] Following the successful conclusion of the negotiations, Adams detoured to Paris. The future American president had heard the stories of Humboldt in Washington from his wife; now he was experiencing Humboldt firsthand.[31] Together he and Humboldt attended lectures at the French Academy of Sciences and socialized with Lafayette and Madame de Staël.[32] This group had a lot to celebrate; as tensions eased across Europe, liberal thinkers could breathe a bit easier in Paris. Adams's immersion in Humboldt's scientific and political world would play a crucial role in 1836 when he advocated strongly for the founding of the Smithsonian Institution, a project that would draw together members of Humboldt's American networks.[33]

Napoleon had represented the most persistent and effective roadblock in the expansion of Humboldt's network. The Corsican's campaigns during the French Revolution had thwarted Humboldt's plans to visit Egypt, and his invasion of Prussia in 1806 had trapped Humboldt in Berlin for close to a year. The international alliance that had coalesced to remove Napoleon from power ushered in a restoration of the Bourbon monarchy and saw the return of political exiles to Paris. One of the most influential of these was Madame de Staël (fig. 47), who had been a prominent supporter of the French Revolution and an equally vocal critic of Napoleon. Napoleon had exiled her in

1803 for her vigorous opposition to his rule. To Klemens von Metternich he had written,

I do not want Madame de Staël in Paris, and for that I have good reasons. . . . If Madame de Staël wanted or knew how to be a royalist or a republican, I would have nothing against her; but she is a machine in movement who stirs the salons.[34]

The salon system was both innocent and subversive, an essential and time-consuming part of European social and intellectual life. Napoleon had been justly wary of hers in particular, as she fomented opposition and provided a safe haven for free speech in the guise of literary and artistic expression. Among other things she maintained a vibrant and candid correspondence with Thomas Jefferson, which Napoleon undoubtedly intercepted and read, in which they shared their views on American and French politics.[35]

In 1814 she resumed her nightly dinner and weekly literary and artistic salons, an international who's who of the brilliant, witty, and popular, at which Humboldt and Lafayette were frequent attendees. The salons were a magnet for Humboldt. His pattern was to spend his days engaged in scholarly pursuits—a fast-paced schedule of listening to and delivering lectures and studying the specimens he had brought back from South America. In the evenings and well into the night, Humboldt emerged from his rooms and made the rounds at dinners, parties, receptions, and salons. Each of these venues afforded him the opportunity to be the center of attention, which added luster to his reputation and attracted a following that, like ripples in a pond, spread his ideas.

Fig. 47
Marie-Éléonore Godefroid, *Germaine Necker, baroness de Staël-Holstein, known as Madame de Staël (1766–1817)*, 1813, oil on canvas, 45 ¹¹⁄₁₆ × 32 ¹¹⁄₁₆ in., Établissement public du château, du musée et du domaine national de Versailles

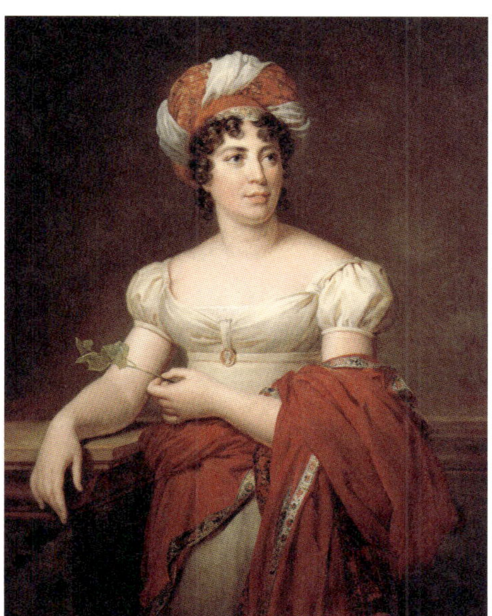

In addition to Madame de Staël's soirées, Humboldt attended those hosted by her close friend and fellow returned exile, Madame Récamier. Juliette Récamier was a legendary beauty whose parties blended politics and literature. She attracted a vibrant coterie that often included Chateaubriand, with whom she became very close. These people openly admired the United States, and each corresponded with or personally knew prominent American statesmen. All of them knew Thomas Jefferson from his tenure in Paris. Humboldt found himself increasingly in the company of pro-American Europeans who encouraged his liberal tendencies. At one Fourth of July gala, Lafayette gave toasts to "Washington, to the republics of South America, to Greece and to American women,"[36] citing his four favorite revolutionary forces.

Although Humboldt never used the salons as a pretext for romantic attachments to the women who ran them, his regard for their intellectual faculties seemed genuine.[37] In 1814 many of the weekly salons in Paris were hosted by women of significant intellectual prowess. This was their avenue into the upper echelon of literary and scientific circles. Gender rules in British and European society dictated that women were barred from membership in the scientific societies but might attend lectures or events as the guest of their father or spouse. To get around this, several women simply moved the conversation from the halls of the Institut to their living rooms. Contributing to Humboldt's regard for intellectually hungry women was his choosing three of them to take on the difficult task of translating his works from French or German into English, the principal avenue by which Humboldt's ideas were starting to permeate American culture. In Paris, Humboldt took his Sunday night dinners with Helen Maria Williams, whose literary salon attracted an international coterie. She engaged Humboldt in discussion of the explosive political situation in South America and had him read from his unpublished manuscripts.[38] She encouraged Humboldt to write up his travels and offered to translate his works from French to English. Her translations of *Views of the Cordilleras* and early volumes of the *Personal Narrative* were for many Americans their introduction to Humboldt's ideas.

The other two translators each had scientific ambitions of their own. Briton Elizabeth Juliana Leeves Sabine assisted her husband, Edward, with his experiments. Her translation of Humboldt's *Aspects of Nature* and the first four volumes of Humboldt's *Cosmos* from German to English remained the authoritative versions for most of Humboldt's lifetime. Elise Otté, whose translation of *Cosmos* supplanted Sabine's and is still in print, was a highly regarded Anglo-Danish translator and historian who maintained her own scientific interests in part by translating other scientists' works. Humboldt worked with all three of these women to varying degrees, fussing over the translations while working on the next volumes. Perhaps the most concrete evidence of Humboldt's regard for his female audience came in 1827, after Humboldt had moved back to Berlin. He scheduled his lectures, which evolved into his magnum opus, *Cosmos*, during the daytime, thereby making them accessible to women. This gesture did not go unnoticed or unappreciated. Humboldt's lectures, like the nightly salons, were well attended and a source of intellectual stimulation.

AMERICANS IN PARIS

With the fall of Napoleon and the Bourbon restoration, Albert Gallatin moved to Paris in 1816 to take up the position of ambassador to France, allowing him to resume his friendship with Humboldt. The two men became quite close, sharing the same social orbit. Travel to the Continent once again became a part of American intellectual life, and in Paris it was now common for American travelers to end up at dinners or salons with a group of Europeans with a great fondness for the United States. The first wave of Americans included a pair of Boston intellectuals, George Ticknor and Edward Everett, who would significantly extend Humboldt's cultural influence in the United States.

Everett would go on to become an influential publisher and orator, advocate for the abolition of slavery, and along with Ticknor, he used his published writings and his teaching position at Harvard to create his own American Humboldtian network. In 1815 the two men had enrolled at the University of Göttingen, Humboldt's alma mater. At Göttingen, Ticknor found himself studying with Humboldt's aging mentor, Johann Friedrich Blumenbach, whom he admired as "the most outstanding of them all."[39] Ticknor had taught himself German by translating Goethe's *Sorrows of Young Werther*, a work immensely popular in its day; and in October 1816 the two Americans made a pilgrimage to meet the author. After visiting Goethe in Weimar, the two young Bostonians arrived in Paris in April 1817.[40] In preparation for the trip, Ticknor had paid a call to the sage of Monticello, where Jefferson wrote him letters of introduction to Gallatin and Lafayette in Paris, which afforded them an opportunity to meet Humboldt as well.

Ticknor (fig. 48) was overwhelmed by Humboldt's curiosity, enthusiasm, and attentiveness to the young American. In his journal he described with fascination Humboldt's philosophy for working, during which

> *night and day form one mass of time which he uses for sleeping, for meals, for labor, without making any arbitrary division of it. . . . But all the rest of the time, when he is not in society, he locks his door and gives himself up to study, rarely receiving visits, but those which have been announced to him the day previous.*[41]

Impressed with Humboldt's physical and intellectual stamina, and equally admiring of his political defiance, Ticknor found him to be one of the "most interesting men in the world, and the idol of Parisian society,"[42] able to get away with a level of idiosyncrasy and political openness due to a combination of his forceful personality and intellectual acumen.

Their friendship deepened over time, forged during this inaugural meeting in Paris. Both men reveled in the social and intellectual swirl of evening society. As Ticknor noted, "Almost everybody who comes to these salons comes to say a few brilliant things, get a reputation for esprit,—the god who serves for Penates in French houses,—and then hasten away to another coterie to produce the same effect." His conclusion was that the French system was "brilliant, graceful, superficial, and hollow."[43] Ticknor, like Humboldt, loved it. He particularly enjoyed running into Humboldt "either by accident or otherwise, nearly every day, and of all the men I have known, he is, in some respects, the most remarkable; the man on whom talent and knowledge have produced their best and most generous effects."[44]

Ticknor and Everett made Humboldt's name familiar to a wide swath of America's literate public. As editor of the *North American Review,* one of the most widely circulated and popular literary magazines in the country, Ticknor printed a hefty number of articles about Humboldt. In 1823 Everett (fig. 49) contributed a lengthy and thoughtful review of Humboldt's most recent publications, providing the first critical assessment

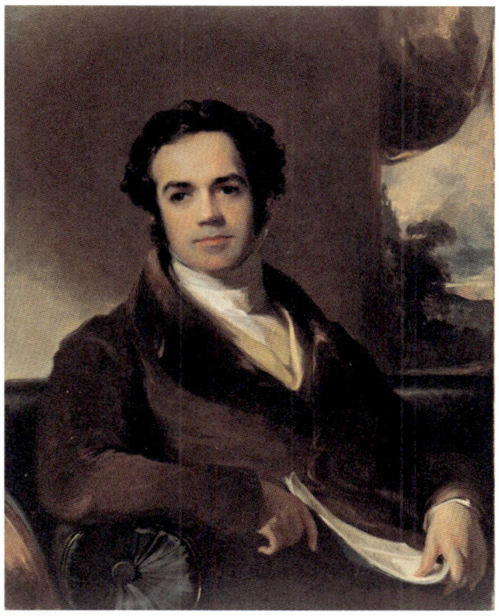

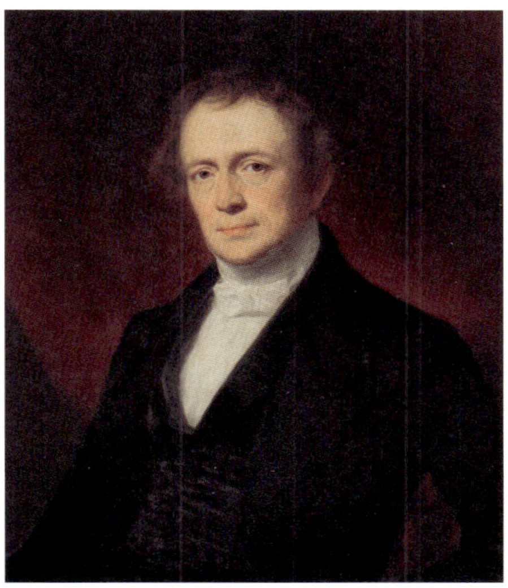

Fig. 48
Thomas Sully, *George Ticknor (1791–1871), Class of 1807*, 1831, oil on canvas, 36 × 28 in., Hood Museum of Art, Dartmouth: Gift of Constance V. R. White, Nathaniel T. Dexter, Philip Dexter, and Mary Ann Streeter

Fig. 49
Unidentified artist, *Edward Everett* (1794–1865), 1845 or 1846, oil on canvas, 29 ¾ × 25 ½ in., Detroit Institute of Arts, Gift of Dexter M. Ferry Jr., 29.193

that described the magnitude of Humboldt's accomplishments and explained their significance.[45] At the time, Everett was Ralph Waldo Emerson's teacher at Harvard. With his encouragement, Emerson began reading Humboldt, making notes in his journals, absorbing an ethos that would permeate his own writing in the decades to come.[46] In time Boston became an epicenter for Humboldtian thinking, encompassing Everett, Ticknor, George Bancroft, ethnographer and archaeologist William Prescott, poet Henry Wadsworth Longfellow, scientists Louis Agassiz and Asa Gray, and politician George Sumner.

Humboldt's ideas made the transatlantic crossing in letters and books, and in a wave of Humboldt's protégés who made America their home. Agassiz and Gray taught Humboldt's scientific theories at Harvard University as Everett promoted his ideas in the humanities. In the fine arts, the transcendentalist authors and Hudson River school artists helped craft a viable wilderness aesthetic, inspired in part by Humboldt. But Humboldt's ideas permeated American culture more broadly as his work was absorbed and incorporated in popular journals and secondary school textbooks. Washington Irving's *Literary Magazine, and American Register*, which was established in Philadelphia in 1803, had been the first American journal to print an article on Humboldt's visit to the United States; in 1804 *The Literary Magazine, and American Register* and the *Weekly Visitor* each had published versions of Humboldt's sixteen-page summary of his South American travels, translated and distributed by John Vaughan, to give Americans a taste of the significance of their recently departed celebrity.[47] In 1810 *The American Register; or, General Repository of History, Politics, and Science* published a seventy-two-page extract from Humboldt's newly published *Political Essay on the Kingdom of New Spain*, which included his compliment to Thomas Jefferson for his enlightened leadership while simultaneously decrying the perpetuation of slavery in the Americas.[48] Reviews and extracts from Humboldt's publications began appearing in American monthly and weekly periodicals beginning in 1818, when passages from Humboldt's third volume of his *Personal Narrative* appeared in several monthly magazines, and in 1821 when *Port Folio* was one of many magazines printing lengthy extracts from Humboldt's recently published fourth volume.[49]

Humboldt's influence in American education extended far beyond professional publications and university curricula. His ideas anchored a generation of textbooks used across the country, most of them written by people who never met him. In 1829 the famous educator Emma Willard co-authored with William Woodbridge an immensely influential book on geography and the study of world history, colloquially known as Woodbridge and Willard's Geography. One of the centerpieces of this book was a large-scale infographic she called The Temple of Time (fig. 50), in which she plotted coordinates between dates, places, and events as a visual means of remembering historical events. Her rubric was a Greek temple structure, in which the ceiling listed the names of individuals important in world history, while the pillars represented the centuries. It was the floor that generated the most interest, as it depicted her "map of time," based on the isothermal lines (fig. 51) Humboldt had created and published in 1817.[50] Those "magnificent lines," as they were called, allowed Willard to overlay them on a world map (fig. 52) making Humboldt's ideas even clearer. Humboldt's revolutionary idea that he could map heat signatures all over the globe and connect equivalent points of temperature in lines that represented similar climates was an extraordinary breakthrough in comprehending his idea of an essential unity of nature. Humboldt had introduced this concept, and his graphic proof, in a series of lectures which Ticknor attended at the Institut National in 1817. Willard and her collaborator found in Humboldt's graphics inspiration for using visual images to teach world history and geography. They were both pioneers in promulgating Humboldt's revolutionary ideas, clear evidence that for the vast majority of Americans, their introduction to Humboldt was filtered through elementary educational materials.[51]

Fig. 50
Emma Willard, *The Temple of Time*, from *Willard's Map of Time: A Companion to the Historic Guide*, 1846, colored print, New York Public Library

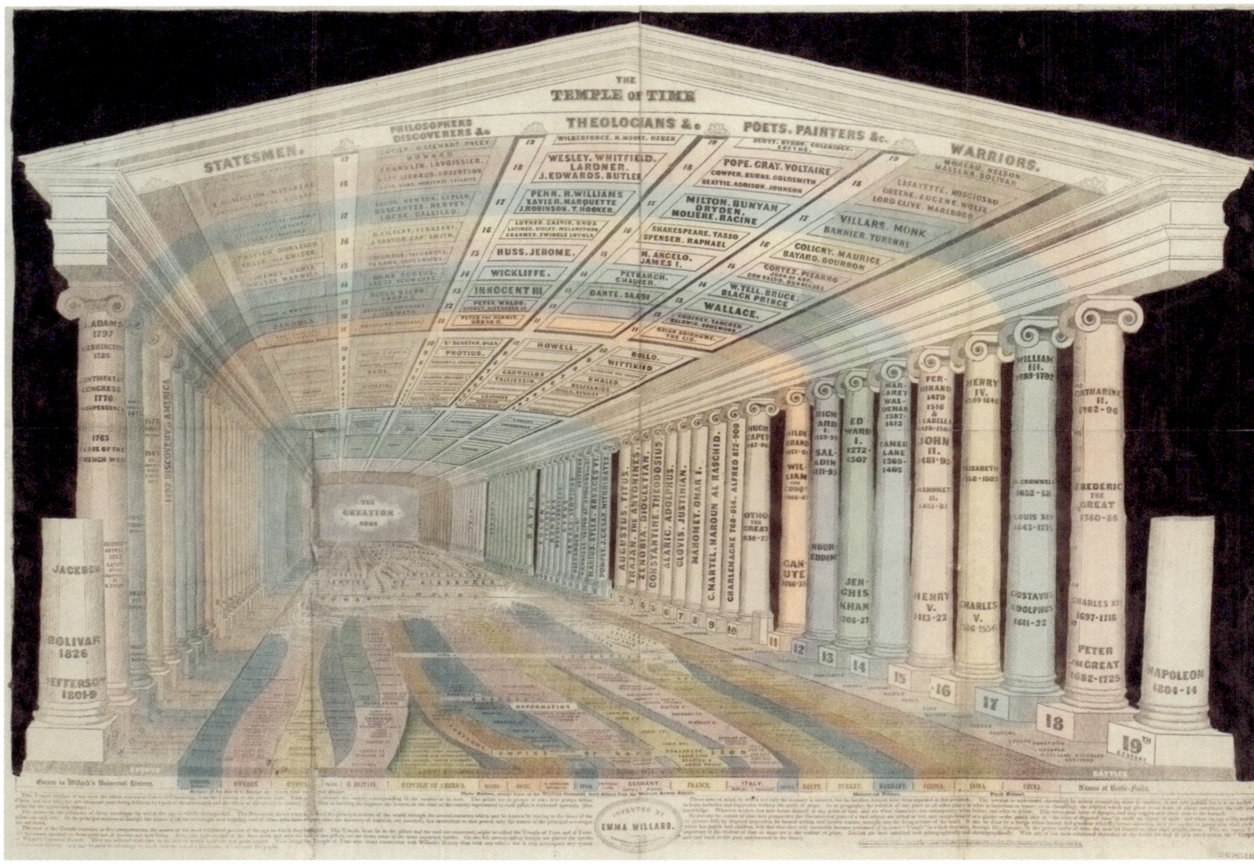

Fig. 51
Alexander von Humboldt, *Isothermal Chart*, 1817, printer's ink on paper, approx. 8 ½ × 9 in., Harvard University Libraries, Widener Library

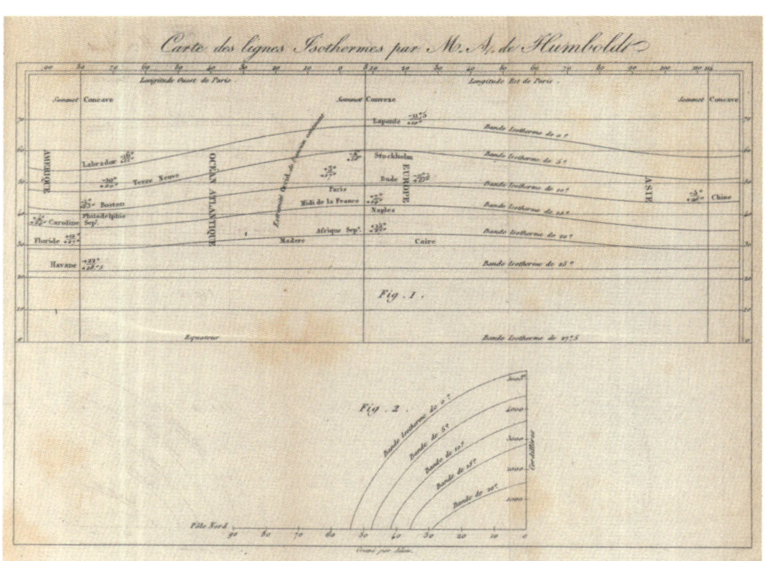

Fig. 52
William C. Woodbridge, *Isothermal Chart, or, View of Climates & Productions; Drawn from the Accounts of Humboldt and Others*, from *Woodbridge's School Atlas*, 1831, hand-colored copperplate print, 7 ½ × 11 in., Library of Congress, Geography and Map Division

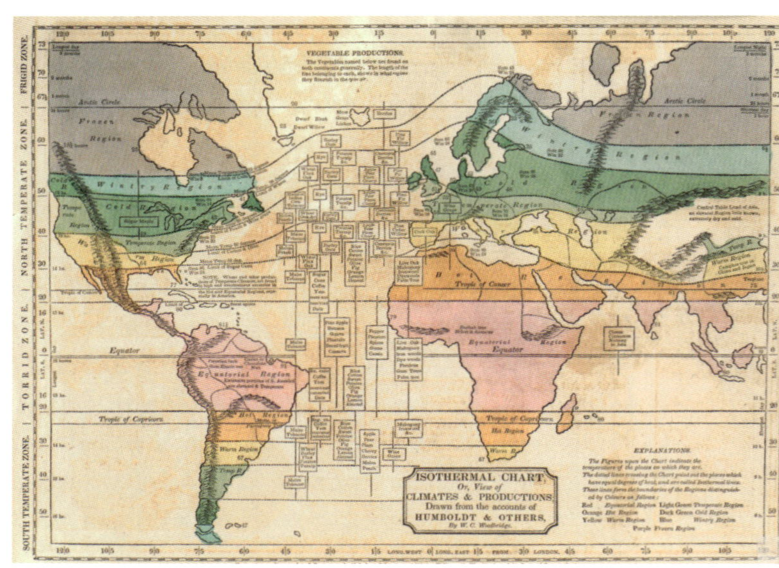

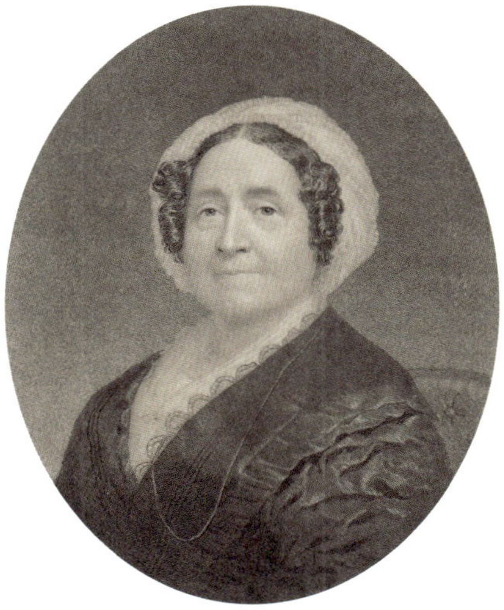

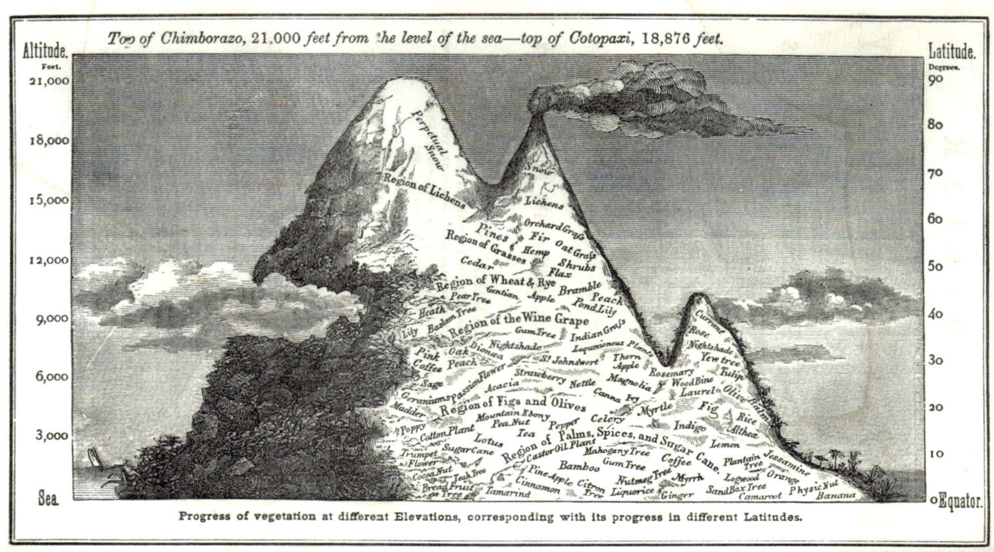

Fig. 53
John Sartain, *Almira Lincoln Phelps* (1793–1884), ca. 1878, engraving, 7 ½ × 5 ¾ in., The Winterthur Library: Joseph Downs Collection of Manuscripts and Printed Ephemera

Fig. 54
Almira Lincoln Phelps, frontispiece to *Familiar Lectures on Botany*, 1829, engraving, 4 ½ × 7 ½ in., Gray Herbarium Library, Harvard University

In 1829 Willard's younger sister Almira Lincoln (fig. 53) published her *Familiar Lectures on Botany*, a textbook used in American classrooms that relied heavily on Humboldt's methods and discoveries.[52] The underlying philosophy was the use of raw data to detect larger systems and in the framework of those systems comprehend the meaning of all that data. The frontispiece of the book (fig. 54) was an English-language adaptation of Humboldt's *Naturgemälde*, his Plant Geography Map of 1805, attesting to the significance accorded to his contributions to the field, and one of the sources for his inspiration for the next generation. Both sisters' books were immensely popular in schools. Lincoln's went through twenty-eight editions over forty-four years, and it may have sold more than 350,000 copies between 1829 and 1872.[53] Sarah Hale, editor of the influential *Godey's Lady's Book*, hailed her with the praise, "No woman in America . . . has made such useful and numerous contributions to the stock of available scientific knowledge."[54] Lincoln was a firm believer in scientific literacy and an enlightened, literate electorate, and in 1859 she became only the second woman elected to the American Association for the Advancement of Science, which counted Humboldt among its members. Her work popularized botany as a classroom subject, spurred the creation of countless herbaria, and was reprinted nearly annually for the next forty years, during which time American botanical and geographical education—for both girls and boys—was taught with a Humboldtian approach to plants, climate, geography, and history.

In 1869 Louis Agassiz delivered an address in Boston celebrating the centennial of Humboldt's birth. In his speech, Agassiz observed that while Humboldt was ubiquitous in American pedagogy, many students were unaware that he was their mentor. Agassiz's eulogy filled fifteen pages in print; it was a masterful celebration of Humboldt's life and legacy. In it he demonstrated how quickly Humboldt's radical ideas had become an integral part of the fabric of American views of nature:

> *To what degree we Americans are indebted to him, no one knows who is not familiar with the history of learning and education in the last century. All the fundamental facts of popular education in physical science, beyond the merest elementary instruction, we owe to him. We are reaping daily in every school throughout the broad land, where education is the heritage even of the poorest child, the intellectual harvest sown by him. See this map of the United States:—all its important features are based on his investigations; for he first recognized the great relations of the earth's physical features, the laws of climate on which the whole system of isothermal lines is based, the relative height of mountain chains and table-lands, the distribution of vegetation on the whole earth. There is not a text-book of geography or a school-atlas in the hands of our children to-day, which does not bear, however blurred and defaced, the impress of his great mind. But for him our geographies would be mere enumerations of localities and statistics. He first suggested the graphic methods of representing natural phenomena which are now universally adopted. The first geological sections, the first sections across an entire continent, the first averages of climate illustrated by lines, were his.*[55]

What Agassiz emphasized was nothing less than the depth of Humboldt's imprint on American exploration and geography. From the 1820s to the 1860s, explorers, authors, and artists would study nature to help construct an American cultural identity.

HUMBOLDT AND THE LITERARY IDENTIFICATION WITH THE AMERICAN WEST

During the 1820s Humboldt came into contact with a small group of American artists and writers who shaped American attitudes toward nature and emphasized the importance of the visual arts. Washington Irving (see fig. 37) was among the first American authors to make Humboldt's acquaintance. He had met Alexander's older brother, Wilhelm, in Rome in 1805, where he and his wife, Caroline, had been stationed as Prussia's emissary to the Vatican. Madame von Humboldt, as George Ticknor recalled, was "called the Mad. de Staël of Germany" and made the traveling American welcome in her social circle.[56] Irving first met Wilhelm's brother, Alexander von Humboldt, at the home of the British Consul in Paris in 1823. Their social circles continued to overlap. On April 6, 1824, Humboldt and Irving attended the salon at the home of Duchess Claire de Duras, whose gatherings Ticknor had described as attracting a literary rather than political set.[57] But most often, the two men could be found together at soirées hosted by Humboldt's favorite artist, Baron François Pascal Símon Gérard.[58] Irving loved Paris and remarked,

> *There is no place in Europe where a young man, who wishes to improve himself, and is determined to act with prudence, can spend a certain space of time to more advantage than at Paris. The doors of knowledge are there thrown open, and the different pursuits, both useful and ornamental, may be prosecuted with more facility and less expense than at any other city in the world.*[59]

Meeting Humboldt had a transformative effect on Irving's life and literary career. For the next three years the two men would pore over old maps and documents in an effort to discover the origin of the name "America," specifically why that name, and not that of Christopher Columbus, was applied to the New World. Together they determined that it was on Martin Waldseemüller's 1507 map that the term "America" was first applied to those two continents. Irving deployed his research as part of his *Life of Columbus*, which he published in 1828.[60] Humboldt was impressed with the American's accomplishment, making reference to it in his *Examen critique de l'histoire de la géographie du nouveau continent* (*Critical examination of the history and geography of the new continent*), published in 1836, commending Irving for having "demonstrated that the power of imagination in the higher intellect does not necessarily preclude the ability to devote oneself with success to the more serious study of the historian."[61] Buoyed by Humboldt's praise, Irving envisioned writing a history of Mexico. This plan, however, came to an awkward end when fellow American and Humboldt devotee William Prescott announced his own interest in writing such a book. Prescott's volume on the history of the conquest of Mexico, published in 1843, also drew Humboldt's praise. Irving turned his attention westward, inspired both by Humboldt and his absorption of the Long Expedition report. His novels of the American West, beginning with *A Tour on the Prairies* and *Astoria* and culminating with the Bonneville stories, established many of the literary conventions that defined this quintessentially American genre.

In July 1826 one of Irving's friends and colleagues, James Fenimore Cooper (see fig. 38), arrived in Paris, accompanied by his wife and their daughter, Susan. They quickly established friendships with Humboldt and Lafayette and enjoyed the company of Benjamin Constant; theirs was a friendship that embraced liberal politics at home and abroad. All of them could be found at Lafayette's Tuesday soirées, where Cooper noted that Humboldt was a frequent guest.[62] Lafayette would invite visiting Americans to see his extensive collection of American memorabilia. Among his prized possessions was a cane made from the wood of George Washington's apple tree, under which the two men breakfasted prior to a battle during the Revolutionary War. Once Napoleon was out of power, Lafayette resumed his role as the chief celebrant at the annual party held to commemorate George Washington's birthday.[63] The celebration on February 22, 1831, was especially noteworthy, as it recognized the centennial of Washington's birth and attracted all of the Americans visiting the city, Cooper foremost among them.

Cooper expanded Humboldt's circle to include the American sculptor Horatio Greenough, who took this opportunity to sculpt a bust of Lafayette. Several of the sittings took place at Lafayette's elegant country estate, La Grange, a five-hour carriage ride outside Paris.[64] Cooper enjoyed adding Americans to this network, clearly relishing the opportunity to introduce his traveling countrymen to Humboldt and Lafayette. He was thus exasperated to learn that Thomas Cole had come through Paris in 1832 without stopping to spend time with this group.[65]

HUMBOLDT AND MORSE'S "LIGHTNING LINES"

Samuel Finley Breese Morse (fig. 55) had arrived in Paris at the end of September 1831, his passport (CAT. 34) signed by Washington Irving, who was serving as a secretary in the U.S. legation in London. Cooper and Morse were old friends; they had met as members of the influential Bread and Cheese Club in New York, along with William Cullen Bryant, poet and editor of the *New York Evening Post*, a newspaper he used as a means of promoting American culture as separate and distinct from its European counterparts.[66] With Cooper's help, Morse settled in a few doors down from Lafayette's Parisian quarters, reuniting two men who had met in New York on the marquis's triumphal return to the United States in 1824. The response to Lafayette's tour of the United States had made it clear just how universally popular the Frenchman still was as a symbol of the American Revolution.[67] New York City Mayor Philip Hone, part of the Lafayette welcoming committee, commissioned Morse to paint a full-length portrait of the returning hero (fig. 56). The Frenchman sat for Morse four times in December 1824 and again in July 1825, during which time the two men had become close friends.[68] Morse's description of his painting captures the emotion of the moment:

> *He is represented standing at the top of a flight of steps, which he has just ascended upon a terrace, the figure coming against a glowing sunset sky, indicative of the glory of his own evening of life. Upon his right, if I remember, are three pedestals, one of which is vacant as if waiting for his bust, while the two others are surmounted by the busts of Washington and Franklin—the two associated eminent historical characters of his own time.*[69]

Now, in 1831, the two men lived just doors apart. Morse joined the group of Americans on their rounds to the various salons, becoming another of the regulars at the soirées held in Gérard's studio.[70] At a dinner hosted by Lafayette shortly after Morse's arrival, Morse and Cooper joined Humboldt and their mutual friend, the Marquis de Marbois, who toasted Lafayette, calling him "the Patron of Americans in Paris" as an affectionate compliment.[71]

Morse's arrival signaled another aspect of Humboldt's extended involvement with the United States. Humboldt had studied Jedidiah Morse's *American Geography* as a student in Hamburg.[72] Now Humboldt tracked down the eminent American geographer's son in the Louvre. Morse was in Paris to begin work on the studies that would culminate in his grandest ambition to date: a massive painting portraying the *Gallery of the Louvre* (CAT. 35), in which he would bring

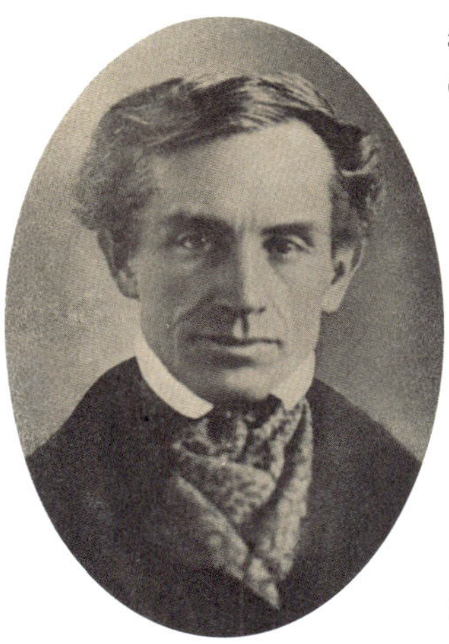

Fig. 55
Possibly Louis-Jacques-Mandé Daguerre, *Portrait of Samuel Finley Breese Morse* (1791–1872), 1840, photographic print on board, 4 ¾ × 3 ⅛ in., Archives of American Art, Smithsonian Institution, Macbeth Gallery Records

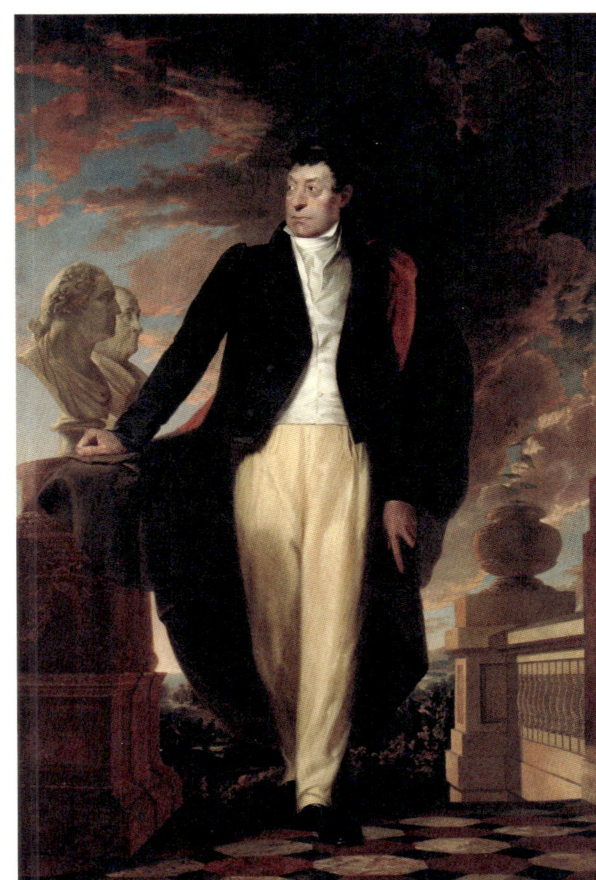

Fig. 56
Samuel F. B. Morse, *Marquis de Lafayette*, 1826, oil on canvas, 96 × 64 in., Collection of the Public Design Commission of the City of New York

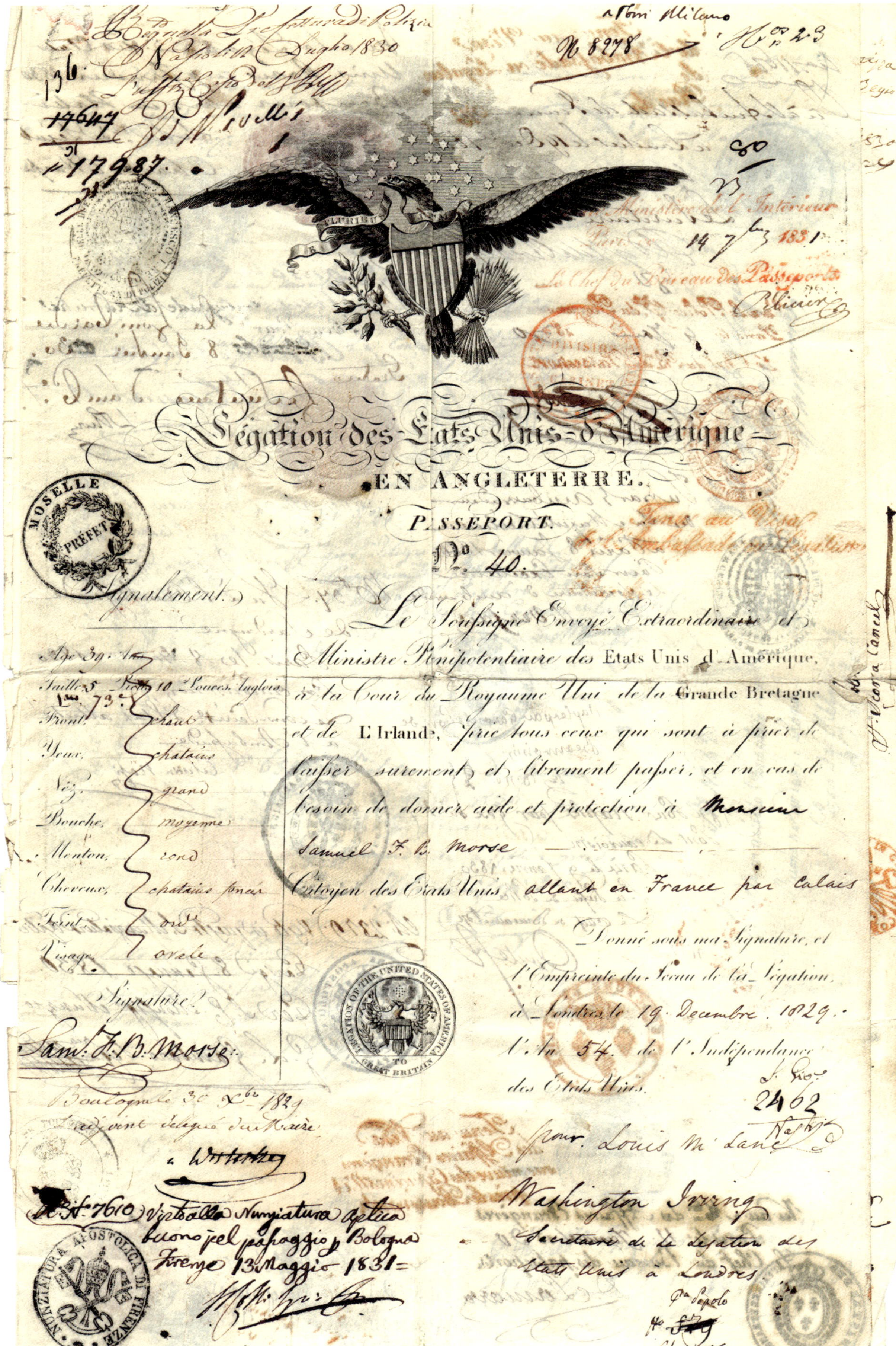

Légation des États Unis d'Amérique
EN ANGLETERRE.
PASSEPORT.
No. 40.

Signalement.
Age 39 Ans
Taille 5 Pieds 10 Pouces Anglois
1m 73c
Front haut
Yeux chatains
Nez grand
Bouche moyenne
Menton rond
Cheveux chatains foncé
Teint ord.
Visage ovale
Signature
Saml. F. B. Morse

Le Soussigné Envoyé Extraordinaire et Ministre Plénipotentiaire des Etats Unis d'Amérique, à la Cour du Royaume Uni de la Grande Bretagne et de L'Irlande, prie tous ceux qui sont à prier de laisser surement et librement passer, et en cas de besoin de donner aide et protection à Monsieur Samuel F. B. Morse Citoyen des Etats Unis allant en France par Calais

Donné sous ma Signature, et l'Empreinte du Sceau de la Légation, à Londres le 19 Decembre 1829. l'An 54. de l'Indépendance des Etats Unis.

pour Louis McLane
Washington Irving
Secretaire de la Legation des Etats Unis à Londres

Boulogne le 30 Xbre 1829

Vº alla Nunziatura Apostolica buono per passaggio per Bologna Firenze 13 Maggio 1831

NUNZIATURA APOSTOLICA DI FIRENZE
MOSELLE PRÉFET
LEGATION OF THE UNITED STATES OF AMERICA TO GREAT BRITAIN

CAT. 34 Samuel F. B. Morse's passport, signed by Washington Irving, issued December 19, 1829, ink on paper, 16 ¼ × 10 ½ in., Smithsonian's National Museum of American History

CAT. 35 Samuel F. B. Morse, *Gallery of the Louvre*, 1831–33, oil on canvas, 73 ¾ × 108 in., Terra Foundation for American Art, Daniel J. Terra Collection, 1992.51

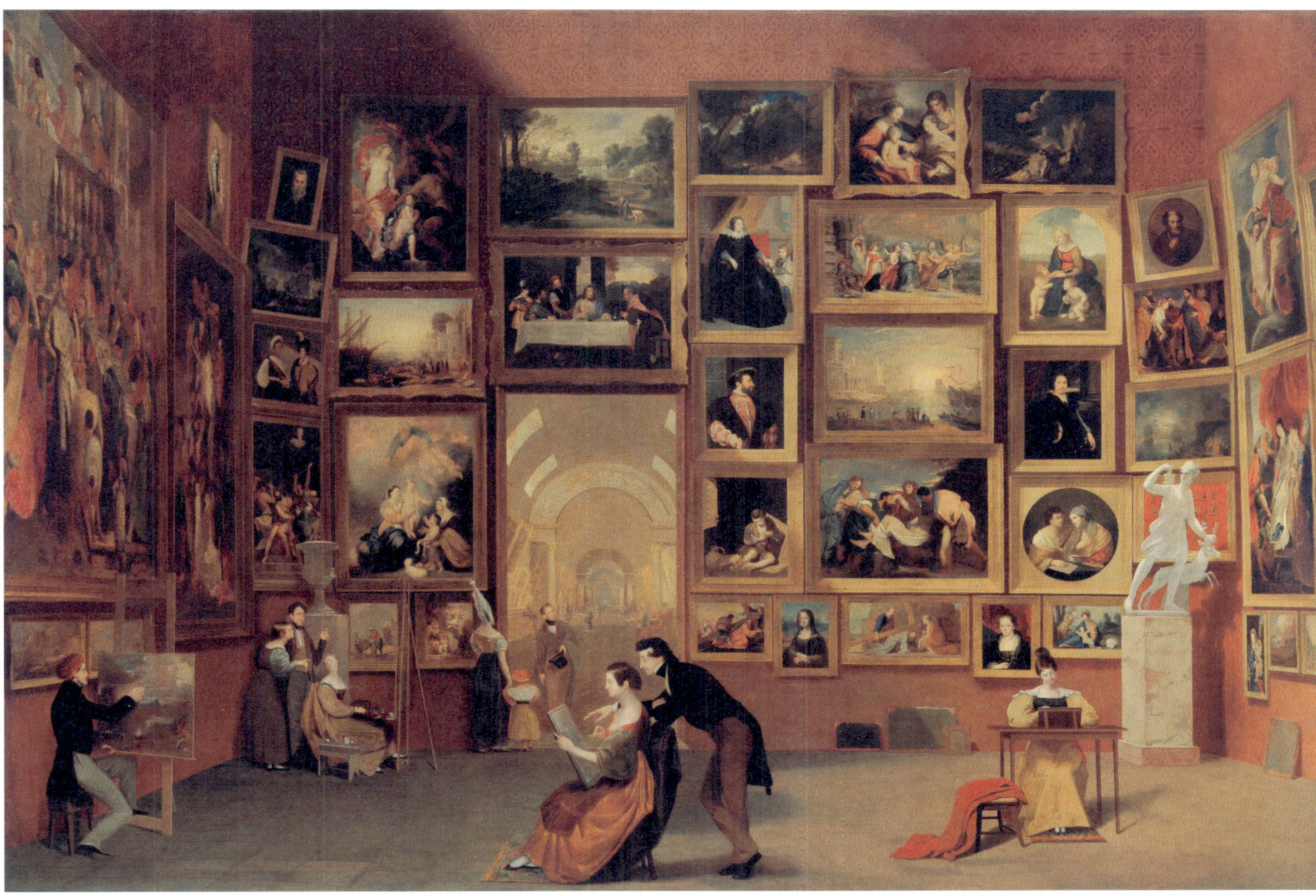

together thirty-eight works of art, now scattered among the galleries of the massive palace, arrayed together in a painting representing the Salon Carré as the pinnacle of achievements of European artistic production, in a setting designed to inspire his American audience to learn from the classics.

Morse's painting, like Humboldt's books, represented the aggregation of massive quantities of data drawn from an entire realm of sources—in this case a tour of the entire Louvre. The goal was to create a singular picture that distilled the history of western European painting. Morse's vision was to bring together works of art that allowed him to synthesize ideas, compare styles, and portray recognizable masterworks as a grand inspirational teaching tool for artists at home. Morse had helped found the National Academy of Design in 1826 and was currently serving as its president. He was intent upon using his position to advance the appreciation of the arts in his home country. Now Morse was touring the Louvre at Humboldt's request, exchanging ideas about art and invention with a man who considered himself half an American.

Morse's practice was to work in the galleries every day, "all day from 8 til dark," making preliminary sketches of each painting on his list.[73] Cooper described to a friend his habit of going "to the Louvre, where I find Morse stuck up on a high working stand, perch myself astraddle of one of the seats, and bore him just as I used to bore you."[74] There Humboldt would join him, climbing up to sit next to the American artist. "Sometimes the great explorer would seat himself beside Morse as he painted at the Louvre, and discourse with the utmost charm from his vast store of observation and thought."[75] Cooper's daughter, Susan, would often accompany her father to the Louvre, where Morse provided art instruction for her. When Morse began painting the finished canvas, he included the Coopers among the visitors to the Salon Carré. Although Susan is shown seated at a large easel, palette in hand, she would soon turn to writing about nature, inspired in part by Humboldt's example.

We can only assume that Humboldt conducted himself as he was long accustomed, dominating the conversation, switching among various topics, occasionally peppering his captive audience with questions. In Morse's case, this led to conversations about his embryonic ideas for developing a telegraph—an invention that captured both of their imaginations for its potential for instantaneous communication across vast distances. As early as 1811, Morse had fantasized about this invention. In a letter to his mother, he had written from London, "I wish that in an instant I could communicate the information: but three thousand miles are not passed over in an instant, and we must wait four long weeks before we can hear from each other." Years later, Morse reread his missive, and wrote along the margins, "Longing for a telegraph even in this letter."[76] Morse and Humboldt clearly enjoyed each other's company. For Morse, Humboldt offered the dual validation of encouraging his art and his enterprise, and pressed the younger American for his insights:

> *Humboldt's custom was to approach the easel of the young artist and engage in conversation, usually upon subjects of art. Frequently he would remark that Mr. Morse must feel fatigued, and would invite him to take a stroll among the paintings, stopping to admire such as attracted his especial notice, and while passing criticisms upon their merits himself, demand of his companion his opinion concerning them.*[77]

Humboldt, apparently, was a welcome distraction.

Although Humboldt courteously and sincerely chatted with Morse about art, he was just as likely to be interested in Morse's ideas for the growth of American visual culture and how this monumental painting would help achieve his goals. Having known of Morse's father, he was unsurprised that the younger Morse's own training had encompassed both the arts and the sciences. Humboldt was attracted to people who held multiple interests and pursued myriad lines of inquiry. He was happiest when he was in the company of others who challenged him to think beyond what he already knew, and further, he was prepared to offer encouragement of his own.

As much as he could respect Morse's commitment to art, deep down Humboldt was far more intrigued by Morse's design for his telegraph. Humboldt and Morse both knew there were competing designs under development in the United States, England, and France; this was Humboldt's chance to evaluate Morse's prospects in their embryonic form. Since his trip to South America,

Humboldt had experienced the frustration of materials lost when ships foundered and missives went astray. Humboldt perceived the telegraph as a network unto itself, capable of connecting people at a great distance, while circumventing the prying eyes of insecure statesmen. Like Lafayette, Humboldt had resorted to sending multiple copies of the same letter to Jefferson, anticipating that one courier might be waylaid or a letter confiscated en route. Under Napoleon, Lafayette had used oblique language to refer to people and events, lest their confidentiality be compromised. The telegraph promised faster and more secure communication, at least overland. Morse believed that the telegraph had the potential to foster national unity, bringing together disparate parts of the country by enabling rapid communication. Edward Everett would be among Morse's supporters, enthusing, "[A]rt and science are, in themselves, progressive and infinite.... Nothing can arrest them which does not plunge the entire order of society into barbarism."[78]

Morse sailed back to the United States in 1832, still thinking about the telegraph but committed to the completion of what he confidently believed would be his greatest painting triumph. At six by nine feet, *Gallery of the Louvre* was his largest work; in scale and scope it was a monumental achievement intended to be an icon for American visual culture. The critics were not impressed, and Morse was devastated by the lukewarm response to his efforts. The reception of this painting convinced him to turn from art to science, and Morse spent the next six years developing his concept for the telegraph. The difficulties he encountered trying to perfect it and establish the patent for it would occupy him for most of the rest of his life.

EXPANDING HUMBOLDT'S NETWORK: "A FOUNTAIN WITH MANY PIPES"

Humboldt's most prodigious accomplishment was writing books, sometimes singly and at other times in collaboration with colleagues, and supervising the publication of every one of the thirty-four volumes relating to his travels. His reputation for working at all hours, fueled on coffee and adrenalin, was borne out by the rapid sequence of publications exploring various aspects of his five-year trip to the Americas, with volumes appearing in print as rapidly as the ideas tumbled out of his head and onto the page.[79] Over the course of the next twenty years he would accrue memberships and honorary degrees from institutions on both sides of the Atlantic, confirming his stature as one of the leading scientists of his age. Humboldt welcomed his American visitors, encouraging the widening sweep of his network. In both Paris and Berlin he encouraged traveling Americans and maintained a robust transatlantic correspondence.

Humboldt had lived in Paris until 1827, delivering lectures, reading scientific papers, and publishing articles in European and English journals. His international renown had earned him the respect of his peers, the undying enmity of Napoleon, and the increasing desire from Prussian King Friedrich Wilhelm III to bring his eminence back to Berlin. Each of Humboldt's published volumes enhanced his reputation and contributed to the depletion of his inheritance. As his personal fortune dwindled further, he finally had no choice but to accept a stipend offered by the king, which required the world-famous explorer to move permanently to Berlin—in an effort to elevate the Prussian city's status in the intellectual community. Humboldt was not happy to leave Paris; it was the center of the scientific world, and a vibrant, cosmopolitan, international hub of activity. Humboldt fretted that Berlin was too quiet and removed to be of much interest, but the need for the king's stipend made that point moot. Cooper echoed Humboldt's frustration with the sleepy city, observing, "We should harness a pair of horses in America in half the time they would harness them in France, and in France, they do it in half the time it is done in Germany."[80]

Now it was up to Humboldt to reinvigorate Berlin. In September of 1828 he founded the Berlin *Naturforscher*, the Society of Naturalists, agreeing to serve as its president.[81] The hunger in Berlin for Humboldt's presence is reflected in the attendance at the inaugural events: 460 members attended the September meeting; 850 attended the dinner afterward, and 1,200 attended the soirée Humboldt hosted that evening.[82] Humboldt threw himself into developing programming at the Berlin Academy of Sciences and delivering public lectures, which became the foundation for *Cosmos*. The German poet Karl von Holtei attended Humboldt's lectures and described the audience in a letter to their mutual friend Goethe,

Such an audience has, I think, never been gathered before the lectern of a man of learning. The king, the whole of the court, the highest members of the administration and the military plus their ladies, all the scholars, the important artists, the whole belle monde.[83]

Holtei concluded, "Eight hundred people barely breathe to hear the One. There is no greater impression than to see the earthly power as it pays homage to the spirit; and that is why Humboldt's current work in Berlin is one of the most uplifting phenomena of the time."[84] Speaking extemporaneously, drawing artful connections among the sciences, Humboldt made his case for the glories of the natural world and reveled in his intellectual prowess in bringing its disparate parts into alignment.

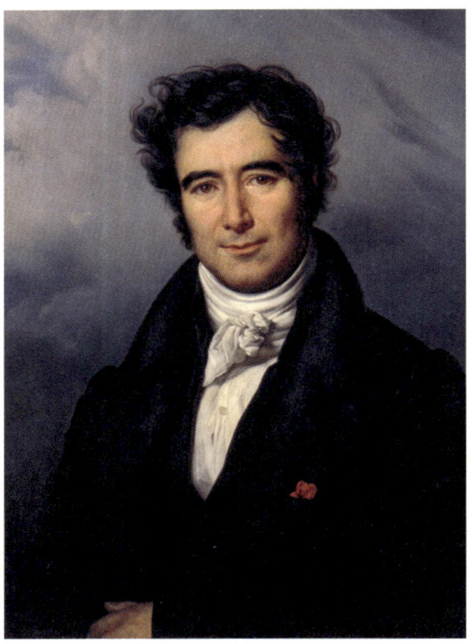

Fig. 57
Charles Auguste de Steuben, *Portrait of François Arago* (1786–1853), 1832, oil on canvas, 43 5/16 x 37 13/16 in., L'Observatoire de Paris

Writing did not come as easily to Humboldt as lecturing. The glowing accounts of his public talks suggest that in his mind's eye, the complex interrelationships that he was discovering were crystal clear, and the relationships between data and ideas were in constant motion. When it came to transforming those dynamic thoughts into a more linear written narrative, Humboldt's impatience to get those complexities on paper resulted in essays and books that were undeniably rich and at times impenetrably convoluted. François Arago (fig. 57) quipped that Humboldt had no idea how to write a book, instead approaching the exercise like a series of interrelated lectures, drawing threads, relating anecdotes, moving swiftly if erratically between topics. Agassiz related the story,

I well remember hearing Arago turning to him, while speaking of composition, and saying, "Humboldt you don't know how to write a book—you write without end, but that is not a book; it is a picture without frame."[85]

Humboldt's working method was evidence of this critique; he frequently chopped up letters and pages from his diaries in order to aggregate information on similar subjects, which he deposited into envelopes marked with the particular topic. His approach threatened to destroy a clear chronology for his discoveries or ideas, but favored a kind of intellectual simultaneity in which "the whole mass of facts collected from different regions of the earth is comprehended in one glance."[86] Humboldt was a magnet in person, an irresistible force, and a compulsory if challenging read.

Goethe waxed poetic about the Prussian's flexible mind and voluble personality:

Alexander von Humboldt has been some hours with me this morning.... What a man he is! Long as I have known him, he ever surprises me anew. One may say he has not his equal in knowledge and living wisdom. Then he has a many-sidedness such as I have found nowhere else. On whatever point you approach him, he is at home, and lavishes upon us his intellectual treasures. He is like a fountain with many pipes, under which you need only hold a vessel, and from which refreshing and inexhaustible streams are ever flowing.[87]

Humboldt's position at the nexus of American and European intellectuals meant that his circles of admirers and colleagues constantly overlapped.[88] Perpetuating that network was Humboldt's ultimate aim.

In some respects, that network perpetuated itself. The Ticknor family returned to Europe in 1835 and the following year visited Humboldt in Berlin. Of all of the visiting Americans, George Ticknor appeared to be one of Humboldt's favorites, someone whose company made the aging Prussian feel closely connected to the United States. Upon hearing of his arrival, Humboldt dashed off a note to his friend Baron Bunsen, then the Prussian envoy in London.

You will remember, my dear friend, the antediluvian times, when two ingenious Americans of classical education, Ticknor and Everett, traveled all over Europe. Ticknor from Boston is now re-appearing on the horizon. Receive him in a friendly way, just

the way you know how to receive guests. Ticknor has been a friend of my family.[89]

Their reunion was warm and genuine, with Humboldt setting aside time to take his old friend through the Bauakademie and the halls of natural history. Ticknor noted the deference he was shown in Humboldt's company, and the respect with which the aging scholar was met. At dinner, Humboldt once again dominated the conversation, as he spoke "with incredible volubility both in French and English, and seems to talk equally well upon all subjects; always, however, I suspect, with a little indulgence of sarcasm towards individuals he does not approve."[90] After catching up in Berlin, Ticknor was off to Austria with a letter of introduction to Klemens von Metternich, now the Austrian empire's chancellor of state, in which Humboldt cheerfully described himself as "Presque Américain moi même," or, "practically an American, myself."[91]

FROM THE TELEGRAPH TO THE TELEGRAPHIC PLATEAU

Humboldt also continued to find reasons to be in Paris. Although he was injecting life into Berlin's intellectual community, Paris was still the center for research and publications. To Ticknor he confided, "'for you must know,' said he, smiling, 'I made my bargain with the King, as the *Cantatrici* do, that I should be allowed to pass three months every year where I like, and that is Paris.'"[92] At first Humboldt was able to make return trips with some frequency, often under the auspices of a diplomatic mission; but as he became more deeply involved with matters at court under King Friedrich Wilhelm IV, he had fewer opportunities to leave Berlin.

One of Humboldt's trips to Paris coincided with Morse's return to the city in September 1838. Arago, who was now the perpetual secretary of the Institut National and director of the Paris Observatory, arranged a formal demonstration of Morse's electromagnetic telegraph at the Institut's meeting, with Humboldt in attendance. Arago translated for Morse. The assembled group of scientists were astonished at the success of his invention and crowded around Morse's telegraph key and register (CAT. 36, CAT. 37), at which point Humboldt "approached Mr. Morse, and shaking him warmly by the hand, congratulated him on his success, at the same time declaring his belief that his invention was destined to produce the most astonishing results."[93] That Paris demonstration resembled the one Morse had successfully conducted a year earlier in Washington, D.C., when he had approached Congress for funding to test his invention at a distance of ten miles between the transmitter and receiver. An observer, who marveled at Morse's ability to communicate at that distance, immediately understood the ramifications of Morse's invention:

This was among the first attempts at communicating intelligence in this manner which had ever been attempted, and no reason could be seen why, after accomplishing a distance of ten miles through a coiled wire, the same might not be done through one extended over that space in a straight line; or why this could not be increased to a hundred or a thousand miles, and thus completely annihilate space in the transmission of intelligence.[94]

Humboldt shared the writer's enthusiasm for the potential of Morse's telegraph. Here was the answer to one of Humboldt's fondest dreams. Instead of waiting months or years in hope of a response, he envisioned conversations without resorting to encoded language to foil eavesdroppers, assured that his messages were received and understood. Humboldt proclaimed Morse's invention was "the best of all the plans that had been devised."[95] Six years later Morse finally had the funding in hand to lay a telegraph wire between Washington and Baltimore. On May 24, 1844, Morse sent the first successful telegraph message to his partner, Alfred Vail: "What hath GOD wrought." Vail replied by sending the same message back to Morse, who carefully transcribed it in his elegant script (CAT. 38). The potential for the telegraph became apparent during the Democratic Convention in Baltimore, where the results of the ballots that eventually nominated James K. Polk for president were relayed to the Washington press by telegraph.[96] Arago pressed his colleagues to support establishing telegraph lines across France after Morse sent him a clipping from the *Baltimore Sun* containing the message from President Polk to

CAT. 36 Samuel F. B. Morse and Alfred Vail, *Telegraph Key*, ca. 1845, wood, brass, and ivory, overall: 3 ½ × 4 ½ × 8 in., Smithsonian's National Museum of American History

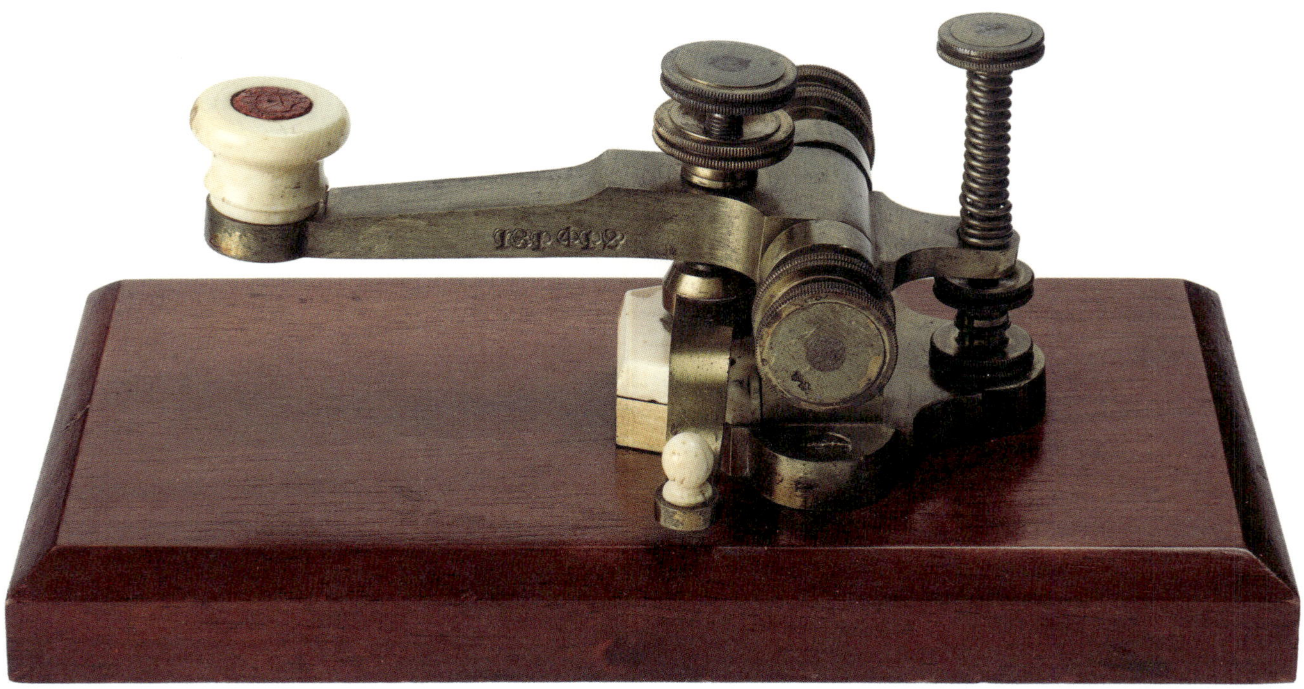

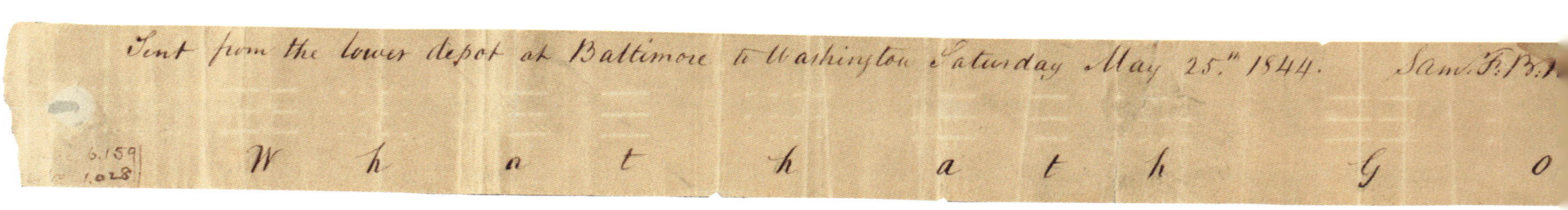

CAT. 37 Samuel F. B. Morse and Alfred Vail, *Telegraph Register*, 1848, wood, brass, iron, and copper, overall: 7 ¼ × 4 × 11 ½ in., Smithsonian's National Museum of American History

CAT. 38 Samuel F. B. Morse, *"What hath GOD wrought" Telegraph Message*, 1844, ink on paper, 1 ½ × 27 ¾ in., Smithsonian's National Museum of American History

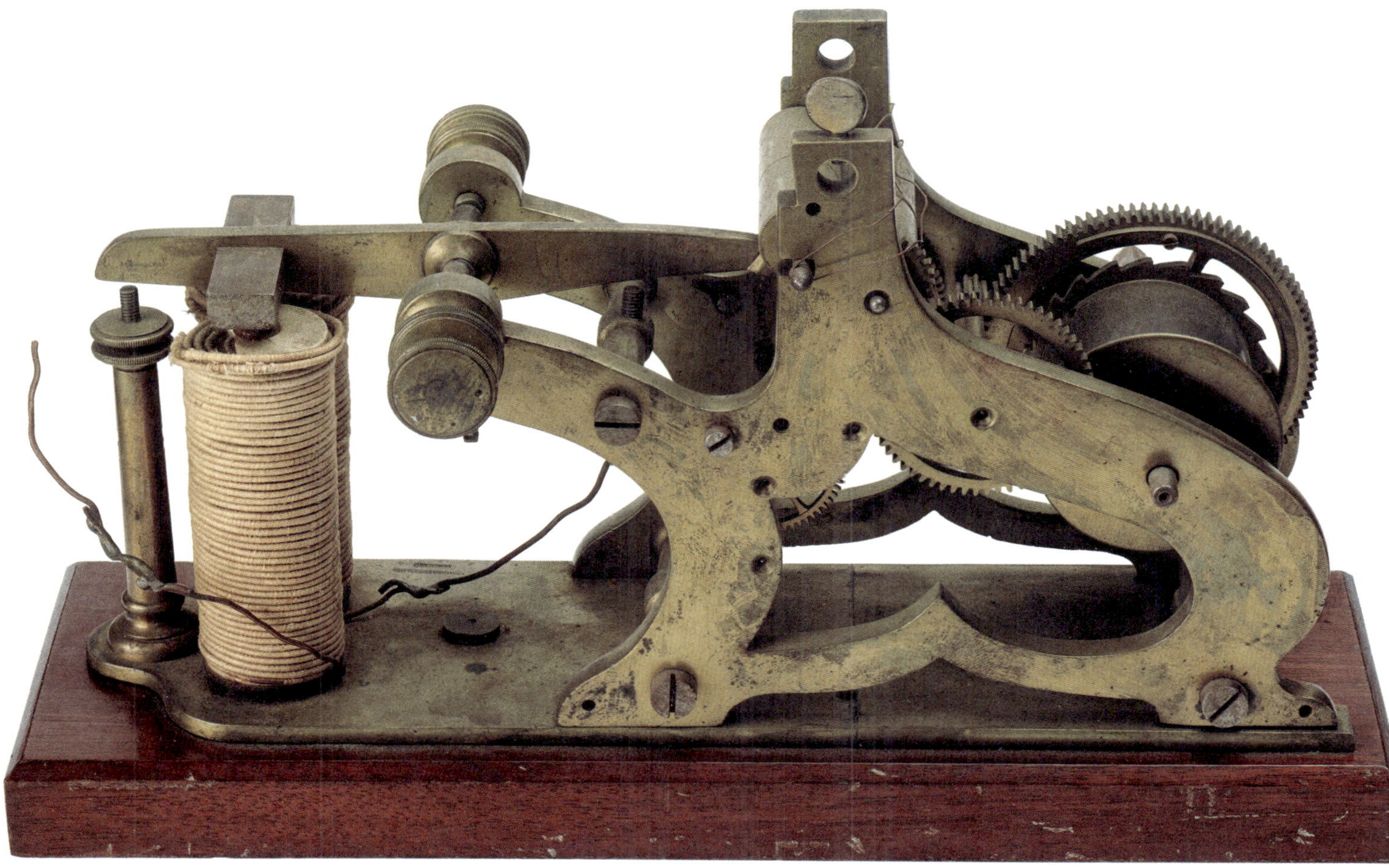

Congress concerning the possibility of a war with Mexico, "transmitted entirely by telegraph in a short space of time." According to a later U.S. magazine article, Arago said, "'If this,' declared the eminent savan, 'is not a fait accompli, I know not what is.'" The article went on to relay, "[T]he appropriation for the erection of the first telegraphic line in France passed by acclimation."[97]

Morse's telegraph transformed communication. In Humboldt's mind, it would revolutionize his entire network, bringing more parts of the globe into faster, more reliable communication. As part of his last ambitious overland trip, Humboldt had traversed more than nine thousand miles across Russia in 1829 at the behest of the tsar. In return for providing the tsar with information regarding the presence of precious metals in the Ural Mountains, Humboldt had extracted his own promise of gold: weather stations to be established every 150 miles across Siberia, each sending him weekly reports with records of temperature, humidity, barometric pressure, and any seismic activity. Shortly before Morse's Paris demonstration of his telegraph, Alexander Dallas Bache, Benjamin Franklin's great-grandson and a U.S. coastal surveyor, visited Humboldt in 1837. That year, several scientific societies in Britain launched the international "magnetic crusade," an effort to establish a network of observatories around the world for the transmission and sharing of meteorological data. Inspired in part by Humboldt's success in persuading the Russian emperor to set up weather stations, the goal was to create a global belt of sites recording terrestrial and atmospheric magnetic data. Bache was instrumental in convincing the U.S. War Department to participate, and he personally supervised the Philadelphia station located at Girard College.[98] Those weather stations were yet another of Humboldt's global networks, providing information that would revolutionize meteorology.

By the time Morse demonstrated his invention in 1838, Humboldt had negotiated similar arrangements spanning close to two-thirds of the populated globe. Humboldt was hailed for having used his study of terrestrial magnetism to foster a kind of scientific international unity impossible through political relationships alone.[99] Still, those readings were arriving mostly through handwritten letters. Electronic communication would compress time, connecting people, data, and ideas into one great network. For Humboldt, the ability to speed up his research and reach across distance was tantalizing. Little wonder the new telegraph wires were called "lightning lines." The success of the overland telegraph gave rise to an even more ambitious conversation concerning the potential for laying a cable on the seabed of the Atlantic Ocean to create a transatlantic cable. Morse understood that such an invention would connect the hemispheres and that it would allow the United States to be a more active participant in world events. As an American invention, the cable would add luster to the country's stature among European scientists. In Humboldt's mind, it would be another way to amass and share more data, feeding his latest enterprise, the culmination of his own career, *Cosmos*. In Berlin Humboldt had begun delivering the lectures that sowed the seeds of this much larger, more ambitious undertaking. In his first book, *Essay on the Geography of Plants*, Humboldt had connected disparate places through the study of similar plants and belts of climate. In his isothermal lines he had expanded those localized temperature readings and understood the way bands of climate wrapped around the globe. Now he wanted to unite all of his research into a single work that would embrace the entire planet—from the outer atmosphere to the bottom of the oceans.

This was not the first time Humboldt had advocated better transit routes. Humboldt had earlier envisioned speeding up travel and lines of communication by creating a canal linking the Atlantic and Pacific Oceans. In South America he had sketched no fewer than seven prospective locations, mostly across Central America. He had broached this topic in Washington with Secretary of State James Madison in 1804, in hopes that the United States' expansionist ambitions might lead to this most worthy construction project. In his *Personal Narrative* he dedicated more than sixty pages to outlining the five points he deemed most likely to provide "the practicability of a water communication between the Atlantic and Pacific Oceans," including a detailed assessment of how to carry out such a venture.[100] In the third edition of *Aspects of Nature*, in 1849, he returned to the topic, pointing out that "For upwards of twenty years

I have been repeatedly consulted on the problem of the Isthmus of Panama, … but in no instance has the simple advice I have given been followed."[101]

With the discovery of gold in California, Humboldt hoped economic expediency might facilitate this goal. He took this opportunity to encourage his "ambassador" to the United States, Johann Gottfried Flügel, to have this matter published in the American press. Flügel served as the Prussian consul from Leipzig and as his country's representative to the Smithsonian Institution. In the June 14, 1850, issue of the *National Intelligencer*, an article appeared printing the letter from Humboldt to Flügel, along with a passage from *Aspects of Nature*:

As the taking possession of a considerable part of the west coast of the New Continent by the United States of North America, and the report of the abundance of gold in New California, (now called Upper California,) have rendered more urgent than ever the formation of a communication between the Atlantic States and the regions of the West through the Isthmus of Panama, I feel it my duty to call attention once again to the circumstance that the shortest way to the shores of the Pacific … is in the eastern part of the Isthmus.[102]

Fig. 58
Mathew B. Brady, *Cyrus West Field* (1819–1892), ca. 1860, salted paper print, 17 13/16 × 14 ½ in., National Portrait Gallery, Smithsonian Institution

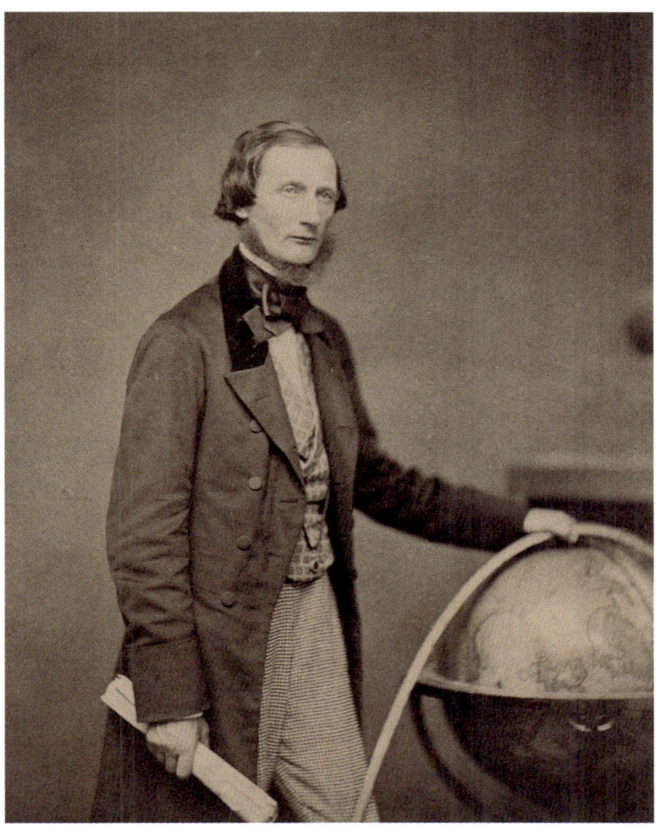

Hoping for added traction, Humboldt expanded on the geopolitical significance of such a canal. Eager for access, hungry for information, Humboldt urged all means of networking the globe during his lifetime. It would be left to another American Humboldt admirer, President Theodore Roosevelt, to make the Panama Canal a reality. Construction finally began in 1904, and the canal opened to ship traffic in 1914.

Transiting the oceans remained a central concern for Humboldt, and not just for letters, books, and other tangible forms of communication. Humboldt had embraced the electromagnetic possibilities for overland communication with the telegraph in 1838 and became involved with envisioning a global version of that network. Morse would spend decades getting control of the patent on his telegraph, but in 1854 he began to forge partnerships with the people who could solve the very real hurdles associated with creating a viable transatlantic cable. The most significant would be determining the length and position of the route. In 1854, fresh from his extended trip across South America with Frederic Edwin Church, Cyrus West Field (fig. 58) became intrigued at the possibility of connecting the Old and New Worlds using Morse's invention. Much as Humboldt had fantasized about a canal to facilitate communication between the Atlantic and Pacific Oceans, Field studied maps to suggest plausible points of connection between North America and Europe. Spinning his globe (CAT. 39) in his study, Field traced the route between Newfoundland and Ireland, the shortest distance across the Atlantic.

Humboldt was deeply enamored of the work being done by the U.S. Coastal Surveys, in part because the explorer had not engaged with those areas. In the introduction to the first volume of his *Personal Narrative*, Humboldt had reasoned, "It is not by sailing along the coast, that we can discover the direction of the chains of mountains, and their geological constitution, the climate of each zone, and its influence on the forms and habits of organized beings."[103] That essential work he left for others to do. It would be Bache's rival, Matthew Fontaine Maury, the "Pathfinder of the Seas," who would map the Atlantic Ocean floor to determine the best place to lay the Transatlantic Cable. He had already come to Humboldt's attention after

CAT. 39 Newton & Son, *Newton's New and Improved Terrestrial Globe*, 1852, wood, paper, brass, and glass, overall: 42 × 26 in., Smithsonian's National Museum of American History

sending him copies of his wind and ocean current charts, adding considerably to Humboldt's understanding of their effect on global weather patterns.[104] It was Humboldt who suggested Maury title his book *The Physical Geography of the Sea*, drawing a further parallel between their respective areas of study. Mapping the Atlantic Ocean floor meant understanding more about ocean currents in order to know where a cable could remain securely in place. Humboldt believed there might be a favorable route for such a cable across the north Atlantic Ocean—what Maury would call the "Telegraphic Plateau." Now Humboldt was especially eager to see this extension of Morse's telegraphy across the ocean.[105] In a letter Humboldt announced, "Morse just sent me a 15-foot long cable," and went on to quote Morse, who claimed "'There will be a possibility, to send you in less than one year a despatch from my home on the Hudson River to Potsdam in less than 5 minutes of the time. I look with sanguine hope to this consummation.'"[106]

Field developed a partnership that included Morse and three others colloquially known as "The Cable Cabinet." On March 10, 1854, the group of investors met at Field's Gramercy Park home in New York. The next decade was spent buying up rival telegraph firms and securing access to the critical points in Canada and Ireland for the terminus of the cable. The Transatlantic Cable project brought together the leading American minds in magnetism, including Bache and Princeton professor, and first secretary of the Smithsonian Institution, Joseph Henry. All of them were motivated by Humboldt's enthusiasm both for the data and its potential for making electromagnetic communication a reality. After several false starts and snapped cables, on August 16, 1858, Queen Victoria was able to send a message using Morse code to President James Buchanan. At the celebratory banquet in New York, Field was reputed to have said, "Maury furnished the brains, England gave the money, and I did the work."[107] The celebration was slightly premature, giving way to frustration as successive attempts to lay a cable that would remain intact and functioning continued until 1866. However, it was simply a matter of time before more reliable hardware could be put in place.

In the winter of 1864 to 1865, Field's close friend Frederic Edwin Church completed the *Aurora Borealis* (CAT. 40), a painting that radiated references to the perils of Arctic exploration and the ongoing conflict of the American Civil War. But in the lurid pulsations of the auroras was a literal manifestation of the electromagnetic pulses that had obsessed Humboldt, Field, and Morse, making them believe there was a way to harness electricity to send messages. Even the landscape in Church's painting suggested a metaphorical resonance with Field's latest enterprise. The mountainous headlands of Labrador along the Canadian coast served as the launching point for the cable in North America. The ocean, frozen solid enough to trap an explorer's ship for the winter, defied human efforts to maintain transatlantic communication. From the distant ocean horizon, an eerie shaft of light streams up like a beacon. Yet overhead the brilliant auroras stretch unfettered from North America across the Atlantic. It is not hard to envision in the broken segments of Church's auroras a metaphor for Field's sequence of snapped cables and broken transmissions.[108]

The Atlantic Cable Projectors (CAT. 41), painted in 1895 at Field's request to commemorate the ultimate success of the project, reunited the extended Cable Cabinet on canvas. Field dominates the right foreground, a commanding presence, while a white-haired Morse is relegated to an ancillary position in the rear of the group. Morse had been a difficult partner during a difficult enterprise; he and Field ended up with a relationship as tense as the span of the cable. A map spread across the table recreates the same oceanic span as Church's *Aurora Borealis*. On the left, one of Field's globes is positioned to show the origin of the cable in Canada; on the far right, Field points to its terminus at Heart's Content in Ireland. The sequence of hands and sheaves of papers passed between the other members of the Cable Cabinet suggests the transmission of information from one side of the painting to the other. A short segment of the cable rests on the table near the globe, a souvenir Field had

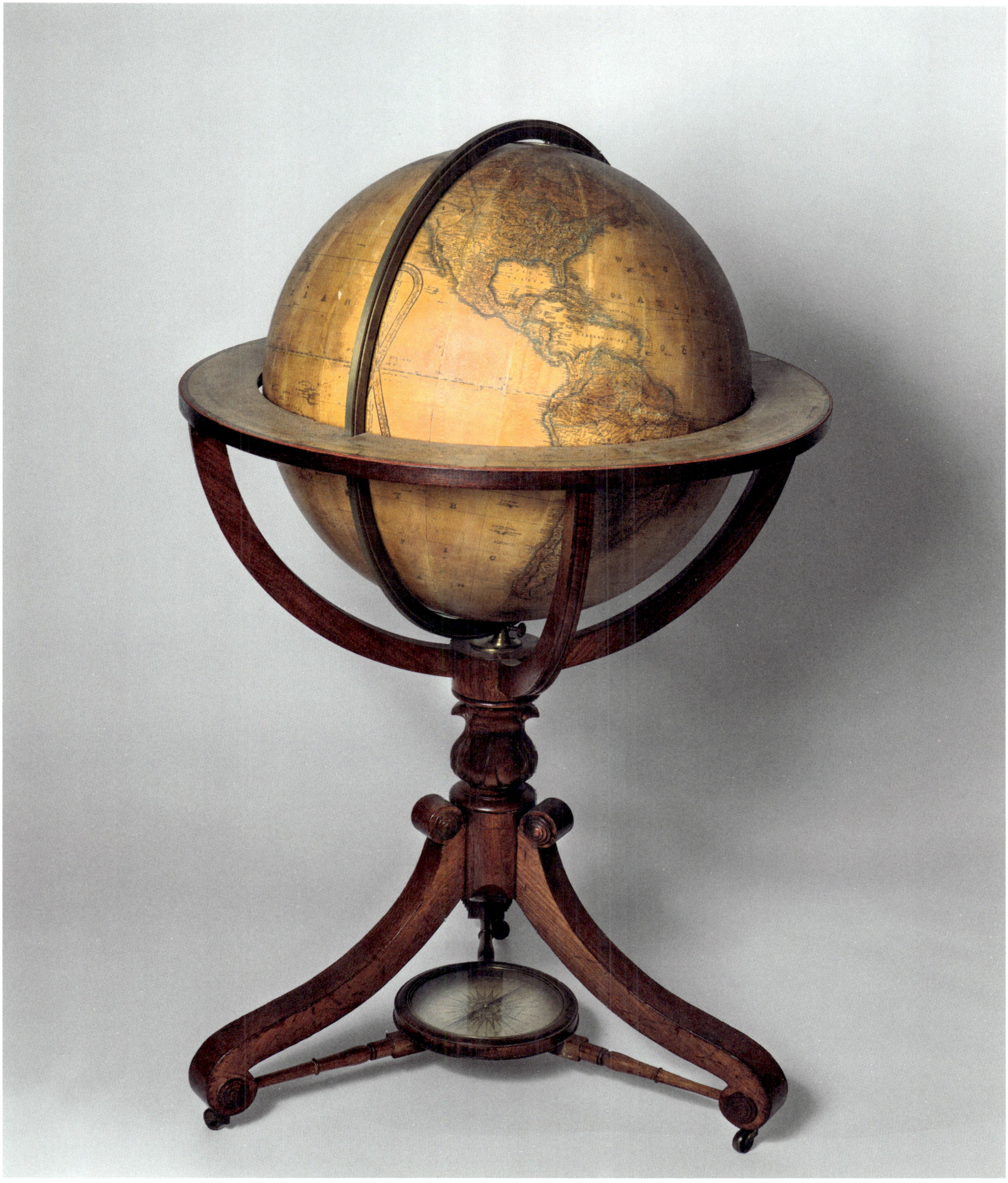

CAT. 40 Frederic Edwin Church, *Aurora Borealis*, 1865, oil on canvas, 56 × 83 ½ in., Smithsonian American Art Museum, Gift of Eleanor Blodgett

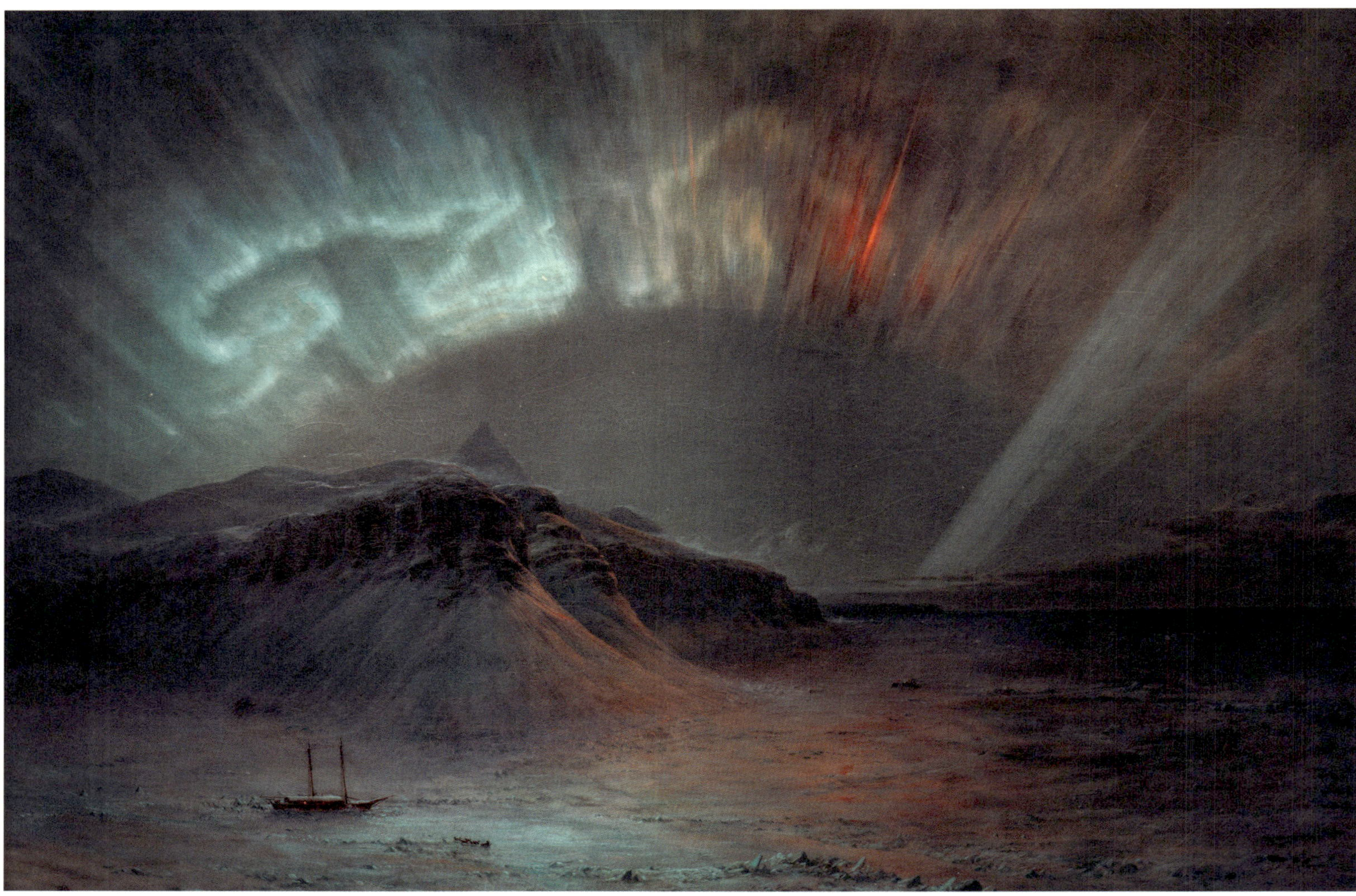

CAT. 41 Daniel Huntington, *The Atlantic Cable Projectors*, 1895, oil on canvas, 87 × 108 ¼ in., New York State Museum, Albany, NY

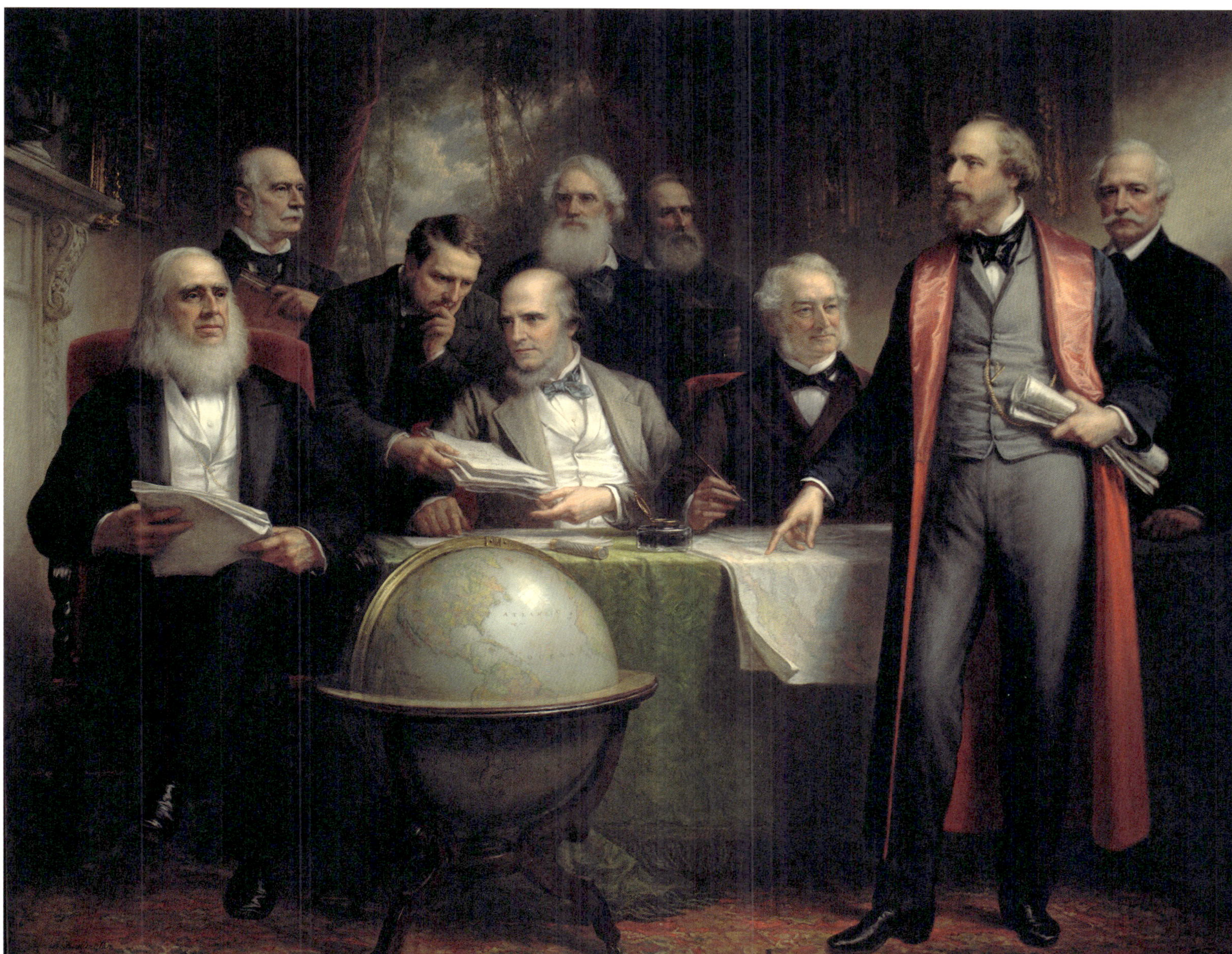

packaged by Tiffany and Co. as a memento of the endeavor (CAT. 42), along with a facsimile of a letter signed by Field attesting to its authenticity (CAT. 43). Humboldt's dream of reliable instant communication across the Atlantic would have to wait until after his death, but he lived long enough to receive his own souvenir segment through his friend and protégé Francis Lieber. Humboldt conveyed his hope that the cable would finally permit the joining of two continents in direct and rapid communication, asking his friend Heinrich Gustav Magnus to show the segment to the members of the Royal Prussian Academy.[109]

Cartoons in both America and Europe extolled the invention (fig. 59). "John Bull" and "Brother Jonathan" clasp hands across the waters, lines of lightning playing around their bodies. The cable itself appears above water, as it was being laid by the two ships, the *Niagara* and the *Agamemnon*. Wags referred to the cartoon as "Morse's History Painting," a mean-spirited jab at his colossal painting, *Gallery of the Louvre,* and Morse's reluctant decision to abandon his career as an artist. Morse paid a final visit to Humboldt in Berlin in 1856. He found the aged explorer still sharp and still working. As a memento of the time they had spent together in Paris, and with regard for Morse's accomplishments, Humboldt gave Morse an engraving of himself, on which he had written, "To Mr. S. F. B. Morse, whose philosophic and useful labors have rendered his name illustrious in two worlds. The homage of the high and affectionate esteem of Alexander von Humboldt. Potsdam, August 1856." Morse took home the engraving and hung it prominently in the library at Locust Hill, his family home.[110] For Morse, who had turned to inventing as solace after feeling like a failure as an artist, Humboldt's approval was doubly sweet.

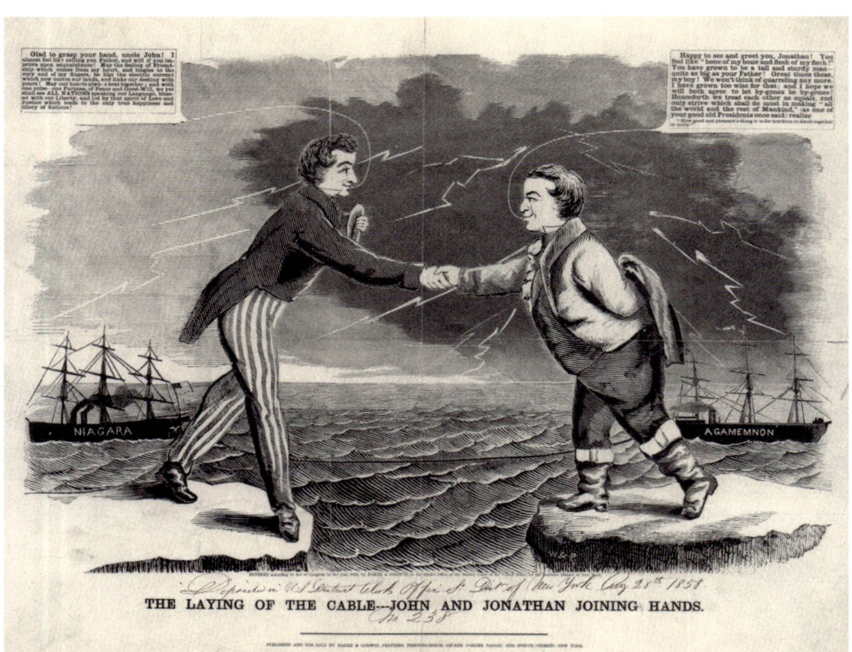

Fig. 59
Baker & Godwin, *Transatlantic cable cartoon*, 1858, woodcut with letterpress on paper, 16 ¾ × 22 ¼ in., Library of Congress, Prints and Photographs Division

THE LIMITS OF APPROBATION

By the 1840s and 1850s, a letter from Humboldt was often just the kind of endorsement an American cherished as evidence of commendable achievement. Humboldt wrote in favor of any exploratory endeavor he felt added to the knowledge of the planet. When they had met in 1837, Bache had been struck by Humboldt's enthusiasm for the United States. At the time he had remarked, "[W]e feel that he was...almost an American."[111] Humboldt's continuing respect for Bache's work became manifestly evident in 1851, when the U.S. government considered cutting Bache's funding. Humboldt came to his defense, writing to urge Congress to continue funding the Coastal Survey, a gesture Bache felt had played a significant role in preserving that endeavor.[112] Authors sent Humboldt inscribed copies of their works, hoping for recognition of their achievements.[113] Both George Ticknor and William Prescott took pride in Humboldt's notice of their publications.[114] In his note to Humboldt, Prescott told him, "I have devoted half a volume to a view of the Aztec civilization; and as in this shadowy field I have been very often guided by the light of your researches, I feel especially indebted to you, and am most desirous that the manner in which my own investigation is conducted may receive your approbation. It will indeed be one of the best and most satisfactory results of my labours."[115]

Not surprisingly, a positive response from Humboldt could validate American cultural achievements. With the founding of the Smithsonian Institution in 1846, Johann Gottfried Flügel became the de facto ambassador between Humboldt and the Smithsonian.[116] Humboldt used Flügel as an intermediary, funneling information to Smithsonian Secretary Joseph Henry. Henry, in return, ensured that every one of the Smithsonian's

CAT. 42 Tiffany and Co., *Tiffany's Transatlantic Cable Souvenir,* 1858, steel, brass, copper, and gutta-percha, 4 × ½ in. diam., Smithsonian's National Museum of American History

CAT. 43 After Cyrus West Field, Facsimile letter accompanying the cable souvenir, 1858, ink on paper, 7 ½ × 4 ⅝ in., Smithsonian's National Museum of American History

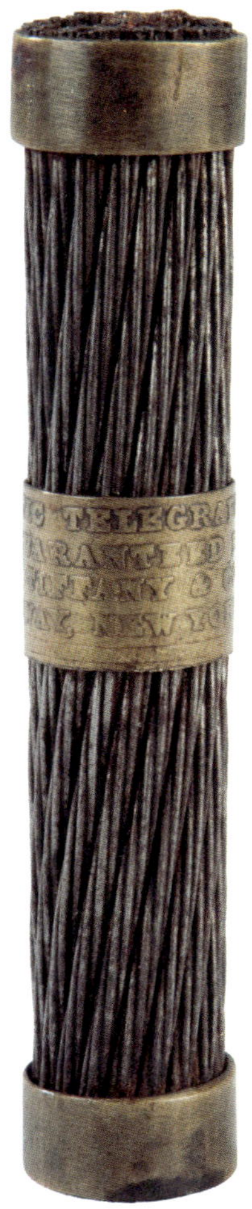

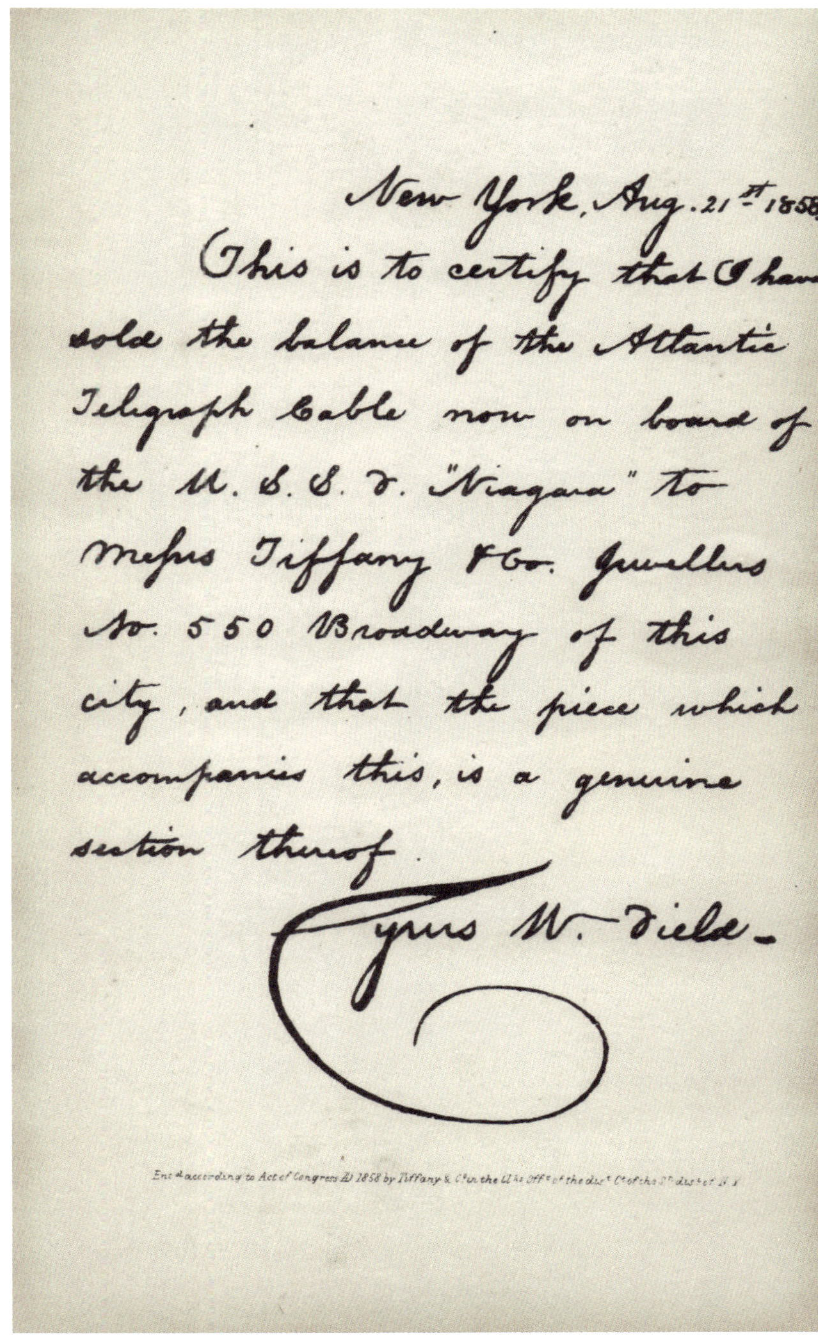

New York, Aug. 21st 1858.

This is to certify that I have sold the balance of the Atlantic Telegraph Cable now on board of the U. S. S. F. "Niagara" to Messrs Tiffany & Co. Jewellers No. 550 Broadway of this city, and that the piece which accompanies this, is a genuine section thereof.

Cyrus W. Field.

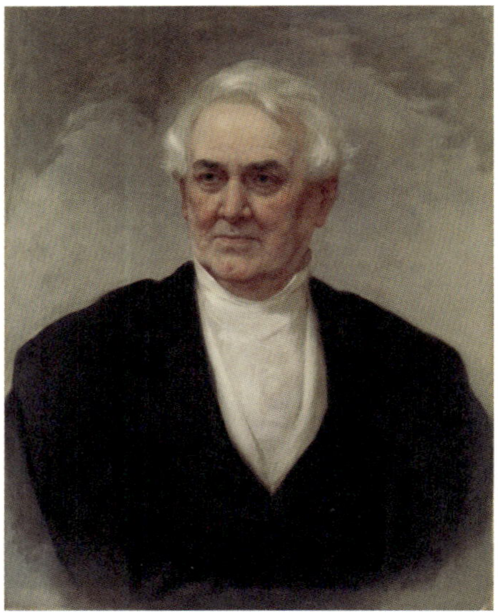

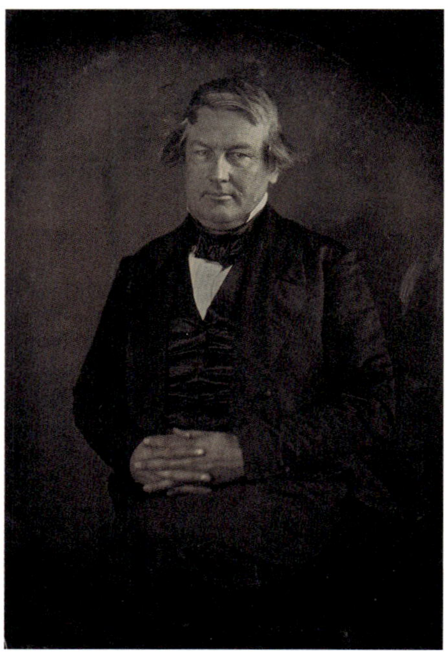

Fig. 60
William Oliver Stone, *William Wilson Corcoran* (1798–1888), ca. 1870, oil on canvas, 28 3/16 × 22 1/8 in., Walters Art Museum, Baltimore

Fig. 61
Mathew B. Brady, *Millard Fillmore* (1800–1874), 1849, half-plate daguerreotype, 7 3/8 × 5 11/16 in., Beinecke Rare Book & Manuscript Library, Yale University

research publications and annual reports made their way to Humboldt's library.[117] Lorin Blodget, a climatologist and statistician working for the Smithsonian, sent Humboldt his "Isothermal and Hyetal, or Rain Charts" in 1856, receiving in return an effusive letter of praise.[118] Blodget had extended Humboldt's research correlating altitude, mean temperature, and characteristic plants across continental North America, making extensive use of Humboldt's published research, from *Aspects of Nature* to the volumes of *Cosmos*. The following year, when Blodget published his findings, he produced a facsimile of Humboldt's letter as part of his promotion for his work.[119]

In 1844 Humboldt suggested that his close friend Baron Friedrich von Gerolt be appointed as extraordinary envoy and minister of the Kingdom of Prussia to the United States, a post he held with short interruptions for twenty-seven years.[120] As ambassador he lived in New York City, effectively becoming Humboldt's eyes and ears in the United States and an advocate for his views in absentia. In his letters, Gerolt occasionally inserted columns from several of the New York newspapers along with his own commentary, keeping Humboldt apprised of events in the United States. This would become especially useful as Humboldt monitored American politics during the 1856 presidential campaign and fretted over the consequences of American slavery.[121]

As Humboldt neared the end of his life, he was overwhelmed with letters and visitors, finally resorting to a newspaper notice published in Europe and the United States, begging people to cease and desist, to allow him to enjoy what he knew was the end stage of his long life. Still, he made himself available to those he admired. Washington art collector and Smithsonian Regent William Wilson Corcoran (fig. 60) arrived in Berlin with former U.S. President Millard Fillmore (fig. 61) in September 1855. Corcoran was carrying a general letter of introduction on Smithsonian letterhead signed by Secretary Joseph Henry, requesting that he be given access to museums and archives.[122] Edward Everett, who knew both men, had helped consult on their itinerary. When Friedrich von Gerolt learned of their arrival, he wrote to Corcoran offering to introduce them to Humboldt, and alerting them that they should anticipate an invitation to meet the king of Prussia.[123] Humboldt dashed off a note to invite the pair to dinner with the king at Sanssouci, the royal palace.[124]

Before leaving Berlin, Corcoran commissioned a marble bust of the explorer by Humboldt's favorite sculptor, Christian Daniel Rauch (fig. 62). Fillmore returned home with a hand-colored lithograph of Humboldt in his library by his friend and long-time illustrator, Eduard Hildebrandt (CAT. 44). The bust, which arrived in Washington months later, has an elegant quality to the features, lending Humboldt a timeless grace. Hildebrandt's lithograph places the elderly Humboldt seated in the midst of his life's work: the room is filled with bookcases and tables overflowing with research materials, portfolios of specimens poking out of trunks on the floor. A stuffed bald eagle visible through the rear doorway recalls the one mounted in Peale's Museum; paintings and sculptures adorn the upper walls and tops of the shelves. And in the foreground is a footed globe stand, the globe positioned to show the Americas, where Humboldt had made his reputation. Thanks to the structure of the stand, though, the continent that is most readily visible is that of North America. Humboldt's library has the air of Peale's Museum, stuffed with art and artifacts, books and maps, its own world in miniature.

Humboldt's imprimatur did not come without a solid dose of opinion. While he was in Europe, Fillmore had been drafted to run against James Buchanan and John C. Frémont in the 1856 presidential election; however, the aging explorer threw his vocal support to Frémont and his abolitionist platform. A year later, Humboldt followed up with Corcoran, inquiring whether the bust had arrived safely, and in the postscript could not resist throwing a little shade on the worsening situation surrounding slavery: "My best wishes to Mr. Fillmore and for inner and outer peace! Also for fewer instances of the sale of negro children who weigh 60 pounds."[125]

Despite Humboldt's criticism, the two men continued to correspond, and in 1857 Humboldt received a letter from Corcoran's wife, Louise: "I cannot resist the desire to offer you my heartfelt congratulations upon this 88th anniversary, and at the same time to express the hope that your useful life may be prolonged many years. I often recur to my interesting visit to Berlin always remembering that it was to your kindness, and attention, that it was made so."[126]

Humboldt's ability to be critical yet kind, curious yet opinionated had intensified with age. Ticknor's last visit to Berlin came in 1856, on the heels of Fillmore's visit. Reporting back to Everett he confirmed,

> *Nobody has said an unkind or unpleasant word to me about our country since I have been in Europe; but I feel, on all sides, that we stand in little favor or respect. Humboldt—whom I have seen every day, or had a note from him—is, I understand, very strong on his remarks sometimes, even to Americans. I cannot say that I am surprised.*[127]

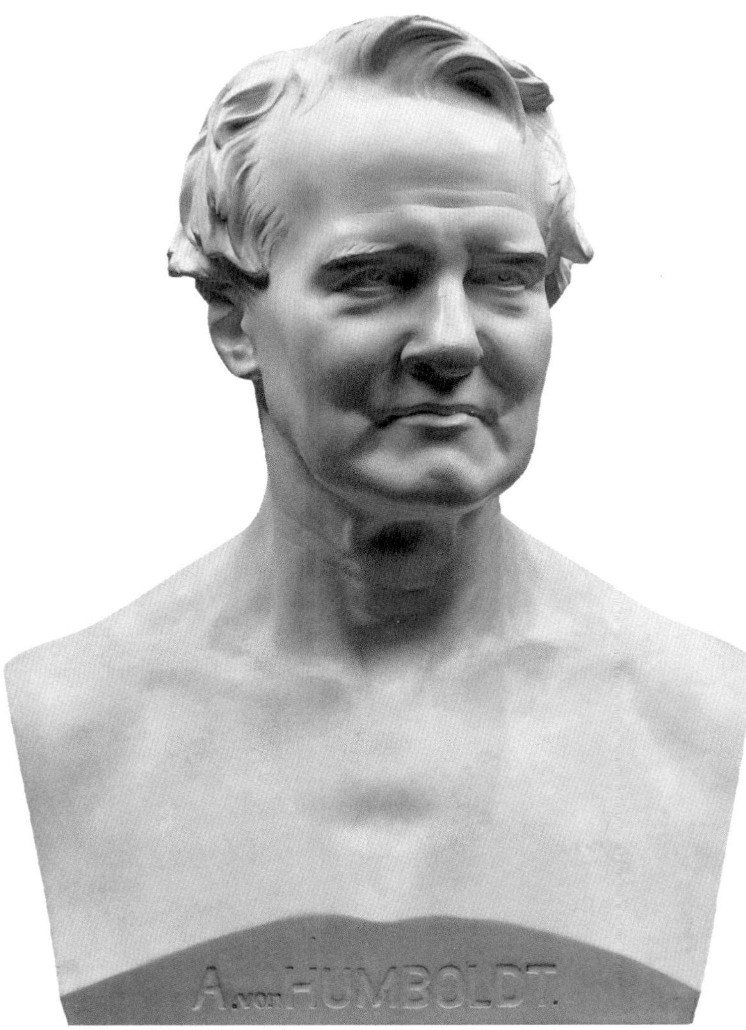

Fig. 62
Christian Daniel Rauch, *Bust of Alexander von Humboldt*, 1857, marble, 24 7/16 × 16 9/16 in., Museum für Naturkunde Berlin

Ticknor admitted to Humboldt that he supported Fillmore in the upcoming presidential election. Even though he did not agree with Ticknor's political views, especially when it came to the prospect of civil war over slavery, Humboldt treated his old friend with respect. Humboldt accompanied Ticknor as he visited the sights in Potsdam, in one case joking about his illegible handwriting serving as a kind of passport. Ticknor recalled, "Once, as we were going along a walk where a cord had been stretched, to signify that the passage was forbidden, he [Humboldt] removed it and told us to go through. I hesitated, and objected on account of the prohibition. 'I should like,' he replied, 'to see anybody, in Potsdam or Berlin, who will stop me when I have these crooked lines that everybody knows'—taking out Humboldt's note—'telling me to go on.'"[128] Humboldt's self-deprecatory remarks about his famously indecipherable handwriting had become a recurring feature in his correspondence. In his first letter to Corcoran, Humboldt dashed off a letter welcoming the travelers, closing with the apology, "Please excuse my illegible handwriting which is a testimony of my antediluvian age."[129]

During Ticknor's final visit to Berlin in 1856, Humboldt set his protégé Balduin Möllhausen to prepare an exhausting if rewarding "Plan stratégique pour Potsdam," through the museums, palaces, libraries, and archives.[130] On 1 November 1856, Ticknor wrote to Prescott,

> *On looking over your letter to see if there is anything to answer, I notice with pleasure what you say of Humboldt. He is, indeed, a man worth knowing, and even more so now, than he was when I was first acquainted with him in 1817–19. His kindliness increases with his years. Every day of the fortnight I was in Berlin he did something for me, and every day I either saw him or had a note from him. The minuteness of his care would have been remarkable in a young man.*[131]

In closing, Ticknor summed up so many similar reactions to Humboldt's generosity with his time, writing, "I do not believe there is another man in Europe who would have taken such trouble for a person of so little consequence, and from whom he could expect only gratitude."[132] That feeling of gratitude was precisely what animated

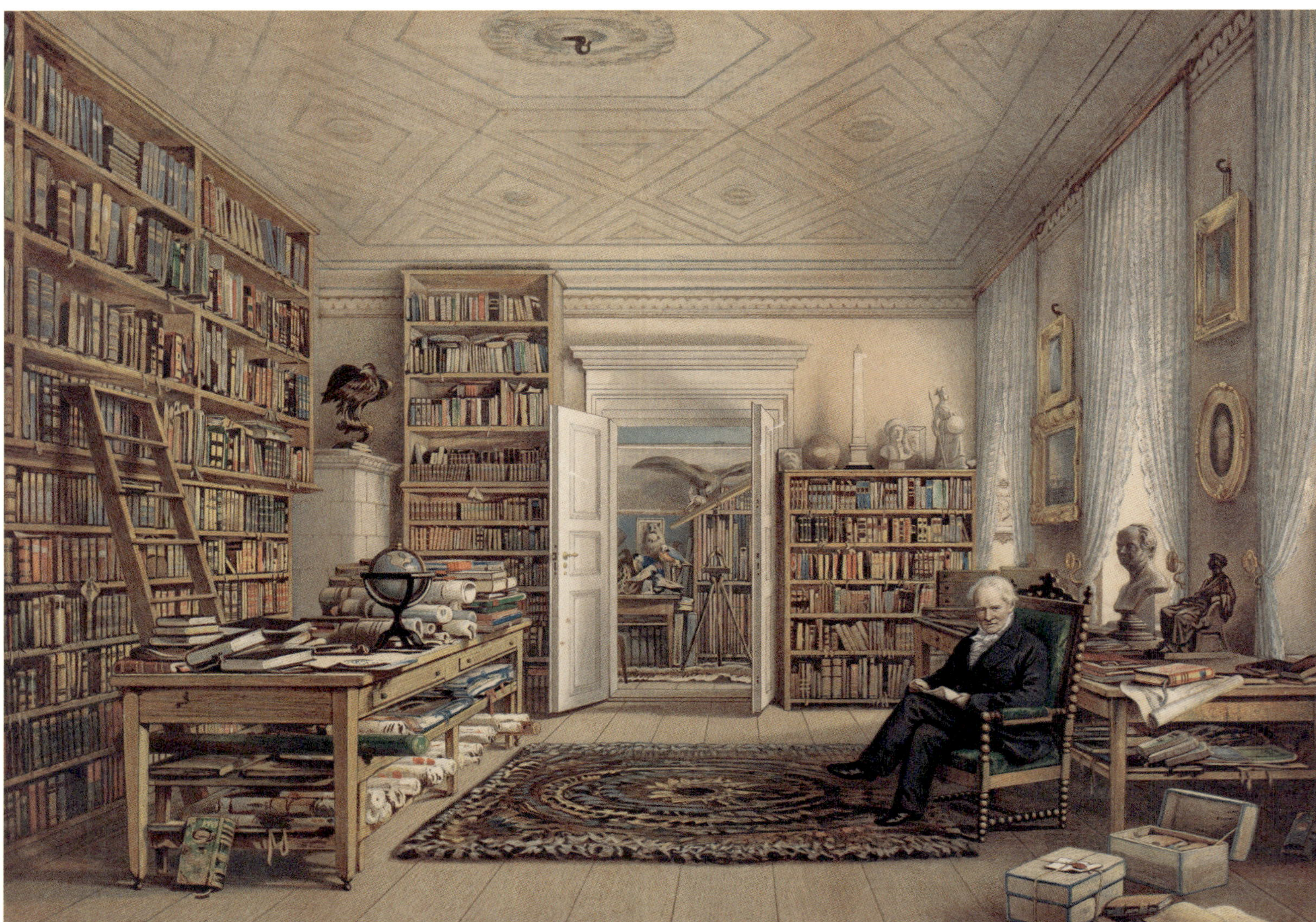

CAT. 44 After Eduard Hildebrandt, *Humboldt in His Library*, 1856, chromolithograph on paper, 18 ⅝ × 26 ⅝ in., Collection of Mr. and Mrs. Robert F. Norfleet Jr.

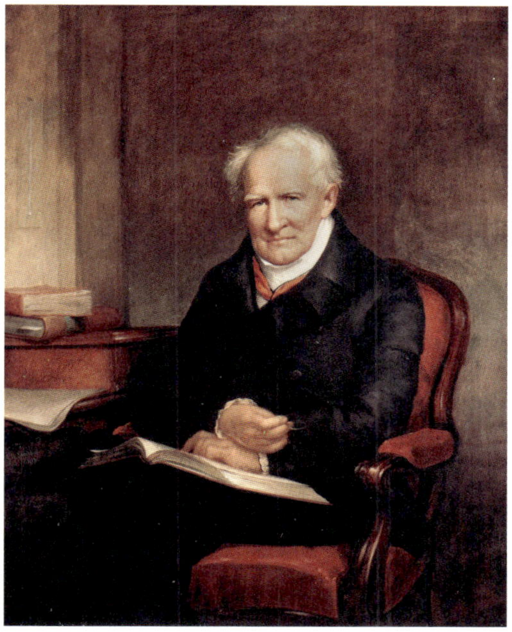

Fig. 63
James Reid Lambdin, *Portrait of von Humboldt*, oil on canvas, ca. 1856–57, 50 × 40 in., American Philosophical Society

so many Americans to strive for Humboldt's approbation and take pride in his description of himself as "half an American."[132]

In private Humboldt was far more vociferous in his disapproval of the way America was expanding, and at what cost. Events during the 1850s that appeared to be leading the United States into a war over slavery prompted the aging traveler to shift the meaning of his approbation. Rossiter Raymond, an American student studying to become a mining engineer, traveled to Berlin, met Humboldt in January 1859, and quoted Humboldt as saying:

> *I am half American; that is, my aspirations and desires are all with you, but I don't like the present position of your politics. The influence of Slavery is increasing, I fear.*[133]

In 1804 Humboldt had been eager to be connected with the promise of American democracy and optimistic about the prospect of his own influence in shaping it. Over five decades he had met with and befriended talented Americans whose continued support for slavery perplexed, angered, and frustrated him. Visitors to Humboldt during the last decade of his life increasingly heard the inference that half of him fully supported what America could be; the other half he withheld pending the abolition of slavery.

At the close of Humboldt's life, the desire to claim Humboldt took on greater urgency. Everett referred to Humboldt as the "American Columbus," echoing Humboldt's colleague, geographer Carl Ritter, who had quipped that Humboldt was the discoverer of America, in a scientific way.[134] Everett took his rhetoric further, proclaiming that Humboldt's "American voyage…was performed on the soil of this continent," a testimony to the depth to which Humboldt's influence on American culture had made him part of the United States' creation story.[135] James Reid Lambdin, who had earlier befriended Prince Maximilian and painted the artist George Catlin, now secured Humboldt's permission to come and paint his portrait in 1856, with the endorsement of Alexander Dallas Bache.[136] Lambdin painted the eighty-seven-year-old Humboldt (fig. 63) not as a highly decorated scientist but as a scholar still actively in pursuit of knowledge, looking up from his books, his gaze direct, and curious. Appropriately this portrait now hangs in the library of the American Philosophical Society, where the eminent naturalist greets every scholar who comes to conduct research of his or her own. In this portrait, as in the one Moses Wight painted the same year, Humboldt expressed a preference for being portrayed "without any ornament whatever."[137]

At the end of his life, Humboldt attended the annual celebrations honoring George Washington, conducted by those Americans who found themselves in Berlin on that anniversary. Richard Henry Stoddard, who would write the first full-length, English-language biography of Humboldt, attended the 1858 celebration of Washington's birthday at the American legation and recalled, "[I]t will be with pride that I can say that I was present upon that occasion in which he honoured the American nation, in his old age, with his presence at a dinner given in remembrance of 'The Father of our Country.'"[138] The following year, three months before he died, Humboldt attended the 1859 meeting, hosted by the American Minister to Prussia Joseph A. Wright, which included a toast to "Washington and Humboldt" accompanied by "three rousing cheers."[139] After Humboldt's death, Wright delivered a eulogy at the Berlin Geographical Society in which he reminisced about this event and declared Humboldt to have been an essential part of the fabric of the American country: "From his uniform kindness to all Americans, his association with our Institutions in their infancy, and the deep interest he ever manifested in our success, we almost feel he was one of us. Two months ago, mingling with seventy of our countrymen on the anniversary of the Father of our common country, with the folds of our lovely flag of our united Republic, the stars and stripes encircling his manly brow, he said: 'I am half an American.'"[140]

NOTES

1 Humboldt to Vaughan, Rome, 10 June 1805: "Vous, vos amis et ce superbe pays que vous habitez et qui présente un si beau tableau moral Pour en être séparé, vous verrez pourtant que, dans mes écrits, je reviens sur les États-Unis. C'est une passion en moi que les louer." Alexander von Humboldt, *Lettres américaines d'Alexandre de Humboldt*, with a preface and notes by Dr. E. T. Hamy (Paris: E. Guilmoto, 1905), 187–90. Also Ingo Schwarz, ed., *Alexander von Humboldt und die Vereinigten Staaten von Amerika: Briefwechsel* (Berlin: Akademie Verlag, 2004), 105–106, letter 19. English translation by the author.

2 Thornton to Vaughan, City of Washington, 6 July 1804, American Philosophical Society Archives.

3 "Vous pourriez aussi insinuer que je m'étais proposé de faire ici une édition pour le Nord-Amérique, où j'ose dire qu'entre le parti anti-fédéralist il règne un certain enthousiasme pour le succès de mon expedition; comme le prouvent toutes les gazettes de ce pays-là. Le débit aux Etats-Unis serait très-grand, et si l'on y voulait des souscripteurs (méthode qui d'ailleurs ne me paraît pas des plus délicate), MM. Jefferson, Madison, Galatin, Whister, Berton, etc., en procureraient un très-grande nombre. Une édition anglaise devrait par conséquent être au moins de 4,000 exemplaires." Humboldt to Pictet, Paris, 3 February 1805, "Lettres d'Alexandre de Humboldt à Marc-Auguste Pictet (1795–1824)," *Le Globe, Journal Géographique* (Genève: Carey Frères, 1868), 7:162–63. [*Le Globe* mistakenly transcribes the letter's year as 1804.] English translation by the author.

4 Humboldt wrote at least 136 letters to his European colleagues from South America and Mexico between 1799 and 1804, as chronicled by Ulrike Moheit in *Alexander von Humboldt: Briefe aus Amerika 1799–1804* (Berlin: Akademie Verlag, 1993).

5 On September 17, 1804, Humboldt delivered the first of a series of lectures at the Institut National covering the basics of his voyage, followed by additional lectures on September 24, October 15, and October 29. In November he lectured twice about the geology and snowcap of the Andean Cordillera. Ingo Schwarz, "Alexander von Humboldt Chronologie, 1804," *Edition Humboldt Digital*, https://edition-humboldt.de/chronologie/index.xql?jahr=1804&l=de.

6 Chimborazo was believed to be the highest peak on earth, at 20,577 feet. On June 23, 1802, Humboldt had climbed to a height of 19,413 feet (3,036 toises), within 1,300 feet of the summit, the highest ascent on any mountain at that time. It was a mountaineering record that would stand for thirty years. Alexander von Humboldt, *Reise auf dem Río Magdalena, durch die Anden und Mexico*, ed. Margot Faak (Berlin: Akademie Verlag, 1986).

7 In this oft-repeated anecdote, when Humboldt was presented to Napoleon at court, the French monarch contemptuously remarked, "You are interested in botany? My wife also studies it." Humboldt confirmed the animosity, writing to Karl Bruhns, "The Emperor Napoleon behaved with icy coldness to Bonpland, and seemed full of hatred to myself." Quoted in Karl Bruhns, ed., *Life of Alexander von Humboldt. Compiled in Commemoration of the Centenary of His Birth, by J. Löwenberg, Robert Avé-Lallemant, and Alfred Dove*, trans. Jane Lassell and Caroline Lassell (Boston: Lee and Shepard, 1873), 1:344. Goethe relayed the anecdote in 1825: "Er ließ—sagte Goethe—die Republic hinter sich, als er nach Amerika zog, und fand einen Dictator als er wiederkehrte, der ihn geringschäßig srug: 'Sie beschäftigten sich mit Botanik? Ich weiß daß auch meine Frau sie treibt.'" Quoted in *Goethes Unterhaltungen mit dem Kanzler Friedrich von Müller*, ed. C. A. H. Burkhardt (Stuttgart: J. G. Cotta, 1870), 101. "Goethe said, he left the Republic behind when he went to America, and returned to find a dictator, who insulted him. 'You are interested in botany? I know that interests my wife, too.'" English translation by the author.

8 The Institut had a fixed number of members; Humboldt's status became official on May 14, 1810, following British chemist Henry Cavendish's death. Maurice Crosland, *The Society of Arcueil: A View of French Science at the Time of Napoleon I* (Cambridge, MA: Harvard University Press, 1967), 166.

9 "Enfin, le célèbre Washington, alors président des États-Unis, m'écrivit deux fois à ce sujet. Dans sa première lettre, il y avait cette phrase remarquable: 'Quoique président des États-Unis, je ne puis m'engager vis-à-vis de vous pour aucun salaire; mais les hommes utiles comme vous ne sont pas délaissés par ma nation. Venez, et nous nous ferons tous un plaisir de vous accueillir.' Toutes ces offres ont été successivement repoussées. Si j'avais voulu m'expatrier, j'aurais préféré la patrie de Washington et de Franklin; mais l'amour de mon pays l'a toujours emporté sur des offres séduisantes de fortune." In English: "Finally, the famous Washington, then President of the United States, wrote me twice about this subject. In his first letter, there was this remarkable phrase: 'Even though President of the United States, I cannot commit myself to you for any payment; but useful men like you are not neglected by my nation. Come, and we will all be happy to welcome you.' All these offers have been successively postponed. If I had wanted to expatriate myself, I would have preferred the fatherland of Washington and Franklin; but the love of my country has always prevailed over attractive offers of fortune." Jean-Antoine-Claude Chaptal, comte de Chanteloup, *Mes souvenirs sur Napoléon* (Paris: E. Plon, Nourrit et Cie, 1893), 40. English translation by the author.

10 Berthollet's estate was described by visiting British scientist Sir John Leslie as "a fine chateau seated on a bank amidst gardens, vineyards, etc." Letter from Sir John Leslie, Arcueil, 1 August 1814; quoted in Crosland, *Society of Arcueil*, 411.

11 "Je sens tous les jours que l'on ne travaille bien que là où d'autres travaillent mieux autour de vous." Humboldt to François Gérard, Berlin, 12 February 1807, in Humboldt, *Lettres américaines d'Alexandre de Humboldt*, 217.

12 Alexander de Humboldt and Gay-Lussac, "Obervations sur l'intensité et l'inclinaison des forces magnétiques, faites en France, en Suisse, en Italie et en Allemagne," in *Mémoires de physique et de chimie, de la Société d'Arcueil* (Paris: J. J. Bernard, 1807), 1:1–22.

13 Humboldt delivered the inaugural address at the Academy of Sciences in Berlin on 21 November 1805. He delivered lectures on 16 and 30 January 1806 at the Royal Prussian Academy of Sciences in Berlin, on 20 March at the Berlin Academy, and on 3 April at the Berlin Philomatic Society.

14 "To embattled minds particularly, these pages are dedicated. '*Who saves himself from life's stormy wave*' will follow me gladly. . ." Alexander von Humboldt, *Views of Nature*, ed. Stephen T. Jackson and Laura Dassow Walls, trans. Mark W. Person (Chicago: University of Chicago Press, 2014), 25.

15 Chaptal recounted: "Il avait toujours regardé le célèbre Humboldt comme un espion de la Prusse. Il lui a demandé cent fois son nom, quoiqu'il le connût parfaitement; il ne lui a jamais parlé." "He had always regarded the famous Humboldt as a Prussian spy. He asked him a hundred times for his name, although he knew him perfectly; he never spoke to him." Chaptal, *Mes souvenirs sur Napoléon*, 382. English translation by the author.

16 There is little question that Bolívar considered himself inspired to lead his native land in revolution after meeting Humboldt. The extent of Humboldt's support for this may be better understood in Humboldt's surprise that Bolívar was able to carry this off, and his acknowledgment that he had perhaps misjudged the younger man. See Maria Arana, *Simón Bolívar: American Liberator* (New York: Simon and Schuster, 2013), 64; and J. Fred Rippy and E. R. Brann, "Alexander von Humboldt and Simón Bolívar," *American Historical Review* 52, no. 4 (July 1947): 697–703.

17 "If Bonaparte had been willing to serve the cause of liberty, I should have been devoted to him. But I can neither approve of an arbitrary government, nor associate myself with it." *Mémoires, correspondence et manuscrits du général Lafayette* (Paris: H. Fournier, 1837), 5:174; quoted in Stanley J. Idzerda, "Lafayette: Apostle of Liberty," in *Lafayette, Hero of Two Worlds: The Art and Pageantry of His Farewell Tour of America, 1824–1825* (Hanover, NH: Queens Museum, distributed by the University Press of New England, 1989), 46. Lafayette and Jefferson worked together to draft the French Declaration of the Rights of Man, which helped trigger the French Revolution and the overthrow of the Bourbon monarchy. Lafayette was imprisoned in the revolution's aftermath, despite intercessions from Washington and Jefferson to have him freed.

18 "Il conspirait en toute sincérité de conscience, acceptant les nécessités de ce rôle, uniquement occupé du but qu'il visait; il aurait conspiré sur la place publique." *Mémoires du Chancelier Pasquier* (Paris, 1894), 4:393; cited in Idzerda, "Lafayette, Apostle of Liberty," 52.

19 Humboldt and Lafayette were hardly alone in this respect. All foreigners and perceived adversaries experienced this level of surveillance. Charles Lyell wrote to his father from Paris that "The seals of many letters going to England, and still more of those into the interior of France, are broken, which Royer tells me 'is very *proper*, for it has been the means of detecting many enemies'!" Charles Lyell to his father, Hôtel Meurice, Paris, 8 July 1823, in Sir Charles Lyell, *Life, Letters, and Journals of Sir Charles Lyell*, ed. K. Lyell (London: J. Murray, 1881), 1:126.

20 Jefferson to Lafayette, Washington, 27 June 1804: "I am in hopes that this will get to Philadelphia in time to be taken charge of by Baron de Humboldt who is embarking there for Paris. He is a Prussian of great science, & the most amiable character who has been five years in S. America, and goes to Paris to publish the result of his labours. He has made a very short stay with us; yet he will be able to give you some account of us, and will be found worthy of any attentions you may bestow on him. Present me respectfully to Me. de la Fayette, M. and Me. de Tessé, and accept yourself assurances of my unchangable affection." In the same letter, Jefferson offered Lafayette the position of first governor of the Louisiana Territory, hoping to coax him back to the United States. Thomas Jefferson, *The Papers of Thomas Jefferson*, vol. 43, *11 March to 30 June 1804*, ed. James P. McClure (Princeton, NJ: Princeton University Press, 2017), 661–62.

21 By 1809 Humboldt is friends with Lafayette's family, and by 1810 Lafayette's letters to Jefferson refer to Humboldt in familiar terms. Lafayette to Jefferson, La Grange, 12 March 1811: "Our friend Humboldt is preparing for His Expedition to Bengal and Thibet—But Will Yet Stay Some time in france—I am under obligations to You for His Acquaintance which Has Become very intimate. as we were Some weeks Ago Speaking of the former Spanish Colonies on the American Continent He gave me an Extract of His Book then under the press, one Copy of which Has Been Sent By mr Russel to Washington, and the other I was, By the Author, desired to present to You." Thomas Jefferson, *The Papers of Thomas Jefferson*, Retirement Series, vol. 3, *12 August 1810 to 17 June 1811*, ed. J. Jefferson Looney (Princeton, NJ: Princeton University Press, 2006), 344–47.

22 Rembrandt Peale to the Mayor of New York, 18 August 1824, Manuscript division of the Museum of the City of New York; cited in Marc H. Miller, "The Farewell Tour and American Art," in Idzerda, *Lafayette, Hero of Two Worlds: The Art and Pageantry of His Farewell Tour of America, 1824–1825* (Hanover, NH: The Queens Museum, distributed by the University Press of New England, 1989), 162.

23 David Brigham observes that these larger portraits were hung in a single row along the west wall of the Long Room and are visible in Titian Ramsay Peale's watercolor of this gallery (cat. 31). David R. Brigham, *Public Culture in the Early Republic: Peale's Museum and Its Audience* (Washington, DC: Smithsonian Institution Press, 1995), 46.

24 To Gallatin he writes at the end of this trip that he can be reached c/o Chaptal, Minister of the Interior in a letter from Philadelphia, 20 June 1804: "Mon Addresse constant à Paris est d'ailleurs le C. Chaptal Ministre de l'Intérieur avec lequel j'ai vécu dans une étroite liaison depuis plusiers années." "My permanent address in Paris is also that of Chaptal, Minister of the Interior, with whom I have been in close cooperation for many years." Schwarz, *Alexander von Humboldt und die Vereinigten Staaten von Amerika*, 95, letter 8. English translation by the author.

25 "M. de Humboldt possède toutes les sciences, et lorsqu'il voyage, c'est toute l'Académie des sciences qui marche." Chaptal, *Mes souvenirs sur Napoléon*, 383.

26 Chaptal relates the entire incident and the exchange between himself and Napoleon in Chaptal, *Mes souvenirs sur Napoléon*, 382–84.

27 "je viendrai finir ma vie dans Votre patrie dont j'aime les habitans, le sol et surtout la sage constitution." Humboldt to Rembrandt Peale, Paris, 15 September 1810, in Schwarz, *Alexander von Humboldt und die Vereinigten Staaten von Amerika*, 116–17, letter 27. Translation from Halina Nelken, *Alexander von Humboldt: His Portraits and Their Artists; A Documentary Iconography* (Berlin: Dietrich Reimer Verlag, 1980), 78.

28 "je préfère Votre Climat à tout autre. Car l'air le plus sain est celui où *l'on respire le plus librement*." Humboldt to Gallatin, Philadelphia, 20 June 1804, in Schwarz, *Alexander von Humboldt und die Vereinigten Staaten von Amerika*, 95, letter 8.

29 W. H. Crawford to the United States Commission, Paris, 13 May 1814, cites Humboldt's intercession and Lafayette's involvement as well. Albert Gallatin, *The Writings of Albert Gallatin*, ed. Henry Adams (Philadelphia: J. B. Lippincott, 1879), 1:614–17; Lafayette to Gallatin, Paris, 3 June 1814, in Gallatin, *Writings of Albert Gallatin*, 1:626–27.

30 Humboldt made a quick trip to London on the Prussian king's behalf in June 1814, where he met with Gallatin. Lafayette alerted Gallatin to his imminent arrival; on June 3, Lafayette wrote (in English): "I have just the time to write a line by my friend M. de Humboldt, who hopes to reach London before you have left it." Gallatin, *Writings of Albert Gallatin*, 1:626. Ingo Schwarz surmises they did see each other while Humboldt was part of Friedrich Wilhelm III's entourage, around June 20—mentioned in a letter from Wilhelm von Humboldt to his wife, Caroline, in Anna von Sydow, ed., *Wilhelm und Caroline von Humboldt in ihren briefen: Federn und schwerter in den freiheitskriegen, briefe von 1812–1815* (Berlin: E. S. Mittler und sohn, 1910), 4:347–65; cited in Schwarz, *Alexander von Humboldt und die Vereinigten Staaten von Amerika*, 133–34, letter 42, n5.

31 John Quincy Adams first met with Humboldt on January 26, 1815. *The Diaries of John Quincy Adams: A Digital Collection*, Massachusetts Historical Society.

32 John Quincy Adams, diary entry, 27 February 1815, 228. For the dinners with Madame de Staël, see John Quincy Adams, diary entries, 15 February 1815, 223; and 20 February 1815, 225. *The Diaries of John Quincy Adams: A Digital Collection*, Massachusetts Historical Society.

33 The founding of the Smithsonian, and Humboldt's influence in that, is the subject of chapter 8 in this volume.

34 Napoleon had a particular dislike for outspoken women. Of de Staël he wrote, "Je ne veux pas de Mme de Staël à Paris, et j'ai pour cela de bonnes raisons.... Si Mme de Staël voulait ou savait être royaliste ou républicaine, je n'aurais rien contre elle; mais elle est une *machine à mouvement*, qui remue les salons." Napoleon to Metternich, in Klemens von Metternich, *Memoirs of Prince Metternich: 1773–1815*, ed., Prince Richard Metternich (London: Spottiswoode, 1880–82) 1:289; quoted in Paul Gautier, *Madame de Stäel et Napoléon* (Paris: Plon-Nourrit, 1903), 255. English translation by the author.

35 In 1814 one of Madame de Staël's contemporaries quipped, "There are three great powers struggling against Napoleon for the soul of Europe: England, Russia, and Madame de Staël." ("Bonaparte l'avait persécutée de manière à ce qu'on dît qu'en Europe il fallait compter trois puissances: l'Angleterre, la Russie et Mme de Staël.") in Victorine de Chastenay-Lanty, *Mémoires de Madame de Chastenay, 1771–1815*, vol. 2, *L'empire. La restauration. Les cent-jours* (Paris: E. Plon, Nourrit et Cie, 1896), 445. English translation by the author.

36 Russell M. Jones, "Flowering of a Legend: Lafayette and the Americans, 1825–1834," *French Historical Studies* 4 (1966): 395; quoted in Idzerda, "Lafayette: Apostle of Liberty," in Idzerda, *Lafayette, Hero of Two Worlds*, 56.

37 There has been a great deal of speculation regarding Humboldt's sexuality. He never married, and his effusive, affectionate letters to favorite male colleagues and traveling companions have raised questions. The extended family destroyed his personal correspondence after his death, further fueling debate. To Wegener, Humboldt wrote: "God! How can one love each other so much and yet be so far apart....I am burning with desire to embrace you again, my dear, my best Wegener." Humboldt to Wegener, 26 June 1788, in Alexander von Humboldt, *Die jugendbriefe Alexander von Humboldt: 1787–1799* (Berlin: Akademie Verlag, 1973), 6:17; translated in Maren Meinhardt, *A Longing for Wide and Unknown Things: The Life of Alexander von Humboldt* (London: Hurst, 2018), 30. In an era of close male friendships and a tendency toward florid prose, such sentiments did not necessarily prove physical passion, but instead foster speculation. Humboldt's brother, Wilhelm, saw in his younger brother an immense capacity for fierce loyalty and on occasion alluded to the potential impropriety of those relationships, writing to his wife, "He'll never be content on his own, because he senses that he won't find fulfilment by himself. Every so often he's even made remarks to me in this vein, though mostly our innermost feelings were concealed as if by a veil, which we both saw, but neither dared lift." "Glücklich wird er schwerlich je sein; er ist nicht ruhig und wird es nie warden, weil ich doch nie glaube, daß irgend ein Interesse sein Herz beschäftigen wird, und er doch gerade für eine solche Existenz Sinn und tiefe Achtung hat. Er wird nie mit sich zufrieden sein, weil er fühlt, daß er sich selbst nicht auszufüllen vermag. Hie und da hat er dies sogar gegen mich geäußert, obgleich meist wie ein Schleier zwischen uns über unsern innersten Gefühlen hing, den jeder sah und keiner aufzuheben wagte." Wilhelm von Humboldt to Caroline von Dacheröden, 3 June 1791, in Sydow, *Wilhelm und Caroline von Humboldt*, 1:477–78, letter 161. Translated in Meinhardt, *Longing for Wide and Unknown Things*, 54. In 1817 while serving as the Prussian ambassador to

London, Wilhelm had planned to host his brother but discovered he was not traveling alone. "You know his passion for always having a person around who, for the moment, is his favorite," he wrote to Caroline. "Now he's got an astronomer, Arago; he refuses to be separated from him; to host both of them, while physically possible, would have been very inconvenient for me." The suggestion of impropriety in a country notably hostile to unconventional living arrangements would indeed have complicated Wilhelm's position as a diplomat. Alexander did not stay in the ambassadorial residence on this trip. "Du kennst seine Passion, immer mit einem Menschen zu sein, der ihm zu jeder Zeit nun eben der liebste ist. Jezt hat er einem Astronomen Arago, von diesem will er sich nicht trennen, und beide zu logieren, wäre wohl physisch möglich, aber mir sehr umständlich gewesen." Wilhelm von Humboldt to Caroline, London, 1 November 1817, in Sydow, ed., *Wilhelm und Caroline von Humboldt in ihren briefen: Federn und schwerter in den freiheitskriegen, briefe von 1812–1815* (Berlin: E. S. Mittler und sohn, 1913), 6:30. Translated in Meinhardt, *Longing for Wide and Unknown Things*, 224–25.

38 May 2, 1817, and Sunday, June 1st, 1817, George Ticknor, *Life, Letters, and Journals of George Ticknor*, ed. George Hillard, Mrs. Anna Eliot Ticknor, and Miss Anna Eliot Ticknor (Boston: Houghton Mifflin Company, 1909), 1:130, 138.

39 "above all from Blumenbach, *ante alios omnes praestantissimus*." Journal entry for Göttingen, March 26, 1817, ibid., 1:121.

40 Jefferson to Gallatin, 19 March 1815. Ticknor presented his letters to Lafayette and Gallatin two years after they had been written. Remarking on the political instability in France, in a letter to Jefferson dated April 25, 1817, from La Grange, Lafayette joked that he was "not a little perplexed Was I to Understand Some parts of it, when I perceived that Since it Had been writen two Revolutions and two dynasties Had Successively Reigned in france." Lafayette to Jefferson, 25 April 1817, in Thomas Jefferson, *The Papers of Thomas Jefferson*, Retirement Series, vol. 11, *19 January to 31 August 1817*, ed. J. Jefferson Looney (Princeton, NJ: Princeton University Press, 2015), 281–83.

41 Journal entry for 26 April 1817, Ticknor, *Life, Letters, and Journals of George Ticknor*, 1:129–30.

42 Ibid.

43 Paris, December 10, 1818, to January 12, 1819, in Ticknor, *Life, Letters, and Journals of George Ticknor*, 1:253.

44 Ibid., 1:257.

45 Edward Everett, "Art. 1.—Humboldt's Works: *Voyage aux regions equinoxiales du Nouveau Continent… avec un Atlas géographique et physique*. Tomes VII & VIII, 8 vo. Paris, 1822," *North American Review*, n.s., 38 (January 1823): 1–30. Up until this time, most articles on Humboldt in American publications were excerpts from Humboldt's writings without any critical analysis.

46 Emerson first mentions Humboldt by name in his journals in 1821, before the English translation of Humboldt's *Personal Narrative* analyzed by Everett was available.

47 *Literary Magazine, and American Register* 2, no. 10 (July 1804): 321–27; *Literary Miscellany* 1, no. 1 (1805): 102–103; *Weekly Visitor/Ladies Miscellany* 2, no. 101 (1804): 387–88, no. 102 (1804): 395–96, no. 103 (1804): 402–403. Interest in this narrative prompted the *Literary Miscellany* to print it again in 1805.

48 "General Considerations on the Extent and Physical Aspects of the Kingdom of New Spain, from Baron de Humboldt's Political Essay on the Kingdom of New Spain," *American Register; or, General Repository of History, Politics, and Science* 7 (2 January 1810): 268–339.

49 "Humboldt's Travels," *New Monthly Magazine and Literary Journal*, American ed. 2, no. 7 (1 January 1821): 314. Other excerpts appeared at this time in "ART. XII.—From Baron Humboldt's Personal Narrative.—Vol. 4," *Port Folio* 12 no. 1 (September 1821): 136; "Humboldt's and Bonland's [*sic*] Travels," *Atheneum; or, Spirit of the English Magazines* 9, no. 12 (15 September 1821): 469; and "Travels of Alexander von Humboldt. Williams, Helen Maria," *Atheneum; or, Spirit of the English Magazines* 10, no. 4 (15 November 1821): 132.

50 Humboldt delivered his lectures on "lignes isothermes" on May 26, and June 2, 9, and 16, 1817. The published article is Humboldt, "Des lignes isothermes et de la distribution de la chaleur sur le globe (On Isothermic Lines and the Distribution of Heat on the Planet)," in *Mémoires de physique et de chimie, de la Société d'Arcueil* (Paris: J. J. Bernard, 1817), 3:462–602. Those lectures were subsequently translated into English for editions of the *Edinburgh Review* in 1820. For Emma Willard, see E. Willard, *Guide to the Temple of Time; and Universal History, for Schools* (New York: A. S. Barnes & Company, 1849).

51 William Woodbridge, *A System of Universal Geography, on the Principles of Comparison and Classification bound together with Emma Willard, Ancient Geography, as Connected with Chronology, and Preparatory to the Study of Ancient History: Accompanied with an Atlas* [also known as Woodbridge and Willard's Geography] (Hartford, CT, 1829).

52 Almira Hart Lincoln Phelps's *Familiar Lectures on Botany* (1829) and *Botany for Beginners* (1833) were widely used in schools and went through many printings and some revisions into the late 1860s. Emanuel D. Rudolph, "Women in Nineteenth Century American Botany; A Generally Unrecognized Constituency," *American Journal of Botany* 69, no. 8 (September 1982): 1347.

53 Emma Lydia Bolzau, *Almira Hart Lincoln Phelps: Her Life and Work* (Philadelphia: University of Pennsylvania Press, 1936), 357; cited in Nina Baym, *American Women of Letters and the Nineteenth-Century Sciences: Styles of Affiliation* (New Brunswick, NJ: Rutgers University Press, 2002), 18.

54 Sarah Hale, *Woman's Record*, 2nd ed. (New York: Harper and Brothers, 1855), 771; quoted and cited in Baym, *American Women of Letters and the Nineteenth-Century Sciences*, 18.

55 "Alexander von Humboldt: Celebration of the Centennial Anniversary of His Birth; Address by Professor Agassiz," *Littell's Living Age*, no 1322, 2 October 1869, 48; reprinted from the *Boston Daily Advertiser*, 15 September 1859.

56 Ticknor, *Life, Letters, and Journals of George Ticknor*, 1:177.

57 Ibid., 1:254. Chateaubriand and Talleyrand were frequent attendees.

58 On August 7, 1825, Humboldt meets Irving at François Gérard's house; on August 25, 1825, Irving is again at Gérard's. Ingo Schwarz, "Alexander von Humboldt Chronologie, 1825," *Edition Humboldt Digital*, https://edition-humboldt.de/chronologie/index.xql?jahr=1825.

59 Irving to his brother, William, Rome, 4 April 1805, in Pierre M. Irving, *The Life and Letters of Washington Irving*, vol. 1 (New York: G. P. Putnam, 1862), 141.

60 Inspired in part by Humboldt's year spent in Mexico, Irving wanted to write a history of Mexico but eventually yielded that project to William Prescott, another of Humboldt's American protégés.

61 "M. Washington Irving a prouvé que dans un esprit supérieur la culture des arts d'imagination n'exclut point la faculté de s'adonner avec fruit aux études sévères de l'historiographe." Alexander von Humboldt, *Examen critique de l'histoire de la géographie du nouveau continent* (Paris: Librairie de Gide, 1836), 1:vxi; quoted in Ingo Schwarz, "The Second Discoverer of the New World and the First American Literary Ambassador to the Old World: Alexander von Humboldt and Washington Irving," *Acta Historica Leopolda* 27 (1997): 93. Translation by Mary Beth Stein in Laura Dassow Walls, *The Passage to Cosmos: Alexander von Humboldt and the Shaping of America* (Chicago: University of Chicago Press, 1979), 118, 344n34.

62 "The other night I met at La Fayette's soirée—in ten feet square—Humboldt, Benjamin Constant, Haslett the Table talker, Capt Sabine, &c &c—….[In] Paris you see people from all quarters of the world—" Earlier in the same letter Cooper had noted, "La Fayette gives a soirée every Tuesday. They are exceedingly well attended….There are, also, a great many Americains. We were there last evening, for the second time—Among the guests were Humboldt, Constant, several savans, artists &c, among the latter Messrs [Ary] Scheffer and [Pierre-Jean] David—The former is making my portrait and the latter my bust, so you see I am likely to live in paint and marble at least." Letter from Cooper to Mrs. Peter Augustus Jay, the wife of John Jay's son, 26 March 1827, in James Fenimore Cooper, *Letters and Journals of James Fenimore Cooper*, ed. James Franklin Beard (Cambridge, MA: Belknap Press of Harvard University Press, 1960), 1:204, 202.

63 Jane Merrill and John Endicott, *Aaron Burr in Exile: A Pariah in Paris, 1810–1811* (Jefferson, NC: McFarland & Company, 2016), 41–42; cited in Henry Levenstein, *Seductive Journey: American Tourists in France from Jefferson to the Jazz Age* (Chicago: University of Chicago Press, 2000), 47n24.

64 Undated letter from James Fenimore Cooper to William Dunlap. William Dunlap, *History of the Rise and Progress of the Arts of Design in the United States* (New York: G. P. Scott, 1834), 2:424–25. Ticknor describes Lafayette's hospitality and estate as "the most venerable castle I have seen in France," in his journal. Ticknor, *Life, Letters, and Journals of George Ticknor*, 1:151–52.

65 Cooper expressed his annoyance to Horatio Greenough when Cole comes through Paris and failed to get in touch, thereby missing his opportunity to meet Humboldt. Cooper to Greenough, Paris, 24 December 1831, in Cooper, *Letters and Journals of James Fenimore Cooper*, 2:167–68. Greenough conveyed Cole's apology, but noted that Cole was prone to temporary hearing loss, rendering him, as he put it, "deaf-mute," a source of embarrassment for the American artist. Horatio Greenough, *Letters of Horatio Greenough, American Sculptor*, ed. Nathalia Wright (Madison: University of Wisconsin Press, 1972), 110.

66 John Dobkin, Nicolai Cikovski Jr., and Paul J. Staiti, *Samuel F. B. Morse: Educator and Champion of the Arts in America* (New York: National Academy of Design, 1982), 47, 75n2.

67 For an assessment of the cultural impact of this visit, and its detailed itinerary, see Izerda, *Lafayette, Hero of Two Worlds*.

68 Miller, "Farewell Tour and American Art," 148.

69 Undated letter from Morse to an unnamed recipient, in Edward Lind Morse, ed., *Samuel F. B. Morse: His Letters and Journals* (Boston: Houghton Mifflin Company, 1914), 1:272.

70 Samuel Iranaeus Prime, *Life of Samuel F. B. Morse, LL.D., Inventor of the Electro-Magnetic Recording Telegraph* (New York: D. Appleton, 1875), 232.

71 Both Cooper and his wife wrote about this dinner. Letter from Cooper's wife to her father, dated 23 March 1832, Paris; see *Correspondence of James Fenimore Cooper*, edited by his grandson (New Haven, CT: Yale University Press, 1922), 1:127. James Fenimore Cooper to Henry Brevoort, Paris, 7–25 March 1832, in Cooper, *Letters and Journals of James Fenimore Cooper*, 2:233, letter 262. It was Marbois who had requested from Jefferson answers to his queries about the nascent United States, answers that Jefferson then expanded to form his *Notes on the State of Virginia*.

72 Alexander von Humboldt's brother, Wilhelm, had corresponded with Samuel F. B. Morse's father, Jedidiah Morse, in 1823 concerning his research on American Indian languages. "Extract of a letter from Baron Wm. von Humboldt, to the Hon. Mr. Pickering, translated from the original, from the National Intelligencer," *Christian Spectator* 5 (September 1823): 495–96. Cited in Walls, *Passage to Cosmos*, 118, 344n36. Humboldt's teacher, Daniel Christoph Ebeling, had also been in contact with the elder Morse in the 1790s.

73 Samuel F. B. Morse to his brothers, Sidney Edwards Morse and Richard Cary Morse, 13 March 1832, Morse Papers, Library of Congress.

74 James Fenimore Cooper to William Dunlap, Paris, 16 March 1832, in Dorothy C. Barck, *Diary of William Dunlap (1766–1839): The Memoirs of a Dramatist, Theatrical Manager, Painter, Critic, Novelist, and Historian* (New York: New-York Historical Society, 1930), 3:608; cited in Elizabeth Kennedy and Olivier Meslay, eds., *American Artists and the Louvre* (Chicago: Terra Foundation, 2006), 66; also in Cooper, *Letters and Journals of James Fenimore Cooper*, 2:239, letter 265.

75 George Iles, "Leading American Inventors," in *Biographies of Leading Americans*, ed. W. P. Trent (New York: Henry Holt, 1912), 137; and letter from Samuel F. B. Morse to his brothers, Sidney Edwards Morse and Richard Cary Morse, 13 March 1832, Morse Papers, Library of Congress; quoted in Paul Staiti, "American Artists and the July Revolution," in *American Artists and the Louvre*, ed. Elizabeth Kennedy and Olivier Meslay (Chicago: The Terra Foundation, 2006), 66.

76 Quoted in Prime, *Life of Samuel F. B. Morse*, 31.

77 James Wynne, "Samuel F. B. Morse," *Harper's New Monthly Magazine* 24, no. 140, January 1862, 228.

78 Edward Everett, *Orations and Speeches on Various Occasions* (Boston: C. C. Little and J. Brown, 1860), 1:275; cited in Paul Staiti, *Samuel F. B. Morse* (New York: Cambridge University Press, 1989), 288n42.

79 François Arago quipped: "'Tu ne sais pas construire; tes livres sont comme des tableaux sans cadre.' (You don't know how to build; your books are like unframed pictures)." Pierre Lazlo, "Humboldt: A French Perspective," in *Alexander von Humboldt: From the Americas to the Cosmos*, ed. Raymond Erickson, Mauricio A. Font, and Brian Schwartz (New York: Bildner Center for Western Hemisphere Studies, 2005), 282. Helmut de Terra recounted this story as: "His friend Arago remarked one day: 'Humboldt, you really don't know how to write a book. You write endlessly, but what comes out of it is not a book, but a portrait without a frame.'" Helmut de Terra, *Humboldt: The Life and Times of Alexander von Humboldt* (New York: Alfred A. Knopf, 1955), 222.

80 James Fenimore Cooper to Peter Augustus Jay, Dresden, 15 July 1830, in Cooper, *Letters and Journals of James Fenimore Cooper*, 1:420–21, letter 165.

81 Humboldt had been an active member of Lorenz Oken's Gesellschaft deutschen Naturforscher und Ärtze since its founding in 1822. This society and Humboldt's Naturforscher would serve as a template for the American Association for the Advancement of Science, founded in 1848.

82 Charles Babbage covered this event, in "Account of the Great Congress of Philosophers at Berlin on the 18th September 1828," *Edinburgh Journal of Science* 10 (October 1828–April 1829): 231–32. Babbage successfully used this example as a rationale for the founding of the British Association for the Advancement of Science. Cited in Susan Faye Cannon, *Science in Culture: The Early Victorian Period* (New York: Dawson and Science History Publications, 1978), 92n53.

83 "Ein solches Publikum ist—glaub' ich—in Deutschland noch nicht vor dem Katheder eines Gelehrten versammelt gewesen. Der König, der ganze Hof, die höchsten Staatsbeamten und Militärpersonen, nebst ihren Damen, alle Gelehrte, Künstler pp. von Bedeutung, die ganze schöne Welt." Karl von Holtei to Goethe, Berlin, 17 December 1827; quoted in Rudolf Borch, ed., *Alexander von Humboldt: Sein Leben in Selbstzeugnissen, Briefen und Berichten* (Berlin: Verlag des Druckhauses Tempelhof, 1948), 242. Translated in Meinhardt, *Longing for Wide and Unknown Things*, 227.

84 "Achthundert Menschen atmen kaum, um den Einen zu hören. Es gibt keinen großartigeren Eindruck, als die irdische Macht zu sehen, wie sie dem Geiste huldigt; und schon deshalb gehört Humboldts jetziges Wirken in Berlin zu den erhebendsten Erscheinungen der Zeit." Karl von Holtei to Goethe, Berlin, 17 December 1827; quoted in Borch, *Alexander von Humboldt: Sein Leben in Selbstzeugnissen*, 242. Translation by the author.

85 Louis Agassiz, "Eulogy by Professor Agassiz upon Baron von Humboldt," delivered at the American Academy of Arts and Sciences, 24 May 1859, from the *Boston Daily Advertiser*," *Littell's Living Age*, no. 785, 11 June 1859, 648. The quotation from Arago has also been cited in Agassiz's memoirs as "Toi, Humboldt! tu ne sais pas ecrire un livre; tu commences, va, va, va, tu ne peux plus t'arrêter; exactement comme tu parles." It translates to "You do not know how to write a book. When you begin a subject, you go, go, go, and cannot stop, just like your never-ending talk." Jules Marcou, *The Life, Letters, and Works of Louis Agassiz* (New York: MacMillan, 1895), 177.

86 Alexander von Humboldt, *Cosmos: Sketch of a Physical Description of the Universe*, trans. Elizabeth J. L. Sabine (London: Longman, Brown, Green, and Longmans, 1849), 1:18.

87 Entry for Monday, 11 December 1826, *Conversations of Goethe, with Eckermann and Soret*, trans. John Oxenford (London: George Bell & Sons, 1874), 177–78.

88 One of Humboldt's surviving address books is peppered with entries representing his expanding American network and is discussed in Ingo Schwarz, "Americans in Alexander von Humboldt's Personal Address Book," in Georgia de Havenon et al., ed., *Unity of Nature: Alexander von Humboldt and the Americas* (Bielefeld: Kerber Verlag, 2014), 29–41.

89 Humboldt to Bunsen, 24 May 1836; quoted and translated in Ingo Schwarz, "Transatlantic Communication in the 19th Century: Aspects of Correspondence between Alexander von Humboldt and George Ticknor," *Asclepio* 56, no. 2 (2004): 26. Original in Alexander von Humboldt and Christian Carl Josias Bunsen, *Briefe von Alexander von Humboldts an Christian Carl Josias Freiherr von Bunsen* (Leipzig: F. A. Brockhaus, 1869), 23–24. "Sie erinnern Sich wohl noch, theurer Freund, der antediluvianischen Zeit, wo zwei geistreiche, classisch gelehrte Amerikaner, Ticknor und Everett, ganz Europa durchreisten. Ticknor aus Boston erscheint wieder am horizont. Nehmen Sie ihn Freundlich auf, wie Sie es verstehen aufzunehmen. Ich werde Ihnen dafür danken. Ticknor ist der Freund unseres Hauses."

90 Entry for 27 May 1836, Ticknor, *Life, Letters, and Journals of George Ticknor*, 1:500–501.

91 Humboldt to Metternich, Berlin, 28 May 1836; quoted in Ingo Schwarz, "Transatlantic Communication in the 19th Century," 31; Schwarz, *Alexander von Humboldt und die Vereinigten Staaten von Amerika*, 196. Ticknor related his visit with Metternich in Ticknor, *Life, Letters, and Journals of George Ticknor*, 2:3–4. Humboldt wrote to Ticknor on September 8, 1837: "Le Prince Metternich, que j'ai vu à Teplitz, a été ravi des entretiens qu'il a eus avec vous. Né dans une république, vous aurez, pourtant, paru plus raisonnable à ses yeux, que ce qu'il appelle mon libéralisme." "Prince Metternich, whom I saw at Teplitz, was delighted with the conversations he had with you. Being born in a Republic, you will have seemed more reasonable to him than what he calls my liberalism." Ticknor, *Life, Letters, and Journals of George Ticknor*, 2:20. English translation by the author.

92 Entry for 26 May 1836, Ticknor, *Life, Letters, and Journals of George Ticknor*, 1:499. A cantatrice is an opera singer.

93 Wynne, "Samuel F. B. Morse," 227.

94 Ibid., 224. Morse's old friend William Cullen Bryant would make similar remarks at a banquet held in 1868 in Morse's honor, at which Bryant pronounced, "Our guest has annihilated both space and time in the transmission of intelligence." William Cullen Bryant, "The Electric Telegraph," in *Prose Writings*, ed. Parke Godwin (New York: Russell and Russell, 1964), 2:258–59; quoted in Staiti, *Samuel F. B. Morse*, 230.

95 Edward Lind Morse, ed., *Samuel F. B. Morse: His Letters and Journals* (1914; repr. New York: Da Capo Press, 1973), 2:105.

96 Wynne, "Samuel F. B. Morse," 226.

97 Ibid., 228.

98 Edward C. Carter II, *"One Grand Pursuit": A Brief History of the American Philosophical Society's First 250 Years, 1743–1993* (Philadelphia: American Philosophical Society, 1993), 28.

99 "It was owing to his representations that the investigations into terrestrial magnetism were first taken up by the Royal Society. Since then, he has seen the British and Russian Governments, urged on by the Royal Society, and more lately by the British Association, also actively supporting this branch of experimental science, till a grand system of observatories has covered both hemispheres with a net-work of stations, which have already furnished the most important data, and on such a scale as, but for such a system of national co-operation, could never have been attainable." Review from *London Morning Chronicle* of *Cosmos* volumes 1 and 2, reprinted in "Humboldt's Cosmos," *Green Mountain Gem; a Monthly Journal of Literature, Science, and the Arts* 7 (1 January 1849): 105.

100 Alexander von Humboldt and Aimé Bonpland, *Personal Narrative*, vol. 6, part 1, trans. Helen Maria Williams (London: Longman, Rees, Orme, Brown, and Green, 1826), 239–300. Quotation is a heading in chapter 26. See Table of Contents, vii.

101 Humboldt, *Aspects of Nature, in Different Lands and Different Climates, with Scientific Elucidations*, trans. Elizabeth J. L. Sabine (London: Longman, Brown, Green, and John Murray, 1849)., 433–35; quoted in Ingo Schwarz, "'Shelter for a Reasonable Freedom' or Cartesian Vortex: Aspects of Alexander von Humboldt's Relation to the United States of America," in "Alejandro de Humboldt y el mundo hispánico," special issue, *Debate y Perspectivas*, no. 1 (2000): 181–82. The published accounts in American newspapers and journals regarding the feasibility and desirability of realizing Humboldt's vision of a canal across Panama are voluminous, and they appear in every decade of Humboldt's life beginning in 1811.

102 *National Intelligencer*, 14 June 1850; quoted in Schwarz, "Americans in Alexander von Humboldt's Personal Address Book," *Unity of Nature: Alexander von Humboldt and the Americas*, 36–37.

103 Humboldt and Bonpland, *Personal Narrative*, vol. 1, trans. Helen Maria Williams (London: Longman, Hurst, Rees, Orme, and Brown, 1814), vii–viii. Also cited in Cannon, *Science in Culture*, 87.

104 Maury received letters from Humboldt praising his data collecting on wind and currents—one dated 11 April 1859 was published as "Compliment from Humboldt to Lieut. Maury," in *New York Times*, 9 May 1859, 5.

105 Maury wrote in his foreword: "Baron Humboldt is of opinion that the results already obtained from this system of research are sufficient to give rise to a new department of science, which he has called the Physical Geography of the Sea." See Matthew Fontaine Maury, *The Physical Geography of the Sea* (New York: Harper & Brothers, 1858), x–xi; quoted in Gerhard Kortum, "M. F. Maury (1806–1873), A. v. Humboldt (1769–1859) und der Mythos des Telegraphen-Plateaus in Nordatlantischen Ozean," in Rolf-Dieter Volmerig, Burkhard Hofmeister, Frithjof Voss, and Michael Brillat, eds., *Geographie der Küsten und Meere. Belträge zum Küstensymposium in Mainz 14.–18. Oktober 1984* (Berlin: Institut für Geographie der Technischen Universität, 1985), 1–2.

106 "Morse schickte mir eben eine 15 Fuss lange Sondierungeskarte." Humboldt went on to quote from Morse's letter to him, dated London, 7 October 1856. For Humboldt's letter, see Gerhard Kortum, "M. F. Maury (1806–1873), A. v. Humboldt (1769–1859) und der Mythos des Telegraphen-Plateaus im Nordatlantischen Ozean," in Rolf-Dieter Volmerig, Burkhard Hofmeister, Frithjof Voss, and Michael Brillat, eds., *Geographie der Küsten and Meere. Beiträge zum Küstensymposium in Mainz, 14.–18. Oktober 1984* (Berlin: Institut für Geographie der Technischen Universität, 1985), 19–20. For Morse's letter to Humboldt, see Schwarz, ed., *Alexander von Humboldt und die Vereinigten Staaten von Amerika: Briefwechsel*, 406–407, letter 264.

107 Diana Fontaine (Maury) Corbin, *A Life of Matthew Fontaine Maury, U.S.N. and C.S.N, Author of 'Physical Geography of the Sea and Its Meteorology'* (London: S. Lowe, Marston, Searle, & Rivington, 1888), 104.

108 Lauren Applebaum has made this argument a central part of one chapter of her dissertation, "Elusive Matter, Material Bodies: American Art in the Age of Electronic Mediation, 1865–1918" (PhD diss., University of Illinois at Urbana-Champaign, 2019). I am grateful to her for conversations concerning her research on this topic. The auroras Church witnessed in 1859 included shafts of light shooting up from the horizon, described in "The Aurora Borealis. The Brilliant Display on Sunday Night. Phenomena Connected with the Event," *New York Times*, 30 August 1859.

109 Ingo Schwarz and Klaus Wenig, eds., *Briefwechsel zwischen Alexander von Humboldt und Emil du Bois-Reymond* (Berlin: Akademie-Verlag, 1997), 154–55; cited in Ingo Schwarz, "Alexander von Humboldt's Visit to Washington and Philadelphia: His Friendship with Jefferson, and His Fascination with the United States," in "Proceedings: Alexander von Humboldt's Natural History Legacy and Its Relevance for Today," special issue, *Northeastern Naturalist* 1 (2001): 55–56n48.

110 Wynne, "Samuel F. B. Morse," 228.

111 Cited in Peter Schoenwalt, "Alexander von Humboldt in the United States of America," in Halina Nelken, ed., *Humboldtiana at Harvard* (Cambridge, MA: Harvard University Widener Library, 1976), 13.

112 "Report of the Secretary of the Treasury, in answer to a resolution of the Senate relative to the transfer of the survey of the coast from the Treasury to the Navy Department," in *U.S. Congress. Senate Executive Documents no. 36, 31st Congress, 2nd Session* (Washington, DC: 1851), 13; cited in Schwarz, "'Shelter for a Reasonable Freedom,'" 181n49.

113 In 1848 Ticknor sent Humboldt an inscribed copy of his three-volume *History of Spanish Literature*; Henry Stevens, *Catalogue of the First Portion of the Humboldt Library* (London: J. Davy & Sons, 1862), 724, entry no. 10208. Prescott sent Humboldt a total of eleven volumes of his writings on Ferdinand and Isabella, Phillip the Second of Spain, the conquest of Mexico, and the conquest of Peru, each a presentation copy with an inscription to Humboldt. Henry Stevens, *Catalogue of the First Portion of the Humboldt Library* (London: J. Davy & Sons, 1862), 572, entry no. 7951. Frémont's inscribed volumes are enumerated in chapter 5, along with his purchases from Humboldt's library after his death.

114 A letter from Humboldt to Flügel, 19 June 1850, from Berlin, praises Ticknor and Prescott's publications and extols great cultural progress in the New World. Ingo Schwarz, "Alexander von Humboldt's Correspondence with Johann Gottfried Flügel," *Yearbook of German-American Studies* 46 (2011): 89.

115 Prescott to Humboldt, Boston, 23 December 1843, *Letters of Alexander von Humboldt, written between the years 1827 and 1858, to Varnhagen von Ense. Together with extracts from Varnhagen's diaries, and letters from Varnhagen and others to Humboldt. Authorized translation from the German with explanatory notes and a full index* (London: Trübner and Co., 1860), 124, letter 86. Not everyone who aspired to meet Humboldt and receive his assessment of his work was successful. John James Audubon secured letters of introduction to Lafayette and Humboldt while he was in England in 1826, hoping to show the eminent naturalist his *Birds of America*. Audubon's "elephant folio" represented his mammoth-scale contribution to the natural history of the United States. Humboldt quite likely would have been pleased to review Audubon's accomplishment, but the American's timing was off by a few months. By the time he arrived in Paris in September 1827, Humboldt had moved to Berlin.

116 Ingo Schwarz, "Alexander von Humboldt's Correspondence with Johann Gottfried Flügel," *Yearbook of German-American Studies* 46 (2011): 87–94.

117 At his death, Humboldt's library included a set of *Smithsonian Contributions to Knowledge*, twelve annual volumes of organizational committee reports, and publications by individual Smithsonian scientists. Stevens, *Catalogue of the First Portion of the Humboldt Library*, 690, entry no. 9701. Humboldt's interest in the founding of the Smithsonian and its research publications is the topic of chapter 8 in this volume.

118 Humboldt to Blodget, Berlin, 9 July 1856, in Schwarz, ed., *Alexander von Humboldt und die Vereinigten Staaten von Amerika*, 398, letter 259. Blodget's inscribed volume was in Humboldt's library at his death. Stevens, *Catalogue of the First Portion of the Humboldt Library*, 77, entry nos. 1010 and 1011.

119 Blodget refers to this letter, dated 7 September 1856, in Lorin Blodget, *Climatology of the United States, and of the Temperate Latitudes of the North American Continent. Embracing a full comparison of these with the climatology of the temperate latitudes of Europe and Asia*...(Philadelphia: J. B. Lippincott, 1857), x. A copy of the facsimile of this letter is in the New-York Historical Society. The original letter is in the American Philosophical Society Archive, Alexander von Humboldt Papers.

120 Friedrich Karl Joseph Freiherr von Gerolt (1797–1879) trained in geology and mining, fields that put him in close proximity with Humboldt. In 1824 he visited Mexico to research silver mines, and his geological map of the region was published in 1828.

121 Humboldt's response to American slavery is the subject of chapter 5 in this volume.

122 "Smithsonian Institution, July, 1855.
The bearer of this letter, Mr. W. W. Corcoran, a gentleman of wealth and influence of this country, a warm friend of the Smithsonian Institution, and to whom has been entrusted the management of a part of its funds, visits Europe as a recreation and on behalf of the regents. I beg to commend him to the attention of the librarians and officers of the societies in correspondence with the Smithsonian Institution which he may visit in the course of his tour.
JOSEPH HENRY, Secretary, Smithsonian Institution.
Washington, July 11, 1855.
To the librarians and officers of the societies in correspondence with the Smithsonian Institution."

Corcoran Papers, LC MSS Division, reprinted in William Wilson Corcoran, *A Grandfather's Legacy; Containing a Sketch of His Life and Obituary Notices of Some Members of His Family, Together with Letters from His Friends* (Washington, DC: Henry Polkinhorn, 1879), 137.

123 "Berlin, 19th September, 1855. W. W. Corcoran, Esq.
My Dear Sir: I leave Berlin to-morrow to return to Washington, where I hope to see you also by and by.
Enclosed you find a letter of introduction to Baron Humboldt, who will be happy to see you with Mr. Fillmore, and will, no doubt, present you to the King./ Mrs. Gazziotti, a distinguished Italian painter, who has made some valuable portraits for the royal family and other people, has also made an excellent portrait of Baron Humboldt, which she intends to send to the United States for sale. Perhaps Baron Humboldt will accompany you to her *atelier*, because she is a fine-looking Italian woman. I regret very much not to have met you here. Adios. Yours, sincerely, FR. GEROLT. / Baron Humboldt will be at Berlin, No. 67 Oranienburger Strasse, till the King returns from the Rhine, the 5th of October. / P. S.—Baron Humboldt will call upon Mr. Fillmore when his arrival will be known to him. You must also see the royal gardens at Sans-Soucis and Potsdam, where the enclosed letter to my friend, Tenné, will guide you and your party," Corcoran, *A Grandfather's Legacy*, 140–41.

124 Humboldt to Corcoran and Millard Fillmore, Potsdam, Thursday October 18, 1855: "Monsieur, Le Roi ayant appris que Vous ne pouvez disposer Vous, mon cher Monsieur et Votre ami Mr. Corcoran que de deux jours, Vous invite Mr. Fillmore et Mr. Corcoran à dîner chez lui demain *Vendredi* à 3 heures à Sanssouci. Je partirai avec Vous à 2h, mais j'espère toujours Vous recevoir chez moi (Oranienburger Str 67) à 11h arrangeant les courses de sorte que Vous serez à Votre hôtel à 1h. /
Mille affectueux hommages / A Humboldt" in Schwarz, *Alexander von Humboldt und die Vereinigten Staaten von Amerika*, 363, letter 237. The king invites them to dine at Sanssouci at 3 p.m. on Friday (the next day); he will meet them at his house at 11 a.m., and they will leave together at 2 p.m. Translation by the author.

125 "Mes très respectueux hommages à Mr. Fillmore et mes voeux pour la paix extérieure et intérieure! aussi des voeux pour la vente un peu moins fréquente 'des négrillons qui pèsent 60 livres.'" Humboldt to Corcoran, Berlin, 9 September 1856, Schwarz, *Alexander von Humboldt und die Vereinigten Staaten von Amerika*, 400–401, letter 260. For an extended discussion of Humboldt's response to American slavery, see chapter 5 in this volume. Translation by the author.

126 Louise Corcoran to Humboldt, Washington, 14 September 1857; quoted in Schwarz, *Alexander von Humboldt und die Vereinigten Staaten von Amerika*, 427, letter 281.

127 Ticknor to Edward Everett, 20 September 1856, in Ticknor, *Life, Letters, and Journals of George Ticknor*, 2:333. Ticknor goes on to report that Fillmore had made a good impression on Humboldt and that while Humboldt supported Frémont in the upcoming presidential election, Ticknor was "of Fillmore's faction."

128 Ibid., 2:340–41; quoted in Schwarz, "Transatlantic Communication in the 19th Century," 34–35.

129 Humboldt to Corcoran, Potsdam, 16 October 1855, "Daignez excuser mon illisible écriture, un témoignage de ma vieillesse antédiluvienne." Corcoran is staying at the Hotel du Nord, Berlin. Schwarz, ed., *Alexander von Humboldt und die Vereinigten Staaten von Amerika*, 361–62, letter 236, with translation.

130 In a letter to Balduin Möllhausen, Berlin, 1 September 1856, Humboldt lists the itinerary of sights he wants Corcoran and Fillmore to see: the Friedenskirche, Sanssouci, the Picture Gallery (only for the hall), the Neues Palais, the statue of the Queen, Charlottenhof, the Marmorpalais, or Babelsberg Castle. The trip is cited and discussed in Schwarz, "Transatlantic Communication in the 19th Century," 34.

131 Ticknor to Prescott, Rome, 24 November 1856, in Ticknor, *Life, Letters, and Journals of George Ticknor*, 2:340.

132 Ticknor, *Life Letters and Journals of George Ticknor,* 2:341; quoted in Schwarz, "Transatlantic Communication in the 19th Century," 35.

133 R.W.R. [Rossiter Worthington Raymond], "A Visit to Humboldt," *New York Times*, 9 June 1859. Reprinted from an address, Saturday, 14 May 1859, from Heidelberg. Schwarz, *Alexander von Humboldt und die Vereinigten Staaten von Amerika*, 571–73, dokument 43.

134 "It has been remarked by Professor Ritter, that America has been twice discovered—actually, objectively by Christopher Columbus; scientifically by Alexander Humboldt." "From the Scientific American. Alexander Humboldt," *Liberator*, no. 53 (31 December 1847): 212.

135 Edward Everett, "Alexander von Humboldt," in *Orations and Speeches on Various Occasions* (Boston: Little, Brown, 1879), 4:170.

136 The Lambdin portrait commission is discussed in Nelken, *Alexander von Humboldt: His Portraits*, 152. Bache's comment quoted in Peter Schoenwalt, "Alexander von Humboldt in the United States of America," in Nelken, *Humboldtiana at Harvard*, 13. For a detailed discussion of Prince Maximilian and George Catlin, see chapter 6 in this volume.

137 Nelken, *Alexander von Humboldt: His Portraits*, 142.

138 Richard Henry Stoddard, *The Life, Travels, and Books of Alexander von Humboldt*, with an introduction by Bayard Taylor (New York: Rudd & Carleton, 1859), 475.

139 "Humboldt and Washington's Birthday," *Littell's Living Age*, no. 775, 2 April 1859, 61.

140 "Joseph Albert Wright on a Session of the Geographical Society, Berlin, May 7, 1859," *Zeitschrift für allgemeine Erdkunde*, n.s., 6 (1859): 415; quoted in Schwarz, "'Shelter for a Reasonable Freedom,'" 182.

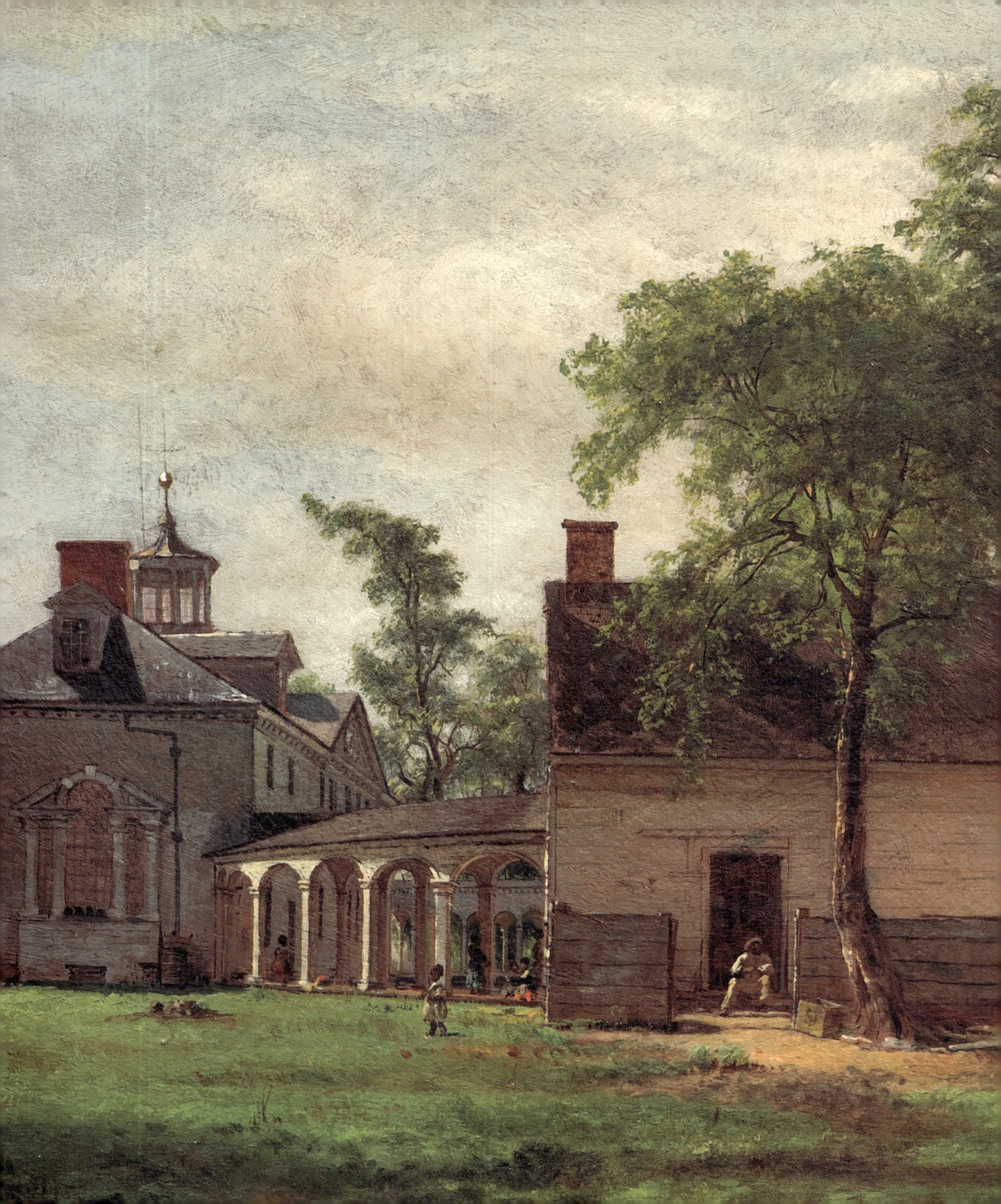

"Slavery is possibly the greatest evil ever to have afflicted humanity"

The one issue on which the United States and Humboldt parted company was slavery. Committed to his stance on the intrinsic equality of all races, Humboldt abhorred colonial rule and enslavement alike. Born and raised during a turbulent period in European history, Humboldt believed America's representational democracy held the best hope for a model government concerned for human rights. In 1804 he witnessed the contradiction between the country's aspirations and the harsh reality of slavery, and for the rest of his life Humboldt would agitate for its abolition. From his earliest conversations with Jefferson, Humboldt had attempted to sway America's attitudes regarding slavery and racial equality. He countered the lingering fear of a Haitian-style rebellion within the United States with arguments emphasizing the possibilities for more peaceful political change. Over time his tone became more strident as he realized the limits of his influence, although his support for what he hoped the United States could be—an example to the rest of the world of a true democracy that lived up to its ideals—never wavered. As his international reputation grew, Humboldt's voice became a powerful factor in the development of America's place in world affairs; he saw no reason his views on slavery should be any less potent.

Eastman Johnson, *The Old Mount Vernon* (detail), 1857. See p. 63.

Humboldt was determined to use his influence but first chose indirect means to advocate for the elimination of slavery. He never broached the subject directly with Jefferson, perhaps not wanting to alienate a man who in other ways seemed like a brother. Instead he used the friendships he made in the United States to drive home his message. Humboldt voiced his opposition first in his letters to American admirers and repeatedly in his published writings, alarmed by the thought that the continuation of slavery might spell the destruction of the United States from within. However, even as Humboldt's clout grew in most other arenas, his inability to gain traction on this issue threatened to rupture those relationships. In the United States, efforts to placate Humboldt strained diplomatic ties with a man who was becoming one of the most influential public figures in Europe.

For close to fifty years Humboldt maintained his concern for racial equality, but it would take until the 1850s for his voice to matter. As abolitionists became an increasingly powerful and polarizing force, they drew attention to his views. With their efforts, Humboldt's reach expanded from the sciences and exploration to human rights. The high-water mark for Humboldt's views on slavery came with his support for John C. Frémont's 1856 presidential bid. Frémont's admiration for Humboldt made him an enthusiastic vessel, and Frémont's support for doing away with slavery made him an attractive candidate for the nascent Republican Party. John C. and Jessie Benton Frémont became the public face of Humboldt's desire to see freedom for all in the United States. Publicly and privately, the Frémonts would assertively use Humboldt's imprimatur to advocate for the goals of abolition and emancipation.

Humboldt's optimism regarding America's future was based on the ideals articulated in the Declaration of Independence. Thomas Jefferson's thinking was rooted in the Enlightenment idea that freedom is innate—a natural law rather than a right conferred by governments. Humboldt's concept of nature developed from the same premise. All human beings enjoyed liberty as a right bestowed by nature; society's role was to secure that right through laws. Humboldt had cultivated his friendship with the American president based on their shared belief in democracy and their mutual interest in the exploration of the expanding country. During his six weeks in the United States in 1804, Humboldt found ways to make his belief in racial equality known, without pushing so hard that he risked rupture with his hosts. When Humboldt arrived in Philadelphia, he had introduced himself to Jefferson by stressing his admiration for his enlightened nature and their shared love of science. To Secretary of State James Madison, Humboldt shifted his focus. He described the extent of his travels through the Americas, countries under the oppressive colonial rule of the kingdom of Spain. After establishing the basis for his global perspective on the human condition, Humboldt mentioned his "intense interest in the human welfare of the United States," which he attributed to the principles of democracy written into the nation's founding documents. He then cut to the heart of his message, stating that he "intended to enjoy the spectacle of a free people worthy of a great destiny."[1] Humboldt chose those words with deliberation, sending an early if oblique signal that slavery did not accord with his expectations for the new republic. He fully expected that his hosts would see past the diplomatic flattery and recognize this was a topic of great importance to their foreign visitor.

Humboldt was a lifelong abolitionist. His study with Johann Friedrich Blumenbach at Göttingen had shaped two things: his liberal political tendencies, and his views on the underlying equality among all races. He had been shocked by the experience of a public slave market in Cumaná (in present-day Venezuela) in 1799.[2] His revulsion upon seeing human beings paraded for examination and sale marked him for life, and his views on racial equality only strengthened over the decades. Humboldt gathered statistics on the human populations and the structures of each society, building an argument for the support of humanitarian ideals over commercial profits. He conducted research during his two trips to Cuba in 1801 and 1804 to create a case study that became his most direct publication on human rights as a moral imperative.[3] Humboldt used this material, combined with his notes from South America, to assert the economic viability of a free labor economy alongside a humanitarian argument for the elimination of race-based slavery.

On June 11, 1804, his last day in Washington, the *National Intelligencer* reprinted a sermon delivered by Dr. David Ramsay in Charleston, South Carolina, wherein he praised the political, cultural, and economic benefits of the acquisition of the Louisiana Territory. Jefferson believed that U.S. expansion would result in a more robust and widespread agrarian economy. For him, that meant the continued use of slave labor in the new territory. Jefferson was hardly alone in this reasoning, and the impulse to tie those two policies together would remain strong until the onset of the Civil War in 1861. However much Humboldt supported the growth and prosperity of the United States, he vigorously opposed enslaved labor as a component of American democracy.

Jefferson's most sympathetic newspaper editor broke with the president and agreed with Humboldt. In an editorial published adjacent to Ramsay's sermon, Samuel Harrison Smith ran a story condemning the continued importation of slaves into South Carolina. That story read in part:

Never has a deadlier blow been levelled at the liberties, the peace, and the happiness of the whole community. It has already fallen with fatal vengeance, and it has sowed the seeds of misery wherever it has touched.... Is there an American, out of the vortex of the little interest that has done all this mischief, not ready to raise his voice in defence of the present and future interests of his country?

Smith made his plea for a public outcry against slavery to redeem the "American character relieved from the darkest cloud that envelopes it."[4]

The unsolved dilemma would remain, that America's economic expansion assumed a commensurate expansion of the slave-based labor force. Humboldt responded to this directly in a June 20 letter to William Thornton, in which he railed,

This abominable law that permits the importation of Negroes in South Carolina is a disgrace for a state in which I know many level-headed people to live. Conforming to the only course of action dictated by humanity, undoubtedly less cotton will be exported at first. But alas! How I detest this politics that measures and evaluates the public welfare simply according to the value of its exports. The wealth of nations is like the wealth of individuals. It is only secondary to our happiness. Before one is free, one must be just, and without justice there is no lasting prosperity.[5]

In an argument based on economic pragmatism, Humboldt pleaded with Thornton to use his influence to accept a short-term economic downturn and take the long view in favor of creating something more important: a just society.

Humboldt's arguments continually returned to his insistence on the need for America to be a free society. He had tied his ideas about nature and human nature together, arguing that "nature is the domain of freedom," a state of existence that replaced the human oppression of colonialism or monarchy with peaceful coexistence. The nascent cultural identity of the United States was bending toward the adoption of the wilderness as a symbol of its democratic ideals. Squaring those ideals with slavery was impossible. The dichotomy at the core of the United States was no more readily apparent than at Mount Vernon. On his day trip there, Humboldt came to understand how revered George Washington was both in life and after his death—and saw firsthand how Americans rationalized the perpetuation of enslaved labor after the Revolution. Humboldt's insistence on a just society remained one of the explorer's core convictions. According to his logic, for the United States to thrive, it must eventually abolish slavery.

In his letters and books, Humboldt decried American slavery as antithetical to the country's founding principles. At first his tone was gentle, like a parent scolding a wayward child; as he grew older, Humboldt's tone grew testier. Humboldt's support of the failed 1848 European revolutions revealed his growing impatience with oppressive political rule. However, his disdain for monarchy paled in comparison to his visceral opposition to the enslavement of human beings. Perhaps the resistance he encountered in the United States fueled Humboldt's interest in making an economic assessment of slavery a central part of one of his earliest books, *Political Essay on the Kingdom of New Spain* (1811). Humboldt took a stand against this practice in the colonies of New Spain, a daring move considering that he dedicated this volume King Carlos IV in gratitude for granting the explorer unfettered access to his colonies in South America.[6] Humboldt decried the enslavement of anyone—on philosophical, humanitarian, and economic grounds.

Humboldt's second daring move was in sending Jefferson the earliest edition of his *Political Essay on the Kingdom of New Spain* and noting in his accompanying letter,

In these works you will find your name cited, a name which must for all time inspire the friends of humanity. May providence keep you alive for the benefit of that world of ours for which I strive to direct my labors and hopes.[7]

In a scathing indictment veiled by diplomatic flattery, Humboldt steered Jefferson directly to the passages that condemned slavery in the United States. The text Humboldt referred to read,

The kingdom of New Spain has one decided advantage over the United States. The number of slaves there, either Africans or of mixed race, is almost nothing; an advantage which the European colonists have only begun rightly to appreciate since the tragical events of the revolution of St. Domingo. So true it is, that the fear of physical evils acts more powerfully than moral considerations on the true interests of society, or the principles of philanthropy and of justice, so often the theme of parliament, the constituent assembly, and the works of the philosophers. The number of African slaves in the United States amounts to more than a million, and constitutes a sixth part of the whole population. The southern states, whose influence is increased since the acquisition of Louisiana, very inconsiderately increase the annual importation of these negroes. It is not yet in the power of Congress, nor the chief of the confederation (a magistrate whose name is dear to the true friends of humanity), to oppose this augmentation, and to spare by that means much distress to the generations to come.*
**The present president, Mr.* Thomas Jefferson, *author of the excellent Essay on Virginia.*[8]

There is no question Jefferson would have understood Humboldt's intent, and there was little room for him to object to the picture Humboldt painted of the United States.

Humboldt was unwilling to confront Jefferson directly with his views on slavery, but he was absolutely willing to proselytize in print. He touched on the connection between national expansion and the increase in slaveholdings and decried the gap between fear of slave revolts and the moral character of the nation. The literary flourish at the end of this passage, commending Jefferson for his ideals while simultaneously berating him for his inaction, left no doubt that Jefferson understood precisely how Humboldt felt.[9] There would be no apologies from Humboldt for applying this pressure. By the time Jefferson received this letter and his personal copy of Humboldt's *Political Essay on the Kingdom of New Spain*, a lengthy excerpt had appeared in an American periodical, including the above quoted material on Jefferson.

Humboldt's was not the only influential voice appealing to Jefferson's higher ideals. Madame de Staël, whose salons Jefferson had attended regularly in Paris, wrote in 1816, "If you should succeed in destroying slavery in the south there would be at least one government in the world as perfect as human mind can conceive."[10] She was well aware of Humboldt's efforts, as Humboldt shared with her his letters from Jefferson. Their mutual friend, the Marquis de Lafayette, was also a committed abolitionist who used his letters to needle Jefferson about slavery, as he had George Washington before him. In 1822 Lafayette challenged Jefferson:

While I feel an inexpressible delight in the progress of Every thing that is Noble minded, Honourable, and Useful throughout the United States, I find, in the Negros Slavery, a great draw-Back Upon My Enjoyments. it Raises a Sigh, or a Blush, according to the Company, American or foreign, Where I Happen to be....this Wide Blot On American Philantropy and Civilisation is Ever thrown in my face When I indulge my Patriotism in Encomiums, otherwise Undisputable. to See that plague Cured, While I live, is here to impossibility, But I Would Like, Before I die, to be assured that progressive and Earnest Measures Have Been adopted to Attain, in due time, So desirable So necessary an object. prudence as Well as Honor Seems to me to Require it.[11]

The three of them formed a powerful nexus of pro-American sentiment in Europe, all of them leaning hard on Jefferson to uphold Humboldt's call for a just society.

THOUGHTS OF HAITI

Jefferson's stubbornness on slavery cannot be adequately explained; however, he was acutely aware of the instability on the island of Hispaniola. In 1791, after years of unrest, the French colony of Saint-Domingue erupted in violence. European colonizers found themselves at war with enslaved blacks from Africa and downtrodden generations of indigenous black and mixed-race inhabitants on the island. Saint-Domingue (modern-day Haiti) had become America's second-largest trading partner and was at that time the wealthiest colony in the New World. Jefferson understood this and in 1793 had noted in a letter, "nature has connected them [the United States and these islands] by the strong link of mutual necessity."[12] By this time the United States counted on the coffee, sugar, and molasses from the island, sending flour, livestock, and dried meats to the island in return.[13] However, in a 1797 exchange with Jefferson over the state of slavery in Virginia, St. George Tucker cautioned, "The recent scenes transacted in the French colonies in the West Indies are enough to make one shudder with the apprehension of realizing similar calamities in this country."[14]

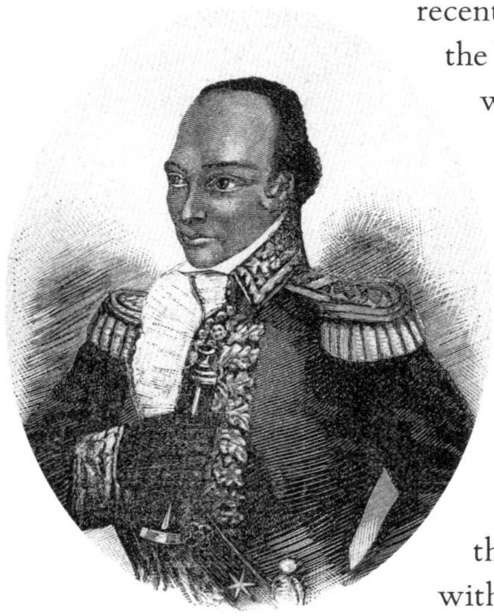

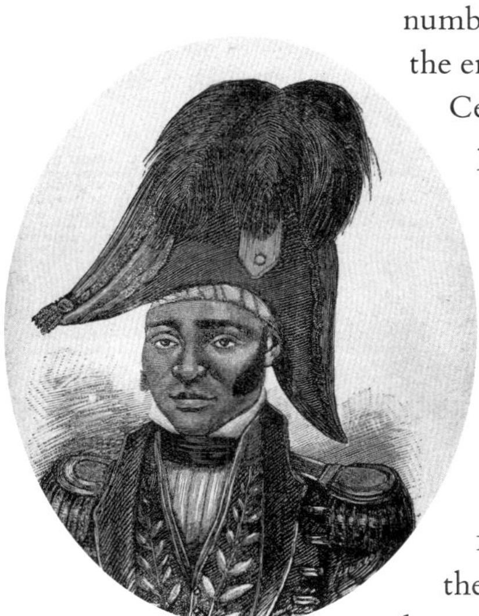

Fig. 64
Toussaint Louverture (1743–1803), from Dantès Fortunat, *Nouvelle géographie de l'île d'Haiti*, 1888, engraving, New York Public Library

Fig. 65
Haitian General Jean-Jacques Dessalines (1758–1806), from Dantès Fortunat, *Nouvelle géographie de l'île d'Haiti*, 1888, engraving, New York Public Library

Toussaint Louverture (fig. 64) and a shifting set of allies spearheaded the successful rebellion on the French half of the island of Hispaniola. From 1791 to 1793 thousands of French refugees from the island emigrated to Philadelphia and New Orleans in the first waves of the bloody revolution.[15] New Orleans absorbed the largest percentage of refugees, along with Charleston, South Carolina. The greatest number of white intellectuals, including nearly the entirety of the scientific community (the Cercle des Philadelphes), migrated to Philadelphia. The American Philosophical Society embraced these émigrés, with whom they had long exchanged letters, publications, and ideas.[16] Following Louverture's imprisonment and death in France, General Jean-Jacques Dessalines (fig. 65), who had a fearsome reputation for massacring whites, announced the first free republic led by a black man.[17] On January 1, 1804, he declared independence, christening the new republic Haiti. Fewer than five months later, Humboldt arrived in Philadelphia, which by that time had absorbed the French colony's scientific community.

The revolution itself and the bloodshed that accompanied it unnerved many in the southern United States who feared that a similar rebellion could take hold and undermine if not destroy the new nation. A report from the *Richmond Enquirer* published while Humboldt was in Washington stated grimly, "between five and six hundred white persons fell under the bloody hatchet of the Haitians, and the warm stream of blood which ran from them, quenched the thirst of their murderers, who went to their knees, to receive it."[18]

Hundreds of such accounts appeared in U.S. newspapers during and after the Haitian revolution. Americans from Virginia to Louisiana saw themselves as part of a pan-Caribbean enterprise now jeopardized by race rebellion. Pro-slavery whites across the South invoked Haiti to instill fear of slave revolts, and newspaper coverage of nearly every slave rebellion cited Haiti's violence as an example of what could happen in the United States. In 1822 Denmark Vesey, who had been brought to the United States from the West Indies, was accused of organizing a state-wide slave revolt in South Carolina, plotting to kill white slave-owners and free enslaved people who would then sail to Haiti for safety. Word leaked of the plan and Vesey was among those who were arrested and executed. The state subsequently clamped down on slave activities and blamed the plot on Haitians. During an address to an agricultural society in Charleston three years later, one orator pronounced,

> *God, in his righteous judgment, will raise up Toussaint, or a Spartacus, or an African Tecumseh, to demand by what authority we hold them in subjection. . . . Our history has verified the melancholy truth, that one educated slave, or colored freeman, with an insinuating address, is capable of infusing the poison of insubordination into the whole body of the black population.*[19]

Trapped in a paradigm that could not envision the stability of the new nation without slave labor in an agrarian system, white slave owners in the United States were unwilling to embrace a change that they believed might topple their fragile country.

In his correspondence with Humboldt, Jefferson stressed the instability of the Caribbean islands and uncertainty over the conditions of slavery in Mexico and South America. Jefferson regarded this as a hemispheric issue, prompting fears about the political stability of the Spanish and French colonies to the south. From Humboldt's perspective, Haiti modeled what a free black republic could look like. By 1814 he optimistically described "the free Africans of Hayti" as a people "who may influence the destiny of the other hemisphere." The European audience understood the Americas to be the "other hemisphere." He went on to write, "Every friend of humanity prays for the development of the civilization which is advancing in so calm and unexpected a manner."[20] The disparity between American views of Haiti as a cauldron of racial violence and Humboldt's perception of a more enlightened society in the making could not have been more pronounced.

In 1826 the political landscape shifted for Humboldt. With Thomas Jefferson's death, he had lost the friendship that mattered most to him in the United States. Subsequent presidents had not cultivated a similar closeness with him, and over the years Humboldt came to recognize that his sway over government policy in the United States was slipping. In 1825 he wrote to his friend, the Prussian geographer Heinrich Berghaus,

> *Should the question of slavery break out one day, I entirely share your opinion that the maintenance of the North American Union as a state is in danger. I do not wish to see this happen. I think highly, very highly of the United States because it is the shelter for a reasonable freedom.*[21]

Despite Humboldt's frustration with the perpetuation of slavery in the United States, he was not willing to give up on its potential as a truly free nation. That phrase, "shelter for a reasonable freedom," reads as a more equivocal statement about the state of affairs in the United States, and represents a grudging acceptance that American democracy—flawed as it was by the continuation of slavery—was still a better political system than the monarchies of Europe.

HUMBOLDT AND THE ISLAND OF CUBA

If appeals to a higher moral ground could not persuade his American audience to abolish slavery, Humboldt could still deploy his economic argument. His next book, titled *Essai politique sur l'ile de Cuba* (*Political Essay on the Island of Cuba*), was published first in French in 1826 and translated into Spanish the following year.[22] Although this was a case study on the economics of the island that had eclipsed Saint-Domingue as the single most prosperous center of trade in the Americas, Humboldt ended with a chapter in which he made his most complete and compelling arguments for abolishing race-based slavery. Returning to the logic he had outlined in his letter to William Thornton, Humboldt rejected the argument that without slaves there could be no large-scale agriculture. Humboldt cited the statistics he had so carefully gathered during his two trips to Cuba, deploying his data to methodically take apart the rationalization that enslaved labor was a requirement for agricultural prosperity. Humboldt wrote that race had no bearing on a man's intelligence or prospects and that enslavement was morally wrong and economically shortsighted, no matter the circumstances. In this book Humboldt did not simply argue for the abolition of slavery, he proposed a gradual emancipation, based on number of years in servitude, the number of children black women brought into the world, and espoused the gradual weaning of enslaved labor, to be replaced with paid positions that would encourage blacks to continue working the land in return for wages or a percentage of the agricultural yield. His attempts to appeal to economic concerns sounded more like a legislative proposal than a moral argument, perhaps in recognition of the market forces he was battling.

For his *Political Essay on the Island of Cuba*, Humboldt deployed an army of statistics and examples to support his humanitarian arguments. Amassing tables of data, he destroyed the argument that slave labor was required to ensure a profitable agricultural enterprise. He demonstrated that the actual number of enslaved people far exceeded the manpower needed to produce the cash crops destined for domestic and foreign consumption. Here, too, Humboldt invoked nature, comparing the inefficiencies of slave labor to "a current of water that one has channeled across a great distance to the colonies, where the water does not reach the soil for which it had been intended."[23] Whether it was the abuse of human beings or the misuse of natural resources, Humboldt was quick to point out the irrationality of human behavior. When it came to slavery, his condemnation was swift and soundly argued.

Humboldt decried the working conditions and mortality rate for the people working on sugar plantations, making a powerful and sophisticated argument for a more enlightened society reliant on free yeoman laborers rather than enslaved people. He reasoned that *if the slave trade ceases altogether, slaves will gradually pass into the class of free men and the society, having rebuilt itself without any of the violent civil upheavals, will return to the paths nature has traced for all societies that have become populous and enlightened.*[24]

Humboldt asserted that all people shared a right to be treated with "sensibility, emotion, intelligence and wisdom." He denounced those who dressed up slavery in terms such as "black peasants" or "black vassalage." These euphemisms he characterized as sophistry designed to minimize its evils. Humboldt decried those who engaged in the perpetuation of slavery as

secret sharecroppers of the slave trade whose intention is to benefit from the misfortune of the black race. In doing so they needed to find a means to reject their own emotional responses which could overcome and surprise them.[25]

He warned of the inevitability of social unrest culminating in violent revolution, taking the example of Haiti. "The horrible catastrophe of St. Domingue" he characterized as "nothing more than the result of the government's incompetence."[26] After making an eloquent and persuasive argument he emphatically stated, "Slavery is possibly the greatest evil ever to have afflicted humanity."[27]

Humboldt's stand-alone *Political Essay on the Island of Cuba* was not translated into English for another three decades. Nevertheless, Humboldt's forceful thoughts on American slavery did not go unnoticed. A popular weekly Quaker journal, *The Friend*, pulled no punches in a review titled "Alexander von Humboldt on Negro Slavery." The review quoted the most pointed passages from the chapter on slavery. *The Friend* assessed that

the views of this celebrated philosopher respecting it [slavery], are in a high degree interesting. They are marked with the same accuracy and minuteness of detail, by the same comprehensive thought, elevated wisdom, and severe impartiality, which distinguish his other writings.[28]

The power of Humboldt's data once again lent gravitas to his arguments. In the second installment of the review, the author quoted Humboldt at length regarding the inequity and barbarism of slavery in the United States in particular:

if... America, between Cape Hatteras and the Missouri, became the principal seat of the light of Christianity, what a spectacle would that centre of civilization offer.... Let us hope that the generous principles which have so long animated the legislatures in the northern parts of the United States, will extend, by degrees, towards the south, and towards those western regions, where, by the effect of an imprudent and fatal law, slavery and its iniquities have passed the chain of the Alleghany, and the banks of the Mississippi: let us hope that the force of public opinion, the progress of knowledge, the softening of manners, the legislation of the new continental republics, and the great and happy event of the recognition of Hayti by the French government, will exert, either by motives of prudence and fear, or by more noble and disinterested sentiments, a happy influence on the amelioration of the state of the blacks.[29]

As Humboldt had feared, slavery was moving westward with the expansion of the nation's borders. He admitted that he had written his *Essay on Cuba* as a cautionary tale for the United States, appealing to America's moral compass by invoking the signature landmarks in the American landscape as inspiration for a high-minded shift in policy. Humboldt drew his audience back to his foundational idea that landscape features served as American monuments, providing an inspirational setting in which to build a just society. Eventually this attitude linking nature to human morality gained wide traction in the United States and later identified the California landscape, in particular, with the concept of a land free of slavery.

THE CARTESIAN VORTEX

Humboldt continued to appeal to America's founding vision of freedom in his first volume of *Cosmos*, published in German in 1845 and almost immediately thereafter in English. Here he ratcheted up the pressure, expanding his rhetoric on the equality of all races, stating,

Whilst we maintain the unity of the human species, we at the same time repel the depressing assumption of superior and inferior races of men. There are nations more susceptible of cultivation, more highly civilized, more ennobled by mental cultivation than others—but none in themselves nobler than others. All are in like degree designed for freedom; a freedom which in the ruder conditions of

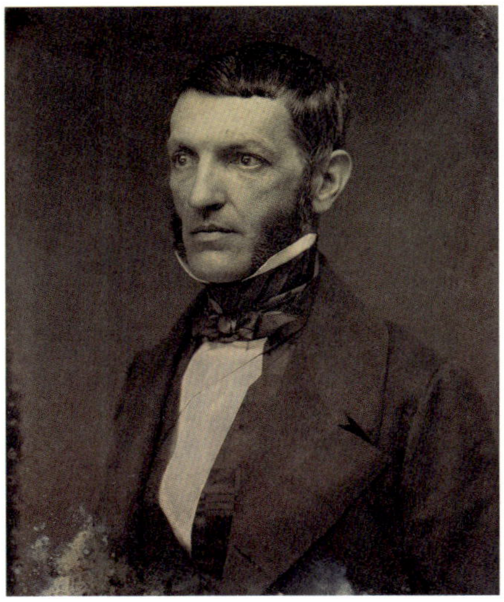

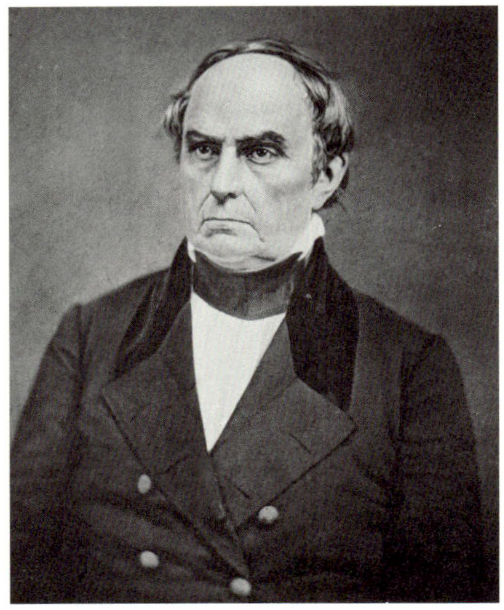

Fig. 66
John Plumbe Jr.,
George Bancroft (1800–1891),
1846, quarter-plate daguerreotype, 4 3/16 × 3 11/16 in.,
National Portrait Gallery,
Smithsonian Institution

Fig. 67
Mathew Brady studio,
Daniel Webster
(1782–before 1852),
printed ca. 1855–65,
wet collodion negative,
Library of Congress,
Brady-Handy Photograph
Collection

society belongs only to the individual, but which in social states enjoying political institutions appertains as a right to the whole body of the community.[30]

Humboldt took the reasoning a step further, confirming human dignity and equality irrespective of race or civilization.

Humboldt's emphatic optimism was tempered by his growing unease that the issue of slavery might tear apart the new republic, a stance he shared with his European colleagues and with many of the Americans who visited him in Europe. In a letter to Bostonian George Sumner, Humboldt urged him to

remain most faithful to the sentiments of Liberty, to which many of your countrymen seem to be growing indifferent, either from the influence of social life in Europe, or under the pretext of exaggerated fears of radicalism. I desire, also, in your work, by the side of a love for the whites, some complaints at the ferocious legislation of the Slave States.[31]

Humboldt's concern over the powerful political bloc of southern states was justified. The war with Mexico from 1846 to 1848 had greatly expanded the national borders, but it also opened new territory for the expansion of slavery. The Polk administration was wary of Humboldt's influence and his penchant for vocal opposition to U.S. policies that he deemed immoral or inhumane. In 1847 President James Polk sent George Bancroft to Berlin to secure the Prussian's blessing for the United States' recent territorial expansion into Mexico. Bancroft (fig. 66) was an astute choice; he had studied at Göttingen and impressed Humboldt when the two men met in Paris. He knew that Humboldt's 1804 map had facilitated U.S. negotiations over the border with Mexico and that Humboldt understood Jefferson's open ambition to expand even farther across the continent. Humboldt apparently was too diplomatic, or Bancroft chose to read him that way, for in his letter to Polk, Bancroft told the president,

In Paris I met Alexander von Humboldt, and he gave me leave to say to you, how greatly he was pleased with it [Treaty of Guadalupe Hidalgo]. The amount of territory you demand [of Mexico], he deemed to be legitimately due to us; and the tone of moderation that prevails through the message won for it his cordial, unhesitating adhesion. His opinion is of value; for having been honored with Mexican citizenship, the bias of his partialities is for Mexico.[32]

In a letter to a friend written earlier that year, Humboldt had expressed far stronger negative feelings about the U.S. incursion, writing, "the ruffians of the American Republic displease me the most. I wish them all the troubles in tropical Mexico. Leave them in the North, where they then spread their wicked slavery."[33]

A series of political events in the early 1850s brought the issue of slavery to a head in the United States, and with them, Humboldt's agitation increased. On March 7, 1850, Daniel Webster (fig. 67), the senior senator from Massachusetts, endorsed the package of bills known as the Compromise of 1850 in a speech called The Constitution and the Union. On the positive side, California would enter the union as a free state, and the slave trade in the nation's capital was banned (although slavery itself was not abolished). But the South blocked implementation of the Wilmot Proviso, which would have banned slavery in the new territories acquired as a result of the war with Mexico. Worst of all, the package included a more draconian version of the Fugitive Slave Law, requiring that it be enforced with strict penalties in the free states. Although U.S. Senators William H. Seward, Salmon P. Chase, and Thomas Hart Benton opposed the Compromise of 1850 for allowing the expansion of slavery, the measures passed in September.

Webster's advocacy for the Fugitive Slave Law in particular horrified abolitionists and antislavery moderates like Ralph Waldo Emerson.[34] In response, Emerson (see fig. 42) delivered a lecture on May 6, 1851, known as An Address to the Citizens of Concord. In it he declared, "Mr. Webster by his personal influence brought the Fugitive Slave law on the country."[35] Emerson's vocal advocacy for abolition only grew stronger. During the debate over the Kansas-Nebraska Act, in hopes of defeating the measure, he delivered an address simply called The Fugitive Slave Law. The date he chose was March 7, 1854—the fourth anniversary of Webster's speech supporting the Compromise of 1850.

The events of 1850 to 1854 prompted seismic shifts in American politics. In 1853 Horace Greeley announced that the *New-York Daily Tribune* would no longer be a Whig paper, but a Republican (abolitionist) one. His decision sent ripples through the North: the *Tribune* boasted more than 100,000 subscribers and probably reached close to 500,000 readers.[36] The *National Anti-Slavery Standard*, the leading abolitionist newspaper, celebrated this announcement, writing that Greeley's influence

> *upon whatever opinion or movement it may see fit to aid can scarcely be overestimated.... The best virtue and the real moral power of the country have rallied to its support. Thousands of eyes moisten with tears of joy and benediction, thousands of hearts thrill with new hope and courage at every defiant word it utters against oppression and wrong. Every* God speed *it pronounces in favour of freedom, temperance and progress is backed by the thunder of a million hearty* amens....All hail, then, *the accession of the New York* Tribune *to the* advancing cause of freedom.[37]

Following the passage of the Kansas-Nebraska Act in 1854, the Whig Party officially splintered, giving rise to the new Republican Party. William Cullen Bryant joined the wave of support, bringing to bear the clout of his own newspaper, the *New York Evening Post*. Emerson was galvanized, and in January of 1855 delivered his American Slavery speech in Boston's Tremont Temple, a lecture he delivered again in New York, Philadelphia, Rochester, and Syracuse during the next thirty days.[38]

Humboldt's outrage over American territorial expansion came to a head in 1854, as the United States pursued Humboldt's nightmare, the annexation of Cuba as a new slave state. Cuba had long been an important trading partner with the United States; the discussions taking place in Ostend, Belgium, represented an expansion of slavery that led Humboldt to denounce the plan, known as the Ostend Manifesto, as "the most outrageous political document ever published."[39] In brief, Secretary of State William L. Marcy ordered three American ministers in Europe to meet and develop a plan to purchase Cuba from Spain, including a rationale for going to war if Spain refused. Pierre Soulé, U.S. minister to Madrid; James Buchanan, U.S. minister to London; and John Y. Mason, U.S. minister to Paris, wrote a circular that justified the seizure of Cuba on the grounds of national security. They met at Ostend in early October to draft a paper that stated up front, "Cuba is as necessary to the North American republic as any of its present members, and that it belongs naturally to that great family of states of which the Union is the providential nursery." The language of diplomacy framed the annexation of Cuba as the addition of another slave state, bringing its prosperous sugar and coffee plantations into the American economy. The trio of diplomats rationalized their position by claiming that unrest on Cuba posed an imminent threat:

> *We should, however, be recreant to our duty, be unworthy of our gallant forefathers, and commit base treason against our posterity, should we permit Cuba to be Africanized and become a second St. Domingo [Haiti], with all its attendant horrors to the white race, and suffer the flames to extend to our own neighboring shores, seriously to endanger or actually to consume the fair fabric of our Union.*[40]

Humboldt fulminated against the Ostend Manifesto, calling the proposal "barbaric."[41] He did not forget this episode, and, when Buchanan ran for president in 1856, he referred to him as "*cubasüchtigen*," or "Cuba-mad Buchanan."[42]

News of the Ostend Manifesto leaked, and negative publicity scuttled it; however, the damage had been done. The timing coincided with the passage of the Kansas-Nebraska Act on May 30, 1854, which repealed the Missouri Compromise of 1820. That statute had mandated that states join the union in pairs, one slave, one free, to maintain the balance of power. It had also prohibited slavery north of the 36°30′ parallel, excluding Missouri. With the Kansas-Nebraska Act, the popular vote of each new state determined its position on slavery. The cumulative impact of these events—Webster's support for the Compromise of 1850 and the Fugitive Slave Law, the attempt to annex Cuba as a slave state, and the passage of the Kansas-Nebraska Act—goaded Humboldt into action. He had cajoled Americans starting with Jefferson and Madison, pleaded with Thornton, and now despaired when his close friends George Ticknor, William Prescott, and others he so admired signed on in support of Webster.

In an interview for Bryant's *New York Evening Post*, the eighty-five-year-old Humboldt expressed his admiration for the American political system but then added,

> *there is one thing in which you stand worse than when I was there.... For thirty years—(and he counted them on his fingers)—you have made no progress about Slavery. You have gone backwards, very far backwards, in many respects, about that—I think especially in your law of 1850, that law by which a man in a Free State, where he ought to be free, can be made a slave of. That I always call the* Webster *law. I always before liked Mr. Webster. He was a great man. I knew him, and always till then liked him. But ever after that I hated him. He was the man who made it. If he had wished to prevent it, he could have done it. That is the reason why I call it the Webster law. And ever after that I hated him.*[43]

The vehemence of Humboldt's language marks the frustrations of a man who had spent decades appealing to reason and saw his influence waning, his time running out.

In July 1854 a despairing Humboldt penned a letter to Karl August Varnhagen von Ense, in which he ruefully noted,

> *In the United States there has, it is true, arisen a great love for me, but the whole there presents to my mind the sad spectacle of liberty reduced to a mere mechanism in the element of utility, exercising little ennobling or elevating influence upon mind and soul, which, after all, should be the aim of political liberty. Hence indifference on the subject of slavery. But the United States are a Cartesian vortex, carrying everything with them, grading everything to the level of monotony.*[44]

In a single metaphor Humboldt expressed his deepest fear—that the United States was misusing liberty. Rather than aspiring to elevate its society, the nation was allowing itself to be ground down to an irredeemable mediocrity. Humboldt's disappointment approached despair.

Humboldt's invocation of a Cartesian vortex freed decades of pent-up frustration with men he viewed as intellectually capable of understanding the evils of slavery but politically unwilling to take concrete action to abolish it. It was a powerful metaphor, less for its scientific nod to Descartes and more for its evocation of lost opportunity and grief over a nation perilously close to dissolution. In 1851 Herman Melville, apparently no stranger to Humboldt himself, had invoked the vortex in a climactic scene in the novel *Moby-Dick*, wherein Ishmael clings to the mast as the *Pequod* is sucked into a whirlpool—a metaphor for Captain Ahab's destructive obsession with the white whale.[45] For Humboldt, America's inability to abolish slavery approached Ahab's willingness to sacrifice everything and everyone in a doomed pursuit. To Humboldt, who had spent a lifetime understanding the downstream consequences of human action on climate and terrain, America's blindness to the consequences of human bondage was especially galling. For someone who had enjoyed significant influence in a wide range of American affairs to be ignored so consistently on the issue of human rights was a devastating failure.

THRASHER'S ESSAY ON CUBA

As if the year couldn't get any worse, in December 1855 Humboldt received a letter from John Sidney Thrasher, an American who had just published an English translation of Humboldt's *Essay on Cuba*.

> *Sir, Will you permit me to offer for your acceptance the accompanying copy of a translation I have attempted to make of your excellent work on 'The Island of Cuba,' and which has been published by J. C. Derby Esq of this city [New York].— I have taken the liberty to differ with you on some of the general principles of social economy in this Continent, but I have endeavored to do so with a proper deference to your great attainments and eminent powers.*[46]

Humboldt pored over Thrasher's version of his text, scribbling his rebuttals in the margins.[47] Whatever clinical appreciation Humboldt had for Thrasher's updated statistics was overshadowed by his shock upon discovering how deeply the American had altered his work. The complete omission of Humboldt's final chapter, titled "Slavery," infuriated the Prussian explorer.[48] What Humboldt had conceived as a manifesto designed to condemn American slavery had been warped into a pro-annexation argument by a man intent on seeing the island of Cuba made a slave-owning state of the United States.

Thrasher was not a disinterested party. He had been raised in Cuba by his mother and earned a college degree on the island. There he also edited a pro-annexation newspaper, and he generally agitated for Cuba's independence from Spain and in favor of its affiliation with the United States. In 1850 he became involved in an internal revolution to establish an independent Republic of Cuba. He was arrested and jailed in Madrid, where he appealed for his release on the grounds that he was an American citizen. The man who came to his defense was none other than Daniel Webster, senior senator from Massachusetts, who took to the floor in 1851 and gave a speech calling attention to Thrasher's plight, leading to his release.[49] Once back in the United States, Thrasher actively promoted the annexation of Cuba as a slave state in order to add vast sugar plantations to the American economy. He further argued for the cessation of the importation of new slaves from Africa—on the grounds that such a move would boost the value of enslaved people already in the United States. The *New-York Daily Times* published his lengthy article espousing these ideas on May 23, 1854, appealing to the proslavery merchants in the city.[50]

Humboldt was having none of it. Why Thrasher thought Humboldt might respond to his book with approval is unclear; instead Humboldt denounced what he saw as outright censorship of his work. He wrote an angry letter published in the *Spenersche Zeitung* in Berlin on July 25, 1856, which was immediately translated and published in numerous newspapers, beginning with the *New York Times* on August 12, 1856, and in the rival *New York Herald* the following day.[51] It also appeared in William Cullen Bryant's *New York Evening Post* and William Lloyd Garrison's abolitionist *Liberator* in Boston. Humboldt's condemnation caused ripples on both sides of the Atlantic.[52] His protest concluded with the stinging rebuke,

> *A steady advocate as I am for the most unfettered expression of opinion in speech or in writing, I should never have thought of complaining if I had been attacked on account of my statements; but I do think I am entitled to demand that, in the free States of the continent of America, people should be allowed to read what has been permitted to circulate from the first year of its appearance in a Spanish translation. Alexander von Humboldt, Berlin, July, 1856.*[53]

Thrasher attempted to defend himself, writing a disingenuous letter published in the *New York Herald* on August 13, 1856, and in the *New-York Daily Times* three days later, protesting his innocence to the charge of mutilating Humboldt's text.[54] From New York, Humboldt emissary Friedrich von Gerolt sent Humboldt a clipping from the *New York Times* carrying Thrasher's rejoinder, calling it "exceedingly lame," and noting the resultant controversy had "excited great attention here, and could not but be welcome to the opponents of slavery, who have made Fremont their candidate."[55] Thrasher's underlying pro-annexation, proslavery views had been unmasked, and Humboldt's denunciation of slavery—and censorship—had gained broad exposure. Humboldt's fury made him as well-known for his views on slavery as for his theories as a natural philosopher.

"FREE MEN, FREE SOIL, FRÉMONT"

The election of 1856 turned into an early referendum on the issue of slavery, and Humboldt played a vocal role in the campaign. The newly created Republican Party planned to nominate John C. Frémont to challenge both the Whigs and the Democrats in the fall election. Frémont had made his reputation during the 1840s as an intrepid explorer, his published narratives painting a picture of a man devoted both to science and American expansion. During his final expedition he had named numerous landscape features in the American West for Humboldt, drawing the explorer's attention. He and Jessie settled in Mariposa, outside Yosemite, where they both

CAT. 45 Unidentified photographer, *John C. Frémont with "Fremont and Freedom!" banner,* 1856, ambrotype housed in half of original leatherette case, 3 ¼ × 3 ¾ in., Collection of Alan V. Weinberg

became involved in territorial politics and the push for California statehood. Their close friends included Thomas Starr King, a fiery abolitionist preacher who had moved from the White Mountains of New Hampshire to San Francisco. King became the most eloquent voice for the abolition of slavery and for California's entering the union as a free state. Jessie, who had ghostwritten her husband's expedition narratives and polished his political rhetoric, provided similar encouragement for King's oratory. When it was time for the crucial vote on California statehood in 1850, it was in large part the power of King's persuasive arguments that swayed the delegates.[56] John C. Frémont was one of the many rising stars who had been in the right place at the right time to influence that decision. His stance on abolition in addition to his contributions to the mapping of the West earned him an accolade from the Prussian king and a glowing letter from Humboldt.[57] Starting in 1850 Frémont served as one of the first U.S. senators from the new state, and although his term was short, it launched him into national politics, setting him up for his presidential bid.

On July 4, 1856, three of the leading pro-Republican newspaper publishers met at James Watson Webb's estate to craft the platform for their preferred candidate. Webb owned the *New York Courier and Enquirer,* Horace Greeley the *New York Tribune,* and William Cullen Bryant the *New York Evening Post.* All three papers became powerful pro-Republican mouthpieces during the upcoming election season. Frémont and his Republican supporters seized on the Thrasher-Humboldt controversy. Frémont hoped that Humboldt's name would make the difference in the upcoming presidential election and in a letter to Humboldt wrote,

In the history of your life and opinions we find abundant reasons for believing that in the struggle in which the friends to liberal progress in this country find themselves engaged, we shall have with us the strength of your name.[58]

Humboldt still fervently desired to see America as an exemplar of democratic ideals. Now, with Frémont challenging Buchanan on an abolitionist platform—one allied with Humboldt's own views—Humboldt threw his vocal support behind Frémont, also known as "the Pathfinder" and "the American Humboldt." Who better to rectify the slavery issue than the explorer whom Humboldt had lauded in 1850 for his contributions to understanding the North American continent, whom he had congratulated in the expanded third edition of *Aspects of Nature,* and upon whom he had arranged to confer the Prussian king's coveted gold medal for scientific endeavors?

Jessie orchestrated much of her husband's political career in the same way she had scripted his expedition narratives. In 1850 Humboldt had written the congratulatory note accompanying the medal presented to Frémont. That letter became the backbone of Frémont's campaign literature. It appeared in the *Republican Scrap Book,* which outlined the new party's platform, in John Bigelow's lengthy biography of Frémont, and in Horace Greeley's pithier pamphlet distributed as campaign literature.[59] The campaign highlighted the final sentence in Humboldt's letter, extolling Frémont's role in bringing California into the union as a free state:

California, which has nobly resisted the introduction of Slavery, will be worthily represented by a friend of liberty and of the progress of intelligence.[60]

Humboldt put the spotlight squarely on Frémont's impassioned support for the abolition of slavery.[61] Although his praise overstated Frémont's role in those negotiations surrounding California statehood, it did accurately capture one strain of the campaign literature promoting his candidacy. Charles Wentworth Upham published a biography of Frémont in time for the campaign, in which he asserted, "His personal history is, in many essential particulars, especially in reference to California affairs, the history of the country."[62] His campaign slogan—"Free men, free soil, Frémont"—was chanted at rallies and printed on small silk banners that could be pinned to a lapel. Versions of it appeared on ambrotype portraits of

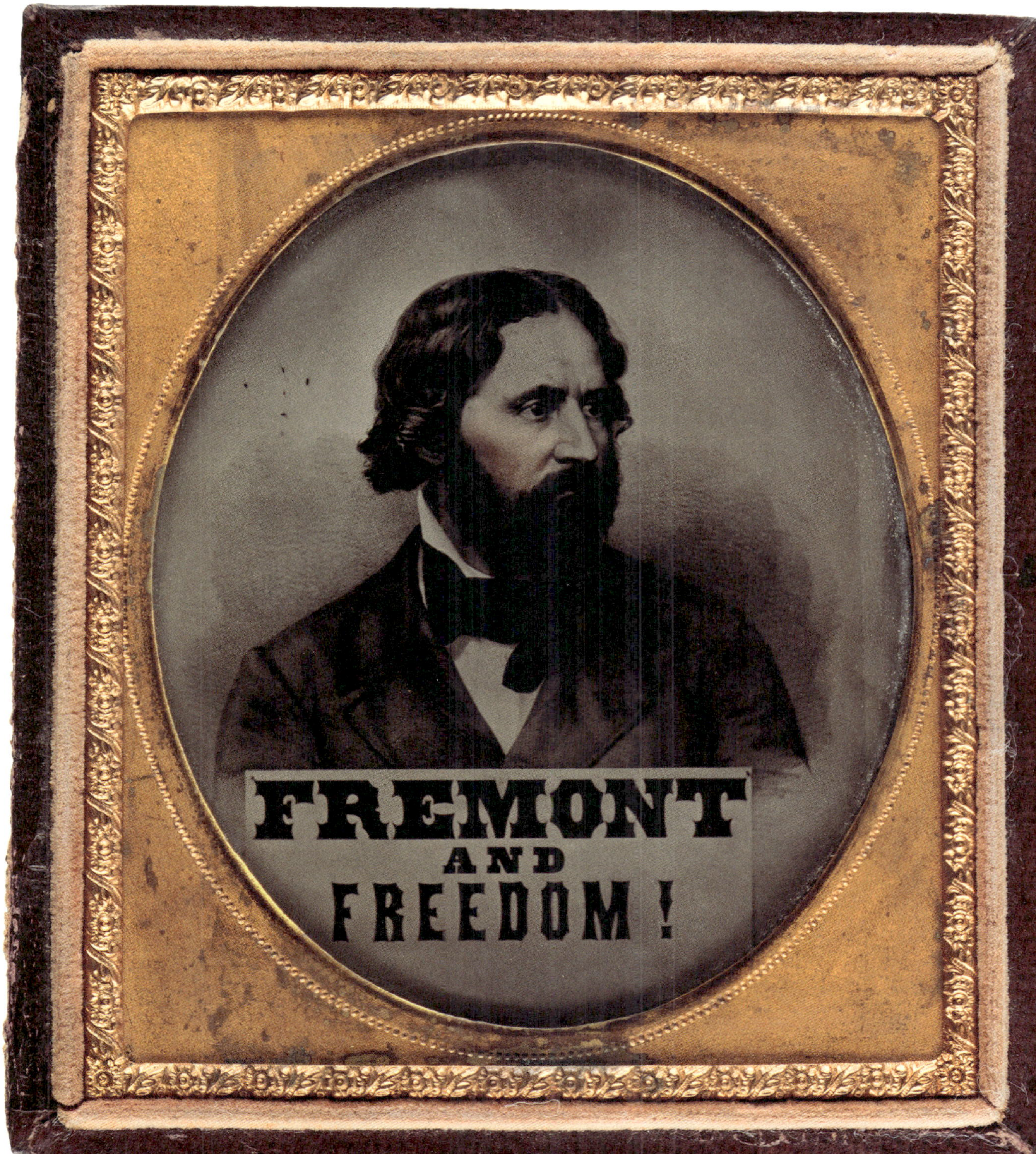
FREMONT
AND
FREEDOM!

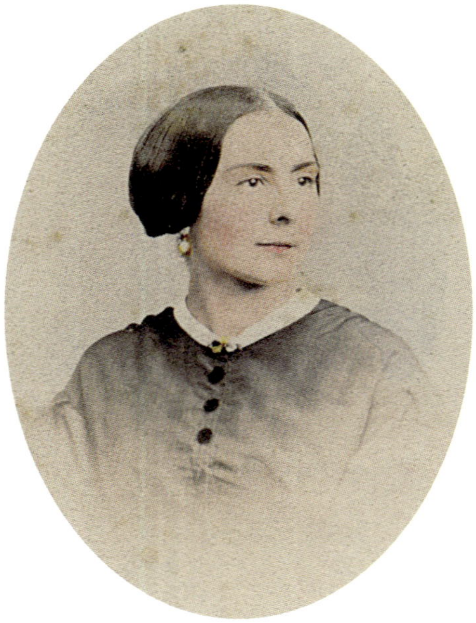

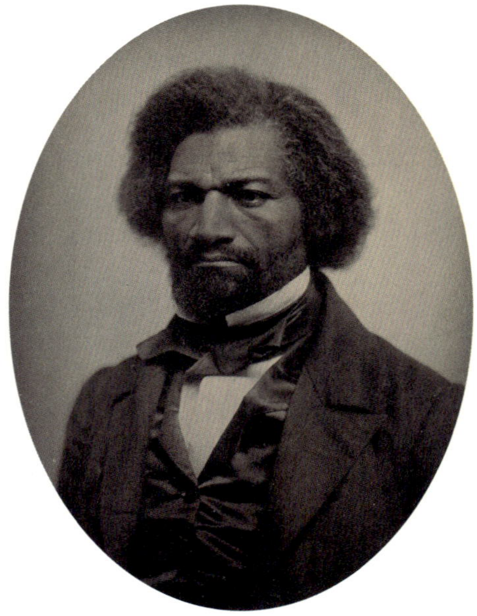

Fig. 68
Unidentified photographer, *Ottilie "Ottie" Assing* (1819–1884), hand-colored photograph, National Park Service, Frederick Douglass National Historic Site

Fig. 69
Unidentified photographer, *Frederick Douglass* (1818?–1895), 1856, quarter-plate ambrotype, 4 3/16 × 3 3/8 in., National Portrait Gallery, Smithsonian Institution; acquired through the generosity of an anonymous donor

the ruggedly handsome candidate (CAT. 45), which deployed the newest photographic techniques to disseminate his likeness. Friends and supporters fanned out to give stump speeches for Frémont, among them Theodore Winthrop, who spent part of the summer in Maine with artist Frederic Edwin Church, delivering outdoor exhortations while Church watched and listened.

Gerolt kept Humboldt apprised of the rising tensions over slavery, remarking, "The slavery question is becoming more alarming from day to day…news is daily coming in from Kansas of bloody conflicts between the free-soilers and the slaveholders." Gerolt had met the Frémonts when he had presented Humboldt's letter and medal to Frémont in 1850. Gerolt also related news of the effectiveness of Frémont's campaign, observing, "his German supporters, many thousands in number, held a mass meeting in his support, and honored him with a splendid torch-light procession in the evening."[63] That support was in some measure due to the network of German-language newspapers scattered across America and one of the most influential writers whose stories were shared among them. Her name was Ottilie Assing (fig. 68), and she was not only a friend of Humboldt's, she was Varnhagen von Ense's niece. Ottilie had honed her journalistic skills in Germany, writing for the influential weekly newspaper *Morgenblatt für gebildete Leser* (literally, Morning paper for educated readers). She arrived in the United States in 1853 on assignment for the paper and the journal *Allgemeine Zeitung*, both of which were owned by Johann Friedrich Cotta's publishing house, which had overseen publication of many of Humboldt's works.[64]

In the spring of 1856, Assing tracked down Frederick Douglass (fig. 69) at his newspaper offices in Rochester, New York. Through Douglass, she met leading American abolitionists, and through her, Douglass met and befriended a number of German American intellectuals in the northern United States.[65] Their professional relationship made them a force to be reckoned with in the wider abolitionist community. That friendship appears to have blossomed into romance, and for the next twenty years their personal and professional lives were intertwined.[66] Assing and Douglass both supported Frémont's candidacy, and the fervor within the German American community likely owes a great deal to more than one hundred articles bearing her byline published during the 1850s in German-language newspapers across the United States. Certainly the extensive German American community around St. Louis was strongly antislavery, their voting record confirming the existence of a small concentration of abolitionists living in a proslavery state.[67]

That support was not enough for the fledgling political party to defeat Buchanan. Nevertheless, Frémont did surprisingly well, and the final numbers were closer than many had predicted.[68] Humboldt's influence appears to have been concentrated among German American communities, which tended to espouse his liberal views, and the pockets of abolitionists who supported the new Republican Party. When Frémont lost the election to "Cuba-mad" Buchanan on November 4, 1856, Humboldt was incensed. Compressing his outrage over the last six months' events in the United States, he wrote to Varnhagen von Ense, "Thus the infamous party, which sells negro children of fifty pounds weight… has triumphed. What a monstrosity!"[69] He had to accept that his hopes for real change in the United States would have to wait even longer.

HUMBOLDT'S MEGAPHONE: THE MASSACHUSETTS ANTI-SLAVERY SOCIETY

Humboldt had caught the attention of an influential group of New England abolitionists who were beginning to contemplate drastic action to push the issue of slavery to a head. For decades the most vocal advocates for doing away with slavery in the eastern United States had been members of the American Anti-Slavery Society, founded by William Lloyd Garrison in 1833. Reverend Theodore Parker had been among the earliest and most fervent among them, delivering his first Sermon on Slavery in 1841. The society and its advocates actively promoted Humboldt's remarks on the abolition of slavery and equality among the races. Beginning in 1844 and throughout the 1850s Garrison's abolitionist newspaper, *The Liberator*, and Frederick Douglass's *North Star* ran articles quoting Humboldt's views mined from his earlier publications, presenting him as an authoritative voice. Humboldt's opinions on Haiti, Webster, the Fugitive Slave Law, and John S. Thrasher's version of *The Island of Cuba* appeared on the pages of these and numerous other abolitionist newspapers and magazines.

In June 1858 Garrison wrote to Parker, asking him to write and deliver an address honoring Humboldt at the Massachusetts Anti-Slavery Society's Fourth of July gathering in Framingham. His intent was to have it printed and sent to Humboldt as a further tribute "to that scientific prodigy and veteran philanthropist."[70] Parker complied, but, because he was unable to deliver the speech in person, it was published in the next issue of *The Liberator* as a "Tribute to Alexander von Humboldt," in which he sketched the accomplishments of his long life. Parker's article claimed,

> *He is the Friend of Mankind, always on the side of Progress, of Humanity. He takes the side of the Indian in North and South America, against his conqueror. He recognizes the natural right of the African, and proclaims it in his early books and his most recent letters.*[71]

Throughout his lengthy address, Parker recounted the history of Humboldt's engagement with American slavery. Finally, he delivered a stirring condemnation of John Thrasher's bowdlerized version of Humboldt's *Political Essay on the Island of Cuba*, enumerating the passages that Thrasher cut from his version and concluding, "Mr. Thrasher has omitted many things of great importance."[72]

Garrison had made one other suggestion in his letter to Parker. "Perhaps you will draw up a resolution, to be adopted by acclamation on the occasion."[73] Parker again acceded. The resolution read,

> *Resolved, That this Society tenders its most hearty thanks to Baron Von Humboldt, for the noble humanity which shines the fairest star in the heaven of his science, and for the faithful testimony he, for fifty years, has always borne against the wrongs of the American slave.*[74]

Parker's public support for Humboldt's position on slavery raises an interesting question. He was a member of the "Secret Six," a group of influential New England abolitionists whose support for John Brown during 1858 and 1859 enabled him to develop his plans for a slave rebellion. Brown's belief in essential human equality begs the question to what extent Brown's support for radical abolition was inspired by Humboldt's views. John Brown (CAT. 46) does not mention Humboldt by name in any of his tracts, but by 1858, he would have had sustained access to Humboldt's views on slavery through multiple channels.[75]

Humboldt did not live to see John Brown's raid on Harper's Ferry, nor the nation's descent into Civil War. But during the 1860s his views on slavery and his discoveries across the natural sciences continued to be published as an authoritative voice, and his standing among those who had revered him in life remained strong. In a tribute to Humboldt, the *New York Tribune*'s Berlin correspondent noted he had "fought against the spread of slavery," maintaining a "strong hope that the great American experiment in self-government would succeed, and mark a new era in the history and progress of man."[76] Following his 1859 death, the *National Anti-Slavery Standard* compiled and published as many notices as possible of Humboldt's remarks on the Fugitive Slave Law, on slavery in general, on Thrasher's book, and Humboldt's letter of support for Frémont.[77]

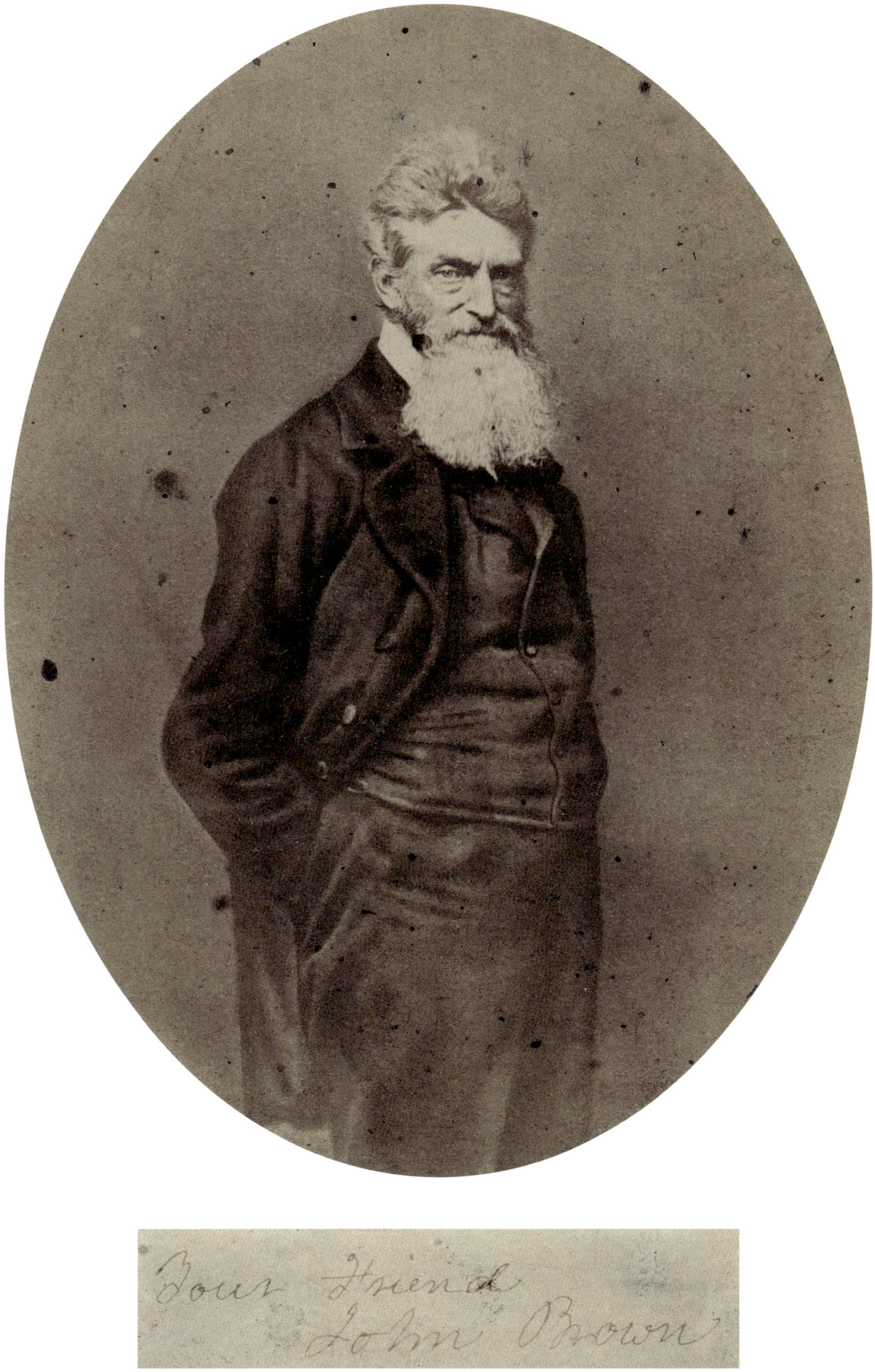
Your Friend
John Brown

CAT. 46 Martin M. Lawrence, *John Brown* (1800–1859), ca. 1858, salted paper print, 7 ⅜ × 5 5⁄16 in., National Portrait Gallery, Smithsonian Institution

FREEDOM AND THE FRÉMONTS' ART COLLECTION

Humboldt's sustained support for John C. Frémont as an explorer and as a political candidate endeared him to both John and Jessie. The Frémonts created a space in their home that became in essence a shrine to Humboldt. The library in their brownstone on West 19th Street in New York City contained a corner étagère on which Jessie Benton Frémont (see cat. 21, ch 2) placed her collection of abolitionist literature, including volumes by her close friend, John Greenleaf Whittier, along with Harriet Beecher Stowe's *Uncle Tom's Cabin*, Wendell Phillips's *Speeches, Letters, and Lectures*, Frederick Law Olmsted's *Journey in the Seaboard Slave States*, Fanny Kemble's *Journal of a Residence on a Georgia Plantation*, and *The Life and Correspondence of Theodore Parker*. In a letter to Whittier, she described the setting, noting,

> *on a projection above the books I put a group of Rogers called the* Slave Auction. *Have you ever seen it? And to finish the idea I have added a piece of a rebel gun taken at Fort Wagner—one of the guns which mowed down the 54th Mass.... Is it not well placed?*[78]

The Frémonts had admired Colonel Robert Gould Shaw, who commanded the Massachusetts 54th Infantry, one of the first regiments of black soldiers. On July 18, 1863, they stormed the battlements of Fort Wagner outside Charleston, South Carolina. Shaw and most of his troops were cut down by Confederate artillery and buried together in a mass grave. The fragment of the cannon served as an emblem of their sacrifice as well as a sober reminder of the war's toll on the country.

The inclusion of John Rogers's *The Slave Auction* (**CAT. 47**) also carried a pointed message. Rogers was best known for his tabletop sculptural vignettes, genre scenes resembling three-dimensional renditions of popular lithographs found in weeklies such as *Harper's* and *Frank Leslie's*, often riffs on bucolic life, or childhood, carrying heartwarming overtones. With the tension mounting over slavery in the 1850s, Rogers turned his attention to more political subjects. In October 1859, the same month as John Brown's raid on Harper's Ferry, Rogers arrived in New York to cast *The Slave Auction*.

> *I think it will be [by] far the most powerful group I have ever made.... I have got a magnificent negro on the stand.... He fairly makes a chill run over me when I look at him.... The auctioneer I have rather idealized and made such a wicked face that Old Nick [Satan] himself might be proud of it—two little quirks of hair give the impression of* horns. *The woman will be as more nearly white and she and the children will come in gracefully. I am entirely satisfied to stake my reputation on it and imagine the present excitement on the subject will give it great popularity.*[79]

Rogers's optimism was short-lived. With Brown's raid, public sentiment turned uneasy, and in Democratic-leaning New York City, there was little enthusiasm for the powerful work. Rogers ruefully noted, "the Slave Auction tells such a strong story that none of the stores will receive it to sell for fear of offending their Southern customers."[80] The sculpture, however, did find a niche among prominent abolitionists; Lewis Tappan bought one of the first copies and spread the word among his friends. Rogers wrote excitedly to his aunt,

> *The Slave Auction is beginning to go like hot cakes.... I have had calls from Dr Cheever & several of the noted abolitionists & others have inquired about me & are going to call—I expect to be in all the papers soon as several have promised to notice it as soon as I send them a group.*[81]

Still, it appears that fewer than thirty castings in plaster were ever made, of which five are known today. That Jessie Frémont secured one for her home and gave it pride of place demonstrates how she displayed art and literature to express her strongly held social and political beliefs as the nation went to war.

The Civil War did not lend itself to direct artistic representation of battles or the evils of

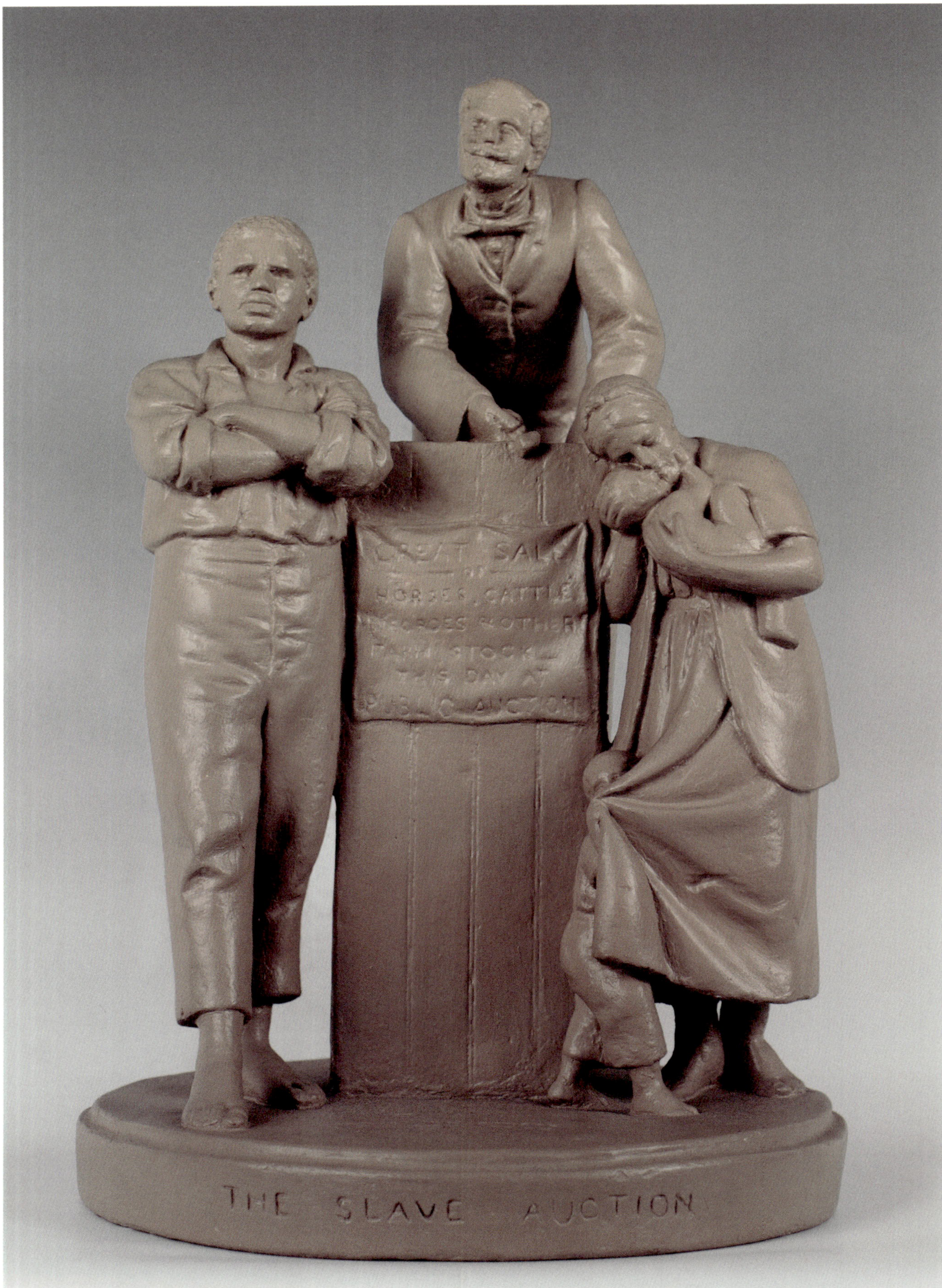
HORSES CATTLE
THIS DAY AT
THE SLAVE AUCTION

CAT. 47 John Rogers, *The Slave Auction*, 1859, painted plaster, 13 ½ × 9 × 8 ½ in., Albany Institute of History & Art, Gift of Mrs. Ledyard Cogswell Jr., from the Benjamin Walworth Arnold Collection

slavery; however, Rogers's groups were among the most widely circulated artworks that addressed the inherent dignity of enslaved blacks. In 1860 the *New York Tribune* asserted, "John Rogers is as familiar a name to Americans as that of the modern martyr, John Brown, and perhaps more so."[82] In February 1864 Rogers debuted another tableau, *The Wounded Scout, a Friend in the Swamp* (**CAT. 48**), which caught Jessie's eye.[83] Rogers described the scene as an escaped slave assisting a wounded soldier, which was inspired by accounts of escaped slaves who found refuge in the Great Dismal Swamp and who were willing to further risk their lives by aiding Union soldiers. The unfolding drama included a snake, which Rogers portrayed "raising its head to strike the negro while he is doing this friendly act."[84] The reference was to the Northern "Copperheads" who supported the Confederacy. Lydia Maria Child, a prominent abolitionist poet and close friend of the Frémonts', admired this work, which she characterized as

> *a stalwart negro guiding a wounded soldier through the swamp. With what brotherly tenderness he holds the fainting Yankee in his arms, and how confidingly the wounded man leans upon him! It brings moisture to my eyes to look upon it; for there is more in that expressive group than the kind negro and the helpless white, put on an equality by danger and suffering; it is a significant lesson of human brotherhood for all the coming ages.*[85]

In June, emboldened by the work's popularity, Rogers sent a copy of the plaster to Henry Ward Beecher and another to President Lincoln, who wrote back,

> *I can not pretend to be a judge in such matters; but the Statuette group 'Wounded Scout'—'Friend in the Swamp' which you did me the honor to present, is very pretty and suggestive, and, I should think, excellent as a piece of art. Thank you for it.*[86]

In 1865 Jessie commissioned a copy of the bronze sculpture, *The Freedman* (**CAT. 49**) from the American artist John Quincy Adams Ward. Ward's sculpture is the single most elegant, compelling work of art commenting on the human condition under slavery. A powerfully built black man, his manacles broken, begins to rise, his eyes uplifted, his expression a strong affirmation of his determination and inherent dignity. Ward had debuted his conception as a two-foot-tall plaster sculpture in the spring of 1863, a few months after the Emancipation Proclamation went into effect. Ward arranged for the plaster to be cast in bronze. The *New York Times* ran an article on the sculpture in which they stated,

> *We welcome with great interest the appearance of Ward's statuette of the "Freedman" in bronze, now on exhibition in Goupil's gallery.... The author of "Art Ideas" says of this work: "It is completely original in itself—a genuine inspiration of American history, noble in thought and lofty in sentiment. It symbolizes the African race in America, the birthday of a new people into the ranks of Christian civilization; we have seen nothing in sculpture more soul-lifting or more comprehensively eloquent. It tells the whole sad tale of slavery and the bright story of emancipation."*[87]

The Frémonts arranged to have their fragment of the rebel cannon melted down and incorporated in the bronze for their cast of *The Freedman*. The artwork thus became a tangible symbol of the sacrifice made by free men of color toward the abolition of slavery in the United States.[88] By the time they installed Ward's *The Freedman* in their home, the Frémonts had moved from the city to an estate near Tarrytown, New York, which they called Pocaho.[89] Hanging alongside the sculpture was a portrait of General Frémont in his military uniform (fig. 70), featuring the sculpture by Ward atop an oak stand, with volumes by ethnologist Henry Rowe Schoolcraft on the shelves below.[90] A pair of portraits by Thomas Buchanan Read also graced the home, painted in 1856 during the presidential campaign (see cats. 20, 21, chapter 2). Jessie had commented at the time that Read "is making a sweet head that resembles me 'as the dew resembles rain'—it is barely founded on fact."[91] Despite Jessie's concern that Read might have idealized her in the portrait, he conveys her resilient charm and

CAT. 48 John Rogers, *The Wounded Scout, a Friend in the Swamp*, 1864, painted plaster, 22 ⅛ × 11 ⅛ × 8 ¼ in., Smithsonian American Art Museum, Gift of John Rogers and Son

CAT. 49 John Quincy Adams Ward, *The Freedman*, 1863, bronze, 19 ½ × 14 ¹¹⁄₁₆ × 9 ⅝ in., Boston Athenæum, gift of Elizabeth Frothingham (Mrs. William L.) Parker, 1922

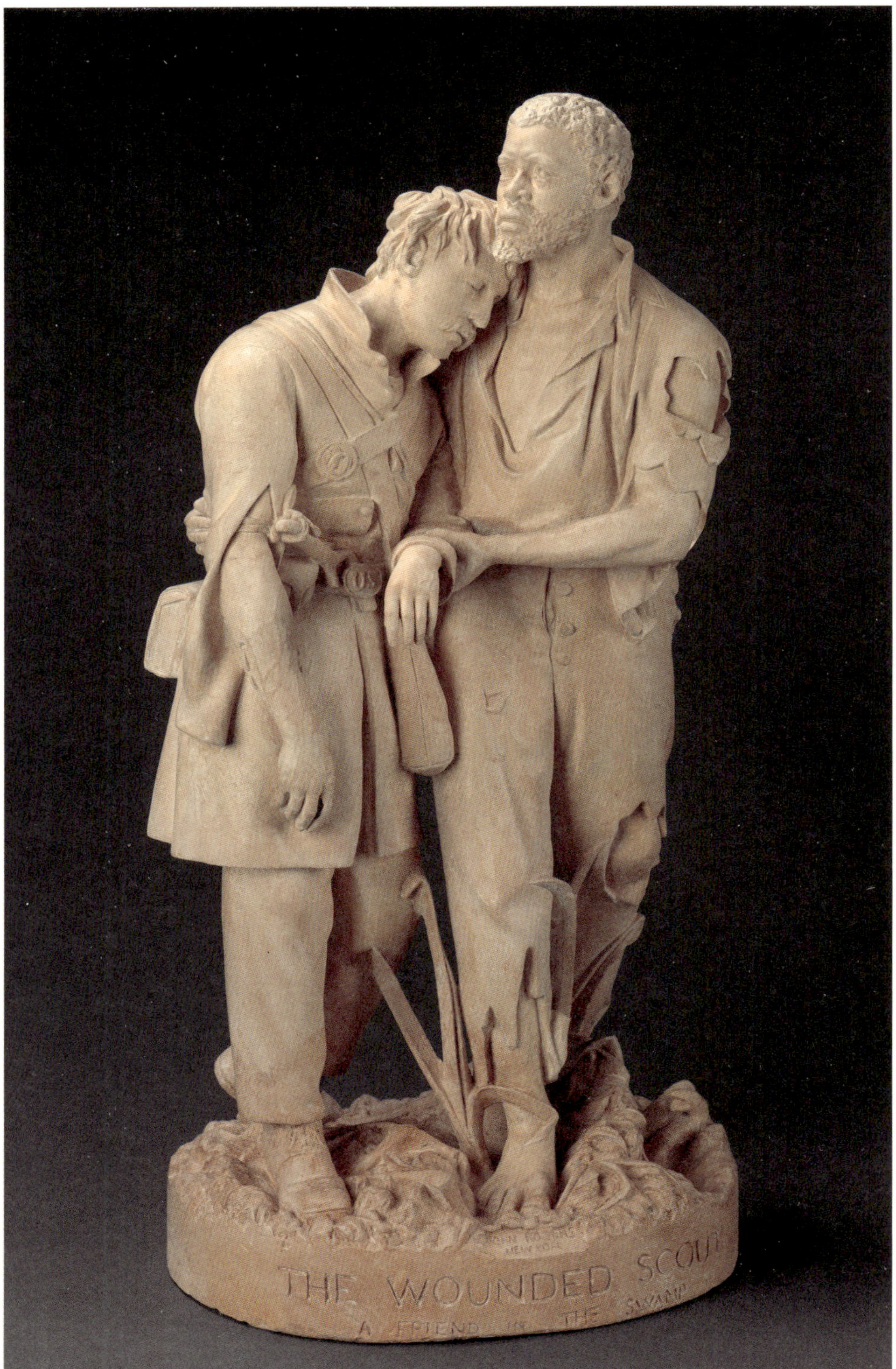

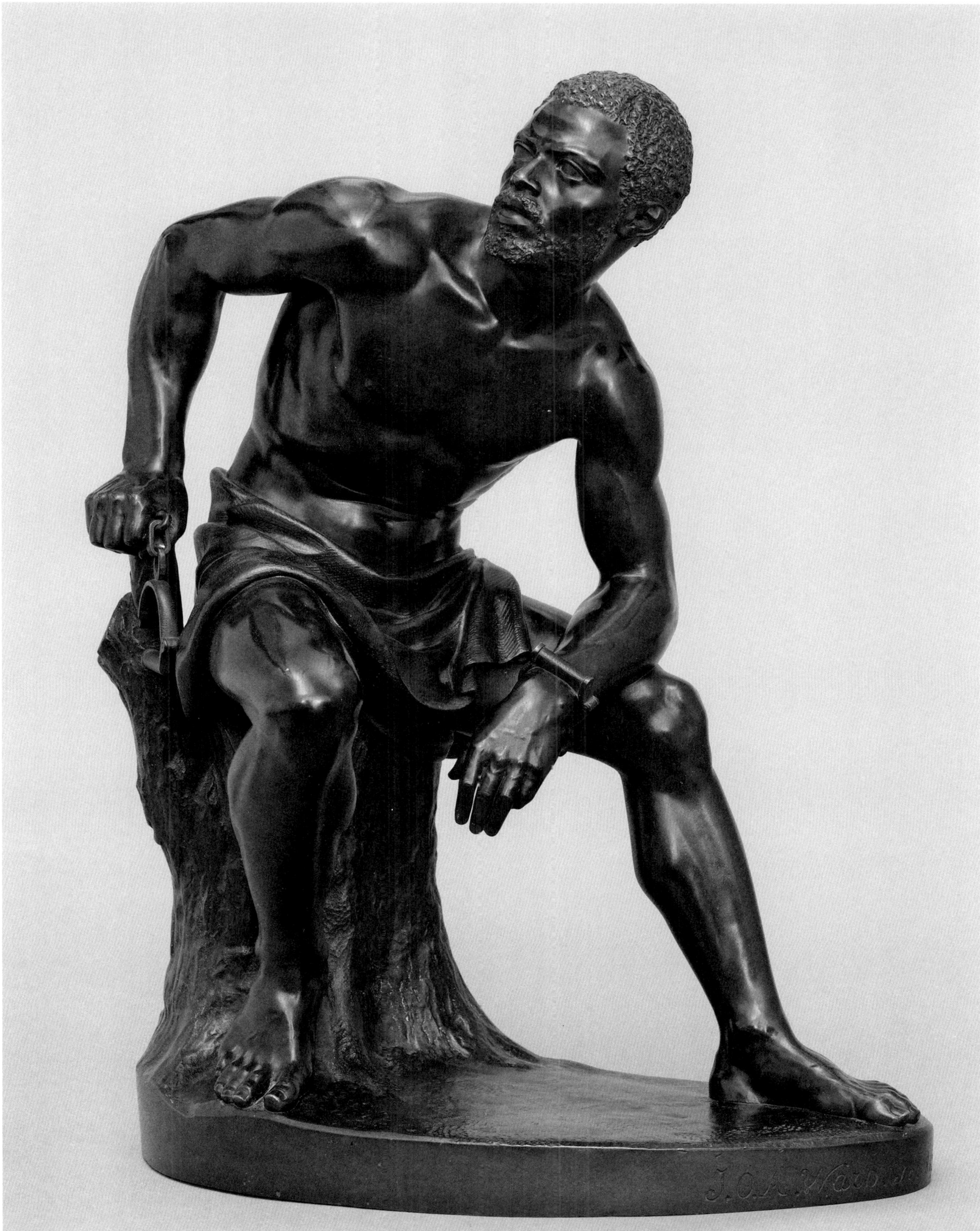
J.Q.A. Ward

CAT. 50 Carleton E. Watkins, *El Capitan*, ca. 1860s, albumen silver print, 25 × 20 in., Smithsonian American Art Museum, Gift of Isaac Lagnado

is equally adept in presenting John in a creditable and positive light. Together, Read's painterly canvases are the most flattering yet accurate portrayals of the power couple.

These antislavery sculptures were the most overt and direct statements of the Frémonts' support for the abolition of slavery. They joined more subtle and metaphorical artistic statements on the quality of freedom: photographs by Carleton Watkins and paintings by Albert Bierstadt, all of which depicted their beloved California. The couple carried these back east with them when they moved to New York. In their minds, California remained a golden dream of prosperity, freedom, and influence. Although explorers and artists had sketched Yosemite's granite features before, it was Watkins who turned a local curiosity into a national icon. Carleton Watkins's views of Yosemite set the itinerary and the expectations for every subsequent artist and visitor. He lugged his large-format camera and glass plates up the steep rock formations and along the valley floor, selecting what would become the canonical views of rock faces, waterfalls, rivers, groves, and meadows.

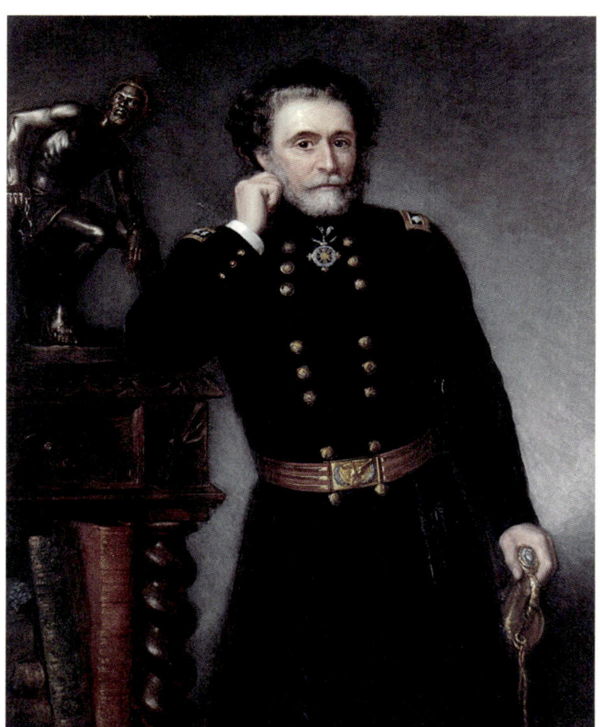

Fig. 70
Giuseppe Fagnani, *Gen. John Charles Frémont*, 1867, oil on canvas, 50 ½ × 41 ½ in., Missouri Historical Society

If Natural Bridge and Niagara Falls had defined eastern natural icons for the new nation, Yosemite offered its own transformative landmarks in the West. The vertiginous face of *El Capitan* (**CAT. 50**) rises straight up from the valley floor like an obelisk, its reflection in the river winding at its base adding height to its already monumental profile. Farther upriver, the plethora of seasonal waterfalls feeding the Merced River made for spectacular vistas. In *Cascade, Nevada Falls, Yosemite, California* (**CAT. 51**) Watkins composed the landscape to feature the ribbon like blur as the water traversed the rocky banks. The giant redwoods proved irresistible yet challenging to photograph. *The Grizzly Giant, Mariposa Grove, Yosemite, California* (**CAT. 52**) provides a steep look up the massive trunk to the gnarled branches and spreading canopy. True to its name, these giant trees conveyed a sense of antiquity not found in any other living form.

Watkins had built his own large-format camera, capable of housing a glass plate measuring eighteen by twenty-two inches, which he dubbed his "mammoth plate" camera. Watkins's mammoth prints echoed Charles Willson Peale's obsession with the presentation of his mammoth skeleton in his Mammoth Room. Watkins's photographs of massive landmarks evoked the impressive and unmatched scale of their fossil namesake, laying claim to the same metaphorical grandeur and importance. The nation had vastly grown in size, but the scale and scope of its ambitions were still writ large in the natural features unique to the country. In competition with large-scale oil paintings, Watkins's large-format photographs asserted both the veracity of their subjects and their claim to national stature.

Albert Bierstadt's appreciation for California had been sparked by seeing Carleton Watkins's mammoth print photographs when they were first shown in 1862 in New York. Like the rest of the nation, he was transfixed by this impressive scenery. For many easterners, this was their first glimpse of the California landscape. The Civil War had made it impossible for him to secure a military escort to the West Coast that year, but in 1863 he arrived

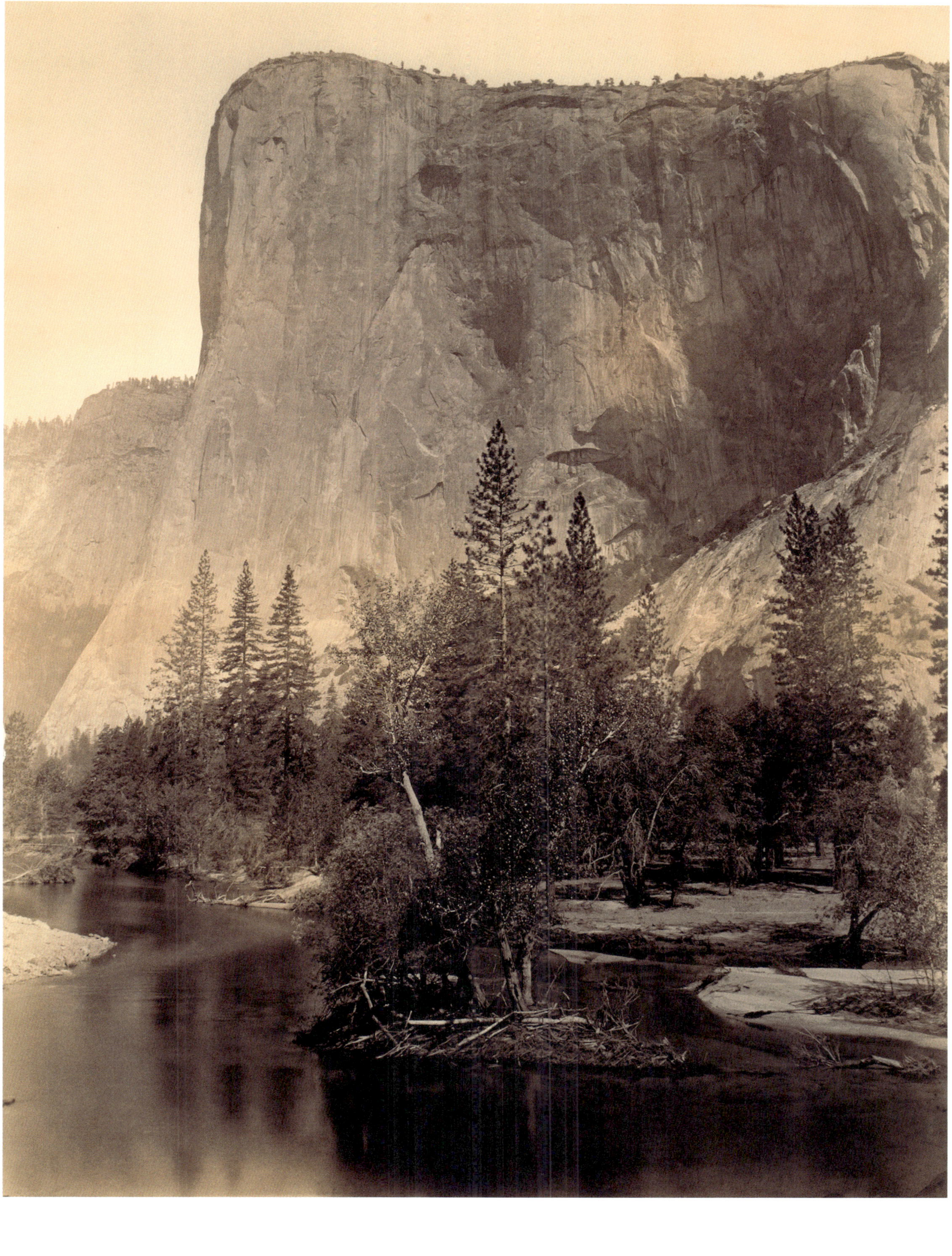

CAT. 51 Carleton E. Watkins, *Cascade, Nevada Falls, Yosemite, California*, ca. 1861, albumen silver print, 15 ⅝ × 20 ⅞ in., Smithsonian American Art Museum, Museum purchase from the Charles Isaacs Collection made possible in part by the Luisita L. and Franz H. Denghausen Endowment

CAT. 52 Carleton E. Watkins, *The Grizzly Giant, Mariposa Grove, Yosemite, California*, 1861, printed after 1875, albumen silver print, 12 × 7 ⅞ in., Smithsonian American Art Museum, Museum purchase from the Charles Isaacs Collection made possible in part by the Luisita L. and Franz H. Denghausen Endowment

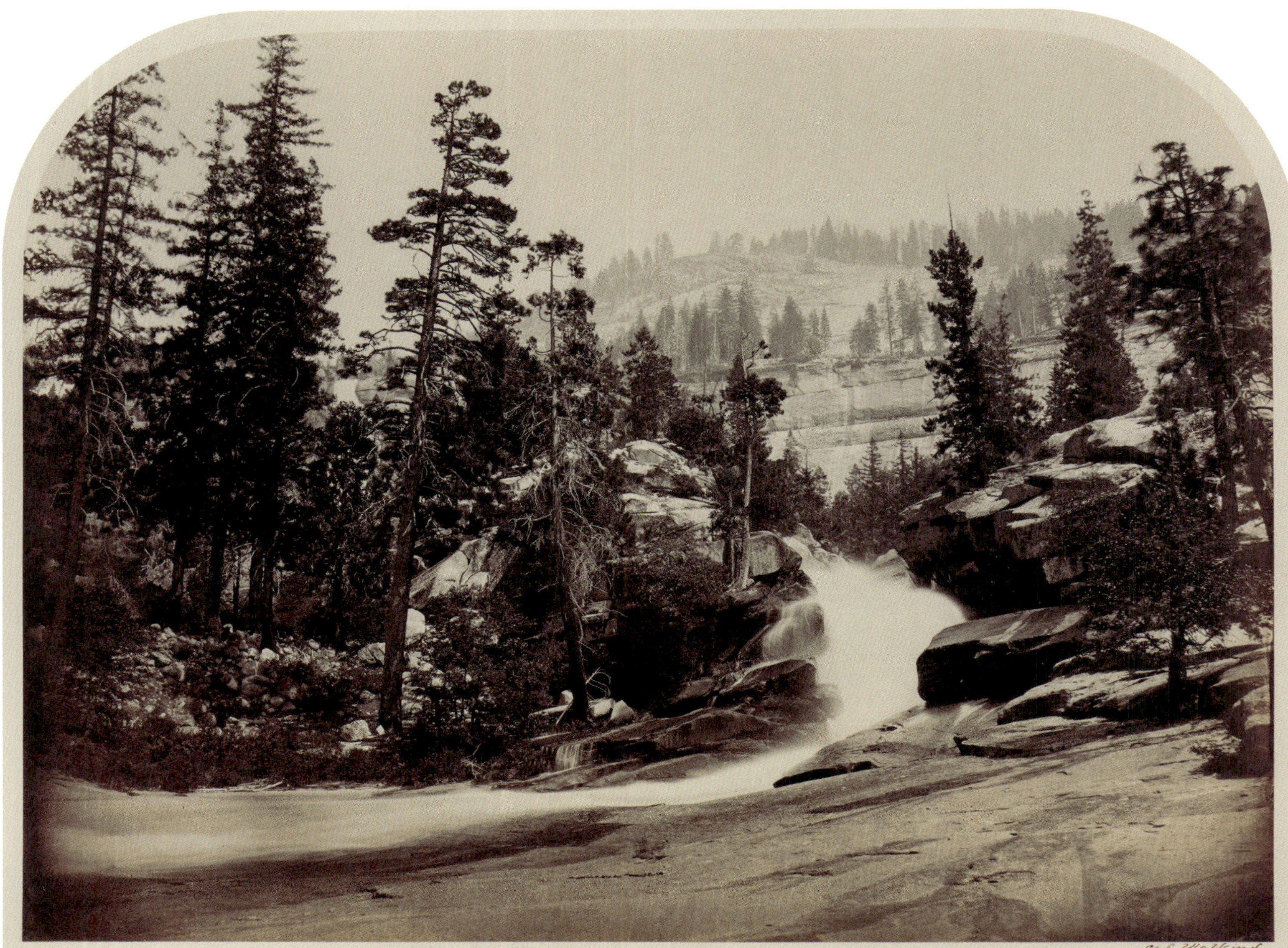

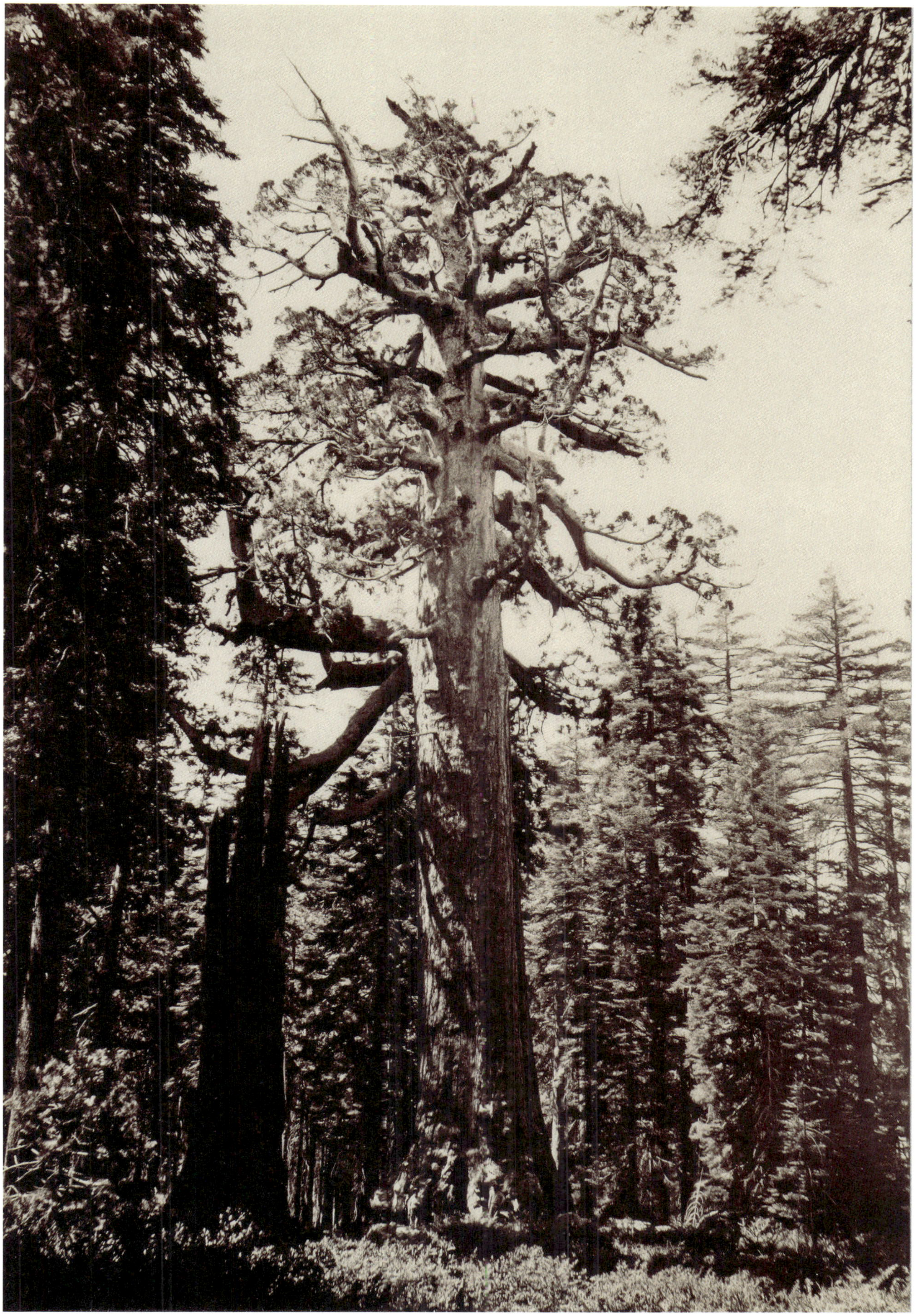

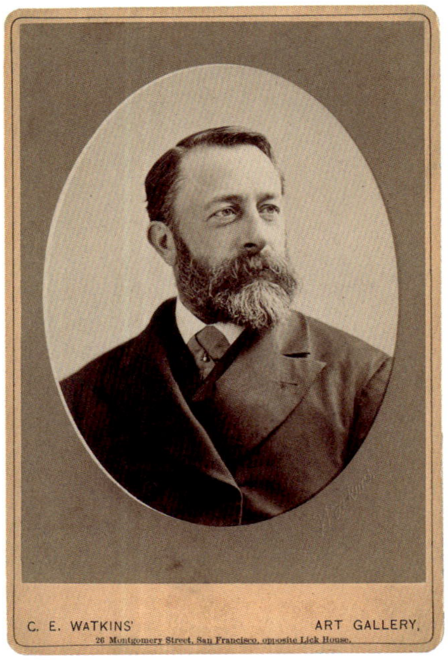

Fig. 71
Watkins-Yosemite Art Gallery, *Photograph of Albert Bierstadt* (1830–1902), 1875–80, carte de visite, $6\frac{5}{16} \times 4\frac{5}{16}$ in., California Historical Society, Carleton E. Watkins photograph collection

and set to work. Bierstadt (fig. 71) recognized in the California landscape an opportunity for him to forge an artistic identity separate from that of the Hudson River school. Among the many sites Bierstadt painted were the iconic views established by Watkins, in several cases hewing closely to the photographer's vantage point and adapting his compositional structure. Watkins's *Yosemite Falls* (CAT. 53) frames the cascade between trees that appear to arch overhead, forming an aperture through which to appreciate the waterfall. The bands of ground cover and grasses in the foreground alternate light and dark, enhancing the sense of distance from camera to rock wall. This quickly became one of Watkins's iconic views and established Yosemite Falls as a primary destination for visitors. Bierstadt's *Cho-Looke, The Yosemite Fall* (CAT. 54) adopts a similar perspective; however, the painter focused on the overwhelming visual effect by adopting a vantage point that emphasizes the monumentality of the rock face and waterfall. The distance between the wisps of clouds hovering near the summit and the stand of trees below makes the height of the cliff seem even higher. On the rocks in the foreground, he included his sketch box and umbrella, lending an air of veracity to what must have seemed like an otherworldly view.

Watkins and Bierstadt portrayed California as a landscape emblematic of freedom. Both men were aware that Thomas Starr King and a small conservation minded group hoped to have Yosemite set aside as a federally protected site. With the nation plunged into a bloody civil war, California became a symbol of the country's future, a potential refuge from the strife and a place in which the nation might bind up its wounds and start anew. Frederick Law Olmsted drafted the proposal, which King promoted, and in 1864, President Abraham Lincoln signed the legislation making Yosemite the precursor to the National Parks. The Frémonts hung several of Watkins's mammoth plate prints and Bierstadt's paintings, first in their California home Las Mariposas, and then in each of their New York homes.

Bierstadt continued to paint scenes of California, making several return trips to Yosemite. Although the artist shared his Prussian birthplace with Humboldt, he seemed less directly affected by the scientist's writings than Frederic Church or contemporary authors like Emerson or Henry David Thoreau. Bierstadt nevertheless channeled one of Humboldt's underlying ideas, that to be able to explore and appreciate nature was a liberating experience, one of unfettered possibility, a kind of freedom denied to those in bondage. In expressing "nature is the domain of freedom," Humboldt had settled on a metaphor that resonated with the artists who depicted the California landscape and claimed it as emphatically American.

THE PATRIOTIC LANDSCAPE

Jessie Benton Frémont had deliberately amassed a socially conscious art collection, understanding that fine art could be a powerful voice for a kind of "cultural unionism" during the Civil War.[92] In a letter to her close friend, King, Jessie wrote that she had been asked to serve on the Arms and Trophies Committee of the Sanitary Commission. The commission, established by a group of civic-minded New Yorkers, set about organizing the largest exhibition of art, artifacts, and other evidence of cultural prowess ever seen in the United States, which was ultimately called the Metropolitan Fair of 1864.[93] The endeavor was enormous, with far-reaching implications for the founding of the Metropolitan Museum of Art in 1870 and the American Museum of Natural History four years later. In the moment, the fair was designed to instill pride in Northern citizens that theirs was a culture worth protecting and celebrating. Furthermore, its various auctions and sales raised more than one million dollars to bolster the war effort.[94]

The Park Avenue Armory exhibition hall, which opened to great fanfare on April 4, 1864, served as the picture gallery and was dominated by three paintings. Emanuel Leutze's massive *Washington Crossing the Delaware*, a painting that had inspired Humboldt as well as countless Prussians to support the 1848 revolutions that swept Europe, anchored the rear wall. It was flanked by Albert Bierstadt's *The Rocky Mountains, Lander's Peak* and Frederic Church's *Heart of the*

Andes (fig. 72), which was displayed draped in velvet and surmounted by portraits of the first three American presidents. This triumvirate of large-scale paintings sent a stirring message of patriotism. Leutze's monumental painting threatened to dwarf everything else in the cavernous hall, the greater-than-life-size presentation of George Washington on his daring river crossing a reminder of how Americans had beaten the odds to establish the nation. Bierstadt's Colorado scene included a mountain named for Colonel Frederick Lander, a recent casualty of the war. Set in the American West, it presented a landscape associated with the nation's future. Church's *Heart of the Andes*, while not a North American view, was the painting Church had created as an homage to Humboldt and was considered among his crowning achievements.[95] Its presentation, surmounted by portraits of the first three presidents, recalled Peale's display of the portraits of eminent Americans in his museum. These were not among the works amassed for sale; they were gathered as a metaphor for the cultural prowess of the northern United States. All three featured the landscape as the means of establishing the setting and the tone for heroic activity—defeating the British army, expanding the country farther west, and taking that exploratory urge global. As metaphorical evidence of patriotism, they presented triumphant visions of hardships overcome. As exemplars of American ambition, they were an inspiration for all.

Fig. 72
Unidentified photographer, *Heart of the Andes*, by Frederic Edwin Church, as exhibited at the Metropolitan Sanitary Fair (detail), 1864, half of a stereograph, New-York Historical Society

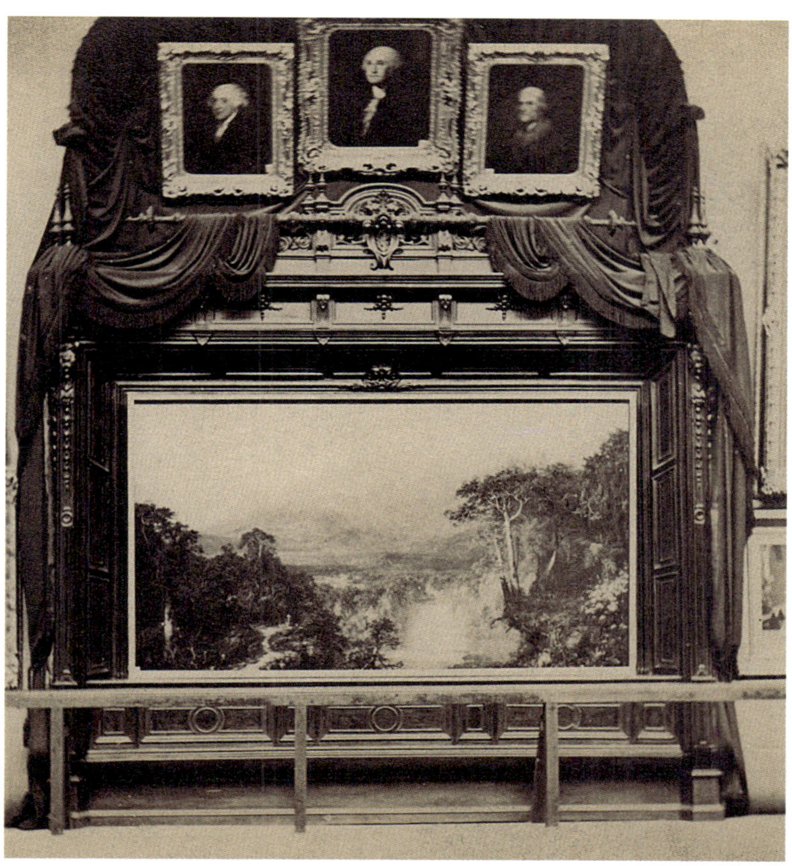

Humboldt's name runs like a thread in the fair's tapestry, extending his humanitarian influence well beyond his lifetime. A photograph of the explorer was one of the items donated for sale, as were a set of "half a dozen long letters of Humboldt."[96] Moreover, the organizers distributed a book titled *The Spirit of the Fair*. The lengthy description of the trauma facing the nation and the reason for the Fair made reference to Humboldt's vision for freedom and human rights in the United States as a primary justification for the war itself. The authors noted, "Long ago Humboldt and other naturalists recognised in the fact that this continent is placed between two oceans, the provision and pledge of a grand destiny," a destiny that was threatened by the slave interests that violated the promise of nature and geography to pursue illegitimate aims.[97] Once again, abolitionists built their support for the war effort around Humboldt's affirmation that American geography was inextricably linked to its cultural destiny. The promise of nature is the promise of liberty—as Humboldt had described it—in both the philosophical and aesthetic senses. In Enlightenment philosophy, natural law confers individual freedom, which should be acknowledged and protected by governments. The Metropolitan Fair promoted the concept that nature is the domain of freedom in both contexts—in its human rights thesis and in its choice of artworks.

Although Jessie's purview included soliciting memoirs of Union soldiers who had died in battle, she played a role in the display of artworks selected as well. She was particularly concerned that Bierstadt's small but exquisite painting *Valley of the Yosemite* (CAT. 55) be appreciated properly, given its small size and how unfamiliar its subject was to an eastern audience. To Horace Greeley she implored,

> *So few people have seen the grand scenery of the Yo Semite that it needs a little explaining. You who have been there and know how much endurance it takes to overcome those weary miles will be among the few who can fully appreciate not only this picture but all that went to its making.*

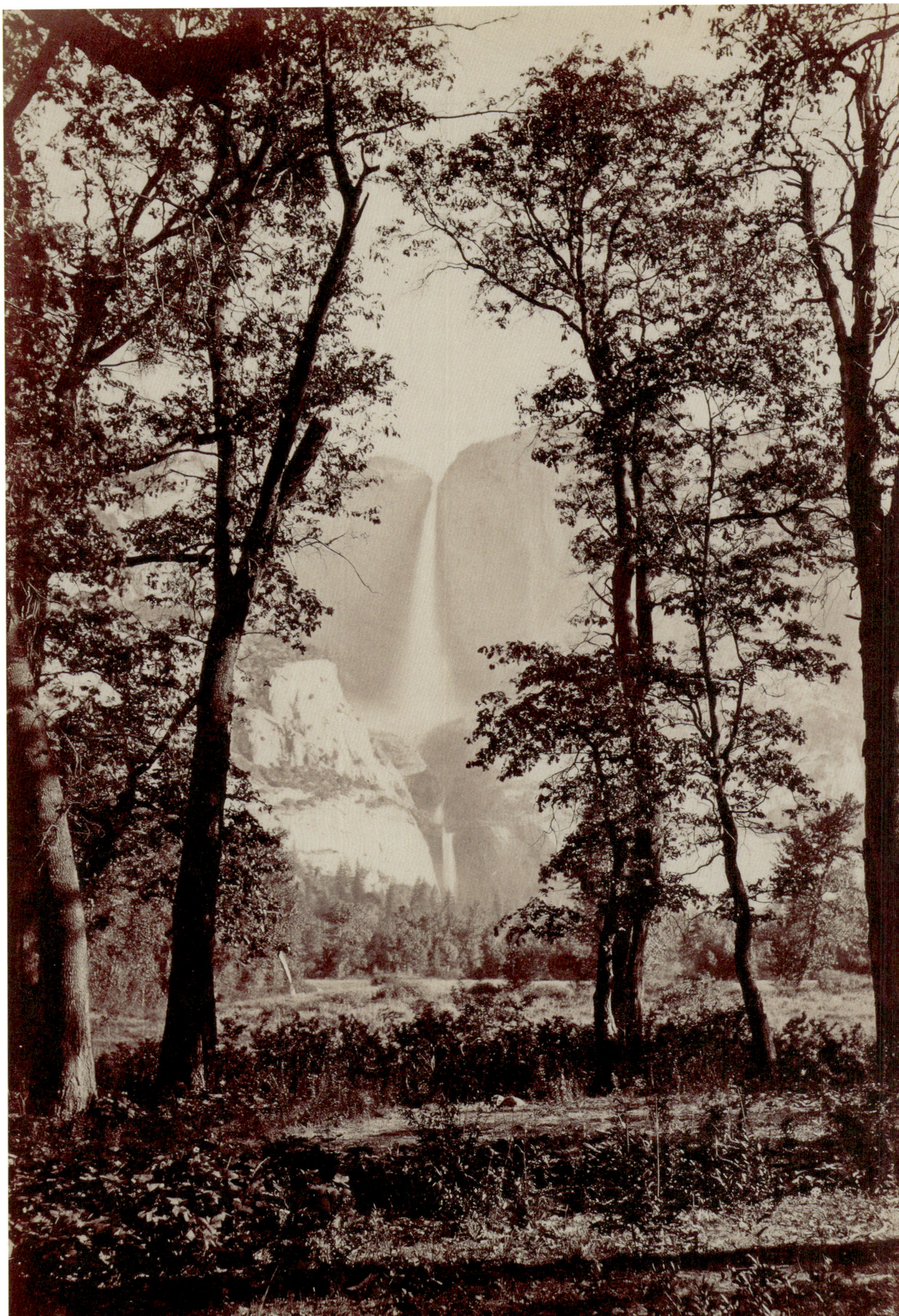

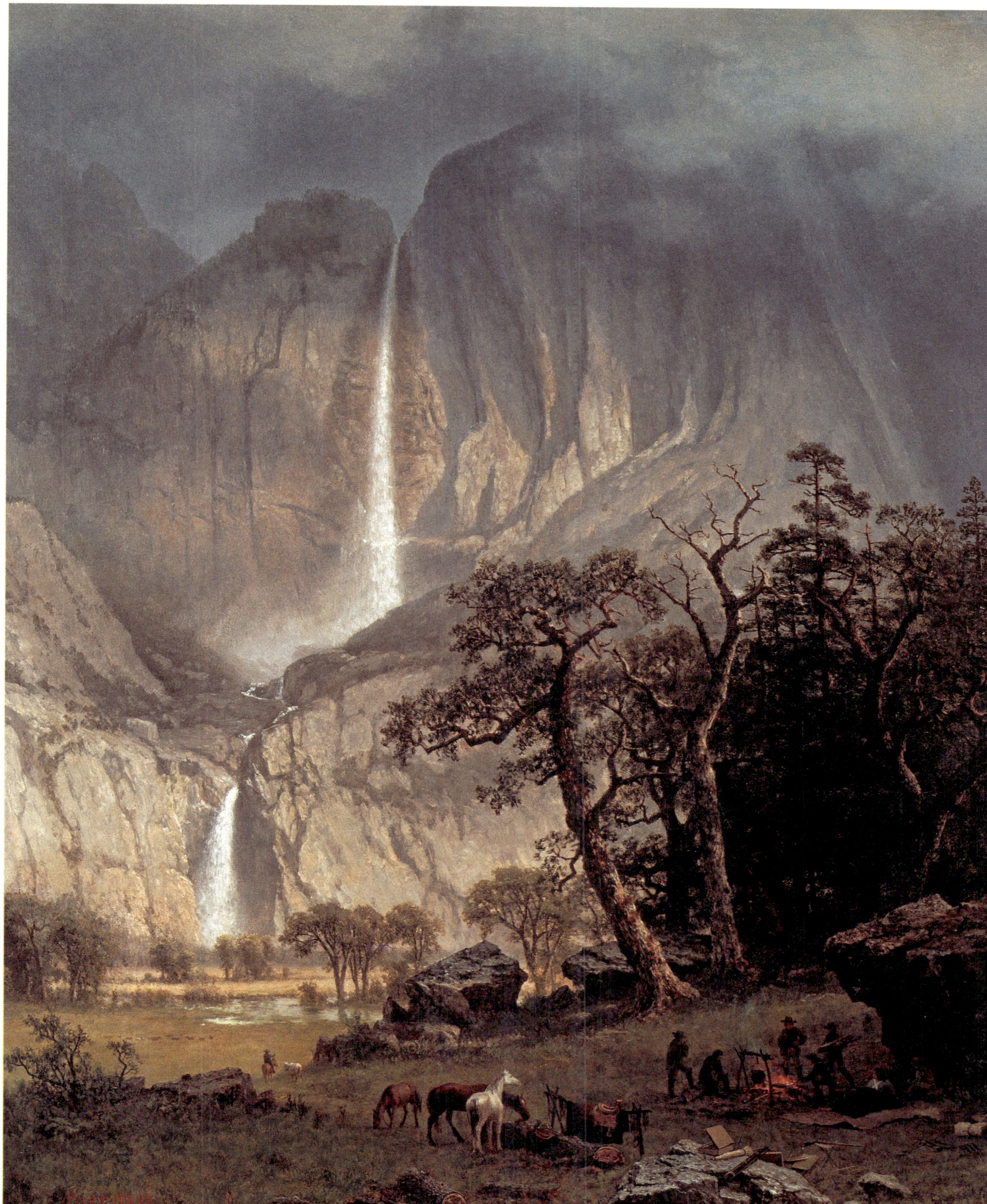

Preceding pages

CAT. 53 Carleton E. Watkins, Isaiah West Taber, *Yosemite Falls*, ca. 1865–66, printed after 1875, albumen silver print, 12 × 8 in., Smithsonian American Art Museum, Museum purchase from the Charles Isaacs Collection made possible in part by the Luisita L. and Franz H. Denghausen Endowment

CAT. 54 Albert Bierstadt, *Cho-Looke, The Yosemite Fall*, 1864, oil on canvas, 34 ¼ × 27 ⅛ in., Timken Museum of Art, Putnam Foundation

Greeley had visited Yosemite in 1851, and had written a popular travelogue published in the *New York Tribune* after his return.[98] Jessie continued, "I wish you would come to the private view—it's, in a manner, my reception, and I shall look for you." In the following sentence, she made absolutely clear what her expectations were: "I shall look too to find such a notice in the Tribune as will bring in money to the object for which it is exhibited & for which I know you are kindly interested." To remove all doubt, she closed, "To give some points, I enclose a rough memoranda which may be of service to some one who does not know the subject so well as I do."[99]

When Jessie set her mind to something, be it politics or fundraising, she did not do things halfway. She and her husband had collected Bierstadt's paintings of California as mementos of a cherished landscape and as symbols of a free state. She made sure that Bierstadt's small oil painting of Yosemite got the attention she desired. Bierstadt's painting was the first painted view of Yosemite to be exhibited on the East Coast; despite its size, it sold for $1,600—the highest price paid for a single work of art sold at the fair.[100] What Bierstadt presented in this sun-drenched landscape was nothing short of a vision of California as a place untouched by war. In 1863 Bierstadt had written to Lincoln's personal secretary John Hay, "we are now here in the Garden of Eden I call it," his prose conveying the awe he felt in a landscape King had asserted had no equal on the planet. Quoting Greeley, King had proclaimed, "I know of no single wonder of nature on earth which can claim a superiority over the Yosemite."[101] In this landscape one could believe it was possible to survive the war and start over. In this sense, California became a postwar opportunity for a New Eden.

As plans for the Metropolitan Fair coalesced, another of Humboldt's advocates developed his own contribution to the cause. In February 1864 New York merchant John Matthews produced a two-sided photographic *carte-de-visite* with the heading "Published for the benefit of the Sick and Wounded Soldiers in the U.S. Armies." Beneath this heading Matthews printed a facsimile of a letter he had received from Humboldt decrying Daniel Webster's support of the Fugitive Slave Law (**CAT. 56**). On the other side, Matthews had typeset an English translation of Humboldt's indecipherable script. Humboldt's letter read

> *I have the warmest attachment to your beautiful and liberal city, New York, but have earnestly and deeply regretted that WEBSTER, whom I long respected, more than favored that* shameful *law which still persecuted colored men after they had regained by flight their natural inborn liberty, of which they had been robbed by Christians.*

Humboldt's message in support of a just society, free of slavery, became part of the American war effort, a message for the common man. Matthews sent a copy of this card in a letter to President Abraham Lincoln. Therein, Matthews wrote,

> *I am aware that the opinions of Humboldt upon slavery cannot be unknown to you but, as there are moments in the life of every one when doubts intrude upon the path of duty,—perhaps oftener when grave affairs of state involving the safety of nations are committed in trust to a single man.—In such a moment I dare venture to hope that this opinion of the good and wise Humboldt may not be without its influence on the progress of human freedom, with which you are honorably identified.*[102]

Matthews's card presented Humboldt's words to inspire Northern men and women of the just nature of the Union cause. Even after his death people invoked Humboldt's name and his humanitarian philosophy to influence an American president. Humboldt would have been pleased to have been mentioned alongside Lincoln in the fight

CAT. 55 Albert Bierstadt, *Valley of the Yosemite*, 1864, oil on paperboard, 11 ⅞ × 19 ¼ in., Museum of Fine Arts, Boston, Gift of Martha C. Karolik for the M. and M. Karolik Collection of American Paintings, 1815–1865

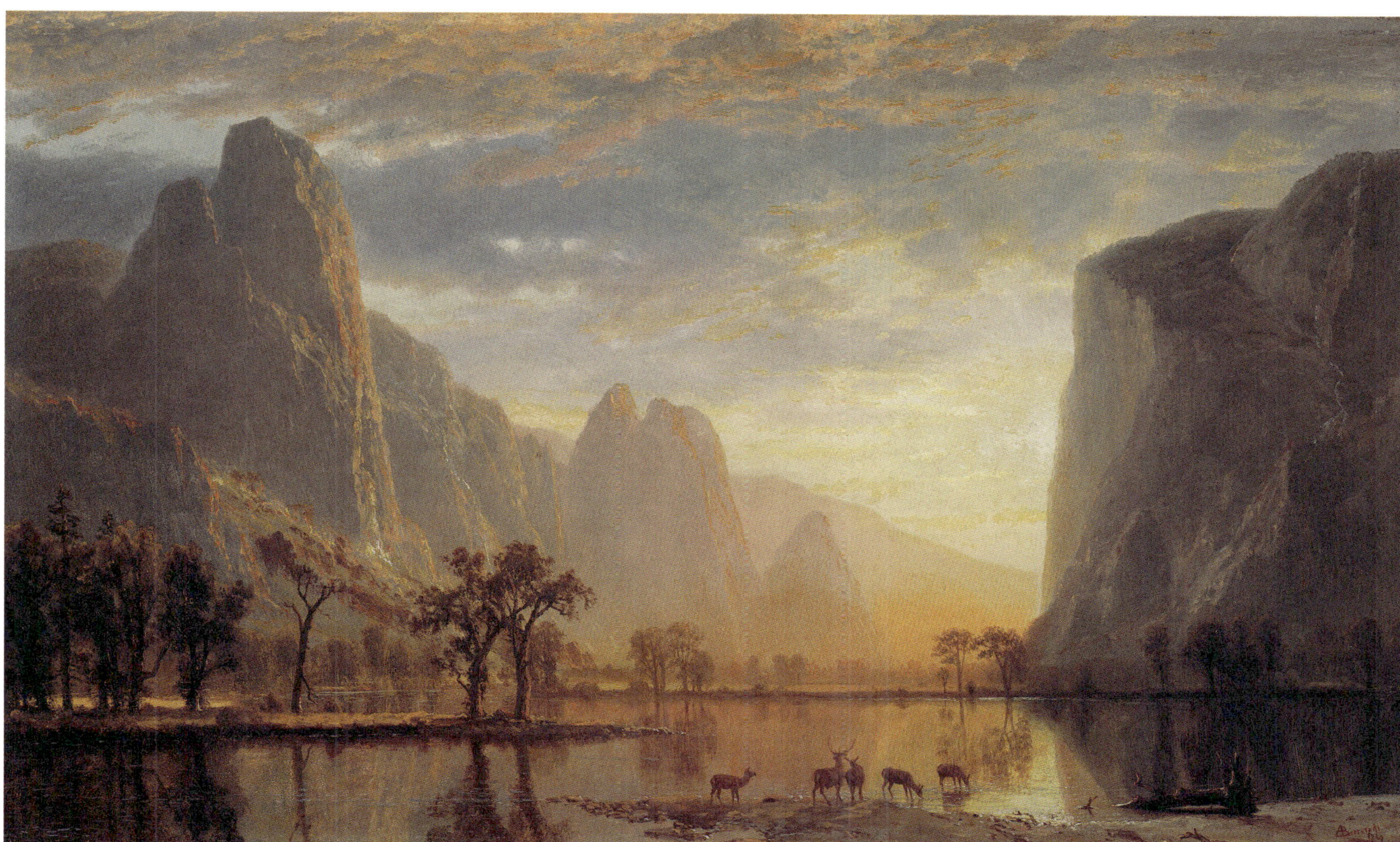

CAT. 56 (recto and verso) John Matthews, *Alexander von Humboldt on the Fugitive Slave Law, Oct. 12, 1858*, 1864, ink on paper, 4 ¾ × 2 ⅝ in., Library of Congress, Manuscript Division, Abraham Lincoln Papers

CAT. 57 E. Haines, *Alexander von Humboldt*, 1858, albumen silver print, 7 ³⁄₁₆ × 5 ⅝ in., National Portrait Gallery, Smithsonian Institution

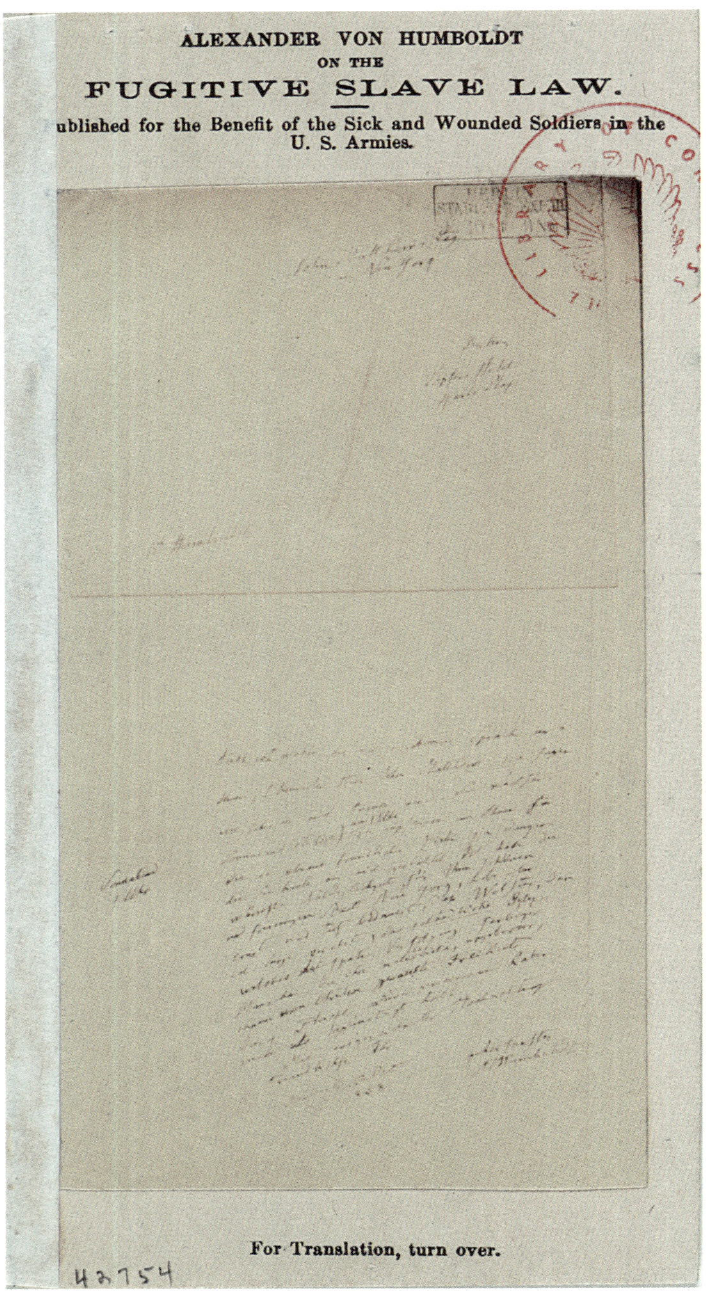

ALEXANDER VON HUMBOLDT
ON THE
FUGITIVE SLAVE LAW.
ublished for the Benefit of the Sick and Wounded Soldiers in the U. S. Armies.

For Translation, turn over.

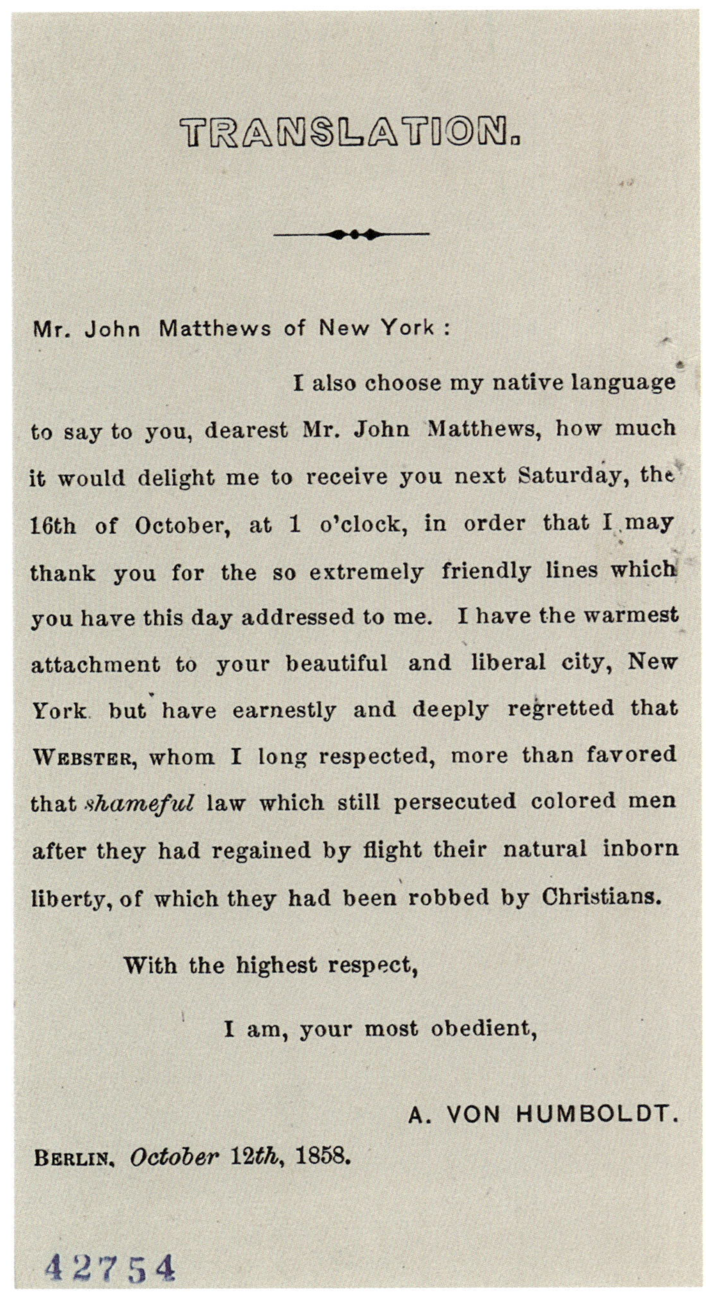

TRANSLATION.

Mr. John Matthews of New York :

I also choose my native language to say to you, dearest Mr. John Matthews, how much it would delight me to receive you next Saturday, the 16th of October, at 1 o'clock, in order that I may thank you for the so extremely friendly lines which you have this day addressed to me. I have the warmest attachment to your beautiful and liberal city, New York. but have earnestly and deeply regretted that WEBSTER, whom I long respected, more than favored that *shameful* law which still persecuted colored men after they had regained by flight their natural inborn liberty, of which they had been robbed by Christians.

With the highest respect,

I am, your most obedient,

A. VON HUMBOLDT.

BERLIN, *October 12th*, 1858.

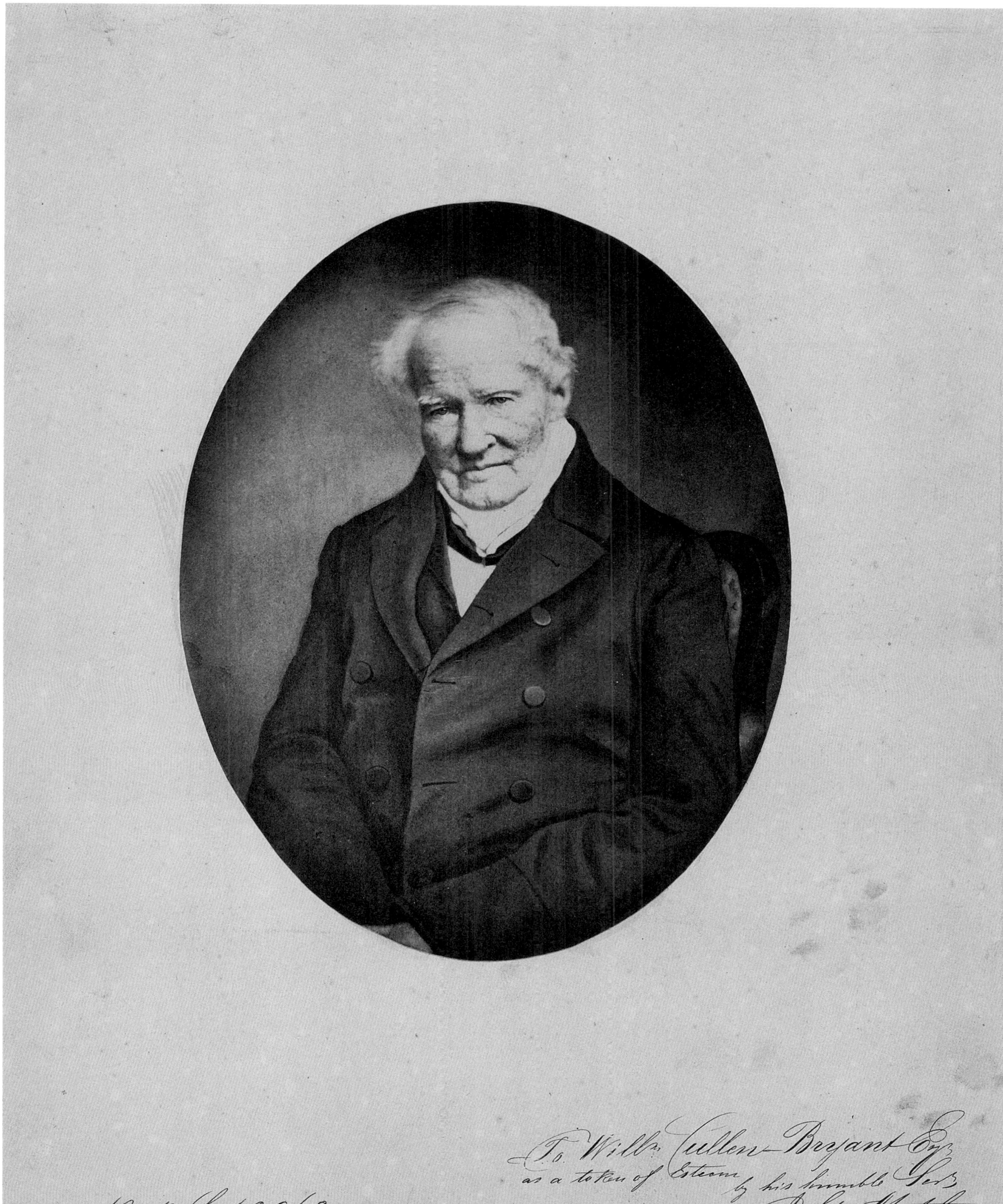
To Willm Cullen Bryant Esq
as a token of Esteem by his humble Servt
N York Sept. 29/58

to abolish slavery and gratified to have his words distributed to inspire the Union troops. Heralded as a champion of liberty for all, Humboldt was more successful than he might have seemed at the end of his life.

Humboldt made a guest appearance at another event celebrating a prominent voice for the abolition of slavery. In November 1864 the cultural elite of New York met at the Century Club to celebrate William Cullen Bryant's seventieth birthday. Bryant's friends had planned for this day months in advance, creating an album of tributes that extolled his contributions to and support for the arts and praised his editorship of a leading newspaper. Poems and letters were interleaved with small sketches by forty-six of Bryant's friends in the New York artistic community. The range of material was broad, but there was a strong current of appreciation for Bryant's solid support for abolition.

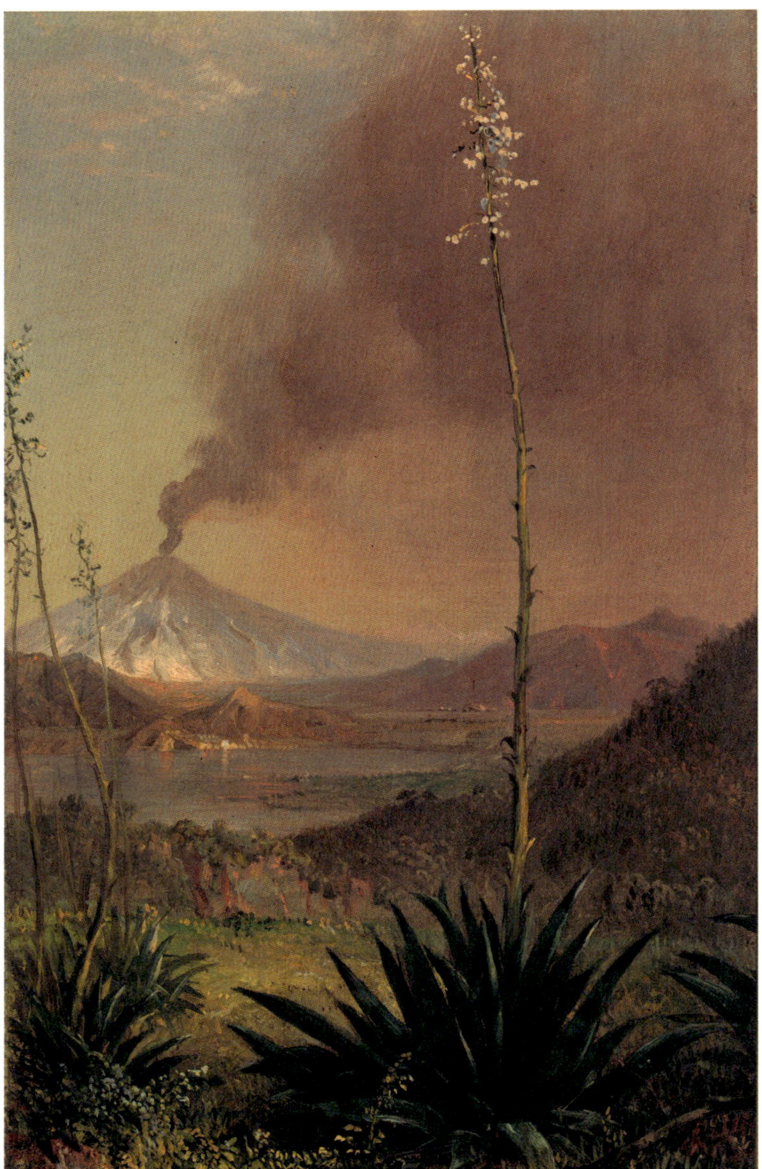

Fig. 73
Frederic Edwin Church, *A Century Plant in Bloom*, ca. 1864, oil on paper mounted on canvas, 7 5/8 × 4 7/8 in., The Century Association, New York, William Cullen Bryant Collection, Gift of the estate of Julia Sands Bryant, 1908

Among the painted sketches in the Bryant Album was a South American landscape by Frederic Edwin Church (fig. 73). A century plant anchors the foreground, a visual pun for the Century Club, of which they were both members. In the background, Cotopaxi smokes away, just as it had when Humboldt visited in 1802, and Church had witnessed five decades later. Humboldt had described the rumbling of volcanoes and the tremors associated with earthquakes as metaphors for social and political upheavals in multiple volumes of his *Personal Narrative*. Simón Bolívar had picked up the reference as a portent for the South American revolutions against Spanish rule, pronouncing

a great volcano lies at our feet.... Who shall restrain the oppressed classes? The yoke of slavery will break, each shade of complexion will seek mastery.

Closer to home, in 1861 Frederick Douglass had warned that American slavery was a "moral volcano" waiting to erupt.[103] Both Bryant and Church knew that Humboldt had admired Cotopaxi for its perfect cinder cone and propensity for erupting. By 1864 the volcano had become so closely associated with throwing off political oppression that implicit in Church's choice of subject was their shared advocacy for abolition. More directly, Bryant received a photograph of Humboldt (CAT. 57), inscribed to him from theatrical writer John Guido Methua, commemorating this milestone with a memento he knew the aging transcendentalist would appreciate. Humboldt's views of nature had been a foundational element in the development of Bryant's landscape poetry, as were his close friendships with many of the leading landscape painters. Like Matthews's letter to Lincoln, Methua's gift of Humboldt's photograph served as an appreciation for Bryant's contributions to that vision of nature and of a united America.

From the moment Humboldt set foot in the United States, he had been agitated about American slavery. He encouraged American adulation of his ideas; he responded to Americans' need for his approbation, but it was on the issue of abolishing slavery that America drew the line. Only by appreciating just how assiduously American politicians and intellectuals had cultivated Humboldt's approval can we begin to appreciate what that rupture meant. Without that perspective, Humboldt seems like just another abolitionist who wished we were different. However, this issue takes on disproportionate significance as we comprehend Humboldt's deep investment in what the nation did and what it stood for. His repeated pleas in American newspapers were intended to sway U.S. policy. In the final months of his life Humboldt did not ease up on this pressure. He continued to give interviews despite the demands on his time in order to intensify his message. His outspoken and principled stance on human rights became as important in the United States as his legacy in the sciences. Eulogists made sure that Humboldt's beliefs in the essential equality of all peoples remained a significant argument for the abolition of slavery, even as the United States descended into a civil war fought over this principle.

For fifty years, abolitionists in the United States retained their stubborn optimism that America could realize its best self by embracing equality and creating the kind of democracy that Humboldt envisioned. For all of Humboldt's sarcasm and frustration with emperors and politicians, he never abandoned his belief in a society that could live up to the ideals expressed by Jefferson and that he cherished himself. That faith in America's potential as an example for the rest of the world might have been idealistic, but it was rooted in his unwavering belief in the essential equality among all races. Over time his tone became more strident as he realized the limits of his influence, although his support for what he believed the United States could be—an example of a true democracy that lived up to its ideals—never wavered. The events leading up to the Missouri Compromise and the Ostend Manifesto, along with John Thrasher's attempts to twist Humboldt's words in support of annexing Cuba as a slave state, tried his patience nearly to the breaking point. His support for Frémont's failed bid for the presidency stands as his final public act; however, despite these setbacks, Humboldt remained an advocate for America's better nature.

NOTES

1 "Je compte sous peu me rendre à Washington moi même et jouir du plaisir de Vous présenter les assurances de ma profonde vénération. Engagé dans une Expédi- [2] tion que j'ai entrepris depuis 5 ans á mes propres fraix, pour le progrès des sciences physiques, après avoir parcouru les Royaumes de la Nouvelle Grénade, et de Quito, la Guayane, les pays Indiens situés entre l'Orénoque, le Rio Negro, et l'Amazone, le Pérou, Venezuela et le Mexique je compte retourner en Europe pour y publier les fruits de mes travaux et ceux de mon ami le Citoyen Bonpland, qui m'accompagne. Pénétré de l'intérêt le plus vif pour la prospérité de l'espèce humaine sur le Sol des États unis, fruit de Votre sage législation et des vertus civiques de Vos Magistrats, j'ai pris la route de ce pays-ci pour me rendre en France. C'est une idée bien consolante pour moi qu'après avoir été témoin des grands phénomènes que présente la Nature magestueuse de la Cordillère des Andes, qu'après avoir vu ce qui est grand dans le monde physique, je puisse jouir du Spectacle moral que présente un peuple libre et digne de sa belle destinée. Vous, Monsieur, qui en homme d'état avez eu une influence si bienfaisante sur cette destinée, recevez les hommages du respectueux attachement avec lequel j'ai l'honneur d'etre Monsieur, A. Humboldt." Humboldt to Madison, Philadelphia, 24 May 1804, in Ingo Schwarz, ed., *Alexander von Humboldt und die Vereinigten Staaten von Amerika: Briefwechsel* (Berlin: Akademie Verlag, 2004), 90–91, letter 4. Translation by Helmut de Terra in "Alexander von Humboldt's Correspondence with Jefferson, Madison, and Gallatin," *Proceedings of the American Philosophical Society* 103, no. 6 (15 December 1959): 796.

2 Alexander von Humboldt and Aimé Bonpland, *Personal Narrative of Travels to the Equinoctial Regions of the New Continent, during the Years 1799–1804*, trans. Helen Maria Williams (London: Longman, Hurst, Rees, Orme, and Brown, 1822), 1:239–40.

3 Humboldt was welcomed by the intellectual community on the island. In 1804 the Sociedad Económica de Amigos del País invited Humboldt to speak on his assessment of the mineral resources of Guanabacoa's highlands. For a thorough and thoughtful analysis of this period in Cuba's history, see Alfonso W. Quiroz, "The Scientist and the Patrician: Reformism in Cuba," in *Alexander von Humboldt: From the Americas to the Cosmos*, ed. Raymond Erickson, Mauricio A. Font, and Brian Schwartz (New York: Bildner Center for Western Hemisphere Studies, City University of New York, 2005), 115–16.

4 *National Intelligencer*, 11 June 1804, 2.

5 "Cette abominable loi qui permet l'importation des Nègres dans la Caroline méridionale est un opprobre pour un État, dans lequel je sais qu'ils existent des têtes très bien organisés. En suivant le seule marche que dicte l'humanité, on exportera sans doute au commencement moins de Coton. Mais hélas! que je déteste cette Politique qui mesure et évalue la félicité publique simplement d'après la valeur des Exportations! Il est de la richesse Nations comme de celle des Individus. Elle n'est que l'accessoire de notre félicité. Avant d'être libre, il faut être juste, et sans

justice il n'y a pas de prospérité durable." Humboldt to Thornton, Philadelphia, 20 June 1804, in Schwarz, *Alexander von Humboldt und die Vereinigten Staaten von Amerika*, 96–97, letter 9. Translated by Frank Baron and Gerhard Casper in Gerhard Casper, "A Young Man from 'ultima Thule' Visits Jefferson: Alexander von Humboldt in Philadelphia and Washington," *Proceedings of the American Philosophical Society* 155, no. 3 (September 2011): 258. Although Humboldt mentions knowing of South Carolinians opposed to slavery, thus far I have not been able to identify any specific individual to whom Humboldt refers.

6 The king of Spain did not react publicly to Humboldt's criticisms; however, it is very likely that the British East India Company repeatedly denied Humboldt entry to India for precisely this reason.

7 "Vous y verrez Votre Nom cité avec cet enthousiasme qu'il a inspiré de tout tems aux amis de l'humanité. Puisse la Providence Vous conserver longtems pour le bonheur d'un monde vers lequel j'aime à diriger mes regards et mes voeux. Je suis, Mr. le Président, avec le plus profound respect / De Votre Excellence / le très-humble et très obéisant serviteur / Humboldt." [I am, Mr President, with profoundest respect / Your Excellency's / very humble and most obedient servant / Humboldt] Humboldt to Jefferson, Paris, 30 May 1808, in Schwarz, *Alexander von Humboldt und die Vereinigten Staaten von Amerika*, 109, letter 21. Translated in de Terra, "Alexander von Humboldt's Correspondence with Jefferson, Madison, and Gallatin," 789.

8 "Le royaume de la Nouvelle-Espagne a un avantage marquant sur les Ètats-Unis. Le nombre des esclaves, soit africains, soit de race mixte, y est presque nul; avantage que les colons européens ne commencent à bien apprécier que depuis les événemens tragiques de la révolution de Saint-Domingue: tant il est vrai que la crainte des maux physiques agit plus puissamment que les considérations morales sur les vrais intérêts de la société, ou les principles de philantropie et de justice, si souvent énoncés au parlement, à l'assemblée constituante et dans les ouvrages de philosophes!

"Le nombre des esclaves africains dans les États-Unis, monte au delà d'un million. Ils font la sixième partie de la population entiére. Les états méridionaux, dont l'influence politique devient plus grande depuis l'acquisition de la Louisiane, augmentent inconsidérément le nombre des nègres qu'ils introduisent tous les ans. Il ne dépend point encore du congrès ni du chef de la confédération (magistrat* dont le nom est cher aux vrais amis de l'humanité), de s'opposer à cette augmentation, et d'épargner par-là de grands malheurs aux races futures. *Le président actuel, M. *Thomas Jefferson*, auteur de l'excellent Essai sur la Virginie." Alexander von Humboldt, *Essai politique sur le royaume de la Nouvelle-Espagne* (Paris: Chez F. Schoell, 1811), 1:221–22. Alexander von Humboldt, *Political Essay on the Kingdom of New Spain,* trans. John Black (London: Longman, Hurst, Rees, Orme, and Brown, 1811), 1:14–15.

9 "Agréez, mon digne et respectable ami, l'hommage de mon admiration et de ma reconnaissance." "Accept, my kind and honored friend, the expression of my admiration and gratitude." Humboldt to Jefferson, Paris, 20 December 1811. In this letter Humboldt also acknowledges receipt of an earlier letter and a copy of Jefferson's *Notes on Virginia*, Schwarz, *Alexander von Humboldt und die Vereinigten Staaten von Amerika*, 122–23, letter 33. Translated in de Terra, "Alexander von Humboldt's Correspondence with Jefferson, Madison, and Gallatin," 792–93.

10 "Si vous parvenez à dètruire l'esclavage dans le midi il y aurait au moins dans le monde un gouvernement aussi parfait que la raison humaine peut le concevoir." Madame de Staël to Jefferson, Pisa, 6 January 1816. Translated in Marie G. Kimball, "Unpublished Correspondence of Mme. De Staël with Thomas Jefferson," *North American Review* 208, no. 752 (July 1918): 69–70, and Thomas Jefferson, *The Papers of Thomas Jefferson*, Retirement Series, vol. 9, *1 September 1815 to 30 April 1816*, ed. J. Jefferson Looney (Princeton, NJ: Princeton University Press, 2012), 326–28.

11 Marquis de Lafayette to Jefferson, La Grange, 1 June 1822, *Founders Online,* National Archives, last modified June 13, 2018, http://founders.archives.gov/documents/Jefferson/98-01-02-2841.

12 Jefferson to Genet, Germantown, 24 November 1793, in Thomas Jefferson, *The Papers of Thomas Jefferson*, vol. 27, *1 September to 31 December 1793*, ed. John Catanzariti (Princeton, NJ: Princeton University Press, 1997), 429–30.

13 Donald R. Hickey, "America's Response to the Slave Revolt in Haiti, 1791–1806," *Journal of the Early Republic* 2, no. 4 (Winter, 1982): 362–63.

14 St. George Tucker, "Of the State of Slavery, in Virginia," in *Blackstone's Commentaries: With Notes of Reference to the Constitution and Laws, of the Federal Government of the United States; and of the Commonwealth of Virginia* (Philadelphia: William Young Birch and Abraham Small, 1803), 75.

15 Edgar F. Smith, *James Woodhouse: A Pioneer in Chemistry (1770–1809)* (Philadelphia: John C. Winston Company, 1918; repr., New York: Arno Press, 1980), 56. Woodhouse noted in 1793 an influx of "several thousand" French refugees from Saint-Domingue. Benjamin Franklin had extended foreign membership in the American Philosophical Society to several of Saint-Domingue's leading scientists who were themselves members of the Société Royale des Sciences et des Arts du Cap François (the Cercle des Philadelphes).

16 James E. McClellan III, *Colonialism and Science: Saint Domingue in the Old Regime* (Chicago: University of Chicago Press, 1992), 4–5, 226–27; and Michael Dettelbach, "Describing the Nation: Local and Universal in Humboldt's Administrative Practice and in Late Eighteenth-Century Cameralism," in *Alexander von Humboldt and the Americas*, ed. Vera M. Kutzinski, Ottmar Ette, and Laura Dassow Walls (Berlin: Verlag Walter Frey, 2012), 166–68. One of these men, botanist Ambroise Marie François Joseph Palisot, Baron de Beauvois, worked for Peale in his museum, helping to catalogue all of his various specimens. Palisot de Beauvois, known as "the unfortunate botanist" for the series of catastrophes that plagued him early in his life, was made a member of the American Philosophical Society in 1792. He counted Peale, Jefferson, and Caspar Wistar among his friends. See Elmer Drew Merrill, "Palisot de Beauvois As an Overlooked American Botanist," *Proceedings of the American Philosophical Society* 76, no. 6 :899–901; and Matthias Schönhofer, *Letters from an American Botanist: The Correspondence of Gotthilf Heinrich Ernst Mühlenberg (1753–1815)* (Stuttgart: Franz Steiner Verlag, 2014), 302–3.

17 Hunt notes that Louverture preferred this spelling to the more customary French spelling, L'Ouverture. Alfred N. Hunt, *Haiti's Influence on Antebellum America: Slumbering Volcano in the Caribbean* (Baton Rouge: Louisiana State University Press, 1988), 3n2.

18 *Richmond Enquirer*, 9 June 1804; quoted and cited in Hunt, *Haiti's Influence on Antebellum America*, 2.

19 Whitemarsh B. Seabrook, *A Concise View of the Critical Situation, and Future Prospects of the Slave-Holding States, in Relation to Their Colored Population*, 2nd ed. (Charleston, SC: A. E. Miller, 1825), 4, 13; quoted in Hunt, *Haiti's Influence on Antebellum America*, 116.

20 Alexander von Humboldt and Aimé Bonpland, *Personal Narrative of Travels to the Equinoctial Regions of America, during the years 1799–1804*, trans. Thomasina Ross (London: Henry G. Bohn, 1853), 3:121.

21 "Ein Anderes ist es, wenn, wie Sie sehr richtig bemerken, die Sklaven-Frage dereinst zum Ausbruch kommen sollte; für den Fall theile ich vollkommen Ihre Ansicht über das Precarium des staatlichen Bestandes der nordamerikanischen Union. Ich wünsche diesen Fall nicht zu erleben. Ich halte viel, sehr viel aus die Vereinigten Staaten, weil sie der Hort einer vernünftigen Freiheit sind!" Alexander von Humboldt, *Briefwechsel: Alexander von Humboldt's mit Heinrich Berghaus aus den Jahren 1825 bis 1858* (Leipzig: Hermann Costenoble, 1863), 16–17. Translation in Ingo Schwarz, "'Shelter for a Reasonable Freedom' or Cartesian Vortex: Aspects of Alexander von Humboldt's Relation to the United States of America," in "Alejandro de Humboldt y el mundo hispánico," ed. Miguel Ángel Puig-Samper, special issue, *Debate y perspectivas*, no. 1 (2000): 170.

22 In 1829 it was translated into English, but it was folded into the much larger seventh volume of his *Personal Narrative of Travels to the Equinoctial Regions of the New Continent*. Alexander von Humboldt and Aimé Bonpland, *Personal Narrative of Travels to the Equinoctial Regions of the New Continent, during the Years 1799–1804*, trans. Helen Maria Williams (London: Longman, Rees, Orme, Brown, and Green, 1829), 7:chapter 28.

23 Alexander von Humboldt, *Political Essay on the Island of Cuba: A Critical Edition*, ed. Vera M. Kutzinski and Ottmar Ette, trans. J. Bradford Anderson, Vera M. Kutzinski, and Anja Becker (Chicago: University of Chicago Press, 2011), 118.

24 Ibid., 131–32.

25 Humboldt writes of those who "ont inventé les mots de *paysans-nègres de Antilles*, de *vassalage noir*" and follows in a note with "Ces rapprochemens ne tranquillisent que ceux qui, partisans secrets de la traite des noirs, cherchent à s'étourdir sur les malheurs de la race noire, et se révoltent, pour ainsi dire, contre toute émotion qui pourroit les surprendre." Alexander von Humboldt, *Essai politique sur l'ile de Cuba* (Paris: Gide Fils, 1826), 1:307–8. Cited and discussed in Irene Prüfer Leske, "Ancient and Modern Forms of Slavery," in Erickson, Font, and Schwartz, eds., *Alexander von Humboldt: From the Americas to the Cosmos*, 471.

26 Humboldt, *Political Essay on the Island of Cuba: A Critical Edition*, ed. Kutzinski and Ette, 68.

27 "L'esclavage est sans doute le plus grand de tous les maux qui ont affligé l'humanité." Humboldt, *Essai politique sur l'ile de Cuba*, 1:309. Translated in Humboldt, *Political Essay on the Island of Cuba: A Critical Edition*, ed. Kutzinski and Ette, 144.

28 "Alexander von Humboldt on Negro Slavery," *The Friend; A Religious and Literary Journal*, 19 February 1831, 145.

29 "Alexander von Humboldt on Negro Slavery (concluded from page 146)," *The Friend; A Religious and Literary Journal*, 26 February 1831, 153.

30 Alexander von Humboldt, *Cosmos: A Sketch of a Physical Description of the Universe*, trans. E. C. Otté (London: Henry G. Bohn, 1849), 1:368. For a smoother translation, see Nicolaas A. Rupke, *Alexander von Humboldt: A Metabiography* (Chicago: University of Chicago Press, 2005), 131–32. "In maintaining the unity of the human race we also reject the disagreeable assumption of superior and inferior peoples. Some peoples are more pliable, more highly educated and ennobled by intellectual culture, but there are no races that are more noble than others. All are equally entitled to freedom; to freedom, which in the state of nature, belongs to the individual and which in civilization belongs as a right to the entire citizenry through political institutions."

31 Humboldt to George Sumner, Paris, 19 May 1845; translated and printed in "Baron Humboldt and the Late Geo. Sumner," *Liberator*, 20 November 1863, 187. Humboldt had invited Sumner to visit him at home in Berlin during December 1839. In a letter to George S. Hillard dated December 25, 1839, Sumner wrote, "Humboldt is very kind to me. He is placed at the head of the conversers of Germany. So far as I can compare conversation in different languages, his reminds me of Judge Story's: it is rapid, continuous, unflagging, lively, various. He has spoken to me in the highest terms of Prescott's book,—which I saw on his table.…In a note to me, he spoke of 'l'excellent et spirituel Gouverneur Everett.'" Edward Lillie Pierce, *Memoir and Letters of Charles Sumner* (Boston: Roberts Brothers, 1878), 2:128.

32 Bancroft to James K. Polk, London, 28 January 1848, in George Bancroft, *The Life and Letters of George Bancroft*, ed. M. A. De Wolfe Howe (New York: Charles Scribner's Sons, 1908), 2:29; also in Helmut de Terra, *Humboldt: The Life and Times of Alexander von Humboldt, 1769–1859* (New York: Alfred A. Knopf, 1955), 347n1, in reference to the Treaty of Guadalupe Hidalgo. In private Bancroft had been more blunt: "For us to come down and take all Mexico he deemed impossible or rather an unwise design, but…against our possessing tropical countries he gave us a warning. Besides he detests slavery and holds the very strongest opinions against its extension. For all this, he regards Cuba as the natural extension of Florida, and that, therefore, one day it must come to the power, of which Florida is a possession." Bancroft to his wife, Paris, 31 December 1847, in Bancroft, *Life and Letters of George Bancroft*, 2:77–78.

33 "Die Groberungen der republicanischen Amerikaner missfallen mir höchlichst. Ich wünsche ihnen alles Unglüd in dem tropischen Mexico. Je leur abandonné le Nord, wo sie dann ihr verruchtes Sklavenwesen verbreiten werden." Humboldt to Bunsen, Sanssouci, 28 July 1847. Alexander von Humboldt and Christian Carl Josias Bunsen, *Briefe von Alexander von Humboldt an Christian Carl Josias Freiherr von Bunsen* (Leipzig: F. A. Brockhaus, 1869), 98, letter 54. English translation by the author.

34 Linck C. Johnson, "'Liberty Is Never Cheap': Emerson, 'The Fugitive Slave Law,' and the Antislavery Lecture Series at the Broadway Tabernacle," *New England Quarterly* 76, no. 4 (December 2003): 552–53.

35 Johnson, "'Liberty Is Never Cheap,'" 552. See also Ralph Waldo Emerson, *Emerson's Antislavery Writings*, ed. Len Gougeon and Joel Myerson (New Haven, CT: Yale University Press, 1995), 74.

36 Johnson, "'Liberty Is Never Cheap,'" 568.

37 "The New York Tribune," *National Anti-Slavery Standard*, 21 January 1854; cited in Johnson, "'Liberty Is Never Cheap,'" 568.

38 Johnson, "'Liberty Is Never Cheap,'" 589, 591.

39 Karl Bruhns, ed., *Life of Alexander von Humboldt. Compiled in Commemoration of the Centenary of His Birth, by J. Löwenberg, Robert Avé-Lallemant, and Alfred Dove*, trans. Jane Lassell and Caroline Lassell (Boston: Lee and Shepard, 1873), 2:253.

40 The Ostend Manifesto, Aix-la-Chapelle, 15 October 1854; cited and partially quoted in David M. Potter, *The Impending Crisis 1848–1861* (New York: Harper & Row, 1976), 190.

41 *New York Tribune*, August 1855 visit to Humboldt, reprinted in the *National Anti-Slavery Standard* commemoration of Humboldt, 2 July 1859; cited in Philip S. Foner, *Alexander von Humboldt über die sklaverei in den USA* (Berlin: Humboldt-Universität 1984), 34.

42 "Cuba-süchtigen," in a letter from Humboldt to Varnhagen von Ense, Berlin, 28 May 1857, in Humboldt, *Briefe von Alexander von Humboldt an Varnhagen von Ense, aus den Jaren 1827 bis 1858*, ed. Ludmilla Assing, 3rd ed. (Leipzig: F. A. Brockhaus, 1860), 368. Translated in Humboldt, *Letters of Alexander von Humboldt to Varnhagen von Ense. From 1827 to 1858. With Extracts from Varnhagen's Diaries, and Letters of Varnhagen and others to Humboldt*, trans. Friedrich Kapp (New York: Rudd and Carleton, 1860), 372–73.

43 "Humboldt's Opinion of Webster and the Fugitive Slave Law," *New York Evening Post*, 1 May 1855; expanded and reprinted in the *Liberator*, 11 May 1855, 73–74; and in the *National Era*, 17 May 1855, 80. The same letter was passed around to numerous papers; it was reprinted nearly annually. An amplified version reprinted from the *Ashtabula Sentinel* appeared as "Humboldt on Webster and Slavery," *The Liberator*, September 1859, 138. It was also included in a pamphlet of anti-slavery letters published by the Boston Anti-Slavery Society in 1860 in commemoration of John Brown's execution. The *National Era* was an abolitionist newspaper (1847–60) published weekly in Washington, D.C.

44 "In den Vereinigten Staaten ist allerdings viel Liebe für mich erwacht, aber das Ganze gewährt mir dort den traurigen Anblick, daß die Freiheit nur ein Mechanismus im Elemente der Nützlichkeit ist, wenig dort veredelnd, das Geistige und Gemüthliche anregend, was doch der Zweck der politischen Freiheit sein soll. Daher Gleichgültigkeit gegen Sklaverei. Uber die V. St. sind ein Cartesianischer Wirbel alles sortreißend, langweilig nivellirend." Humboldt to Varnhagen von Ense, Berlin, 31 July 1854, in Assing, *Briefe von Alexander von Humboldt an Varnhagen von Ense*, 295. Translated in *Letters of Alexander von Humboldt to Varnhagen von Ense*, trans. Kapp (New York, 1860), 305.

45 Both Aaron Sachs and Laura Dassow Walls consider the debt Melville owes to Humboldt. See Aaron Sachs, *The Humboldt Current: Nineteenth-Century Exploration and the Roots of American Environmentalism* (New York: Viking, 2006); and Laura Dassow Walls, *Passage to Cosmos: Alexander von Humboldt and the Shaping of America* (Chicago: University of Chicago Press, 2009). For a deeper consideration of Melville's use of this idiom, see David Charles Leonard, "The Cartesian Vortex in Moby-Dick," *American Literature* 51, no. 1 (March 1979): 105–109.

46 Thrasher to Humboldt, New York, 14 December 1855, Staatsbibliothek Berlin, Nachl, Alexander von Humboldt, gr. Kasten 2, Folder Nr. 3.

47 The copy of Thrasher's translation annotated by Humboldt is now in the Biblioteca Nacional José Martí, Havana, Cuba. For an incisive discussion of this controversy, see Vera M. Kutzinski, "Translations of Cuba: Fernando Ortiz, Alexander von Humboldt, and the Curious Case of John Sidney Thrasher," in *Alexander von Humboldt's Transatlantic Personae*, ed. Vera M. Kutzinski (London: Routledge, 2012), 116–17.

48 Vera Kutzinski and Ottmar Ette note that Thrasher had translated the Spanish edition of Humboldt's *Political Essay on Cuba* into English. This version had a separate chapter titled "Esclavos" (slaves); the French edition did not give this section a separate heading. Humboldt, *Political Essay on the Island of Cuba: A Critical Edition*, xxiii.

49 Obituary for Thrasher in *Appleton's Annual Cyclopedia and Register of Important Events of the Year 1879* (New York: D. Appleton, 1886), 696–97.

50 "Cuban Annexation: Influence of Annexation on the Sugar Interest—Letter from John S. Thrasher," 4 May 1854, New Orleans, published in the *New-York Daily Times*, 23 May 1854. Thrasher knew that New York City was a democratic, proslavery enclave in an otherwise neutral or anti-slavery leaning state.

51 "Baron von Humboldt's Political Essay on Cuba—Letter from the Author on the Omission of a Chapter by the Translator," *New York Times*, 12 August 1856, 2; "Humboldt Corrects Thrasher," *New York Tribune*, 12 August 1856, 3; "The Works of Humboldt and J. S. Thrasher on Cuba," *New York Herald*, 13 August 1856, 8. Variations appeared as "What Slavery Does to Our Literature," *Lowell (Mass.) Daily Citizen and News*, 14 August 1856; "A Thrasher Thrashed," *Boston Daily Atlas*, 16 August 1856; "Baron von Humboldt on Slavery in the United States and Cuba," *Liberator*, 22 August 1856, 138; and "Humboldt's Protest," *Friends' Review; A Religious, Literary and Miscellaneous Journal*, 20 September 1856, 28.

52 Vera M. Kutzinski astutely notes that Thrasher did more than just omit a chapter. He changed the underlying political message entirely, downplaying the racial disparity between blacks and whites and twisting Humboldt's use of statistics away from flagging injustices and toward an argument for making Cuba a U.S. state and divorcing it from Spain altogether. Kutzinski, "Translations of Cuba," 119. As Fernando Ortiz notes, a Cuban proclamation from 1848 made it clear there were those in Cuba who also sought to align themselves with the United States. Therefore, annexation was not an external land grab; it was a careful political calculation with advocates and dissenters in both places. Ortiz quotes Cuban historian Vidal Morales y Morales: "Uniting Cuba with this strong and respected nation,

whose interests in the South will be identical with our own, means to strengthen her peacefulness and future fate; it will increase her wealth, doubling the value of her plantations and her slaves, tripling the value of her land; it will give liberty to individual action and get rid of this hateful and pernicious system of restrictions that paralyzes trade and agriculture." Morales y Morales also wrote, "We would prefer above all the protection of Great Britain, if it were not for the law that abolishes slavery which will take hold here before long and ruin our sugar factories." These quotations reveal that the core issue behind the maintenance of slavery was to fuel the sugar plantation system. Vidal Morales y Morales, *Iniciadores y Primeros Mártires de la Revolución Cubana* (Havana, 1931), 226, 95; both cited in Fernando Ortiz, "Humboldt's Translator in the Context of Cuban History," in *Alexander von Humboldt's Transatlantic Personae*, ed. and trans. Vera M. Kutzinski (London: Routledge, 2012), 94, 95.

53 "Humboldt's Protest," *Friends' Review; A Religious, Literary and Miscellaneous Journal*, 20 September 1856, 28. This final paragraph appeared in all of the printings of Humboldt's letter of protest.

54 "Humboldt and Thrasher in Cuba: To the Editor of the Herald," *New York Herald*, 17 August 1856, 2; "Baron Humboldt and Mr. Thrasher," *New-York Daily Times*, 17 August 1856.

55 Gerolt to Humboldt, New York, 25 August 1856; translated in Humboldt, *Letters of Alexander von Humboldt to Varnhagen von Ense*, trans. Kapp (New York, 1860), 326, letter 177. The letter included a clipping with Thrasher's rebuttal printed in the *New York Herald*.

56 Tyler Green has done a masterful job of describing the political landscape in California as part of the essential backdrop to statehood in *Carleton Watkins: Making the West American* (Berkeley: University of California Press, 2018).

57 Frémont's career as an explorer and his emulation of Humboldt is addressed in chapter 2 of this volume.

58 Frémont to Humboldt, 16 August 1856; quoted in Bruhns, ed., *Life of Alexander von Humboldt. Compiled in Commemoration of the Centenary of His Birth*, 2:253. Also see Schwarz, *Alexander von Humboldt und die Vereinigten Staaten von Amerika*, 387, letter 252.

59 After the election, Frémont sent Humboldt an inscribed copy of Bigelow's biography. Ever quick with a buoyant (if catty) bon mot, Humboldt wrote of the biography, "Fremont's portrait reminds one vividly of Chateaubriand. A biography of the former has just appeared in New York, dedicated to me—'Memoirs of the Life and Public Services of John Charles Fremont, by John Bigelow.' The dedication says: 'To Alexander von Humboldt this memoir of one whose genius he was among the first to discover and acknowledge, is respectfully inscribed by the author.' Delicate words, a little artificially combined. There is a copy of the letter written to him from Sans Souci, in the King's name, in 1850, accompanying the great prize medal for science and art, upon his having projected the most extensive barometrical level ever executed, from Missouri to the South Sea [Pacific Ocean]. It closes with the words of which Sans Souci has no reason to be ashamed: '*La Californie, qui a* NOBLEMENT *résisté* à l'introduction *de l'esclavage*, sera dignement représentée par *un ami de la liberté et des progrès de l'intelligence*.'* [*California, which has nobly resisted the introduction of slavery, will be worthily represented by a friend of liberty and of the progress of intelligence.] The biography has passages of a strange romantic interest. At one time cold and hunger having driven a party to fury and almost phrensy, when they all pray and sing, and then an oath from Fremont that there shall not in any case be a resort to cannibalism." Humboldt to Varnhagen von Ense, Berlin, 3 December 1856, in Humboldt, *Letters of Alexander von Humboldt to Varnhagen von Ense*, trans. Kapp (New York, 1860), 348–49.

60 This letter was printed in three separate biographies of Frémont published during his presidential campaign: *The Republican Scrap Book; Containing the platforms, and a choice selection of extracts, setting forth the real questions in issue, the opinions of the candidates, the nature and designs of the slave oligarchy*... (Boston: John P. Jewett, 1856), 19–20; John Bigelow, *Memoir of the Life and Public Services of John Charles Fremont, including an account of his explorations, discoveries and adventures on five successive expeditions across the North American continent*... (New York: Derby & Jackson, 1856), 327–28; and Horace Greeley, ed., *Life of Col. Frémont* (New York: Greeley & McElrath, 1856), 31.

61 For a thorough and insightful discussion of the politics of California statehood and the major players, see Green, *Carleton Watkins: Making the West American (2018)*.

62 Charles Wentworth Upham, *Life, Explorations and Public Services of John Charles Fremont* (Boston: Ticknor and Fields, 1856), 354.

63 Gerolt to Humboldt, New York, 25 August 1856, in Humboldt, *Letters of Alexander von Humboldt to Varnhagen von Ense*, trans. Kapp (New York, 1860), 326–27. Although the Frémonts counted Gerolt among their close friends, in his opinion, "Mr. Fillmore would be the best President; but he appears to have little hope of succeeding against Frémont and Buchanan; and the Knownothings have lost all credit."

64 T. H. Pickett, "The Friendship of Frederick Douglass with the German, Ottilie Assing," *Georgia Historical Quarterly* 73 no. 1 (Spring 1989): 88–105.

65 Hartmut Keil, "Race and Ethnicity: Slavery and the German Radical Tradition," talk, Madison, WI, sponsored by the Max Kade Institute for German-American Studies and the Center for the History of Print Culture, 3 February 1999, text available at https://mki.wisc.edu/sites/mki.wisc.edu/files/graphics/Keil-RACE_AND_ETHNICITY.pdf

66 Scholars disagree as to whether the two became lovers; certainly Assing's letters to her sister, Ludmilla, suggest theirs was a committed extramarital relationship. In a letter to Ludmilla, January 29, 1874, Ottilie wrote, "The last seventeen years being not married and still living in a union of the deepest mutual affection, more firmly found than many who are married, without the faintest hope that it might be different, and kept apart by being incapable of valuing or giving love; it is fate." Quoted in Pickett, "Friendship of Frederick Douglass with the German, Ottilie Assing," 99–100; see also William S. McFeeley, *Frederick Douglass* (New York: W. W. Norton, 1991), 183–86. Two months after Douglass remarried, Ottilie Assing took her own life.

67 Although accurate figures for 1856 are not readily available, 80 percent of German Americans in St. Louis voted for Lincoln in 1860. See Walter D. Kamphoefner, "St. Louis Germans and the Republican Party, 1848–1860," *Mid-America: An Historical Review* 57, no. 2 (April 1975): 69–88; cited in Kirsten L. Anderson, "German Americans, African Americans, and the Republican Party in St. Louis, 1865–1872," *Journal of American Ethnic History* 28, no. 1 (Fall 2008): 34, 47n2.

68 The Whig Party had disintegrated over the issue of slavery, splitting into the Republican Party and the American, or "Know-Nothing," Party. In the end, Buchanan earned 174 electoral votes to Frémont's 114, with Fillmore earning 8 electoral votes.

69 "Und die schändliche Parthei, die fünfzigfündige Negerkinder verkauft...die erweist, daß alle weiße Arbeiter auch besser Sklaven als Freie wären,—hat gesiegt. Welche Unthat!" Humboldt to Varnhagen von Ense, Berlin, 21 November 1856, in Assing, *Briefe von Alexander von Humboldt an Varnhagen von Ense*, 332, letter 187. Translated in *Letters of Alexander von Humboldt Written between the Years 1827 and 1858, to Varnhagen von Ense, together with Extracts from Varnhagen's Diaries, and Letters from Varnhagen and Others to Humboldt* (London: Trübner, 1860), 263–64. Schwarz notes that Humboldt had a copy in his library of Preston Brooks's *Speech Delivered at Columbia South Carolina* (the inscription noted his being "presented with a pitcher, goblet, and cane for caning Senator the Hon. Charles Sumner") 29 August 1856, Boston. Henry Stevens, *Catalogue of the First Portion of the Humboldt Library. First Day's Sale*. (London: J. Davy & Sons, 1862), 101, entry 1353. Schwarz, "'Shelter for a Reasonable Freedom' or Cartesian Vortex: Aspects of Alexander von Humboldt's Relation to the United States of America," 170.

70 Garrison to Theodore Parker, Boston, 17 June 1858, in Walter M. Merrill, *Letters of William Lloyd Garrison*, v. 4, no. 222, Boston Public Library, Rare Books Department, Anti-Slavery Collection, https://digitalcommonwealth.org/search/commonwealth:cv43rj68p.

71 Rev. Theodore Parker, "Tribute to Humboldt," *Liberator*, 9 July 1858, 110. Parker wove Humboldt into several of his sermons attacking slavery.

72 Ibid. In an earlier issue of the *Liberator*, Humboldt had granted permission for them to print a letter to Julius Froebel, in which he vented, "Continue to brand the shameful devotion to slavery, the treacherous importation of negroes, under the pretense of their becoming free—a means to stimulate the hunting of negroes in the interior of Africa. What atrocities have been witnessed by one who had the misfortune to live from 1789 to 1858! My book against slavery (Political Essay on the Island of Cuba) is not prohibited in Madrid, but cannot be published in the United States, which you call 'The Republic of distinguished people,' except with the omission of everything that relates to the sufferings of our fellow-men, who, according to my political views, are entitled to the same enjoyment of freedom as ourselves. Add to this, the anathema on other races of men, forgetting that the most ancient cultivation of humanity, before that of the white Hellenic race in Assyria, in Babylon, in the valley of the Nile, in Iran, in China, was the work of colored men, though not wooly haired." Humboldt signed this letter "I am ever your illegible Al. Humboldt." Humboldt to Froebel, Berlin, 11 January 1858. Published with his permission in the *New York Tribune*; and then in "Humboldt on American Slavery," *Liberator*, 11 June 1858, 96; and *Friends' Review; A Religious, Literary and Miscellaneous Journal*, 3 July 1858, 697.

73 Garrison to Parker, Boston, 17 June 1858.

74 "Anti-Slavery Celebration of Independence Day," *Liberator*, 9 July 1858, 110.

75 It may have been pure coincidence that following John Brown's execution, his widow and three of their children left the East Coast and resettled in Humboldt County, California, named for the explorer. "It may not be generally known that the widow of John Brown is a resident of Humboldt County living with her son Salmon and her daughters Sarah and Ellen." *West Coast Signal* (Eureka, CA), 10 May 1871; cited in "The Family of John Brown, Abolitionist," *SunnyFortuna,* updated March 2010, https://sunny-fortuna.com/history/rohnerville/browns.htm. A fourth son, John Brown Jr., enlisted in the Union Army. Letters to his wife written in 1862 were posted from Camp Wendell Phillips, near Lawrence, and from Humboldt, Kansas, an abolitionist community founded by German émigrés in Allen County. See Letters from John Brown Jr. to his wife, Wealthy Brown, written between February and May 1862. Kansas Historical Society, John Brown Jr. Coll. #833, unit ID 216668, https://kshs.org/index.php?url=archives/216668.

76 *New York Tribune*, notice of Humboldt's death posted 10 May 1859 by the paper's Berlin correspondent, printed in the 31 May 1859 issue; cited in Philip S. Foner, ed., *Alexander von Humboldt on Slavery in the United States* (Berlin: Humboldt-Universität, 1984), 12, and note 5.

77 "Humboldt on Slavery in the United States," *National Anti-Slavery Standard*, 2 July 1859. Reprinted in Foner, ed., *Alexander von Humboldt on Slavery in the United States*, 31–33.

78 Frémont to John Greenleaf Whittier, New York, 14 February 1864, in Jessie Benton Frémont, *The Letters of Jessie Benton Frémont*, ed. Pamela Herr and Mary Lee Spence (Urbana: University of Illinois Press, 1993), 372.

79 John Rogers, October 1859, Rogers Family Papers, New-York Historical Society; quoted in "For Sale to the Highest Bidder: *The Slave Auction*, Sculpture by John Rogers, 1859," in "The American Companion to *A History of the World in 100 Objects*, a fresh, visual perspective on the Civil War," *Erenow*, last modified 2019, https://erenow.net/ww/the-civil-war-in-50-objects/5.php.

80 Rogers to his mother, Sarah Ellen Derby Rogers, New York, 24 December 1859, Rogers Family Papers 1614–1950, MS 523, New-York Historical Society.

81 Rogers to "Aunt" Mary Jane Derby Peabody, New York, 31 January 1860, Rogers Family Papers 1614–1950, MS 523, New-York Historical Society. Rogers had difficulty placing this work in New York shops and settled on the strategy of hiring "a good looking negro to carry them around on a sort of tray, with an appropriate notice printed on the front." Rogers to Sarah Ellen Derby Rogers, New York, 16 December 1859 and 20 January 1860; quoted in "For Sale to the Highest Bidder: *The Slave Auction*, Sculpture by John Rogers, 1859," in "The American Companion to *A History of the World in 100 Objects*." In this manner, Tappan discovered and purchased his copy. See also Kimberly Orcutt, ed., *John Rogers: American Stories* (New York: New-York Historical Society, 2010), 156.

82 *New York Tribune*, 8 September 1860, 4.

83 Rogers noted in a letter dated 28 February 1864 that Jessie had admired *The Wounded Scout* and asked if she could bring "the General," her husband, to see it. Cited in Frémont, *Letters of Jessie Benton Frémont*, 372n7; and David H. Wallace, *John Rogers: The People's Sculptor* (Middletown, CT: Wesleyan University Press, 1967), 100–101.

84 Wallace, *John Rogers: The People's Sculptor*, 100.

85 Lydia Maria Child, "To the Editor of the Independent," *Independent*, 21 December 1865, 1; quoted in Lydia Maria Child and Carolyn L. Karcher, *A Lydia Maria Child Reader* (Durham, NC: Duke University Press, 1997), 279–80.

86 Lincoln to Rogers, 13 June 1864, Rogers Family Papers 1614–1950, MS 523, New-York Historical Society; quoted in "For Sale to the Highest Bidder: *The Slave Auction*, Sculpture by John Rogers, 1859," in "The American Companion to *A History of the World in 100 Objects*."

87 "Ward's Statuette of the 'Freedman,'" *New York Times*, 19 November 1864, 2.

88 Jessie Benton Frémont corresponded with Ward concerning the casting of this sculpture using that gun fragment: Frémont to Ward, 23 December 1865, and two undated letters (probably December 1865–January 1866) in the collection of the Albany Institute; cited in Frémont, *Letters of Jessie Benton Frémont*, 558n2.

89 The Frémonts purchased the home near Tarrytown and roughly one hundred acres from James Watson Webb, owner of the *New York Courier and Enquirer*, who helped found the Republican Party and had been instrumental in planning Frémont's 1856 campaign. Frémont, *Letters of Jessie Benton Frémont*, 387.

90 Jessie described this portrait in a letter to Emma Hill Ballard, Los Angeles, 26 June 1902, in which she wrote, "The statue is by Ward—the books below it are Schoolcraft's History of the American Indians. It is all part of our old home—." The Ballards had purchased the painting as part of the Frémont's household sale in 1877. In 1909 the Ballards donated the painting to the Missouri Historical Society. See Frémont, *Letters of Jessie Benton Frémont*, 557–58.

91 Ibid., 113.

92 "Cultural unionism" is a term coined by Green, in *Carleton Watkins: Making the West American*, 55–68, 120, 175.

93 Jessie Benton Frémont to King, January? 1854. Here Jessie informs King that she and George McClellan's wife were assigned to the same committee, wryly noting, "After that let your lambs & lions go to work together too." The two women held opposing political views, but by March, Jessie wrote to another close friend, "we get on very civilly—even amicably." Frémont to Peabody, New York, 20 March 1864; both letters published in Frémont, *Letters of Jessie Benton Frémont*, 360, 374–75.

94 A complete discussion of the Metropolitan Fair exceeds the scope of this project. For an itemization of the Humboldt materials, see "List of Contributors and Articles Contributed," and "Autograph Collection of the Metropolitan Fair," *Spirit of the Fair*, 5 April 1864, 10, 99.

95 Frederic Church's debt to Humboldt is discussed at length in chapter 7 in this volume.

96 "List of Contributors and Articles Contributed," and "Autograph Collection of the Metropolitan Fair," *Spirit of the Fair*, 5 April 1864, 10, 99. Among the items donated for display and sale were "1 photograph of Humboldt, Pierre Seroude," and "half a dozen long letters of Humboldt." Rosaly Torna Kurth suggests that the Coopers might have donated the Humboldt letters, noting their generosity in donating autographs to that part of the fair. Rosaly Torna Kurth, *Susan Fenimore Cooper: New Perspectives on Her Works* (Bloomington, IN: iUniverse, 2016), 341n6.

97 Henry T. Tuckerman, "English Abuse of America," in *Spirit of the Fair*, 15 April 1864, 116.

98 Horace Greeley, *An Overland Journey, from New York to San Francisco in the Summer of 1859* (New York: C. M. Saxton, Barker, 1860).

99 Frémont to Greeley, March/April 1864?, in Frémont, *Letters of Jessie Benton Frémont*, 378.

100 "Sale of Pictures at the Fair," *New York Evening Post*, 21 April 1864, 2.

101 Albert Bierstadt to Hay, 3 August 1863, John Hay Papers, Brown University; quoted in Eleanor Jones Harvey, *Civil War and American Art* (New Haven, CT: Yale University Press in association with the Smithsonian American Art Museum, 2012), 62. Thomas Starr King, *A Vacation among the Sierras: Yosemite in 1860*, ed. John A. Hussey (San Francisco: Book Club of California, 1962), xxiv; quoted in Harvey, *Civil War and American Art*, 60.

102 Matthews to Lincoln, 26 February 1864, Library of Congress, Abraham Lincoln papers, Series 2, General Correspondence, 1858–1864.

103 Simón Bolívar to General José Antonio Páez, 8 August 1826; Frederick Douglass, "The American Apocalypse," delivered in Rochester, NY, 16 June 1861. Bolívar and Douglass, quoted in Harvey, *Civil War and American Art*, 45, 43. For a thought-provoking argument about Humboldt's use of volcanoes and earthquakes as metaphors for political upheavals, see Carlos Fonseca, "Shaky Grounds: Bolívar, Humboldt, and the Birth of Catastrophe Politics," *Revista Hispánica Moderna* 67, no. 2 (December 2014): 163–81.

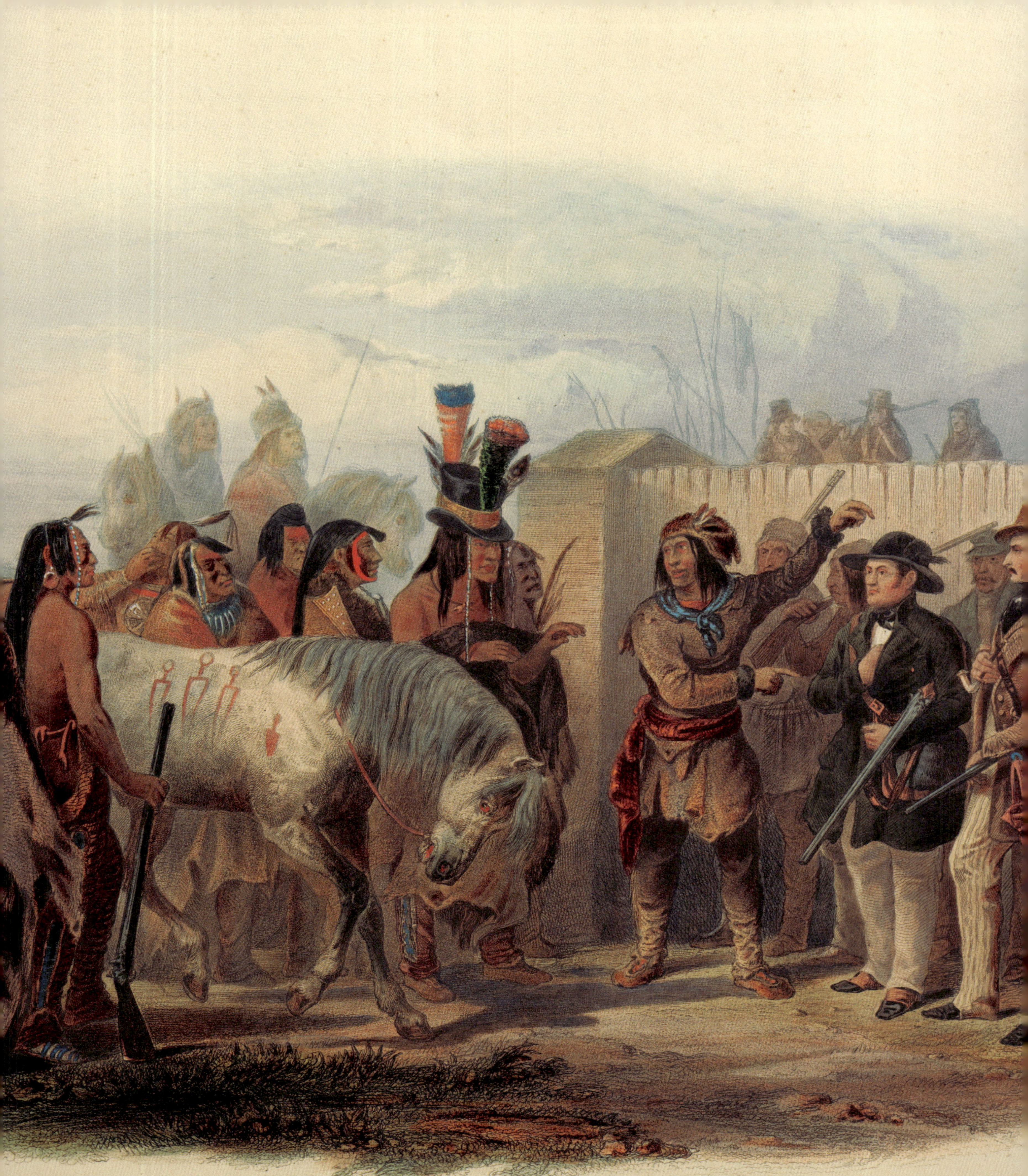

Imp. de Bougeard

CHAPTER SIX | HUMBOLDT AND NATIVE AMERICAN ETHNOGRAPHY

"We maintain the unity of the human species"

Humboldt's belief in a just society extended to declarations of respect for all cultures. Theodore Parker, a leading New England abolitionist, praised the breadth of his humanitarian concerns, noting, "He takes the side of the Indian in North and South America, against his conqueror."[1] On this Humboldt was consistent: a moral compass was the most important instrument in his equipage. The naturalist decried race-based slavery and found equal fault in the mistreatment of indigenous cultures by European colonizers. Five years in the Americas had taught him that local inhabitants often knew more about an environment than government officials. As a result, when Humboldt sought to understand the landscapes through which he traveled, he often relied on indigenous guides.[2] These relationships provided him with greater access to different cultures. Trained at Göttingen by Johann Friedrich Blumenbach to believe that all races were essentially equal, Humboldt used his clout to urge American politicians and explorers to follow his lead. In his view, the information exchanged during encounters with native people enhanced global knowledge of the earth and encouraged mutual respect among its peoples.

Humboldt's experiences convinced him that the indigenous people he encountered were descendants of advanced pre-Columbian civilizations who were oppressed by generations of Spanish colonial rule. His conclusion was a radical departure from the prevailing European ethnocentrism. European naturalist Cornelius de Pauw had targeted Native Americans in his sweeping indictment of all aspects of life in the American hemisphere. Humboldt's data undermined the argument of Georges Louis Leclerc, comte de Buffon, exposing as false any correlation between

After Karl Bodmer, *Travellers Meeting Minatarre Indians near Fort Clark* (detail), 1842. See p. 266.

geography and physical stature or mental ability. Having studied an Aztec calendar stone in Mexico (fig. 74), Humboldt pointedly observed:

A people whose festivals were arranged according to the stars and whose calendar was engraved in a public monument probably had a higher level of civilization than it is granted by those clever historians who have taken America as their subject. These authors perceived as barbaric such human condition as was remote from a type of culture formed in accordance with certain systematic ideas they had. These sharp distinctions between barbaric and civilized nations are unacceptable.[3]

Humboldt understood that expansion and exploration carried ethical considerations that too often became casualties of government preoccupation with political boundaries. He had seen this play out in expeditions launched from Europe to locations all over the world. He firmly believed that colonization destroyed valuable cultural knowledge—and sometimes the culture itself. His concern for the place of native tribes and cultures within the burgeoning United States mirrored his opinions regarding the destruction of indigenous civilizations in South America stemming from European colonial rule. When he arrived in the United States in 1804, he hoped that Jefferson's more enlightened attitude toward the Native Americans, as expressed in his *Notes on the State of Virginia* in 1785, would be reflected in official government policies. Jefferson had argued that Native Americans were capable of being coeval in intelligence and potential with whites. The Marquis de Chastellux, a French philosopher and general who had fought with the Americans alongside the comte de Rochambeau at Yorktown, queried Jefferson on this point.[4] In response, Jefferson wrote, "The proofs of genius given by the Indians of North America place them on a level with Whites in the same uncultivated state...in body and mind equal."[5] Humboldt intended to add his own data to hold Jefferson to that standard.

However, by the time of Humboldt's visit, Jefferson's idealistic views were being strained by the realities of two fundamentally different belief systems regarding landownership. Jefferson had long hoped that the United States could develop meaningful treaties with the tribes to allow for peaceful coexistence, and in 1786 he had written:

It may be taken for a certainty that not a foot of land will ever be taken from the Indians without their own consent. The sacredness of their right is felt by all thinking persons in America as much as in Europe.[6]

Representatives from Indian tribes had been visiting the U.S. capital to meet the president and negotiate peaceful relations, starting with George Washington and continuing under Jefferson. However, westward migration of Anglo settlers was proving uncontrollable, with little regard for sites sacred to each native community. Publicly, Jefferson's stance was to find ways to coexist peacefully. Privately Jefferson could adopt a harder line. In a letter to Ohio territorial governor William Henry Harrison in 1803, Jefferson wrote that the Indians could "either incorporate with us as citizens of the United States or remove beyond the Mississippi.... We have only to shut our hand to crush them."[7] Jefferson's harsh assessment may have been phrased to placate a nervous territorial governor

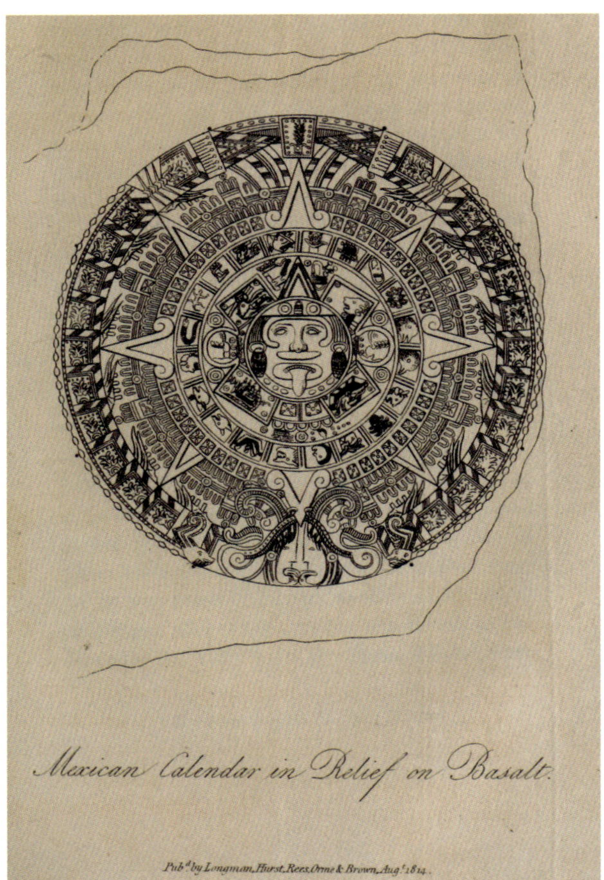

Fig. 74
Alexander von Humboldt, *Mexican Calendar in Relief on Basalt* from *Researches, Concerning the Institutions & Monuments of the Ancient Inhabitants of America*, 1814, engraving, 9 × 6 in., Getty Research Institute, Los Angeles

on a newly expanded international border with Spain, but as president, Jefferson's relations with Native American tribes were clearly shifting toward U.S. governmental control.

Humboldt kept trying to persuade Jefferson to change his approach, but in the wake of the War of 1812 Jefferson found it more convenient to blame the British for the deteriorating relationship between the U.S. government and native communities. In a letter to Humboldt, Jefferson complained:

> *You know, my friend, the benevolent plan we were pursuing here for the happiness of the Aboriginal inhabitants in our vicinities. . . . but the interested and unprincipled policy of England has defeated all our labors for the salvation of these unfortunate people. They have seduced the greater part of the tribes, within our neighborhood, to take up the hatchet against us, and the cruel massacres they have committed on the women and children of our frontiers taken by surprise, will oblige us now to pursue them to extermination, or drive them to new seats beyond our reach* [8]

Knowing that Humboldt would disapprove of the policy the United States was pursuing, Jefferson framed this as a geopolitical issue rather than a humanitarian one. The fur trade—which was dependent on a complex set of relationships among British, French, and native trappers, guides, and customers—essentially owned the Missouri and Mississippi waterways. Lewis and Clark had relied heavily on their French-Canadian guide, Toussaint Charbonneau, and one of his wives, a Shoshone woman named Sacagawea (Bird Woman). By 1809, when Clark succeeded Lewis as the territorial governor of the northern part of the Louisiana Territory, he used his headquarters in St. Louis to establish himself as a powerful and effective interlocutor for all of the participants in that community, Indians and Euro-Americans alike. In 1818 Stephen Harriman Long discovered quickly that his expedition's survival depended on the willingness of the Otoe, Kansa, and Pawnee to provide assistance.[9] For Humboldt, these were the kinds of relationships that he hoped would foster peaceful coexistence.

As had been the case with his argument for abolition, Humboldt found that his humanitarian concern for Native Americans fell on deaf ears in the halls of government. His frustration with Jefferson was nothing compared to his shock upon learning of President Andrew Jackson's Indian Removal Act. On May 26, 1830, Jackson signed into law "an act to provide for an exchange of lands with the Indians residing in any of the States or Territories and for their removal West of the river Mississippi."[10] The Indian Removal Act ushered in decades of contested negotiations with the various tribes occupying the trans-Mississippi west, forever changing the political relationships between the government and these numerous nations. In explaining the need for this policy, Jackson reasoned that Indian civilization was destined to disappear in the face of more advanced civilization. Moving Native American tribes farther west might hasten that scenario, but Jackson's primary goal was to facilitate white expansion. Jackson's Indian policies ran roughshod over the rights of the individual tribes, breaking existing treaties and forcing the mass relocation of entire communities to make way for those new settlements.

Jackson's draconian approach directly affected two travelers who focused their efforts on documenting the North American Indian tribes along the trading routes of the upper Mississippi and Missouri Rivers. One was self-taught artist and amateur ethnographer George Catlin; the other was professionally trained Prussian cultural anthropologist Prince Maximilian zu Wied-Neuwied, accompanied by Swiss artist Karl Bodmer. In 1832 Catlin made an extended trip upriver from St. Louis deep into the Missouri River watershed with the express intent of studying the local tribes. He returned with more than one hundred painted sketches: portraits of individuals, genre scenes, and landscapes that would form the foundation of his artistic career. Catlin was determined to make multiple trips and visit as many native communities as possible in order to create a gallery of painted portraits and genre scenes, artifacts, and regalia that would resemble Peale's Museum in Philadelphia—a kind of national institute for commemorating North America's indigenous peoples before, as he believed, Jackson's policies destroyed their cultures.

The following year, Prince Max and Bodmer traveled much the same route to study the same tribes with the assistance of many of the same people. Although Catlin and Max did not meet in the field, their paths would cross a decade later in Europe. Prince Maximilian's underlying reason for coming to the United States was to fulfill Humboldt's desire to travel across the trans-Mississippi west, gathering information about North American natural history and learning from encounters with native communities. He was already a well-respected ethnographer based on his earlier travels in Brazil; now he aimed to publish a work devoted to the indigenous tribes on the North American continent.

Their motives were as different as their training; however, what they shared was an appreciation for Humboldt's respect for indigenous cultures and his concern for the detrimental impact of expansionist policies. Both Catlin and Max believed the U.S. government was willing to allow the wholesale obliteration of multiple native nations. Max had seen the devastation caused by colonial rule in Brazil and had conversed with Humboldt about his own experiences in New Spain. Catlin wanted to generate public outrage in the United States on behalf of native communities, but deep down he did not see himself saving these people so much as preserving traces of their culture.

GEORGE CATLIN'S SELF-APPOINTED MISSION

George Catlin trained to be a lawyer but, after a scant three years, abandoned the practice, sold his lawbooks to buy paints, and decided to become an artist in 1821. He moved first to Philadelphia, where he cultivated a friendship with Rembrandt Peale. Contact with the Peale family provided Catlin with his early exposure to Humboldt's ideas. There Catlin would have heard Rembrandt's accounts of his recent visit with Humboldt in Paris and the elder Peale's stories of traveling with the baron on his trip to the United States, and he could sense Humboldt's influence in the organization of Peale's Museum. The museum distinguished itself in Catlin's mind in one other arena: it featured the largest and most comprehensive collection of Native American artifacts on the East Coast. In 1804 Charles Willson Peale had installed ten wax figures representing different Indian tribes, each "dressed in their proper habiliments; and sundry arms, utensils, &c,"[11] and in 1809 Peale had received a tremendous amount of material gathered by the Lewis and Clark Expedition. He further augmented the collection with materials his son, Titian, brought back as part of the Long Expedition in 1821 and through the additional purchase in 1826 of a collection of Indian garments and artifacts representing several midwestern tribes.[12]

Catlin actively constructed his own identity as an artist. When he took up painting, it was with the intent to be a portrait painter; however, he recalled witnessing a visiting delegation of Indians in Philadelphia during the 1820s, which inspired him to change course. Years later his friend and fellow artist George Harvey described a studio visit from Catlin in 1826 during which Catlin "disclosed the ambition of his life to become the historian and limner of the aborigines of the vast continent of North America."[13] In doing so, Catlin allied his artistic practice with a socio-political agenda. He envisioned himself as the artist who would use his paintings and his collection of Indian artifacts to educate his audience in much the way the Peale family had structured their museum. As he planned his trips out West to meet and paint his subjects, he also indulged in the same dream as the Peales: to build a collection and devise an educational program that would become a national institute. Catlin's Indian Gallery would be the result: a public installation of more than five hundred of his paintings and an impressive collection of Native American artifacts, a forum in which he would teach visitors to respect Indian culture in hopes of instigating opposition to Jackson's Indian Removal Act.

By the mid-1820s Humboldt's influence had permeated American educational texts and the study of natural history. His *Personal Narrative* was widely read in the Peale family circle and had been among the volumes Stephen Long had taken with him on his eponymous expedition to the Rocky Mountains from 1818 to 1820. Humboldt's descriptions of his travels had helped shape American exploration protocols, and Long had found—as had Humboldt before him—that cultivating friendly relations with the Indian tribes helped ensure his survival in unfamiliar territory or adverse conditions. Long's ethnographic observations were among the most valuable aspects of the record of the expedition team. His published narrative became required reading for future explorers but also provided inspiration for a decade of novels by James Fenimore Cooper in the 1820s and Washington Irving beginning in the 1830s, both of whom used Long's settings and his descriptions of native communities as backstories for their fictional characters. Cooper's Leatherstocking novels, in particular, became the inspiration for paintings by Thomas Cole and multiple stage performances in the United States and Europe.

The Peale family's respect for Humboldt and their involvement with the Long Expedition provided a rich matrix for Catlin's developing interests in Native American culture and natural history. However, it was Jackson's Indian Removal Act that spurred Catlin to forgo painting members of Philadelphia society and instead portray individual Indians. He adopted a position similar to Humboldt's, asserting that

> *[t]he Indians of North America are . . . human beings, with features, thoughts, reason, and sympathies like our own; but few yet know how they* live, *how they* dress, *how they* worship, *what are their actions, their customs, their religion, their amusements, &c., as they practise them.*[14]

In a letter to Major James Hook in Washington, D.C., Catlin expressed his desire to study the Indians, proposing that he undertake a self-funded enterprise outside of governmental control:

> *If my life be spared, nothing shall stop me short of visiting every nation of Indians on the Continent of N. America. I go entirely at my own expense, and though as poor as Lazarus I would still go.*[15]

Humboldt would have approved. However, Catlin was not trained as an ethnographer nor as a cultural anthropologist. Catlin's was a well-intentioned amateur enterprise, fueled by his sincere desire to chronicle tribal customs and portray prominent individuals he believed were doomed to extinction by Jackson's misguided policies.

By the spring of 1830, Catlin had left the East Coast for St. Louis to establish a studio and concentrate on painting Native Americans. He explained his rationale for devoting his career to painting Indians:

> *I have flown to their rescue,—not of their lives or their race (for they are "*doomed*" and must perish), but to the rescue of their looks and their modes, at which the acquisitive world may hurl their poison and every besom of destruction, and trample them down and crush them to death; yet, phoenix-like, they may rise from the "stain on a painter's palette" and live again upon the canvass, and stand forth centuries yet to come, the living monuments of a noble race.*[16]

Catlin clearly articulated the fundamental conundrum in his practice. His interest in Native American cultures was genuine, but his outlook was relentlessly fatalistic. His aim was not, in fact, to rescue Native Americans as individuals or communities, nor to preserve the complexities of native cultures. Catlin was racing the clock to capture the likeness of Native Americans before their inevitable—to him—demise. When among the tribes, Catlin could be, at times, an astute observer allowed to witness major ceremonies, and, at others, a completely tone-deaf outsider who believed he could disregard cultural taboos with impunity. His belief that the disappearance of native culture was assured relied on the same ethnocentric assumptions behind Jackson's Indian Removal Act. Ironically, while Jackson viewed this as justification for his atrocities, Catlin saw it as a mandate for action. His moral outrage took the form of recording likenesses, regalia, and tribal ways as a cultural eulogy, not a rescue mission.[17] However, stripped of its melodrama, Catlin's belief in the value of preserving what he could of Native American culture would have resonated with Humboldt, as would the underlying sentiment that European colonization and intemperate expansion had created this dire circumstance. Catlin's determination was fixed in his belief that his art would change people's minds. If the Peales had imbued their museum with a civics lesson, Catlin would teach his audience to appreciate native culture through his Indian Gallery.

For access Catlin enjoyed the support of William Clark, who provided passports to enter Indian territory. Clark and his nephew, Benjamin O'Fallon, maintained their own collections of Indian artifacts, rivaling the display in Peale's Museum, which Catlin had become familiar with during his years in Philadelphia. It is possible that the

CAT. 58 George Catlin, *Máh-to-tóh-pa, Four Bears, Second Chief, in Full Dress*, 1832, oil on canvas, 29 × 24 in., Smithsonian American Art Museum, Gift of Mrs. Joseph Harrison Jr.

Peale family members provided Catlin with an introduction to Clark, helping to foster the frontiersman's early support for Catlin's endeavors. In 1831 Clark became one of Catlin's earliest patrons, for which the artist was profusely grateful.[18] Catlin eventually made five trips among native communities to amass his detailed accounts of the Indian tribes, but it was his first trip up the Missouri River in 1832 that formed the backbone of what would become Catlin's Indian Gallery.

On March 26, 1832, Catlin embarked on the inaugural run of the steamer *Yellowstone*, carrying a letter of introduction from Secretary of War Lewis Cass to his Indian agents along the Missouri River.[19] His itinerary took him up the Mississippi to the Missouri River, with lengthy stopovers at Fort Pierre and Fort Union on the way upriver and at Fort Clark on the return trip. At Fort Pierre, Catlin witnessed a Dog Feast, an important tribal ceremony, thanks to an invitation from Pierre Chouteau Jr. from the American Fur Company, and Indian Agent John Sanford, who was assigned to the upper Missouri River. At Fort Union, about two thousand miles farther upriver on the present-day border between North Dakota and Montana, Catlin stayed at the American Fur Company trading post from mid-June to mid-July. His garrulous host was Scotsman Alexander McKenzie, the larger-than-life fort manager. There he sketched and painted Blackfeet, Crow, Cree, and Ojibwe Indians.

Catlin's introduction to individuals from various tribes came through the trading posts of the international fur trading empire, the individual Indian agents stationed at the forts and trading posts, and the native leaders themselves. Mandan (Nų́mą́khų́·ki) Chief Mató-Tópe provided the critical approval on this trip. As was the custom, Catlin arrived bearing gifts to exchange for the opportunity to meet him and to negotiate painting his portrait (**CAT. 58**). Mató-Tópe (Four Bears) chose how he intended to be portrayed, his stance and his regalia asserting his authority. He did more than barter with Catlin for permission to paint his likeness. He granted Catlin extensive access to the village, specific members of the tribe, and even important ceremonies. The cordial relationship between the two men meant that Catlin's stay at Fort Clark, adjacent to the Mandan village, was the longest on this trip. After departing, Catlin passed more quickly through the territory of the adjacent Arikara (Sahnish), with most of his stops between there and St. Louis affording him time for painting quick oil sketches of the landscape.

This first trip upriver was a marathon, during which the artist covered roughly four thousand miles and created as many as 170 paintings.[20] Carrying precut lengths of canvas and a limited palette, he sketched as many Indians at each fort as he could. For his portraits he concentrated on the sitter's head and distinguishing characteristics of dress. To record ceremonies, tribal games, and hunts, Catlin relied on pencil sketches and his memory, working on his paintings after the fact. It appears that many of his small, dynamically painted landscapes were made *en plein air* as he stopped along the Missouri River bluffs. His technique allowed his paintings to dry quickly for ease of transport; afterward he intended to work them to a higher degree of finish in his studio. By October 20, Catlin was back in St. Louis, where he accompanied Clark to the Jefferson Barracks, a prison holding Ma-ka-tai-me-she-kia-kiak (Black Hawk). In St. Louis he painted a haunting and dignified portrait of this prominent Sauk leader, one of the earliest victims of Jackson's Indian Removal Act. He had been held prisoner there since August, following his surrender at Prairie du Chien. Catlin painted him in a matter of hours in late October before taking 140 of his paintings with him to display in Cincinnati en route to Pennsylvania.[21] Ultimately Catlin went on to paint more than five hundred works representing forty-eight native communities.

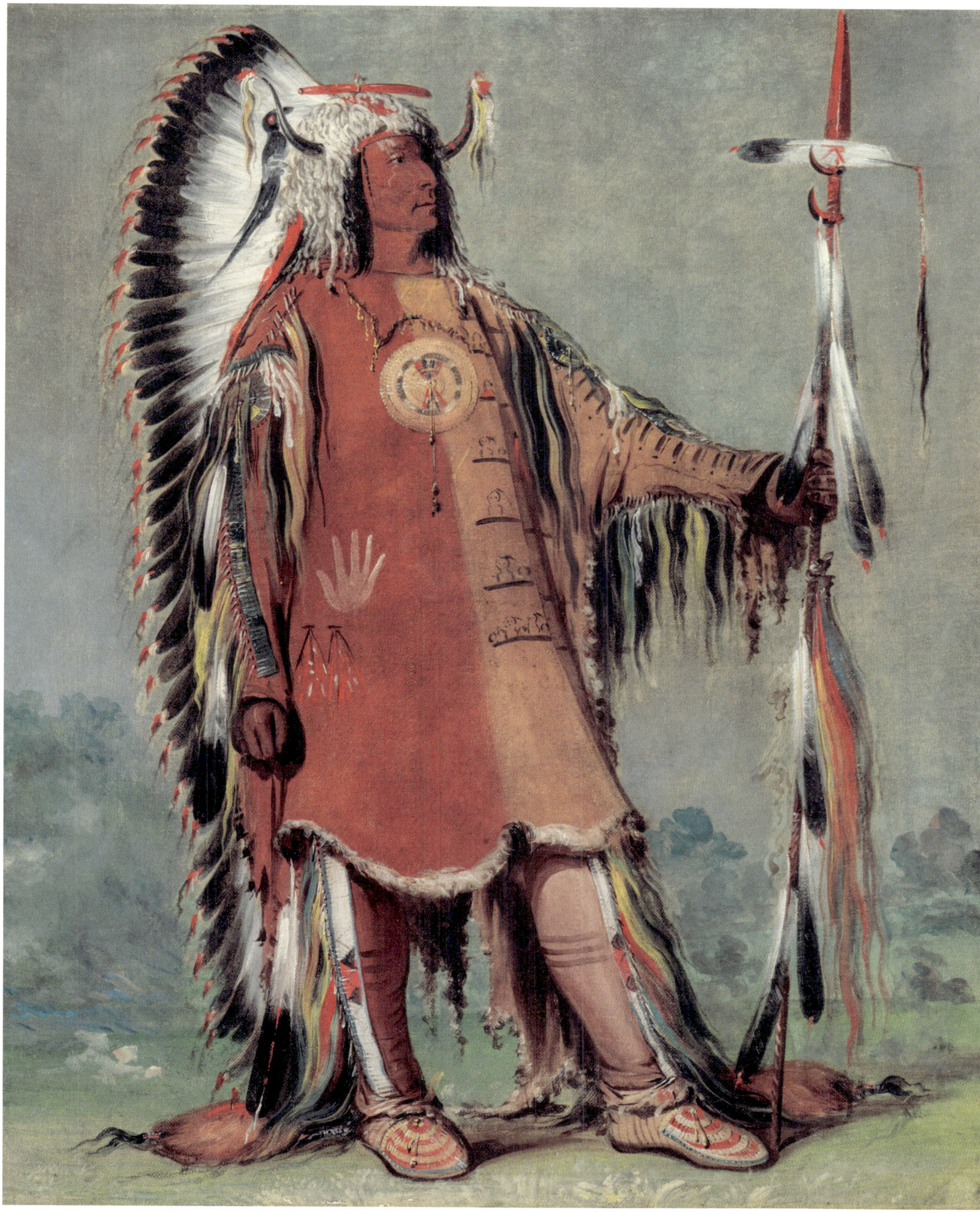

While Catlin was making his way along the Missouri River, Humboldt's protégé Prince Maximilian, accompanied by artist Karl Bodmer and the Wied family hunter and taxidermist, David Dreidoppel, arrived in Philadelphia.[22] Max's formal training in cultural anthropology and the broader array of natural sciences, like Humboldt's, was conducted under Blumenbach's auspices. At Göttingen he had learned about linguistics, tribal relationships, and field collecting.

Between 1814 and 1815 Max deployed those skills in Brazil, a trip he undertook with Humboldt's support and encouragement. He spent much of his time among the Botocudo (Aimoré), learning their lifeways and studying the region's physical geography.[23] His published account, translated into English in 1843, was widely heralded as a sensitive and perceptive study of their culture and environment.[24] Max then set his sights on traveling to North America to pick up where Humboldt had left off. He would spend the next ten years preparing for this trip, which was designed to study the indigenous cultures of North America as a corollary to his South American work. As he arrived in Boston Harbor on July 4, 1832, he was already watching everything with the trained eye of a cultural anthropologist. Max's gift for ethnographic observation manifested on his first day in the United States. Landing in Boston, Max and Bodmer were swept up in Fourth of July celebrations. Boston was in some ways as foreign to him as Brazil. Max's diary contains detailed descriptions of the houses and yards and the dress and manners of the local inhabitants, comparing them with his experiences in Europe and South America.

Max had read extensively about Native Americans before leaving Europe and understood the social and political tensions created by U.S. government Indian policy. He kept daily journals with detailed notes about many subjects, one of which was American attitudes toward the Indian tribes. Hoping to purchase copies of American books describing Indian cultures, he noted in his journal and in the published account of his trip:

How much I was astonished, that I could not find, in all the towns of this country, one good, that is, characteristic representation of them, but only some bad or very indifferent copper-plates, which are in books of travels! It is incredible how much the original American race is hated and neglected by the foreign usurpers.[25]

His choice of words—calling the European settlers "usurpers"—sent an early and clear indication of his sympathies with Humboldt's humanitarian views.

Prince Max's eastern itinerary, and the winter he spent with Thomas Say and Charles Lesueur in New Harmony, Indiana, are considered elsewhere in this book.[26] For this leg of his journey, he carried a letter of recommendation from Duke Bernhard of Saxe-Weimar to William Clark; additional letters had introduced him to influential members of the scientific community in and around Philadelphia. He generated goodwill every place he stopped on his journey, which meant he often came away from meeting one person armed with a letter introducing him to another valuable contact at his next destination. In Philadelphia, Dr. Richard Harlan, a physician and naturalist who had been affiliated with the Long Expedition, wrote a letter of introduction for Max to present to James Reid Lambdin, an artist who had been inspired by the Peales to open his own museum in Pittsburgh.[27] There Max saw Indian artifacts borrowed from William Clark, along with natural history specimens.[28] Lambdin took a liking to Max and willingly wrote letters to introduce him to friends who could show him the local Indian mounds. Lambdin composed another letter to his brother-in-law, Alfred Cochrane, who shared Max's interest in amphibians and reptiles.[29]

Prince Max was eager to meet Native Americans and quickly made the connection between their absence on the East Coast and current Indian policy. In New Harmony during the winter of 1833, he noted in his diary:

Unfortunately, during my entire North American journey, I have not yet seen one native Indian or any of their descendants. Now one already has to cross the Mississippi to find them. During the course of this winter, they even want to drive across the Mississippi the last settled remnants of these nations, the Cherokees, Choctaws, and other peoples—an act of sheer brutality! Such is the vaunted liberty of America![30]

Max's anger at the inhumanity of American Indian policy resonates throughout his journals. Like Humboldt, he had difficulty squaring proclamations of liberty and justice that, in practice, applied to only a fraction of the country's inhabitants. Like Catlin, he was not in a position to do more than record what he experienced. However, Max documented in his journals his impressions of every race he encountered. His disgust with American Indian policy was matched by his revulsion at the perpetuation of slavery.

Max's itinerary resembled Humboldt's more than Catlin's in that the first half was devoted primarily to observing the German American communities dotting the countryside and taking notes on the natural history of the landscape. Although all of his letters and journals were written in German, he appeared to speak English, and possibly French, well enough to communicate. Both languages were useful as he and Bodmer ascended the Missouri River. But first Max had to decide which route he would take once he arrived in St. Louis.[31]

Both Catlin and Prince Max relied heavily on Clark's assistance in securing guides to accompany them up the Missouri River. They also needed his help obtaining permission to visit specific tribes and individuals as they traveled. Clark warmly welcomed Max and Bodmer, providing letters of introduction to the Indian agents at Fort Leavenworth, Prairie du Chien, and St. Peters along the Missouri River.[32] Beyond that, Clark arranged for the travelers to ride the steamboat *Yellowstone* up to Fort Union, escorted there by agents from St. Louis. Max also met some of the same people who had aided Catlin the year before, notably Pierre Chouteau Jr., of the American Fur Company, who graciously allowed Max and his companions to ride the company's boats and stay in their trading posts. Max's connection with the Astors probably influenced Chouteau. One of Max's classmates at Göttingen was William Backhouse Astor, son of John Jacob Astor, founder of the American Fur Company. Max clearly hoped the Astor name would facilitate his access to the Indians living on the upper Missouri River.[33]

Max's diplomatic skills and network of references also made a favorable impression on Clark and O'Fallon. The two maintained collections of Native American artifacts, to which each gave Prince Max and Bodmer ready access. Clark in particular made sure Max was carrying appropriate gifts to present to the leaders of each community he would meet. O'Fallon introduced Max to Charbonneau, who accompanied the Europeans upriver, and his skills as a translator were immensely helpful when Max and Bodmer reached the Mandan and Arikara villages.

Perhaps even more valuable than the advice and the gifts was Clark's extraordinary gesture of respect in providing Prince Max with a complete set of the route maps he had drawn during the Corps of Discovery's travels up the Missouri River (see cat. 22). The reciprocal generosity Humboldt had practiced in Philadelphia and Washington in sharing his map of Louisiana now carried forward to Clark's sharing his own hand-drawn maps.[34] As they ascended the river, Max studied and annotated this atlas of maps and read from the Long Expedition report, which he considered the best account of that portion of their route.[35] Max, however, had done the bulk of his preparation in Germany and reinforced it during his extended stay in New Harmony. On his journey, he worked as he had in Brazil, from memory, using his journals as a daily diary, vocabulary log, and running list of plants and animals, with room left along the margins and at the bottom of most pages for additional or comparative information that could be added at a later date (see cat. 59).

ARTISTIC PRACTICE IN THE FIELD

Before they left St. Louis, Max finally met his first Native Americans. Clark invited Max and Bodmer to attend a meeting concerning the continued imprisonment of Black Hawk. Max was impressed. He wrote, "The first sight of them, which did not a little surprise me, convinced me at once of their great affinity to the Brazilians, so that I cannot hesitate to consider them as belonging to the same race." In a footnote, Max continued,

> *In confirmation of the similarity of the Americans to each other, we may quote the authority of Humboldt, and other travelers.*[36]

Throughout his journals, Max evaluated Native American physiognomy, manners, dress, language,

CAT. 59 Prince Maximilian zu Wied-Neuwied, *Indian Ponies, Manuscript Journal of Prince Maximilian's Travels in North America, Parts 7–14 (Vol. 2)*, 1833, ink and watercolor on paper, single page: 12 ¾ × 7 ¾ in., Joslyn Art Museum, Omaha, Nebraska, Gift of the Enron Art Foundation

and culture. He recorded separate vocabularies for each tribe and, in most cases, appeared to be effective in making his subjects comfortable with his questions. In numerous places Max extended his comparative assessment of indigenous peoples from Brazil, the Andes, Mexico, and the United States, enumerating distinctive and commonly held physical traits. But Max invoked Humboldt as well—explicitly channeling his mentor and colleague.

The previous winter, Prince Max had contracted what might have been a mild case of cholera, forcing the travelers to remain in New Harmony. While Max was convalescing, Bodmer had taken a side trip to New Orleans, where he made several tentative sketches of local Indians. He realized that he needed to master the details of Indian facial features, body language, ornamentation, and dress as the central focus of his work. Bodmer used this first encounter with the Indians to make significant adjustments to his working method in order to make the most of his limited time sketching and painting the Indians. Because everything about North America seemed different, every impression was a first impression, a situation he found taxing at first but critical for the success of his participation in Prince Max's venture.

Bodmer acknowledged to Max that his "lack of expertise" in sketching and painting Indians from life meant that initially he had "to copy with unusual eagerness and attention every aspect realistically...in an almost ridiculously fussy manner but also with as much authenticity and truth."[37] Later the artist observed to a colleague the importance of practicing his desired subjects ahead of time in order to be able to reflexively compose as much as possible, rather than having to learn every facet of his subject and surroundings on the ground.

In New Harmony, Max had also heard that an artist named Catlin had spent the previous year traveling among the Indians along the same route he was now contemplating.[38] In St. Louis he had a chance to inspect a group of Catlin's paintings from the 1832 trip to Fort Union that were in O'Fallon's collection.[39] The Europeans were not particularly impressed with Catlin's work. The stark contrast between Catlin's large-scale oil sketches and Bodmer's smaller, more precise watercolors was pronounced, so much that when they returned from their year among the Indians and viewed Catlin's paintings a second time, Max cryptically, if diplomatically, stated, "Here we saw a collection of Indian portraits and scenery by Mr. Catlin, a painter from New York, of which we were able to form an opinion after our recent travels in the country."[40] Their working methods in the field were significantly different, as were their ultimate aims. Bodmer knew his works would be translated into delicate and highly detailed prints for a book; Catlin envisioned a room covered wall to wall with his portraits to create an immersive experience. Both men succeeded admirably in their chosen goals. And both men could capture a distinctive likeness, and having made their field paintings only a year or two apart, they often ended up painting the same individuals.

Bodmer also worked exclusively on paper, either in pencil or in the unforgiving medium of watercolor. His watercolors made in the eastern United States, around Bordentown and in the Pennsylvania wilderness, demonstrated his sure hand and feel for landscape features. Although he apparently had little formal training in portraiture, Bodmer turned out to be remarkably well suited for this role and quickly learned how to observe and capture the specifics of features, dress, and regalia for his Indian sitters. He inscribed most of his field sketches with the name of the individual, tribal affiliation, date, and location. These drawings and field watercolors, in essence, formed Bodmer's record of their travels, a visually rich complement to Max's detailed journals.

The importance of Bodmer's contributions to Max's expedition cannot be overstated. Bodmer's skill is most evident in his watercolors, whether they are landscapes or portraits. Anchored by a sensitive, fluid pencil line, his compositions are

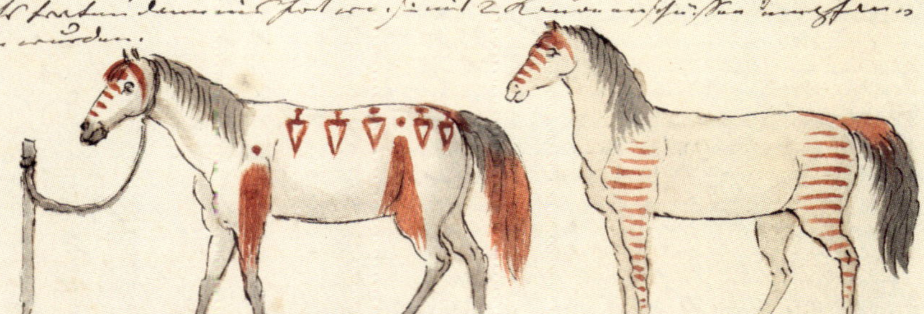

CAT. 60 After Karl Bodmer, *Travellers Meeting Minatarre Indians near Fort Clark*, 1842, hand-colored aquatint, plate size: 11 15⁄16 × 13 ½ in., Joslyn Art Museum, Omaha, Nebraska, Gift of the Enron Art Foundation

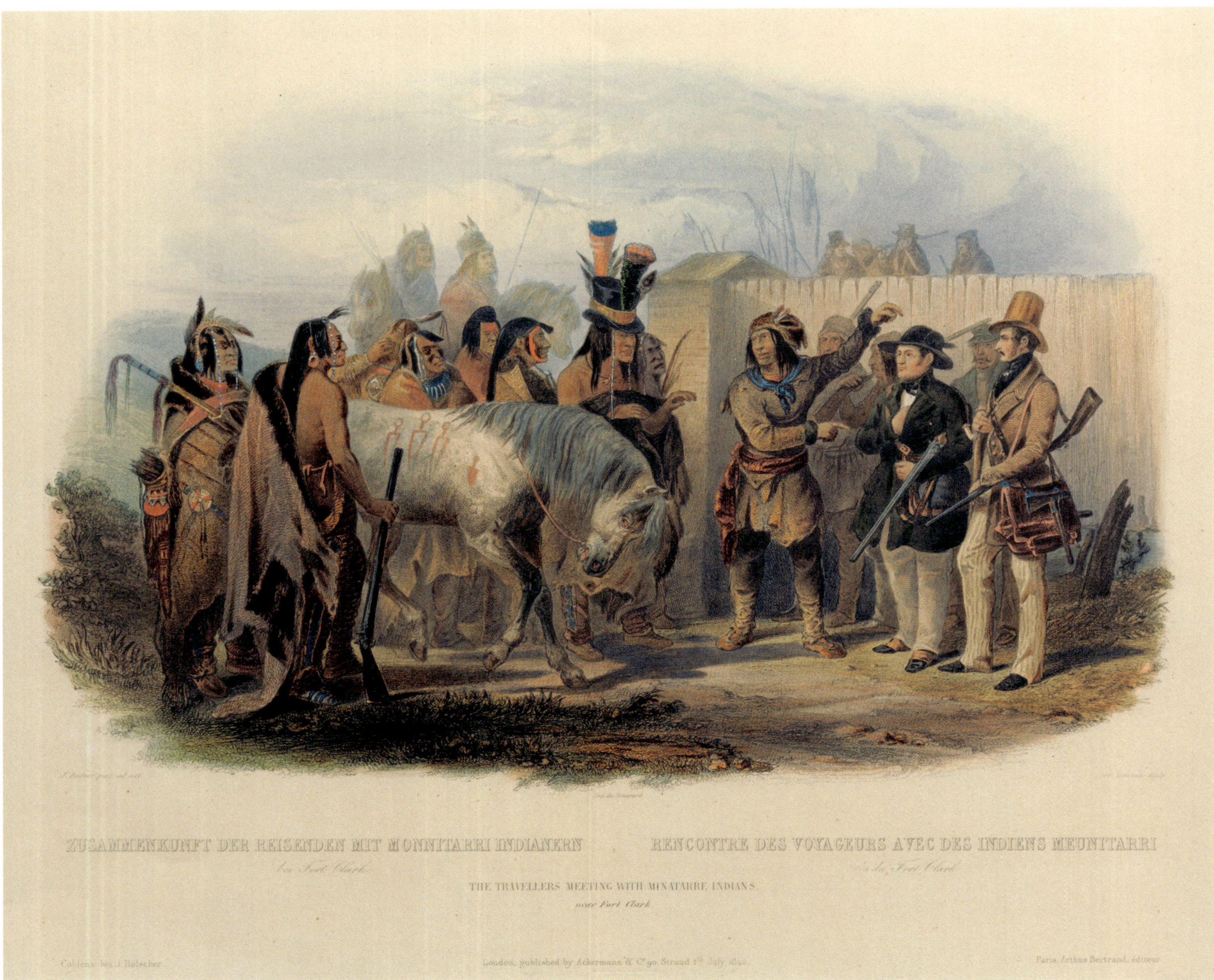

indicative of his academic training; in watercolor, he handles the difficult medium with innate grace and nuance. As a draftsman, Max had decent talent for rendering artifacts and structures, and although these drawings (like those of the reptiles and amphibians he had sketched in Brazil) demonstrate an acute level of observation, his repertoire was limited. A side-by-side comparison of their respective renderings of the painted horses they saw when they met the Minnetaree makes it abundantly clear why Humboldt had held firm in his advice to Max to pay a trained artist to accompany him. Max's drawing (CAT. 59) has the charm of an amateur, more concerned with recording the painted markings than the correct anatomy of the horses, while Bodmer's finished print (CAT. 60) stages the encounter as a complex, detailed narrative. The now-fiery painted pony occupies center stage in the formal meeting among Max (in dark green), Bodmer (far right), Dreidoppel (background), and the group of Minnetaree. The visual record with which Bodmer returned to Paris became the most highly accomplished body of work depicting the Indian tribes of the upper Missouri, especially the Mandan. Paired with Max's acute written observations about the culture and appearance of these tribes, Bodmer's hand-colored aquatints based on his fieldwork resulted in a book every bit as important as that Max had produced following his trip to Brazil.

MATÓ-TÓPE

Both Catlin and Max spent considerable time among the Mandan Indians, in part due to the tribe's willingness to engage with white artists and travelers. The Mandan lived in two permanent villages outside the trading post at Fort Clark. Mató-Tópe, second chief of the Mandan, got to know each of these visitors well, and it was he who determined their access to his villages, their dwellings, and their ceremonies. In 1832 Catlin's access to the Mandan was a function of Mató-Tópe's willingness to entertain Catlin's questions and satisfy his curiosity, which sent a signal to his tribe that Catlin could be trusted. Catlin's regard for the Mandan is apparent in his books. Nearly half of his first volume was devoted to describing them. Max and Bodmer arrived at Fort Clark months later, stopping briefly on their way farther north to Fort Union. A violent battle between the Blackfeet and Assiniboine outside of Fort McKenzie so unnerved Prince Max that he decided to return to Fort Clark to spend the winter of 1833 to 1834 with the Mandan. The extended stay provided Max with a rich vein of ethnographic information and allowed Bodmer to make some of his finest watercolors of individual Indians. Images of the Mandan account for fully a quarter of Bodmer's portfolio from this trip.[41]

Mató-Tópe knew how he wished to be seen, understood the power of his stance and appearance, and negotiated the terms of his depiction. The similarities between the formal, full-length portraits by Catlin (see cat. 58) and Bodmer (CAT. 61) suggest that the chief exerted a high degree of control, making sure his pose and his regalia conveyed his importance to the artists and within his tribe. In a letter to his brother, August, Max relayed an anecdote from the chief:

Just now, the Mandan chief Mató-tope (translated: the 4 bears) was here, an excellent warrior and a man of noble behavior. Only a few features about his character. A traveller from New York, Painter C. [Catlin] was here during the former years, and he asked Mató-tope to paint him a few buffalo-hides, in which he is particularly skilled. It is usual to paint on such hides the warlike deeds of the heroes in figures with bright colors. The Indian bought some beautiful hides, which are rather expensive here, painted them in a fine way, and brought them to C. The latter gave him some unimportant small things, much below the value of what he had got, and Mató-tope's revenge was that he told C; "that he must be very poor if he is unable to pay according to the value, so he wants to give it as a present."[42]

Mató-Tópe's preference for Max's gifts and his dismissal of Catlin's may have been a diplomatic ploy to ensure that the Europeans were grateful for his approval, but it also undoubtedly signaled the respect developing between the prince and the chief. Mató-Tópe sold Max several of his painted robes, as he had sold them to Catlin the year before, and he continued to exchange them with fur traders and Indian agents. Mató-Tópe recognized that these artifacts were highly prized by his white visitors. Bison robes painted with pictographic scenes of valor were made by the men; geometric

Tab. 13.

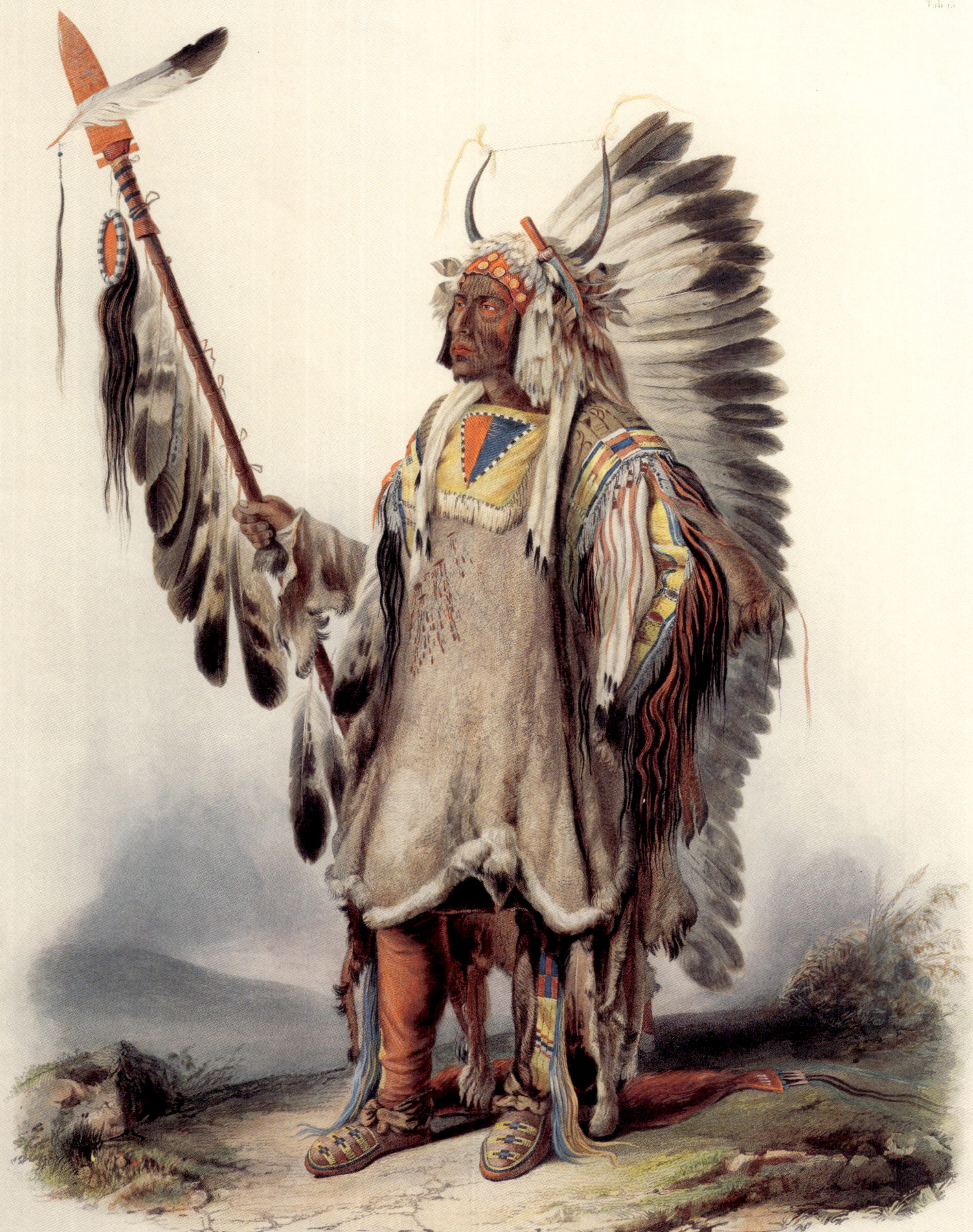

MATÓ-TOPE

Mandan chef *Chef Mandan*

A MANDAN CHIEF

CAT. 61 After Karl Bodmer, *Mató-Tópe, a Mandan Chief*, 1839, hand-colored aquatint, plate size: 20 ¼ × 14 ⅝ in., Joslyn Art Museum, Omaha, Nebraska, Gift of the Enron Art Foundation

designs were often painted by the women. Max was particularly interested in a robe that depicted Mató-Tópe's battle exploits. In a letter to his brother, he wrote:

In a fight against the Chayennes, he had marched off with the Manitaries [Minnetarees]. When they were riding against the enemy, a Chayenne-chief came and shouted: "Is Mató-tope among you?" The latter rode fastly ahead, they fired at each other and fought, but the powder-horn of the Mandan was shot into pieces and he could not load any more. Now they sprang from their horses and went one against the other, The Chayenne pulled his knife and wanted to stab Mató-tope, but the latter caught the knife, whereby he was severely wounded and stabbed the Chayenne.[43]

Max eventually purchased a painted robe from Mató-Tópe that depicted this encounter, and Bodmer made a watercolor of this robe that served as the basis for an aquatint reproduced in the book of their journey (**CAT. 62**). Max was amassing an extensive collection of Indian artifacts, which he packed into trunks to be shipped back to Germany from points along the way. Mató-Tópe provided much of the material Max acquired from the Mandan, and in a letter to his sister, Max mentioned some of these objects. Mató-Tópe showed Max some of his own prized possessions, which Max described as well:

He owns quite curious things, feather-caps and ornaments, drums, everything painted and decorated in a beautiful manner, objects of great value in the eyes of these people, as these are considered to be medicine, i.e. a protective device or magic objects, because superstition and sacrifices for their guardian spirits are still at their day's programme.... Mr. Bodmer will continue...to make pictures of all these objects and I describe them afterwards, and this affords to me an agreeable occupation for the long and sad winter.[44]

At some point during Max's stay, the chief used Bodmer's watercolor paints to make a drawing (**CAT. 63**) that depicted the episode wherein he defeated the Cheyenne chief. In Europe, Bodmer arranged to have Mató-Tópe's watercolor copied as a print for the same volume, where these two works appeared side by side. Max published

a fac-simile of a representation of one of his exploits, painted by himself, of which he frequently gave me an account.... Mató-Topé's drawing of the scene in the above-named plate, shows the guns which they had discharged and thrown aside, the blood flowing from the wounded hand of the Mandan chief, the footsteps of the two warriors, and the wolf's tail at their heels—the Chayenne being distinguished by the fillet of otter skin on his forehead.[45]

Over the winter that Max and Bodmer spent with the Mandan, the level of trust between them continued to grow. Sharing Bodmer's watercolor paints with the Mandan was its own form of cultural exchange; including one of the Mandan chief's own paintings in the published account of this journey became another sign of the mutual respect that developed between them.

As a result of their extended stay at Fort Clark, Mató-Tópe was not the only chief to command Max and Bodmer's attention. On March 7, 1834, Péhriska-Rúhpa, a Minnetaree chief also known as Two Ravens, allowed the Europeans to witness the dog dance. Mató-Tópe had not been pleased when Péhriska-Rúhpa had started spending time with Max and Bodmer. Max sensed the two men did not get along and were in competition for the travelers' attention; however, gradually the iciness thawed, and they tolerated each other's company to a far greater degree.[46] Mató-Tópe attended the dog dance, which Bodmer painted as *Pehriska-Ruhpa, Moennitarri Warrior in the Costume of the Dog Dance* (**CAT. 64**).

Péhriska-Rúhpa was one of the chiefs of the Dog Society. The dog dancers were considered the bravest warriors in the tribe's Dog Society, tasked with fighting to protect the tribe. The dance featured war maneuvers and was performed to bring success in battle. In his journal Max made an elegant line drawing of Two Ravens in his elaborate regalia (fig. 75) and described his headdress:

CAT. 62 After Karl Bodmer, *Indian Utensils and Arms*, 1840, hand-colored aquatint, plate size: 16 × 20 ¾ in., Joslyn Art Museum, Omaha, Nebraska, Gift of the Enron Art Foundation

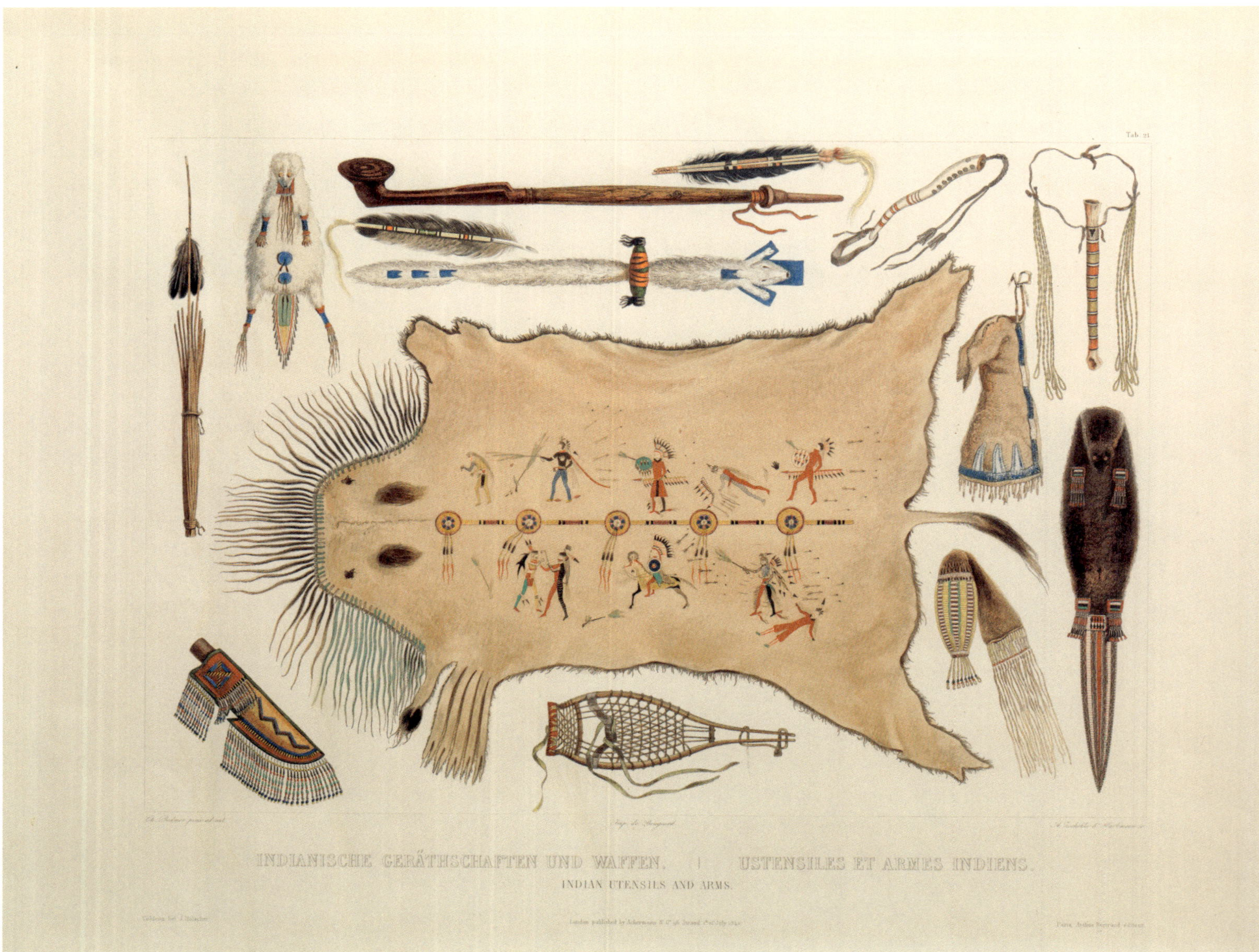

CAT. 63 Mató-Tópe, *Battle with a Cheyenne Chief*, 1834, watercolor and pencil on paper, 12 ⅜ × 15 ⅜ in., Joslyn Art Museum, Omaha, Nebraska, Gift of the Enron Art Foundation

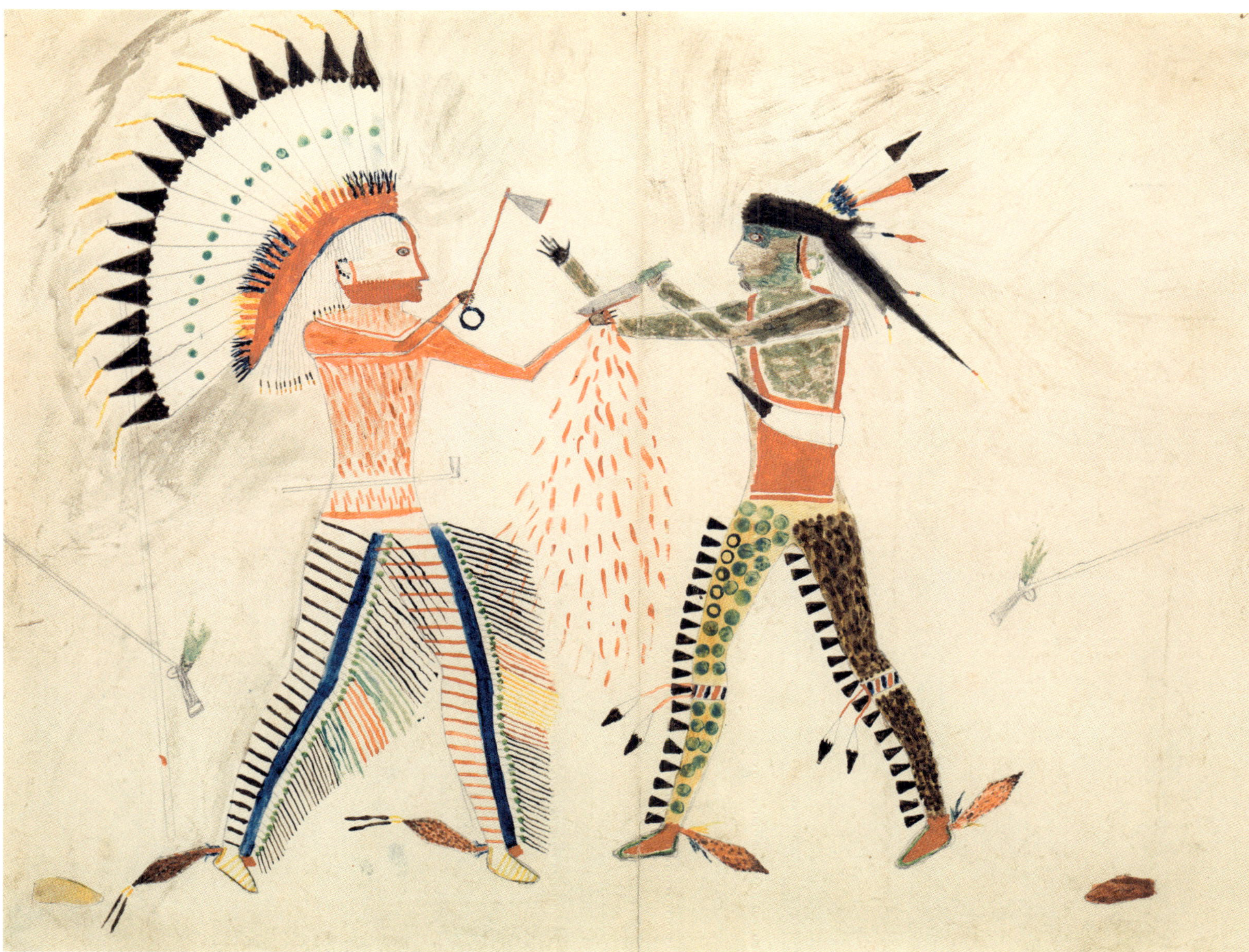

Tab. 29.

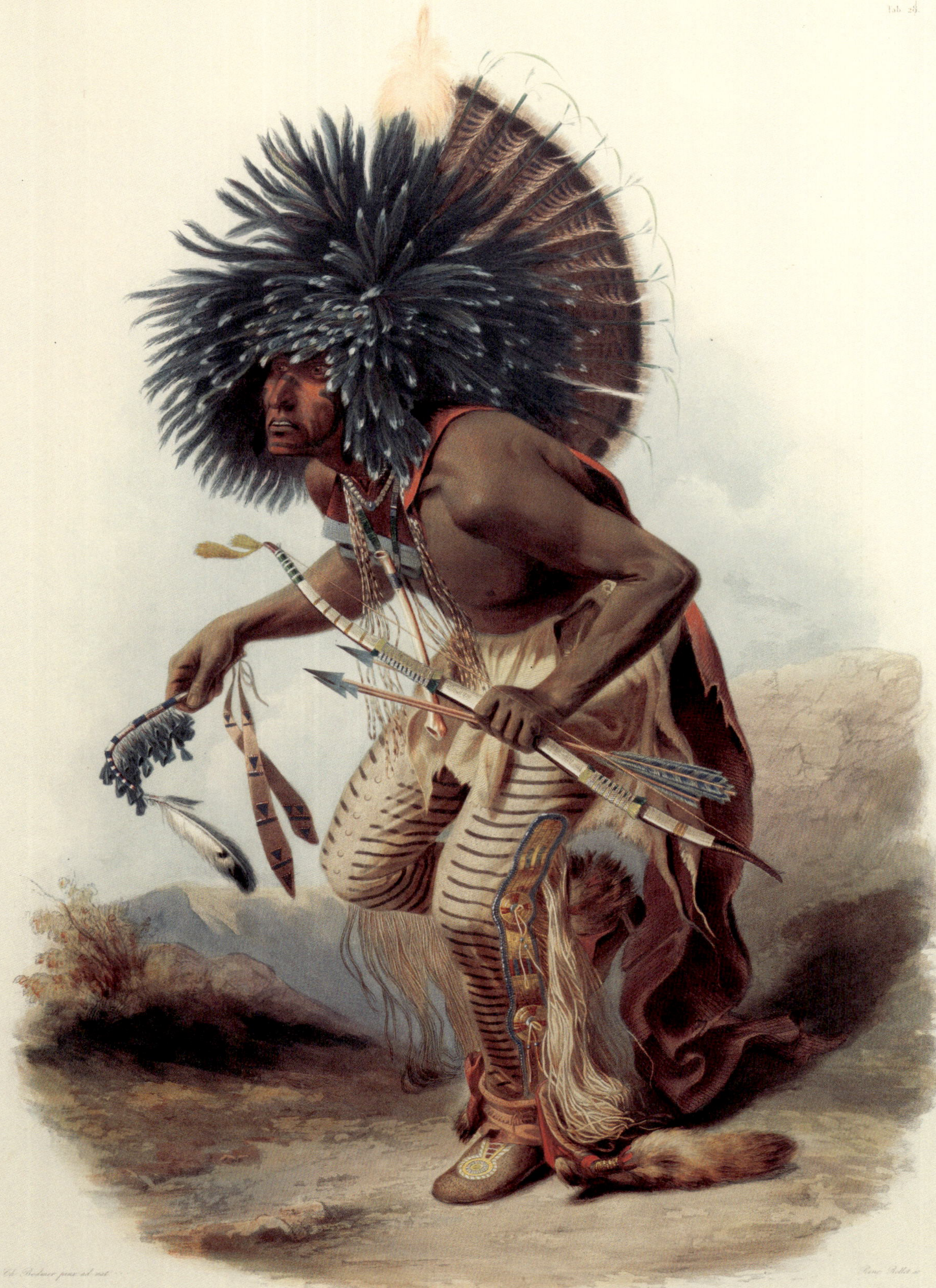

Ch. Bodmer pinx. ad nat. Réné Rollet sc.

Imp. de Bougeard

PEHRISKA-RUHPA.

Mönnitarri Krieger im Anzuge des Hundetanzes. Guerrier Mönnitarri costumé pour la Danse du Chien

MŒNNITARRI WARRIOR IN THE COSTUME OF THE DOG DANSE.

Coblenz, bei J. Hölscher — London, published by Ackermann & Co. 96, Strand February 1st 1841. — Paris, Arthus Bertrand, éditeur

CAT. 64 After Karl Bodmer, *Pehriska-Ruhpa, Moennitarri Warrior in the Costume of the Dog Dance*, 1841, aquatint and engraving on paper, plate size: 21 ¹⁄₁₆ × 15 ¼ in., Joslyn Art Museum, Omaha, Nebraska, Gift of the Enron Art Foundation

Four of them—these are the leading, or real, Dogs—wore enormous bonnets or caps on their heads, [which hung down] far below their shoulders: giant tufts of densely arranged raven tail feathers [with] small down feathers attached [to each raven feather] tip. In the middle of [each] mass of feathers, the beautiful white and black tail of a war eagle stood upright, attached in a line reaching from front to back. When they dance, all these feathers rock up and down.[47]

In his color aquatint, Bodmer also includes a war whistle and a rattle made from a deer hoof attached to a beaded stick. Max followed this with his own watercolor of Péhriska-Rúhpa, seen from behind (**CAT. 65**). Perhaps in recognition of his more limited ability to handle the watercolor medium, Max chose to focus on the regalia rather than the individual. A week later, Max wrote:

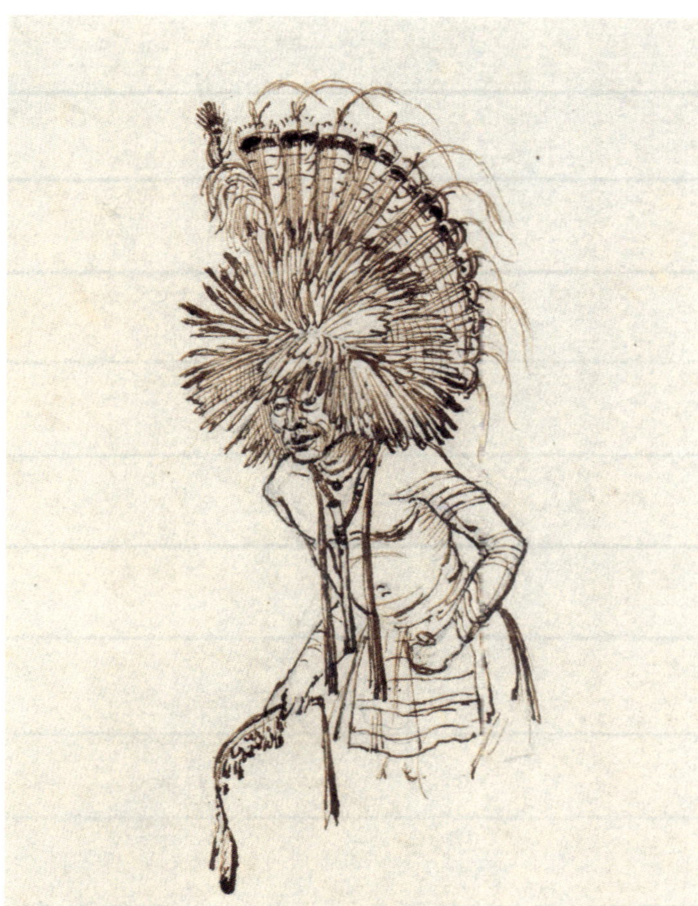

Fig. 75 Maximilian zu Wied-Neuwied, *Line drawing of Pehriska-Ruhpa, Manuscript Journal of Prince Maximilian's Travels in North America, Parts 15–29 (Vol. 3)* (detail), 1834, ink on paper, full page: 12 ¾ × 7 ¾ in., Joslyn Art Museum, Omaha, Nebraska, Gift of the Enron Art Foundation

Pehriska-Ruhpa spent several days with us, in order to have his portrait taken in his dress of one of the chiefs of the dog band. When the sitting was over, he always took off his ponderous feather cap, and rubbed it twice on each side of his head, a charm or precaution which he never neglected. He then seated himself with his friend, Mato-Topé, by the fire-side, when both took their pipes, the latter, however, always turning around first, and making everybody in the room sit down. . . . I was enlivened by the frequent visits of the Indians.[48]

Max was also able to transcribe the Mandan vocabulary. He spent winter evenings interviewing the chiefs of the Mandan and Minnetaree with the help of Charbonneau and James Kipp, the Indian agent assigned to Fort Clark who was in charge of trade with the Mandan. Max was grateful, noting,

Mató-Tópe gave me very accurate information respecting his own language, and that of the neighboring Indian nations, and took great pleasure in communicating to me some words of the Mandan and Arikkara languages, the latter of which he spoke fluently.[49]

During the winter, Max began writing a detailed ethnology of the Mandan, covering their appearance, health, diet, family life, and tribal structure. He also sketched their dwellings and described their internal hierarchies. He commented favorably on their morals and intelligence and described their spirit animals, medicine, ceremonies, hunting methods, techniques of warfare, and rituals surrounding sickness and death. Dipäuch, a prominent Mandan, contributed greatly to Max's study of the tribe's religious beliefs and material culture. With translation help from Kipp, Max transcribed the Mandan's stories of ceremonies. Dipäuch invited Max and Bodmer to enter his lodge, a privilege not often granted. Throughout the winter, Bodmer had several opportunities to enter the lodge, allowing him to compile an impressive portfolio of pencil sketches and watercolors. His finished print conveys an astonishing level of detail (**CAT. 66**).

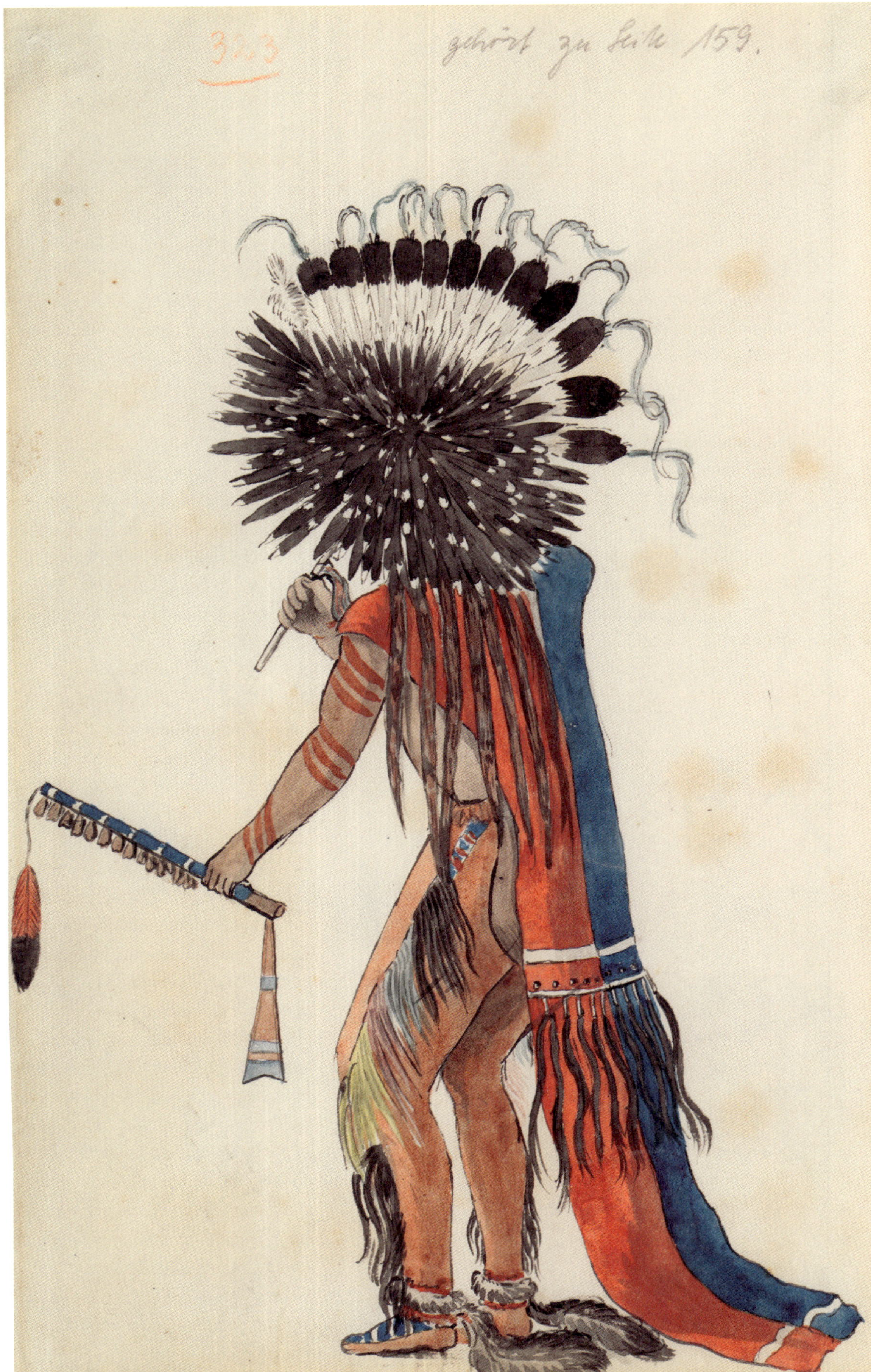
gehört zu Seite 159.

CAT. 65 Maximilian zu Wied-Neuwied, *Pehriska-Ruhpa in Costume of Dog Dancer*, 1834, watercolor and pencil on paper, 11 ⅝ × 7 ½ in., Joslyn Art Museum, Omaha, Nebraska, Gift of the Enron Art Foundation

When Catlin had visited Fort Clark in the spring of 1832, Mató-Tópe had granted the artist permission to witness part of the O-Kee-Pa ceremony, which lasted four days and was the central religious rite for the tribe. This complex ceremony related to the creation of the earth and involved a series of ordeals through which warriors proved their physical strength and mental endurance. Preparation for the ceremony could take the better part of a year, and those who completed the grueling challenges were honored by the tribe. Access to this rite was severely restricted, another unusual honor that suggests Mató-Tópe's ease with Catlin's presence. With Kipp's help as an interpreter, Catlin watched the ceremony, making sketches from memory afterward, followed by a series of small paintings.

Catlin's painting of the Bull Dance (or Bison Dance) represents the opening event of the ceremony, designed to call the buffalo to the Mandan village to ensure a good hunt (**CAT. 67**). The following week, Catlin took the time to write a more detailed description of the entire sequence of the O-Kee-Pa ceremony.[50] Max and Bodmer were not with the Mandan in the spring or fall, the usual times for the ceremony to be held; instead, Dipäuch provided the description that Max recorded in his journal, down to the meaning behind all of the participants' roles and ceremonial regalia. The records Catlin, Bodmer, and Max accrued took on added significance after 1837, when smallpox ravaged the tribe, nearly wiping it out.[51] More than twenty years later, the aftermath of that tragic epidemic would bring their lives into alignment.

Curiosity about what Bodmer was doing with his paints extended beyond Mató-Tópe to "a handsome tall Mandan-Indian, his name is Sih-chi-dä, the yellow feather."[52] Sih-Chidä (**CAT. 68**) asked Bodmer to paint his portrait and then to teach him to use watercolor. Max described him as a close friend, who "came almost every evening, when his favourite employment was drawing, for which he had some talent."[53] Bodmer ended up teaching several of the Mandan how to paint in watercolor, which could be a challenge during the worst of the winter, when temperatures dropped far enough to freeze the artist's paints, even indoors.[54] Max began to collect paintings made by the Mandan themselves and sent several home in a letter to his brother:

> *I own highly curious drawings from several Indians representing figures, fights etc., which will interest you very much, in general, my collection of Indian drawings will be one of the greatest interest and fill out many hours with pleasant entertainment.*[55]

Max's fascination with North American Indians prompted him to consider bringing a member of another tribe back to Europe with him, although in the end he decided the risks were too high. As he concluded in a letter to his sister:

> *I would have been glad also to bring an Indian from here, but they do not like to go away and it costs too much. A young Dacota* [sic] *told me repeatedly he would accompany me, however one does not know if he would not become homesick later, and this would be bad.*[56]

CULTURAL TRANSGRESSIONS

That Max even considered returning to Europe with a Dahkóta is a reminder of the limits of cultural awareness of the era. Respect for native peoples had gained Max and Catlin extraordinary access to the Mandan through Mató-Tópe; however, both Max and Catlin were also very much products of their age, believing that the acquisition of knowledge at times excused cultural insensitivity. One reason Max had traveled in Brazil was the opportunity to do the research denied to his mentor, Humboldt; the other was to contribute information to Blumenbach's ongoing research on human origins. Blumenbach believed that the indigenous races of both American continents derived from the same common ancestor.[57] This larger issue of human origins would divide scientists and ethnographers throughout the nineteenth century. Monogenists hewed to the belief that all humans descended from a single ancestor; polygenists believed there were many different races

CAT. 66 After Karl Bodmer, *Interior of a Hut of a Mandan Chief*, 1841, hand-colored aquatint, plate size: 16 ⅜ × 21 $^{5}/_{16}$ in., Joslyn Art Museum, Omaha, Nebraska, Gift of the Enron Art Foundation

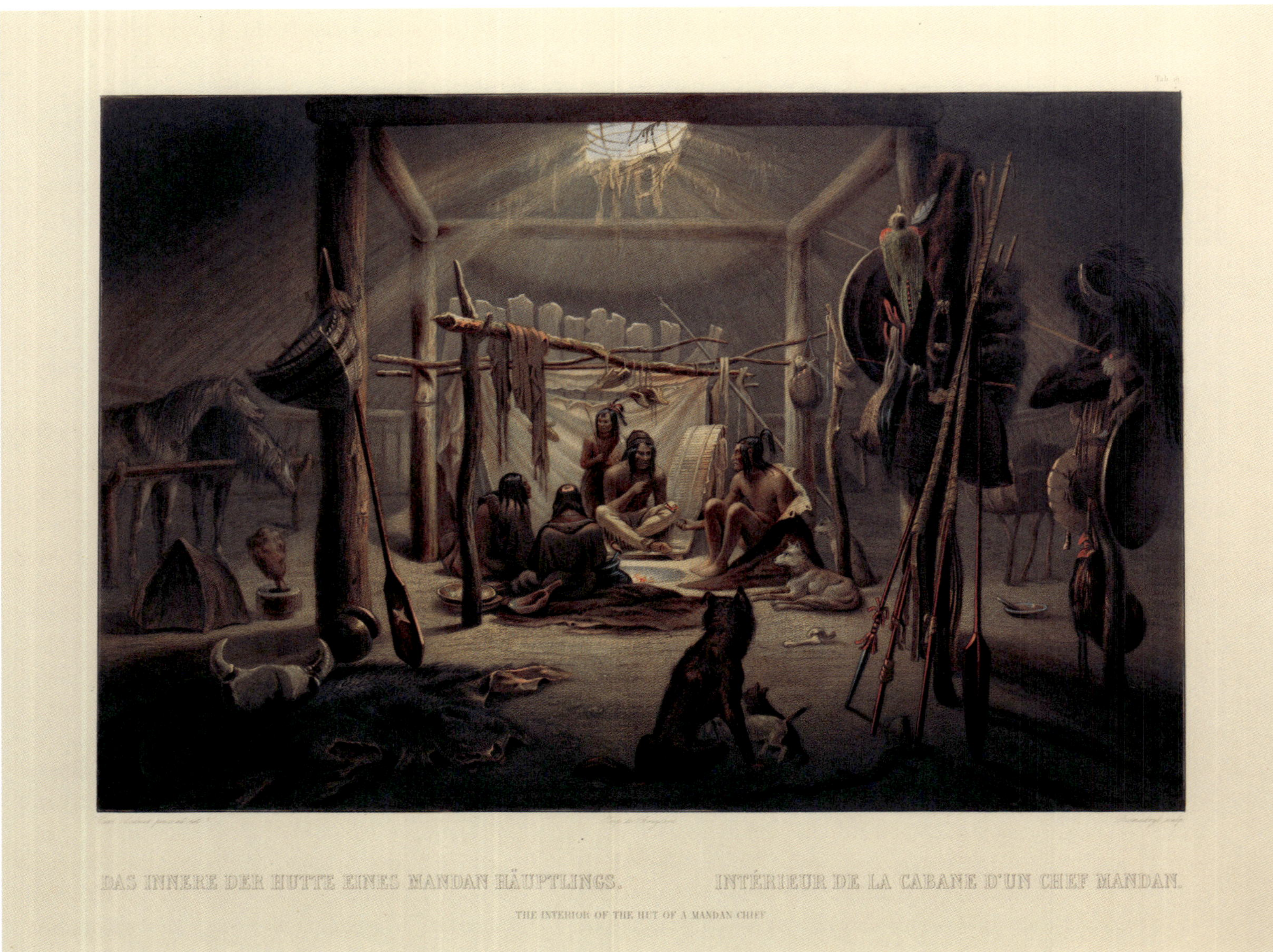

CAT. 67 George Catlin, *Bull Dance, Mandan O-kee-pa Ceremony*, 1832, oil on canvas, 23 ¼ × 28 in., Smithsonian American Art Museum, Gift of Mrs. Joseph Harrison Jr.

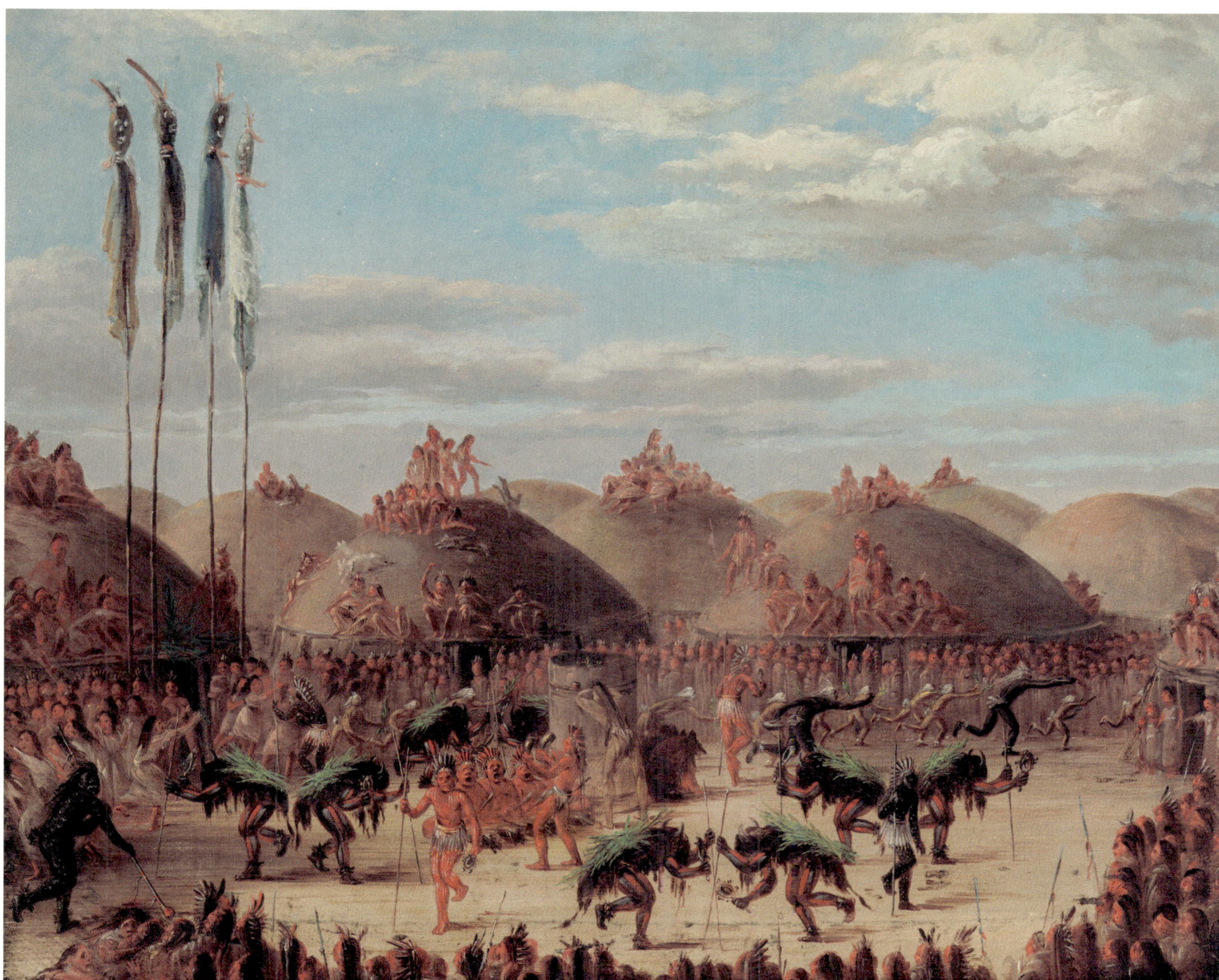

Tab. 20

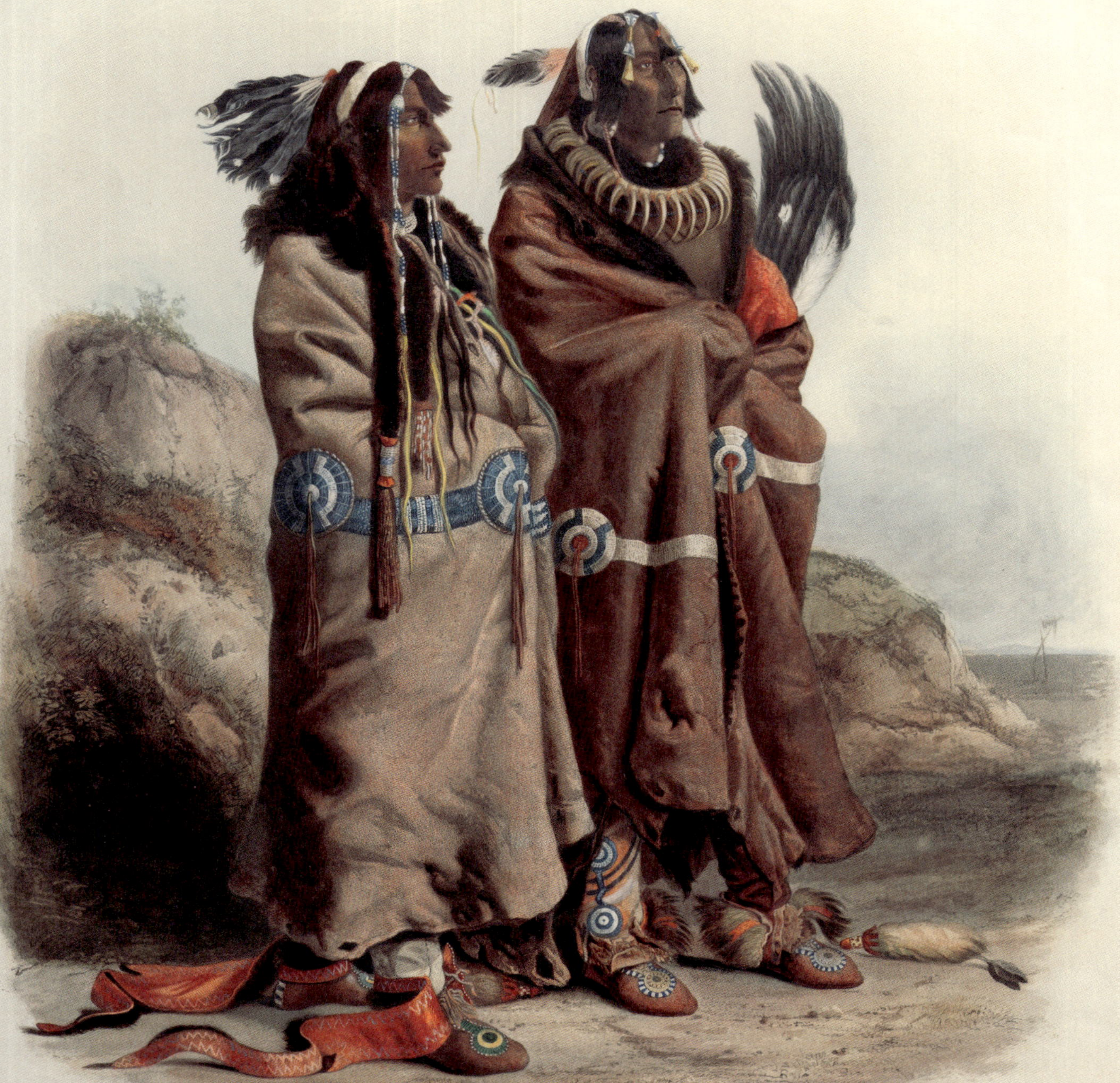

SIH-CHIDA & MAHCHSI-KAREHDE.

Mandan Indianer — *Indiens Mandans*

MANDAN INDIANS

Coblenz bei J. Hölscher — London, published by Ackermann & C.º 96, Strand — Paris, Arthus Bertrand éditeur

CAT. 68 After Karl Bodmer, *Sih-Chida and Mahchsi-Karehde, Mandan Indians*, 1841, hand-colored aquatint, plate size: 19 13⁄16 × 16 ½ in., Joslyn Art Museum, Omaha, Nebraska, Gift of the Enron Art Foundation

with separate ancestors whose descendants populated the earth. Blumenbach, based on his comparative study of the architecture of the human skull, was a monogenist who believed there were five races of man and that physical features, including skin color and cranial shape, were a product of environment. Both Humboldt and Prince Max were among his many protégés who sent him skulls to add to his collection.[58] His office in Göttingen, ringed with skulls, was affectionately called "Golgotha" by his students, with no sense of irony or revulsion.

Blumenbach's belief in a common ancestor was the foundational principle that shaped both Prince Max's and Humboldt's belief in the essential equality of all human beings. Humboldt himself had occasionally collected skulls in South America for the same reason but endeavored to remain within the bounds of cultural propriety. Where he ultimately transgressed was along the Orinoco, at the cavern of Ataruipe. The cave served as an ossuary for the Ature people, whose bones were carefully preserved in baskets. Humboldt insisted on bringing back several skulls and complete skeletons for scientific study. His horrified indigenous guides protested and refused to handle the donkeys carrying their contraband load. One guide quit and refused to assist Humboldt's party as a result. For a man who had such respect for indigenous peoples, his decision to raid burial sites seems out of character. However, Humboldt believed the knowledge gained for scientific progress outweighed cultural taboos. Ironically the shipment that contained the remains foundered off the coast of Africa, taking the remains to the bottom of the Atlantic Ocean.[59]

Fig. 76
George Catlin, *Pipestone Quarry on the Coteau des Prairies*, 1836–37, oil on canvas, 19 ½ × 27 ¼ in., Smithsonian American Art Museum, Gift of Mrs. Joseph Harrison Jr.

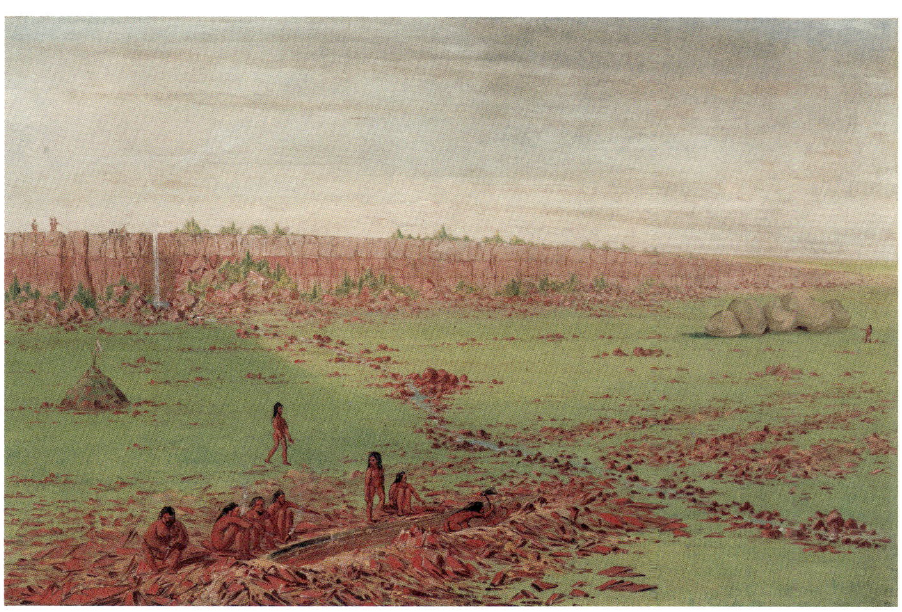

Although Max was also careful to respect the customs of the communities he and Bodmer encountered, he did collect human skulls to ship back to Blumenbach. In New Harmony he also collected bones of Swabian peasants buried in tumuli in the German fashion alongside Indian mounds.[60] After coming across numerous unexcavated mounds outside St. Louis, Max lamented that the U.S. government had not seen fit to systematically excavate them.[61] Along the Missouri River, Max encountered a medicine site with human and bison skulls arranged in a circle. He never learned exactly what all of the components and their decorations meant; however, he expressed a desire to return by night "to commit some thefts" in the name of science.[62] Max's phrasing makes it abundantly clear that he understood that his quest for scientific knowledge did not, in fact, outweigh cultural proprieties. Despite describing his impulse as "theft," he acknowledged his strong desire to excuse his actions. In this case, Max resisted that urge and left the burial site intact.

Catlin, too, participated in collecting materials that violated tribal protocols. In 1830 when he first arrived in St. Louis, Catlin was struck by Clark's collection of Indian pipes.[63] In 1836, on his last trip into Indian territory, Catlin insisted—over the strong objections of the Dahkóta—on visiting the pipestone quarry sacred to their people.[64] He collected specimens of the sacred stone, along with carved pipes made from it, and made paintings depicting the quarry and the process of extracting and shaping the stone (fig. 76). Having returned east with his prizes, Catlin then advertised his samples as "Catlinite." The whole episode reveals a level of cultural insensitivity or tone deafness that was invasive and self-serving. Catlin excused his behavior in his own mind, focusing on his interest in documenting and collecting artifacts associated with sacred spaces and rituals of cultures he believed were on the verge of destruction or extinction. Like Humboldt and Max, Catlin rationalized that the ends justified the means.

CATLIN'S INDIAN GALLERY

As Max and Bodmer headed home in the summer of 1834, Catlin engaged in the first of several public protests regarding Andrew Jackson's Indian Removal Act. Catlin's open hostility to this policy was genuine. In Florida Catlin met John Howard Payne, an actor who had spent time with the Cherokee, and together they protested the removal of that tribe. Payne was arrested, and a warrant was issued for Catlin, which prompted the artist to travel to Washington and take his grievances to the office of President Andrew Jackson. Many years later Catlin recalled:

I... went to the "White House" and was admitted to the presence of Genl Jackson who was in his office alone. I asked him for what cause an order had been issued for our arrest—to which he replied, "it is none of your business, Sir." Of course my interview ended there.[65]

Catlin's vocal objections helped promote his achievements. They also became an obstacle to his ultimate goal, seeing his Indian Gallery purchased by the government.[66]

After his fifth and final trip among the Indians in 1836, Catlin summarized his accomplishments:

I have visited forty-eight different tribes, the greater part of which I found speaking different languages, and containing in all 400,000 souls. I have brought home safe, and in good order, 310 portraits in oil, all painted in their native dress, and in their own wigwams; and also 200 other paintings in oil, containing views of their villages—their wigwams—their games and religious ceremonies—their dances—their ball plays—their buffalo hunting, and other amusements (containing in all over 3000 full-length figures); and the landscapes of the country they live in, as well as a very extensive and curious collection of their costumes, and all their other manufactures, from the size of a wigwam down to the size of a quill or rattle.[67]

His numbers were legitimately impressive and were intended to convey authority over his subject matter, both in artifacts and art. He understood from Humboldt's works that the quantity of accurate data guided the significance of the insights gleaned from them. His numbers made their own case for the seriousness of Catlin's purpose and the importance of its outcomes.

Catlin's attention shifted following his final trip out West. Thereafter, his preferred form of advocacy was through his Indian Gallery—a display of his collection of painted portraits, landscapes, and genre scenes installed along with the artist's sizable collection of Indian artifacts collected on his travels. Catlin pronounced that his intent was to "perpetuate them in a Gallery unique, for the use and instruction of future ages."[68] In this regard his vision resembled Charles Willson Peale's aspirations for his museum as a world in miniature that would teach Americans their own natural and cultural history. Catlin couched his concerns in the rhetoric of Manifest Destiny: "I have seen thus, in all its forms and features, the grand and irresistible march of civilization.... this splendid Juggernaut rolling on, and beheld its sweeping desolation." The artist compared the fate of Native Americans to that of "Caius Marius on the ruins of Carthage," invoking the fatalism woven into Thomas Cole's *The Course of Empire* but here superimposed on Native American civilization.[69] For the next decade he maintained the stubborn belief that he could use his art and his performances to change government policy.

That stubbornness extended to the artist's business model, as Catlin needed to make his enterprise profitable as well. To that end, he shaped the display of the paintings and artifacts into more than a static museological display. He himself often appeared in buckskins or Indian garb and used the opportunity to promote the sale of his paintings and deliver lectures on the need for the preservation of America's Indians. In 1833 he had visited James Reid Lambdin's studio in Louisville, Kentucky, and had Lambdin paint him in the guise of a Sioux, dressed in a shirt decorated with painted quillwork (fig. 77). When Catlin inaugurated his performances as part of his Indian Gallery in 1837, William Dunlap, a pioneer in American theater and a founder of the National Academy of Design in New York, attended several of his lectures and

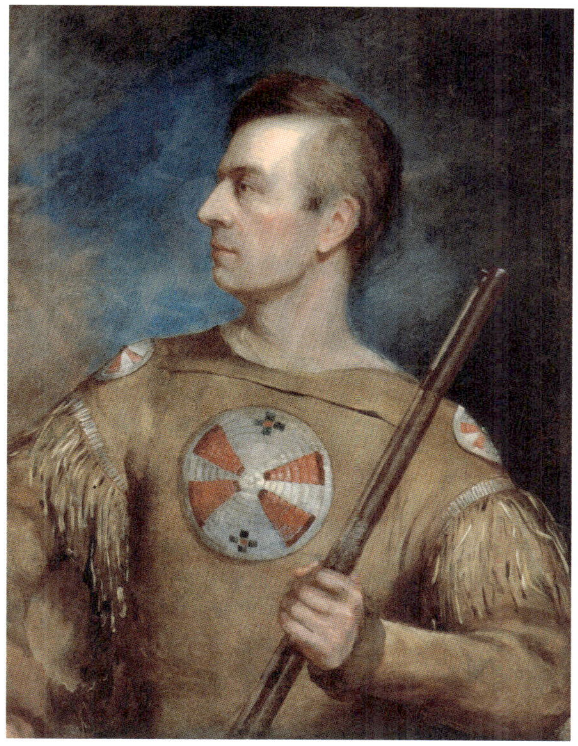

Fig. 77
James Reid Lambdin,
Portrait of George Catlin in the Dress of a Sioux Warrior, 1833,
oil on canvas, 32 ⅞ × 25 in.,
Private Collection

admired them.[70] Philip Hone, who as mayor of New York City had orchestrated Lafayette's triumphal tour of the United States in 1824, wrote in his diary:

I went this morning by invitation of the proprietor, Mr. Catlin, to see his great collection of paintings… and implements of husbandry, and the chase, weapons of war, costumes, etc.… The enthusiasm, zeal, and perseverance with which he has followed up this pursuit are admirable. I have seldom witnessed so interesting an exhibition. Among the invited guests were Mr. [Daniel] Webster, some of the members of the Common Council, the mayor, and some of the newspaper editors. We had a collation of buffaloes' tongues, and venison and the waters of the great spring, and smoked the calumet of peace under an Indian tent formed of buffalo skins.[71]

Catlin's fete for the mayor, the senator, and the rest of the dignitaries went beyond the usual theatrical performance in an effort to garner their support for his enterprise.

Hone noted approvingly what he perceived as the growing interest in Native American ways of life. In his entry for March 28, 1839, he wrote:

Indian stories are now all the vogue in the United States and Canada. Col. McKinney's "Indian Portraits" is a stupendous work, by which, I have been told, the publishers will realize a profit of $100,000. Catlin is delighting everybody by an exhibition of his own paintings and of curiosities and relics collected during a residence of five years among the Indians. Washington Irving's Tour of the Prairies *and subsequently his* Astoria *have aided in directing public attention to the annals of the red men.*[72]

Irving and Cooper had created a literary genre that introduced sympathetic portrayals of Native Americans faced with unalterable changes in their ways of life as white settlements spread westward. In his Leatherstocking novels, Cooper used the destruction of nature to foretell the demise of native cultures. In *The Last of the Mohicans*, the portent was the clear-cutting of old-growth forests; in *The Pioneers*, it was the mass slaughter of the passenger pigeon. The Indian Removal Act of 1830, followed by a massive outbreak of smallpox among the Plains Indian tribes, had encouraged a wave of nostalgia as part of that cultural eulogy for people who had not disappeared and would indeed persist.

The literary vogue for stories of Native Americans might have been thriving, but the market for scholarly ethnographic depictions of Native Americans was in decline. The amount of time, expertise, and labor required for such books did not command a robust audience willing to pay for them. Thomas McKinney and James Hall's three-volume tome on the North American Indians was an impressive achievement that bankrupted its authors. As both Max and Catlin would discover, very few illustrated accounts actually turned a profit for their artists or publishers. For Catlin, this trend did not augur well for the viability of his Indian Gallery as a scholarly enterprise.

By the end of 1839, Catlin had exhausted the interest in his Indian Gallery on the East Coast and had failed in his earliest attempts to entice the federal government to buy his paintings and artifacts for a national collection. As Peale had discovered decades earlier, the desire for national collections far outstripped congressional interest in actually spending money to acquire and house such things. Neither Peale nor Catlin was in a position to make the materials of his respective enterprise a gift. Catlin's open opposition to Jackson's Indian Removal Act soured many of his would-be supporters in Congress who might privately have been sympathetic to the acquisition of the collection but could not afford politically to cast a vote to actually do so.

Catlin's investment in recording Native Americans outstripped his market, and he was in near-constant financial distress. He decided to take his paintings and his artifacts to London. On November 25, 1839, he sailed to Liverpool with eight tons of materials, including a Crow tepee, his collection of artifacts, and 507 paintings.[73] In London, his agent, Charles Augustus Murray, had already booked space in Egyptian Hall for Catlin's Indian Gallery. In short order, pamphlets describing each painting were made available for distribution. Murray arranged for Catlin's publicity, including presentations at several British scientific societies, invitations to society dinners, and three visits with Queen Victoria.[74] At first, Catlin and his nephew, Theodore Burr Catlin (fig. 78), dressed as Indians and delivered the lectures to the public. Their most memorable publicity stunt was the night they and Murray dressed in costume, complete with face paint, and crashed the exclusive Royal Caledonian Ball. It was not until the heat inside the building began to melt their greasepaint that the ruse was unmasked. Their escapade was covered in the morning papers.[75]

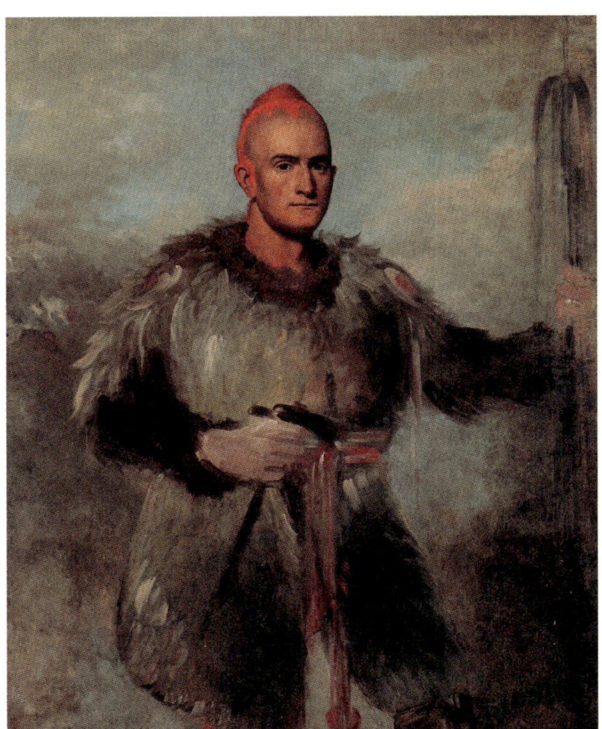

Fig. 78
George Catlin, *Theodore Burr Catlin, in Indian Costume*, 1840–41, oil on canvas, 50 ⅛ × 40 in., Smithsonian American Art Museum, Gift of Mrs. Joseph Harrison Jr.

That charade was just the beginning. Catlin then hired twenty Englishmen to dress and act the part of Indians, creating a full-blown theatrical event to accompany his display of paintings and artifacts. Catlin worked up a script and explained the specific dances being performed; his interest in educating his audiences was still integral to his presentations.[76] This was not the first time whites had performed as Native Americans. During the 1820s, troops of English performers had enacted scenes from Cooper's *The Last of the Mohicans*. For the next two decades Cooper's novels fueled British interest in portrayals of American Indians. During this same period, Native Americans had occasionally toured and performed in England and in Europe as well.[77]

Amid the publicity, in 1840 Bodmer and Max became aware of Catlin's arrival in England. They hoped Catlin's Indian Gallery might "rouse interest" in their own book.[78] In 1841 Bodmer traveled to London and visited Catlin's gallery. He had heard that Catlin was also planning to publish a volume and calculated that his own work with Max might benefit from the attention surrounding Catlin's enterprise. Bodmer was horrified to find white performers pretending to be Indians and, in a letter to Max, excoriated Catlin as a "charlatan."[79] The following year, when Catlin published his *Letters and Notes on the Manners, Customs, and Condition of the North American Indians*, a two-volume illustrated travelogue, Max annotated his copy, writing a scathing, page-by-page critique. His approach was that of a professional ethnographer marking up the work of an amateur. He took Catlin to task for every conceivable error, down to the way Catlin had phonetically spelled Indian names. Max published his review in the German press. He was contemplating a French version to broaden his reach but was dissuaded from doing so by Bodmer, who felt the negative publicity could only hurt their own plans.[80]

MAX AND BODMER'S *REISE*

Max and Bodmer had returned to Europe by the end of 1834, and for the next four years, they set about organizing Max's voluminous journals and Bodmer's portfolios into what would become the most lavishly illustrated travel volume of its kind. From his estate in Wied, Max wrote the narrative.

In it he incorporated references to the work of Americans he most admired, from Stephen H. Long to Thomas Say, even Washington Irving's *A Tour of the Prairies*. But the largest number of references was to Humboldt. Max explicitly confirmed that he had shared the ethnographic materials he brought back from the United States and from Brazil with Humboldt before publishing his book.[81] In Paris, Bodmer hired around twenty of the best engravers to assist with the translation of his watercolors into aquatints. Max's two-volume set chronicling his travels in Brazil was still considered the most significant anthropological assessment of the Botocudo culture; Max envisioned something similar for his North American publication. Of paramount importance to Max was the ethnographic accuracy of both text and image. To this end, the two collaborators exchanged letters in which they held each other to a high standard and queried each other's recollections.[82] For the individual portrayals of the Indians, Max insisted on accuracy. For the *Bison Dance of the Mandan Indians in front of Their Medicine Lodge in Mih-Tutta-Hankush* (CAT. 69), both Bodmer and Max were relying on Dipäuch's description of the dance from the O-Kee-Pa ceremony. Bodmer, who preferred painting from life, ended up hiring models to help block in the composition.

The perfection both artist and explorer demanded from this enterprise meant that it took nearly five years for the first portions of the text to be published. In 1839 publisher Jakob Hölscher of Koblenz began selling the text in installments, leaving it to the individual buyer to determine the binding. In 1841 the entire text and the accompanying atlas of eighty-one of Bodmer's illustrations were advertised as a two-volume set for purchase.[83] *Reise in das innere Nord-America in den Jahren 1832 bis 1834* was as ruinously expensive as Humboldt's most lavish volumes. Like Humboldt, Max bore all of the associated costs personally.[84] The purchase price for the set was staggering: more than a year's wages for a skilled laborer.[85] The publisher balked at committing to a large run even in German; the English language edition of 1843 was a much smaller single volume with the prints bound in.[86] Their efforts, however, resulted in an ethnographically significant tome worthy of Humboldt's praise.

When the two-volume *Reise* finally appeared in print in 1843, both Max and Bodmer were eager for Humboldt's imprimatur. In a flurry of letters back and forth, they fretted that Humboldt had not yet opined as to the quality of Bodmer's illustrations or the veracity of Max's ethnographic observations. Finally, in September 1843, Humboldt set them at ease, congratulating Max as follows: "There is no other travel-book written in our language, which might be compared with this publication that is so perfect in all its details."[87] This was high praise indeed, given the accolades for Humboldt's own publications. In describing Max's observations, Humboldt continued,

> *We may find in all this book the portrayal of a rather important section of mankind in its condition, language and customs and this human race is about to get extinct soon; so you have produced a most important and attractive book for the history of mankind.*

Like Catlin, Humboldt assumed that the fate of North America's indigenous peoples was sealed by Jackson's policies and that Max's volumes would stand as a lasting chronicle of their cultures.

Humboldt continued, complimenting Bodmer on "the large Atlas and the excellent copper engravings," and pronounced,

> *I can't find anything like that in other literary achievements, with regard to beauty and reality.... How poor are, compared to others, the recent books of the French about travels around the world, which had been published at the expense of the government!*

Bodmer, he pronounced, "is a great artist, who is able to perceive nature as something alive."[88] Humboldt's admiration for Bodmer's depictions of Native Americans and the North American landscape is a reminder that Humboldt had seen neither in person and was absorbing every detail of this visual record as the closest he would come to that experience.

THE OJIBWE AND IOWA IN EUROPE

When Humboldt penned his effusive praise of Max and Bodmer's achievement, Catlin received a letter from promoter Arthur Rankin offering to lease a group of Ojibwe (Anishinaabe) Indians to him as performers.[89] The two quickly worked out a deal, and in December 1843, the *Illustrated London News*

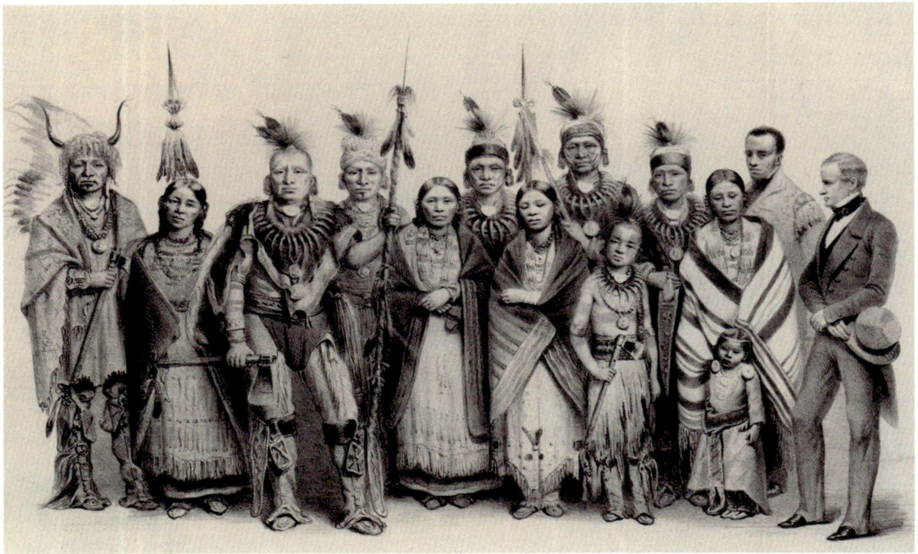

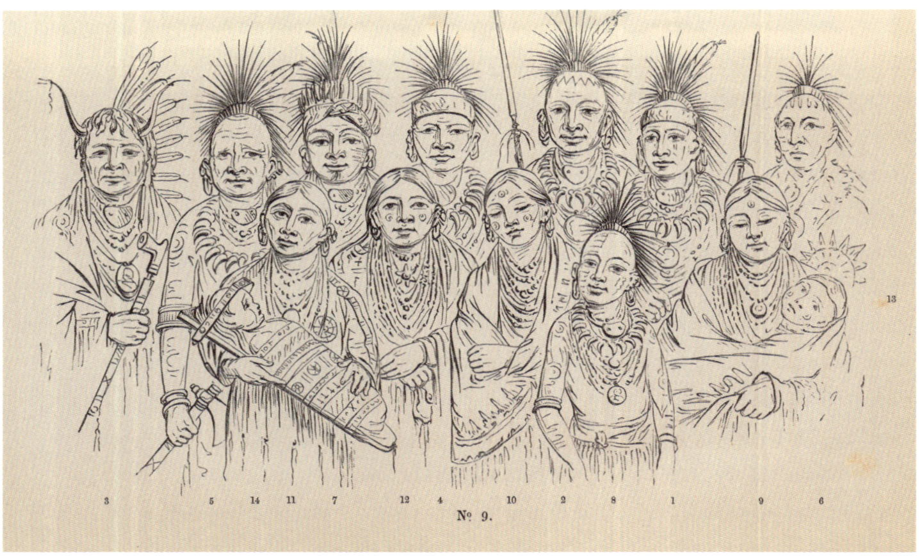

Fig. 79
After George Catlin, *Catlin and Iowa Indians with Jeffrey Deroine*, lithograph on chine collé, 1845, 9 ⅞ × 17 ⁷⁄₁₆ in., The British Museum

Fig. 80
George Catlin, *Line drawing of Iowa Indians as a group* (plate 9) in Catlin's *Notes of Eight Years' Travels and Residence in Europe, with His North American Indian Collection (Vol. 2)*, 1848, engraving, 6 × 9 in., National Gallery of Art Library, David K. E. Bruce Fund

announced their arrival.[90] The Ojibwe replaced the English performers, providing a much-needed jolt for flagging ticket sales. Catlin and Rankin toured the Indians to Windsor Castle, where they met the queen, and took them to see the Thames Tunnel, which was under construction.[91] The London press followed their actions in some detail, fascinated to learn what the Indians thought of British society. Relations between Catlin and Rankin deteriorated over time, and in June 1844, the Ojibwe left. However, before the Ojibwe had even arrived in London, Secretary of War J. M. Porter had approved P. T. Barnum's request for his associate, George Curzon Melody, to transport a group of Iowa (Báxoje) Indians to London.[92] Catlin signed a contract splitting the costs and the proceeds. While Melody made those arrangements, Barnum sailed to London with his premiere attraction, General Tom Thumb, to join Catlin's entourage. In June 1844, Catlin and his Indians shared the stage with Barnum's dwarf at Egyptian Hall.[93] The performances were kept separate, but the shared setting was its own commentary on the triumph of theatrics over substance. Even with that backing, Catlin could not make the Indian Gallery profitable, and the alliance with Barnum undercut his scholarly and humanitarian credibility.

Catlin was delighted when the group of fourteen Iowa Indians replaced the Obijwe. These were people he knew, whose villages he had visited a decade earlier. They were accompanied by their mixed-race translator, Jeffrey Deroine, who had served as one of Catlin's translators before. Their paths had continued to cross afterward in St. Louis and in New York.[94] The artist set about touring the Iowa as he had the Ojibwe. They performed on Lord's Cricket Ground and at Vauxhall in addition to Egyptian Hall.[95] All of this was documented in the London press, no doubt orchestrated by Barnum, Melody, and Catlin.[96] They arranged to have a photograph taken of the group, including Deroine, and printed in the *Illustrated London News* (fig. 79), along with a story in the London *Sunday Times* that extolled the Iowa as "the Indians which we read about in Cooper's novels."[97] Cooper's Leatherstocking novels had ushered in the romanticized trope of Native Americans as being one with nature and not corrupted by Eurocentric civilization. *The Last of the Mohicans* had been a staple of the British stage for close to twenty years. However, despite all the fanfare, the appeal of Catlin's enterprise was fading.

Catlin made a line drawing of the group in London (fig. 80) as an illustration for his book chronicling the eight years he ultimately spent in Europe. However, within a year of their arrival, disease claimed two of the Iowa. The ten-month-old baby, Corsair, named for the sailing vessel on which he had been born, succumbed in London before Catlin could paint his portrait. He had attracted a great deal of attention during the group's dances, strapped to his mother, Oke-we-me (Female Bear That Walks on the Back of Another), in a decorated cradleboard. The youngest adult in the group, No-ho-mun-ya (One Who Gives No Attention) (CAT. 70), also called Roman Nose, had

CAT. 69 After Karl Bodmer, *Bison Dance of the Mandan Indians in front of Their Medicine Lodge in Mih-Tutta-Hankush*, 1842, hand-colored aquatint, plate size: 16 ½ × 21 ⁵⁄₁₆ in., Joslyn Art Museum, Omaha, Nebraska, Gift of the Enron Art Foundation

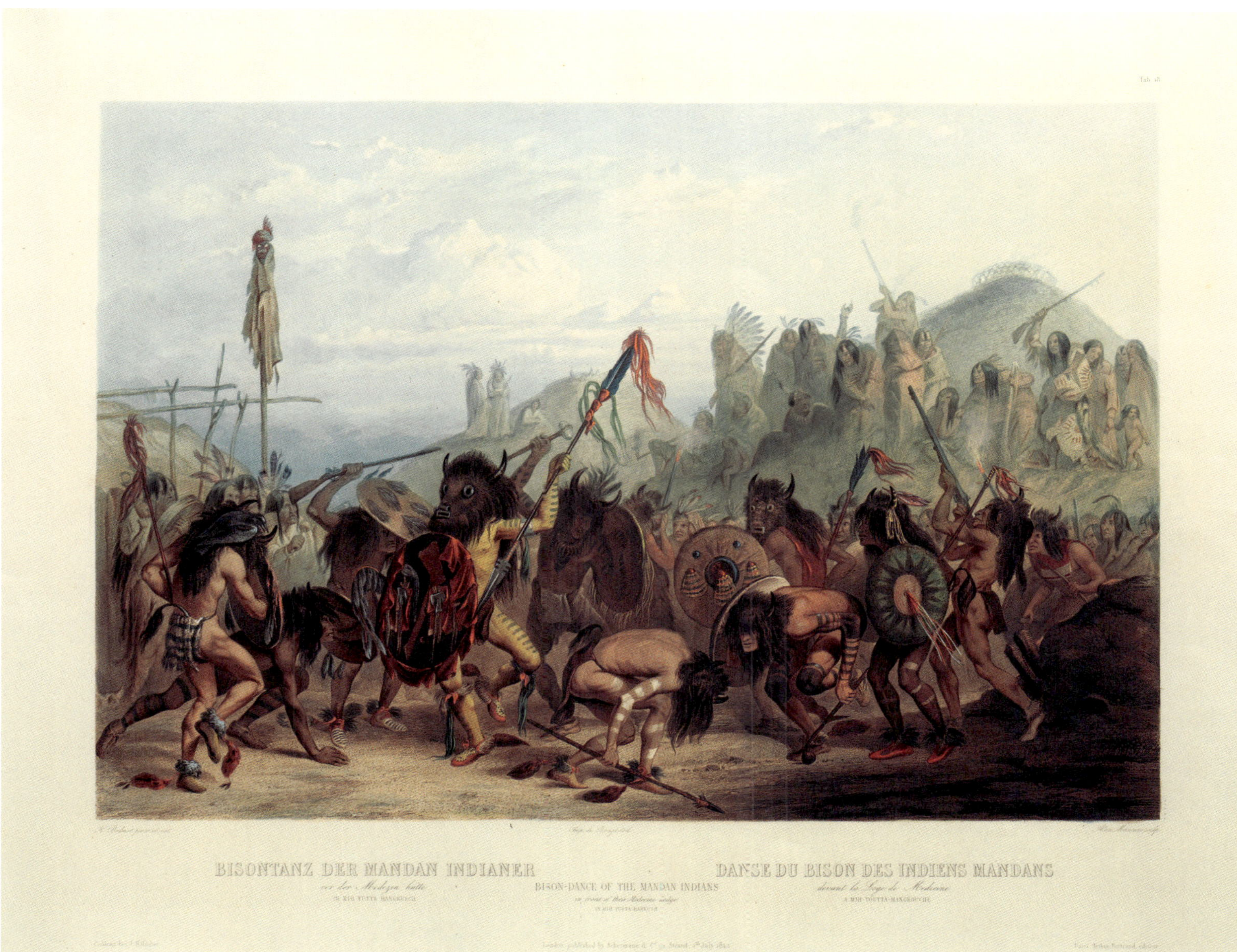

CAT. 70 George Catlin, *No-ho-mun-ya, One Who Gives No Attention*, 1844, oil on canvas, 29 × 24 in., Smithsonian American Art Museum, Gift of Mrs. Joseph Harrison Jr.

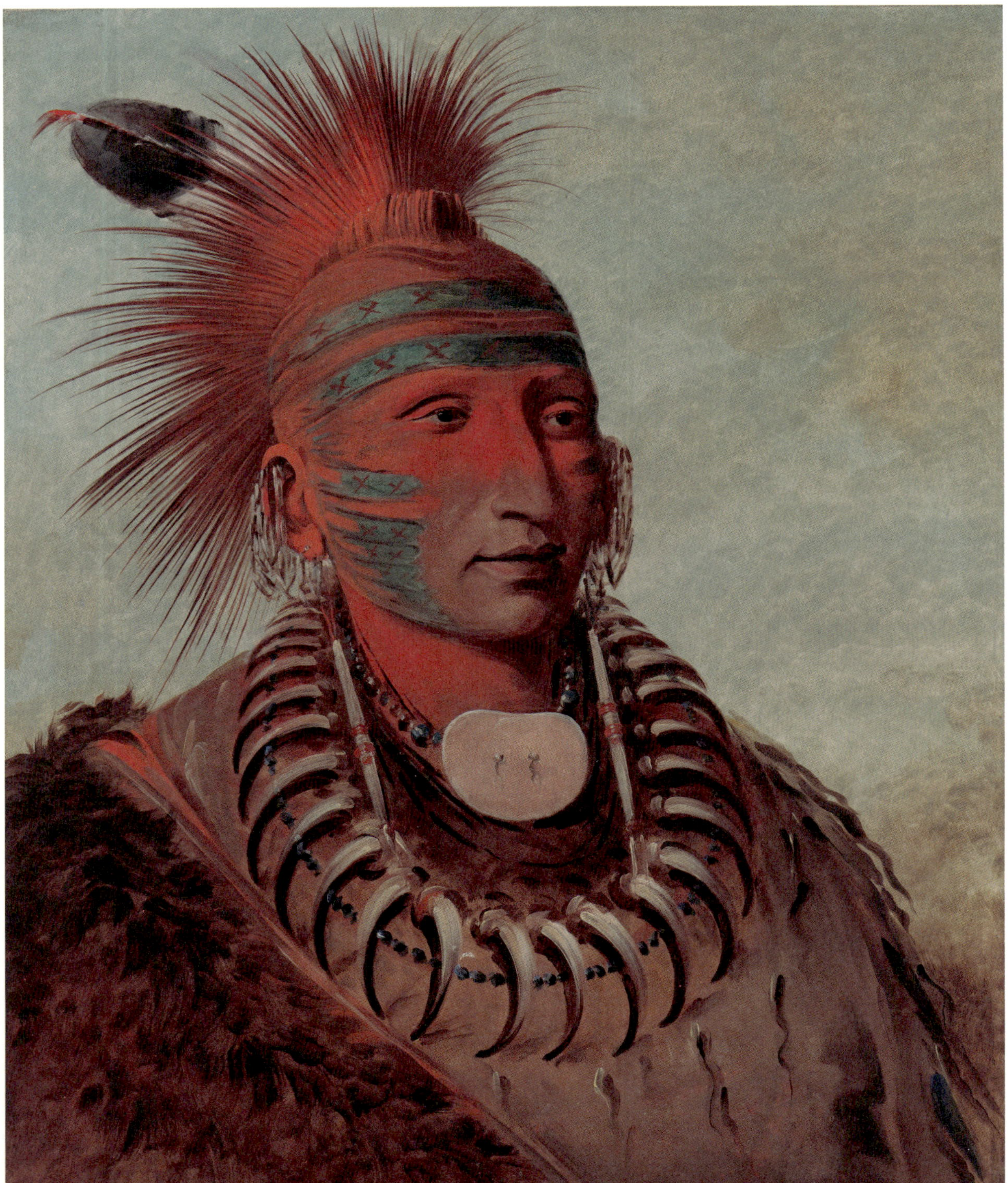

also been immensely popular with audiences in London, admired for his handsome features. The illness he contracted in London proved fatal, and he died while the group was touring the British Isles.[98] Catlin painted his portrait in London before he died. His appearance is that of a warrior, his face painted, and his scalp shaved to create a central scalp lock typical of the Iowa Indians. A bison robe encircles his shoulders, and he wears a bear claw necklace. A shell gorget rests at his throat on a string of glass beads. His ears contain multiple piercings along the outer ridge with a single piercing in the lobe. A single eagle feather is woven into his hair. After touring the surviving Iowa Indians around England, Scotland, and Ireland, Catlin booked passage to Paris for the group.

CATLIN AND THE IOWA IN PARIS

Catlin's trip to Paris in 1845 represents, in some respects, the high-water mark of his career and certainly was the last moment he believed everything might turn out in his favor. The French monarch expressed interest in seeing the Iowa perform and later commissioned Catlin to paint fifteen works for the royal collection. Catlin was still hoping he could cover his expenses and find a buyer for his entire gallery. After arriving in Paris and moving everything into the Salon Valentino, Catlin and the Iowa toured the city in carriages, visiting the major landmarks. They were accompanied by Alexandre Vattemare, who served as Catlin's manager in Paris, orchestrating publicity and handling ticket sales. While the group waited for their invitation to meet King Louis-Philippe and the royal family, Catlin began the process of installing the Indian Gallery. Meanwhile, the Iowa met with groups of curious visitors.[99]

Fig. 81
George Catlin, *The White Cloud, Head Chief of the Iowas*, 1844–45, oil on canvas, 27 15/16 × 22 13/16 in., National Gallery of Art, Paul Mellon Collection

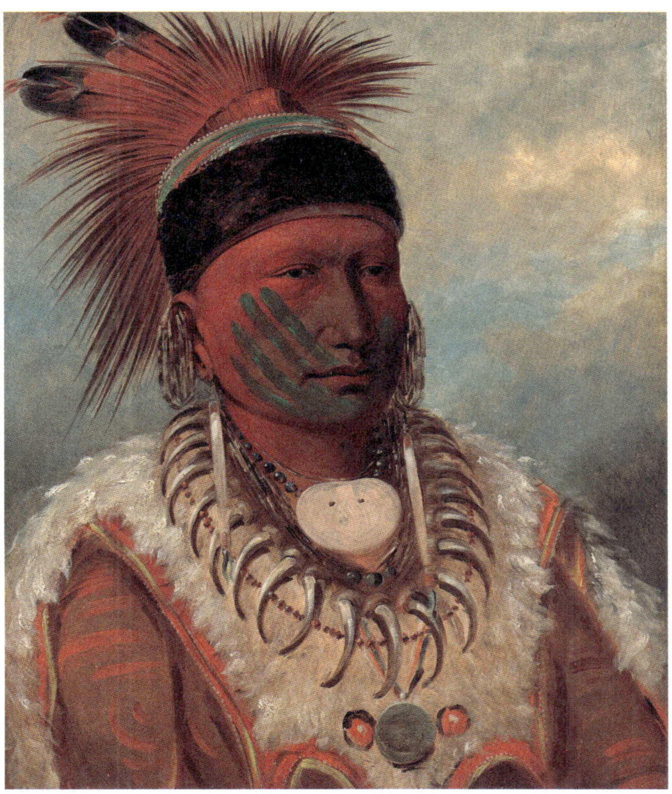

The small, full-length portraits Catlin had painted of the Iowa in the 1830s were hanging as part of the Indian Gallery already. In England Catlin had painted bust-length portraits of each of the Iowa, representing their position within their tribe as indicated by their dress and accoutrements. In 1845 Catlin added to several of these portraits the medals the men received from French King Louis-Philippe, which hung from a tricolor ribbon in the French national colors of blue, white, and red. These diplomatic gifts joined the peace medals that had been given to them by the John Tyler administration during treaty negotiations over landownership.

Their leader was Mew-hu-she-kaw (The White Cloud) (fig. 81), first chief of the Iowa. He had chosen the group that would travel with him to Europe. The White Cloud was only thirty-five years old and already a highly regarded chief. He is shown dressed in his white wolf cloak and bear claw necklace, a blue handprint painted across his face, and an otter skin turban around his forehead. His hair is shaved in a traditional Iowa roached scalp lock, painted red and adorned with eagle feathers. His ears are pierced in multiple places with strands of wampum woven through the upper cartilage and long pendant hair-pipe beads hanging from the lobes. His war chief Neu-mon-ya (Walking Rain) (CAT. 71) was fifty-four years old and the third chief of the Iowa. In person he towered over everyone, standing well over six feet tall. Like The White Cloud, he wears a bear claw necklace and similar ear ornaments. The fur of a bison robe wraps around his torso, and his turban is of woven cloth instead of fur. He wears a concave shell gorget at his throat and a gold medal from the French king around his neck, below a larger silver medal bearing the likeness of President Tyler.

CAT. 71 George Catlin, *Neu-mon-ya, Walking Rain, War Chief*, 1844–45, oil on canvas, 29 × 24 in., Smithsonian American Art Museum, Gift of Mrs. Joseph Harrison Jr.

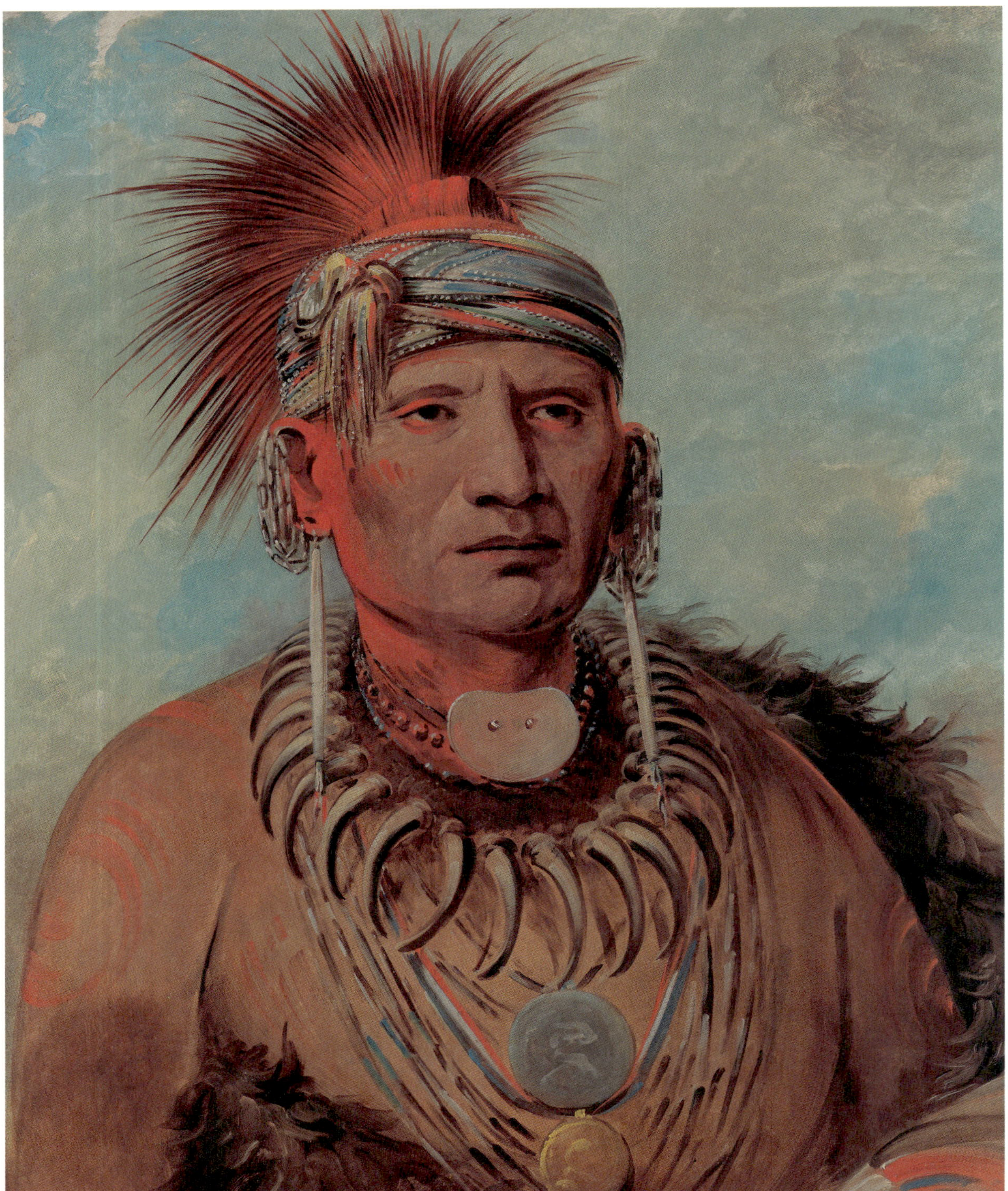

The third man whom Catlin had met before was Wash-ka-mon-ya (Fast Dancer) (CAT. 72), who turned out to be among the most curious about their European hosts. He among the Indians could communicate in English and kept a diary of the trip.[100] Known by his nickname, Jim, Wash-ka-mon-ya is presented as a warrior, the blue handprint—like that of The White Cloud—possibly indicating he had killed an enemy in hand-to-hand combat. The shaved areas of his scalp are exposed, instead of being covered by a turban, his forehead painted with a red zig-zag that denoted speed and agility. Like Walking Rain, Jim wears a shell gorget at his throat and two medals: the larger one from President Tyler, the smaller one a silver version of the medal presented by the French king.

The rest of the group were people Catlin had not painted during his visit to the Iowa community. Catlin's riveting portrait of the group's religious leader, See-non-ty-a (Blister Feet) (CAT. 73), conveys the artist's respect for and unease with his strong personality. Known as The Doctor, See-non-ty-a struck Catlin as a powerful, if slightly terrifying, member of the group, prone to disappearing in order to ascend to the roof and observe his surroundings. Catlin noted that in many tribes the doctor was also deemed a soothsayer and considered to have oracular powers—attributes that Catlin respected but privately dismissed. See-non-ty-a's full frontal portrayal and hooded gaze challenge the viewer in a manner unlike the rest of Catlin's portraits of the Iowa. His accoutrements resemble those of the warriors in the party, although he does not wear a medal from Louis-Philippe. His bare chest, shoulders, and arms are painted, his torso wrapped in a red and white woven blanket.

Shon-ta-yi-ga (Little Wolf) (CAT. 74) was a highly esteemed Iowa warrior. Like The Doctor, Little Wolf appears bare-chested, his upper torso and arms painted, a bison robe thrown around his shoulders. His neck and shoulders are covered in multiple beaded necklaces, ornaments, and a bear claw necklace. The shell gorget is joined by two medals, a silver medal awarded him by the Tyler administration for his bravery in defending his tribe and a gold medal from the French king. His ears are heavily pierced and threaded with wampum from top to bottom, adding to the density of ornament that would have moved and made noise with his every step. A metal armband also includes rings made of a different metal. Louis-Philippe admired this portrait enough to ask that Catlin include a copy of it in the group of paintings he commissioned from the artist. Little Wolf's gaze carries a touch of the sorrow he experienced during this trip; it was his son, Corsair, who had died in London, followed by his wife, Oke-we-me (Female Bear). She succumbed to what might have been tuberculosis in Paris, a tragedy that would prompt the end of the Iowa's visit to Europe.

Although Catlin rarely mentions them in his book, three other Iowa women were part of the touring group. They were active participants in the performances and very popular with audiences in London and Paris, but Catlin's portraits of these women are as vague as his written descriptions. He was more successful in depicting their regalia than conveying their personalities. White Cloud's wife, Ru-ton-ye-wee-ma (Strutting Pigeon), is portrayed along with her young daughter, Ta-pa-ta-me (Wisdom) (CAT. 75). The young girl looks straight at the viewer, while her mother's gaze is deliberately oblique. Both are dressed in buckskin shirts decorated with dyed porcupine quills. Strutting Pigeon's shirt is further enhanced by a line of small shells stitched to the yoke. Both have ear piercings that display multiple small sprays of silver ornaments. Ta-pa-ta-me reaches for the strands of hair-pipe beads around her mother's neck, which are interwoven with strands of glass bead necklaces. A delicate silver cross on a thin silver chain encircles her throat.

Another was Ru-ton-wee-me (Pigeon on the Wing) (CAT. 76), who faces the viewer but does not make eye contact. Her thoughtful gaze is distant, much like the presence of the women in Catlin's narrative. She wears her hair in two braids, which can suggest that she was unmarried, as many married women wore their hair in a single braid. Koon-za-ya-me (Female War Eagle) (CAT. 77) wears her hair loose around her shoulders. While Catlin took pains with the articulation of the other women's shirts, in this case the garment seems like a loose drape, without the same sense of structure. The necklaces are similarly abstracted, as though the artist was under some time pressure to complete the portrait. Catlin put far more effort into her facial features, calling attention to the red

CAT. 72 George Catlin, *Wash-ka-mon-ya, Fast Dancer, a Warrior*, 1844–45, oil on canvas, 29 × 24 in., Smithsonian American Art Museum, Gift of Mrs. Joseph Harrison Jr.

CAT. 73 George Catlin, *See-non-ty-a, an Iowa Medicine Man*, 1844–45, oil on canvas, 27 15⁄16 × 22 13⁄16 in., National Gallery of Art, Paul Mellon Collection

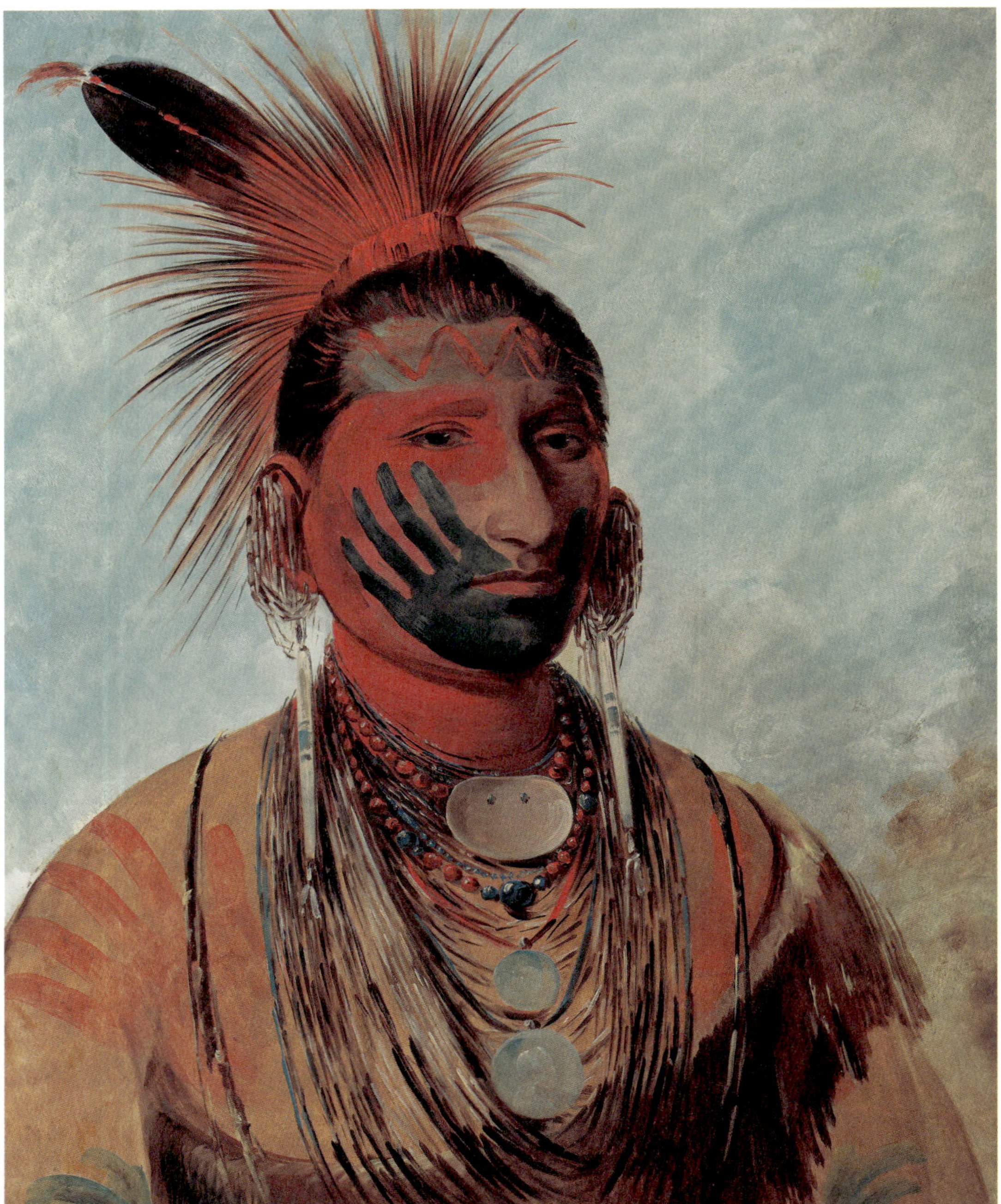

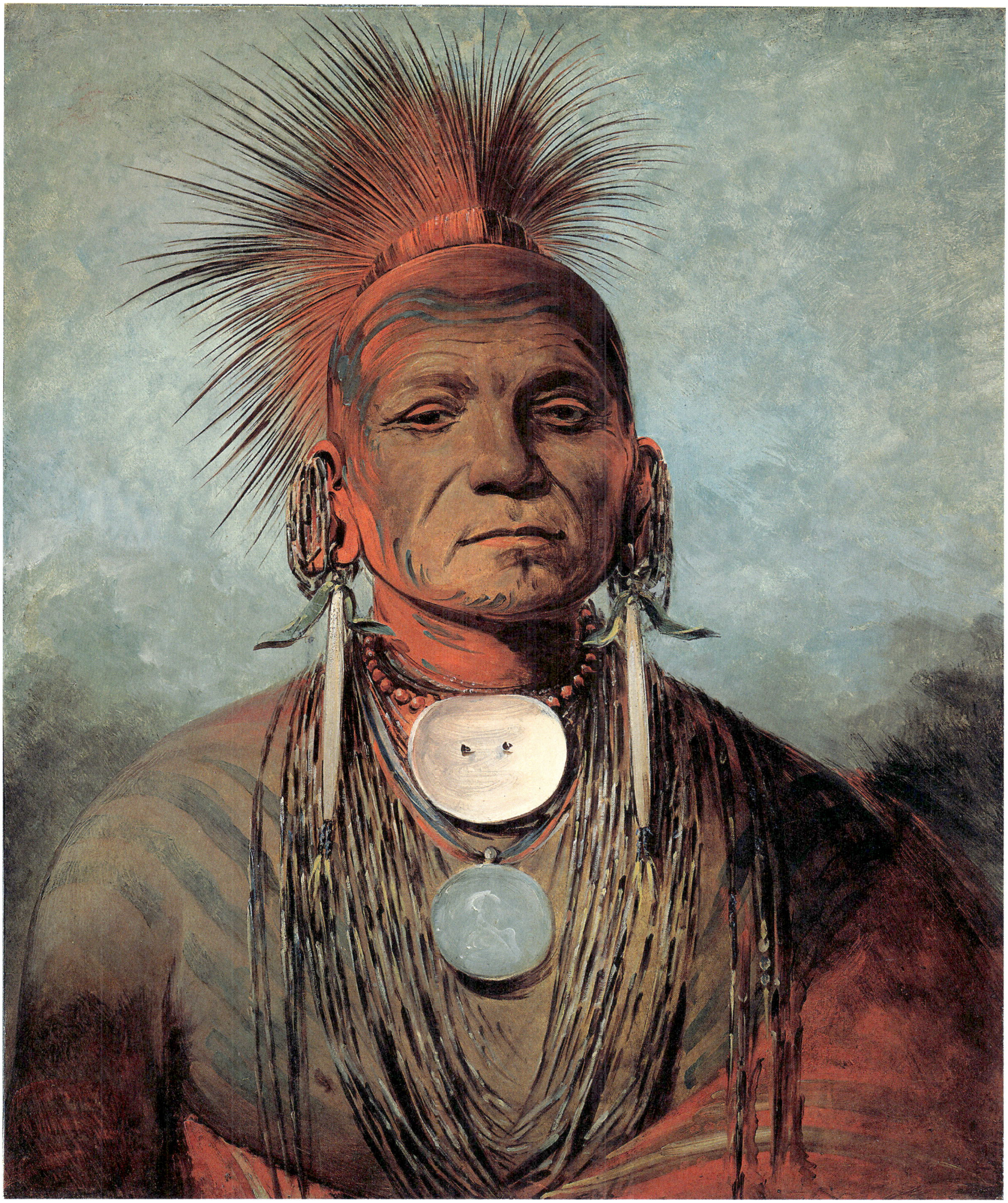

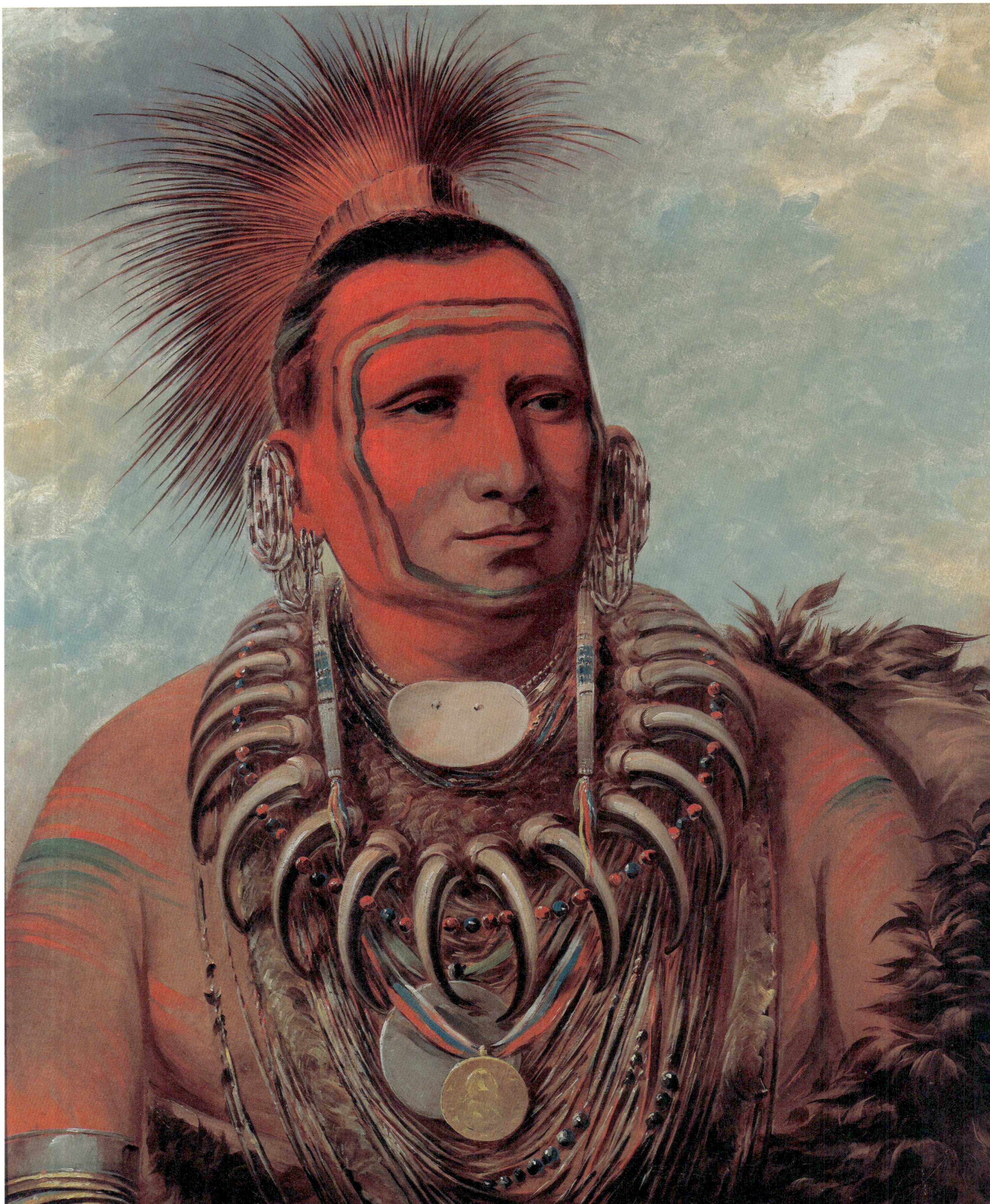

CAT. 74 George Catlin, *Shon-ta-yi-ga, Little Wolf, a Famous Warrior*, 1844–45, oil on canvas, 29 × 24 in., Smithsonian American Art Museum, Gift of Mrs. Joseph Harrison Jr.

CAT. 75 George Catlin, *Ru-ton-ye-wee-ma, Strutting Pigeon, Wife of White Cloud*, 1844, oil on canvas, 29 × 24 in., Smithsonian American Art Museum, Gift of Mrs. Joseph Harrison Jr.

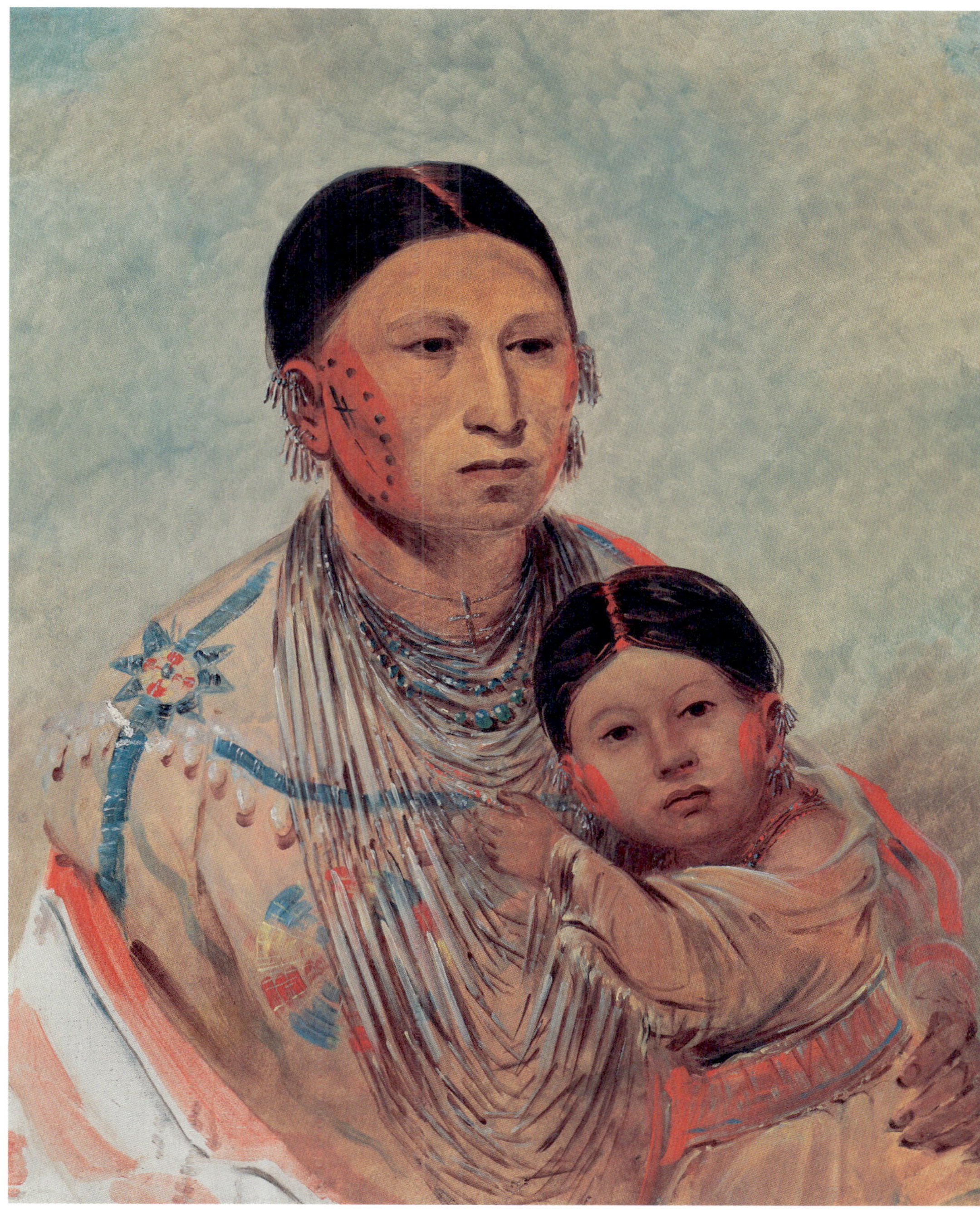

CAT. 76 George Catlin, *Ru-ton-wee-me, Pigeon on the Wing*, 1844, oil on canvas, 29 × 24 in., Smithsonian American Art Museum, Gift of Mrs. Joseph Harrison Jr.

CAT. 77 George Catlin, *Koon-za-ya-me, Female War Eagle*, 1844, oil on canvas, 29 × 24 in., Smithsonian American Art Museum, Gift of Mrs. Joseph Harrison Jr.

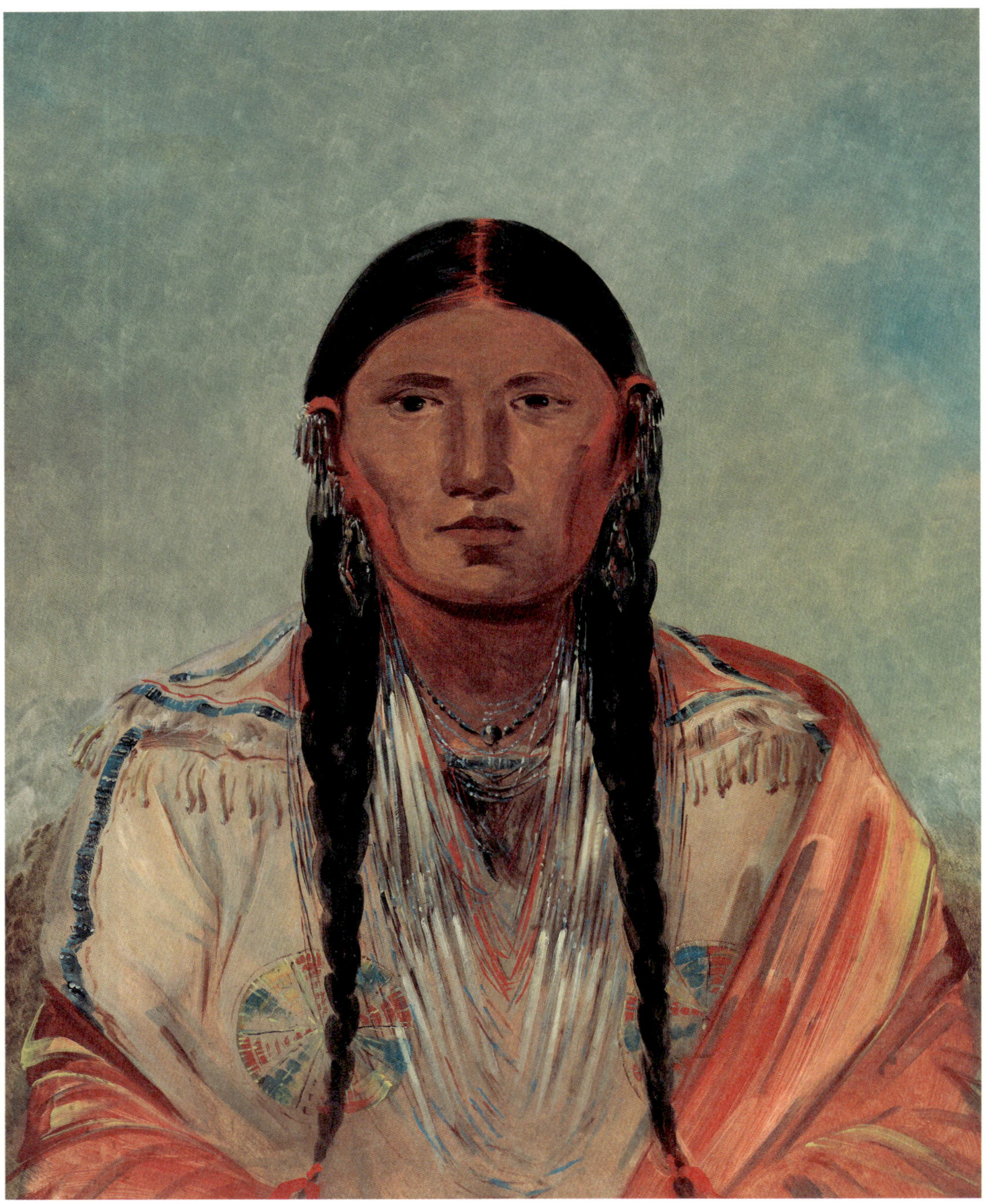

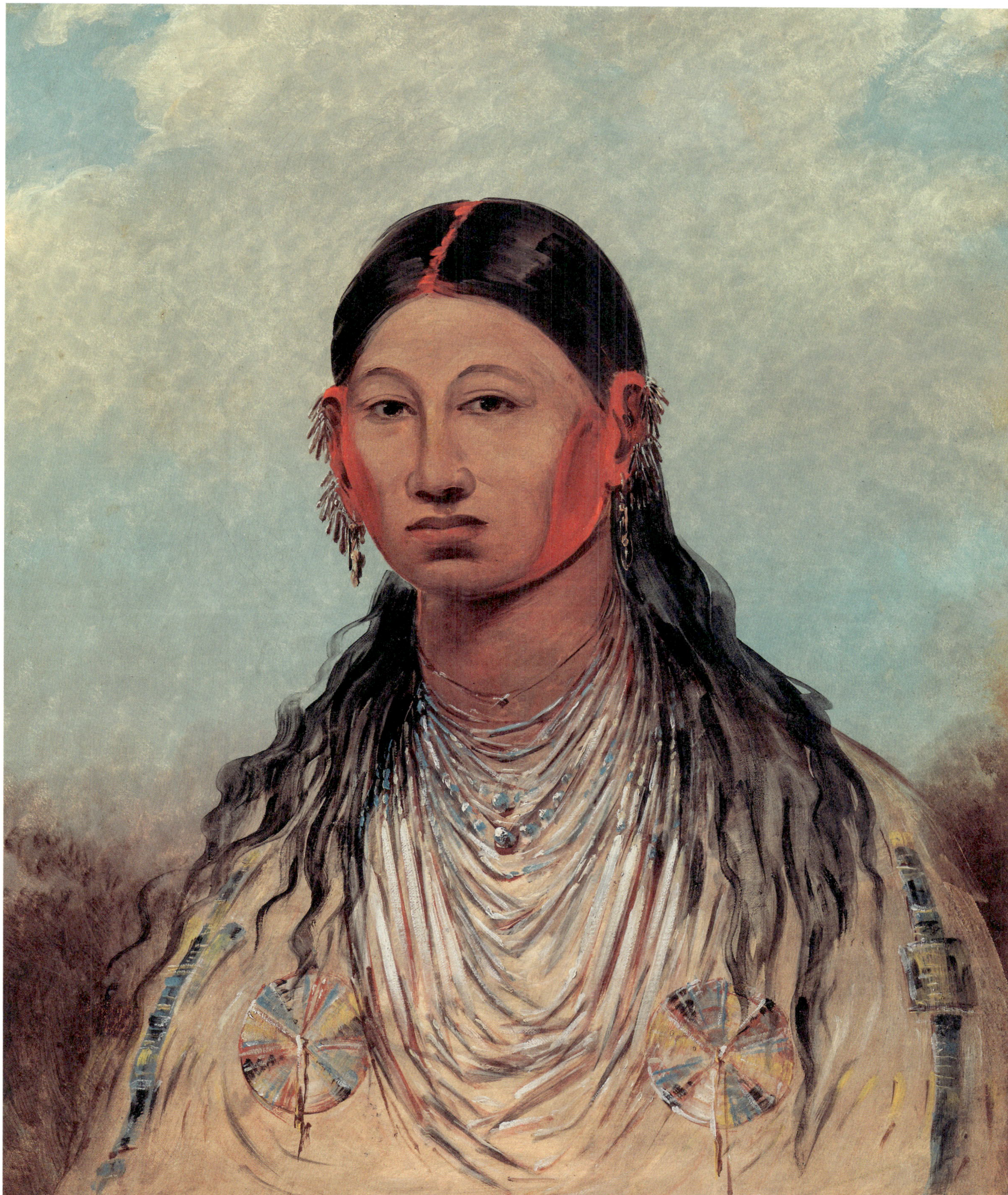

paint that was typically applied to the sides of the women's faces and applied along the center part of their hair. Her gaze is direct and unforgiving, her features a mask to her emotions.

Catlin painted the young boy in the group, Wa-ta-we-bu-ka-na (Commanding General), who was Walking Rain's ten-year-old son. This painting remained in France and became part of Louis-Philippe's commission of fifteen paintings for the royal collection. No painting is known of the final member of the group, Wa-tan-ye (Foremost Man), outside of his appearance in Catlin's line drawing of the entire group.

On May 3, the London newspapers covered the visit of the Iowa Indians to the Tuileries, where they had danced for the king and his court on April 21.[101] Catlin wrote an extensive and detailed description of the pageantry and costuming:

Each one came out from his toilet, in a full blaze of colour of various tints, all with their wampum and medals on, with their necklaces of grizly bears' claws, their shields, and bows, and quivers, their lances, and war clubs, and tomahawks, and scalping knives. In this way, in full dress, with their painted buffalo robes wrapped around them, they stepped into the several carriages prepared for them, and all were wheeled into the Place Carousel, *and put down at the entrance to the Palace. We were met on the steps by half a dozen huge and splendid looking porters, in flaming scarlet livery and powdered wigs, who conducted us in, and being met by one of the King's* aides-de-camp, *we were conducted by him into His Majesty's presence, in the reception hall of the* Tuileries.[102]

Catlin was in his element. He had cultivated the attention of royalty in two countries, far outstripping the attention he had received at home.

Catlin, Melody, Deroine, and the Iowa Indians were presented to the king, queen, and the royal court.[103] Louis-Philippe addressed the Indians, as Catlin recalled:

His Majesty in the most free and familiar manner (which showed that he had been accustomed to the modes and feelings of Indians) conversed with the chiefs, and said to Jeffrey, "Tell these good fellows that I am glad to see them; that I have been in many of the wigwams of the Indians in America when I was a young man, and they treated me every where kindly, and I love them for it.—Tell them I was amongst the Senecas near Buffalo, and the Oneidas—that I slept in the wigwams of the chiefs—that I was amongst the Shawnees and Delawares on the Ohio; and also amongst the Cherokees and Creeks in Georgia and Tennessee, and saw many other tribes as I descended the Ohio river the whole length, and also the Mississippi to New Orleans, in a small boat, more than fifty years ago."[104]

The king's speech conveyed the most sincere appreciation for Native American culture that Catlin and the Iowa had encountered in Europe.

When the group began to dance in the Salle de Bal, the queen asked Catlin to explain to her the meaning of each dance. Karl Girardet, a Swiss-born court painter, captured the scene in a painting (CAT. 78). Catlin described one of the dances as follows:

The Doctor led off first in the character (as he called it) of a soaring eagle, sounding his eagle whistle, which he carried in his left hand, with his fan of the eagle's tail, while he was brandishing his lance in the other. At the first pause he instantly stopped, and, in the attitude of an orator, made his boast of an instance where he killed an enemy in single combat, and took his scalp. The Little Wolf, and Wash-ka-mon-ya and others, then sprang upon their feet, and sounding their chattering whistles, [*An ingenious whistle made to imitate the chattering of the soaring eagle, and used in the eagle dance.] and brandishing their polished weapons, gave an indescribable wildness and spirit to the scene. When the dance was finished, the Indians had the pleasure of receiving their Majesties' applause, by the violent clapping of their hands, and afterwards by expressions of their pleasure and admiration, conveyed to them through the interpreter.*[105]

Girardet's painting, commissioned by Louis-Philippe to commemorate the event, captures the gulf between the ornate palace interior and the raw energy of the performance. The French monarch sits on the far right, his wife turning to Catlin, who provides a commentary on the dance. The Doctor, The White Cloud, and Jim are the principal dancers, along with the young boy, Commanding General. The remaining Iowa Indians are seated on the floor, providing the music and vocals for the performance. Catlin had sought out

CAT. 78 Karl Girardet, *Danse d'indiens Iowas devant le roi Louis-Philippe aux Tuileries (Dance of the Iowa Indians before the King Louis-Philippe at the Tuileries)*, 1845, oil on canvas, 15 $^{9}/_{16}$ × 21 $^{1}/_{16}$ in., Établissement public du château, du musée et du domaine national de Versailles

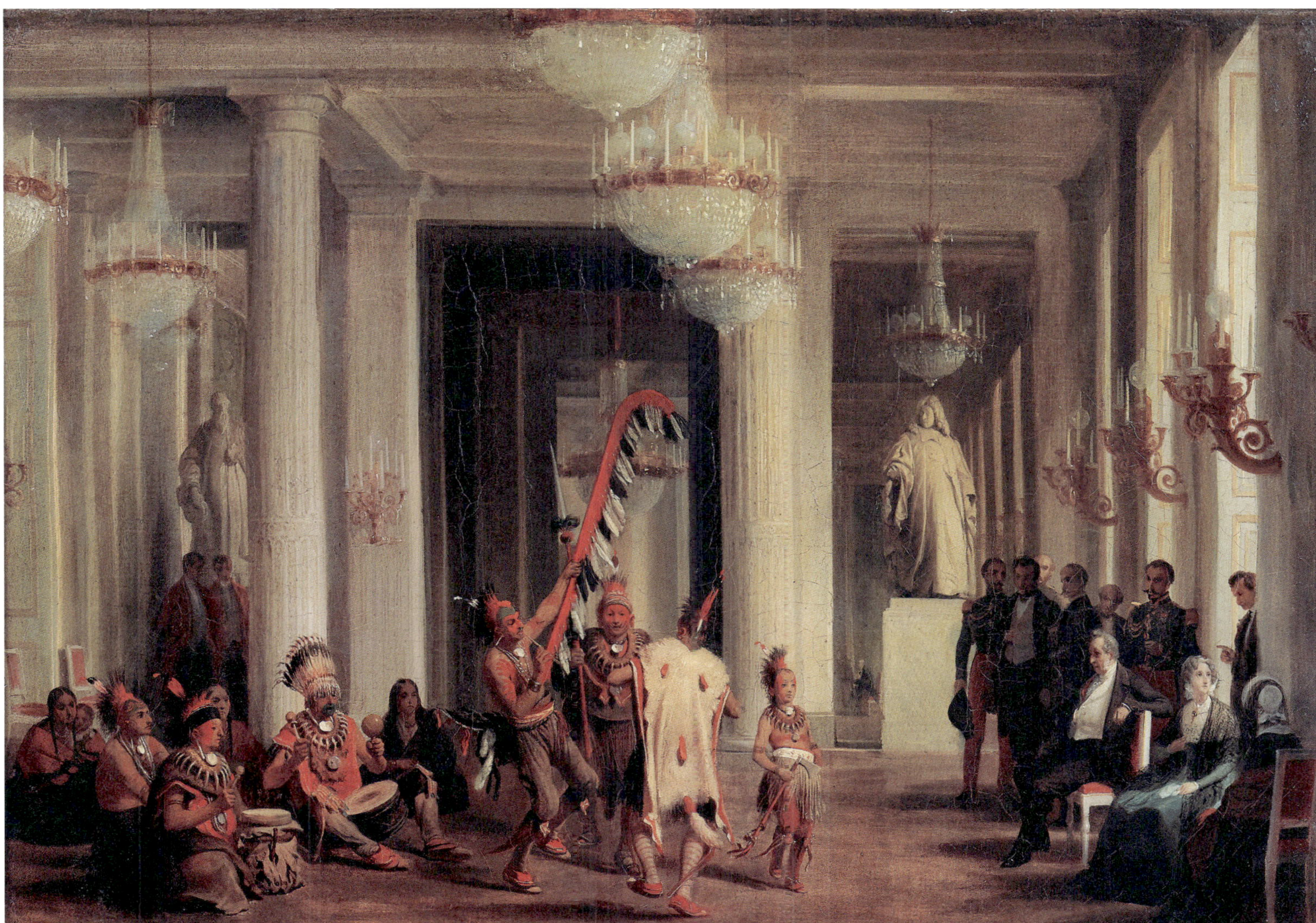

the highest level of acknowledgment and approval of his Indian Gallery and performers from the U.S. government, the queen of England, and now the French king. In each case Catlin had looked to profit from his labors and the Indians' willingness to perform. At moments like this, Catlin's original desire to draw attention to the plight of Indian tribes yielded to his increasingly dire financial straits. No matter how he tried, he could not generate meaningful respect for indigenous cultures, nor could he afford to maintain his Indian Gallery.

To promote public interest, Catlin and Vattemare engaged in a campaign to have the Indian Gallery and Indian performers reviewed in the press. Théophile Gautier, the leading art critic, penned a mixed review on the front page of *La Presse* for May 19, 1845, noting of Catlin's paintings:

In the absence of talent, sincerity shines through in distinguishing features. Most of the time, these are studies where the background is sketched in. But nothing characteristic is omitted, even with this rapid handling, [and] a more talented artist might not have done as well. When a civilization or barbaric culture is disappearing, an historian inevitably emerges whose portrayals memorialize his subject.

Gautier compared Catlin's paintings to Cooper's novels, noting that the artist presents "the most exact *facsimile* of the wild life at the moment when these unfortunate tribes are being decimated, several annihilated, with no traces remaining except for the sketches of Mr. Catlin."[106] Catlin coached his reviewers to pick up the theme of the Indians' imminent demise, still hoping for traction. Gautier described the performance in his review, noting the differences between the Approach Dance, the Scalp Dance, and the Eagle Dance, and concluding that there was

Nothing more fantastic and more frightening than these dancers. The Ioways will bring an energy and extraordinary entertainment: they wiggle, they jump, they wave their tomahawks, brandishing their spears, shouting, rolling their eyes.[107]

Gautier's written reaction to the dancers and the dances echoes Girardet's painted response—in both cases expressing discomfiture with a foreign culture they did not understand.

When Catlin was ready to begin regular performances, Vattemare distributed tickets to the cultural influencers in Paris, among them authors Charles Baudelaire, Victor Hugo, and George Sand. Sand attended a performance on May 29 and was so struck that she secured tickets for artist Eugène Delacroix and composer Frédéric Chopin to accompany her again the following day.[108] She asked Catlin for permission to interview the Iowa, and on June 3, with their consent, she spent the day talking with them through Deroine.[109] She fashioned her notes into an article, "Relation d'un voyage chez sauvages de Paris," in which she described a fictionalized trip to the United States to see the Iowa.[110] The attendance of these literary and artistic luminaries provided Catlin with the cultural company he longed to keep, raising his hopes that their interest would make this venture a financial and performative success.

CATLIN, HUMBOLDT, AND PRINCE MAX

For Catlin, one of the high points of his European experience occurred during a private tour of the Louvre, arranged on a day the museum was closed. The director had invited several guests for the tour, among them Alexander von Humboldt. Catlin was thrilled and proudly wrote:

Amongst the distinguished persons who were present, and to whom I was introduced on the occasion, was the Baron de Humboldt. He accompanied us quite through the rooms of the Louvre, and took a great deal of interest in the Indians, having seen and dealt with so many in the course of his travels. I had much conversation with him, and in a few days after was honoured by him with a private visit to my rooms, when I took great pleasure in explaining the extent and objects in my collection.[111]

Catlin was correct that Humboldt had spent considerable time among the indigenous people of South America, but the Iowa may have been the only North American Indians Humboldt ever met. North American Indians had toured England and France before, but not when Humboldt could have seen them.

Three years earlier, in 1842, Humboldt had had the opportunity to see Catlin's Indian Gallery, minus its proprietor and the white performers then

CAT. 79 George Catlin, *Catlin's Notes of Eight Years' Travels and Residence in Europe, with His North American Indian Collection (2 vols.)* (frontispiece and cover), 1848, embossed and gilt cloth covers, engraving and letterpress printing, closed: 9 × 6 × 1 in., National Gallery of Art Library, David K. E. Bruce Fund

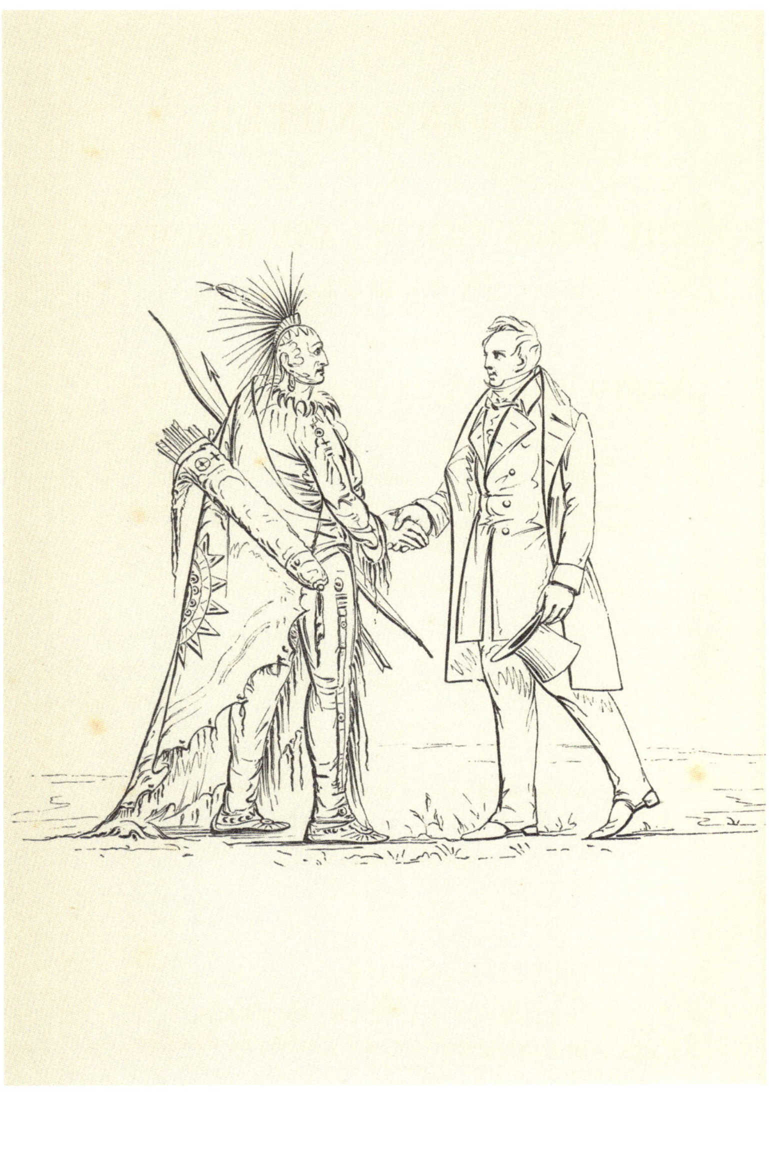

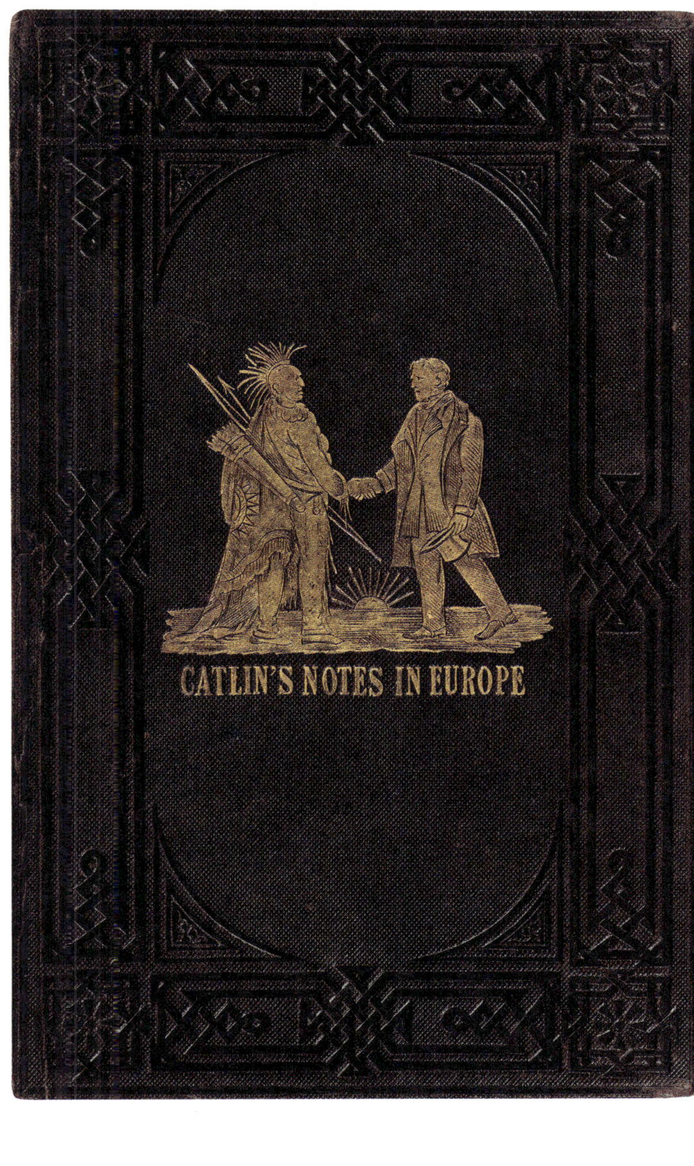

portraying Indians. He had made a brief visit to London in February with King Friedrich Wilhelm III for the baptism of the infant who would be the future King Edward VII.[112] Catlin knew of Humboldt's arrival and wrote excitedly to his father that he expected the explorer to pay a call the following day. Catlin makes no further mention of this, implying that the artist and the baron did not meet that day. Now, in 1845, Humboldt finally had the opportunity to speak with North American Indians. Deroine most likely translated Humboldt's French and accented English and the Iowa language.[113]

Catlin sought Humboldt's approval for his enterprise and used it to market himself and the Indian Gallery. In 1848 Catlin published his two-volume book chronicling this European tour, *Catlin's Notes of Eight Years' Travels and Residence in Europe with His North American Indian Collection* (CAT. 79). The line drawing that appears as the frontispiece to the first volume shows a smiling Humboldt, who doffs his hat and extends his hand in friendship to one of the Iowa Indians. Catlin took this stamp of approval one step further, embossing this vignette in gold leaf on the covers of both books in the two-volume set. For Catlin, Humboldt's imprimatur validated his efforts to depict American Indian ways of life and educate the public. Humboldt did not appear to be put off by Catlin's self-promotion, and in fact the two men established a friendship that would deepen over the next decade.

Catlin mentioned in his book that French artists had been among his Parisian visitors to the Indian Gallery, several of whom complimented him on his work. Delacroix asked to sketch both the Iowa and the second group of Ojibwe who replaced them at the end of the summer. However, even though Bodmer was in Paris, there is no record of his either attending a performance or introducing himself to Catlin.[114] In April 1845, Dr. Al Donné folded coverage of the live Indians and Bodmer's paintings into a single review. The article opened with a description of the Iowa Indians in their native dress—"thirteen individuals, men, women, and children...presented at the Academy of Sciences after visiting the royal family"—and then included a lengthy review of Max and Bodmer's publication: "We are perfectly well acquainted with Prince Max's description of them, in his account of his journey to their homeland, and in the fine drawings of M. Bodmer, taken from life." Bodmer's portraits, according to Donné,

> *were so realistic that we could recognize the savages easily. When we saw the real people it was almost as if we were transported to the heart of the Mandan tribe. We expected to start out dancing the scalp dance or another strange or wild ceremony that Prince Maximilian described so well.*[115]

King Louis-Philippe certainly knew of Prince Max and Catlin's overlapping, competing interests. He had met with Bodmer the year before, graciously committing to the purchase of three full sets of Max and Bodmer's publication, in which Max had described Catlin as essentially accurate but prone to exaggeration.[116] On May 20, 1845, Bodmer wrote to Max that the king had sent him a ring, probably shortly after the Iowa's performance at the Tuileries.[117] Whether that was to console him or encourage him, Bodmer seemed satisfied with the king's notice. Meanwhile, at Louis-Philippe's invitation, Catlin moved his entire operation to the Salle de Séance at the Louvre. Catlin then provided a tour for the royal family, with Vattemare providing translation as needed. The king was most taken by the Crow wigwam and scenes of ball play that reminded him of his own experiences in the United States. At the end of the visit, Catlin noted,

> *His next object was to designate the paintings he wished me to copy and somewhat enlarge, and soon pointed out the number of fifteen, which I was commanded to paint for the palace at Versailles.*[118]

Catlin completed his commissions in Paris.[119]

At Baudelaire's invitation, Catlin entered two of his paintings in the 1846 Salon. One was the portrait of *Little Wolf* that he had just completed for Louis-Philippe, and the other was of Blood Chief Stu-mick-o-súcks (Buffalo Bull's Back Fat) that he had painted from life in 1832 (fig. 82). Baudelaire wrote:

> *When M. Catlin came to Paris, with his Museum and his Ioways, the word went round that he was a good fellow who could neither paint nor draw, and that if he has produced some*

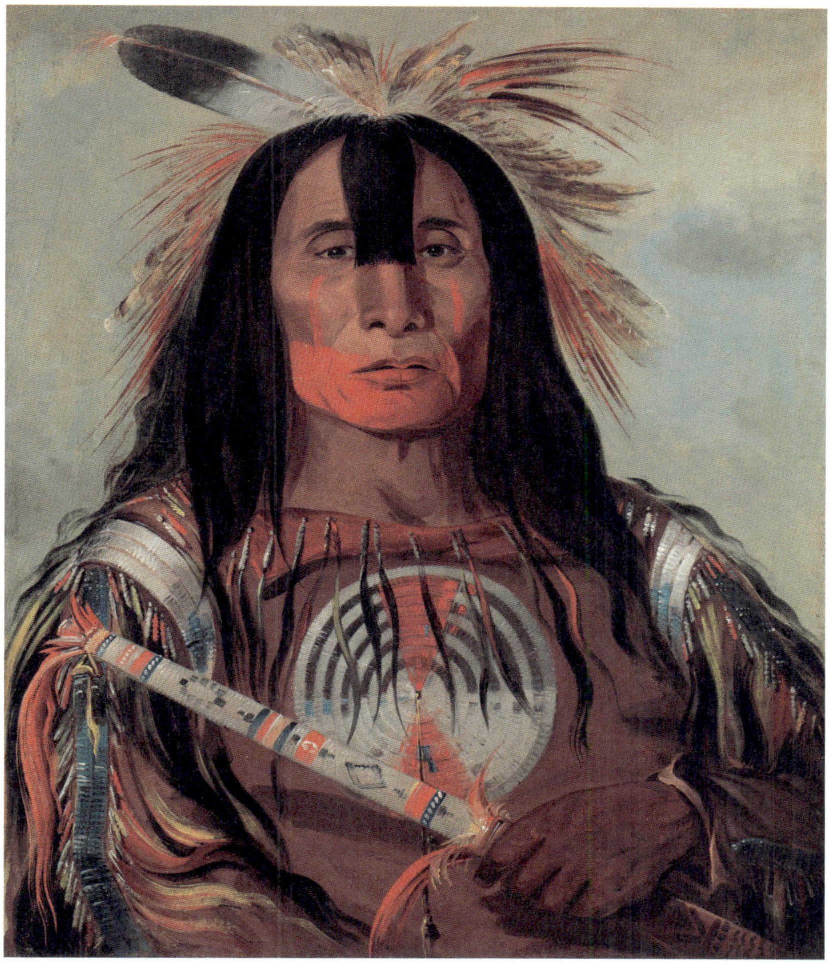

Fig. 82
George Catlin, *Stu-mick-o-súcks, Buffalo Bull's Back Fat, Head Chief, Blood Tribe*, 1832, oil on canvas, 29 × 24 in., Smithsonian American Art Museum, Gift of Mrs. Joseph Harrison Jr.

tolerable studies, it was thanks only to his courage and his patience. Was this an innocent trick of M. Catlin's, or a blunder on the part of the journalists? For today it is established that M. Catlin can paint and draw very well indeed.... M. Catlin has captured the proud, free character and the noble expression of these splendid fellows in a masterly way.[120]

Ten years earlier Bodmer had arranged to exhibit his watercolors from North America at the Salon, where the criticism had been far harsher. The critic for *L'Artiste* had damned his work with faint praise, concluding,

Finally, the Indian characters *of M. Bodmer have excited our curiosity thanks to the bizarreness of their garb, and we must at least thank the designer for the kind of artlessness with which he has collected the details of this savage attire.*[121]

Bodmer's work had been critiqued by people who had never seen a Native American. With the arrival of Catlin's Indian Gallery and the Iowa Indians, the reviewer's perspective had changed.

SCHOOLCRAFT'S OPPROBRIUM

Although there was no evidence of affection among Catlin, Max, and Bodmer, their shared respect for Humboldt played a noteworthy role the following decade. Catlin had made a powerful enemy in the person of American ethnographer Henry Rowe Schoolcraft. Schoolcraft and Catlin had begun their respective careers as friends, and Schoolcraft endorsed the authenticity of Catlin's portraits and portrayal of Native American life in the early Indian Gallery promotional brochures. The friendship began to sour, however, when Schoolcraft approached Catlin to secure the artist's illustrations for his own volumes on Native American ethnography. Catlin refused, as he had his eye on publishing his own accounts of his travels and did not want to dilute the impact of his artworks. Catlin's vocal disapproval of official Indian policy widened the gulf between them, as Schoolcraft positioned himself as the expert, stationed in Indian territory in an official capacity. From that position, Schoolcraft sought and received government sponsorship of his publications on the North American Indians.

Humboldt did not become deeply involved in Native American policy after Jefferson's death in 1826. Instead he championed the work of specific ethnographic scholars. His support encouraged Albert Gallatin to produce his landmark treatise on North American Indians in 1836 and, in 1842, to found the American Ethnological Society in New York. Although Schoolcraft traveled to Europe in 1841 carrying a letter of introduction from Gallatin, his much-desired friendship with Humboldt did not materialize and, in fact, foundered on what appeared to be a rare personal and professional dislike for the American Indian agent and ethnographer. In 1853 Schoolcraft wrote to Humboldt, sending him the first volume of his treatise, seeking his approbation.[122] Humboldt responded, but not in the way that Schoolcraft had hoped. The *North American Review* devoted eighteen pages to its less-than-glowing assessment, concluding with the coup de grâce:

Baron Humboldt, having had occasion to examine the work, expressed in strong terms his opinion that it was a crude and worthless compilation, and his great surprise that it

should be allowed to appear with the sanction and at the expense of the government of the United States.[123]

Schoolcraft wrote an intemperate, blistering response in *Literary World*, under the pseudonym "Curtius," which he certainly was not. Schoolcraft proclaimed that America put too much stock in "foreign modes and authors" and slammed Humboldt personally, asserting,

However eminent this learned gentleman is on his favorite topics . . . [on subjects] which he is not acquainted with, his opinions are worth just nothing at all.[124]

If the back and forth had stopped there, Humboldt might never have seen the need to ask Max to defend Catlin, but in the fifth volume of his book, Schoolcraft attacked Catlin directly. He deliberately accused Catlin of being as wrong about the Mandan O-Kee-Pa ceremony as he had been in speculating that the Mandan descended from Madoc, prince of Wales.[125] That story had long been discredited, but in linking the two examples, Schoolcraft called all of Catlin's ethnographic work a fabrication and Catlin a fraud. Humboldt received his copy of this volume in the spring of 1856 and, upon reading it, informed Catlin that he needed to defend himself. After expressing his outrage that Schoolcraft's work carried the endorsement of the American government, Humboldt invoked a powerful supporter of his own:

I have often conversed with our illustrious traveler in America, the Prince Maximilian, of Neuwied, who spent a winter with the Mandans subsequent to your visit to them, and gained information from the chiefs entirely corroborating your descriptions. You should write to the prince at once, and, getting a letter from him (with your other proofs), lay it before the Government of your country, which cannot fail, by some legislative act, to do you justice.[126]

Schoolcraft might have succeeded in discrediting Catlin had he not also gone after Humboldt. Humboldt knew that, regardless of Max's opinions of Catlin, their accounts of the Mandan, and the O-Kee-Pa ceremony in particular, lined up.[127] Catlin had witnessed it firsthand; Max had transcribed Dipäuch's firsthand account. Mató-Tópe could no longer speak for himself: smallpox had nearly wiped out the entire tribe in 1837, and Mató-Tópe had succumbed that July.[128] The surviving tribe members banded with the Arikara and Hidatsa, becoming the Three Affiliated Tribes. Catlin and Max were two of the remaining authorities on Mandan culture.

By the time Humboldt sent his letter, Catlin was dodging creditors in South America.[129] In 1852 he lost control of his Indian Gallery after yet another failed effort to urge the U.S. government to purchase it for the nation. Joseph Harrison had paid many of the artist's debts and took possession of the entire Indian Gallery. The paintings and artifacts remained in Harrison's family until after Catlin's death. Catlin simply disappeared, and over the next three years he apparently traveled through South America and up the west coast of both American continents. Then, in September 1855, he renewed his acquaintance with Humboldt. Humboldt welcomed Catlin to Berlin, where the two men talked about Indians, South America, and Catlin's interest in geology. Catlin suggested, and Humboldt eagerly agreed, that the facial features of the Crow Indians resembled those of the Toltec, which would suggest that they descended from a common ancestor.[130] Although their reasoning was flawed, the similarities implied they were peer civilizations.

While Catlin was there, Humboldt arranged for King Friedrich Wilhelm IV to see the American artist's work and for Catlin to paint replicas of some of his paintings for the Prussian king's art collection.[131] The selection of paintings Humboldt commissioned for the royal collection focused on genre scenes and groupings of several specific tribes as ethnographic vignettes. The genre scenes were copies of earlier works Catlin had made during and after his trips among the Indians during the 1830s, including his *Buffalo Hunt under the Wolf-Skin Mask* (**CAT. 80**), a demonstration of a clever hunting practice designed to use the wolf scent to mask the Indians' own. The tribal groups were in turn copies of the "cartoon collection" of paintings Catlin had begun after having to sell his primary Indian Gallery. Mató-Tópe appears in the group featuring the Mandan (fig. 83), striking the pose he had favored when Catlin and Bodmer had painted him from life. Catlin was planning to return to South America, a prospect the aging

CAT. 80 George Catlin, *Buffalo Hunt under the Wolf-Skin Mask*, 1832–33, oil on canvas, 24 × 29 in., Smithsonian American Art Museum, Gift of Mrs. Joseph Harrison Jr.

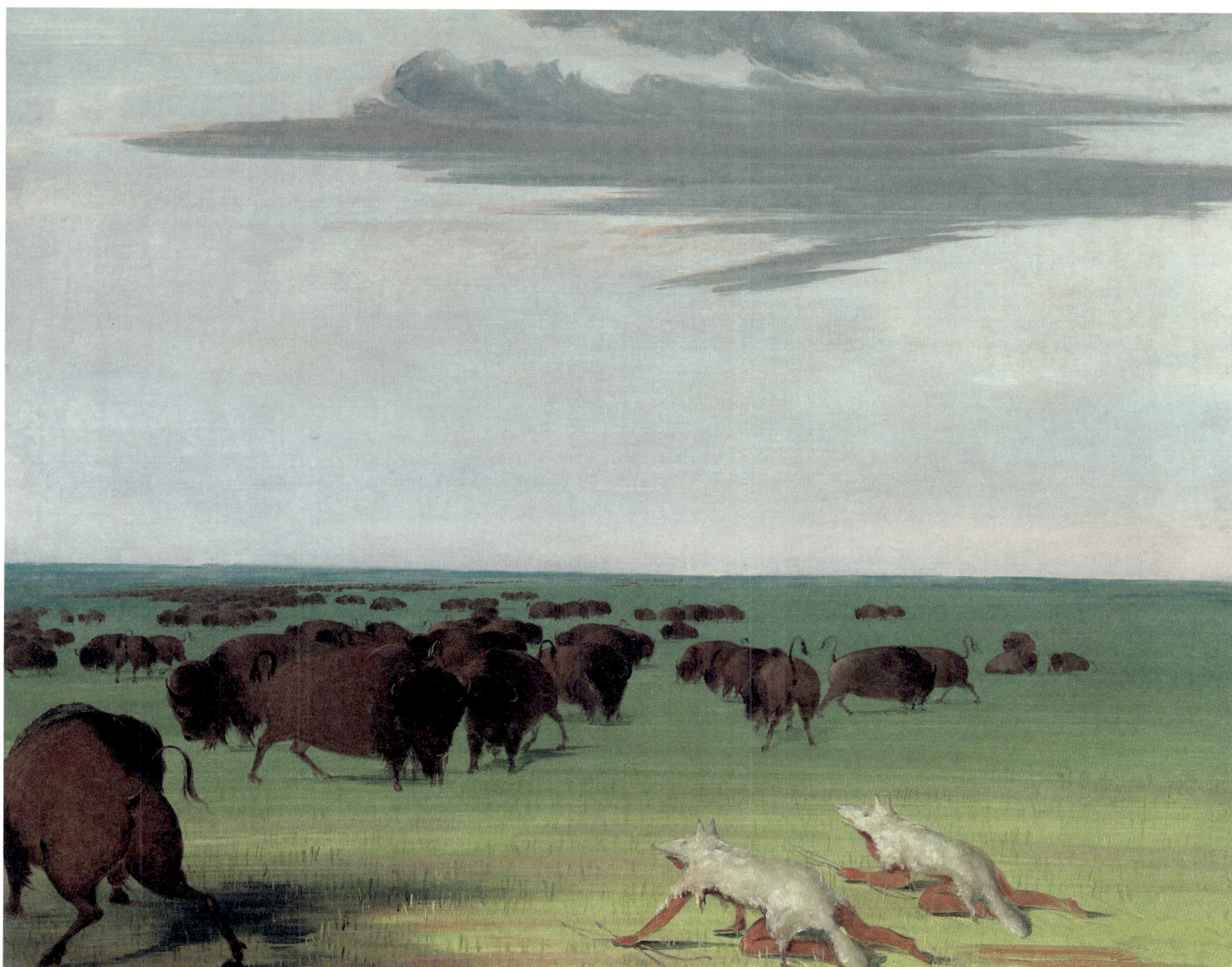

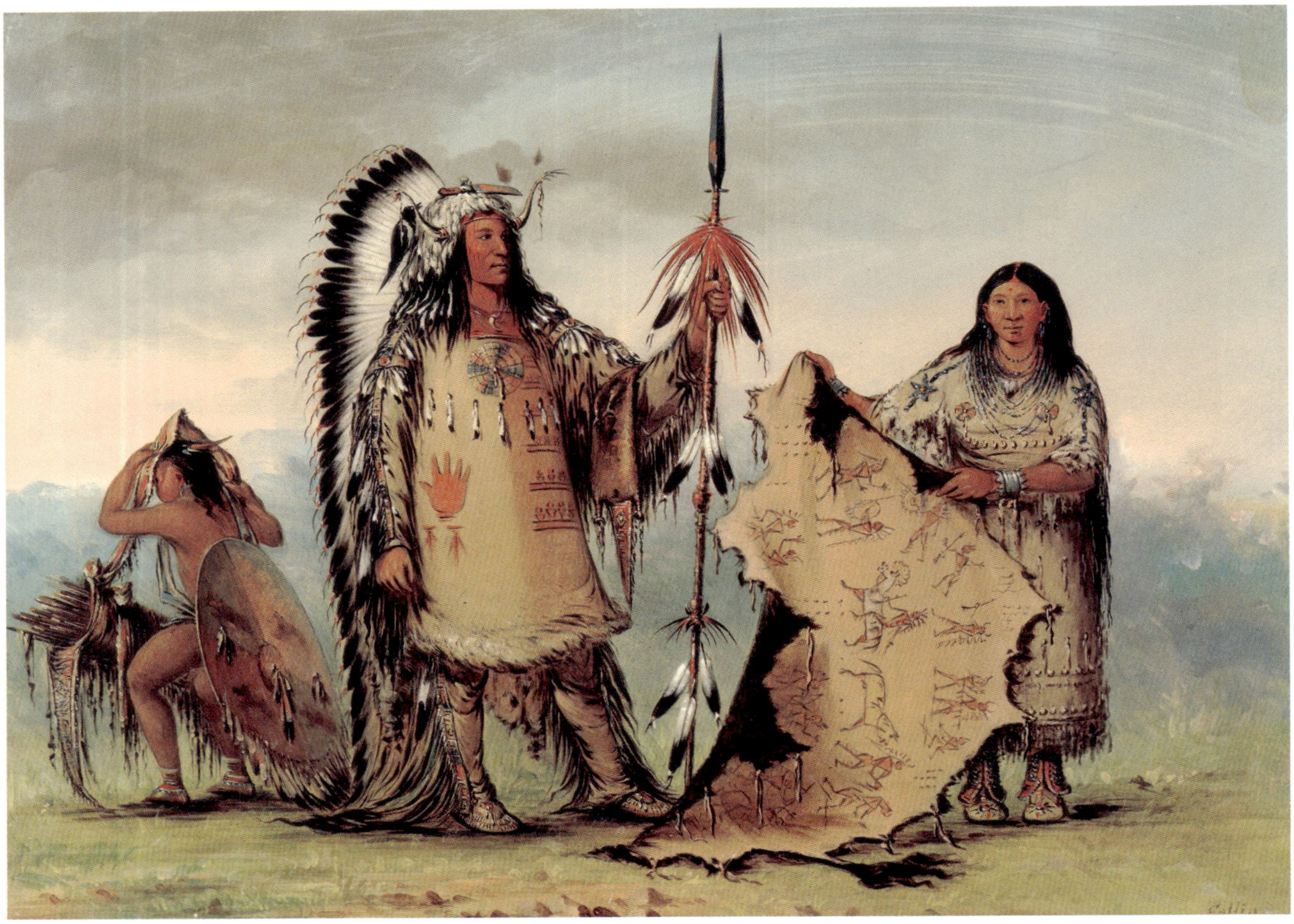

Fig. 83
George Catlin,
Mandan Indians, ca. 1854,
oil on canvas, 18 ⅞ × 25 ¾ in.,
Ethnologisches Museum,
Staatliche Museen, Berlin,
Germany

Humboldt still found tempting. Before Catlin had left Berlin to return to South America, he was carrying a letter from Humboldt to give to his old friend Bonpland.[132]

Catlin did not write to Max immediately. Instead, after spending several years in South America, he returned to Europe in 1860 and began writing his own authoritative essay on the O-Kee-Pa ceremony, intent on defending his honor. He sent his essay to Max in 1866, along with four oil paintings he had made on site depicting moments during the ceremony. He requested that Max

inform me and the world how far my descriptions of those scenes will be supported by information gathered by yourself from Mr. Kipp and others during the winter which you spent in the Mandan village, and for which I feel deeply indebted.[133]

Max replied immediately, asserting,

I can attest your relation of it to be a correct one, after all that I heard and observed myself. In my description of my voyage in North America (English edition) I gave a very detailed description of the O-kee-pa, *as it was reported to me by all the chiefs and Mr. Kipp, and it is about the same that you told, and nobody would doubt our veracity, I hope.... If he [Schoolcraft] should doubt what we have both told in our works of the great medicine festivities of the* O-kee-pa, *he would be wrong, certainly. If my statement, as that of a witness, could be of use to you, I should be very pleased.*[134]

Catlin published Max's statement in the front matter of his 1867 book, *O-Kee-Pa: A Religious Ceremony; and Other Customs of the Mandans*. Max's letter, confirming Catlin's written account, became Catlin's most powerful weapon, which he deployed for years after Humboldt's death. Max's respect for Humboldt must have played a role in Max's agreeing to defend a man he viewed as less than a worthy competitor—but for Humboldt, truth mattered more than defects of personality. His entire life Humboldt had sought to gather as much evidence as he could to support what he deemed the underlying laws of nature. Nothing stood in the way of this pursuit. He sought out local wisdom, but he also supported the exploration that put those very civilizations in jeopardy. He defended the rights of native peoples, yet he also defended Catlin. The irony was not lost on men like Schoolcraft, but in truth none of their motives were pure, and their outlook was an accurate reflection of the conflicted motivations of their era. Long after Humboldt's death, Catlin treasured the letters connecting him to Max and to Humboldt, letters of approbation from two of the most influential European voices on American ethnography. Thus, even after his death, Humboldt's name carried a particular weight, especially in the United States.

Humboldt's ability to create and extend networks of like-minded people—artists, explorers, ethnographers alike—remained an integral part of his life well beyond his years as an active traveler. His opinion retained the ability to make or shape the careers of those who sought his advice. For Max and Bodmer, Humboldt's praise reinforced the significance of their published accounts within the broad European scientific community. Max was deeply influenced by the Humboldtian worldview: gather the data, derive observations, connect the dots. The more he could see and experience, the more acute his understanding. Max, like Humboldt, measured the success of his venture in the production of the published account. For Catlin, Humboldt's support validated the sincerity at the core of his enterprise. Humboldt could look beyond the more exploitative aspects of Catlin's traveling with Native Americans to make ends meet; he also saw Catlin's unspoken plea to honor and admire a culture he believed doomed, whether through disease or government policy. Catlin's efforts were more circumscribed, focusing intently on the plight of the nations he saw as vulnerable to white encroachment and fighting a battle he believed they would ultimately lose. On ethical grounds, Humboldt believed he could have things both ways: rather than seeing the traveling Indian performances as the exploitation of race to satisfy the curiosity of "civilized" Europeans, he believed that access to traveling Indians could augment knowledge accrued from others' accounts.

For Humboldt, questions about the essential humanity and equality among all races touched on the fine arts, literature, and the emerging fields of archaeology and ethnography in the United States. He was not a quiet bystander. He got to know American artists and writers, along with statesmen and ethnographers, and he had a hand in shaping their attitudes as well as their reputations. During the 1830s and 1840s, Humboldt's growing unease with America's expansionist policies were becoming clear, as emissaries seeking his approval for a territorial war with Mexico to further expand the United States were met with diplomatic ambivalence masking deeper dismay. Humboldt's desire for corollary information on North American Indians fueled his interest in Prince Maximilian's trip and fostered a lasting friendship with artist George Catlin. But Humboldt's relationships with explorers—whether their goals were geopolitical, scientific, or artistic—were firmly grounded in his unceasing desire for more information to fuel his crowning achievement: the publication of *Cosmos*, the series of books that would make Alexander von Humboldt perhaps the best known public intellectual figure anywhere on the globe.

NOTES

EPIGRAPH
Alexander von Humboldt, *Cosmos: A Sketch of a Physical Description of the Universe*, trans. E. C. Otté (London: Henry G. Bohn, 1849), 1:368.

1 Rev. Theodore Parker, "Tribute to Humboldt," *Liberator* 28, no. 28 (9 July 1858): 110. Parker wove Humboldt into several of his sermons attacking slavery.

2 Humboldt engaged with his guides and with the communities he visited. In 1800 he arrived at Lake Valencia, in Venezuela, where the local inhabitants asked his advice on their failing agriculture. Humboldt discovered that all of the mature trees had been felled years earlier. As a result, the land's ability to hold moisture had dropped and, with it, the lake's water level. Runoff from the rains accelerated soil erosion, all of which had contributed to a change in the local climate. His logic provided perspective on this community's situation and a case study for Humboldt's interest in the effects of deforestation. See Alexander von Humboldt and Aimé Bonpland, *Personal Narrative of Travels to the Equinoctial Regions of the New Continent, during the Years 1799–1804*, trans. Helen Maria Williams (London: Longman, Hurst, Rees, Orme, and Brown, 1819), 4:143–44.

3 Modern translation in Ottmar Etté, "Alexander von Humboldt: The American Hemisphere and Trans-Area Studies," *Iberoamericana* 5, no. 20 (December 2005): 101. Original translation in Alexander von Humboldt, *Researches, Concerning the Institutions and Monuments of the Ancient Inhabitants of America, with Descriptions and Views of Some of the Most Striking Scenes in the Cordilleras!*, trans. Helen Maria Williams (London: Longman, Hurst, Rees, Orme, and Brown, 1814), 1:408. The calendar stone, also known as the sun stone, is an elaborate solar disk rather than a true calendar.

4 For a transcript of Chastellux's letter to Jefferson, see Thomas Jefferson, *The Papers of Thomas Jefferson*, vol. 8, *25 February to 31 October 1785*, ed. Julian P. Boyd (Princeton, NJ: Princeton University Press, 1953), 174–75. For a larger consideration of Jefferson's evolving views, see Nicholas W. Gentile, "Philosophical Sympathy and 'Seeds of Extinction': Jefferson's French Correspondence and the Roots of His Presidential Indian Policy, 1785–1789," *We Proceeded On* 36, no. 2 (May 2010): 18–27.

5 Jefferson, *Papers of Thomas Jefferson*, 8:185–86.

6 Thomas Jefferson, *The Papers of Thomas Jefferson*, vol. 10, *22 June to 31 December 1786*, ed. Julian P. Boyd (Princeton, NJ: Princeton University Press, 1954), 44.

7 Joseph J. Ellis, *American Sphinx: The Character of Thomas Jefferson* (New York: Alfred A. Knopf, 1997), 201.

8 Jefferson to Humboldt, Montpelier, 6 December 1813, Heineman Collection (V), The Pierpont Morgan Library, New York, in Ingo Schwarz, ed. *Alexander von Humboldt und die Vereinigten Staaten von Amerika: Briefwechsel* (Berlin: Akademie Verlag, 2004), 130, letter 41.

9 An extended discussion of the Long Expedition occurs in chapter 2 in this volume.

10 *Register of Debates in Congress, 21st Congress, 1st Session* (Washington, DC: US Government Printing Office, 1830), 382; cited in Kathryn S. Hight, "'Doomed to Perish': George Catlin's Depictions of the Mandan," in "Depictions of the Dispossessed," special issue, *Art Journal* 49, no. 2 (Summer 1990): 119. In his 1830 address, Jackson asserted, "What good man would prefer a country covered with forests and ranged by a few thousand savages to our extensive Republic, studded with cities, towns, and prosperous farms, embellished with all the improvements which art can devise or industry execute...?" Andrew Jackson, Second Annual Message, December 6, 1830, *The American Presidency Project*, http://presidency.ucsb.edu/node/.

11 "Account of the Philadelphia Museum," *National Intelligencer and Washington Advertiser*, 24 December 1804. Among those were portraits of Captain Joseph Brant (Thayendanegea) of the Mohawk and wax figures of Blue Jacket and Red Pole, both of the Shawnee. See Charles Coleman Sellers, "'Good Chiefs and Wise Men': Indians as Symbols of Peace in the Art of Charles Willson Peale," *American Art Journal* 7, no. 2 (November 1975): 10–18.

12 Charles Coleman Sellers, *Mr. Peale's Museum: Charles Willson Peale and the First Popular Museum of Natural Science and Art* (New York: W. W. Norton, 1980), 426.

13 George Catlin, *Letters and Notes on the Manners, Customs, and Condition of the North American Indians* (London: Tilt and Bogue 1842; repr., Minneapolis: Ross & Haines, 1965), 1:2–3. For George Harvey's description, see Thomas Donaldson, *The George Catlin Indian Gallery in the U.S. National Museum (Smithsonian Institution), with Memoir and Statistics*. (Washington, DC: US Government Printing Office, 1887), 715. Catlin did see a delegation of Winnebago in Washington, D.C., during the winter of 1828. It is worth bearing in mind that Catlin can be quite vague in both recalling his early chronology and specific whereabouts once he arrived in St. Louis.

14 Catlin, *Letters and Notes*, 1:5.

15 Catlin to Major James H. Hook, Commissary General's Office, DC, 20 December 1832, National Archives, Record Group 92; quoted in Brian W. Dippie, *Catlin and His Contemporaries: The Politics of Patronage* (Lincoln: University of Nebraska Press, 1990), 445n36.

16 Catlin, *Letters and Notes*, 1:16.

17 W. Richard West provides a thoughtful assessment of Catlin's conflicted aims in his introduction to *George Catlin and His Indian Gallery*, ed. George Gurney and Therese Thau Heyman (Washington, DC: Smithsonian American Art Museum in association with W. W. Norton, 2002), esp. p. 11. John Hausdoerffer astutely coined the term "cultural eulogist" to describe Catlin's approach. See John Hausdoerffer, *Catlin's Lament: Indians, Manifest Destiny, and the Ethics of Nature* (Lawrence: University Press of Kansas, 2009).

18 Catlin extended his thanks for Clark's support: "My works and my design have been warmly approved and applauded by this excellent patriarch of the Western World; and kindly recommended by him in such ways as have been of great service to me." Catlin, *Letters and Notes*, 2:30.

19 In 1832 Catlin carried with him a letter of introduction from Secretary of War Lewis Cass that read in part: "[S]o far as your object in visiting the Indian Villages is explained in your communication, I think it laudable, and every facility should be afforded you in its promotion. You can show this letter to the Indian Agents you may meet in your tour who I am confident will render you every assistance in their power. Very respectfully &c., Lewis Cass. Department of War, February 18, 1832." Letters Sent by the Office of Indian Affairs 1824–1881, Records of the Bureau of Indian Affairs, Record Group 75. It is reproduced on National Archives Microfilm Publication M21, Letters Sent by the Office of Indian Affairs, 1824–1881, roll 8, p. 118.

20 Estimates vary for concrete numbers. John C. Ewers, a Smithsonian anthropologist, estimated 135 paintings in eighty-six days; Bill Truettner makes a strong case for 170 paintings in five months. See William H. Truettner, *The Natural Man Observed: A Study of Catlin's Indian Gallery* (Washington, DC: Smithsonian Institution Press in cooperation with the Amon Carter Museum of Western Art and the National Collection of Fine Arts, Smithsonian Institution, 1979), 23.

21 Truettner, *Natural Man Observed*, 143–49; Dippie, *Catlin and His Contemporaries*, 30.

22 The first part of Prince Maximilian and Karl Bodmer's journey through the United States is covered in chapter 2 in this volume.

23 Botocudo is the name the Portuguese gave for the Aimoré, derived from the word "botoque" (plug), referring to the wooden discs worn in their ears and lips.

24 For a discussion of this publication, please see chapter 2 in this volume.

25 Prince Maximilian of Wied, *Travels in the Interior of North America, by Maximilian, Prince of Wied. With Numerous Engravings on Wood, and a Large Map*, trans. H. Evans Lloyd (London: Ackermann, 1843), 20. Max knew that Thomas McKenney's and James Hall's *History of the Indian Tribes of North America* was in production, but he would not see it during his travels. Maximilian, Prince of Wied, *The North American Journals of Prince Maximilian of Wied*, ed. Marsha V. Gallagher and Stephen S. Witte, trans. William J. Orr, Paul Schach, and Dieter Karch (Norman: University of Oklahoma Press, 2012), 1:72n98.

26 See chapter 2 in this volume for an extended discussion of Prince Max's background and the eastern part of his trip in the United States.

27 "Dr. Harlan of Philadelphia had also given me a letter to the owner of the museum, a portrait painter, Mr. Lambdin; and through this I had been able to familiarize myself with those interesting animals." Entry for 23 August 1832, in *North American Journals of Prince Maximilian of Wied*, 1:171.

28 Christian F. Feest, "James Reid Lambdin, George Catlin, and the 'Costume of a Sioux Warrior,'" *European Review of Native American Studies* 17, no. 2 (2003): 23.

29 James Reid Lambdin to Alex. Wilson, Esq., Pittsburgh, 30 September 1832: "Dear Sir I beg leave to introduce to your friendly notice Mon'r le Baron of Braunsburg [*sic*] of Germany, who is now making a scientific tour through the United States. —He will probably spend a day or two in Wheeling principally for the purpose of visiting the Indian mounds at Grove creek. An introduction to Dr. [James] Clemens will probably be of service— which with all other attentions bestowed will be cheerfully reciprocated by very truly your obedient servant J. R. Lambdin." Lambdin also wrote to Alfred Cochran, Pittsburgh, 28 September 1832: "Any information you can render him regarding the Reptilae of your neighborhood— will be cheerfully received by him and esteemed as a particular mark of favour conferred on, very truly your friend & brother, J. R. Lambdan." Both letters, Maximilian-Bodmer Collection, Margre H. Durham Center for Western Studies, Joslyn Art Museum, Omaha, Nebraska.

30 Prince Max collected newspaper clippings that contained stories about Indian populations, including one that noted the number of Indian tribes still living in the territory of the United States: "1) Choctaws 20,000. 2) Snakes [Shoshones] 20,000. 3) Crees 20,000. 4) Cherokees 15,000. 5) Blackfeet 15,000. 6) Chippewas 15,000. 7) Sioux 15,000. 8) Pawnees 12,000. 9) Assiniboines 8,000. 10) Potawatomis 6,500. 11) Winnebagos 5,800. 12) Sauks 6,800. 13) Osages 5,000. 14) Menominees 5,200. 15) Crows 4,500. 16) Kansa ——." Cited in *North American Journals of Prince Maximilian of Wied*, 1:235n M39.

31 In the annals of Humboldtian what-might-have-beens, Max considered joining up with Washington Irving after receiving a letter from a German-American who alerted him that "[t]he American poet, Irving, as I heard, is about to start a journey toward the West, and as they say, he wants to go from St. Louis to the Rocky Mountains, he might be perhaps also a pleasant fellow-traveller for you, Baron von Braunsberg, in the primeval forests of America." Fred. V. Gebhard to Baron de Braunsberg, New York, 5 October 1832, Maximilian-Bodmer Collection, Margre H. Durham Center for Western Studies, Joslyn Art Museum, Omaha, Nebraska. Max had other options as well. In St. Louis he met with William Stewart Drummond, who was also looking for a companion with an artist to accompany him out West. Drummond would end up traveling with Alfred Jacob Miller, establishing that artist's career interest in Western subjects.

32 Prince Max was carrying letters from E. Broghan to four Indian agents: Major William Davenport, U.S. Army Commandant at Fort Leavenworth; Col. Zachary Taylor, 1st Infantry, Fort Crawford, Prairie du Chien; General Joseph Montfort Street, Indian Agent, Prairie du Chien; and Major Lawrence Taliaferro, Indian Agent, St. Peters Agency, Minnesota. Maximilian-Bodmer Collection, Margre H. Durham Center for Western Studies, Joslyn Art Museum, Omaha, Nebraska.

33 In his journal entry for 10 July 1832, Prince Max noted their shared German heritage and wrote: "The most extensive fur trade is conducted by Mr. Astor. Originally from Mannheim on the Rhine, he...came to the United States as a poor man and is now said to be the richest person in the United States....It was therefore very important for me to receive from young Astor, who lived in Göttingen when I did in 1811 and 1812, a letter of recommendation to a certain Mr. Abbott in Detroit, which is quite close to the border of civilization and near several Indian nations. Unfortunately, the state is now at war with part of the Indian population of those regions." *North American Journals of Prince Maximilian of Wied*, 1:60–61. At the end of his voyage, Max went to great lengths to thank the American Fur Company for its support and thank by name each of the Indian agents who facilitated his and Bodmer's contact with various Indian tribes.

34 For the significance of the Lewis and Clark route maps, see chapter 2 in this volume.

35 Prince Maximilian of Wied, *Travels in the Interior of North America*, see pp. 20, 74, 103, 119–20, 132, 336, 389, 474.

36 Prince Maximilian of Wied, *Travels in the Interior of North America*, 102. Max cites Humboldt in a footnote, referring his readers to Humboldt's *Political Essay on the Kingdom of New Spain*, trans John Black (London: Longman, 1811), 1:115.

37 Bodmer to Maximilian, 18 July 1840, Maximilian-Bodmer Collection, Margre H. Durham Center for Western Studies, Joslyn Art Museum, Omaha, Nebraska; cited in Ron Tyler, "Karl Bodmer and the American West," *Studii și cercetări de istoria artei, Artă Plastică* 5, no. 49 (2015): 14.

38 "In the evening a stranger arrived who gave me some very interesting news from St. Louis. A painter, Mr. Catlin, who had lived for a long time among the Indians high up on the Missouri above the Yellowstone River [and] had painted many portraits of them and made interesting collections, was staying there, but had now left for New York to describe his journey." Entry for 22 January 1833, in *North American Journals of Prince Maximilian of Wied*, 1:274.

39 "We found at his house an interesting collection of Indian articles, and a great number of Indian scenes by Catlin, a painter from New York, who had travelled in 1831 to Fort Union." Prince Maximilian of Wied, *Travels in the Interior of North America*, 111.

40 Prince Maximilian of Wied, *Travels in the Interior of North America* 475.

41 Fully 25 percent of Bodmer's works depict the Mandan. See Kimberly Minor, "Stealing Horses and Hostile Conflict: 1833–1834 Drawings of Mato-Tope and Shi-Chida," (M.A. Thesis, University of Nebraska, 2011), 17.

42 Maximilian to August, begun 18 November 1833, from Fort Clark, Maximilian-Bodmer Collection, Margre H. Durham Center for Western Studies, Joslyn Art Museum, Omaha, Nebraska.

43 Ibid.

44 Ibid.

45 "Plate XXII. gives a fac-simile of a representation of one of his exploits, painted by himself, of which he frequently gave me an account. He was, on that occasion, on foot, on a military expedition, with a few Mandans, when they encountered four Chayennes, their most virulent foes, on horseback. The chief of the latter, seeing that their enemies were on foot, and that the combat would thereby be unequal, dismounted, and the two parties attacked each other. The two chiefs fired, missed, threw away their guns, and seized their naked weapons; the Chayenne, a tall, powerful man, drew his knife, while Mato-Topé, who was lighter and more agile, took his battle-axe. The former attempted to stab Mato-Topé, who laid hold of the blade of the knife, by which he, indeed, wounded his hand, but wrested the weapon from his enemy, and stabbed him with it, on which the Chayennes took to flight. Mato-Topé's drawing of the scene in the above-named plate, shows the guns which they had discharged and thrown aside, the blood flowing from the wounded hand of the Mandan chief, the footsteps of the two warriors, and the wolf's tail at their heels—the Chayenne being distinguished by the fillet of otter skin on his forehead. The buffalo robe,...painted by Mato-Topé himself, and which I have fortunately brought to Europe, represents several exploits of this chief, and, among others, in the lower figure on the left hand, the above-mentioned adventure with the Chayenne chief." Journal entry for 9 April 1834, in Prince Maximilian of Wied, *Travels in the Interior of North America*, 450.

46 In his journal entry for 25 December 1833, Max describes Mató-Tópe and Péhriska-Rúhpa as not being on good terms when Mató-Tópe arrives. Péhriska-Rúhpa leaves but promises to let Bodmer paint him in his finery at another time. On 10 January 1834, Max wrote: "He [Mató-Tópe] had become reconciled with Péhriska-Rúhpa, for whom he bought a green blanket as a present." *North American Journals of Prince Maximilian of Wied*, 3:107, 239.

47 Journal entry for 7 March 1834, in *North American Journals of Prince Maximilian of Wied*, 3:269.

48 Max contracted scurvy during this winter and remained bedridden. Journal entry for 14 March 1834, in Prince Maximilian of Wied, *Travels in the Interior of North America*, 447–48

49 Journal entry for 6 March 1834, in Prince Maximilian of Wied, *Travels in the Interior of North America*, 446.

50 Catlin, *Letters and Notes*, 1:155.

51 In 1838 Joseph Nicollet, accompanied by John C. Frémont, vaccinated as many of the Mandan as he could in an effort to save them from the ravages of smallpox, calling it "[a]n act of justice and an act of kindness." Edmund Bray and Martha Coleman Bray, eds., *Joseph N. Nicollet on the Plains and Prairies: The Expeditions of 1838–39 with Journals, Letters, and Notes on the Dakota Indians* (St. Paul: Minnesota Historical Society Press, 1993), 78; cited in Laura Dassow Walls, *The Passage to Cosmos: Alexander von Humboldt and the Shaping of America* (Chicago: University of Chicago Press, 2009), 136.

52 Maximilian to Louise, begun 5 December 1833, Maximilian-Bodmer Collection, Margre H. Durham Center for Western Studies, Joslyn Art Museum, Omaha, Nebraska.

53 Journal entry for 13 November 1833, in Prince Maximilian of Wied, *Travels in the Interior of North America*, 414.

54 "Mr. Bodmer's colours and pencils froze, so that he could not use them without hot water. Writing, too, was very difficult, because our ink was congealed." Journal entry for 30 December 1833, in Prince Maximilian of Wied, *Travels in the Interior of North America*, 435.

55 "For Carl as a busy portrait-painter, I enclose Bodmer's portrait and also mine, which were made by the young Mandan-Indian Sihchidä (the yellow feather)." Maximilian to August, begun 18 November 1833, from Fort Clark, Maximilian-Bodmer Collection, Margre H. Durham Center for Western Studies, Joslyn Art Museum, Omaha, Nebraska. Two watercolor self-portraits by Sih-Chidä are in the collection of the Joslyn Art Museum.

56 Maximilian to Louise, begun 5 December 1833, Maximilian-Bodmer Collection, Margre H. Durham Center for Western Studies, Joslyn Art Museum, Omaha, Nebraska.

57 Of Max's trip to Brazil: "One stated purpose of his journey was to continue his zoological and botanical studies, but he also intended to compare the indigenous cultures of North America with those of southeastern Brazil in order to verify Blumenbach's thesis that the peoples of South and North America were, in fact, members of the same race." Michael G. Noll, "Prince Maximilian's America: The Narrated Landscapes of a German Explorer and Naturalist" (PhD diss., University of Kansas, 2000), 39.

58 Anatomist and anthropologist Johann Friedrich Blumenbach fathered physical anthropology at the University of Göttingen. Blumenbach (1752–1840), best remembered for his classification of humans into five "races," sparked Maximilian's interest in native peoples and ethnology. His doctoral thesis, *On the Natural Variety of Mankind*, which was published in 1775, had tried to determine whether the known human races were varieties of one species or several distinct species. Blumenbach assumed there were five races: American, Caucasian, Ethiopian, Malayan, and Mongolian.

59 In May 1801, Humboldt visited the cave. There were more than six hundred skeletons, each placed in a basket made of palm fronds: "We took several skulls, the skeleton of a child of six or seven years old, and two of full-grown men of the nation of the Atures, from the cavern of Ataruipe." Humboldt delivered one of the skulls to Blumenbach, but the rest of the remains were lost in a shipwreck off the coast of Africa along with their collections. Alexander von Humboldt and Aimé Bonpland, *Personal Narrative of Travels to the Equinoctial Regions of America, during the Years 1799–1804*, ed. and trans. Thomasina Ross (London:Henry G. Bohn, 1852), 2:481–85.

60 Prince Maximilian of Wied, *Travels in the Interior of North America*, 79, 81.

61 When he published the account of his travels, Max expressed a hope that these would become archaeological sites and referred his interested readers to Humboldt's writings on indigenous barrows. Ibid., 475–76.

62 *North American Journals of Prince Maximilian of Wied*, 3:62.

63 Max, too, admired these artifacts. In his diary he noted, "The pipes of the Dakotas are very beautiful, in truth the most beautiful of all the North American Indians, which they make, in various forms, of the red indurated clay, or stone. The pipe has a long, flat, broad wooden tube, which is ornamented with tufts of horse-hair, dyed red, yellow, or green, and wound round with strings of porcupine quills of divers colours." Prince Maximilian of Wied, *Travels in the Interior of North America*, 157–58.

64 Dippie, *Catlin and His Contemporaries*, 22, 42.

65 Catlin to California geologist William H. Brewer, Brussels, 18 January 1869, William Henry Brewer Papers, Correspondence, Box 2, Folder 47, Manuscripts and Archives, Sterling Memorial Library, Yale University. Quoted and cited in William H. Goetzmann, *Exploration and Empire: The Explorer and the Scientist in the Winning of the American West* (New York: Alfred A. Knopf, 1971), 190. Partially quoted in Dippie, *Catlin and His Contemporaries*, 362.

66 In April 1838, Catlin showed the Indian Gallery in the Old Theater in Washington, D.C., to force the issue of buying his gallery, but Congress did not vote to approve the purchase. House Journal, 25th Congress, 2nd session, p. 974, 28 May 1838. In 1839 Catlin pleaded again with Congress to buy his gallery, but the resolution died without action. House Journal, 25th Congress, 3rd session, pp. 518–19, 11 February 1839; recounted in Dippie, *Catlin and His Contemporaries*, 64–65.

67 Catlin, *Letters and Notes*, 1:4.

68 Ibid., 1:3.

69 Ibid., 2:156–57.

70 Dunlap likes his lectures—"Mr. Catlin's Lectures," *New York Mirror*, 14 October 1837, 126; cited in Dippie, *Catlin and His Contemporaries*, 10n12.

71 Entry for Wednesday, 6 December 1837, in Allan Nevins, ed., *The Diary of Philip Hone, 1828–1851* (New York: Dodd, Mead, 1927), 1:290–91.

72 Diary entry for Thursday, 28 March 1839, in Nevins, *Diary of Philip Hone, 1828–1851*, 1:385–86.

73 George Catlin to Alexandre Vattemare, 24 November 1839, Indiana Historical Society, Indianapolis; cited in Dippie, *Catlin and His Contemporaries*, 67.

74 Murray was a Scotsman who had American ties and had met Cooper, Goethe, and Sir Walter Scott. He served as Catlin's literary agent in London. Murray had visited the United States from 1834 to 1836, living for a time with the Pawnee near St. Louis. He and Catlin first met there in 1835. His book, C. A. Murray, *Travels in North America during the Years 1834, 1835 & 1836, Including a Summer Residence with the Pawnee Tribe of Indians and a Visit to Cuba and the Azore Islands*, 2 vols. (London: Richard Bentley, 1839), made him the reigning expert on North American Indians in Britain.

75 George Catlin, *Catlin's Notes on Eight Years' Travels and Residence in Europe with His North American Indian Collection*, 2 vols. (New York: Burgess, Stringer, 1848), 1:69–78.

76 Herring notes that the March 1843 performance in Sunderland, Scotland, included dancing by twenty whites in Indian clothing performing the "scalp dance" and reenacting battle scenes. The local paper was enthusiastic despite the fictive quality of the whole event: "It was a rich treat to all, especially to those acquainted—through their reading—with Indian life.... The figures of these genuine sons of the forest were striking...their costumes elegant and appropriate, and their statue-like attitudes; and varied quick, and simultaneous movements, chained, as it were, the attention of the audience, and drew from them loud and repeated acclamation." *Northern Times* (Sunderland, Scotland), 24 March 1843; quoted in Joseph B. Herring, "Selling the 'Noble Savage' Myth: George Catlin and the Iowa Indians in Europe, 1843–1845," *Kansas History: A Journal of the Central Plains* 29 (Winter 2006/2007): 231. See also Therese Thau Heyman, "George Catlin and the Smithsonian," in George Gurney and Therese Thau Heyman, eds., *George Catlin and His Indian Gallery* (New York: W. W. Norton, 2002), 251–52.

77 In 1817 a Seneca chief and six braves arrived in England to perform tribal dances for audiences in Liverpool, Manchester, Leeds, and London; in 1819 a group of Oneida traveled to France; and in July 1827, six Osage arrived in Le Havre. In Paris they were presented to Charles X, and the chief ascended to view the city from a balloon. Their portraits were engraved, and they subsequently toured Belgium and Germany before returning to the United States in 1820. See Hugh Honour, *The New Golden Land: European Images of America from the Discoveries to the Present Time* (New York: Pantheon Books, 1975), 231.

78 Bodmer to Prince Maximilian, 9 April 1840, Maximilian-Bodmer Collection, Margre H. Durham Center for Western Studies, Joslyn Art Museum, Omaha, Nebraska; cited in Ron Tyler, "Karl Bodmer and the American West," in Brandon K. Ruud, *Karl Bodmer's North American Prints* (Lincoln: University of Nebraska Press, 2004), 19.

79 Bodmer to Maximilian, 20 May 1845, Maximilian-Bodmer Correspondence, Maximilian-Bodmer Collection, Margre H. Durham Center for Western Studies, Joslyn Art Museum, Omaha, Nebraska.

80 Prince Maximilian zu Wied, "Einige Bemerkungen über Geo. Catlins Werk: 'Lettres and Notes on the Manners, Customs, and Condition of the North-American Indians," *Isis* (1842): 726–41; Bodmer to Maximilian, 9 April 1842, Maximilian-Bodmer Collection, Margre H. Durham Center for Western Studies, Joslyn Art Museum, Omaha, Nebraska.

81 Max appears to have channeled Humboldt's dislike of Zebulon Pike, as he noted: "Pike, and some other travellers, pretend to have found Mongol physiognomy among the North American Indians, especially the Pawnee and the Sioux; but I can affirm that I met with no such physiognomy, though I saw a few instances of it in Brazil. Mr. Von Humboldt very justly observes, on this head, that not merely bodily conformation, but likewise the mode of living, of the two races, are entirely different." Prince Maximilian of Wied, *Travels in the Interior of North America*, 104.

82 Bodmer to Maximilian, 22 February 1840, re: substituting one Indian for another because no one will notice; Bodmer to Maximilian, 6 November 1840, re: re- engrave Périskha-Rúhpa with his war bonnet; cited in Tyler, "Karl Bodmer and the American West," in Ruud, *Karl Bodmer's North American Prints*, 19.

83 Max's German edition of his travels ran nearly 300,000 words; Max's manuscript journals are longer—more than 500,000 words with daily diary, longer essays, tables of statistics, and multiple Indian vocabularies. The English edition of 1843, translated by H. Evans

Lloyd in England, omitted a lot of the scientific data and vocabularies. Lloyd's version was reprinted in 1906 by Reuben Gold Thwaites in *Early Western Travels, 1748–1846* (Cleveland, OH: A.H. Clark, 1904) as volumes 22–24.

84 See Ron Tyler, "Illustrated Government Publications Related to the American West, 1843–1863," in Edward C. Carter II, ed., *Surveying the Record: North American Scientific Exploration to 1830* (Philadelphia: American Philosophical Society, 1999), 147–72.

85 Orr estimated that the cost was 300–850 francs, more than a skilled worker made in a year. William J. Orr, "Karl Bodmer: The Artist's Life," in *Karl Bodmer's America*, ed. David C. Hunt and Marsha V. Gallagher (Lincoln: Joslyn Art Museum and University of Nebraska Press, 1984), 363.

86 See Bernhard Gondorf, "Die Expedition von Maximilian Prinz zu Wied und Karl Bodmer in das innere Nord-amerika," in *Prärie- und Plainsindianer: Die Reise in das Nord-America von Maximilian zu Wied und Karl Bodmer* (Mainz: Verlag Hermann Schmidt, 1993), 42, 45.

87 Humboldt to Maximilian, 18 September 1843, Maximilian-Bodmer Collection, Margre H. Durham Center for Western Studies, Joslyn Art Museum, Omaha, Nebraska.

88 Ibid.

89 The Ojibwe came to seek legal redress from the British government in the form of annuities. Their arrangement with Catlin provided them an income and added authenticity to Catlin's presentation of his Indian Gallery. See Herring, "Selling the 'Noble Savage' Myth: George Catlin and the Iowa Indians in Europe, 1843–1845," 231, 231n18. See also L. G. Moses, *Wild West Shows and the Images of American Indians, 1883–1933* (Albuquerque: University of New Mexico Press, 1996), 17.

90 *Morning Chronicle* (London), 25 and 26 December 1843; cited in Herring, "Selling the 'Noble Savage' Myth," 231n17.

91 The *Illustrated London News* provided steady coverage of the Ojibwe. "Court and Haut Ton," 23 December 1843, 401–407, carried extensive notice of their arrival, including an engraving of an Ojibwe dancing and noted that the Ojibwe have just visited the queen at Winsor Castle. An advertisement in the 13 January 1844 issue affirmed that the Indians would love to return to Windsor Castle again (p. 30). The 16 March 1844 issue recounted the visit by the Ojibwe to the Thames Tunnel (p. 167), and the year-end "Chronology of Remarkable Events, January 1–June 29, 1844" listed this as a notable visit in the Supplement to the *Illustrated London News*, 6 July 1844, 426. An advertisement in the 6 April 1844 issue stated they are performing at Piccadilly (p. 223).

92 On Melody, see Ray V. Denslow, *Out of the Past: The Story of George Henry Curzon Melody* (St. Louis: Royal Arch Masons, 1942), 3, 14–18. For an account of Melody's initial dealings with the Iowa Indians, see Barry, *Beginning of the West*, 519–20. Letter from Secretary of War J. M. Porter, War Department, Washington City, 14 September 1843, to recipient unknown, approving petition on Melody's behalf to bring the Iowa to London; quoted in *Catlin's Notes on Eight Years' Travels and Residence in Europe*, 2:2.

93 An advertisement in the June 1, 1844 issue of the *Illustrated London News* states that Catlin has teamed up with Tom Thumb to perform jointly at Piccadilly (p. 358). In the June 29, 1844 issue, Catlin announces that he and Tom Thumb are reopening together (p. 422).

94 Catlin recalls his acquaintance with Deroine, whom he calls Doraway, in *Catlin's Notes on Eight Years' Travels and Residence in Europe*, 2:7.

95 The *Illustrated London News* also covered the Iowa. The 27 July 1844 issue reports the Iowa are on a boat heading to England to replace the Ojibwe (p. 64). The first advertisement for the new group appears in the 10 August 1844 issue (p. 86), with an advertisement and an article on the Iowa and their interpreter, including an engraving of the group (pp. 91, 93). An advertisement in the 24 August issue states they are performing at Lord's Cricket Ground (p. 126), and in the 31 August issue there is an engraving of the group at that location (pp. 129–30). An advertisement in the 7 September issue indicates they have moved to Vauxhall Gardens and cites additional coverage in the London *Morning Chronicle* and *Sunday Times* (p. 158). For descriptions of Iowa performances at Vauxhall Gardens, see the London *Morning Chronicle*, 7 and 9 September 1844.

96 On January 18, 1844, Barnum and Charles "General Tom Thumb" Stratton sailed for England—the Iowa were to be "exhibited by Mr. Catlin on our joint account." Melody recognized that Catlin was better at working with native performers: "I find him a very kind, sociable and excellent gentleman." Catlin left Barnum with the impression that he was "making a fortune" in London. Philip B. Kunhardt Jr., Philip B. Kunhardt III, and Peter W. Kunhardt, *P. T. Barnum: America's Greatest Showman* (New York: Alfred A. Knopf, 1995), 61–62; quoted in Herring, "Selling the 'Noble Savage' Myth," 237. For their first night in a London hotel, Catlin and Melody agreed to split all costs and receipts. See *Catlin's Notes of Eight Years' Travels and Residence in Europe*, 2:1–7.

97 The *Sunday Times* added that the dances and songs of Indian men were "singularly wild and interesting, while the contented air of the women, with their little children, are those of the Indians which we read of in Cooper's novels." *Illustrated London News*, 27 July and 10 and 24 August 1844; quoted and cited in Herring, "Selling the 'Noble Savage' Myth," 239, 239n40.

98 Catlin relates the story of Roman Nose's death in *Catlin's Notes of Eight Years' Travels and Residence in Europe*, 2:200–201.

99 Ibid., 2:206–212.

100 Catlin mentions the diary frequently in *Catlin's Notes of Eight Years' Travels and Residence in Europe*, 2:115, 153–54. The whereabouts of that diary are unknown at this writing.

101 *Illustrated London News* describes the performance for Louis-Philippe [vol. 6, issue 157, p. 274]; this event is then cited in "Chronology of Remarkable Events, January to June 1845" [vol. 6, issue, 165, p. 417]. Catlin describes it in *Catlin's Notes of Eight Years' Travels and Residence in Europe*, 2:210–14.

102 *Catlin's Notes of Eight Years' Travels and Residence in Europe*, 2:211–12.

103 "The royal party…were—their Majesties the *King* and the *Queen*, the *Duchess of Orleans* and *Count de Paris*, the *Princess Adelaide*, the *Prince* and *Princess de Joinville*, the *Duke* and *Duchess d'Aumale*, and his *Royal Highness* the *Duke de Brabant*." Ibid., 2:211.

104 Ibid., 2:211–12.

105 Ibid., 2:214.

106 "Quand une civilisation ou une barbarie va disparaître, il naît fatalement un historien qui en fait le portrait et en conserve ainsi le souvenir. » Ce que Fenimore Cooper a fait par le roman, Catlin le fait par l'image, il donne à voir, écrit Gautier, « le *fac-similé* le plus exact de la vie sauvage » au moment où ces « malheureuses tribus » sont décimées, quelques-unes anéanties « sans laisser d'autres traces que les croquis de M. Catlin." Gautier, *La Presse*, 19 May 1845; quoted in Daniel Fabre, "The Catlin Effect: Paris, 1845–1846," *Gradhiva, Anthropology Review and History of Arts* 3 (2006): 7. I am grateful to Mary Lublin for the translation.

107 "Rien de plus fantastique et plus effrayant que ces danseurs. Les Ioways y mettent une énergie, une animation extraordinaires : ils se trémoussent, ils sautent, ils agitent leurs tomahawks, brandissent leurs lances en poussant des cris, en roulant des yeux." Gautier, *La Presse*, 25 August 1845; quoted in Daniel Fabre, "The Catlin Effect: Paris, 1845–1846," *Gradhiva, Anthropology Review and History of Arts* 3 (2006): 9; translated by the *Gradhiva* website from French to English.

108 On May 20, 1845, Sand sent Delacroix tickets to the performance. George Sand, *Correspondance*, ed. Georges Lubin (Paris: Garnier frères, 1969), 6:870–71, 873–74. They appear to have gone together on the 30th, possibly with Chopin. According to Herring, Sand wrote to Delacroix encouraging him to attend, and wrote that she bought a large painting of the Iowas. The whereabouts of this painting are unknown. See Herring, "Selling the 'Noble Savage' Myth," 240. On the 31st, Sand wrote to Vattemare: "I found great interest in the work of this traveler and a great talent in his paintings. The savages impressed me with the luxury and strangeness of their costumes, with the beauty of their race, with the contrasts of their sweet and affectionate countenance, and suddenly energetic and even ferocious in mimic dance." ["J'ai trouvé un grand intérêt dans l'oeuvre de ce voyageur et un grand talent dans ses peintures. Les sauvages m'ont vivement frappée par le luxe et l'étrangeté de leurs costumes, par la beauté de leur race, par les contrastes de leur physionomie douce et affecteuse, et tout à coup énergique et même féroce dans la danse mimique."] George Sand, *Correspondance*, ed. Georges Lubin (Paris: Garnier Frères, 1969), 6:873–76; quoted and cited in Sharon L. Fairchild, "George Sand and George Catlin—Masking Indian Realities," *Nineteenth-Century French Studies* 22, no. 3/4 (Spring/Summer 1994): 439–40.

109 Fairchild, "George Sand and George Catlin—Masking Indian Realities," 440.

110 Written as a letter to a friend, she describes what Catlin endured to make his paintings and bemoans the hardships of the Iowa Indians and the fate of the fourteen individuals who traveled to Europe. Sand's description of the Indians hews so closely to language Catlin used to compare the Iowa to Greek gods and Olympic athletes that it is hard to believe that Catlin did not have a hand in suggesting how she might best describe them. Sand's article appeared in *La Diable à Paris*, "Relation d'un voyage chez sauvages de Paris," and was republished later the same month in three issues of *L'Eclaireur* (June 14, 21, 28). Quoted from *Oeuvres Illustrées*, ed. Hetzel-Blanchard Maresq (Paris, 1852–1856), 44. See esp. pp. 41–48. Also cited in Fairchild, "George Sand and George Catlin—Masking Indian Realities," 440.

111 *Catlin's Notes of Eight Years' Travels and Residence in Europe*, 2:246.

112 Humboldt was in England as of January 15 and arrived in London on February 5. He was back in Berlin on February 11. Alexander von Humboldt Chronologie: Basierend auf Alexander von Humboldt, Chronologische Übersicht über wichtige Daten seines Lebens, Bearbeitet von Ingo Schwarz, Kurt-R. Biermann, Ilse John und Fritz G. Lange (Berlin: Berlin-Brandenburgische Akademie der Wissenschaften), http://avh.bbaw.de.

113 Our understanding of this encounter would be much enriched by the appearance of a diary or other account. As of this writing, Catlin's is the only record of this encounter between Humboldt and the Iowa Indians at the Louvre.

114 Bodmer and Prince Max had met some of the Iowa Indians in 1833, but they had been unimpressed with them, especially compared with their fondness for the Mandan. The Iowa Indians decided to leave Paris and journey home after three members of their group died from tuberculosis, including Little Wolf's wife, Oke-we-me, who died June 21, 1845.

115 "Une troupe d'Indiens sauvages, composée de treize individus, hommes, femmes et enfans...est arrivé à Paris et a été présentée hier à l'Académie des Sciences, après avoir rendu visite à la famille royale...nous étaient parfaitement connus par la description qu'en fait le prince Maximilien de Wied-Neuwied, dans le récit de son voyage chez ces peuplades, et par les beaux dessins de M. Bodmer, pris sur nature. Les portraits sont si resemblans qu'il n'était pas possible de les méconnaître; il nous semblait, en voyant les originaux, que nous étions transportés au milieu de la tribu des Mandans.... à voir commencer la danse du scalpe ou quelque autre cérémonie bizarre et féroce si bien décrite par le prince Maximilien." Dr. Al Donné, "Les Sauvages du Missouri," *Journal des Débats*, 23 April 1845, https://gallica.bnf.fr/ark:/12148/bpt6k446903b/f1.image. Partial translation in Tyler, "Karl Bodmer and the American West," *Studii şi cercetări de istoria artei, Artă Plastică*, 25. Translation also by the author.

116 On May 14, 1844, Bodmer had a thirty-minute audience with the French king, who purchased three sets of his prints. In several places in his book, Max acknowledges Catlin's accomplishments and then corrects his errors. "auch einige Irrthümer sich eingeschlichen haben, welche bei einer so schnell vor den Sinnen vorbei eilenden Aufhäufung verschiedenartiger sonderbarer Gestalten und Gebräuche sehr verzeihlich ist, auch die indianischen Worte häufig nicht richtig ausgesprochen und geschrieben sind." Prince Maximilian zu Wied, *Reise in das innere Nord-America in den Jahren 1832 bis 1834* (Koblenz: Hoelscher, 1839–41), 2:658. Catlin allowed "some erroneous figures to creep in, which are very pervasive in the piling-up of various curious forms that rush so quickly past the senses, and the Indian words are often not properly pronounced and written." Translation by the author.

117 Louis-Philippe sent Bodmer a ring at about the same time the king met with Catlin. Bodmer to Maximilian, 25 March and 20 May 1845, Maximilian-Bodmer Collection, Margre H. Durham Center for Western Studies, Joslyn Art Museum, Omaha, Nebraska. Läng surmises that when the Iowa performed for the king, Louis-Philippe recalled having seen Bodmer's paintings and afterward sent the artist a precious ring in recognition. "Bei einer Vorführung am Hofe erinnerte sich Louis-Philippe der sprechend änlichen Bilder, die er von Karl Bodmer gesehen hatte, und sandte dem Künstler als nachträgliche Anerkennung einen kostbaren Ring." Hans Läng, *Indianer waren meine Freunde: Leben und Werk Karl Bodmers 1809–1893* (Bern and Stuttgart: Hallwag Verlag, 1976), 142. Translation by the author.

118 *Catlin's Notes of Eight Years' Travels and Residence in Europe*, 2:291, 293.

119 "When I had completed the pictures ordered by the King, his Majesty graciously granted me an audience in the Palace of the Tuileries to deliver them, on which occasion he met me with great cheerfulness, and, having received from me a verbal description of each picture, he complimented me on the spirit of their execution, and expressed the highest satisfaction with them, and desired me to attach to the back of each a full written description. The dimensions of these paintings were 30 by 36 inches, and the subjects as follow:—

No. 1. An Indian ball-play.
2. A Sioux Council of War.
3. Buffalo-hunt on snow-shoes.
4. Mah-to-toh-pa (the Four Bears), a Mandan chief, full length.
5. A Buffalo-hunt, Sioux.
6. Eagle-dance, and view of loway village.
7. Mah-to-he-ha (the Old Bear), a medicine-man of the Mandans.
8. Wan-ee-ton, one of the most distinguished chiefs of the Sioux.
9. Ee-ah-sa-pa (the Black Rock), a Sioux chief, full length.
10. Mu-hu-shee-kaw (the White Cloud), loway chief.
11. Shon-ta-ye-ee-ga (the Little Wolf), loway warrior.
12. Wa-tah-we-buck-a-nah (the Commanding General), Ioway boy.
13. Maun-gua-daus, Ojibbeway chief.
14. Say-say-gon (the Hail Storm), Ojibbeway warrior.
15. Ah-wun-ne-wa-be (the Thunder-bird), Ojibbeway warrior."

Catlin's Notes on Eight Years' Travels and Residence in Europe, 2:316–17.

120 "Quand M. Catlin vint à Paris, avec ses Ioways et son musée, le bruit se répandit que c'était un brave homme qui ne savait ni peindre ni dessiner, et que s'il avait fait quelques ébauches passables, c'était grâce à son courage et à sa patience. Était-ce ruse innocente de M. Catlin ou bêtise des journalistes? —Il est aujourd'hui avéré que M. Catlin sait fort bien peindre et fort bien dessiner. Ces deux portraits suffiraient pour me le prouver, si ma mémoire ne me rappelait beaucoup d'autres morceaux également beaux. Ses ciels surtout m'avaient frappé à cause de leur transparence et de leur légèreté." Charles Baudelaire, *Salon de 1846* (Paris, 1846), 124. Translated in Honour, *The New Golden Land*, 238. See also Robert Beetam, "George Catlin in France: His Relationship to Delacroix and Baudelaire," *Art Quarterly* 24 (1961): 129–45.

121 "Enfin, *les Personnages indiens* de M. Bodmer ont excité notre curiosité par la bizarrerie de leurs accoutremens, et nous devons au moins remercier le dessinateur de l'espèce de naïveté avec laquelle il a recueilli les détails de cette toilette sauvage." "Salon de 1836: IXe. Article. Peinture," *L'Artiste* 11 (1836): 166–70; quote is on p. 170. Translation by the author.

122 Schoolcraft to Humboldt, Washington, 30 July 1853: "I have directed the first volume to be forwarded to you (without being apprized, of their reception) and to send the future volumes. I have the honor to transmit this, for its greater security, through the Baron von Geroldt [sic]." Schwarz, *Alexander von Humboldt und die Vereinigten Staaten von Amerika*, 323–24, letter 203.

123 Francis Bowen, "Schoolcraft on the Indian Tribes," *North American Review* 77 (July 1853): 262.

124 Curtius (Schoolcraft), "Examination of an Article in the North American Review," *Literary World* 13 (20 August 1853): 51–54. Marcus Curtius was a legendary Roman hero who sacrificed himself to save the empire.

125 Madoc was a prince from Welsh folklore who was said to have sailed across the Atlantic Ocean in 1170, thereby establishing a British claim to North America. Madoc and his followers were said to have settled along the upper Missouri and were occasionally surmised to be the Mandan Indians. Proponents of this story pointed to similarities in Mandan village architecture and Welsh beehive construction, parallels between Mandan "bull boats" and Welsh coracle boats, and similarities in the "ll" diphthong distinctive to both languages. Catlin initially supported the identification of the Mandan as "Welsh Indians" but later disavowed it.

126 On June 9, 1856, Humboldt wrote to Catlin in care of Aimé Bonpland, from Potsdam to Uruguay, warning him about Schoolcraft, urging him to contact Max regarding his O-Kee-Pa story. Schwarz, *Alexander von Humboldt und die Vereinigten Staaten von Amerika*, 377, letter 249. “My Dear Friend: An immense scrap-book on the North American Indians, written by Schoolcraft for the Government of the United States, in three huge volumes, has been sent to me as a present, and I find, in looking into it, that he denies the truth of your descriptions of the ‘*Mandan religious ceremonies*,’ distinctly saying that they are contrary to facts, and that they are works of your imagination, &c. Now my dear and esteemed friend, this charge, made by such a man as Schoolcraft, and ‘*under the authority of the Government of the United States*,’ to stand in the libraries of the scientific institutions of the whole civilized world, to which they are being sent as presents from your Government, is calculated not only to injure your hard-earned good name, but to destroy the value of your precious works through all ages, unless you take immediate steps with the Government of your country to counter its effects. I have often conversed with our illustrious traveler in America, the Prince Maximilian, of Neuwied, who spent a winter with the Mandans subsequent to your visit to them, and gained information from the chiefs entirely corroborating your descriptions. You should write to the prince at once, and, getting a letter from him (with your other proofs), lay it before the Government of your country, which cannot fail, by some legislative act, to do you justice. Your sincere friend, Humboldt.”

127 Max had commented on Catlin’s description of the O-Kee-Pa ceremony [Catlin, *Letters and Notes*, vol. 2 (1841): 127], finding the truncated description Catlin had published in an American newspaper to be “incomplete,” based on the description Kipp had given him. *North American Journals of Prince Maximilian of Wied*, 3:185.

128 At the conclusion of the smallpox epidemic, the Mandan tribe lost all but about 120 members. Mató-Tópe contracted the disease and died July 30. W. Raymond Wood, Joseph C. Porter, and David C. Hunt, *Karl Bodmer’s Studio Art* (Chicago: University of Illinois Press, 2002), 81.

129 Catlin’s first trip to South America was underwritten by his traveling companion, Samuel Colt. The firearms manufacturer paid Catlin to create hunting images featuring his rifles.

130 Catlin to Humboldt, Berlin, 10 September 1855, in Schwarz, *Alexander von Humboldt und die Vereinigten Staaten von Amerika*, 358–59, letter 232. “Is it not possible that the Crows were the builders of the Mexican Monuments, or that they are the existing descendants of the people who did build them? There is a *facial outline so peculiarly* and *strikingly* like the Crow in those monuments you showed me, that the image has been constantly in my mind since, both day and night. I recollect distinctly that those people told me they had migrated from a great way south, and that their ancestors had lived in great houses of stone. Were not these people the *Toltecs*? Do not Mexican historians say the Toltecs came from the North West?”

131 *Briefe Alexander v. Humboldt’s an Ignaz v. Olfers, Generaldirektor der Kgl. Museen in Berlin*, ed. E. W. M. von Olfers (Nürnberg and Leipzig: A. E. Sebald, 1913), 191, 193; cited in Ingo Schwarz, “Americans in Alexander von Humboldt’s Personal Address Book,” in *Unity of Nature: Alexander von Humboldt and the Americas*, ed. Georgia Riley deHavenon, Christina De Léon, Alicia Lubowski-Jahn, and Gabriela Rangel (Bielefield Kerber Verlag, 2014), 35.

132 Humboldt encouraged Catlin to visit Bonpland in South America. In a letter from Humboldt to Catlin from Potsdam, 12 September 1855, he wrote: “My dear sir: The bearer of these lines Mr. Möllhausen, whom I recommend to your kindness, is a man of talent, for whom I have much regard. He has been a long time among the Indians; first alone, on the Upper Missouri, and afterward in the employ of the United States Government as topographer and draughtsman, in the expedition of Lieutenant Whipple to California, across New Mexico (lat[itude] 36°), and back via Panama. He has married the daughter of Mr. Seifert, the hunter and valet de chamber, who accompanied me to Siberia. He brings you a few lines for my friend, M. Bonpland. A thousand kind regards.” Schwarz, *Alexander von Humboldt und die Vereinigten Staaten von Amerika*, 360, letter 234.

133 Catlin to Maximilian, Bruxelles, 2 December 1866. “Dear Prince: Since we traveled together on the Upper Missouri, Mr. Schoolcraft, who has published a large work on the North American Indians for the United States Government, and who never had the industry or the courage to go within one thousand miles of the Mandans, has endeavored to impeach my descriptions of the Mandan religious ceremonies, which, as the tribe has become extinct, he has supposed rested on my testimony alone.... Your highness spent the winter with the Mandans subsequent to the summer season in which I witnessed those ceremonies, and, of course, lived in the constant society of Mr. Kipp, the fur-trader at that post, who witnessed, in company with me, the whole of those four days’ ceremonies, and interpreted everything for me, and from whom you no doubt drew a detailed account of those scenes as we saw them together. I send you with this letter my four oil paintings of those four days’ ceremonies...*seen and approved by the chiefs and the whole tribe*, and having attached to their backs the certificates of Mr. Kipp and two other men who were with us, that ‘those paintings represent strictly what we saw, and without exaggeration.’ I send you also herewith the manuscript of a work (‘*O-kee-pa*’), descriptive of those ceremonies, which I am about to publish, and on reading this and examining my paintings you will be able to inform me and the world how far my descriptions of those scenes will be supported by information gathered by yourself from Mr. Kipp and others during the winter which you spent in the Mandan village, and for which I shall feel deeply indebted. Your highness’ obedient servant, Geo. Catlin.” Thomas Donaldson, *The George Catlin Indian Gallery in the U.S. National Museum (Smithsonian Institution), with Memoir and Statistics* (Washington, DC: Government Printing Office, 1887), 377.

134 Maximilian to Catlin, Neuwied, Prussia, 20 December 1866, “To Mr. George Catlin: Dear Sir: your letter of 2d December came safely to hand, and revived the quite forgotten recollections of my stay among the Indian tribes of the Missouri, now thirty-three years past....Not having been, like yourself, an eye-witness of those remarkable starvations and tortures of the *O-kee-pa*, but having arrived later, and spent the whole of a winter with the Mandans, I received from all the distinguished chiefs, and from Mr. Kipp (at that time director of Fort Clarke, at the Mandan village, and an excellent interpreter of the Mandan language) the most detailed and complete record and description of the *O-kee-pa* festival, where the young men suffered a great deal; and I can attest your relation of it to be a correct one, after all that I heard and observed myself. In my description of my voyage in North America (English edition) I gave a very detailed description of the *O-kee-pa*, as it was reported to me by all the chiefs and Mr. Kipp, and it is about the same that you told, and nobody would doubt our veracity, I hope.... If [Schoolcraft] should doubt what we have both told in our works of the great medicine festivities of the *O-kee-pa*, he would be wrong, certainly. If my statement, as that of a witness, could be of use to you, I should be very pleased. Your obedient, Max, Prince of Neuwied.” Donaldson, *The George Catlin Indian Gallery in the U.S. National Museum (Smithsonian Institution), with Memoir and Statistics*, 378. Prince Max died two months later, on February 3, 1867.

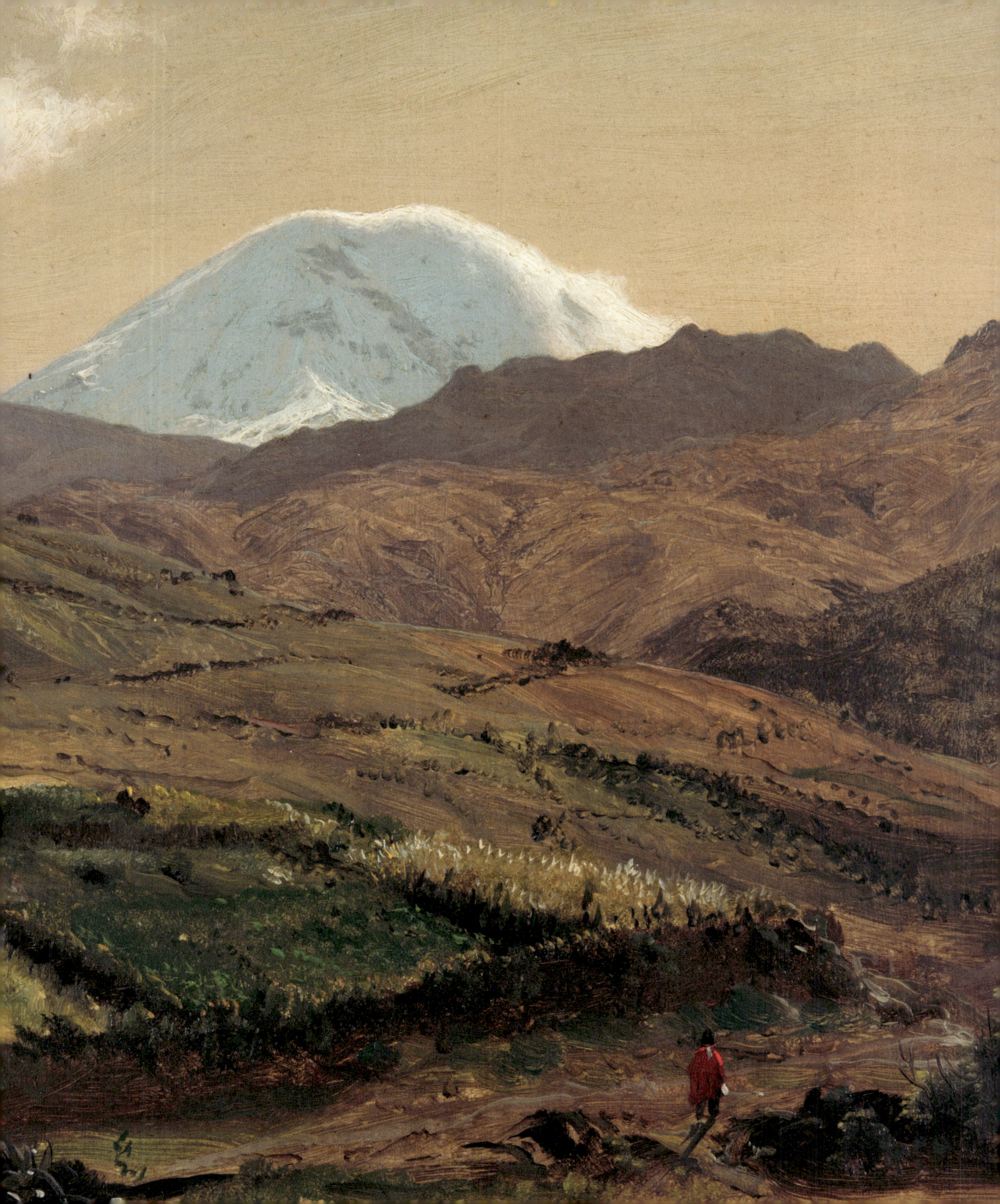

CHAPTER SEVEN

EMBODYING COSMOS: FREDERIC EDWIN CHURCH

"One great whole, animated by the breath of life"

In April of 1845 Alexander von Humboldt's first volume of *Cosmos* appeared in print in German. The response was electric, and the book became an instant best-seller. The sixty-page-long introduction explained that all of the years Humboldt had spent traveling, taking notes, asking questions, querying colleagues, exchanging and sharing specimens and ideas had culminated in a comprehensive vision of the underlying natural laws governing the planet. *Cosmos* attracted so much attention for many reasons—its unparalleled scope, its ambition, and Humboldt's amplification of his insistent theme of the interconnectivity of everything from outer space to the bottom of the oceans and into the earth's molten core. His intent, as he stated it, was to describe nature as "a unity in diversity of phenomena; a harmony, blending together all created things, however dissimilar in form and attributes; one great whole animated by the breath of life."[1] This is the book that made Humboldt an international celebrity. It extended his fame beyond the scientific institutes, the halls of academia, and the intellectual salons, and into a broad range of literary journals and the ranks of the weekly and daily newspapers. Between 1845 and his death in 1859, Humboldt published four volumes of *Cosmos* (the fifth was published posthumously), each one well-received, distributed internationally, and translated rapidly into numerous languages.

The cumulative impact of Humboldt's works had so permeated American cultural nationalism that it seemed sui generis—natural, obvious, and inspired by the perceived synergy between the American landscape and the American experiment. The cultural identity of the United States

Frederic Edwin Church, *Mount Chimborazo, Ecuador* (detail), 1857. See p. 347.

was tied closely to an appreciation of nature, which had by 1845 infused the fine arts and literature in the form of the Hudson River school of painters and the transcendentalist writers. Humboldt inspired several notable creative men to change the scope of their work and adjust its content as they consciously absorbed something of the breadth and depth of his interests. In the arena of American art, it would be Frederic Edwin Church, proclaimed as "the American Humboldt," who most assiduously adopted aspects of a Humboldtian persona as part of his own. Church pursued the study of art and natural sciences, and he invested each of his Great Pictures with elements of Humboldt's vision. His own travels to South America, his drive to continually reinvent the rules governing landscape painting, and the creation of his personal universe at Olana, the home he built above the Hudson River, were inflected by his exposure to Humboldt, and altogether they provide a window into understanding Humboldt's reach at midcentury.[2]

In American literature, Ralph Waldo Emerson, Henry David Thoreau, and Walt Whitman fell under Humboldt's spell, and each author's signature works bear Humboldt's fingerprints. Emerson came to Humboldt earliest, and he invested in his journals and his essay *Nature* (1836) an express desire to be a naturalist as well as a poet and essayist. Emerson in turn introduced Thoreau to Humboldt's writings. Thoreau's obsession with measurements stands as a key component of *Walden,* which became a kind of *Cosmos* in miniature. For Whitman, Humboldt's inspiration is most evident in the evolutionary process within his groundbreaking poem, *Leaves of Grass,* with the poet's declaration of himself as "a Kosmos." What was it about *Cosmos* that inspired such deliberate infusion of the spirit of Humboldt's work in their own? How did these creative giants come to embody *Cosmos* in their way of encountering and distilling the natural world? And what did it mean to aspire to be an American Humboldt?

HUMBOLDT'S *COSMOS*

For forty years American readers had been absorbing Humboldt's key ideas regarding the unity of nature, the relationship between nature and culture, and the essential humanity of all races. In *Cosmos*, Humboldt wove many of the ideas put forth in his earlier works into a more ambitious framework. What distinguished Humboldt as a thinker was his ability to hold multiple layers of data and interconnected implications in his head and to make those ideas legible on the pages of his book. He organized the first volume of *Cosmos* in three sections. He first took the reader into the outer atmosphere, addressing celestial phenomena. Then he brought his acute sensibilities down to Earth, focusing on the forces acting on the planet's surface, including earthquakes and geomagnetic forces. Finally he addressed flora and fauna, including the human race. For Humboldt, the concept of the unity of nature appeared as "one fair harmonious whole" that accounted for everything that constituted planet Earth and its environments.[3]

In the United States, Humboldt's *Cosmos* was widely reviewed, those stories often excerpting lengthy passages that inspired a new wave of interest in Humboldt's ideas. One of the earliest in English appeared in the November issue of the *North British Review*, which devoted more than fifty pages to recounting Humboldt's travels, praising his ideas, and heralding the book as a superlative accomplishment. In his opening lines the reviewer asserted:

> *Were the republic of letters to alter its constitution, and choose a sovereign, the intellectual sceptre would be offered to Alexander Humboldt.... The title of our author to such distinguished pre-eminence will be readily acknowledged by all who have followed him in his brilliant career as a traveller and a naturalist; and those who have known him only by the voice of fame, will recognize in the present work a mind... exercising the highest powers of combination and analysis, intensely alive to the beauty and grandeur of the material world, and thus qualifying its possessor to be the historian and the interpreter of inorganic nature,—the expounder of her phenomena and laws,—the high-priest of her holiest mysteries, and the most enthusiastic yet humblest worshipper at her shrine.*[4]

Reviews like this ensured that Humboldt's publisher would have to scramble to keep up with demand, reprinting additional editions after the first printing of 20,000 copies had sold out within a matter of weeks. A bootleg translation in English appeared almost immediately; Charles Darwin borrowed English botanist Joseph Dalton Hooker's copy, lamenting the book's stilted prose and awkward phrasing of Humboldt's concepts, but Darwin was happy to have even a glimpse into what Humboldt had written. This unauthorized version made its way across the Atlantic and into the hands of Ralph Waldo Emerson and Edgar Allan Poe, each of whom began incorporating Humboldt's ideas into works of his own. Reviews in the American press appeared during 1846, often with lengthy excerpts from the introduction and a more cursory assessment of the rest of the book.

The demand for Humboldt's *Cosmos* circled the globe. By 1847, the authorized translations of this first volume had appeared in English, French, Italian, Spanish, Russian, Dutch, Polish, Swedish, Danish, and Hungarian.[5] By this time, Humboldt was ready to release his second volume of *Cosmos*, which meant that in the United States, his English-speaking audience frequently absorbed both authorized volumes at once. The second volume offered a different experience from the first, opening with a history of human thought that brought together elements of history, philosophy, and religion in the service of a "poetic description of nature."[6] Those descriptions, Humboldt noted, were a shared trait among all civilizations around the globe and across time. From literary evocations of nature, Humboldt then turned to the work of landscape painters from all epochs, validating the genre far above its lowly rank on the accepted hierarchy of artistic achievements. He followed with a section on tropical plants as a means of understanding how he had arrived at his ideas. The second section of the book comprised an extended discourse chronicling the history of human attitudes toward nature, from the ancient Greeks to the modern era. Humboldt's disquisition was an "extension of the idea of the Cosmos as a natural unity."[7]

This second volume of *Cosmos* provided a template for American landscape painters to take inspiration in the scientific and poetic facets of their subjects and invest their works with a dual love of science and art. Here the preeminent intellectual of the age helped readers comprehend the universe by studying every aspect of the landscape in some detail. He then synthesized those impressions and wrote lyrically about the attendant feelings of poetic wonder and scientific curiosity derived from the study of nature, and how that formed the foundation of landscape painting. Humboldt did not privilege any specific country's landscape: his was a call to all human beings to open their eyes to the natural world and see nature for both its scientific and aesthetic appeal.

Humboldt's approach to aesthetics had been shaped by Immanuel Kant and his close friend Johann Wolfgang von Goethe. Starting with his Plant Geography Map, which he titled "*Naturgemälde*," or Picture of Nature, Humboldt had blended scientific rigor with aesthetic joy, both fueled by his curiosity. He began his first volume of *Cosmos* with the observation that "'Nature,' as Schelling remarks in his poetic discourse on art, 'is not an inert mass; and to him, who can comprehend her vast sublimity, she reveals herself as the creative force of the universe—before all time, eternal, ever active, she calls to life all things, whether perishable or imperishable.'"[8] Now Humboldt encouraged his readers to emulate that holistic embrace of a wide spectrum of experiences and stimuli. He did not hesitate to differ from the leading aesthetic philosophers when he had reason to disagree. On Sir Edmund Burke's concept of the sublime, which ascribed the power and intensity of its effects to fear of the unknown, Humboldt pronounced,

> *The vault of heaven, studded with nebulae and stars, and the rich vegetable mantle that covers the soil in the climate of palms, cannot surely fail to produce on the minds of these laborious observers of nature, an impression more imposing and more worthy of the majesty of creation, than on those who are unaccustomed to investigate the great mutual relations of phenomena. I cannot, therefore, agree with Burke when he says, "it is our ignorance of natural things that causes all our admiration, and chiefly excites our passions."*[9]

For Humboldt, the sublime was not an extension of fear but of exhilaration. The emotional rush associated with seeing, learning, knowing, and synthesizing every aspect of nature inspired Humboldt.

By 1849 *Cosmos* was so popular that there were multiple, competing English translations from which to choose, all published in London, a situation Humboldt described as the "sheer madness about *Cosmos* in England."[10] His exuberant German prose proved a significant challenge to his translators, but their versions were what the vast majority of his American audience read, absorbed, and understood. One of those readers was Frederic Edwin Church (fig. 84), a young man who had decided despite his family's reservations, to pursue a career as an artist. Through a combination of raw talent and influential supporters, Church secured his father's somewhat reluctant permission to study with Thomas Cole, the leading landscape painter in the United States. Cole recognized his pupil's immense potential, proclaiming that he had "the finest eye for drawing in the world."[11] That skilled draftsmanship was coupled with a prodigious memory for color and texture. Church excelled at sketching under the variable conditions in the field and then returning to the studio to synthesize those immediate impressions from nature. Between 1844 and 1846, Cole helped shape his pupil's approach, a combination of sketches made out of doors and paintings crafted in the studio.

Fig. 84
Mathew Brady studio, *Frederic Edwin Church* (1826–1900), ca. 1855–65, wet collodion negative, 3 13⁄16 × 2 3⁄8 in., Library of Congress, Brady-Handy Photograph Collection

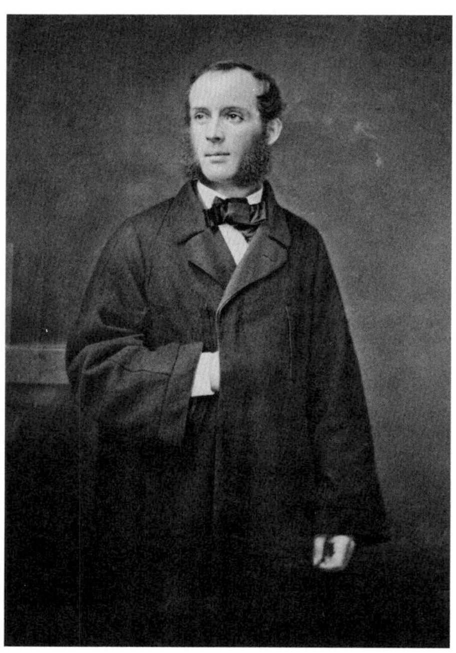

Pinpointing the origin of Church's introduction to Humboldt's books may remain elusive. Born in 1826, Church would have had the opportunity to begin absorbing Humboldt's ideas in his early youth. By the time he embarked on his own as a landscape painter, the Prussian naturalist's enthusiasm for the United States already had deep roots. Church came of age as Humboldt's reputation had already made him an oft-quoted authority on numerous topics, and his early publications could be found in various editions in American libraries. American scientific societies in Philadelphia, New York, and Boston featured Humboldt's writings, as did wealthy collectors in Church's orbit. Cyrus West Field, Robert L. Stuart, William Blodgett, and James Lenox each acquired Humboldt's volumes along with Church's major paintings.

By 1849, when Humboldt's first two volumes of *Cosmos* had sold more than forty thousand copies in English alone, Church had absorbed Humboldt's ethos, and most likely he had begun to absorb his words. Church's copies of the first two volumes of *Cosmos* were published in London in 1849.[12] The Prussian's guidance to landscape painters resonated with Church most especially:

a distinction must be made in landscape painting, as in every other branch of art, between the elements generated by the more limited field of contemplation and direct observation, and those which spring from the boundless depth of feeling and from the force of idealizing mental power.[13]

That idea of direct observation feeding the intellect harmonized with Cole's approach to landscape. Plein air impressions became the inspiration for what Cole called "a higher style of landscape."

By this time Humboldt's ideas had also permeated the works of the leading art critic of his generation, John Ruskin. The consonances between Humboldt's and Ruskin's approaches to studying physical geography and the aesthetics of landscape painting were so strong that Ruskin felt obliged to address the issue in print. In an appendix to the third volume of his influential book *Modern Painters*, Ruskin wrote,

Many people will suppose that for several ideas in the chapters on Landscape I was indebted to Humboldt's Kosmos. . . . *Of Humboldt's* Kosmos *I heard much talk when it first came out, and looked through it cursorily; but thinking it contained no material (connected with my subject) which I had not already possessed myself of, I have never since referred to the work. I may be mistaken in my estimate of it, but certainly owe it absolutely nothing.*[14]

His defensiveness suggests the extent to which Humboldt's ideas had saturated the fabric of landscape aesthetics and its ties to natural history. As poet John Greenleaf Whittier would later note, Humboldt's reputation had outgrown "the limits of nationality and breaking down distinction of race and language, has become universally acclimated, and common property of Science, Civilization and Human Progress."[15] Church clearly absorbed Humboldt's ideas as well as Ruskin's. Reading Ruskin would have confirmed the importance of blending art and science in the pursuit of a career as a landscape painter.

Church's eye for botanical detail and the nuances of cloud forms and weather phenomena became hallmarks of his works. His drawings began to include significant annotations regarding the weather and the artist's impressions of the colors of everything, from the sky and the clouds to the silvery undersides of leaves fluttering in the breeze. Humboldt had described landscape painting as a kind of visual poetry, requiring a heightened sense of awareness and aptitude on the part of the artist. Church marked one such passage in his copy of *Cosmos* (CAT. 81):

The azure of the sky, the form of the clouds, the vapoury mist resting in the distance, the luxuriant development of plants, the beauty of the foliage, and the outline of the mountains, are the elements which determine the total impression produced by the aspect of any particular region. To apprehend these characteristics, and to reproduce them visibly, is the province of landscape painting; while it is permitted to the artist, by analysing the various groups, to resolve beneath his touch the great enchantment of nature—if I may venture on so metaphorical an expression—as the written words of men are resolved into a few simple characters.[16]

Humboldt envisioned the artist as a scientist and a poet, capable of distilling the essence of nature while retaining the specific character of the local landscape.

By the end of the 1840s Humboldt's writings and his ideas permeated American landscape aesthetics, travel narratives, and scientific investigative methods. The wilderness aesthetic that fueled both the prominence of the Hudson River school and more broadly America's cultural identification with nature was well established. Church's acute powers of observation were a perfect match for Humboldt's exultation in observing and marveling at these details in nature. It was Humboldt's second recommendation that rang true for Church: that the artist should also see the world as a scientist—well-read in natural philosophy and observant of nature and its processes, so that he could invest his works with scientific truth as well as aesthetic merit. Church studied each facet of the physical sciences—whether it was optics, botany, or geological strata—as an enthusiastic amateur, a trait shared by some of his landscape painter colleagues and some of his scientifically minded patrons. Humboldt's *Cosmos* was perfectly timed for the launch of Church's artistic career. The book's combination of scientific and aesthetic appreciation of the landscape—including the North American landscape—gave Church the framework for contributing to an American cultural identity grounded in that synthesis of art and science.

CHANNELING HUMBOLDT IN SOUTH AMERICA

Frederic Church's two trips to South America in 1853 and 1857 were directly inspired by his appreciation for Humboldt. Church's South American paintings were upheld then, just as they are now, as evocations of Humboldt's ideas; however, this observation has often obscured the larger truth that Humboldt influenced more than the artist's interest in South America. When Henry T. Tuckerman wrote a lengthy article on Humboldt in 1850, he described Humboldt's work in South America as pivotal to the traveler's larger understanding of the unity of nature, but he expounded much more on Humboldt's ideas than the geography that had inspired them. Tuckerman praised Humboldt for the "scope and impulse" of his inquiries, the goal being to understand the myriad forces at work in nature and convey a sense of how they were interconnected.[17] The same can be said of Church's most ambitious compositions.

Every work of art Church produced, every voyage he undertook, beginning with his first trip to South America in 1853, was inflected by Humboldt's works. Although Church and Humboldt never met or corresponded directly, Church was surrounded by those who did. Using Humboldt's early writings as his inspiration, Church fashioned Humboldt as an absent mentor. Later, in a profile published in 1866, Tuckerman extolled Church as the "American Humboldt."[18] His accolade was not merely in recognition of the specific subjects of his paintings. Tuckerman understood that Humboldt's ideas had permeated Church's way of envisioning the world. In the artist's intensive study of the physical properties of nature, from

CAT. 81 Alexander von Humboldt, *Cosmos: A Sketch of a Physical Description of the Universe*, 1849, ink on paper, open: 7 ½ × 10 in., Olana State Historic Site, New York State Office of Parks, Recreation and Historic Preservation

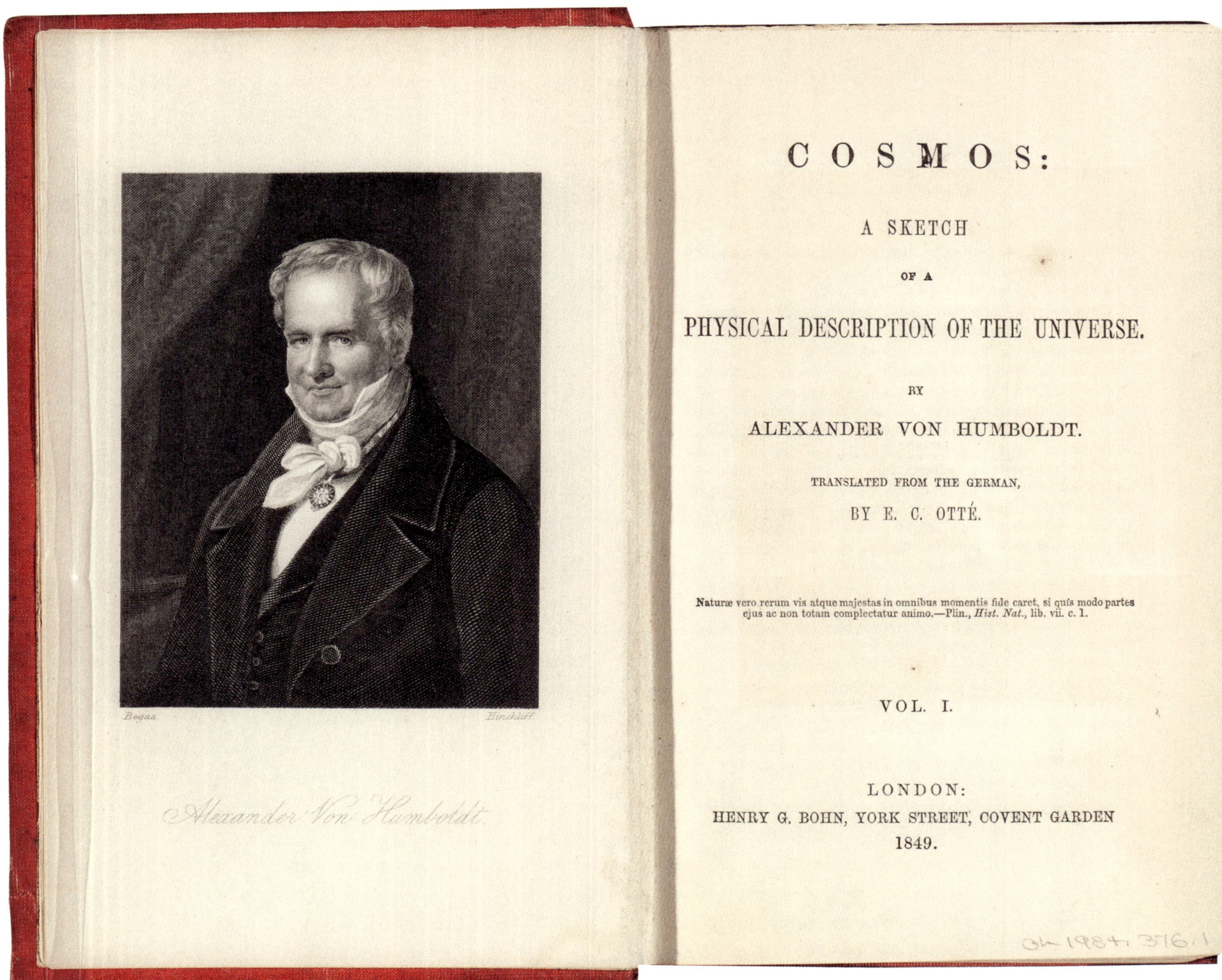

Begas Hinchliff

Alexander Von Humboldt

COSMOS:

A SKETCH

OF A

PHYSICAL DESCRIPTION OF THE UNIVERSE.

BY

ALEXANDER VON HUMBOLDT.

TRANSLATED FROM THE GERMAN,

BY E. C. OTTÉ.

Naturæ vero rerum vis atque majestas in omnibus momentis fide caret, si quis modo partes ejus ac non totam complectatur animo.—Plin., *Hist. Nat.*, lib. vii. c. 1.

VOL. I.

LONDON:
HENRY G. BOHN, YORK STREET, COVENT GARDEN
1849.

clouds and water to rocks and trees, Church emulated his distant mentor. In elevating landscape as a genre, he invested each of his paintings with a scientist's eye for accuracy and an artist's eye for emotional impact. Each of his Great Pictures was an attempt to surpass the last, a refinement of technique and composition that found a ready parallel in Humboldt's own published works. In doing so, Church pushed landscape painting to greater prominence as the genre most capable of conveying America's cultural ambitions.

Church's friendship with Cyrus Field developed over several decades and a shared appreciation for Humboldt's ideas. Field was a Springfield, Massachusetts, paper manufacturer whose business acumen allowed him to retire with considerable wealth by the time he was thirty-three years old. The two men belonged to the same American scientific societies, including the American Geographical and Statistical Society. Church followed Field's decade-long effort to successfully complete the transatlantic cable. Moreover, in 1864 they helped found a New York group called The Travellers, bringing together twenty-five men of arts, letters, and science who shared a passion for natural history—many of whom were shaped by Humboldt's works. The two took sketching excursions together in New England, and Field purchased Church's *West Rock, New Haven* (fig. 85) in 1849, the painting that vaulted the artist to the forefront of American landscape painting.[19] *West Rock* portrays a contemporary vista imbued with historical overtones and has a personal layer of meaning with the artist's inclusion of a small white-steepled church, a visual pun on his name that would find its way into a number of his paintings.[20] The painting impressed his fellow artists sufficiently to vote him a full member of the National Academy of Design (NAD) the same year.

Church and Field began making plans for an extended trip to South America that ultimately lasted seven and a half months. For Field, it was a chance to scout possible investments and indulge in the search for a brother who had last been seen somewhere in South America two decades earlier. For Church, it was the chance to follow in Humboldt's footsteps and experience the continent through Humboldt's words and engraved images.

Before Church and Field set out for South America, they planned a shorter sojourn to Virginia, a trip that may have confirmed that they were amiable traveling companions. Two

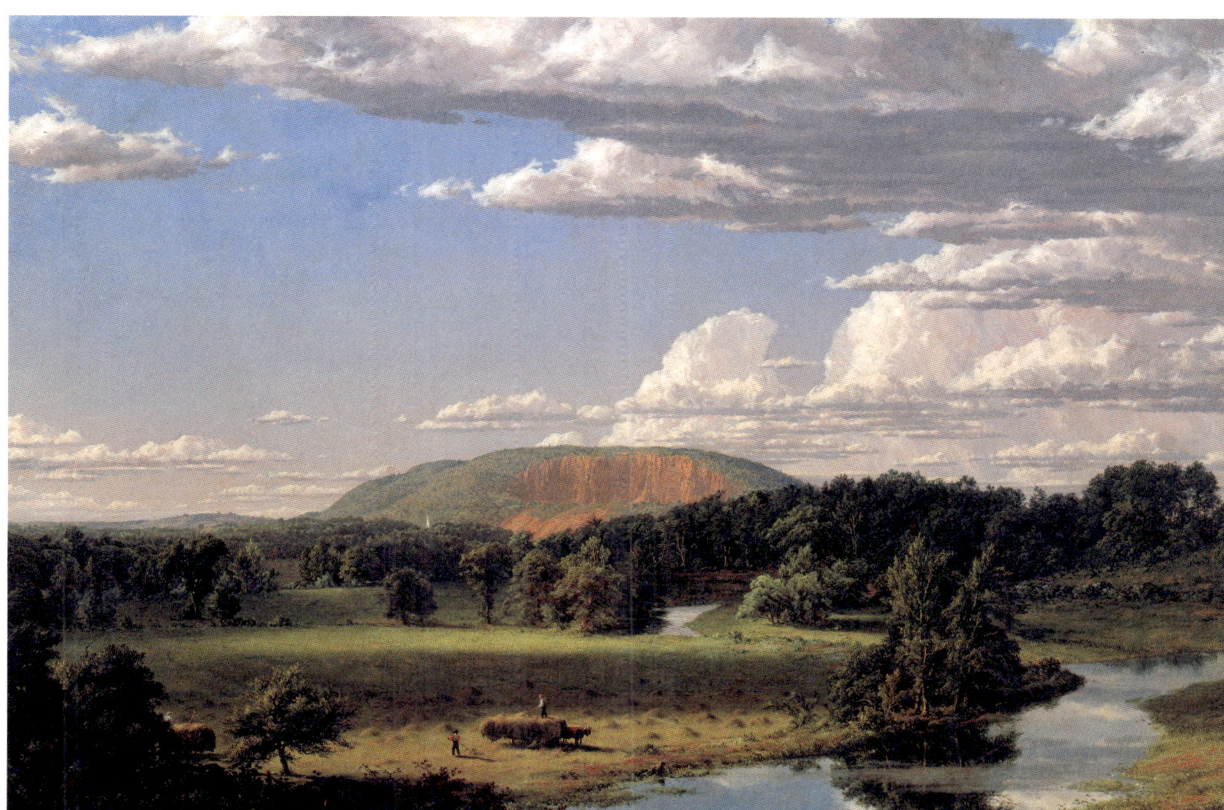

Fig. 85
Frederic Edwin Church, *West Rock, New Haven*, 1849, oil on canvas, 26 ½ × 40 in., New Britain Museum of American Art, John Butler Talcott Fund

CAT. 82 Frederic Edwin Church, *The Natural Bridge, Virginia*, 1852, oil on canvas, 28 × 23 in., Courtesy of The Fralin Museum of Art at the University of Virginia, Gift of Thomas Fortune Ryan

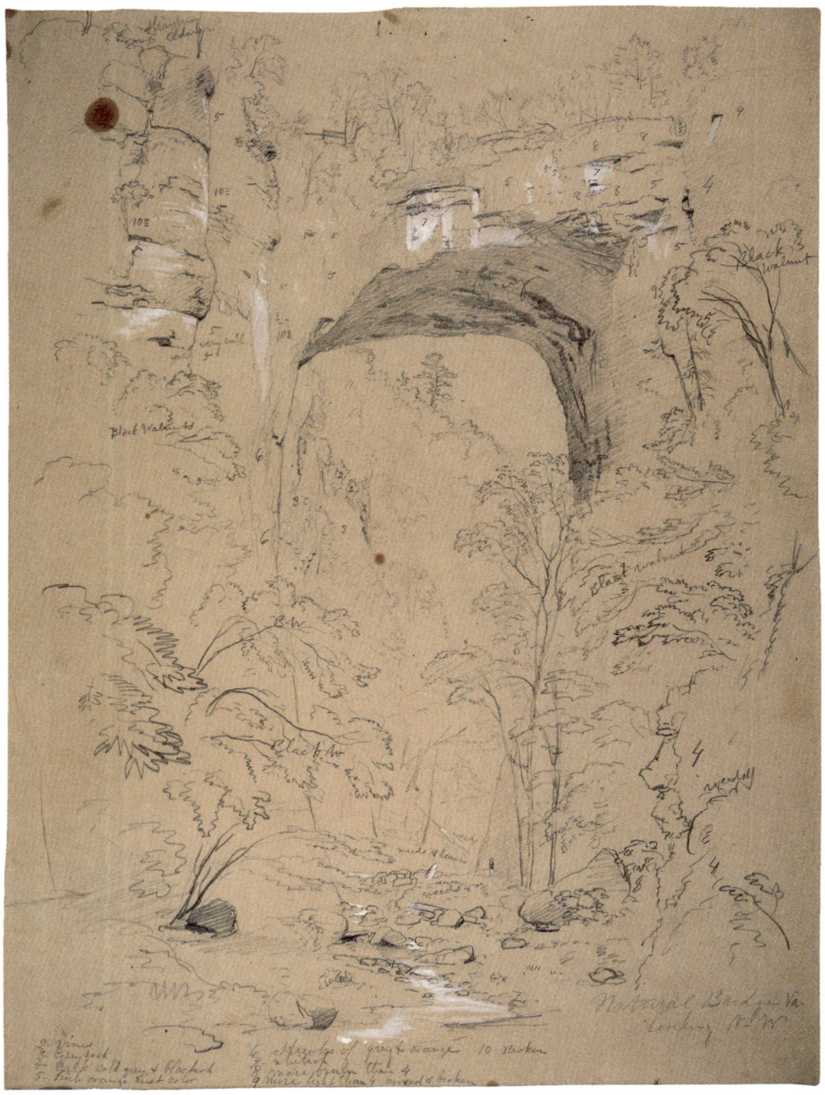

Fig. 86
Frederic Edwin Church, *Natural Bridge, Virginia*, June 1851, graphite, pencil, and gouache on paper, 15 × 10 ⅞ in., Olana State Historic Site, New York State Office of Parks, Recreation and Historic Preservation

of their stops were at Mount Vernon and Natural Bridge. Both sites held associations with George Washington and blended American history with a compelling landscape view, dovetailing with Field's interest in early American history. However, when they reached Natural Bridge, they knew that Thomas Jefferson had featured the landmark in *Notes on the State of Virginia,* which in turn had spawned two decades of written, engraved, and painted descriptions of the landmark. What they also most likely knew was that Humboldt had written about it as well in *Views of the Cordilleras.* In South America, Humboldt had been struck by a natural bridge formation on the Icononzo, enough to have included an engraving of the geological wonder (see cat. 32) to accompany the two-page description of his response to its impressive appearance. It reminded him of the Virginia formation. He wrote,

> *A phenomenon, similar to the upper bridge... exists in the mountains of Virginia, in the county of Rockbridge. This Mr. Jefferson has examined with an attention, that distinguishes all the observations of that excellent naturalist.*[21]

Field and Church were also aware of the story published in Nathaniel Parker Willis's *American Scenery* in 1840: "Under the arch, thirty feet from the water, the lower part of the letters G.W. may be seen, carved in the rock. When Washington was a young man, he climbed up hither, to leave this record of his visit."[22] This was a site replete with references to the founding fathers, Humboldt, and the establishment of natural wonders as icons of American identity.

At Natural Bridge, Church sketched the arch in pencil, annotating the sheet with color notes designed to prompt his memory back in the studio (fig. 86). Field chose to collect small rock samples that displayed the vibrant reddish hues of the stained limestone formations. As the two men discussed the painting Church intended to make for Field after their return to New York, Field offered to let Church use his rocks as a guide. Church demurred and painted the scene from memory and his penciled annotations. Field enjoyed telling the story that when he arrived to see Church's finished painting of *The Natural Bridge* (CAT. 82), he pulled the rocks from his pockets. The colors matched.[23] This was proof, in both men's minds, of Church's

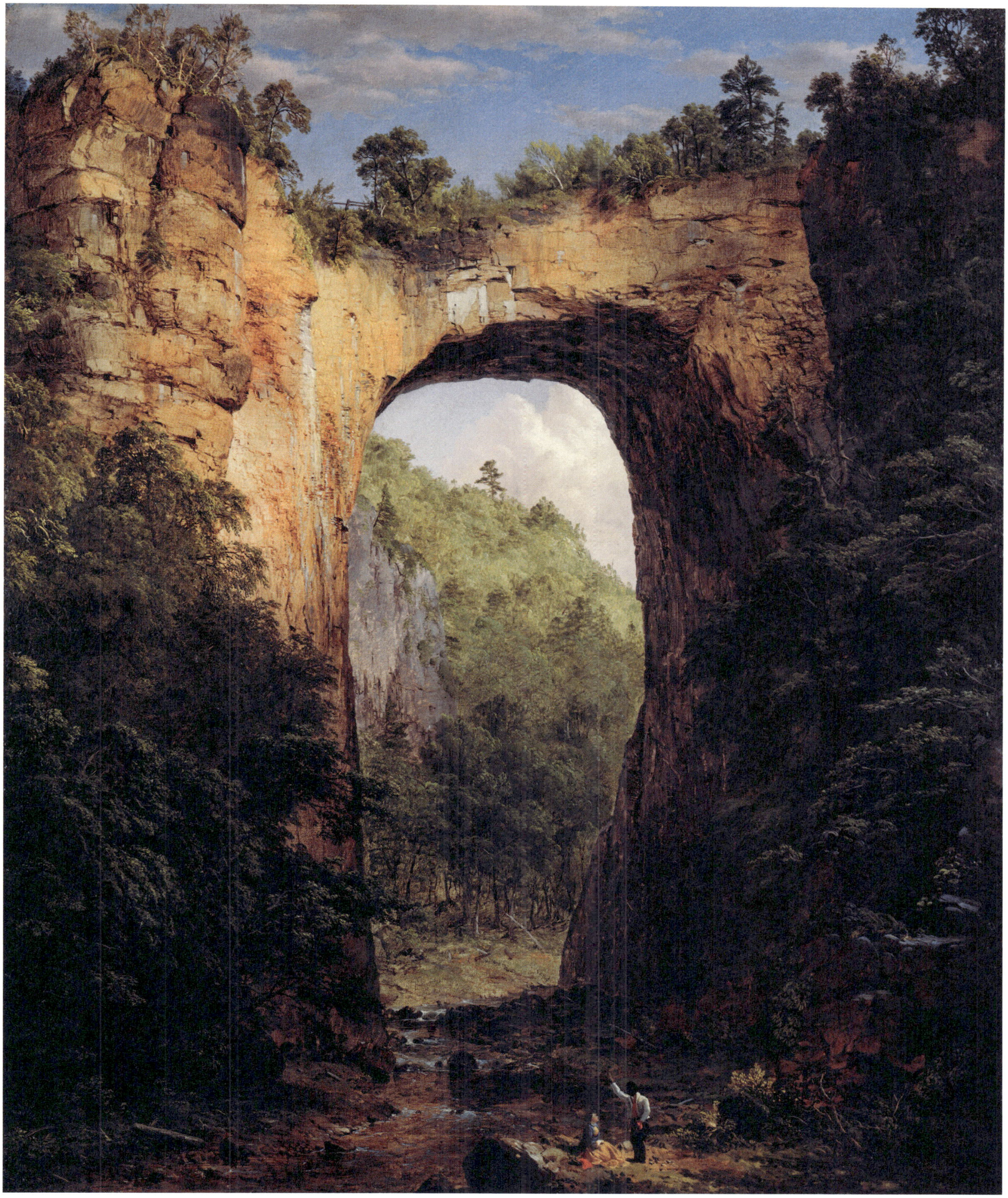

prodigious memory for color, his ability to rely on his annotations, and his acute observations in the field. All were attributes Humboldt would have admired. Church was prepared to be the landscape painter Humboldt had called for, one who could approach nature as a blend of scientist and artist, marrying close observation with an eye for composition and visual effect.

This successful trial encouraged the two men to develop ambitious plans for an extended stay in South America. Field underwrote the cost for the trip, which they planned meticulously. They decided to concentrate on Colombia and Ecuador, following a route Humboldt had taken fifty years earlier across the Andes. *Cosmos* was not the only book that inspired Church. Humboldt had long urged his readers to travel to South America to appreciate the differences and the underlying unity of nature to be found around the globe. They had easy access to the folio edition of Humboldt's *Atlas géographique et physique* in the Astor Library in New York, which Church frequented.[24] This deluxe volume included large-scale maps of the entire route and color engravings of the key features. Church owned a copy of Humboldt's *Personal Narrative,* one volume of which was also separately published in English as *Views of the Cordilleras* in 1814. It provided Church and Field with a travelogue that described the route and the major attractions for their trip to South America. The frontispiece, with its exuberant typography and multiple exclamation points, captured the enthusiasm of Humboldt's prose (see cat. 90), while the engraved view of the smoking cinder cone of Cotopaxi illustrated one of their destinations. *Views of the Cordilleras* was structured as a narrative journey across South America and served as the pair's Baedeker during their travels in 1853.

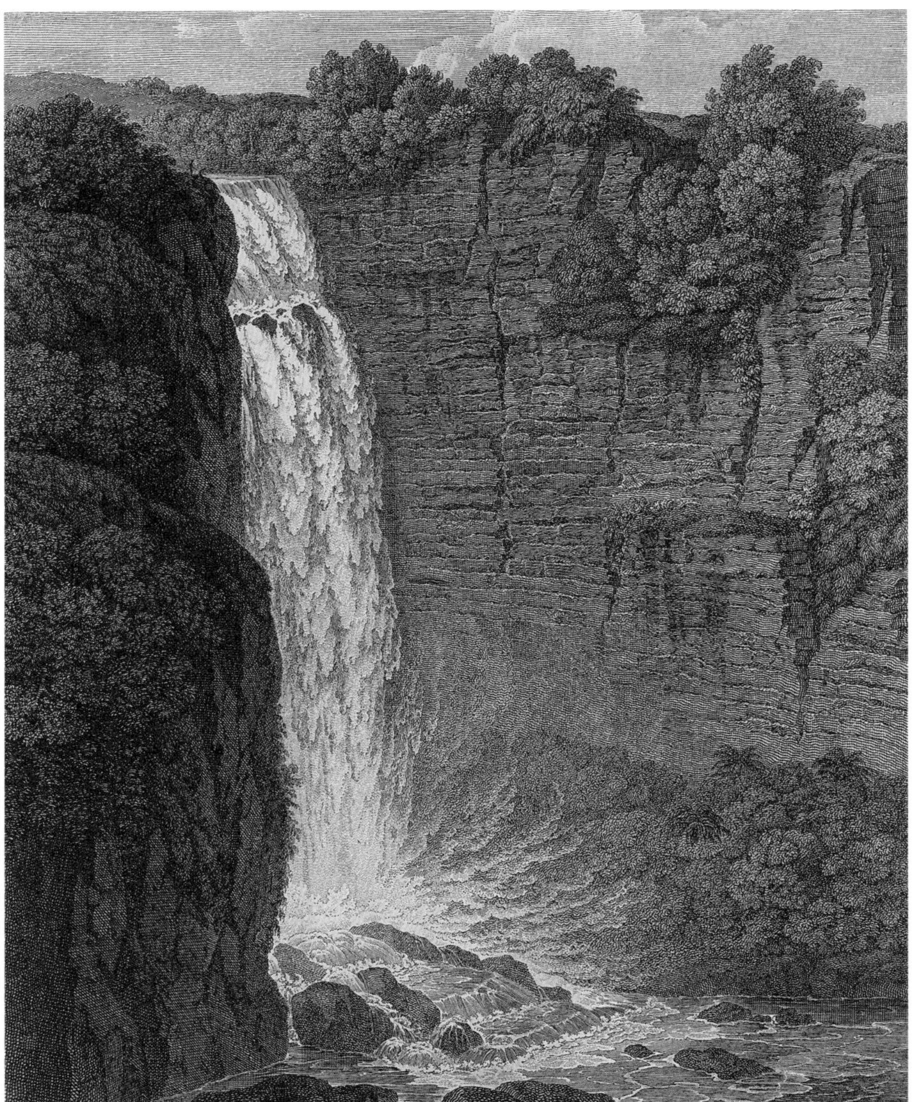

Fig. 87
Alexander von Humboldt, *Tequendama Falls*, from *Vues des Cordillères et monuments des peuples indigènes de l'Amérique*, 1810, engraving, single page: 9 × 6 in., Smithsonian Libraries

Field and Church sailed from New York on April 8, arriving at the mouth of the Magdalena River near Barranquilla twenty days later. Church's lengthy letters to his mother and sisters convey both his apprehensiveness and excitement, while his pocket journal—which he endeavored to maintain in Spanish—indicates his willingness to immerse himself in the local environment, much as Humboldt had done wherever he went. Their itinerary steered them inland across Colombia, along the Magdalena River toward the Andean Cordillera, where they traveled south and west along the high altitude valley between the two parallel ridges of volcanic peaks in Ecuador. Along the way Church and Field stopped at the major landscape features Humboldt had described and illustrated in his book. At each of the sites Church scouted the area for scenery worthy of a sketch or finished painting.

Tequendama Falls outside of Bogotá was the first scenic stop on their route. There they stayed for ten days, affording Church plenty of time to survey the waterfall, first from head-on, then from a distance, and finally from below the falls. Humboldt had devoted eight pages to his description of the waterfall, including an engraving based on his own field sketches (fig. 87). Humboldt wrote of his feelings of anticipation as the roar from the cataract presaged the view, recounting the difficulties he and his entourage experienced trying to find a viable path—first along the rim, and then carrying his instruments for three hours to reach the base of the falls. There, Humboldt proclaimed,

The solitude of the place, the richness of the vegetation, and the dreadful roar that strikes upon the ear, contribute to render the foot of the cataract of Tequendama one of the wildest scenes, that can be found in the Cordilleras.[25]

Church and Field followed the same progression. On June 20 Church wrote in his diary, "After depositing our baggage we advanced to Tequendama and quickly we saw the waterfall. I am terrified by the grandeur and beauty of the deluge [torrent?]."[26] The artist spent the next two days scouting additional sketching locations along the rim. Church clearly relished the view from the top of the falls and made a detailed sketch in pencil and gouache, locating it within a panoramic setting. One of the vertical views adopts a vantage point similar to the one Humboldt chose in his engraving; again Church relied on numbered observations to prompt his memory of the atmospheric effects (fig. 88). Much as Humboldt had done in 1801, Church arranged for six local men to assist with clearing a path to reach the base of the falls. Bracing for the day's trek, Church wrote on June 29, "Tomorrow morning we intend to undertake the arduous descent to the bottom of Tequendama." The next morning, accompanied by a group of local men, Church recounted,

Today we carried out our destiny and had the satisfaction of a view of the Salto *(falls) in a new and very beautiful view. We had 6 peons who chopped the tree branches which interrupted the view. I made the sketches in pencil. The view is admirable.*[27]

The detailed pencil and gouache drawing captures the difficult terrain and emphasizes the power of the water as it tumbled from the rim above and ran furiously along the path downstream.

Humboldt had observed, "the cataract of the Tequendama is so high, that the water falls in one bound, from a cold *(tierra fria)* into a warm country."[28] In a letter to his mother, Church elaborated on the experience and confirmed the pronounced change in the climate at the top and bottom of the falls:

I have been twice to the Salto or falls of Tequendama which is a most wonderful cataract. The river Bogotá…falls in one unbroken sheet into a terrific chasm of 670 feet and then descends in a series of waterfalls and cascades about as much more. At the top of the fall you are in what is called the cold country with the trees, plants and fruits of temperate climates: at the bottom grow palms, oranges, etc. The clouds formed by evaporation from the falls sometimes cover the mountains for miles and entirely prevent a view of the falls.

Church described their descent into the "warm country" at the base of the falls. Church continued,

At some little distance I made a sketch in pencil and also one as near as the spray would allow me to see. Near the cataract below a perpetual and heavy rain falls

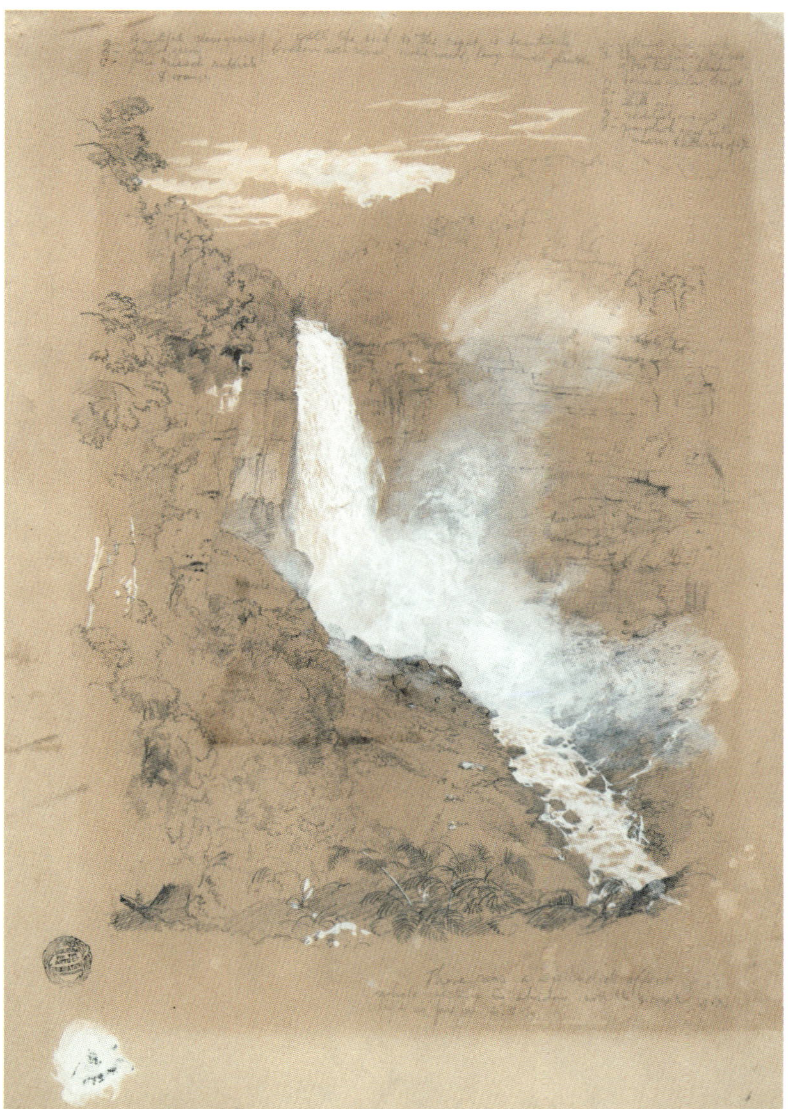

Fig. 88
Frederic Edwin Church, *Tequendama Falls near Bogotá, Colombia*, July 1853, gouache and pencil on paper, 17 ½ × 11 15⁄16 in., Cooper Hewitt, Smithsonian Design Museum, Gift of Louis P. Church, 1917-4-261

CAT. 83 Frederic Edwin Church, *Tequendama Falls near Bogotá, Colombia*, 1853, pencil and gouache on paper, 18 ⅛ × 12 ⅜ in., Cooper Hewitt, Smithsonian Design Museum, Gift of Louis P. Church, 1917-4-265

CAT. 84 Frederic Edwin Church, *Tequendama Falls near Bogotá, Colombia*, July 1853, pencil and gouache on paper, 18 ⅛ × 12 ½ in., Cooper Hewitt, Smithsonian Design Museum, Gift of Louis P. Church, 1917-4-260

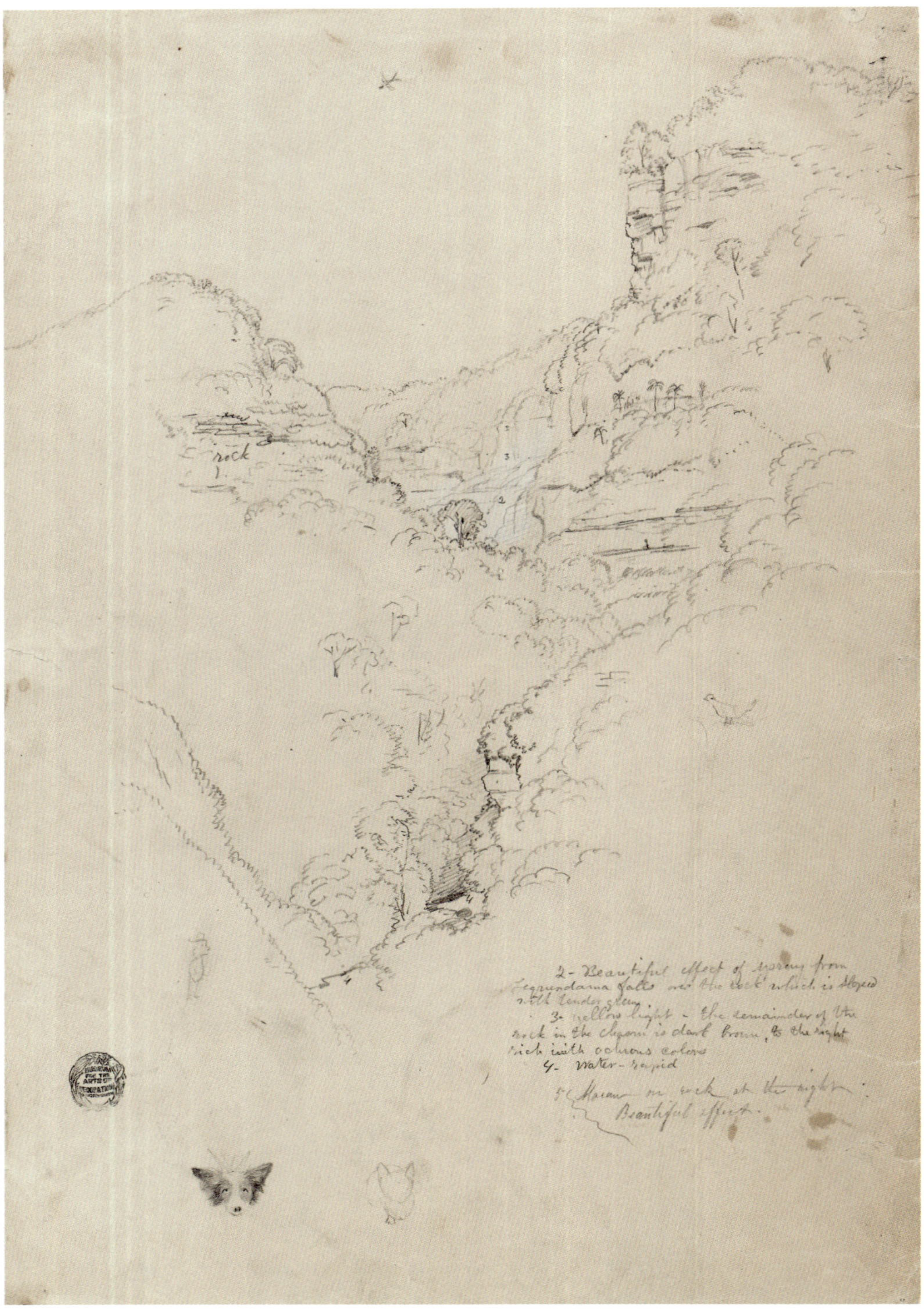

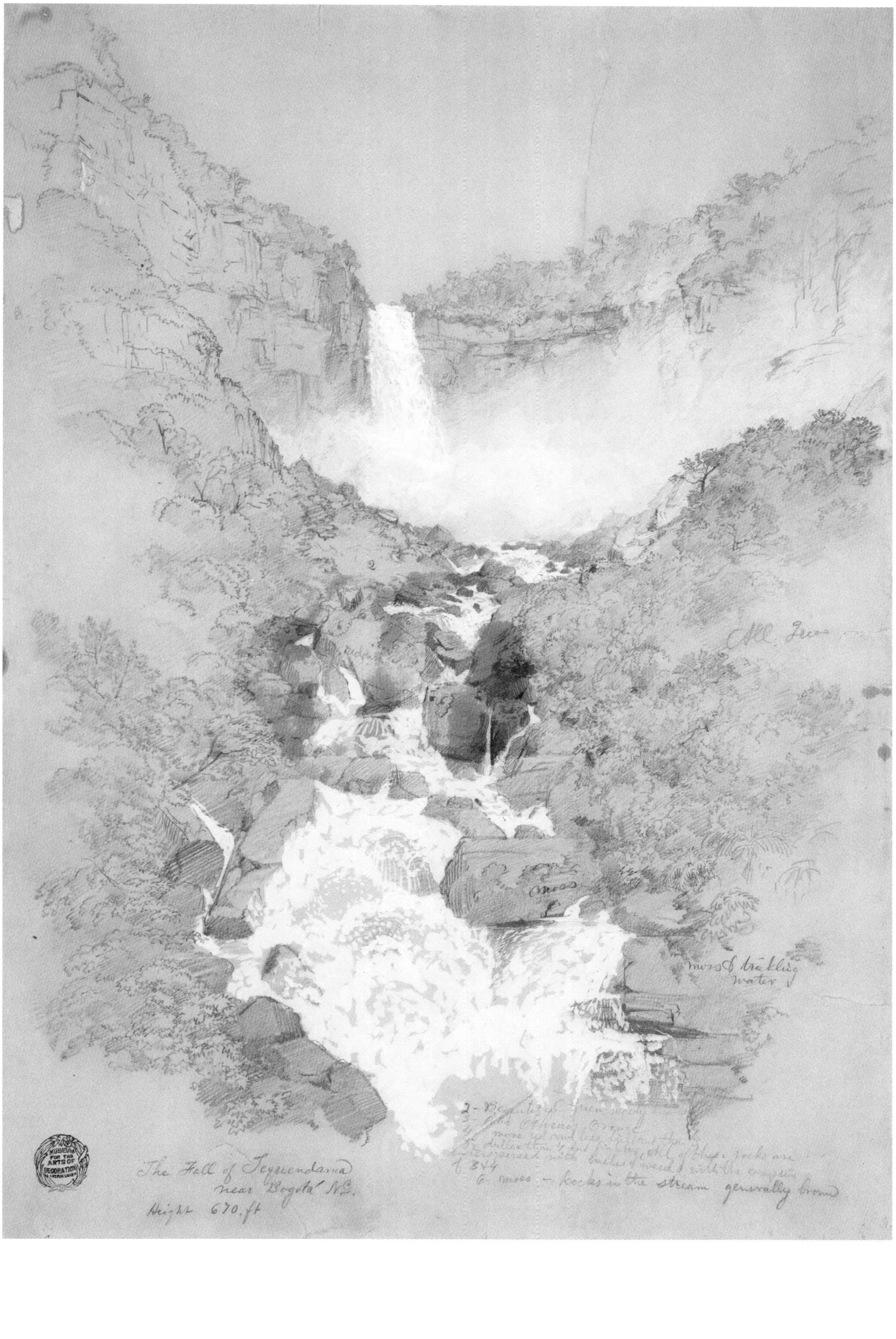
moss
moss & trickling water
6 moss – Rocks in the stream generally brown
MUSEUM FOR THE ARTS OF DECORATION
The Fall of Tequendama
near Bogotá N.G.
Height 670 ft

CAT. 85 Frederic Edwin Church, *The Falls of Tequendama, near Bogotá, New Grenada*, 1854, oil on canvas, 60 7⁄16 × 48 1⁄16 in., Cincinnati Art Museum, The Edwin and Virginia Irwin Memorial

which causes a number of rivulets, which run among the rocks into the main river. Owing to the tremendous evaporation the river below is not more than half as large above.[29]

Another vertical view adopts a light touch, with numbered annotations guiding his memory for the play of light on the cascade (**CAT. 83**). In the selection of the waterfall, the scouting of optimal sketching sites, and his description of the ecosystem, Church relied on Humboldt as a mentor and guide who encouraged the artist to see the landscape through scientific as well as artistic eyes.

Church's second trip to the base of the falls yielded a sketch taken from that lower vantage point (**CAT. 84**) that establishes the composition for the finished painting. Here his focus shifted slightly, from previously emphasizing the height of the falls to this time re-creating the roar of the river as it continued downhill. In both sketches Church drew on his adrenaline and Humboldt's inspiring prose to convey the raw energy and aesthetic power of the cascade. Here, though, the artist's numbered notations shifted to record observations on the foliage. Ever the botanists, Humboldt and Aimé Bonpland had taken delight in accumulating more than four thousand specimens of plants they deemed new to the study of botany. Humboldt's published travelogues convey that excitement to the reader, and Church, who had also studied the local flora under Thomas Cole's tutelage, reveled in the sheer variety of unfamiliar flowering plants. Church took note of these plants, seeing the South American landscape through a Humboldtian lens. En route to Bogotá, he remarked,

The flowers by the roadside would have enchanted a Botanist. I have saved such seeds as I could pick up by the wayside among them specimens of the resisitive plant which is a perfect marvel.[30]

His botanical sketches responded to Humboldt's suggestion that

An artist of distinction would find it worthwhile to study, not in greenhouses or in botany books but in nature itself, the physiognomy of the plant groups that I have enumerated.[31]

Church mused about building a greenhouse upon his return to the United States, for propagating these seeds, which would give him living examples of the plants he would incorporate in his paintings. In some cases he had no idea what kind of plant he was collecting:

if all the seeds which I have got are planted although a large part of them I have picked without knowing whether the flowers are desirable or not, having made it a principle to gather any dry seeds or pods that I saw.[32]

Like Humboldt, Church remained curious about the plants he sketched, whose seeds he gathered, accumulating aesthetic and scientific records of the specific flora that would populate his South American landscape paintings.

CHURCH'S FIRST HOMAGE TO HUMBOLDT: *THE FALLS OF TEQUENDAMA*

Within a year of his return to New York, Church painted *The Falls of Tequendama, near Bogotá, New Grenada* (**CAT. 85**), measuring five feet tall and four feet wide, his first full-height waterfall, capturing the roar of the torrent as it cascaded down the cliffs, the foaming Bogotá River running quite literally underfoot in the immediate foreground. This painting represents the artist's first painted homage to Humboldt.[33] Church had very specific goals in mind. Humboldt had admired the cataract from its boulders,

which seem to have been cut by the hand of man; to that narrow gulf, into which falls headlong the mass of waters that issue from the valley of Bogotá; to those rainbows reflecting the most vivid colours, and of which the forms vary every instant; to that column of vapour, rising like a thick cloud.[34]

Taking his cues from Humboldt, Church trusted his instincts and training to establish his composition, adapting Humboldt's perspective and making it his own. *The Falls of Tequendama* was

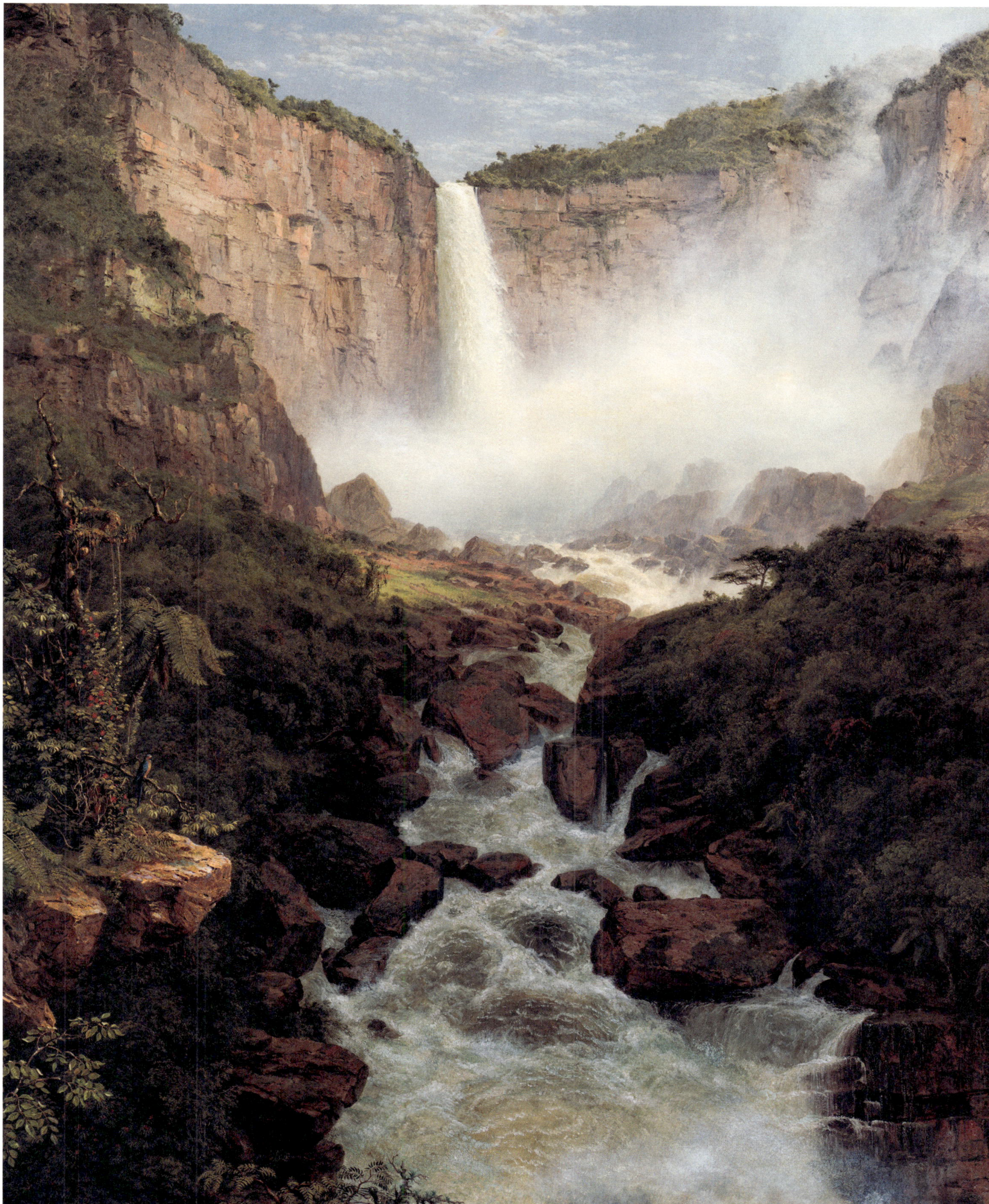

a departure for the artist, a clear signal that he had ramped up his ambitions in keeping with the enlarged scale and unusual vertical format of the painting. Water is the central dramatic element. Church places the viewer in the path of the frothing water, looking up toward the parabolic sweep of the precipice. However, Church has also gone to great lengths to incorporate all of the botanical, geological, and climatic information he gleaned from reading Humboldt and observing the same landscape in person. Referring to his color notes, Church captures the shadows created by the canyon and the pale reddish color of the exposed rock. Mindful of Humboldt's observations regarding the change in temperature and plants, Church consulted the notes he had made in his own diary, confirming and reinforcing the Prussian traveler's experiences. The vegetation at the bottom of the fall includes an impressive variety of tropical plants and vines, drawn with great botanical fidelity. Church's suite of drawings demonstrates how the artist developed his strategy for composing a painting of the cataract. His detailed pencil sketches from this trip (CAT. 86) (CAT. 87) (CAT. 88) provided the data he needed to populate his foreground and middle distance with identifiable plants, much as Humboldt had described them in his own writings. The blue-and-yellow parrot eying the viewer resembles a bird featured in another of Church's sketches (CAT. 89), further animating the scene.

Church's intention with this painting was to convey the exhilaration and logistical difficulties he had to overcome to reach this vantage point, combining Humboldt's dramatic travel narrative with the wealth of visual observations across multiple disciplines of natural history. Of all of the paintings Church created in the immediate wake of this trip, *The Falls of Tequendama* represents the capstone experience. While Church depended on scale to announce its grandeur, it was the artist's use of a multipoint perspective that conveyed its distinctiveness. The curved rim, canyon walls, and the steep drop into the valley below were each rendered in sharp focus, brought together in a massive vertical composition that emphasizes the vertiginous sweep of the terrain. Here Church had taken up Humboldt's challenge to convey both the grandeur of the scenery and the specific scientific insights into the links among terrain, altitude, and plant life. With *The Falls of Tequendama,* Church demonstrates how Humboldt's writings had made a material impact on the way the artist approached nature—distilling the experiences of both explorer and artist on canvas. Cyrus Field appreciated Church's impressive achievement as well. After Church completed it in New York in 1854, Field purchased the monumental painting for his collection.

Humboldt's words were like a voice whispering in Church's ear, or a hand on his shoulder, pointing out characteristic aspects of the South American landscape, highlighting its singular features while noting details about the local inhabitants and their culture. Church also sketched the local villages, paying close attention to the people, architecture, and scenery—details of quotidian life along the Magdalena River. Like Humboldt he maintained a vocabulary in the back of his diary, noting the words for specific plants and foodstuffs. Back in New York, Church mined those drawings and the notes in his diary to faithfully render details in his paintings. And like Humboldt, Church created two parallel bodies of work from his explorations: on the one hand, paintings like *The Falls of Tequendama* that bridged art and science to create a Great Picture with a naturalist's accuracy; and, on the other hand, compositions that synthesized elements from various moments on their journey in an effort to paint a quintessential scene that might or might not be any specific location along their route.

Church's record of his travel experiences acquired a Humboldtian slant as he noted events that paralleled those Humboldt had endured. After leaving Bogotá on July 8, 1853, he and Field made their way toward the Andes to Quito; from there they crossed the Magdalena River and began their climb into the Cordillera Central, through Quindio Pass, down into the Cauca Valley to Cartago. Church wrote of the extreme swings in weather and climate as they ascended. In a letter to his sister, Elizabeth, Church described the freezing conditions on the summit of Quindio Mountain, which echoed Humboldt's recollection of his ascent of Chimborazo.[35] The shortness of breath Humboldt experienced on that epic climb mirrored Church's exhaustion in scaling the Colombian volcano Puracé (he called it Puresé), whose

CAT. 86 Frederic Edwin Church, *Studies of Trees and Plants seen from the Río Magdalena, Colombia*, 1853, pencil on paper, 8 7/16 × 10 7/8 in., Cooper Hewitt, Smithsonian Design Museum, Gift of Louis P. Church, 1917-4-135

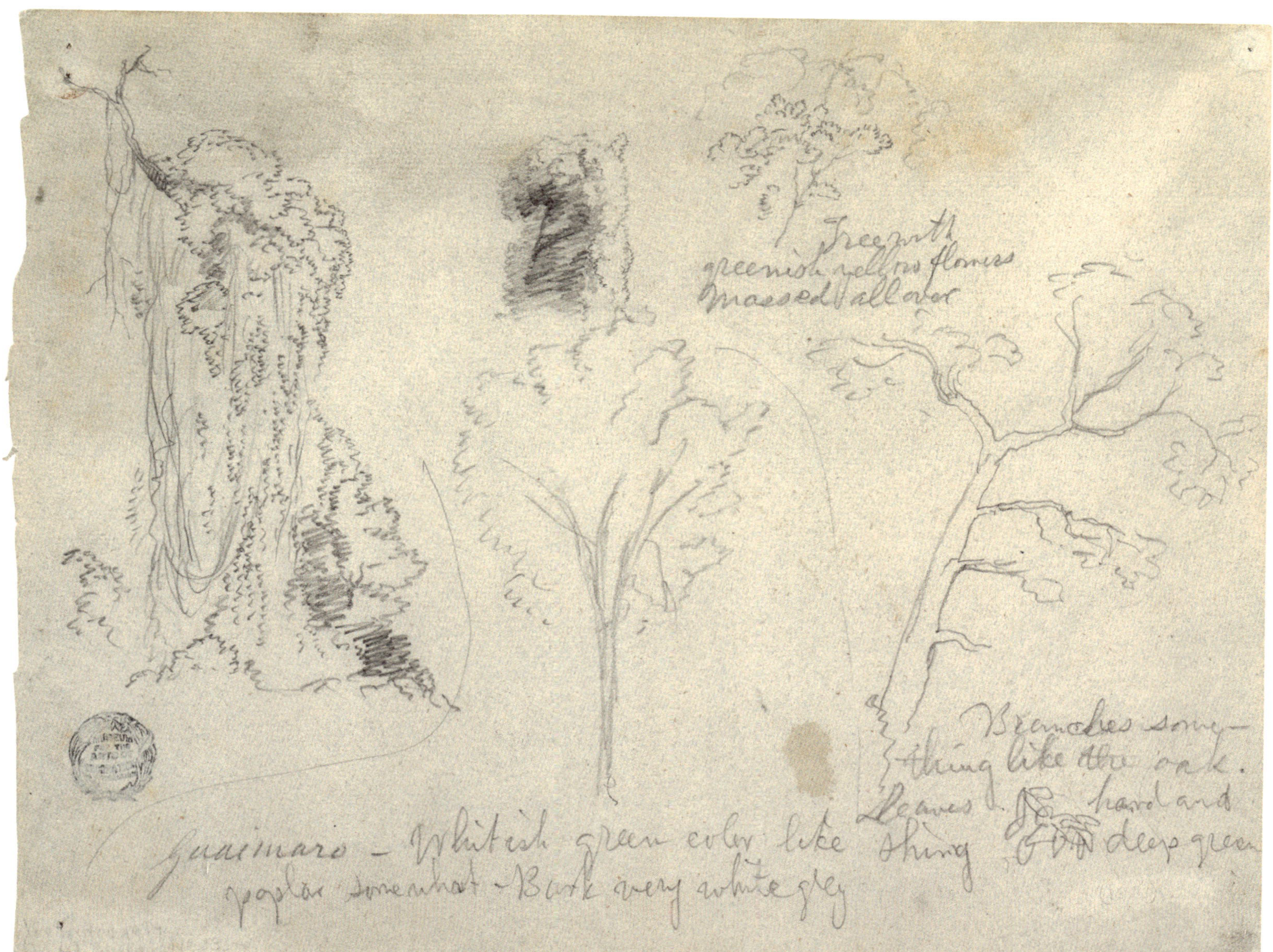

CAT. 87 Frederic Edwin Church, *Botanical Sketches from Colombia: Bamboo Trees, Trunks and details*, July 1853, pencil on paper, 4 ⅞ × 7 ¹³⁄₁₆ in., Cooper Hewitt, Smithsonian Design Museum, Gift of Louis P. Church, 1917-4-85

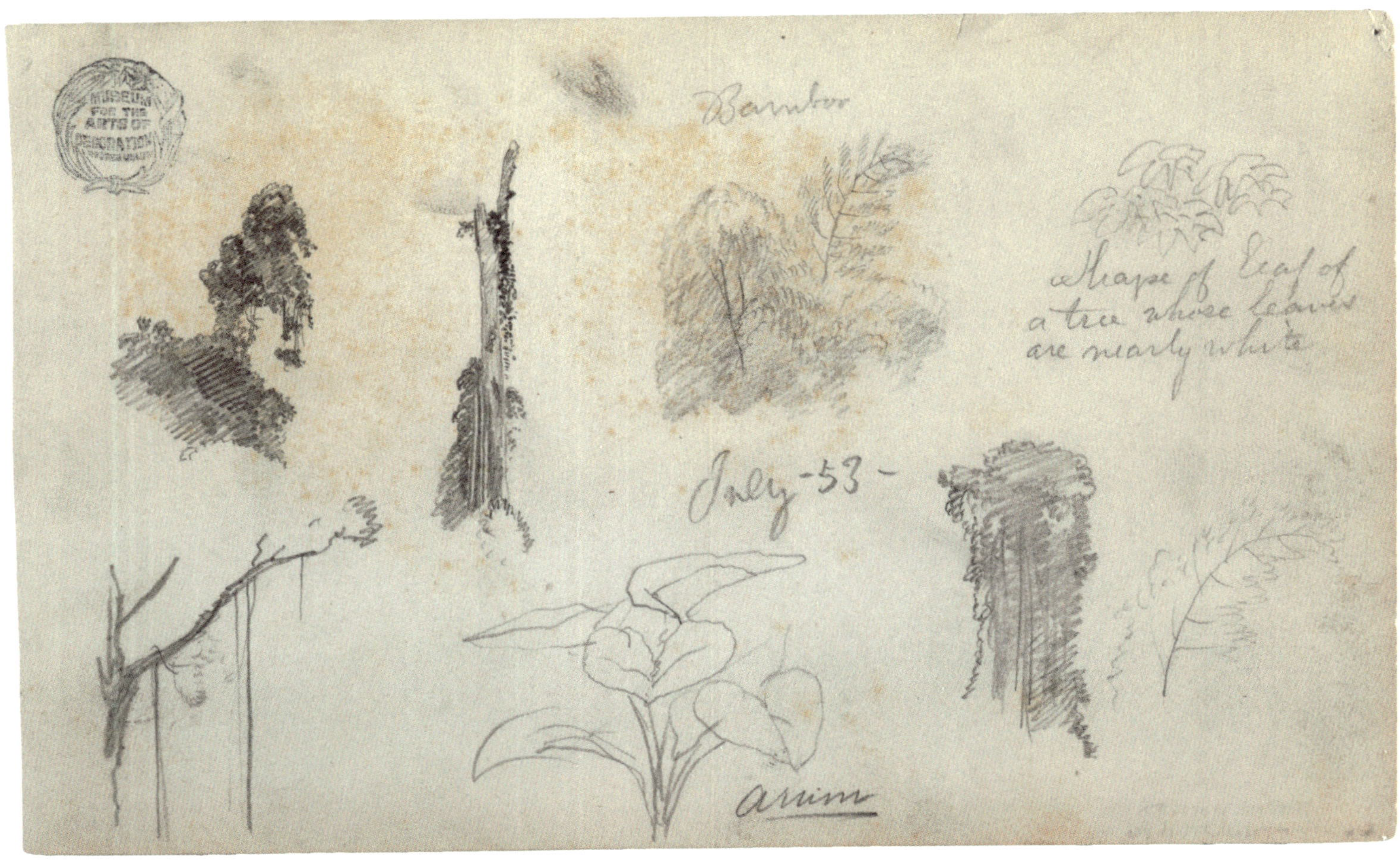

CAT. 88 Frederic Edwin Church, *A Group of Thirteen Botanical Studies from Colombia or Ecuador*, 1853, pencil on paper, 10 13/16 × 18 1/16 in., Cooper Hewitt, Smithsonian Design Museum, Gift of Louis P. Church, 1917-4-879

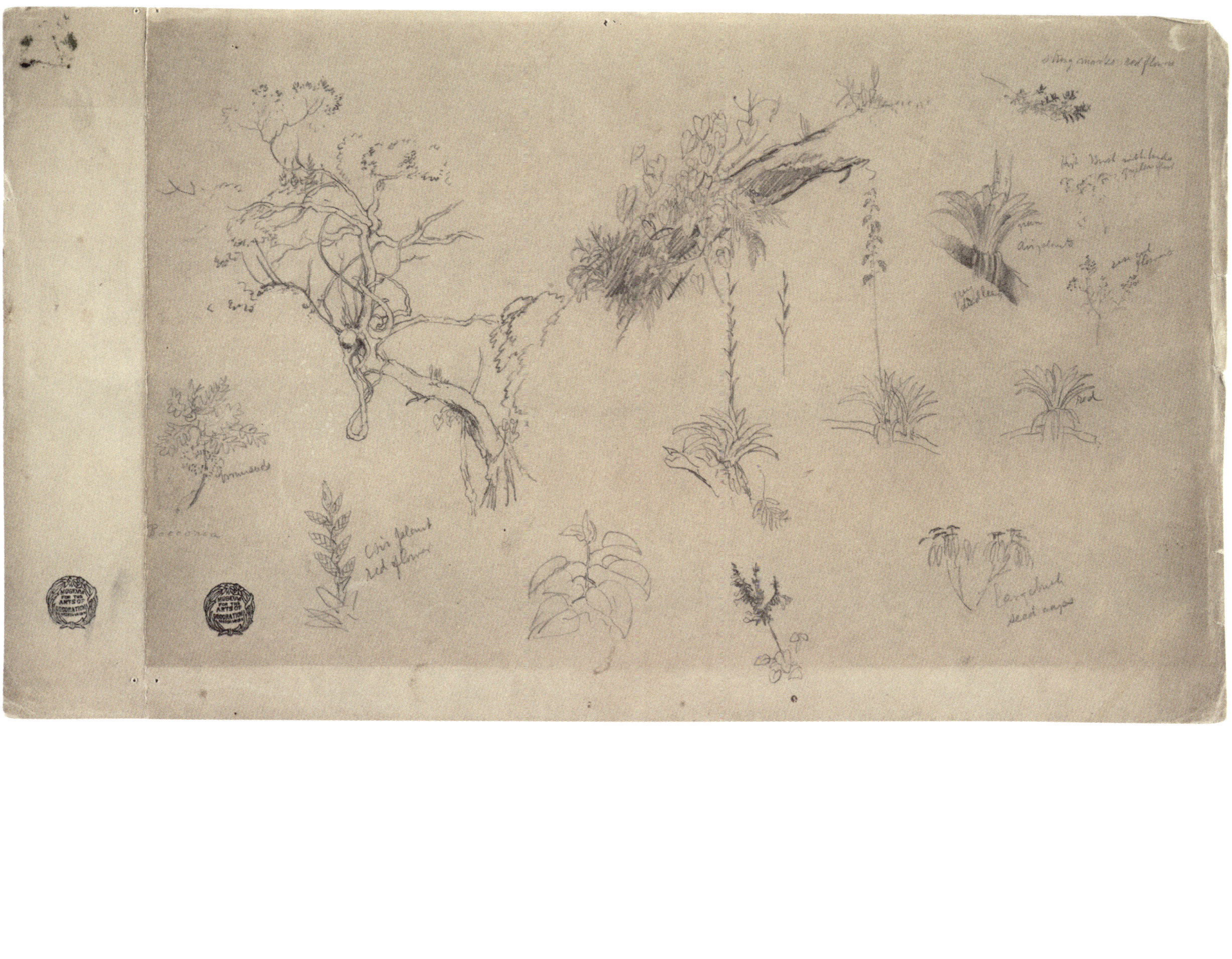

CAT. 89 Frederic Edwin Church, *Studies from Guacamaya, Bogotá, Colombia*, June–July 1853, pencil and watercolor on paper, 10 ¾ × 14 ⅝ in., Cooper Hewitt, Smithsonian Design Museum, Gift of Louis P. Church, 1917-4-877

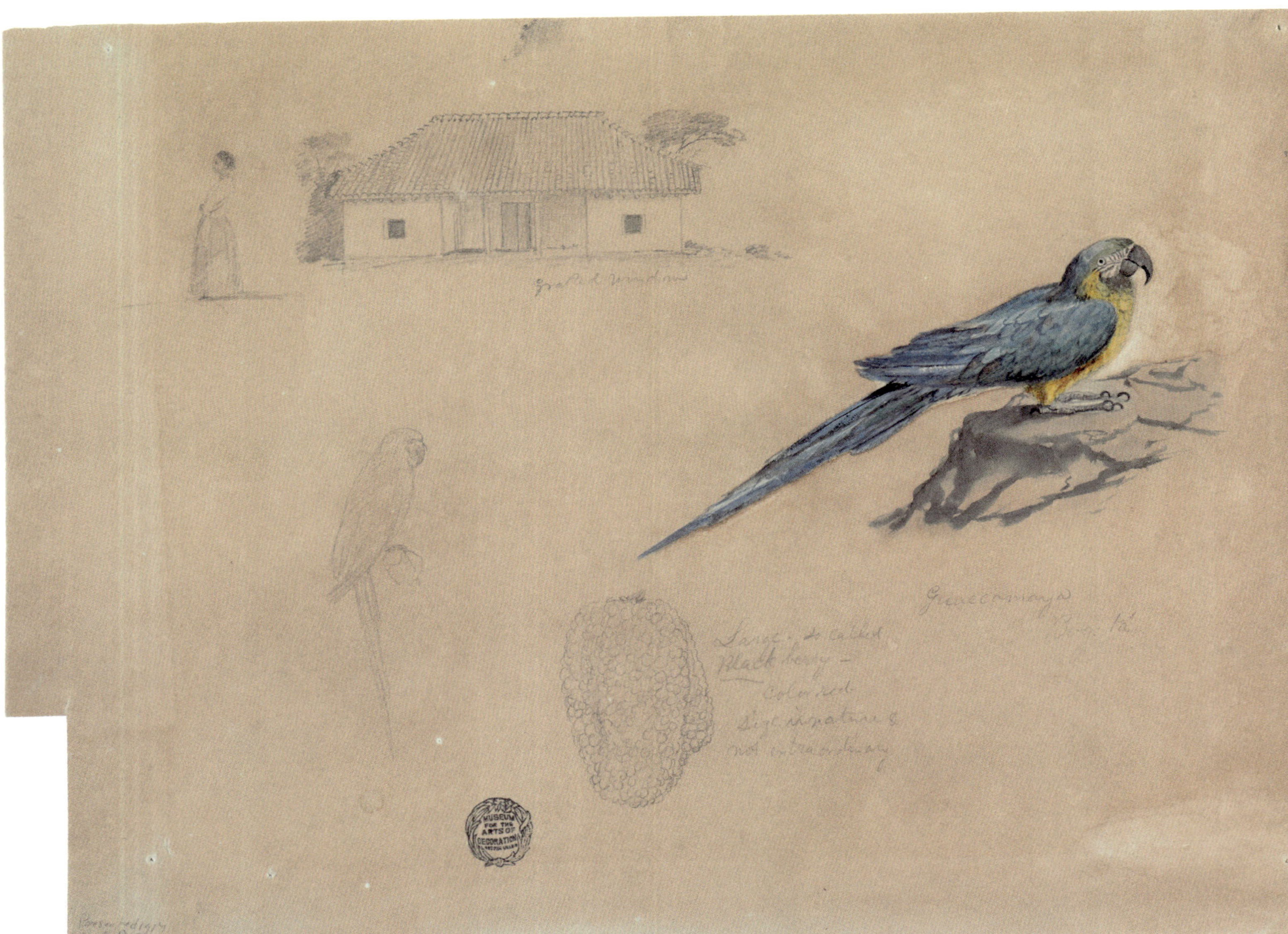

summit reached fifteen thousand feet. As the temperature dropped, the conditions changed from hail to snow. Church kept climbing until the horses gave out, at which point he continued on foot. He wrote, "You can scarcely imagine the strange effect that so rare an atmosphere has upon the body ten steps will put a strong man out of breath."[36] Despite the conditions, Church was eager to reach the summit, continuing his climb after Field gave up. Finally, Church reluctantly abandoned his ascent due to adverse weather conditions, but he vowed, like Humboldt, to return for another attempt.

As the two travelers crossed into Ecuador, they read from Humboldt's travelogue by nightfall about the treks they would make in the following days. Humboldt had lavished tremendous attention on this volcanic chain of mountains, extrapolating from his observations a connection between seismic and volcanic activity and beginning to discern what would later become the theory of plate tectonics. His exuberant descriptions of smoking and rumbling volcanoes lent an air of excitement to his scientific observations. Church felt the same exhilaration as did Humboldt. Faced with these smoking giants, Church recorded the contours of the volcanoes, climbed extensively as the weather permitted, and, not surprisingly, made these volcanoes central to his South American subjects for the rest of his career.

COTOPAXI: HUMBOLDT'S PERFECT VOLCANO

Church's letters home provide a sense of the travelers' experiences after leaving Bogotá. However, some of his observations were lost when, on August 24, a donkey slipped off an elevated trail. The animal perished and with it part of the artist's diary and presumably some of his sketches.[37] Whatever dismay Church felt in the loss of nearly six weeks of notes, his enthusiasm upon reaching the Andes is evident in his diary entry for August 26:

After a disagreeable journey across an elevated plain with a cold piercing wind and a sprinkling of rain we finally came to the edge of an eminence [?] which overlooks the valley of Chota And a view of such [?] unparalleled magnificence presented itself that I must pronounce it one of the great wonders of Nature. I made a couple of feeble sketches this evening in recollection of the scene. My ideal of the Cordilleras is realised.[38]

Church's language here is another echo of Humboldt's prose. In the preface to *Views of the Cordilleras*, Humboldt stated,

I shall think I have accomplished my purpose, if the feeble sketches contained in this work should lead other travellers, friends of the arts, to visit the regions which I traversed, and to retrace accurately those stupendous scenes, to which the Old Continent offers no resemblance.[39]

Church was never one to describe his own efforts as feeble; here he channeled Humboldt's words, drawn from the volume that had inspired this trip.

Church also channeled Humboldt's excitement. Embedded in a typographic word cloud of irrepressible enthusiasm, Humboldt's frontispiece, with its engraving of Cotopaxi (CAT. 90), alludes to one of the more dramatic encounters between Humboldt and a smoking volcano.[40] Humboldt admired the volcano's profile, noting,

The form of Cotopaxi is the most beautiful and regular of the colossal summits of the high Andes. It is a perfect cone, which, covered with an enormous layer of snow, shines with dazzling splendor at the setting of the sun, and detaches itself in the most picturesque manner from the azure vault of Heaven.[41]

Humboldt sketched Cotopaxi and the side vent, which is known as the "head of the Inca" and is visible in Church's painting from a nearby hacienda, exclaiming, "This is one of the most majestic and most awful views I ever beheld in either hemisphere."[42]

Church, too, visited the region at a relatively peaceful moment. In his diary Church noted, "After a pleasant journey of seven leagues we reached this place (the hacienda of Señor Larrea) which has a commanding view of Cotopaxi."[43] While they were based there, Church sketched from multiple vantage points. In one pencil sketch, Church drew the volcano from an angle that hid the prominent side vent, and shows the prevailing wind coming from a different direction. His difficulty was in the vagaries of the weather. Church's entry for September 12 reads,

RESEARCHES,

Concerning

THE INSTITUTIONS & MONUMENTS

OF

the Ancient Inhabitants

OF

AMERICA,

with Descriptions & Views

OF SOME OF THE MOST

Striking Scenes

in the

CORDILLERAS!

Written in French by

ALEXANDER DE HUMBOLDT,

& Translated into English by

Helen Maria Williams.

Vol. I.

Plate 10.

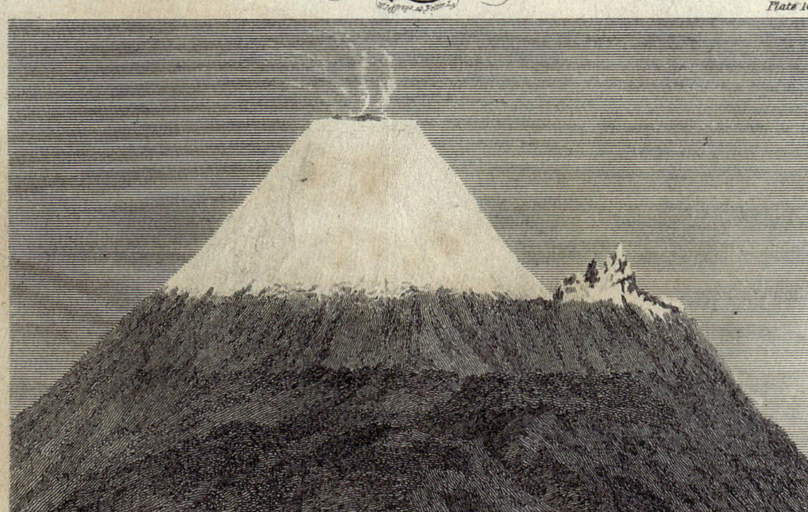

Page 118. Vol. 1.

VIEW OF COTOPAXI.

LONDON:

Published by Longman, Hurst, Rees, Orme & Brown, J. Murray & H. Colburn.

CAT. 90 Alexander von Humboldt, Frontispiece to *Researches, Concerning the Institutions & Monuments of the Ancient Inhabitants of America*, 1814, engraving, closed: 9 × 6 × 1 in., Smithsonian Libraries

> *Waited patiently all day for the clouds to disappear from the Summit of Cotopaxi and about Sun-set had the satisfaction of a partial view which was magnificent. [—?] a slight sketch.*[44]

The following day stretched Church's patience, as he wrote with some exasperation,

> *Left the hacienda of San Juan early this morning in the rain but the sun broke through the clouds before noon. Reached this village before sunset. There must be magnificent scenery on the road which the clouds today prevented me from witnessing.*[45]

Their itinerary took them close to Chimborazo, but here again the weather and the topography prevented Church from spending the time required to make a significant group of sketches of Humboldt's favorite peak. He managed what he called a "slight sketch of Chimborazo" on September 15, while visiting a hacienda near Riobamba with a particularly good view.[46] Two days later, Church attempted a plein-air oil sketch of Chimborazo, but "a thunder storm drove me down the hill to a house for shelter after-wards the mountain was totally obscured."[47] The artist's diary is peppered with references to getting out of the weather and not being able to see or reach the summits of individual volcanoes. Church and Field departed for Guayaquil shortly thereafter, ending their Andean sojourn. The warmer climate and lower elevation there made for a pleasant end to the trip, and the two men departed on the steamer *Bogotá* on September 30, arriving in New York a month later.

Fig. 89 Thomas Cole, *Mount Aetna from Taormina*, 1843, oil on canvas, 78 ⅝ × 120 ⅝ in., Wadsworth Atheneum Museum of Art, Hartford, CT, Purchased by Daniel Wadsworth

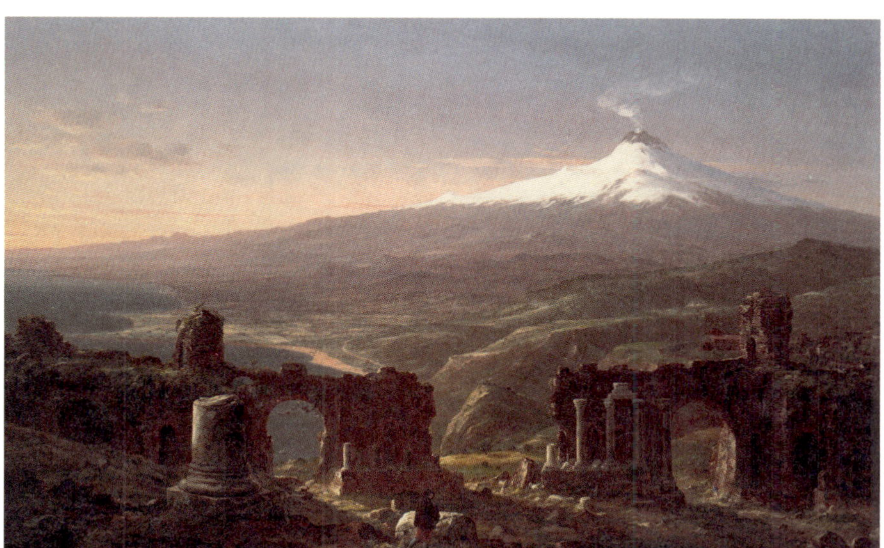

While weather conditions in early September hindered Church's sketching opportunities, once back in New York, the artist again relied on his prodigious memory to re-create scenes he viewed while in transit or glimpsed during fleeting moments of clearer skies. Not surprisingly, Church also depended on Humboldt's engraved scenes of Andean landscapes to supplement his firsthand experiences. This is particularly true of Church's first painting featuring Cotopaxi (**CAT. 91**). Here the artist brought together a group of his pencil and oil sketches of the smoking cinder cone, placing the volcano in a rolling landscape that is true to its setting, but also one that bears a strong resemblance to Humboldt's hand-colored view of *Chimborazo from the Plains of Tapia* (**CAT. 92**) in structure and palette. The orientation of the volcanic cone matches that found in the smaller engraving embedded in the frontispiece of Humboldt's *Views of the Cordilleras*. The distinctive cinder cone with its side vent, visible in both Humboldt's engraving and Church's painting, reminds us of Church's efforts to find similar vantage points for his own studies of nature.

In Church's painted landscape, two diminutive travelers make their way to a hacienda, very likely one of the same ones Humboldt had stayed in fifty years earlier. Church has taken artistic liberties with the location of the dwelling vis-à-vis the volcano, all to provide a sweeping vista that encompasses the terrain from the vast fertile plains to the volcanic foothills and up to the snow-clad summit. The clear sky and peaceful plume of steam rising from the caldera convey a sense of stability and timeless calm. Small, blooming plants along the foreground recall passages in Church's letters about his admiration for the superabundance of flowering plants native to South America (**CAT. 93**). As an encapsulation of their South American journey, this view of Cotopaxi brought together the best features of their travels in a single frame. Like *West Rock*, it provides a broad vista and a compelling geological formation. In portraying a classically

CAT. 91 Frederic Edwin Church, *Cotopaxi*, 1855, oil on canvas, 28 × 42 in., Smithsonian American Art Museum, Gift of Mrs. Frank R. McCoy

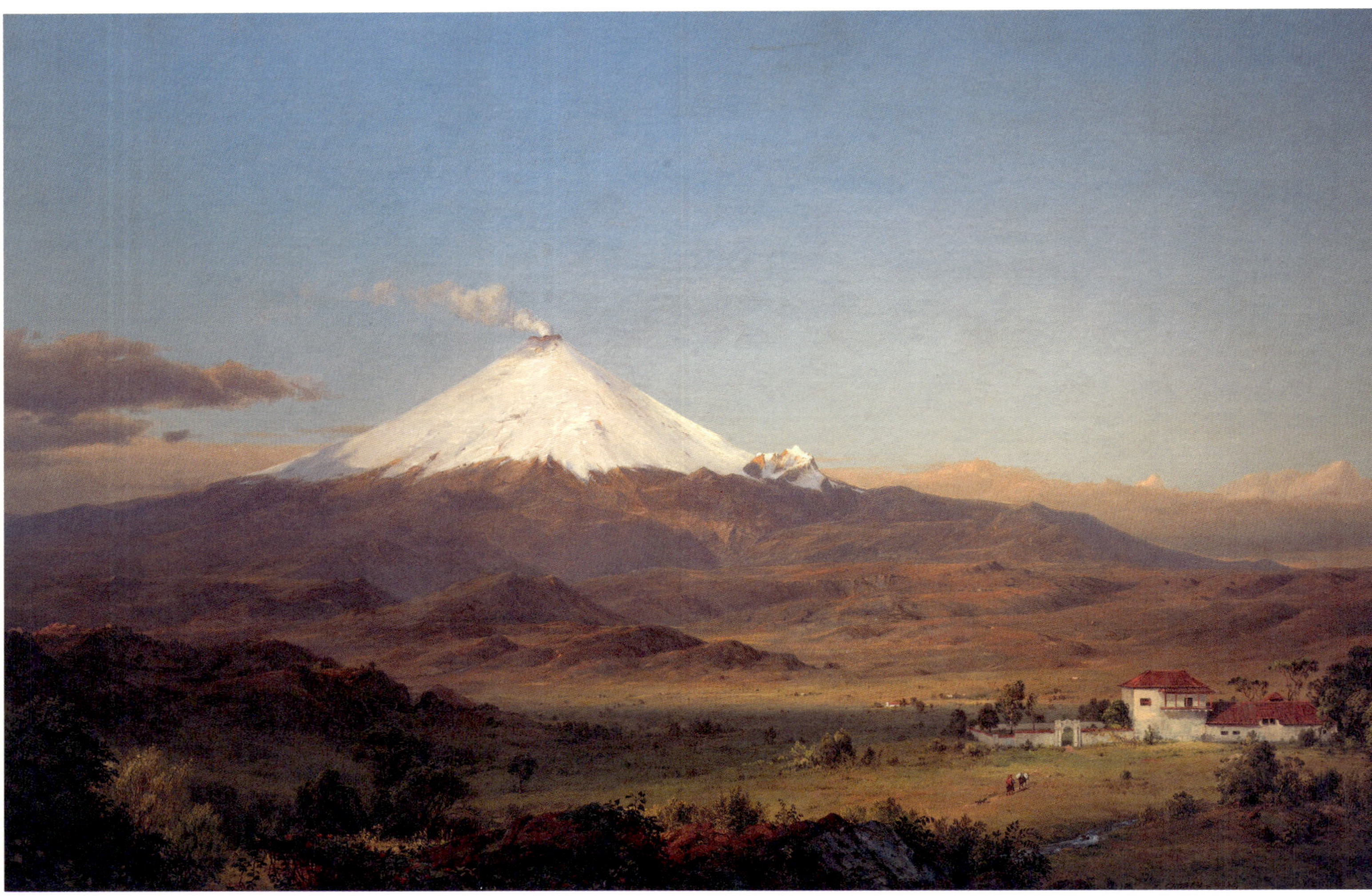

CAT. 92 Alexander Von Humboldt, *Chimborazo from the Plains of Tapia*, from *Vues des Cordillères et monuments des peuples indigènes de l'Amérique*, plate 25, 1810, color engraving, open: 23 × 34 in., Smithsonian Libraries, Gift of the Burndy Library

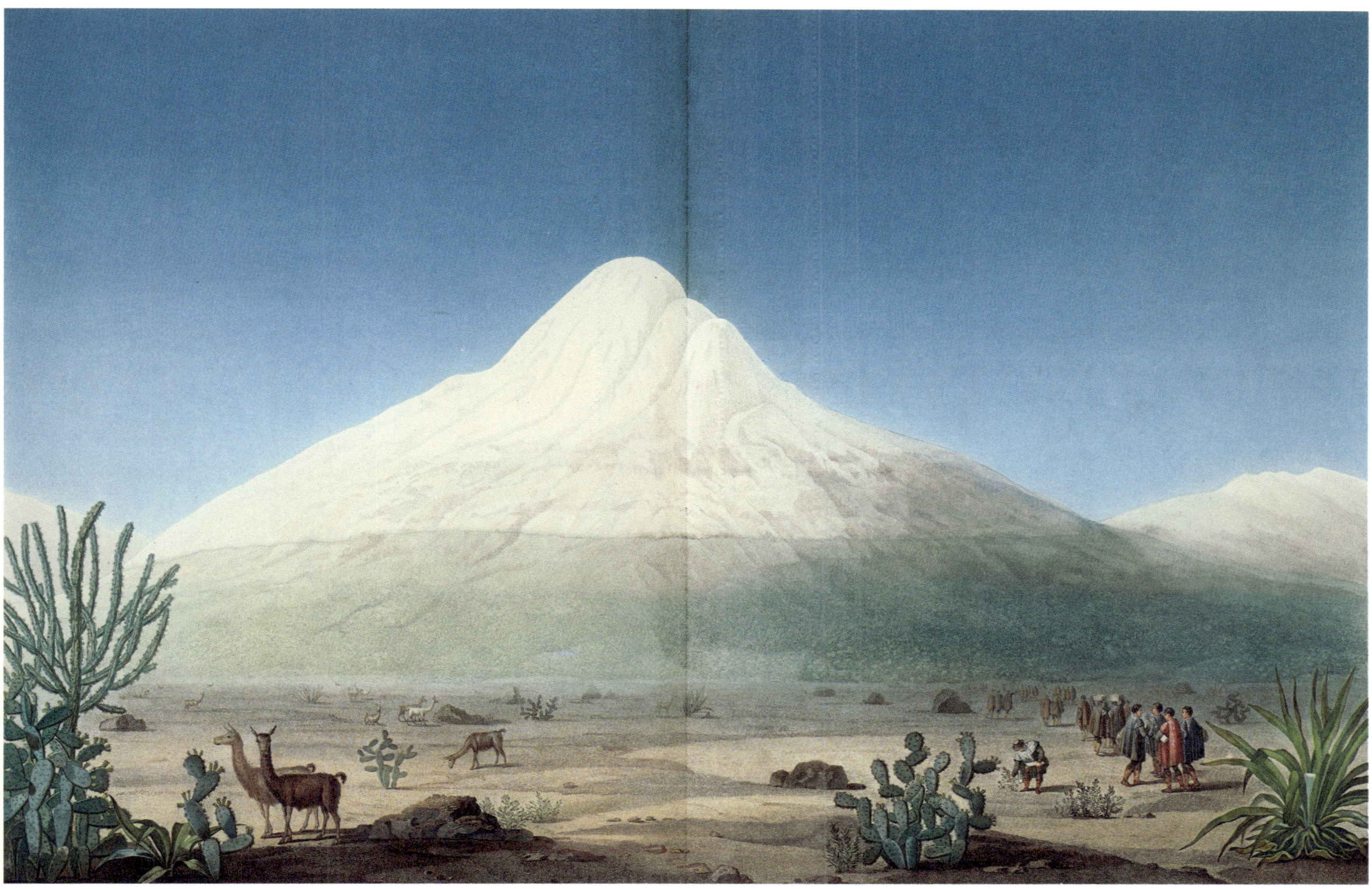

clear day, Church adapts a composition similar to Cole's views of *Mount Aetna from Taormina* (fig. 89), with the smoking cinder cone a benign presence on the horizon. If Humboldt yearned for an artist who could approach South American scenery to rival that of the Mediterranean, Church accomplishes that in *Cotopaxi*.[48] Inspired and encouraged by Humboldt's guidance, this painting is simultaneously a glance back to Humboldt's pioneering travels and a souvenir of Church and Field's more recent sojourn. Not surprisingly, then, it joined *West Rock, Natural Bridge,* and *The Falls of Tequendama* in Cyrus Field's home once it was completed.

In 1855 Church displayed *The Falls of Tequendama* at the NAD. The critics marveled at the dramatic composition but were less impressed with the artist's rendering of the turbulent water, with one acerbically noting, "He should not paint falling water—for he cannot."[49] Undaunted, or perhaps annoyed, Church took steps to correct that deficiency by going to Niagara Falls in 1856. There he made close to one hundred sketches in pencil and in oil. The primary focus was not as much on the plunge of the waterfall as a detailed, incisive understanding of the play of water in motion. The body of work dating from this trip includes detailed sketches of foaming rapids above the cataract, swirling eddies, and whitecaps—all revealing a keen eye for movement and color. Church's sketches from the Niagara River above the falls are exuberant, the result of days spent in close observation of his subject. Church's approach to understanding Niagara demonstrates the depth to which Humboldt's influence had permeated the artist's working method. Thus, Humboldt's influence was not confined to Church's South American subject matter.[50] Instead, Humboldt's ideas—about the relevance of landscape painting, the need for a scientific rigor linked to an aesthetic sensibility, and the call for an artist capable of creating a higher style of landscape—became Church's inspiration for focusing his own pursuits on the intersection of science and art.

NIAGARA AS A HUMBOLDTIAN PAINTING

The year 1857 was a turning point for Church. *The Falls of Tequendama* had demonstrated how ambitious Church could be and how willing he was to work outside of his usual parameters of composition and subject matter. It had also revealed that Church's command of nature's visual or artistic vocabulary still required refinement. He later used Niagara Falls to push beyond all established conventions of landscape painting, composing the work that would redefine his career as it announced his embrace of Humboldt's immersive approach to understanding nature. *Niagara* (CAT. 94) broke with the traditional proportion and size of an American landscape painting. The painting stands out for more than its masterful rendering of moving water or the evanescent rainbow shimmering over the gorge. Church destroyed the conventional perspective accorded to views of the falls, stretching the canvas into a panoramic rectangle. He dropped out the entire foreground, suspending the viewer over moving water at the very edge of the cataract, in danger of being swept over the falls. The singular focus here is on the movement of water: precisely the element from *The Falls of Tequendama* the critics had found wanting. By radically changing the expected format of a landscape painting, Church amplified its immersive qualities and the tendency toward synesthesia. As the reviewer in the *Home Journal* proclaimed, "if the object of painting be to render faithfully, and yet poetically, what an artist's eye discerns, this *is* Niagara, *with the roar left out!*"[51] Church used an entire suite of pencil and oil sketches to fix his data points and invest the scene with as much scientific accuracy as possible while setting a higher goal—to dramatize the grandeur of the falls and, by extension, to ennoble the country's cultural identity. His final oil sketch, painted across two sheets of prepared paper, signals the development of his daring composition (CAT. 95). The same critic termed Church's painting "the eighth wonder of the world!," eliding the reputation of the falls with Church's rendering of the cascade.[52]

Niagara is every bit as much a Humboldtian painting as was *The Falls of Tequendama*. The artist's virtuosity answered his critics who had found fault with his rendering of moving water; his unorthodox format literally expands his viewer's perception of the rapidly flowing water just before it descends over the lip of the falls. Humboldt approved of the panorama as the ideal format for conveying the scale and scope of nature and

CAT. 93 Frederic Edwin Church, *Botanical sketches*, May 1857, pencil on paper, 4 ¼ × 7 ⅛ in., Cooper Hewitt, Smithsonian Design Museum, Gift of Louis P. Church, 1917-4-835-A

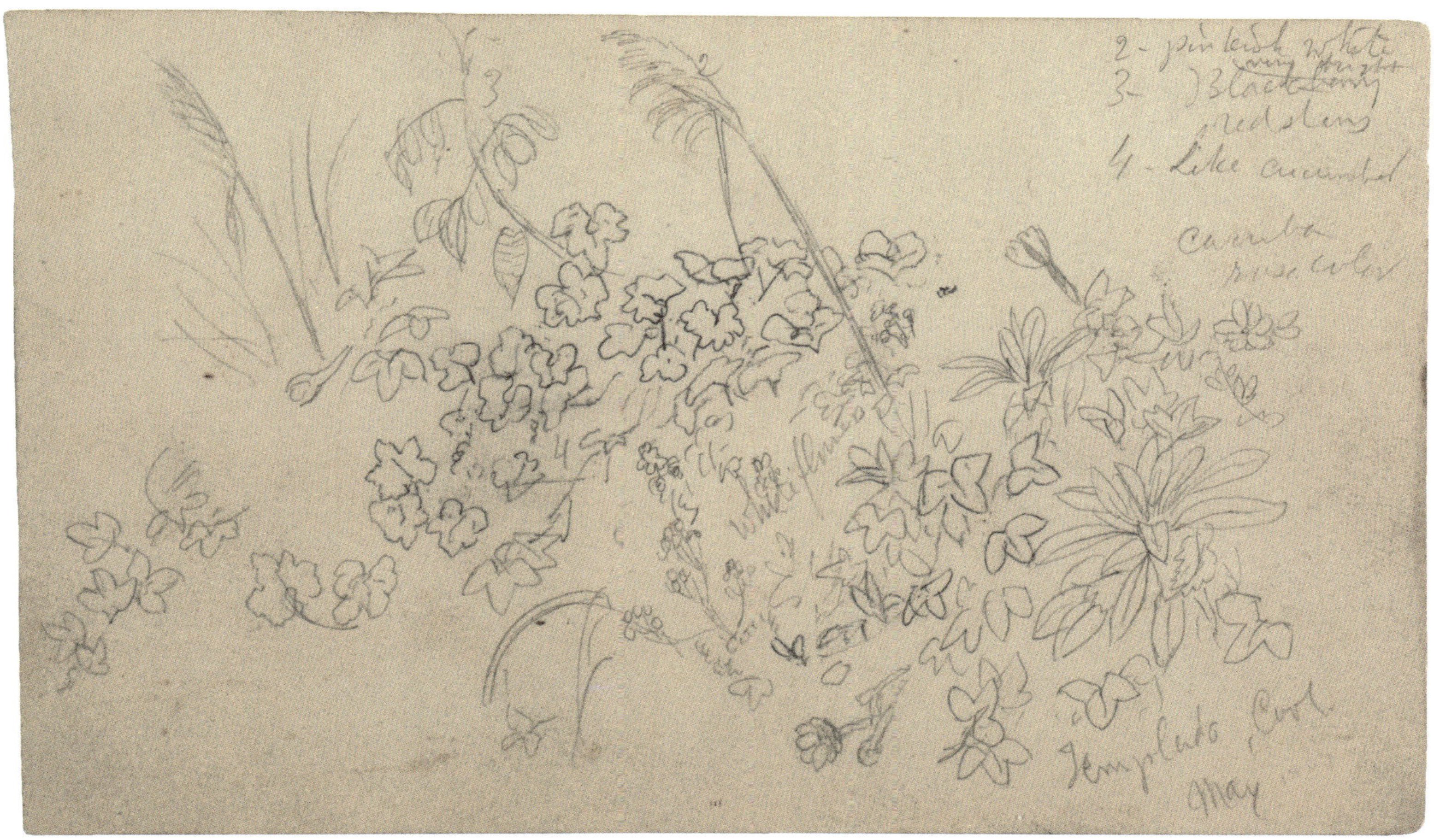

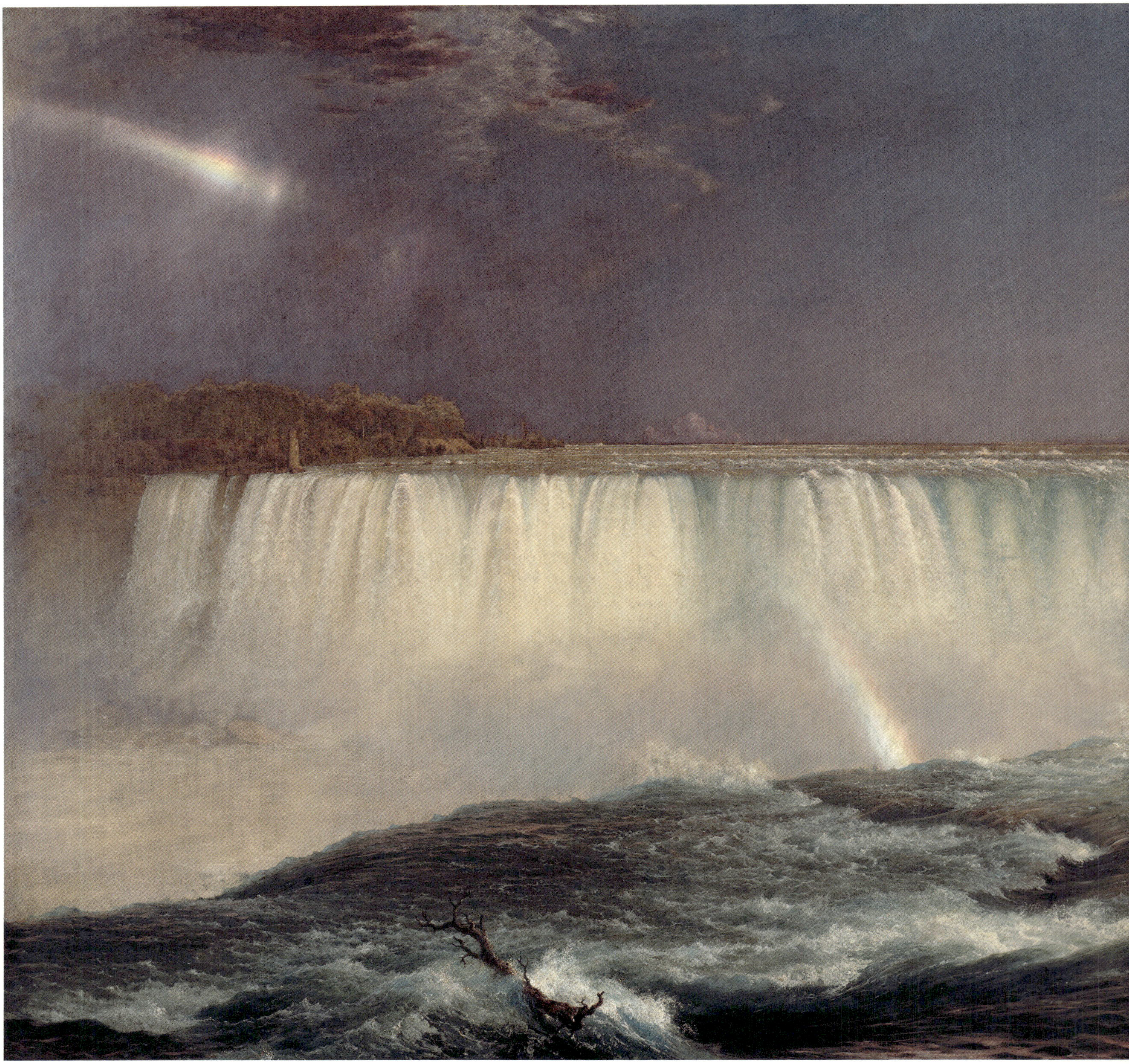

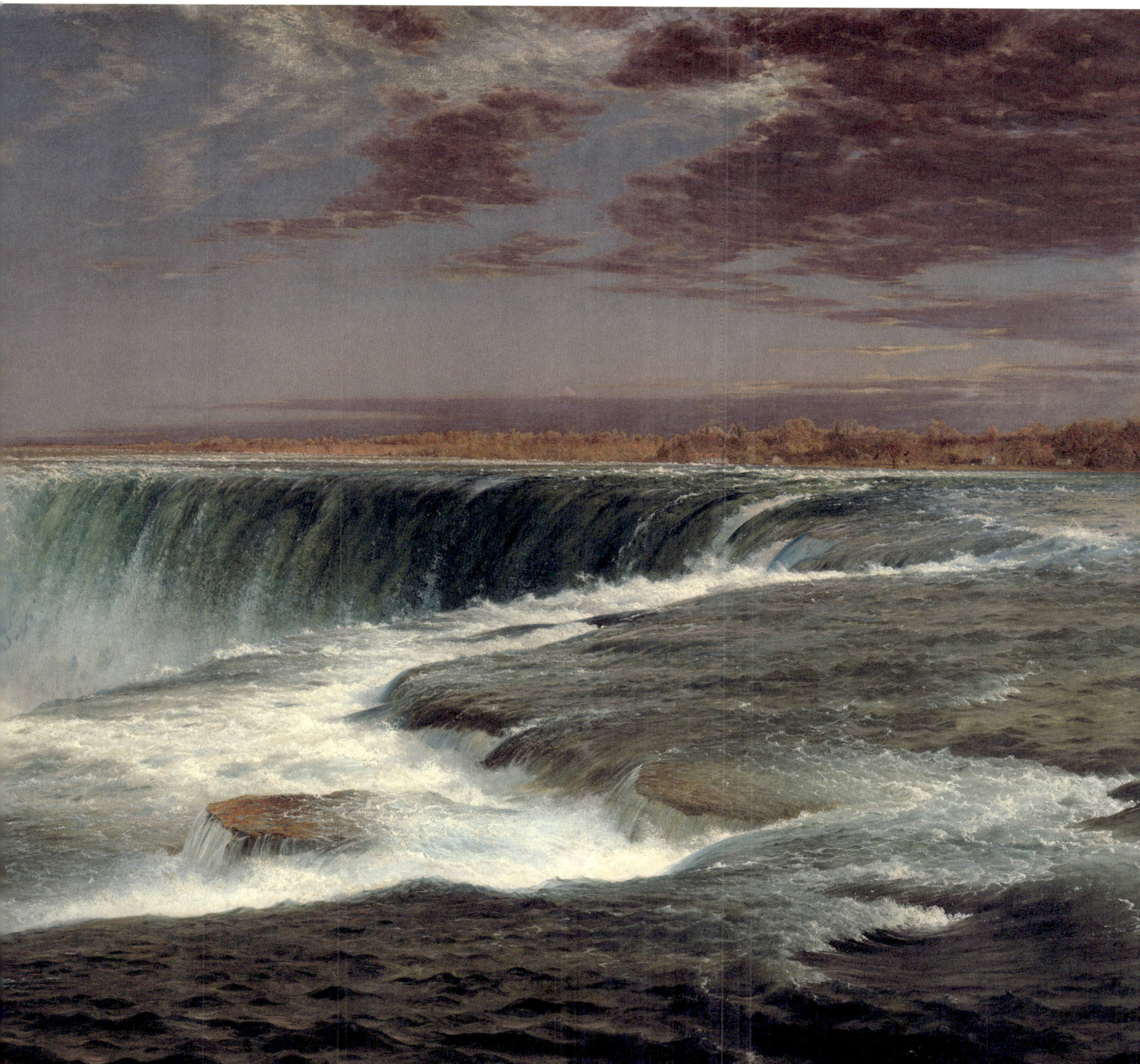

Preceding pages

CAT. 94 Frederic Edwin Church, *Niagara*, 1857, oil on canvas, 40 × 90 ½ in., National Gallery of Art, Corcoran Collection (Museum Purchase, Gallery Fund)

Below

CAT. 95 Frederic Edwin Church, *Horseshoe Falls*, December 1856–January 1857, oil on paper mounted to canvas, 11 ½ × 35 ⅝ in., Olana State Historic Site, New York State Office of Parks, Recreation and Historic Preservation

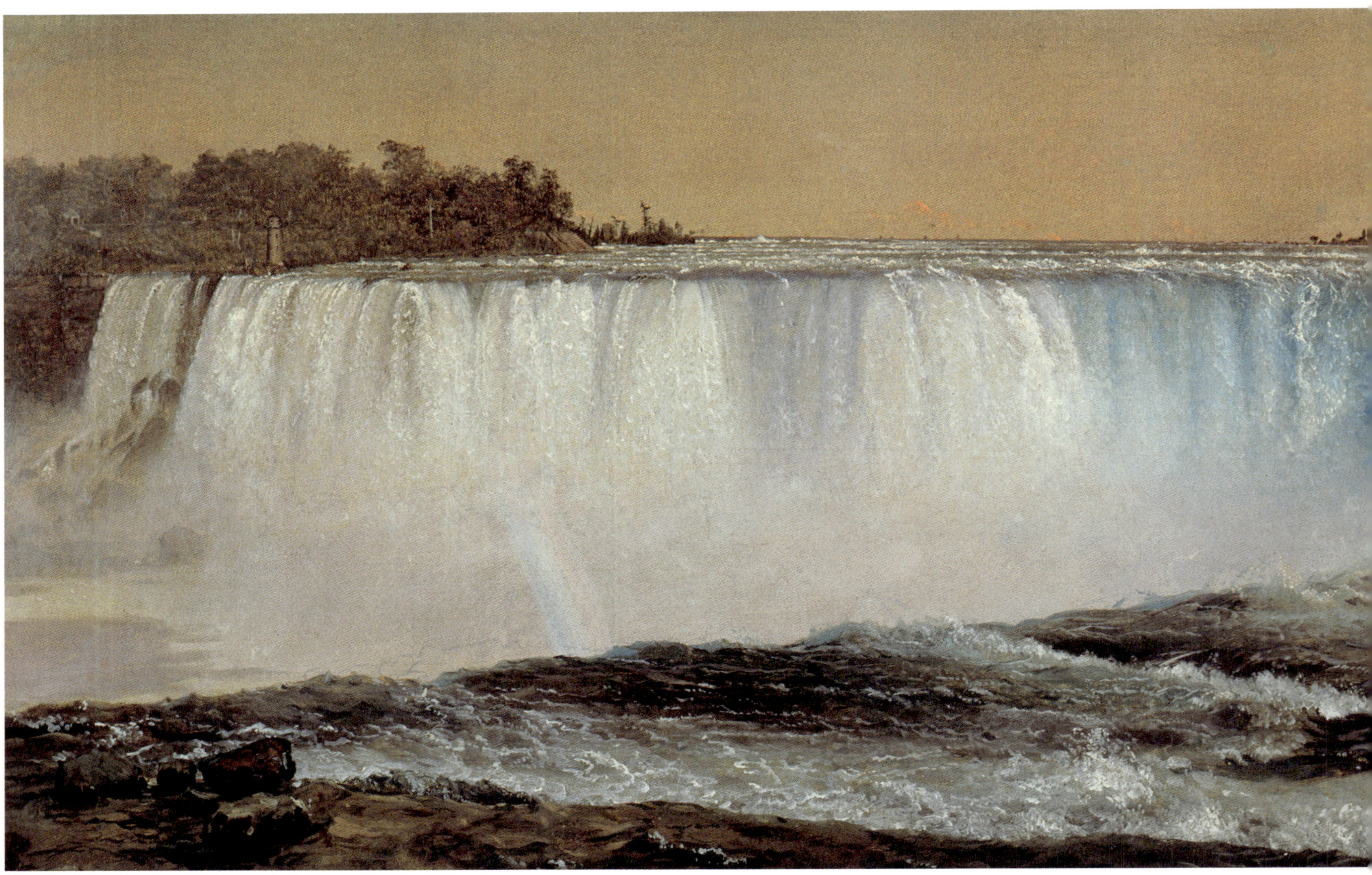

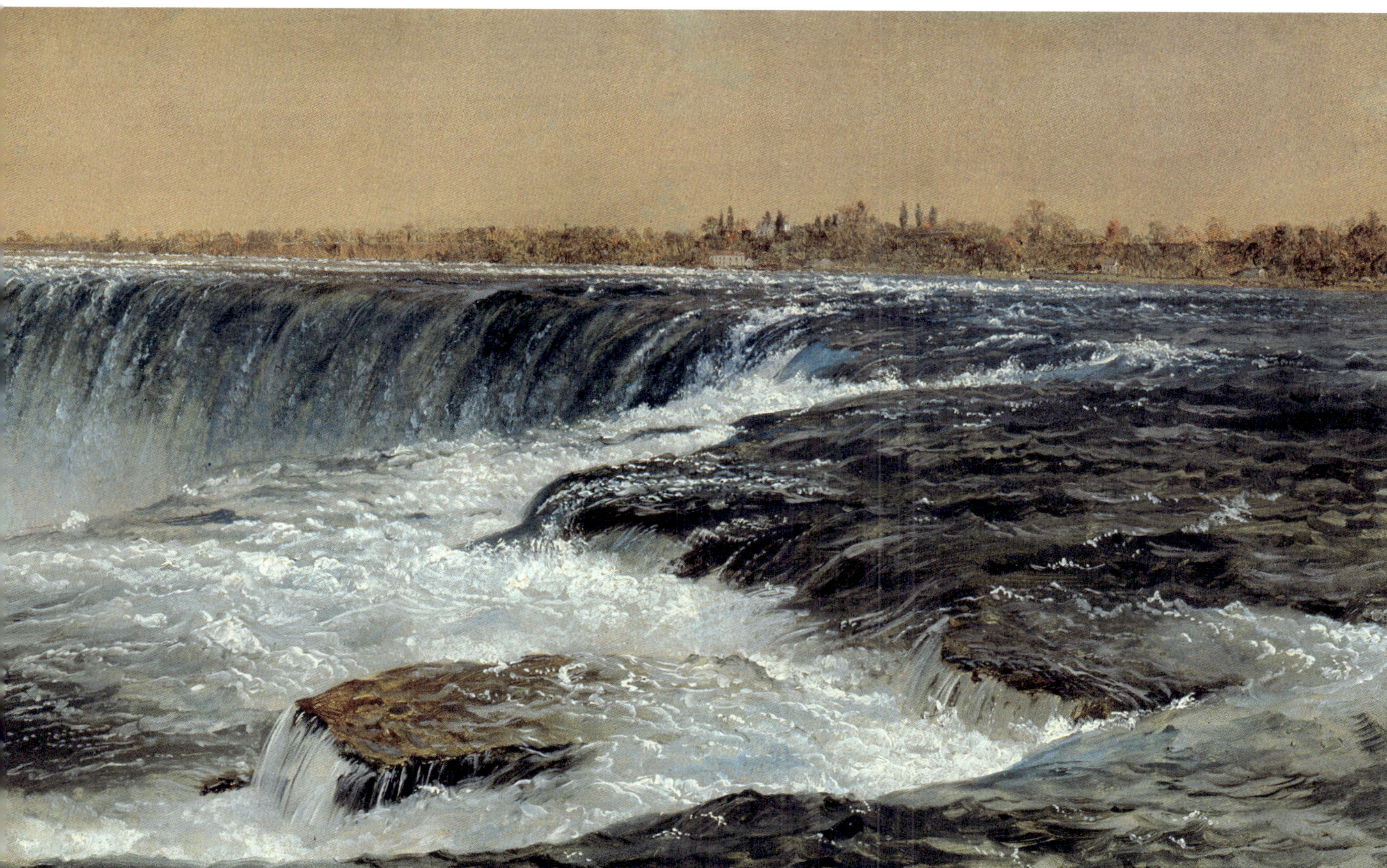

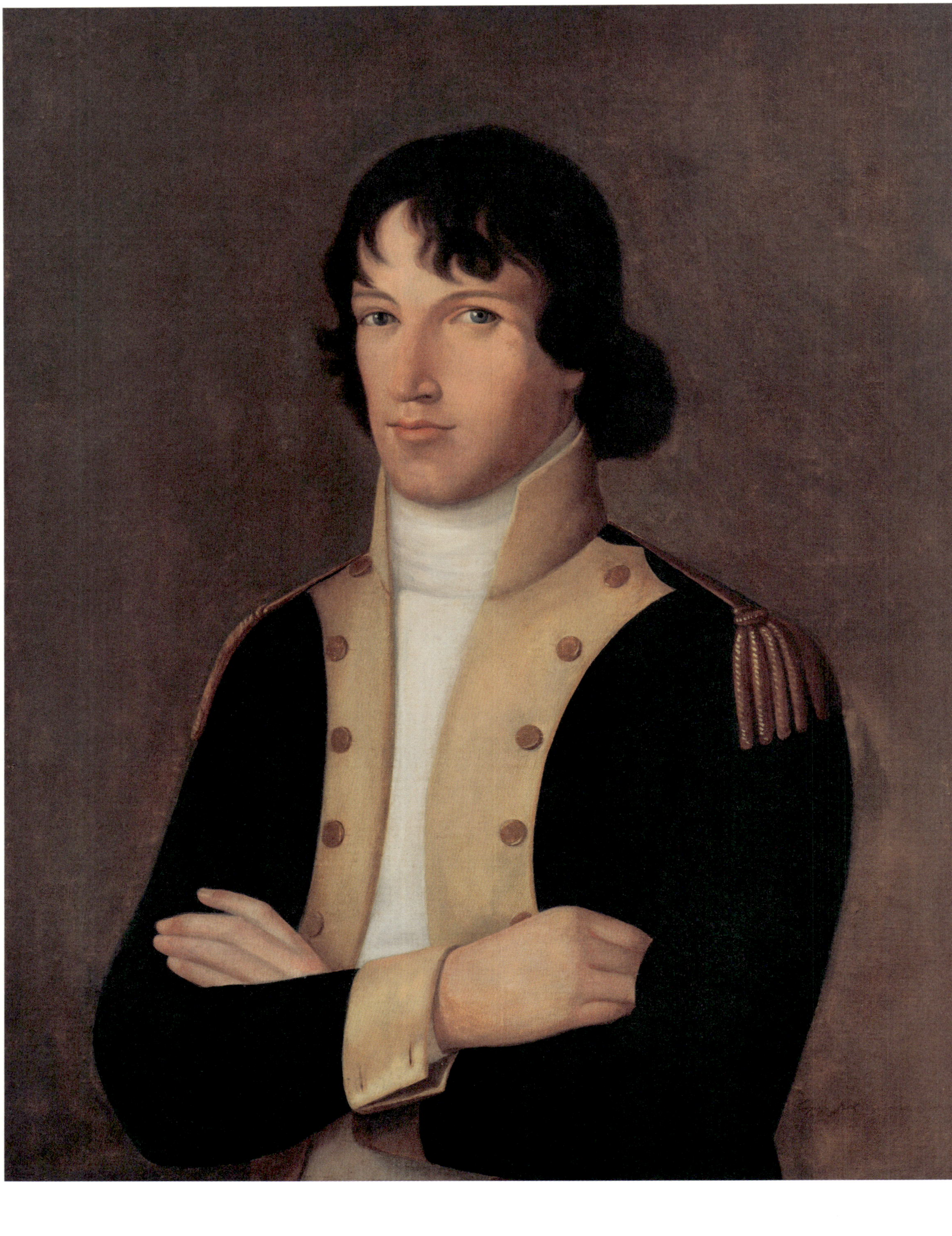

CAT. 96 Rafael Salas, *Portrait of Humboldt*, 1857, oil on canvas, 27 ¾ × 22 in., Private Collection

encouraged large-scale paintings as a means of "increase[ing]...the force of these impressions."[53] With each successive painting, Church pushed his understanding of Humboldt's synthesis of nature further, experimenting with the scale and proportions of his canvas and adjusting his focal point to absorb a wider range of terrestrial and atmospheric phenomena to convey the range of his travels, observations, and insights. What Humboldt accomplished in writing served as inspiration for what Church would essay in paint.

In 1857, as *Niagara* began its triumphal tour across America and London, the artist embarked on his second, shorter trip to South America. Although Field underwrote the trip, he did not accompany Church, who traveled this time with fellow artist Louis Rémy Mignot. The four-month foray was intended to reacquaint Church with key features in Humboldt's narratives, notably the explorer's favorite volcano, Chimborazo. Church's goal was to develop a suite of sketches of Chimborazo, at that point considered the highest peak on the earth, a dormant volcanic massif that had earned the affectionate appellation of "Humboldt's mountain." Chimborazo had been socked in by low cloud cover during Church's first trip, and his itinerary did not afford him the luxury of waiting for the weather to clear. This time he could stay near the mountain in hopes of capturing sketches of the summit. The pencil sketches Church made in the field once again took in the welter of botanical detail Church envisioned for the foreground of his next Great Picture that would become *Heart of the Andes* (see fig. 90). However, the bulk of his pencil and oil sketches, both in the field and back in his studio, present the surrounding terrain and the snow-covered summit in all weather. They form a body of work similar in feel to the detailed studies of water Church had made at Niagara Falls. Church's journal confirms the difficulties the artist and his party encountered trying to approach the summit, and his stories parallel Humboldt's frustration in dealing with the unpredictable weather.

TACKLING "HUMBOLDT'S MOUNTAIN," CHIMBORAZO

About Chimborazo, Humboldt had exulted, "Travellers who have approached the summits of Mont Blanc and Mont Rose are alone capable of feeling the character of this calm, majestic, and solemn scenery."[54] The massive snow-capped summit of Chimborazo towered over the rest of the Andean Cordillera, prompting Humboldt to compare it to St. Peter's in Rome: "like that majestic dome, produced by the genius of Michel Angelo, over the antique monuments, which surround the Capitol."[55] Humboldt's equivalences between Old World architectural monuments and New World natural features spurred and reinforced the rapid development of a nature-based aesthetic linking American scenery with the nation's cultural ambitions. For Church, it made his South American sojourn seem equivalent to the European Grand Tour for landscape painters. It also gave him a worthy focal point for this second trip to the Andes. In the field, Church created a suite of drawings and oil sketches of Chimborazo and the surrounding Andean plateau. As the painting took shape in his mind, these sketches cemented the link between his own artistic production and Humboldt's grand vision.

Church and Mignot traversed Panama and arrived on the west coast at Guayaquil on May 23 and spent ten days getting acclimated and sketching the Guaranda highlands. From there they made their way to the outskirts of Quito, the countryside inspiring the middle distance in Church's *Heart of the Andes*. In the Chillo Valley, twenty miles southeast of Quito, the two artists stayed at a hacienda that was still in the family of Carlos Montúfar, who had accompanied Humboldt and Bonpland on their ascent of Chimborazo in 1802. Montúfar had then traveled with them to the United States and Europe on the final leg of their own South American journey. Fifty years earlier, the Ecuadorian artist José Cortés de Alcocer had painted a portrait of Humboldt, which still hung in the Montúfar hacienda. During their stay, Church arranged to have it copied by Rafael Salas (**CAT. 96**). The portrait presents the young Humboldt in a South American military jacket, his crossed arms and direct gaze exuding

CAT. 97 Frederic Edwin Church, *Mount Chimborazo, Ecuador*, 1857, oil and pencil on paperboard, 13 7⁄16 × 20 9⁄16 in., Cooper Hewitt, Smithsonian Design Museum, Gift of Louis P. Church, 1917-4-1296-B

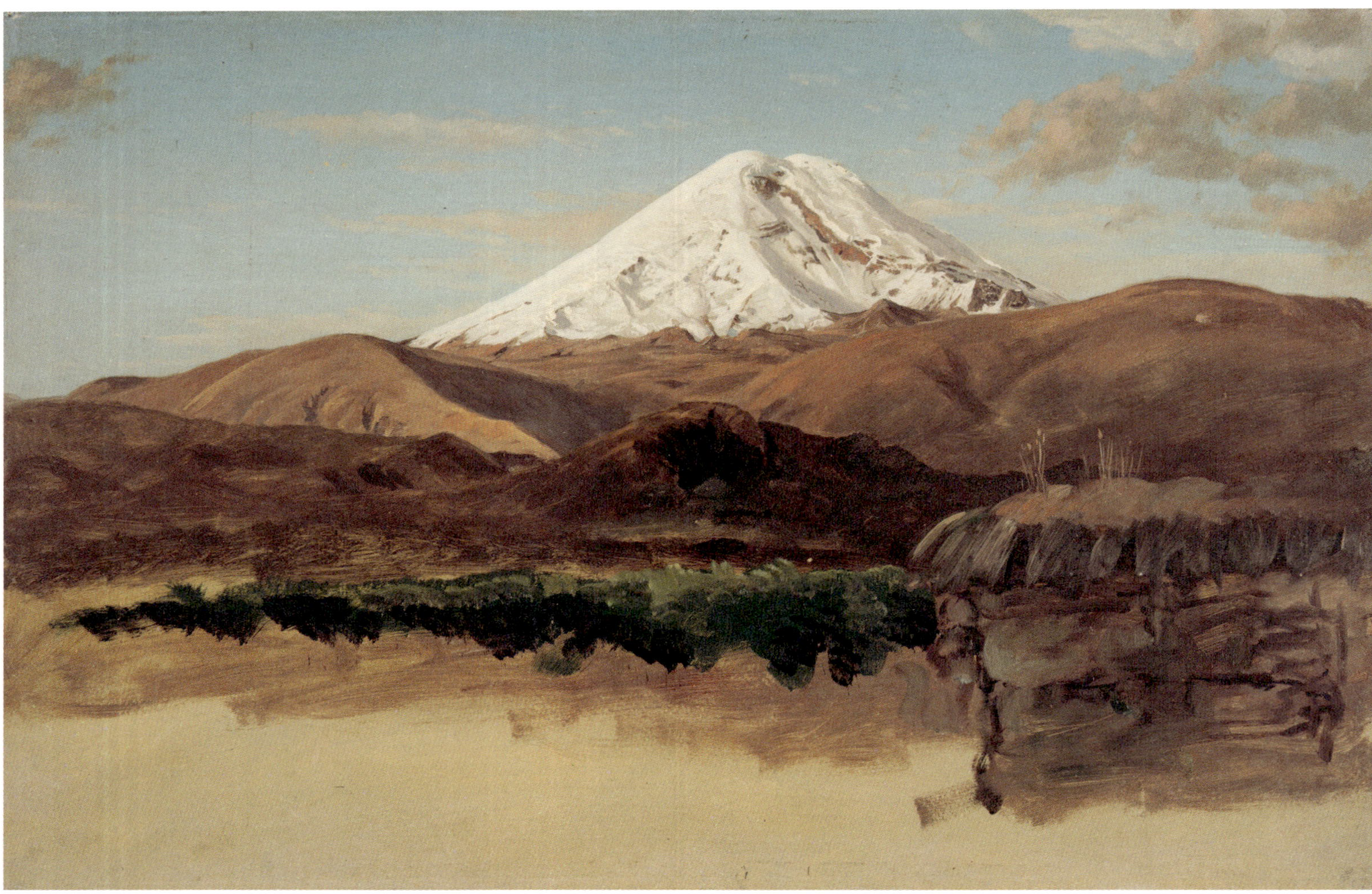

CAT. 98 Frederic Edwin Church, *Mount Chimborazo, Ecuador*, 1857, oil and pencil on paperboard, 9 ¼ × 17 ½ in., Cooper Hewitt, Smithsonian Design Museum, Gift of Louis P. Church, 1917-4-1296-A

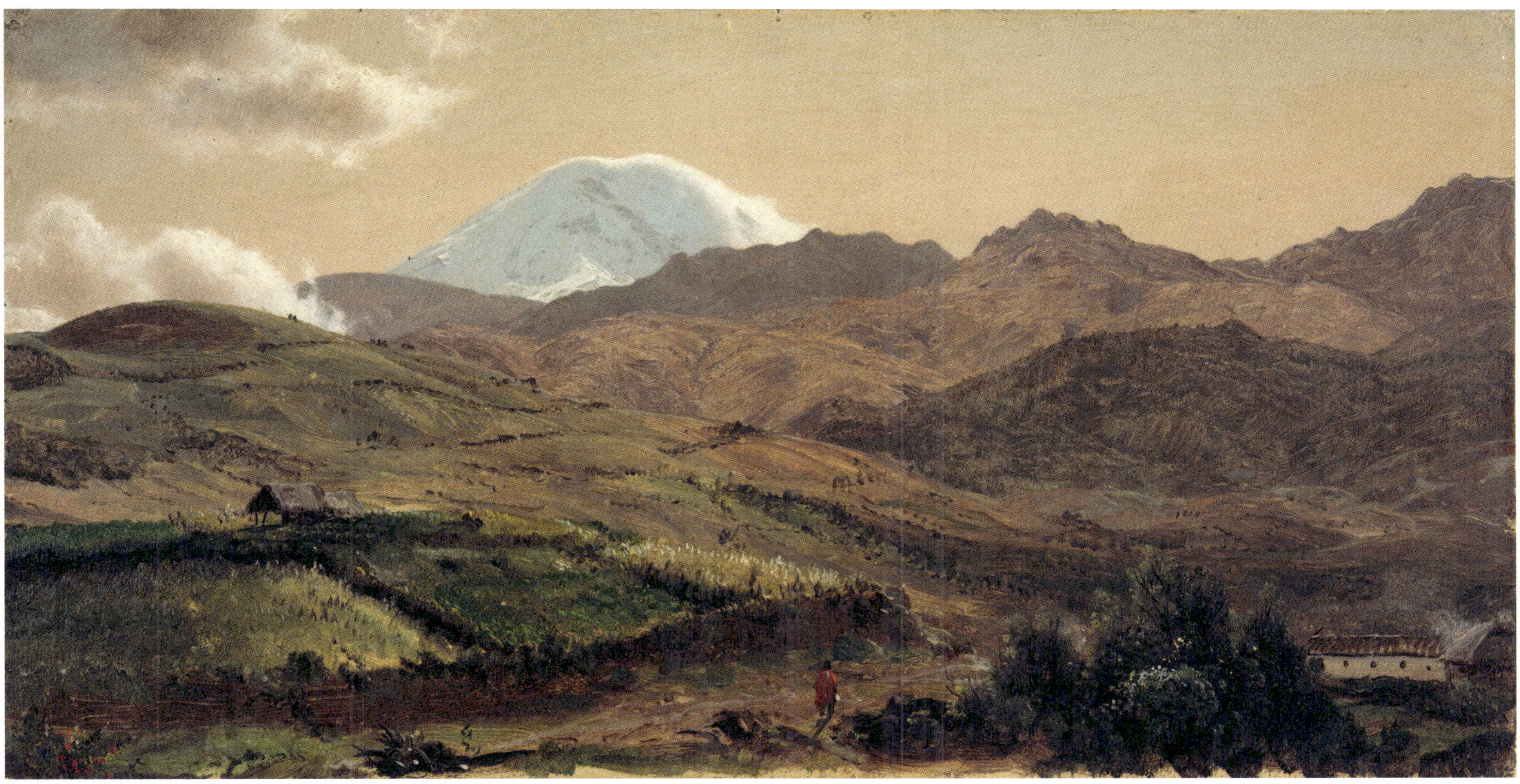

CAT. 99 Frederic Edwin Church, *Mount Chimborazo at Sunset*, 1857, oil on academy board, 12 × 21 ⅔ in., Olana State Historic Site, New York State Office of Parks, Recreation and Historic Preservation

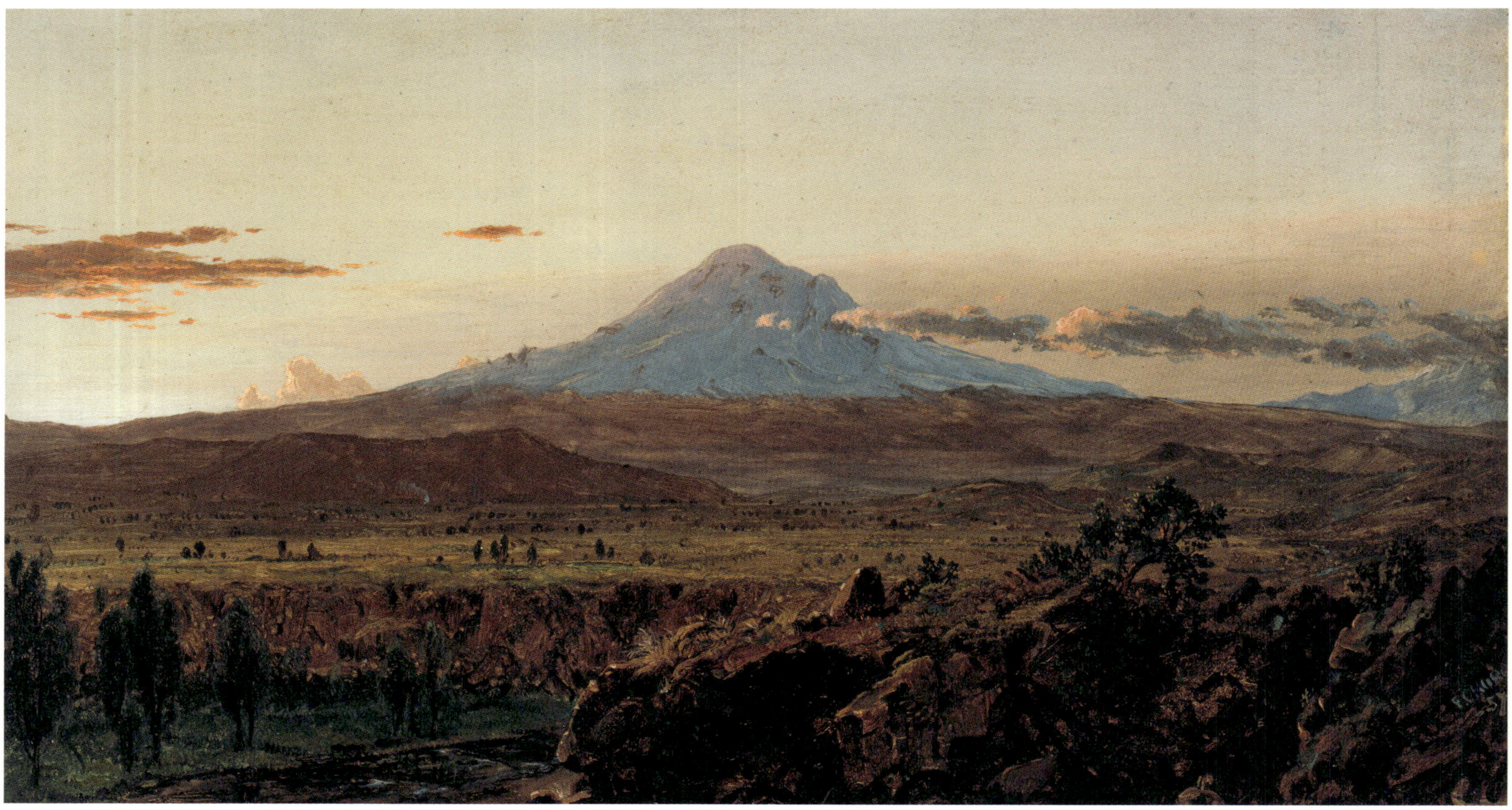

CAT. 100 Frederic Church, *Study for "The Heart of the Andes,"* 1858, oil on canvas, 10 ¼ × 18 ¼ in., Olana State Historic Site, New York State Office of Parks, Recreation and Historic Preservation

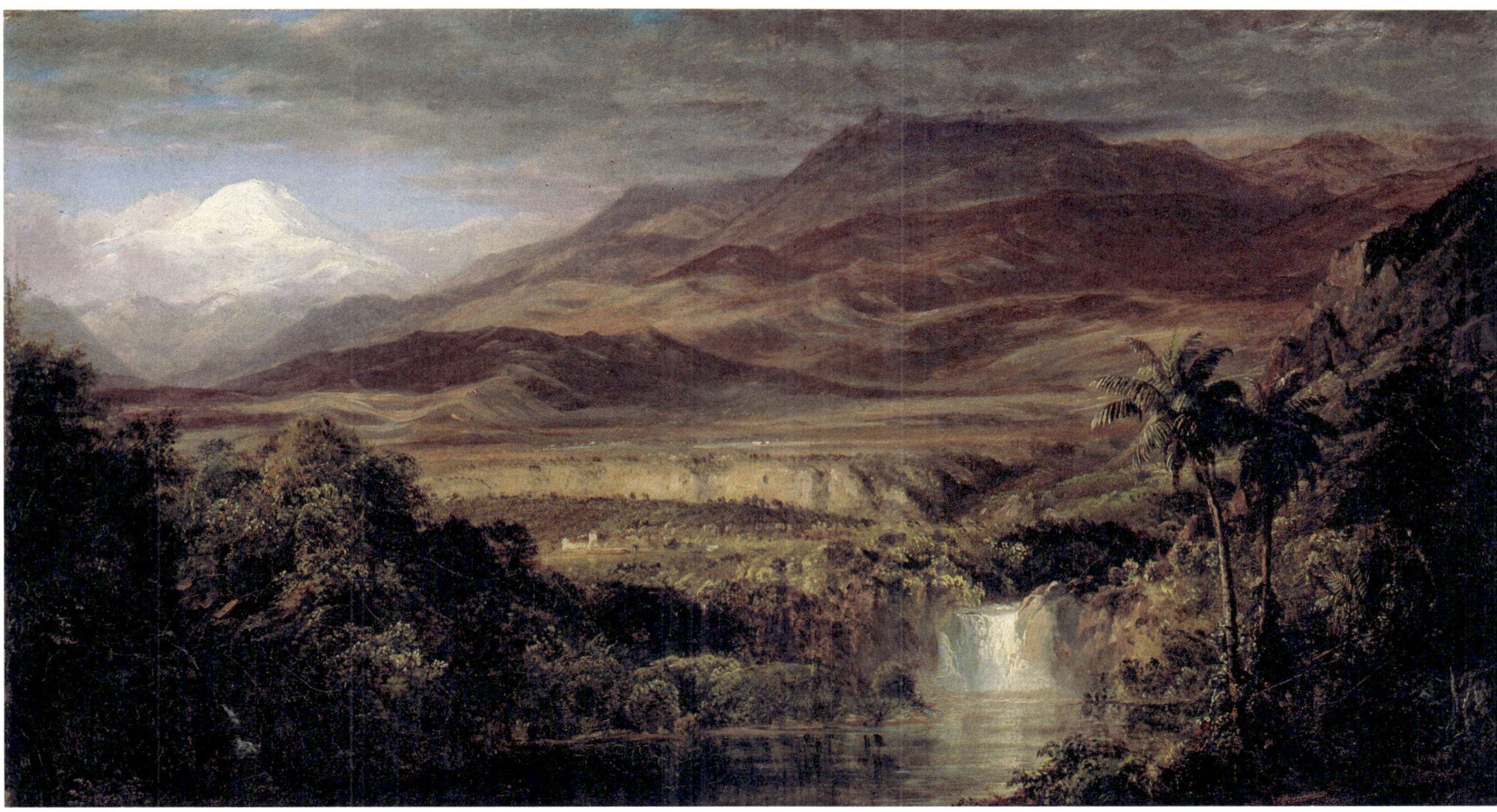

CAT. 101 Alexander von Humboldt, *View of Cayambe in the Region of Quito, Ecuador*, from *Vues des Cordillères et monuments des peuples indigènes de l'Amérique*, plate 42, 1810, color engraving, open: 23 × 34 in., Smithsonian Libraries, Gift of the Burndy Library

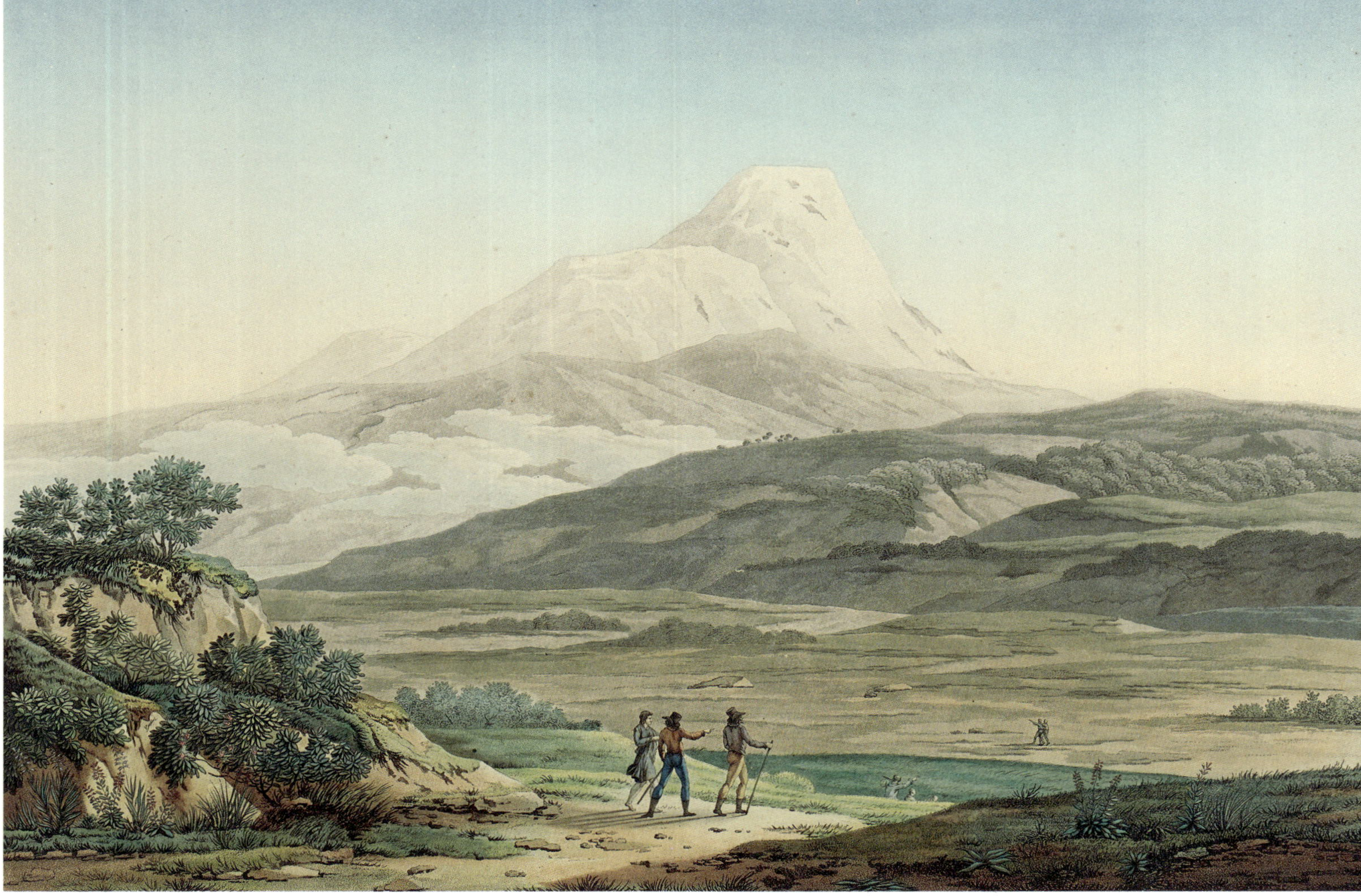

CAT. 102 Frederic Edwin Church, *Cayambe*, 1858, oil on canvas, 30 × 48 ⅛ in., New-York Historical Society, The Robert L. Stuart Collection, the gift of his widow Mrs. Mary Stuart

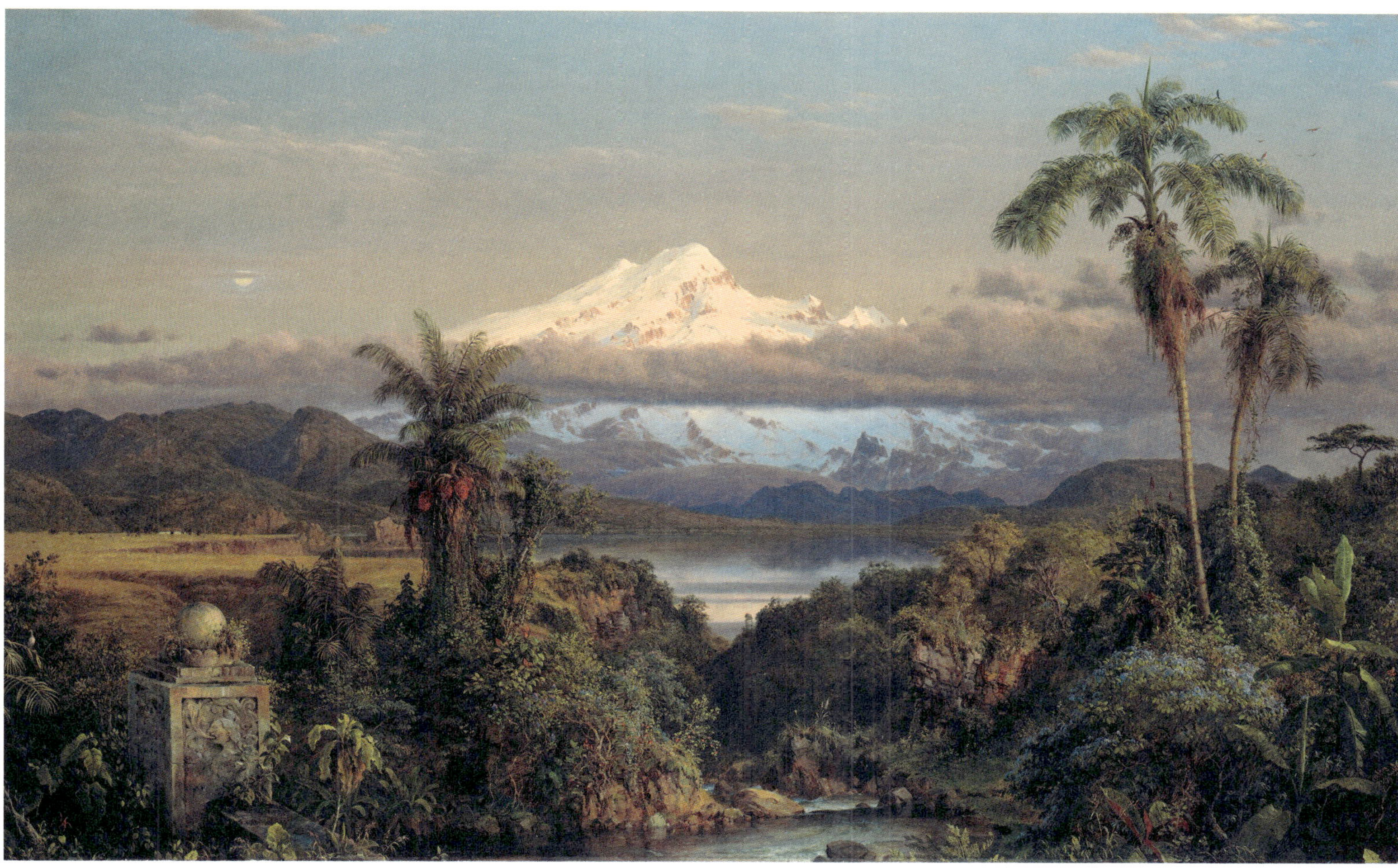

CAT. 103 Frederic Edwin Church, *Mount Chimborazo through Rising Mist and Clouds*, 1857, oil and pencil on paperboard, 13 $\frac{9}{16}$ × 21 $\frac{1}{8}$ in., Cooper Hewitt, Smithsonian Design Museum, Gift of Louis P. Church, 1917-4-824

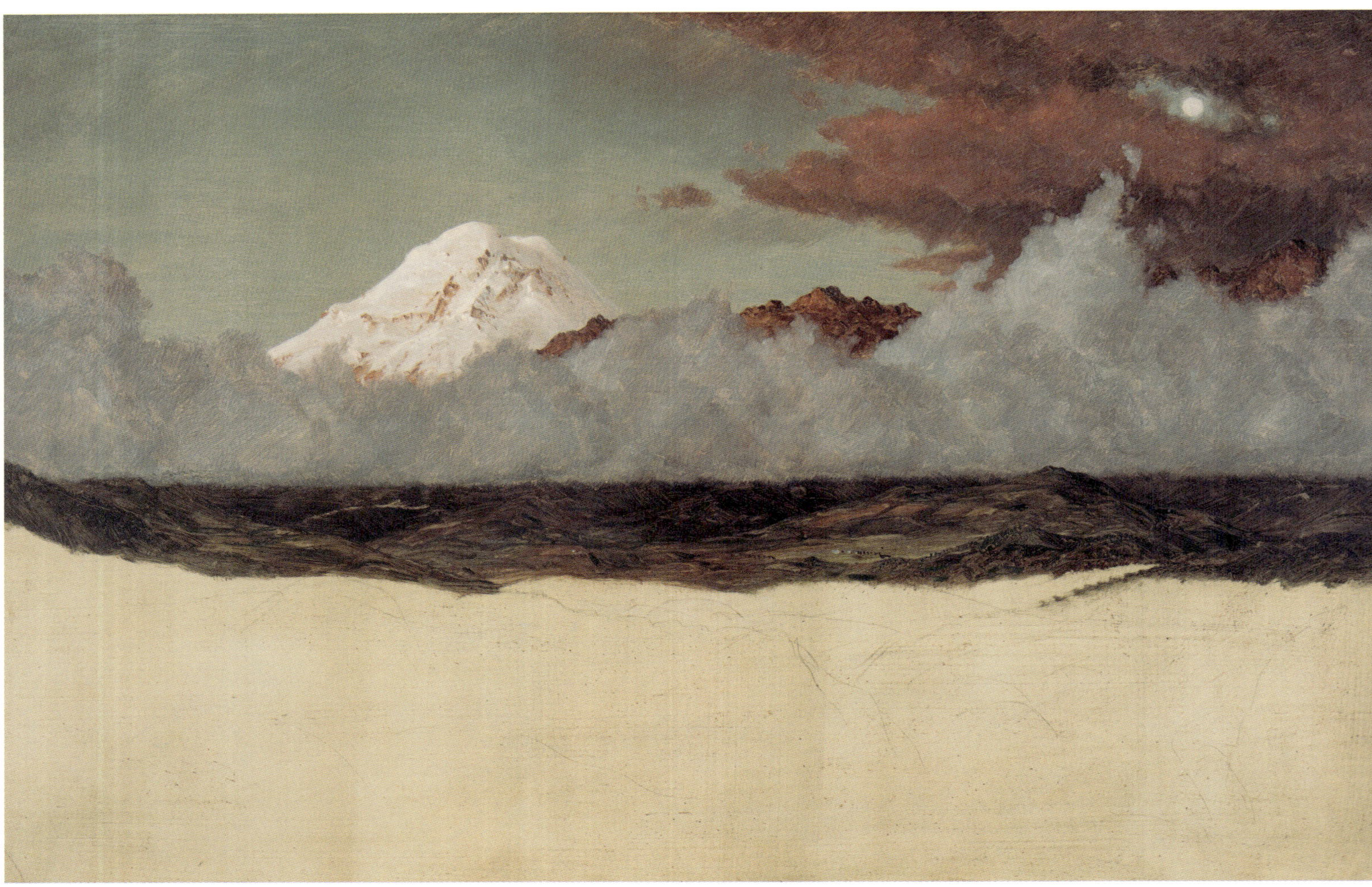

confidence. The smile playing about his lips suggests the sitter's serene happiness in the Andean environment, a man at home in his surroundings. Church brought the portrait of Humboldt home with him and eventually hung it at Olana, signaling his fond remembrance of this trip and his regard for Humboldt.

Church's oil sketches of Chimborazo reveal the artist's compositional experiments in how to feature the mountain that meant so much to Humboldt. Studies of the dormant volcano's cone made at different times of the day caught the light on the snow and the crevasses that had prevented Humboldt from completing his ascent. At first, Church centered the volcanic dome as the visual anchor of the painted sketch (CAT. 97). Gradually he moved the summit off-center, sketching in the suggestion of multiple dwellings in the unpainted foreground. As he moved the dome farther back behind the foreground ridge, Church painted a rolling agricultural landscape across the foreground (CAT. 98). In a small finished sketch, Church took the measure of Chimborazo at sunset, the light glowing across the Andean landscape (CAT. 99), offering a glimpse into Church's first thoughts about composing a larger painting. His final preliminary oil sketch (CAT. 100) sets the composition for *Heart of the Andes* under a lowering sky reflective of the changeable weather he had experienced on both trips to South America. Palm trees in the foreground eventually gave way to massive oaks as Church redefined the botanical parameters of his vista.

Chimborazo was not the sole focus of this trip. Five years earlier Church and Field had passed by the most active volcano in the range, Sangay, but had not stopped to spend much time there. Now Church and Mignot made a five-day detour with the hope of watching it erupt. Church's fragmentary diary from this trip focuses on this episode and the drama of watching it perform.[56] The two also spent time admiring Cayambe, another extinct volcano that Humboldt considered one of the most beautiful volcanoes in the Andes. Humboldt had published a full color engraving of Cayambe in his *Relation historique, atlas pittoresque* (CAT. 101), the same volume that included his view of Chimborazo. As Humboldt noted, Cayambe

> *never ceases to excite admiration at sunset, when the volcano of Guagua-Pichincha, situate to the west, or toward the Pacific Ocean, throws its shadow over the vast plain, which forms the foreground of the landscape.... We may consider this colossal mountain as one of those eternal monuments, by which nature has marked the great divisions of the terrestrial Globe.*[57]

On June 24, 1857, Church had taken careful note of the time of day, atmospheric conditions, and color observations to aid his memory. On his drawing, delicately rendered in pencil with a thin layer of gouache to give substance to the perpetual snow on the summit, Church wrote, "Cayambe, Morning, from the Temple of the Sun (June 24, 1857). snow...lying on the summits of the lesser mountains and paramos [high plateaus]... beautiful bluish haze." The haunting electric-blue shadows on the snow distinguish this work, although Church adapted the time of day to resemble evening when he made Cayambe the subject of one of the first paintings from this second trip (CAT. 102).

Cayambe represents Church at an inventive moment, blending the actual profile of the mountain with elements drawn from an oil sketch of a cloud bank below Chimborazo (CAT. 103). The foreground llano with its lake and small waterfall appears to have been drawn from the artist's imagination or another place he had seen along the way. More puzzling is Church's inclusion of what appears to be a Mexican stele in the lower left foreground, an artifact of an earlier civilization, but not one native to the Andes.[58] It is a reminder that Church was less interested in ethnological and archaeological purity, and more focused on painting a work that would please his patron. In this case, Robert L. Stuart, a New York sugar merchant who maintained a lifelong interest in Central and South America, commissioned *Cayambe*. Stuart had long admired Humboldt's writings and maintained a longstanding interest in pre-Columbian civilizations.[59] Stuart's library, which he made available to friends, boasted several of the deluxe folio volumes with their maps and hand-colored engravings that Humboldt had bankrupted himself producing, including seven volumes of Humboldt's *Personal Narrative,* and his *Essai sur la géographie des plantes* with its *Naturgemälde* map.[60]

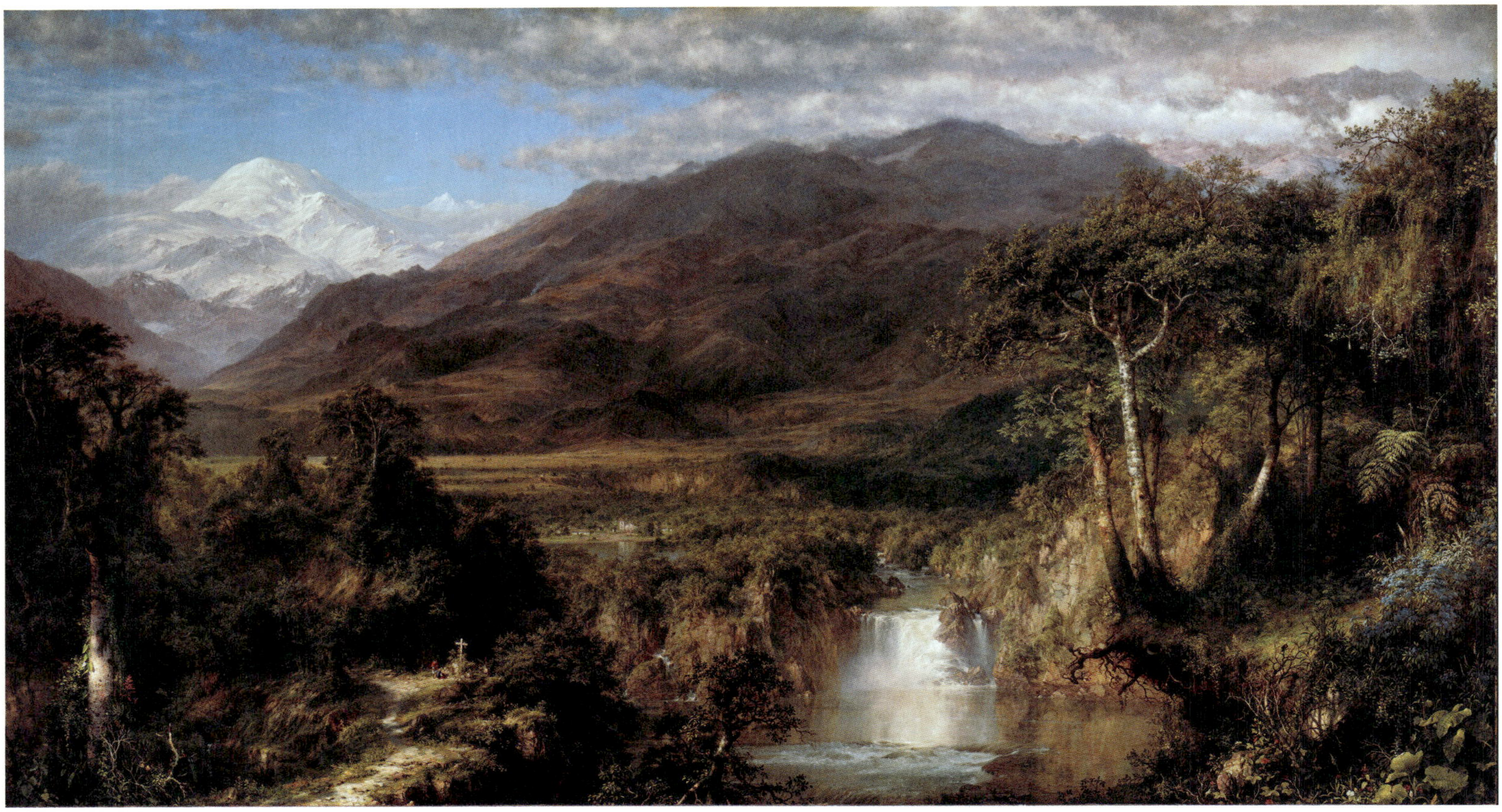

Fig. 90
Frederic Edwin Church, *Heart of the Andes*, 1859, oil on canvas, 66 ⅛ × 120 3/16 in., Metropolitan Museum of Art, NY, Bequest of Margaret E. Dows, 1909

HEART OF THE ANDES AND HUMBOLDT'S *NATURGEMÄLDE*

When *Niagara* made its debut in 1857, a critic issued a challenge that was also a warning, that Church "must go on to greater and still greater attainments, or meet this picture in condemnation."[61] Upon his return to New York, Church let it be known publicly that he intended to outdo his earlier efforts and surpass *Niagara* in scale, scope, size, and attendant drama. Unlike *The Falls of Tequendama,* there was no optimal angle from which to compose a painting featuring Chimborazo. Instead Church composed a painting that became a summation of what he had learned from Humboldt.

When Church turned his attention to *Heart of the Andes* (fig. 90), Humboldt was in the forefront of his mind.[62] In his lectures that preceded *Cosmos*, Humboldt sketched what his sister-in-law, Caroline, marveled at as his "entire great *Naturgemälde*."[62] If *Cosmos* was the summation of Humboldt's life, the *Naturgemälde* was where everything had begun. In that Plant Geography Map, Humboldt had compressed a staggering amount of information onto a single sheet of paper (see cats. 15–16). Chimborazo was the standard against which Humboldt calibrated all of the corollary data collected on European mountains. The whole was a picture far greater than its fragmentary parts: what Humboldt had visualized was an argument that altitude and geology on each continent could predict the appearance of particular families of plants. Humboldt's Picture of Nature was a half-verbal, half-pictorial essay on what he had observed climbing from the tropical lowlands to 19,600 feet above sea level.

Views of the Cordilleras had provided the descriptions of the rugged Andean range and an analysis of the underlying geology that created high-altitude valleys between volcanic peaks. Humboldt had amassed vast quantities of plant and rock specimens, written descriptions, and atmospheric measurements on his trip, noting details as exact as where he ceased to see insect life as he ascended Chimborazo's snowy slopes, and the altitude at which mosses and lichens gave way to bare rock. Church made extensive use of sketches he had made on his two trips to compose the picture, responding to Humboldt's observation that

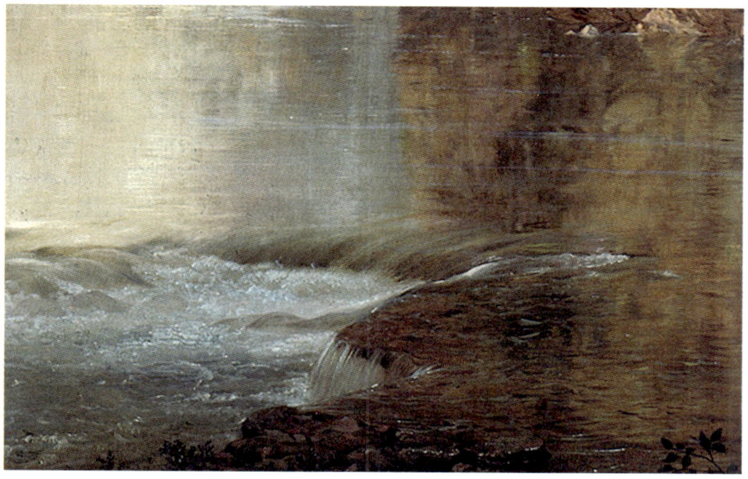

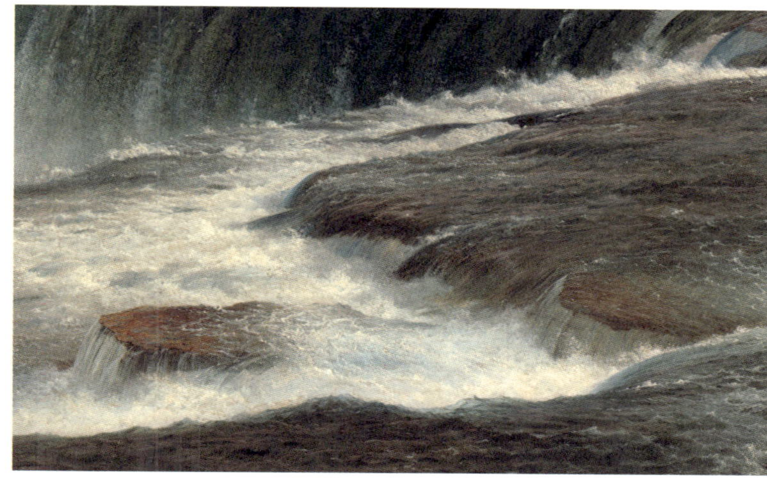

Frederic Edwin Church, *Heart of the Andes*, 1859 (detail of fig. 90, opposite)

Frederic Edwin Church, *Niagara*, 1857 (detail of cat. 94, p. 340–41)

> *Colored sketches, taken directly from nature, are the only means by which the artist, on his return, may reproduce the character of distant regions in the more elaborately finished pictures; and this object will be the more fully attained where the painter has, at the same time, drawn or painted directly from nature a large number of separate studies of the foliage of trees; of leafy, flowering, or fruit-bearing stems; of prostrate trunks, overgrown with Pothos and Orchideae.*[63]

The superabundance of plants and birds in the foreground of *Heart of the Andes* paid dual homage to Humboldt's advice to artists and his passion for studying every possible feature in nature.

Heart of the Andes was Church's extrapolation of Humboldt's *Naturgemälde*. On one six-foot-by-ten-foot canvas, the artist covered the same range of territory, incorporating the same density of detail, and invoking a similar acuity of observation. Critic Henry Tuckerman had admired Humboldt's ability to absorb massive quantities of information and then synthesize it into a comprehensible idea. He noted that Humboldt's "great distinction is the comprehensive view he takes of the laws and facts of the physical world. No naturalist ever so united minute observation with the ability to generalize. The smallest trait of material form or action did not evade his curious eye; and the grandest hypothesis could not subdue his intelligent soul." Tuckerman went on to praise how Humboldt "intuitively recognized the unity of nature, and understood the relative worth of details far better than those who were satisfied with grasping them in an isolated way."[64] Tuckerman recognized that Church's methodical approach to painting *Heart of the Andes* invoked Humboldt's philosophical system.

Heart of the Andes accounts for plants that Humboldt had collected and chronicled from the near-tropical equatorial valley floor up to the arctic summit, a span of more than ten thousand vertical feet. Church added the third dimension to Humboldt's flattened template, allowing the visitor's eye to float over the foreground and into the vast middle distance, across the broad llano and up the rumpled sides of the mountain chain. The recognizable summit of Chimborazo, instead of being the focal point, now became the destination, just out of reach, the most distant point in the landscape. This was not a peak to be summited so much as the capstone to a world to be understood. Church, like Humboldt, combined observations from different geographic sources. He repeated numerous specific plants that had populated the foreground of *The Falls of Tequendama*, a reminder that his first monumental-scale South American landscape had functioned as his earliest deliberate attempt to put Humboldt's ideas into a single painting.

In *Heart of the Andes,* the artist quoted from another of his own works, *Niagara*. In the center foreground, the South American river makes a sharp bend to the viewer's left (detail, fig. 90), and over the lip of a protruding shelf of rock, the water that falls is a direct echo of the same formation in the foreground of Church's *Niagara* (detail, cat. 94). Church's oeuvre, like Humboldt's publications, became interconnected over time, bringing insights from one place to bear on another, each work indebted to the last. Church had completed *Niagara* in a burst of activity lasting two months; *Heart of the Andes* required more than a full year.[65]

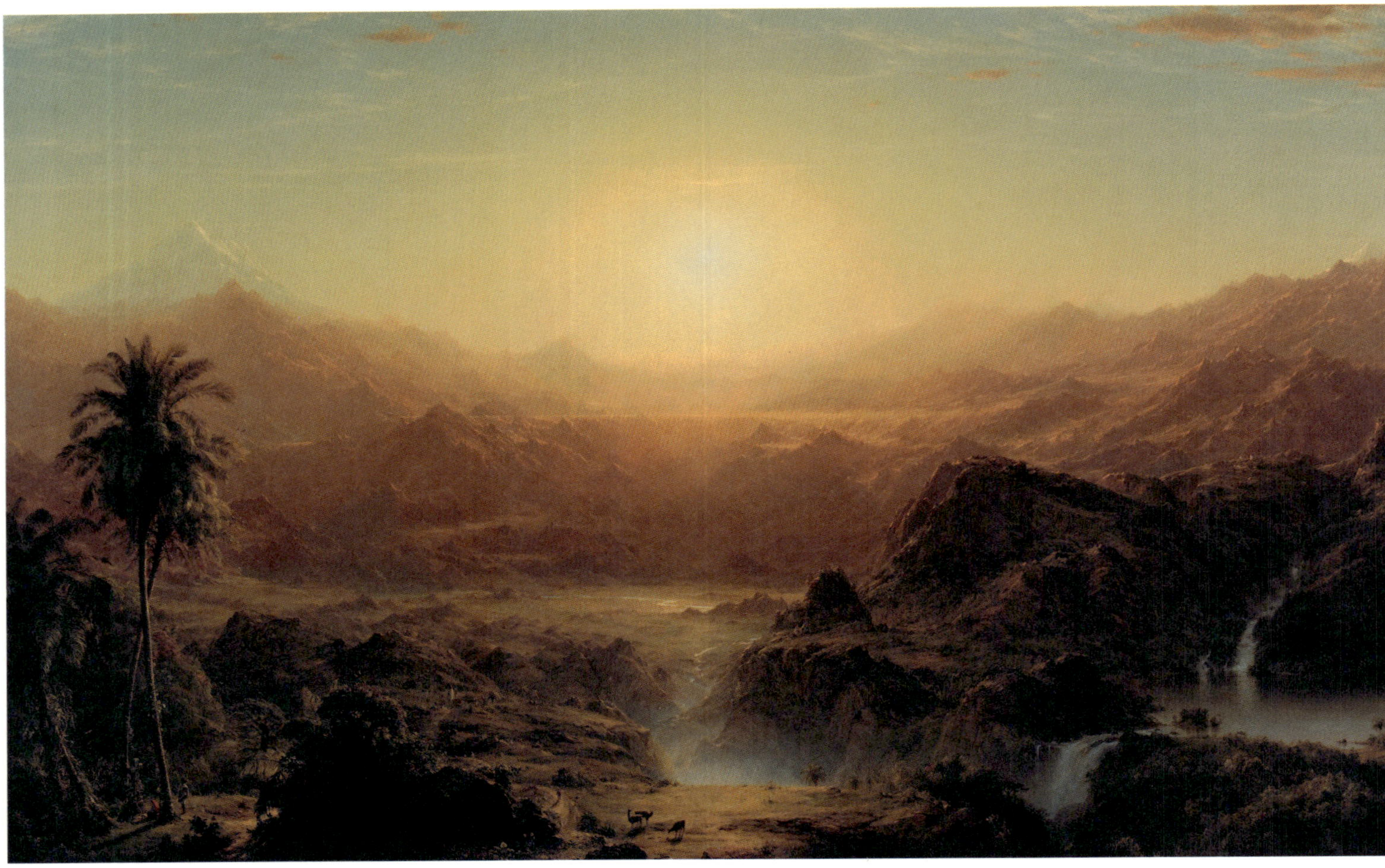

Fig. 91
Frederic Edwin Church, *The Andes of Ecuador*, 1855, oil on canvas, 48 × 76 ½ in., Reynolda House Museum of American Art, Original Purchase Fund from Mary Reynolds Babcock Foundation, Z. Smith Reynolds Foundation, ARCA, and Anne Cannon Forsyth

There was nothing accidental or inadvertent about the way Church composed *Heart of the Andes*. He intended this painting to be, like Humboldt's *Cosmos*, a singular work that relied on a lifetime's accumulated experiences and observations brought together to form a larger expression of the capacity for the natural world to be overwhelming and satisfying all at once. In Humboldt's third volume of *Cosmos*, published before Church had taken his second trip to South America, the traveler had stated his case even more emphatically:

> *I have…considered Nature in a two-fold point of view. In the first place, I have endeavored to present her in the pure objectiveness of external phenomena; and, secondly, as the reflection of the image impressed by the senses upon the inner man, that is, upon his ideas and feelings.*[66]

With *Heart of the Andes*, Church strove to achieve the synthesis of Humboldt's dual objectives. Church sketched in pencil and in oil, both on-site and back in his studio, developing a body of work that provided him with the armature for his composition. *Heart of the Andes* required Church to draw on the totality of his accumulated sketches from nature: whether they were made in New England, at Niagara Falls, or in South America. Close observation informed his method, and Humboldt's quest to find the common threads tying together all of creation inspired the means by which he created the largest and arguably most synthetic composition of his career.

It is worth pausing to understand what Church intended with this painting and how he constructed it. First consider how this painting differs from the paintings Church produced following his first trip to South America. The work that stood out to the art critics of the day was *The Andes of Ecuador*, extolled for the warm humid light that suffuses the landscape (fig. 91). That quality of saturation, of landforms bathed in a harmonious glow, appealed to North American viewers, as it presented a landscape that was both lovely and unfamiliar. Although it seems to be the visual evocation of the unity of nature promised by Humboldt, it is in fact more similar to Humboldt's lectures, riffing on a singular idea, portraying the sense of a place in its ideal form rather than the concrete reality of the place itself. The lyrical qualities of *The Andes of Ecuador* thus stand in pronounced contrast to the crispness of each individual plant and landform in *Heart of the Andes*. If *The Andes of Ecuador* captures the mesmerizing quality of Humboldt's lectures, *Heart of the Andes* wrestles with the impossibility of getting every piece of information into a single book. The progression from *The Andes of Ecuador* to *Heart of the Andes* may be read in the titles, as Church moved from a generalized and distilled overview to portraying the overwhelming amount of specific information found in his sketches. In doing so, he followed Humboldt's own progression, visible in his cutaway landscape maps, moving from the accumulation of raw data to a synthesis of its meaning, and from there, diving back in to arrive at an even more nuanced and more complete understanding of natural law. If *The Andes of Ecuador* visualizes the moment of synthesis, *Heart of the Andes* revels in the restlessness associated with endless discovery of new data.

In the latter painting, Church's reach, like Humboldt's, exceeds his grasp. But neither man ceased striving for that monumental achievement. Humboldt's public lectures beginning in 1827 sketched the framework for *Cosmos*, but in lecture form, Humboldt was able to hold in his mind the totality of his argument, bringing selective evidence to bear on the distillation of his ideas. When forced to put pen to paper, he found it far more difficult to organize the vast quantities of data he had amassed and absorbed over the years. His travel journals bore annotations accrued over several decades; he cannibalized letters and notes and sorted them by topic in envelopes bearing single words like "mountains" or "slavery." These aggregations of evidence complicated his arguments that rang most clearly in distilled form—whether it was a single line in a letter to a friend or the free flow of an extemporaneous lecture. Humboldt would spend the better part of fourteen years organizing data stored in his head and scattered through thousands of pages of notes. As a result, reading *Cosmos* is like reading well-organized stream of consciousness, with one brilliant, coruscating idea segueing into another, the trajectory loosely structured but prone to diversions and byways en route to the next intersection of evidence and ideas. Like Humboldt's books, Church's painting is a massive, overwhelming, exuberant, impassioned compendium of all that Church had learned from his absent mentor.

Church orchestrated the display of *Heart of the Andes* at Lyrique Hall on Broadway beginning April 27, 1859. The colossal painting appeared even larger thanks to an elaborate frame designed to resemble a window casement; parted draperies intensified that illusion. That frame serves mostly as a reminder that what we see is still but a fragment of the whole, a synecdoche perhaps of all of South America. *Heart of the Andes* presents the essence of the artist's metaphorical journey of scientific understanding and the revelations accruing from it, given back as a singular summation rather than a poetic synthesis. Compositionally it barely contains the superabundance of data drawn from two trips to South America, and it incorporates close to a decade of intense study of Humboldt's writings.

Church soon moved the painting into the gallery of the Tenth Street Studio Building, where it was lit by gas jets concealed behind silver reflectors in a darkened space, enhancing the drama even further. Visitors were encouraged to bring opera glasses in order to take in the fine degree of detail in the oversized painting.[67] He must have been pleased with the pronouncement made by one critic, who wrote:

> *It is seldom the fortune of any artist to achieve a triumph so complete as that accorded to Mr. Church for his fine picture of* The Heart of the Andes. *Some twelve months ago we were told that he had commenced a composition picture, designed to illustrate the general idea he received of tropical mountain scenery, and it was further stated, that he intended to surpass his* picture *of Niagara. We deemed the idea a chimera of an overenthusiastic artist, but Mr. Church has accomplished his determinate prophecy, and the picture of Niagara has unquestionably paled before the more glowing and sentient merits of the piled up mountains of our Southern Continent. A moment's absorbed examination, and apparently the* real scene *is before you—the medium of its expression disappears.*[68]

Humboldt's influence was widely understood and appreciated, as fellow artist Robert Walter Weir noted in a letter published in the *New York Post* pronouncing,

Humboldt has given us the word-painting of the magnificent scene of this picture, and now comes the true magician, with his harp of a thousand strings, to present the subtle tones that no words can describe.[69]

The public response to *Heart of the Andes* was as overwhelming as the experience of viewing the painting itself.[70] Lines formed along lower Broadway as more than five hundred people per day stood, sometimes for hours, to see this singular artistic achievement.[71] Among those who came were a number of Humboldt's American friends, notably Edward Everett and Washington Irving. Irving, the elder statesman of American literature, arrived the last day of the exhibition, the crowds parting to allow the frail man to approach Church's painting, which he pronounced "glorious—magnificent!"[72] Collector and fellow Humboldt enthusiast William T. Blodgett had arranged to purchase the painting for the unheard of sum of $10,000, with the understanding that he could not take possession of the painting until it had completed its two-year, international tour.[73]

No one doubted the power behind Church's pictorial accomplishment, but the sheer volume of information coupled with the lack of a single, fixed focal point made viewing it as much a challenge as visiting the actual volcano. Church likely encouraged two of his close friends and supporters as they each prepared a lengthy essay intended to elucidate the highlights of the massive painting. These were travelogues, designed to take the reader on a virtual journey through Church's painting the same way Humboldt's prose conveyed Church and Field through the actual South American landscape. Louis Legrand Noble had been Thomas Cole's pastor. Now he deployed his considerable oratory in a book-length disquisition titled *Church's Painting, The Heart of the Andes*. Theodore Winthrop had been one of John C. Frémont's most fervent campaigners in 1856, traveling extensively from Pennsylvania to Maine giving stump speeches on Frémont's behalf.[74] Winthrop's *A Companion to The Heart of the Andes* and Noble's essay approached the painting from opposite starting points. Noble worked the viewer from foreground to summit, beginning with physical geological forces, while Winthrop arranged his tour in ten sections, starting with the summit and descending to the foreground. Winthrop's divisions, reminiscent of Humboldt's *Aspects of Nature*, gave visual coherency to each aspect of the painting, rather than trying to treat it as a seamless achievement. His section headings were: The Sky; The Snow Dome; The Llano, or central plain; The Cordillera; The Clouds, their shadows and the atmosphere; The Hamlet; The Montana or central forest; The Cataract and its Basin; The Glade on the right foreground; and The Road and left foreground.[75]

The painting does not promise visual coherence. Instead it holds out the rewards of close looking and aesthetic and scientific inquiry, and the revelation that the mountain is the apex of an ecosystem that extended from the Amazon River basin near the equator, across the llano, up through the foothills, and ultimately among the peaks of the Andean Cordillera. That vertical progression from a tropical foreground up to an arctic summit accomplished another of Humboldt's goals, that of extrapolating the changes in climate and habitats if one were to travel from the equator to the poles, moving laterally from a tropical environment to a polar one. Whether working in a latitudinal or altitudinal vector, the results were similar: plants and climate were linked together, their appearance and location predictable by type as one moved around the globe and compared similar landforms.

The brilliance of Humboldt's epiphany was that he could correlate information from various mountain chains on multiple continents and use that comparative data to develop higher laws of nature. This approach eventually animated multiple disciplines that branched out to become all of the physical sciences. Church's painting captures that overwhelming moment of understanding, that without massive quantities of data, without as nearly complete an understanding of a place as possible, all we are capable of comprehending would be isolated fragments. Humboldt's Plant Geography Map, published in 1805, laid the groundwork for the Prussian explorer's ideas, and forty years later acted as a template for *Cosmos*.

The reviews of *Heart of the Andes* resembled those of Humboldt's books. Audiences marveled at the ambition and the command of the subject matter but often came away overwhelmed by the experience. Mark Twain paid three separate visits to the painting in March 1861 when it was on view in St. Louis. He articulated the difficulty in taking in the enormous size of Church's painting, combining admiration with a mental gasping for breath that seemed appropriate to the high altitude of the Andean chain. He advised bringing opera glasses, the better to marvel at the artist's accomplishment.

> *There is no slurring of perspective effect about it—the most distant—the minutest object in it has a marked and distinct* personality*—so that you may count the very leaves on the trees.*

That lack of a single focal point contributed to Twain's fatigue, as the viewer's eyes restlessly canvass the scene looking for a place to alight. Twain continued, "your third visit will find your brain *gasping* and straining with futile efforts to take all the wonder in...and understand how such a miracle could have been conceived and executed by human brain and human hands."[76]

Humboldt had ruefully admitted the impossibility of encapsulating all of South America in a single, linear narrative, a perspective that applies equally to Church's *Heart of the Andes*:

> *It is a difficult task, to describe those sensations, which act with so much the more force as they have something undefined, produced by the immensity of the space as well as by the greatness, the novelty, and the multitude of the objects, amidst which we find ourselves transported. When a traveller attempts to furnish descriptions of the loftiest summits of the globe, the cataracts of the great rivers, the tortuous vallies [*sic*] of the Andes, he is exposed to the danger of fatiguing his readers by the monotonous expression of his admiration.*[77]

Church encountered a similar dilemma in his own work: the fragments he accrued through his sketches, diary entries, and letters, when brought together in the service of a single Humboldtian homage, threatened to overwhelm rather than enlighten his viewer. The reviewer for *The Crayon* expressed this as a shortcoming, noting that the artist had sacrificed pictorial unity and repose. Instead, "every square inch of the canvas [is] covered with nature's statistics."[78] This, the reviewer concluded, meant that instead of being absorbed by the scene, the viewer's mind was "employed upon an infinity of details."[79] Church's painting provides a visual simulacrum to the experience of dipping in and out of Humboldt's oeuvre, coming up for air and diving in once again during successive encounters with the painting.

As Church painted this massive work, he was also planning to place it in front of his distant mentor as an expression of how deeply the American artist had absorbed and could reflect Humboldt's philosophy of nature. To accomplish this he worked with his dealer, John McClure, who was making arrangements to crate the painting and ship it to Berlin. Bayard Taylor, who had met and spent time with Humboldt several times, had agreed to write a letter of introduction to present to Humboldt. Church expressed his desire in a letter to Taylor, as his

> *principal motive in taking the picture to Berlin is to have the satisfaction of placing before Humboldt a transcript of the scenery which delighted his eyes sixty years ago—and which he has pronounced to be the finest in the world.*[80]

Taylor then wrote a letter in German to the aging explorer, expressing his hope that were Humboldt to

> *recognize its truthfulness, and take delight in it, the painter will count himself fortunate, for he feels that you have shown him the way he has taken with such great success.*[81]

Taylor sent the letter to Church, along with a cover note, cautioning him to check on Humboldt's whereabouts. But Humboldt had died on May 6, 1859, a week after *Heart of the Andes* first went on public view in New York. When Church heard the news, he wrote of it to Taylor, "It touched me as if I had lost a friend."[82] Newspapers noted Church's aspirations to present the work to Humboldt and were encouraged to mourn with Church the loss of that opportunity. Plans to ship the painting to Berlin were scrapped in favor of a different public tour, which certainly made the painting and Frederic Edwin Church famous.[83] Church, who may have planned to accompany the painting to Europe, changed his own plans as well.[84]

The *New York Times* anointed Church the "artistic Humboldt of the new world."[85] What did it mean to so fully absorb a Humboldtian ethos that it consumed an artist's way of working and permeated his subject matter, regardless of where he traveled? Like Humboldt, Church sought out more distant and challenging subject matter. He

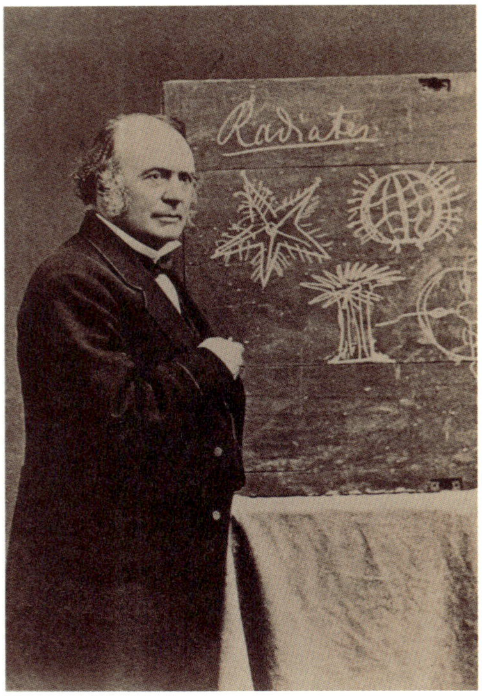

Fig. 92
Pach Brothers Studio, *Louis Agassiz*, ca. 1850, cabinet card, 6 ½ × 4 ⅓ in., Harvard University Archives

followed up *Heart of the Andes* with a trip to Labrador, where having read extensively ahead of time about the properties of ice, refraction, and water, he was able to draw and paint sketches of icebergs while fending off persistent seasickness.[86] In doing so, Church himself moved from the vertical climb up Chimborazo to the lateral voyage to Labrador. His treks from the equator to perpetual ice fulfilled Humboldt's desire to observe the changes in the landscape, recording the characteristic features at both ends of this spectrum. *The Icebergs* (Dallas Museum of Art) would, if anything, present an even greater challenge to Church's viewers, who initially struggled to comprehend the internal scale of the painting while marveling at the effects of compressed ice and glacial erratics, boulders deposited by the movements of ice sheets to locations far from their points of origin.

Each successive painting Church produced seemed calculated to push those boundaries even further, moving from documentation into the delivery of an emotional experience grounded in the landscape, as *Cotopaxi* (Detroit Institute of Arts) belched forth hot ash that occluded the sun, and *Rainy Season in the Tropics* (Fine Arts Museums of San Francisco) pushed atmospheric phenomena into a science lesson on hope.

HUMBOLDT AND THE CONCORD TRANSCENDENTALISTS

Ralph Waldo Emerson experienced a similarly sustained immersion in Humboldt's ideas. Emerson began writing about Humboldt under Edward Everett's tutelage at Harvard; and by 1821 references to the Prussian naturalist began appearing in his journals. His assertion that he would become a naturalist spoke not just to his growing appreciation of what he could learn from the study of nature, but what that study would do to shape his life and his writing. His influential essay *Nature,* published in 1836, would in turn inspire Henry David Thoreau. The two men became the leading literary exponents for Humboldt's ideas, creating a response to nature in their works that pushed beyond the earlier narrative threads of Cooper and Irving, just as landscape painters like Church had moved beyond expedition sketches to a cultural identity grounded in landscape painting.[87] The epistolary stamina of Humboldt's ideas, spanning the Knickerbocker writers of the 1820s and '30s to the Concord transcendentalists of the 1840s and '50s, resembles the relays that developed in the realms of exploration and the appreciation of the American wilderness.

In August 1858 Emerson found himself in the Adirondack wilderness at Follensby Pond, part of a group affectionately known as the Adirondack Philosophers' Club. There two scientists, two poets, two lawyers, two doctors, a writer, and an artist gathered to hunt, fish, explore the surrounding wilderness, and commune with nature.[88] William James Stillman, the artist of the group, painted *The Philosophers' Camp in the Adirondacks* (CAT. 104) to commemorate the event. Camp Maple, so named for the large trees, is dappled with sunlight that filters through the canopy and provides the setting for what Stillman described as an outdoor symposium, where nature is the locus for higher education.[89]

Although the affectionate name bestowed on the gathering may have arisen from the unlikelihood of finding a learned group of Bostonians camping in the wilderness, it is simultaneously a nod to the study of natural philosophy from which all of the sciences derived.[90] On the left, the celebrated Harvard professor and comparative anatomist Louis Agassiz (fig. 92) dissects a fish, his lab table replaced by a giant tree stump.[91] On the right, a group of the company takes target practice. In the center, standing apart from both groups, is Emerson, rapt in his immersion in nature. Stillman observed of Emerson that he "in his single interview with a true and uncontaminated Nature saw all the relations between her and not merely one individuality, narrow or large, but one humanity."[92] It was a worldview deeply inflected by decades of studying Humboldt's works. Stillman described the Concord philosopher looking for meditative moments: "a pilgrim's staff in hand [he] took a place alone and between the two

CAT. 104 William James Stillman, *The Philosophers' Camp in the Adirondacks*, 1858, oil on canvas, 20 × 30 in., Courtesy William Munroe Special Collections–Concord Free Public Library Corporation

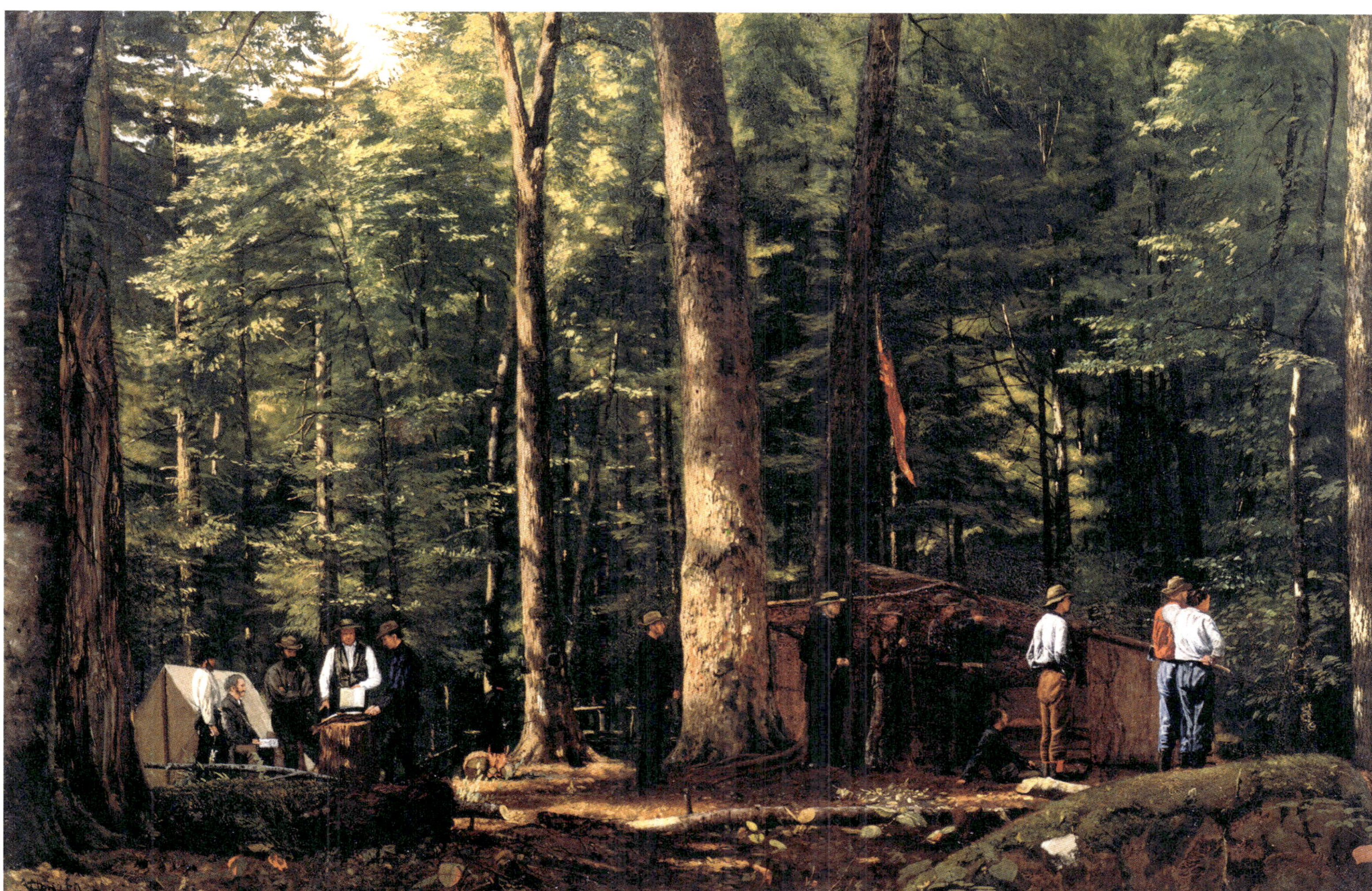

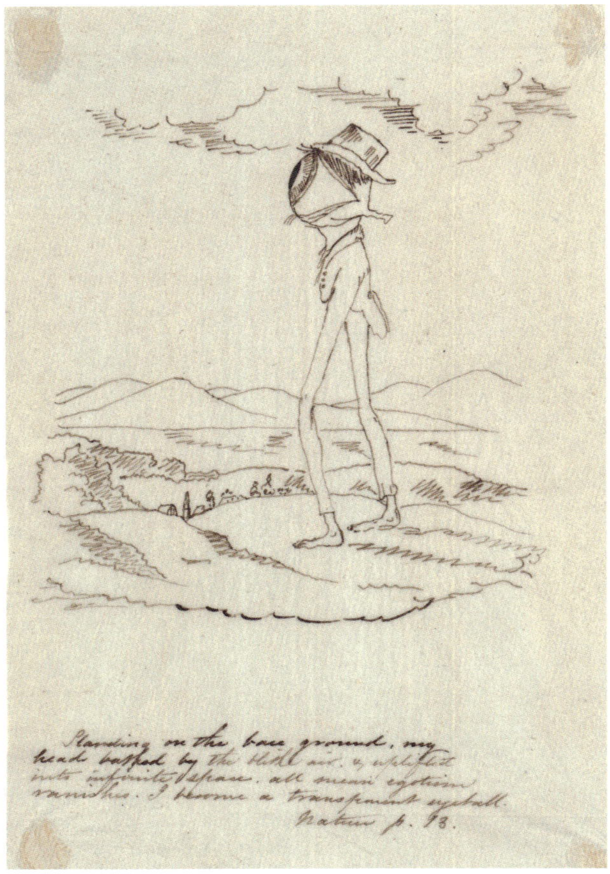

Fig. 93
Christopher Pearse Cranch, *Standing on the Base Ground… I Become a Transparent Eyeball (Illustration for Ralph Waldo Emerson's "Nature")*, 1830–92, pen and ink, 8 3/8 × 5 11/16 in., The Metropolitan Museum of Art, Gift of Whitney Dall, Jr., in Memory of Emily Dall, 1976

groups, with an intentional symbolism of his position in the world."[93] That position, as a poet and philosopher, is at the fulcrum between the arts and sciences, and between looking intently inward and gazing expansively outward.

Emerson admired in Humboldt the vast reach of his ideas. Thoreau, by contrast, turned his acute gaze toward the minute and particular aspects of the world around him. If the aphorism about transcendentalism is that one must look down in order to look up, Thoreau and Emerson epitomized the two sides of that duality. Humboldt embodied both aspects of it. In a letter to his friend Marc-Auguste Pictet in 1805, Humboldt sketched his plan for the multivolume *Personal Narrative*, already envisioning eleven volumes. Several would appeal to the public eagerness for an engaging travel narrative, while others would satisfy Humboldt's preference for a "comprehensive view of Nature." But he singled out the proposed *Essay on the Geography of Plants* as a volume devoted to a specific topic in which he could dive deep in an effort to prove the larger truth of his unity of nature. As Humboldt explained, "The world likes to see, and I there exhibit a microcosm in a leaf."[94] In *Cosmos* he expanded on this duality, understanding that in science as well as the fine arts, one needed massive quantities of data in order to extrapolate a larger truth, but that there was also merit in the suppression of detail in order for the "great masses" to be visible.

Emerson's *Nature* adopted as its premise that "the whole of nature is a metaphor of the human mind."[95] It was an ethos that perfectly summed up the American predisposition to see the nation's cultural values and individual identity bound up in the features of the landscape. Emerson's visual paradigm for this immersive experience was the transparent eyeball, capable of absorbing all that nature has to offer. Christopher Pearse Cranch's playful cartoon (fig. 93) captures that poetic merging of eye and brain as vision becomes the faculty of cosmic understanding. For Emerson, looking up and seeing far was his visual paradigm for enlightenment in nature. In 1843 he returned to this metaphor in describing Humboldt as being "magnificent, too, as a distributing eye. His glance is stratification; geography of plants, etc."[96] In *Essay on the Geography of Plants*, Humboldt's consistent use of botany to understand how the individual plant, scrutinized up close, could be a window to a cosmic understanding of planetary forces remained a revelation for Emerson, as it was for Church and Thoreau.

Conversely, Humboldt inspired Thoreau (fig. 94) to examine the world through the other end of the telescope. By 1842 Thoreau, who was at that time living with Emerson and his family, also began citing Humboldt, embracing his insatiable desire for detail based on close observation. In 1842 he invoked Humboldt in his essay "A Walk in Wachusett" and read Humboldt's *Cosmos* in English as soon as he could borrow Emerson's unofficial translation.[97] The same year, Emerson encouraged Thoreau to publish his "Natural History of Massachusetts," a work that responded both to Humboldt and to Jefferson.[98] "A Walk to Wachusett," part of Thoreau's *Week on the Concord and Merrimac Rivers,* opens with the lines,

> *Concord, July, 19, 1842. Summer and winter our eyes had rested on the dim outline of the mountains, to which distance and indistinctness lent a grandeur not their own, so that they served equally to interpret all the allusions of poets and travellers; whether with Homer, on a spring*

morning, we sat down on the many-peaked Olympus, or, with Virgil, and his compeers, roamed the Etrurian and Thessalian hills, or with Humboldt measured the more modern Andes and Teneriffe. Thus we spoke our mind to them, standing on the Concord cliffs.[99]

Instruments were just as important to Thoreau as they had been to Humboldt. Thoreau was trained as a land surveyor, taking measurements and recording changes with the same obsessiveness as Humboldt, fretting over the accuracy of his equipment as the explorer had, and delving into terrestrial magnetism, one of Humboldt's favorite topics.

Thoreau's notebooks, journals, and maps carried extensive annotations, some of them in keeping with his surveyor's training and his obsessive need for minute detail. Other notations convey the sensory appreciation of nature, the quality of light, or the texture of sand and grass. He kept detailed notes of which plants grew where and treated his local surveying tasks as a kind of expedition in miniature. When it was published in 1845, Thoreau read Frémont's 1842 expedition report, which contained the explorer's Humboldtian narrative of braving inclement weather in pursuit of measurements and naming landscape features for the Prussian traveler and his French botanist.[100] Thoreau spent his life learning the Concord River, developing such a loving obsession for it that Emerson noted with annoyance that his friend "occupies himself with the history of the river, measures it, weighs it, strains it through a colander to all eternity."[101]

Thoreau was building to a work that would rival Church's *Heart of The Andes* and Humboldt's accrued data for its visual and statistical density. *Walden* turns Humboldt's cosmic telescope around, aiming it like a microscope on the perimeter of Walden Pond. Thoreau absorbed all of Humboldt's charts, graphs, tables of data, and accumulated observations, and dedicated himself to making a humble Massachusetts pond a *Cosmos* in miniature. When Thoreau looks down into the depths of the water, he sees stars. Grasping this as the essence of transcendentalism, Thoreau wrote, "A lake is the landscape's most beautiful and expressive feature. It is earth's eye; looking into which the beholder measures the depth of his own nature."[102] While Humboldt gravitated to the mountains, Thoreau was drawn to water. His contemplation of the gravel in a streambed launched his celestial meditation, "Time is but the stream I go a-fishing in."[103] He drew his own map of Walden Pond (fig. 95), designed to sound its depths and locate its deepest spot, that visualizes this poetic paradigm. Each notation of depth is a point on a line, collectively resembling the plotting of individual stars and constellations. Together they form their own network, interconnected lines of information and communication resonant of Humboldt's concept of cosmos.[104]

Like Humboldt's *Naturgemälde*, Thoreau's map of Walden Pond embodies multiple points of view designed to make the connections across a landscape visible. The map is not oriented toward magnetic north; the arrow in the lower right corner instead points to the true meridian. The resulting

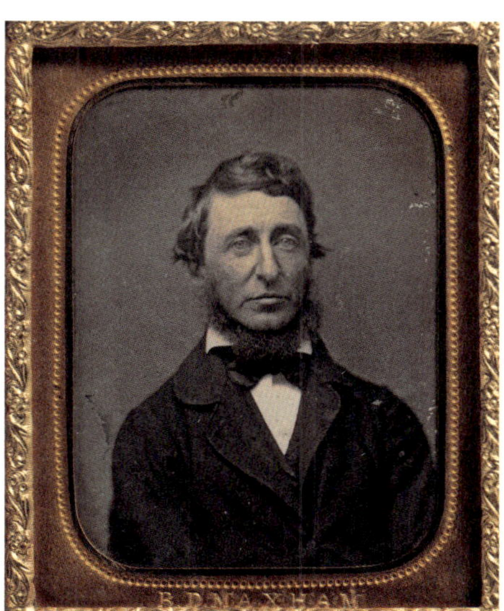

Fig. 94
Benjamin D. Maxham, *Henry David Thoreau* (1817–62), 1856, ninth-plate daguerreotype, 2 ½ × 1 ⅞ in., National Portrait Gallery, Smithsonian Institution; gift of anonymous donor

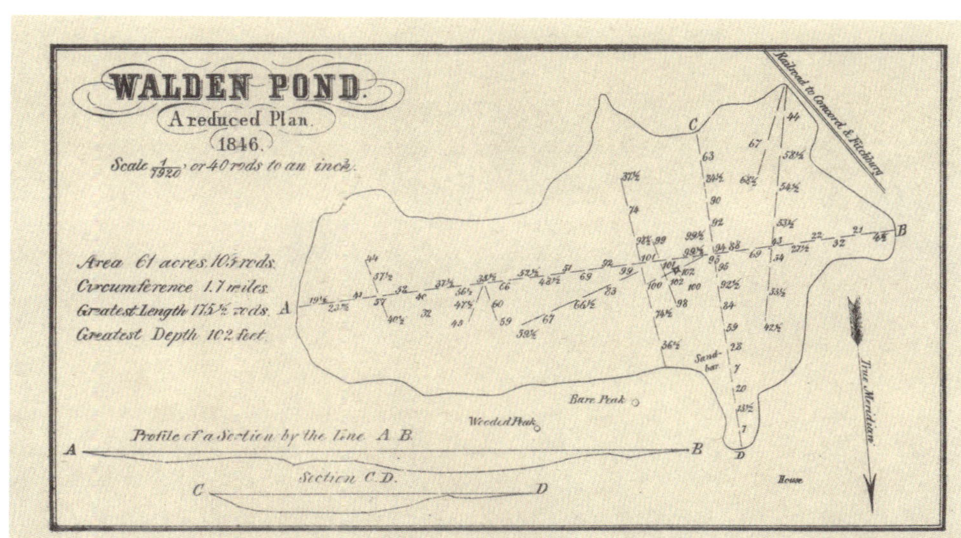

Fig. 95
Henry David Thoreau, *Walden Pond, A Reduced Plan* from *Walden: or, Life in the Woods*, engraving, 3 ½ × 6 ⅓ in., The New York Public Library

shift in perspective presents the contour of Walden Pond as resembling the outline of the United States. In his own words, *Walden* represented what the poet called "a reduced plan," a distillation of Thoreau's years of accumulated information, perceptions, and ideas. As Humboldt had synthesized a life's worth of observations and ideas in providing what he called a physical description of the universe, Thoreau studied every aspect of Walden Pond as his own exercise in understanding the depths and the contours of his personal universe. Taken together, Thoreau's evolving text and his carefully structured map dive deep to reveal his conceptual framework and reflect his own thought process.

The overwhelming statistical information Thoreau accrues is what threatens to make *Walden* impenetrable, as he delineates every inch of his immediate surroundings. Mark Twain's review of *Heart of the Andes* described a viewer overcome, smothered by Church's enumeration of detail on a single canvas; Thoreau's *Walden* is a similarly dense literary experience. Both men understood and reiterated Humboldt's central tenet, that more information makes for a more comprehensive understanding of a subject and fosters the ability to synthesize the experience and find its inner truth. But for Church and Thoreau, that mass of distilled material threatened to behave like a black hole, sucking in everything around it rather than illuminating the visible world through exploration. Viewing *Heart of the Andes* and reading *Walden* underscores the difficulty any other person encountered in attempting to replicate Humboldt's grand achievement in *Cosmos*.

After Louis Agassiz's arrival in Boston in October 1846, Thoreau briefly became part of the scientist's network of collectors, eagerly sending him pressed plants, only to realize after sending him close to 1,500 plants that the naturalist was neither acknowledging nor returning them.[105] When Thoreau stopped collecting for Agassiz, he turned to writing his own botany based on his detailed analysis of his local landscape. As Church was composing *Heart of the Andes* as an index to Humboldt's South American botany, Thoreau climbed Mount Washington in July 1858, making detailed notes of every plant he saw as he ascended and descended, compiling the most thorough list by zones, a list that remained unsurpassed well into the twentieth century.[106] For the last decade of his life he kept a daily "Kalendar," the spelling an acknowledgment of Humboldt's original German title *Kosmos*. In this Thoreau expanded the kind of record-keeping he had made of the Concord River and Mount Washington, applying it to the entirety of his existence. Like Humboldt's work, it was intended to be a summation and cross-referencing of the observations Thoreau had accrued in his diaries. Setting aside time each morning, Thoreau strove to be present in each moment.[107] Like Humboldt's *Naturgemälde*, Thoreau's Kalendar incorporated tables of comparative statistics; unlike Humboldt, Thoreau's observations remained resolutely local.

Thoreau did not need to seek out a remote wilderness. He treated local travel as if it were an expedition. *Walden* was Thoreau's embodiment of *Cosmos*—and like Humboldt's book, it was simultaneously a manuscript in endless revision and a reality that never was the same from one day to the next. Thoreau's revisions of *Walden* incorporated more and more scientific observations between 1852 and 1854, adding layers of data beyond his original literary aspirations.[108] Walden Pond became Thoreau's index of measurement from which he extrapolated a whole universe; like the femur bone in Peale's mastodon, it is a synecdoche of Humboldt's *Cosmos*, wherein Thoreau has turned Humboldt's wide-angle lens into a macro lens, aimed at discovering every bit of data he can amass about the ecosystem of a single pond.

In 1853 Spencer Baird, then assistant secretary at the Smithsonian Institution and a member of the American Association for the Advancement of Science, wrote to Thoreau that he had placed his name in nomination for membership. Thoreau demurred but did send back the membership questionnaire, which included questions about his formative scientific influences. In the remarks, he noted,

> *I may add that I am an observer of nature generally, and the character of my observations, so far as they are scientific, may be inferred from the fact that I am especially attracted by such books of science as White's* Selborne *and Humboldt's* Aspects of Nature.[109]

Thoreau absorbed more than Humboldt's views on nature. Humboldt had written "Nature is the domain of freedom" as an ethos that embraced nature as much as it decried slavery and the strictures of civilization. Thoreau asserted that wildness is a landscape in which "no fugitive slave laws are passed,"[110] because the whole point of being "an inhabitant, or a part and parcel of Nature"[111] was to be part of that wild condition. Thoreau understood an essential aspect of Humboldt's delight in the cosmos, observing,

I [am] reminded of the advantage to the poet, and philosopher, and naturalist, and whomsoever, of pursuing from time to time some other business than his chosen one,—seeing with the side of the eye. The poet will so get visions which no deliberate abandonment can secure. The philosopher is so forced to recognize principles which long study might not detect. And the naturalist even will stumble upon some new and unexpected flower or animal.[112]

Thoreau values peripheral vision as a means of widening his appreciation of nature, taking in more than what is directly in his path. His fascination with oblique observation recalls the sidelong glance Charles Willson Peale had captured in his portrait of Humboldt from 1804, as the Prussian naturalist, philosopher, and poet first assessed the United States.

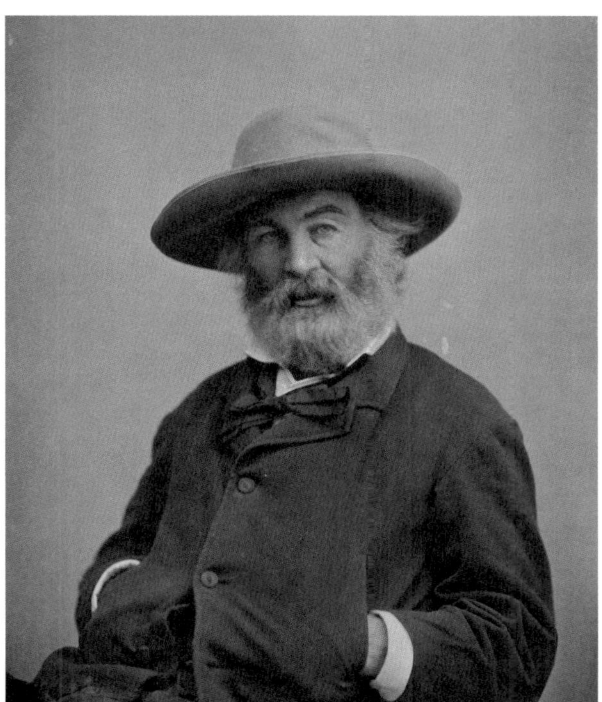

Fig. 96
Mathew Brady Studio, *Walt Whitman* (1819–92), ca. 1866–67, wet plate negative, National Archives, Mathew Brady Photographs of Civil War-Era Personalities and Scenes

WALT WHITMAN AS A KOSMOS: ANOTHER EMBODIMENT OF HUMBOLDT

The American literary giant who came closest to absorbing and matching Humboldt's *Cosmos* was Walt Whitman (fig. 96), whose endless revision and expansion of *Leaves of Grass* resembled Humboldt's continual revision and expansion of each successive edition of each of his books. Whitman paid close attention to Humboldt's life and his writings, developing what has been called a cosmological geography.[113] His friend and biographer Horace Traubel recounted Whitman's admiration for Humboldt's habit of working at all hours:

"I have no rule: I live, move, just as the spirit directs." He had read somewhere of Humboldt's informal mode of life in Paris—eating, sleeping, &c., not by hours but by instinct. W.[hitman] liked the idea. Spoke of it.[114]

He is reported to have written *Leaves of Grass* with a copy of Humboldt's *Cosmos* on his desk. He kept clippings from magazines about Humboldt, and in 1849, as he was reading an article in a magazine, scribbled a line from the just-published *Cosmos*, citing Schiller's observations on the relationship between landscape and human activity.[115] By 1855 Whitman saw that the inspiration he took from the sciences was foundational to his poetry. In the first edition of *Leaves of Grass,* Whitman proclaimed,

The American bards shall be marked for generosity and affection, and for encouraging competitors... They shall be Kosmos...glad to pass any thing to any one.... Exact science and its practical movements are no checks on the greatest poet, but always his encouragement and support....The sailor and traveller...the anatomist, chemist, astronomer, geologist, phrenologist, spiritualist, mathematician, historian and lexicographer, are not poets, but they are the lawgivers of poets, and their construction underlies the structure of every perfect poem.[116]

The American bard, in Whitman's universe, owed his art to a Humboldtian range of inspirations.

Whitman relished the same reach to which Humboldt aspired. The cumulative power of Whitman's volume, which expanded and pulsated with each new iteration, presented the poet as, quite literally, "A Kosmos." As Whitman's sense of himself

CAT. 105 Frederic Edwin Church, *Clouds Over Olana*, August 1872, oil on paper, 8 ⅔ × 12 ⅛ in., Olana State Historic Site, New York State Office of Parks, Recreation and Historic Preservation

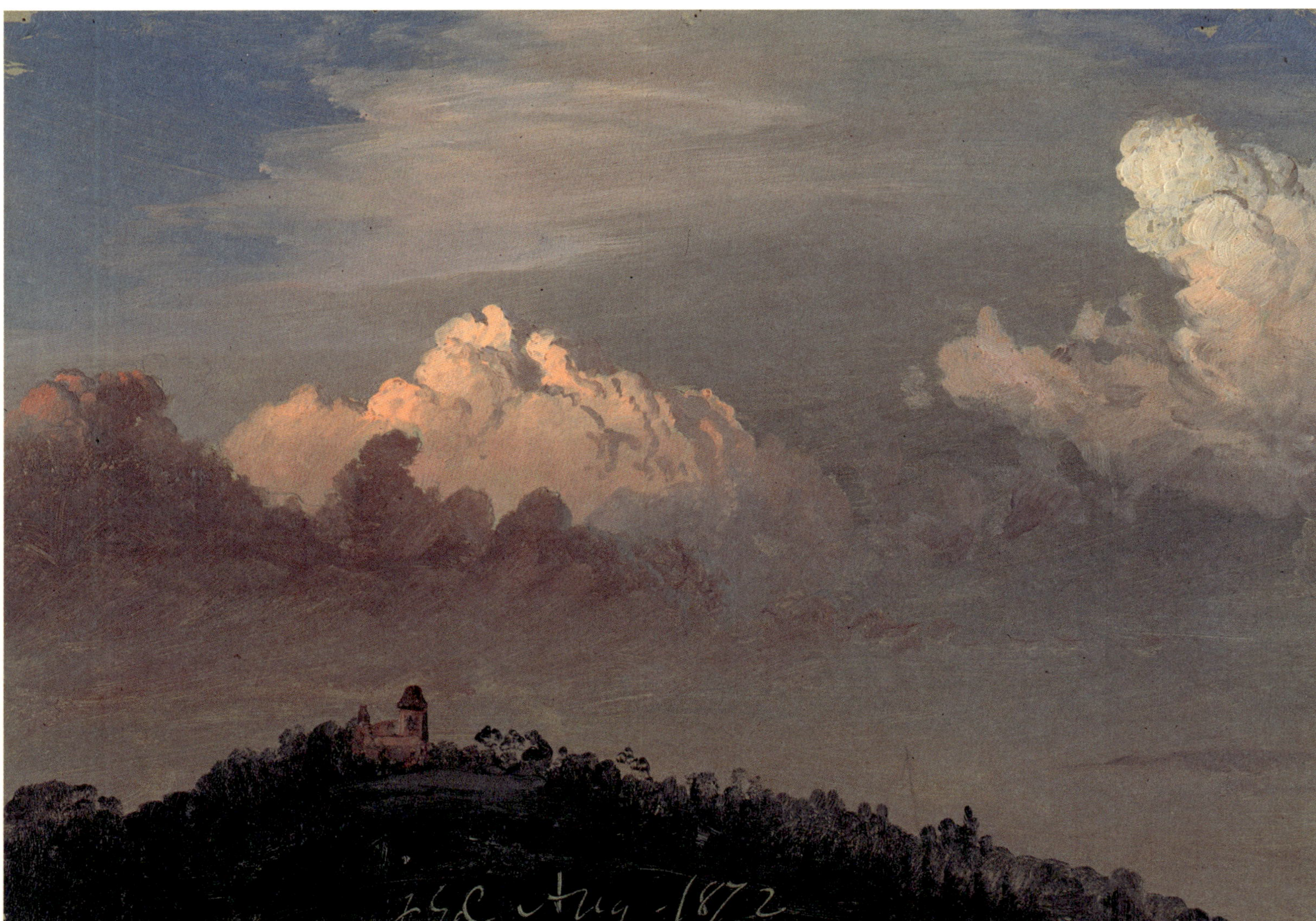

came into focus in these poems, so too did the feeling that he was, like Humboldt, attempting to invoke the cosmos. In 1866 his admiring reviewer William Douglas O'Connor wrote,

Mr. Whitman's book may not be understood at all for a long time. Nature and Art alike often seem chaotic and incomprehensible. . . . The physical cosmos cooled its heels a good while before HUMBOLDT arrived.[117]

The consonance is appropriate: *Leaves of Grass* is also an evolutionary project, a never-ending set of revisions and reappraisals, an expansion of Whitman's worldview centered on his own being as an avatar for the transparent eyeball postulated by Humboldt, envisioned by Church, embodied by Emerson, and examined by Thoreau. Emerson was awed by Humboldt's ability to simultaneously look to the entirety of the cosmos and yet discern its individual elements, concluding his "eyes are natural telescopes & microscopes."[118] That ability to look wide and still look close, to amass fragments and unify them into a cosmic whole, mirrored the whole of transcendentalism, described Walt Whitman's restlessness, and summarized Frederic Church's working method.

EMBODYING HUMBOLDT'S COSMOS

The 1850s turned out to be a decade in which the leading landscape painter, the leading transcendentalists, and the singularity that is Walt Whitman each strove to embody Humboldt's *Cosmos* in the capstone works for their careers. For Whitman, it involved absorbing Humboldt's confident command of a wide swath of scientific observation transformed by literary metaphor. Emerson's centennial eulogy for Humboldt, delivered in 1869, visualizes that reach, describing him as "a universal man." Of *Cosmos,* he averred,

There is no book like it; none indicating such a battalion of powers. You could not put him on any sea or shore but his instant recollection of every other sea or shore illuminated this.[119]

Thoreau's debt to Humboldt—so evident in his meticulous keeping of records, his desire to know his local universe, and in his close proximity to Emerson—recasted his own trajectory as a writer along Humboldtian lines.[120] For Frederic Edwin Church, his encounters with Humboldt—on the pages of his books, on his travels to South America, and in his paintings—redefined his career.

In 1883 Church wrote in frustration, "I wish science would take a holiday for ten years so I could catch up."[121] He expressed the impossibility of absorbing it all, grasping it all, synthesizing it all. It is exactly the conundrum Humboldt faced in writing *Cosmos:* how do you convey all that you know, how it all fits together, when the very concept is dynamic, in flux? François Arago had quipped that Humboldt did not know how to write a book—that he simply put words on a page like unframed pictures. When writing a book, that became a fatal flaw, but in Church's case, by building Olana, he had created his own personal cosmos. As he built his house, he sketched its setting in works like *Clouds Over Olana* (CAT. 105), and afterwards, he painted nearly daily from its prospect. Every view, every season, the daily weather, added another page to the book, another layer to the experience. Church might not ever catch up on paper or on canvas, but in constructing his "center of the world," he had come as close as he could to embodying what Humboldt had aspired to: a summation, a point from which he could view the surrounding world and understand in his mind how it all interconnected, without the need to write it all down and make it concrete.

Out of the singularity of Humboldt's *Essay on the Geography of Plants* to the narrative cohesion of *Aspects of Nature,* Humboldt developed the serial template for his multivolume *Personal Narrative* and then attempted the comprehensiveness of *Cosmos.* The American response to Humboldt was to embody that apprehension of both the forest and the trees: to revel in every detail, all the while envisioning the underlying natural laws that accounted for every one of them. Humboldt's influence on American views of nature and culture ultimately became so fully absorbed they seemed invisible, ubiquitous. For Humboldt, the path to *Cosmos* was a lifelong journey, incomplete at the finish, but he pursued it with the reassurance that others were taking up various strands of ideas and theories—ultimately extending them to a point where Humboldt's fingerprints had faded away, replacing them with those of future generations of thinkers and creators.

NOTES

1 Alexander von Humboldt, *Cosmos: A Sketch of a Physical Description of the Universe*, trans. E. C. Otté (London: Henry G. Bohn, 1848), 1:2–3.

2 See Eleanor Jones Harvey, "Capturing the Cosmos," in *Frederic Church's Olana on the Hudson: Art, Landscape, Architecture*, ed. Karen Zukowski and Julia B. Rosenbaum (New York: Rizzoli, 2018), 101.

3 Alexander von Humboldt, *Cosmos: Sketch of a Physical Description of the Universe*, trans. Elizabeth Sabine (London: Longman, Brown, Green, and Longmans, 1846), 1:5, 279.

4 The author of the review begins with a sketch of Humboldt's life and travels, including the detour to the United States: "they set sail for the United States, where they spent six weeks in Philadelphia and Washington, studying the commercial and political relations of that interesting Republic." "Art. VIII.—*Kosmos. A General Survey of the Physical Phenomena of the Universe*. By Baron Alexander von Humboldt. London, 1845," *North British Review* 4 (November 1845): 202, 219.

5 The most complete accounting of Humboldt's publications is Horst Fiedler and Ulrike Leitner, *Alexander von Humboldts Schriften: Bibliographie der selbständig erschienenen Werke* (Berlin: Akademie Verlag, 2000).

6 Alexander von Humboldt, *Cosmos: A Sketch of a Physical Description of the Universe*, trans. E. C. Otté (New York: Harper & Brothers, 1850), 2:iii.

7 Ibid., 2:106.

8 Humboldt, *Cosmos*, trans. E. C. Otté (London, 1848), 1:36.

9 Ibid., 1:19; quoted in Edmunds V. Bunkše, "Humboldt and an Aesthetic Tradition in Geography," *Geographical Review* 71, no. 2 (April 1981): 138.

10 Humboldt to his publisher, Johann Georg von Cotta, 16 March 1849; quoted and cited in Andrea Wulf, *The Invention of Nature: Alexander von Humboldt's New World* (New York: Alfred A. Knopf, 2015), 248. Quoted from Alexander von Humboldt, *Alexander von Humboldt und Cotta. Briefwechsel*, ed. Ulrike Leitner (Berlin: Akademie Verlag, 2009), 359.

11 Thomas Cole, quoted in Louis Legrand Noble, *The Course of Empire, Voyage of Life, and Other Pictures of Thomas Cole* (New York: Cornish, Lamport, 1853), 364.

12 Church's library contains several of Humboldt's books: Alexander von Humboldt and Aimé Bonpland, *Personal Narrative of Travels to the Equinoctial Regions of America, during the Years 1799–1804*, 3 vols., trans. and ed. Thomasina Ross (London: Henry G. Bohn, 1852); Alexander von Humboldt, *Aspects of Nature*, 3rd ed., trans. Elizabeth Sabine (London: Longman, Brown, Green, and Longmans, 1849); and all five volumes of Alexander von Humboldt, *Cosmos: A Sketch of a Physical Description of the Universe*, trans. E. C. Otté (London: Henry G. Bohn, 1848–59). I am grateful to Valerie Balint, Evelyn Trebilcock, and Ida Brier at Olana State Historic Site for answering my incessant queries about Church's connections to Humboldt.

13 Humboldt, *Cosmos*, trans. Otté (New York, 1850), 2:95.

14 "Appendix III: Plagiarism," in John Ruskin, *Modern Painters* (Boston: Estes and Lauriat, 1856), 3:371. An earlier edition of this volume is in the library at Frederic Church's Olana, Hudson, New York. My thanks to Valerie Balint and Ida Brier for providing me with a complete list of the volumes in the library at Church's home. The most astute study of this issue is Alicia Lubowski-Jahn, *"A Comparative Analysis of the Landscape Aesthetics of* Alexander von Humboldt and John Ruskin," *British Journal of Aesthetics* 51, no. 3 (July 2011): 321–33.

15 John Greenleaf Whittier, "J. G. Whittier on Humboldt, Amesbury, 9th mo. 6th, 1869," *Littell's Living Age*, 16 October 1869, 148.

16 Alexander von Humboldt, *Cosmos: A Sketch of a Physical Description of the Universe*, trans. E. C. Otté (London: Henry G. Bohn, 1849), 2:456.

17 Henry T. Tuckerman, "Alexander von Humboldt," *Godey's Lady's Book (1848–1854)*, September 1850, 134.

18 Henry T. Tuckerman, "Frederic Edwin Church," *Galaxy* 1, no. 5 (1 July 1866): 422. This essay is reprinted in expanded form in Tuckerman's *Book of the Artists. American Artist Life, Comprising Biographical and Critical Sketches of American Artists: Preceded by an Historical Account of the Rise and Progress of Art in America* (New York: G. P. Putnam and Son, 1867), 370–86.

19 *West Rock* was credited with establishing Church's career. He showed it at the National Academy of Design, where the reviewer proclaimed, "Church has taken his place, at a single leap, among the great masters of landscape." "Fine-Art Gossip," *Bulletin of the American Art-Union* 2, no. 1 (April 1849): 20.

20 The hayfield and the white steeple, which do not appear in Church's initial drawing, can be read as puns on Field's and Church's surnames. Franklin Kelly notes the dual punning potential in his entry for *West Rock, New Haven* in *New Britain Museum of American Art: Highlights of the Collection*, ed. Pamela T. Barr (New York: New Britain Museum of Art in association with Prestel, 1999), 1:51.

21 Alexander von Humboldt, *Researches, Concerning the Institutions and Monuments of the Ancient Inhabitants of America, with Descriptions and Views of Some of the Most Striking Scenes in the Cordilleras!*, trans. Helen Maria Williams (London: Longman, Hurst, Rees, Orme, and Browne, 1814), 1:59. (The short form of this title is referred to as *Views of the Cordilleras*.) Humboldt includes the page reference to Jefferson's *Notes on the State of Virginia* and provides the comparative heights and geological composition of both natural bridges.

22 Nathaniel Parker Willis, *American Scenery; or, Land, Lake, and River Illustrations of Transatlantic Nature* (London: George Virtue, 1840), 2:60.

23 Isabella Field Judson, ed., *Cyrus W. Field: His Life and Work (1819–1892)*, (New York: Harper and Brothers, 1896), 39.

24 David Huntington notes that Church was a frequent visitor to Astor's library. "Landscape and Diaries: The South American Trips of F. E. Church," *Brooklyn Museum Annual* 5 (1963–64): 76n37.

25 Humboldt, *Views of the Cordilleras*, trans. Williams, 1:80.

26 Entry for Monday, 20 June 1853, Frederic Church's South American Diary, May–October 1853, Olana's David C. Huntington Archive and Olana Research Collection, Olana State Historic Site, Hudson, New York, OL.1980.27.

27 Entries for Wednesday, 29 June and Thursday, 30 June 1853, Frederic Church's South American Diary, May–October 1853.

28 Humboldt, *Views of the Cordilleras*, trans. Williams, 1: 79.

29 Church to his mother, Eliza, 7 July 1853, Henry Francis du Pont Winterthur Museum, Joseph E. Downes Manuscript Collection, no. 67 x 18.36; quoted in John K. Howat, *Frederic Church* (New Haven, CT: Yale University Press, 2005), 51.

30 Church to his father, Joseph, Bogotá, 9 June 1853, Henry Francis Du Pont Winterthur Museum, Joseph E. Downes Manuscript Collection, no. 57 x 18.35.

31 Alexander von Humboldt and Aimé Bonpland, *Essay on the Geography of Plants*, ed. Stephen T. Jackson, trans. Sylvie Romanowski (Chicago: University of Chicago Press, 2009), 74.

32 Church to his mother, Eliza, 7 July 1853, Henry Francis Du Pont Winterthur Museum, Joseph E. Downes Manuscript Collection, no. 67 x 18.36; quoted in Howat, *Frederic Church*, 51.

33 I am grateful to Bethany Palkovitz for suggesting this avenue of interpretation for *The Falls of Tequendama*.

34 Humboldt, *Views of the Cordilleras*, trans. Williams, 1: 76.

35 Church to his sister, Elizabeth, Cartago, 23 July 1853, Henry Francis Du Pont Winterthur Museum, Joseph E. Downes Manuscript Collection. One of Humboldt's proudest accomplishments was ascending Chimborazo. He made it close its summit (20,577 feet) but couldn't cross a final chasm. Humboldt's ascent to 19,413 feet above sea level stood as a record for thirty years. See Kevin J. Avery, *Church's Great Picture:* The Heart of the Andes (New York: Metropolitan Museum of Art, 1993), 13.

36 Church to his sister, Charlotte, Popayan Volcano Puresé, 8 August 1853, Henry Francis Du Pont Winterthur Museum, Joseph E. Downes Manuscript Collection.

37 On Wednesday, August 24, Church noted in his diary that they had "Left Tuquerres to day but had the misfortune to lose my notebook containing a part of my diary and some valuable information relating to our further journey here[?]," which appears to account for the pages missing between July 10, shortly after leaving Bogotá, and August 23. Frederic Church's South American Diary, May–October 1853.

38 Entry for 26 August 1853, Frederic Church's South American Diary, May–October 1853.

39 Humboldt, *Views of the Cordilleras,* trans. Williams, 1:41–42.

40 Of Cotopaxi, Humboldt wrote: "The explosion which took place in the month of January, 1803, was preceded by a dreadful phenomenon, the sudden melting of the snows that covered the mountain. For twenty years before, no smoke or vapour, that could be perceived, had issued from the crater; and in a single night the subterraneous fire became so active, that at sunrise the external walls of the cone…appeared naked, and of the dark colour….we heard, day and night, the noises of the volcano, like continued discharges of a battery." Ibid., 1:119.

41 Ibid., 1:120–21.

42 Ibid., 1:122, 125.

43 Entry for Saturday, 10 September 1853, San Juan, Frederic Church's South American Diary, May–October 1853.

44 Entry for Monday, 12 September 1853, San Juan, Frederic Church's South American Diary, May–October 1853.

45 Entry for Tuesday, 13 September 1853, Ambato, Frederic Church's South American Diary, May–October 1853.

46 Entry for Thursday, 15 September 1853, Frederic Church's South American Diary, May–October 1853.

47 Entry for Saturday, 17 September 1853, Frederic Church's South American Diary, May–October 1853.

48 "Are we not justified in hoping that landscape painting will flourish with a new and hitherto unknown brilliancy when artists of merit shall more frequently pass the narrow limits of the Mediterranean, and when they shall be enabled, far in the interior of continents, in the humid mountain valleys of the tropical world, to seize, with the genuine freshness of a pure and youthful spirit, on the true image of the varied forms of nature?" Humboldt, *Cosmos*, trans. Otté (New York, 1850), 2:93.

49 "Editor's Table: Exhibition of the National Academy of Design," *Knickerbocker* 45, no. 5 (May 1855): 532.

50 Franklin Kelly made this observation in his book *The Early Landscapes of Frederic Edwin Church*, coauthored with Gerald L. Carr (Fort Worth, TX: Amon Carter Museum, 1987), 65, 75.

51 "Our Private Correspondence," *Home Journal*, 9 May 1857, 2; quoted and cited in Jeremy Elwell Adamson et al., *Niagara: Two Centuries of Changing Attitudes, 1697–1901* (Washington, DC: Corcoran Gallery of Art, 1985), 15, 76n19.

52 "Our Private Correspondence," 2. *The New-York Daily Times* compared Church's painting favorably to the Great Pyramids in a review published Thursday, 21 May 1857.

53 Humboldt, *Cosmos*, trans. Otté (New York, 1850), 2:97–98.

54 Humboldt, *Views of the Cordilleras,* trans. Williams, 2:11.

55 Ibid., 1:239.

56 Gerald L. Carr noted that Church had folded down the corner of the page in the fifth volume of his copy of *Cosmos*, describing Sangay. Gerald L. Carr, *Frederic Edwin Church: Catalogue Raisonné of Works of Art at Olana State Historic Site* (New York: Cambridge University Press, 1994), 1:244n3. Humboldt wrote, "The most active of the South American volcanoes…is the *Sangay*, which is also called the *Volcan de Macas*….I myself heard it thunder for months together, especiallyin the early morning, in Chillo, the pleasant country seat of the Marquis de Selvalegre near Quito." Alexander von Humboldt, *Cosmos: A Sketch of a Physical Description of the Universe*, trans. by E. C. Otté and W. S. Dallas (New York: Harper and Brothers, 1860), 5:249. Church would not have read this prior to his 1857 trip; volume 5 was published posthumously from Humboldt's notes in 1859. Frederic Church's Sangay Diary 1857, Olana's David C. Huntington Archive and Olana Research Collection, Olana State Historic Site, Hudson, New York, OL.1980.28.

57 Humboldt, *Views of the Cordilleras,* trans. Williams, 2:100.

58 Kevin J. Avery has argued convincingly that Church was possibly adapting a stela retrieved by Frederick Catherwood and John L. Stevens from Kabah, in the Yucatan, not far from Palenque. Those steles were installed in a folly on an island in the Hudson near Olana and subsequently given to the American Museum of Natural History in New York. See Kevin J. Avery, "Maya on the Hudson: Church's *Cayambe* and Cruger's 'Folly,'" *Hudson River Valley Review: A Journal of Regional Studies* 31, no. 1 (Autumn 2014): 50–61; and AMNH Anthropology Department accession records, nos. 4773–4776, from Miss Cornelia Kruger, purchased 1919–55.

59 *Lenox Library Catalogue of the Paintings in the Robert L. Stuart Collection* (New York: Printed by Order of the Trustees, 1893), 19, no. 135. Cayambe. The description with the entry states: "Mountain in Ecuador, on the line of the equator, about 20,000 feet high. Painted to order." Stuart owned all of John L. Stephens's books, illustrated by Catherwood; both men had been inspired to explore Mexico and Central America from reading Humboldt's work. In *Incidents of Travel in Central America, Chiapas and Yucatan* (1841) and *Incidents of Travel in Yucatan* (1843), complemented by Catherwood's illustrations, Stephens built on Humboldt's radical proposition that pre-Columbian civilizations like the Maya were not primitive but had been advanced cultures. Albert Gallatin introduced Stephens to Humboldt in a letter dated August 1842 from New York, in Ingo Schwarz, ed., *Alexander von Humboldt und die Vereinigten Staaten von Amerika: Briefwechsel* (Berlin: Akademie Verlag, 2004), 225–28, letter 128. Stephens recalled his visit in "An Hour with Alexander von Humboldt," *Littell's Living Age* 15, no. 180, 23 October 1847, 151–53.

60 Stuart also owned Humboldt's *Political Essay on the Kingdom of New Spain* (1811) and the 1850 third edition of *Aspects of Nature*. He owned two more of Humboldt's books printed in the original French: the sixth volume of Humboldt and Karl Sigismund Kunth's *Voyage… botanique* (1818 folio edition) and *Vues des Cordillères* (1810 folio edition). Although Church was not an especially talented linguist, he was attempting to teach himself French. One possible reason for his interest in this language might have been to get more out of his perusal of these volumes as he planned his trips to South America. Stuart served as a founding member of the American Museum of Natural History in 1870 and its second president. He made his extensive library of more than 11,000 volumes and his natural history collections available to visitors to his home. See *Catalogue of the Library of Robert L. Stuart* (New York: J. J. Little, 1886); and Arnold Turner Lewis, James Turner, and Steven McQuillen, *The Opulent Interiors of the Gilded Age* (New York: Courier Corporation, 2013). His books were eventually donated to the Lenox Library, now part of the New York Public Library. *Bulletin of the New York Public Library* 20, no. 7 (July 1916): 702.

61 "Our Private Correspondence," 2.

62 "ein bild des Ganzen großen Naturgemäldes." Karoline von Humboldt to Alexander von Rennenkampff, 28 January 1828, in Albrecht Stauffer, ed., *Karoline von Humboldt in Ihren Briefen an Alexander von Rennenkampff* (Berlin: Mittler und sohn, 1904), 215. Translated in Wulf, *Invention of Nature,* 194.

63 Humboldt, *Cosmos*, trans. Otté (New York, 1850), 2:94. Humboldt had mentioned frequently in his writings of the need to create sketches "at the spot" and "frequently under very afflicting circumstances" rather than relying on memory. Alexander von Humboldt and Aimé Bonpland, *Voyage de Humboldt et Bonpland: Relation historique, atlas pittoresque*, 3 vols. and 2 atlases (Paris: Librairie de Gide, 1813–34), 1:v. Translation in Barbara Maria Stafford, *Voyage into Substance: Art, Science, Nature, and the Illustrated Travel Account, 1760–1840* (Cambridge, MA: MIT Press, 1984), 422.

64 Tuckerman, "Alexander von Humboldt," 133.

65 "Our Private Correspondence," 2.

66 Alexander von Humboldt, *Cosmos: A Sketch of a Physical Description of the Universe*, trans. E. C. Otté (New York: Harper & Brothers, 1855), 3:5.

67 The use of opera glasses was first recommended to viewers in a notice for the exhibition printed in the *New York Herald*, 19 May 1859, 4. The writer advised that with opera glasses "the details of perspective will then be all clearly brought out." Kevin J. Avery compiled an extensive list of newspapers in New York and Boston making similar recommendations. Kevin J. Avery, "*The Heart of The Andes* Exhibited: Frederic E. Church's Window on the Equatorial World," *American Art* 18, no. 1 (Winter 1986): 59, 70n45.

68 "The Heart of the Andes," *Spirit of the Times*, 14 May 1859, 157. This account noted that this information was circulating twelve months earlier (i.e., in the spring of 1858).

69 Robert Walter Weir, "Fine Arts. Church's New Picture," *New York Post*, 20 April 1859, 2; quoted and cited in Avery, *Church's Great Picture*, 37, 61n110.

70 Several scholars have thoroughly documented the response to *Heart of the Andes*, notably David C. Huntington in "Frederic Edwin Church, 1826–1900: Painter of the New World Adamic Myth" (PhD diss., Yale University, 1960), 1–12; David C. Huntington, *The Landscapes of Frederic Edwin Church: Vision of an American Era* (New York: George Braziller, 1966), 5–9; Gerald L. Carr, "American Art in Great Britain: The National Gallery Watercolor of *The Heart of the Andes*," *Studies in the History of Art* 12 (1982): 83–86; and Avery, "*The Heart of the Andes* Exhibited," 52–72.

71 Kevin J. Avery has estimated that attendance on the final day reached 2,000 people, based on the receipts reported in "The Heart of the Andes," *New York Herald*, 24 May 1859, 5. Avery, *Church's Great Picture*, 36, 61n102.

72 Pierre M. Irving, *The Life and Letters of Washington Irving* (New York: G. P. Putnam, 1864), 4:288–89.

73 The contract between Church and Blodgett is dated 6 June 1859 and is in the archive at Olana. That contract is discussed in detail in Carr, "American Art in Great Britain," 88–89.

74 Winthrop and Church had taken time from their canoe trip in Maine that summer so that Winthrop could wax eloquent in favor of his antislavery candidate, delivering speeches that could last over two hours and at which Church was in attendance. See George William Curtis, "Theodore Winthrop," *Atlantic*, August 1861, 244.

75 See Louis Legrand Noble, *Church's Painting: The Heart of the Andes* (New York: D. Appleton, 1859); and Theodore Winthrop, *A Companion to the Heart of the Andes* (New York: D. Appleton, 1859), 13. By 1863 another of Church's friends, composer George William Warren, wrote a libretto titled "The Heart of the Andes," which was set to music by another friend, the composer and pianist Louis Gottschalk. The painting itself ended up in the collection of William T. Blodgett, another of Church's friends and patrons. This entire group of people surrounding *Heart of the Andes* had close ties to Humboldt, and their collective contributions form a kind of rousing send-off to the late naturalist. In 1864 *Heart of the Andes* made a public appearance as part of the Metropolitan Sanitary Fair to benefit the Union troops during the Civil War. The exhibition was trumpeted as proof of the superiority of Northern artistic prowess as well as its military might, and *Heart of the Andes* was surmounted by portraits of three American presidents (two of whom had met Humboldt when he visited Washington, D.C., in 1804). Jessie Benton Frémont was one of the fair's organizers, charged with soliciting items for the auction to benefit Union soldiers. Among the items listed for auction in the program were six long Humboldt letters. *Heart of the Andes* remained a public feature long after it was first painted and displayed. For a discussion of the Metropolitan Fair, see chapter 5 in this volume.

76 Samuel Clemens to Orion Clemens, St. Louis, 18 March 1860; quoted in *Mark Twain's Letters*, ed. Albert Bigelow Paine (New York: Harper and Brothers, 1917), 1:45–46.

77 Alexander von Humboldt and Aimé Bonpland, *Personal Narrative of Travels to the Equinoctial Regions of the New Continent, during the Years 1799–1804*, trans. Helen Maria Williams (London: Longman, Hurst, Rees, Orme, and Brown, 1822), 1:180–81.

78 "The Heart of the Andes," *Crayon* 6, no. 6 (June 1859), 193.

79 Ibid., 193–94. The *New York Times* reviewer encountered the same struggle but concluded, "What seems, at first, a chaos of chords or colors gradually rises upon the enchanted mind a rich and orderly creation, full of familiar objects, yet wholly new in its combinations and its significance." "Mr. Church's New Picture," *New York Times*, 28 April 1859, 4.

80 Church to Bayard Taylor, New York, 9 May 1859, Bayard Taylor Correspondence, Letters to Taylor, Box A–Cr., letter 1, Cornell Regional Archives, Cornell University, Ithaca, NY.

81 "Wenn Sie seine Wahrheit erkennen, und Freude daran empfinden, wird der Maler sich glücklich preisen, den er füllt dass Sie ihm den Weg gezeigt haben, denn er mit so reichen Erfolg gegangen ist." Bayard Taylor to Humboldt, Rockford, Illinois, 16 May 1859. Translation by Kevin J. Avery, Frederic Church Archive, Olana State Historic Site, OL.1998.1.35.1. Quoted in Avery, *Church's Great Picture*, 61n123.

82 Church to Bayard Taylor, Hartford, 13 June 1859, Bayard Taylor Correspondence, Letters to Taylor, Box A–Cr., Cornell Regional Archives, Cornell University, Ithaca, NY.

83 The new tour began in Boston in December 1859. *Heart of the Andes* then traveled to Philadelphia, Baltimore, Cincinnati, Chicago, and St. Louis before arriving at its final destination, Brooklyn, in April 1861. I am grateful to Valerie Balint for sharing the chart created by Moriah Mudd-Kelly tracking the tour of this painting. The tour is also discussed in Avery, *Church's Great Picture*, 40–44.

84 "Domestic Art Gossip," *Cosmopolitan Art Journal* 3, no. 2 (March 1859): 87, noted that "Church and the 'Albany Sculptor,' Palmer, go to Europe this spring for study and recreation.... Mr. Church is busy upon his great canvas of South American scenery."

85 "Church's Cotopaxi," *New York Times*, 17 March 1863, 4.

86 Church and Noble ran into Louis Agassiz on the *Great Republic* sailing from Boston to Nova Scotia. The Harvard scientist was headed back to Switzerland for the first time since his arrival in the United States in 1846. Agassiz, whose early work on glaciation had prompted him to theorize that the planet had undergone an Ice Age, had never seen an iceberg and envied Church's trip.

87 I am grateful to Peter Betjemann and Dan Peck for their observations on the longevity of Humboldt's influence on American literature.

88 Attendees were Agassiz and Jeffries Wyman, scientists; Emerson and James Russell Lowell, poets; Ebenezer Hoar and Horatio Walker, lawyers; Amos Binney and Estes Howe, doctors; John Holmes (Oliver Wendell Holmes's younger brother) and William James Stillman. Stillman had founded *The Crayon*, the first magazine devoted to the fine arts in America, which had published Asher B. Durand's "Letters on Landscape Painting," extensive articles on John Ruskin's aesthetics, and excerpts from Humboldt's *Cosmos*.

89 James Schlett has written a thoughtful and detailed history of the Adirondack Philosophers' Club and Stillman's painting in *A Not Too Greatly Changed Eden: The Story of the Philosophers' Camp in the Adirondacks* (Ithaca, NY: Cornell University Press, 2015).

90 Two of the group's guides bestowed the nickname in an article they wrote. See F. S. Stallknecht and C. E. Whitehead, "Sporting Tour in August, 1858," *Frank Leslie's Illustrated Newspaper*, 13 November 1858, 379; cited and discussed in Schlett, *A Not Too Greatly Changed Eden*, 88.

91 Agassiz had come to Humboldt's attention in 1837, when his work on glaciation led to his radical hypothesis that the earth had experienced an Ice Age, which had shaped the earth's surface more than wind or water. When Agassiz's personal and professional lives collapsed in 1846, Humboldt helped him secure a position at Harvard and obtained funds from the Prussian government that enabled the Swiss-born naturalist to immigrate to the United States and rebuild his life and career.

92 William James Stillman, "The Philosophers' Camp: Emerson, Agassiz, Lowell, and Others in the Andirondacks," *Century Magazine*, August 1893, 604.

93 Ibid., 603.

94 "Les hommes veulent *voir*, et je leur montre un microcosme sur une feuille." Humboldt to Marc-Auguste Pictet, Bern, 3 February 1805; "Lettres d'Alexandre de Humboldt à Marc-Auguste Pictet (1795–1824)," *Le Globe, Journal Géographique de la Société de Géographie de Genève* 7 (1868): 162. English translation by the author.

95 Ralph Waldo Emerson, *Nature* (Boston: James Munroe, 1836), 41.

96 Ralph Waldo Emerson, *Journals of Ralph Waldo Emerson, 1841–1844, with annotations*, ed. Edward Waldo Emerson and Waldo Emerson Forbes (Boston: Houghton Mifflin, 1911), 400–401.

97 Emerson read the same bootleg English translation as Darwin; Thoreau likely read it as well, but preferred to wait for the complete, official English translation before acquiring a copy for himself. Thoreau also owned three volumes of his *Personal Narrative* reprinted in 1852–53, along with Klencke's *Lives of the Brothers Humboldt* (1853). Laura Dassow Walls notes Thoreau would not settle for abridged versions of Humboldt's books; he waited for the full-length translations, which is why his edition of the three-volume *Personal Narrative* is the 1852 edition. Laura Dassow Walls, *Seeing New Worlds: Henry David Thoreau and Nineteenth-Century Natural Science* (Madison: University of Wisconsin Press, 1995), 135–36.

98 "Natural History of Massachusetts," *Dial* 3, no. 1 (July 1842): 19–40. https://walden.org/web/viewer.html?file=https://www.walden.org/wp-content/uploads/2016/03/NaturalHistoryMassachusetts.pdf. The *Dial* was the only journal by the New England transcendentalists as a group. Margaret Fuller served as editor from 1840 to 1842. Ralph Waldo Emerson edited the magazine from 1842 to 1844, with the exception of the April 1843 issue, for which Henry David Thoreau served as editor. Laura Dassow Walls, *Henry David Thoreau: A Life* (Chicago: University of Chicago Press, 2017), 131–32.

99 Henry David Thoreau, "A Walk to Wachusett," in *Excursions* (Boston: Ticknor and Fields, 1863), 73.

100 Walls, *Seeing New Worlds*, 110.

101 Emerson to Elizabeth Hoar, 3 August 1859, J. Pierpont Morgan Library, New York; quoted and cited in Daegan Miller, "A Map of Radical Bewilderment: On the Liberation Cartography of Henry David Thoreau," *Places: The Journal of Public Scholarship on Architecture, Landscape, and Urbanism* (March 2018): n17. https://placesjournal.org/article/a-map-of-radical-bewilderment/.

102 Henry David Thoreau, *Walden*, with an introduction by Bradford Torrey (New York: Riverside Press, 1897), 2:291–92.

103 Ibid., 1:155.

104 I am grateful to Peter Betjemann for sharing his insights on Thoreau's work, his epistolary goals, and in particular his map of Walden.

105 Agassiz arrived in New England in October 1846. By April 1847 he was in touch with fellow scientists to create a collecting network. Among those who sent him specimens were Spencer Baird and Thoreau, who sent Agassiz an impressive number of botanical specimens. Walls, *Seeing New Worlds*, 113–15.

106 Ray Angelo described Thoreau's detailed list of plants by climate zones as having been inspired by Humboldt, and that it remained the most comprehensive for this site until the twentieth century. Ray Angelo, "Botanical Index to Thoreau's Journal: Thoreau as Botanist," *Botanical Index to Thoreau's Journal*, at http://ray-a.com/ThoreauBotIdx/ThAsBot.html.

107 H. Daniel Peck, *Thoreau's Morning Work: Memory and Perception in "A Week on the Concord and Merrimack Rivers," the Journal, and "Walden"* (New Haven, CT: Yale University Press, 1990), 79–114, 163–66. I am grateful to Dan for sharing his insights on Thoreau's Kalendar. For images of Thoreau's Kalendar, see http://thoreauskalendar.org.

108 Walls, *Seeing New Worlds*, 156.

109 "Concord Dec. 19th 1853
Spencer F. Baird,
Dear Sir,
I wish hereby to convey my thanks to the one who so kindly proposed me as a member of the Association for the Advancement of Science, and also to express my interest in the Association itself. Nevertheless, for the same reason that I should not be able to attend the meetings, unless held in my immediate vicinity, I am compelled to decline the membership.
Yrs, with hearty thanks,
Henry D. Thoreau
(*To be returned to S. F. Baird, Washington, with the blanks filled.*)
Name Henry D(avid) Thoreau
Occupation (Professional, or otherwise). Literary and Scientific, combined with Land-surveying
Post-office address Henry D. Thoreau Concord Mass.
Branches of science in which especial interest is felt The Manners & Customs of the Indians of the Algonquin Group previous to contact with the civilized man.
Remarks I may add that I am an observer of nature generally, and the character of my observations, so far as they are scientific, may be inferred from the fact that I am especially attracted by such books of science as White's Selborne and Humboldt's "Aspects of Nature."
With thanks for your "Directions," received long since I remain
Yrs &c
Henry D. Thoreau"

Henry David Thoreau, *The Correspondence of Henry David Thoreau*, ed. Walter Harding and Carl Bode (New York: New York University Press, 1958), 309–10.

110 Henry David Thoreau, "Walking (1862)," in Thoreau, *Excursions*, 212.

111 Ibid., 161.

112 Journal entry for 28 April 28 1856, in Henry David Thoreau, *The Writings of Henry David Thoreau: Journal*, ed. Bradford Torrey (Boston: Houghton Mifflin, 1906), 8:314. In his journal entry for 23 July 1851, Thoreau wrote, "But this habit of close observation,—in Humboldt, Darwin, and others. Is it to be kept up long, this science? Do not tread on the heels of your experience. Be impressed without making a minute of it. Poetry puts an interval between the impression and the expression,—waits till the seed germinates naturally." Thoreau, *Writings of Henry David Thoreau: Journal*, 1:341.

113 Rebecca Ann Walsh, "The 'Terraqueous' Globe: Walt Whitman and the Cosmological Geography of Humboldt and Somerville," *Geopolitics of Modernism* (2015): 59.

114 Horace Traubel, *Walt Whitman in Camden (March 28–July 14, 1888) (November 1, 1888–January 20, 1889)* (New York: Mitchell Kennerley, 1914), 3:446; quoted in Walsh, "'Terraqueous' Globe," 61.

115 In the margin of his copy of the *Edinburgh Review* from April 1849, Whitman wrote, "Humboldt, in his *Kosmos*, citing Schiller, has observed of the Greeks: 'With them the landscape is always the mere background, of a picture, in the foreground of which the human figures are moving.'" This journal is part of the Trent Collection of Whitmania, David M. Rubinstein Rare Book and Manuscript Library, Duke University. Cited in David S. Reynolds, *Walt Whitman's America: A Cultural Biography* (New York: Alfred A. Knopf, 1995), 245, 614.

116 Walt Whitman, *Leaves of Grass: Preface to the Original Edition, 1855* (London: Trübner, 1881), 16–17.

117 William Douglas O'Connor, "Current Literature: Walt Whitman," *New York Times*, 2 December 1866, 2; quoted in Walsh, "The 'Terraqueous' Globe," 65.

118 Ralph Waldo Emerson, notes for his centennial Humboldt address of 14 September 1869, in *The Journals and Miscellaneous Notebooks of Ralph Waldo Emerson*, ed. Ronald A. Bosco and Glen M. Johnson (Cambridge, MA: Belknap Press of Harvard University Press, 1982), 16:160.

119 Ralph Waldo Emerson, "XXIV. Humboldt," in *The Complete Works of Ralph Waldo Emerson: Miscellanies* (Boston: Houghton Mifflin, 1878), 11:457–58.

120 Laura Dassow Walls has explored Thoreau's debt to Humboldt in a masterful survey of his work, reading each of Thoreau's works and his voluminous diaries and journals in chronological order to chart the development of his ideas in *Seeing New Worlds* and *Henry David Thoreau: A Life*.

121 Quoted in Huntington, *Landscapes of Frederic Edwin Church*, 108.

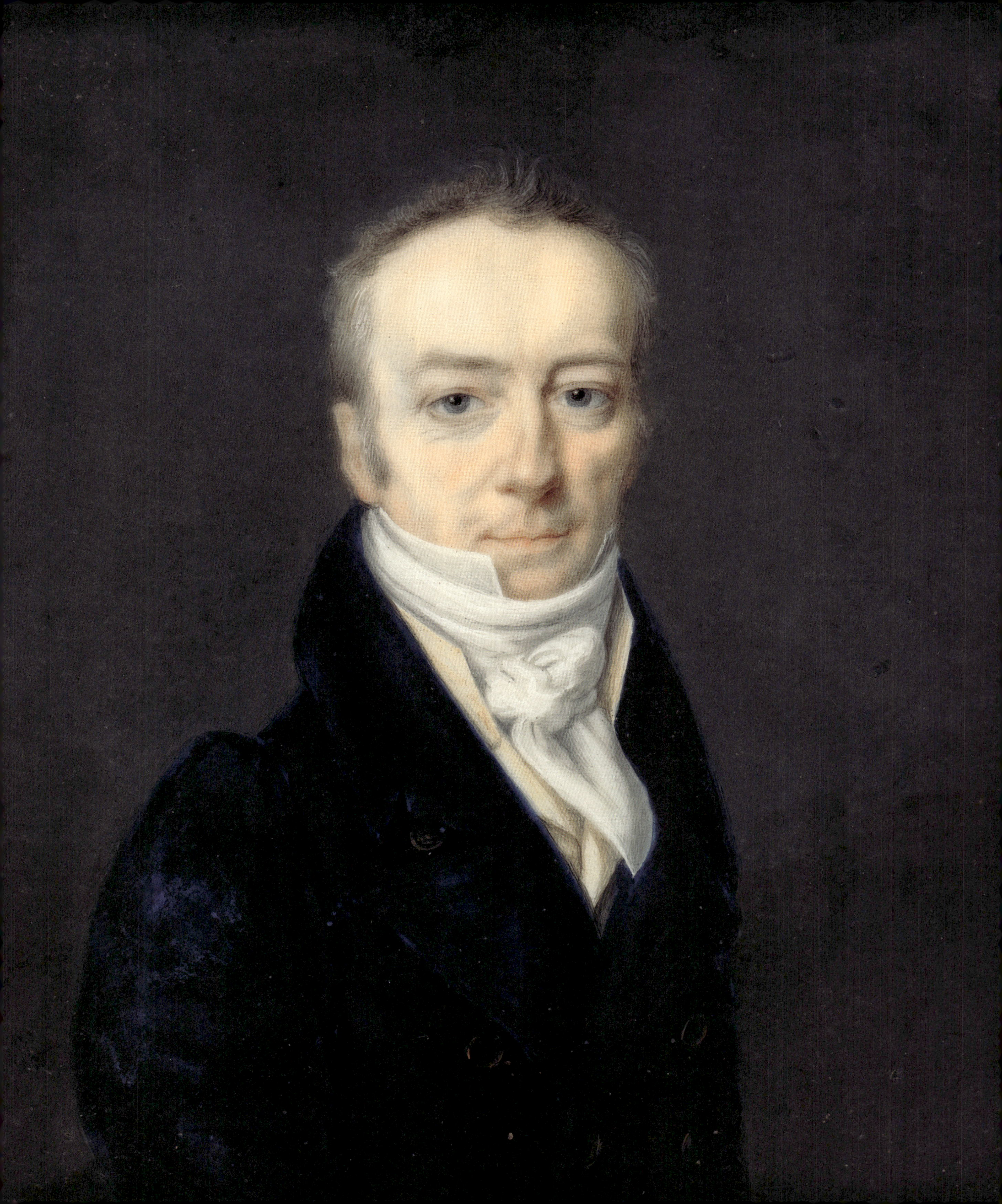

CHAPTER EIGHT | HUMBOLDT AND SMITHSON

"The increase and diffusion of knowledge"

Alexander von Humboldt envisioned America as a practical laboratory for his ideas—on science, exploration, politics, and the human condition. America's territorial expansion fed the twin engines of scientific discovery and geopolitical engagement. Humboldt was enthusiastic in the establishment and development of his American cultural network, creating lines of communication and exchange that he relied on to fuel his own work and that benefited all involved. As that expansionist network turned into an expansionist geopolitical project, Humboldt expressed his unease, first over the U.S. war with Mexico, then with the push for Cuban annexation, and the growing internal tensions that led to the Civil War. The displacement of Native Americans and the perpetuation of slavery fueled Humboldt's worries for the stability of the American experiment. As he had with Jefferson in 1804, Humboldt remained outwardly diplomatic while privately fuming about the outrages perpetuated in the name of democracy; however, he never stopped appealing to Americans' better nature.

From the outset, Americans had believed that Humboldt could provide a measurable influence on the country's sense of identity and its cultural development. Charles Willson Peale had hoped his museum would be adopted as a national institute to fulfill that mission. He enlisted Humboldt to help persuade Jefferson to purchase his museum and move it to Washington. Peale envisioned his collections as proof of the country's tangible assets, and his educational program as a persuasive case for the nation's cultural identity. George Catlin never ceased hoping that a national institute in Washington could absorb his Indian Gallery,

Preceding spread

CAT. 106 Henri-Joseph Johns, *James Smithson*, 1816, gouache on ivory, 3 × 2 ¾ in., National Portrait Gallery, Smithsonian Institution; transfer from the National Museum of American History, Conserved with funds from the Smithsonian Women's Committee

although his protests against Andrew Jackson's Indian Removal Act appear to have dampened any Congressional enthusiasm for his plan. Margaret Fuller had been among Catlin's prominent advocates, hoping Congress would vote to found a "national institute, containing all the remains of the Indians, all that has been preserved by official intercourse at Washington, Catlin's collection, and a picture-gallery as complete as can be made."[1]

Discussion of a national institute had been a constant refrain since the founding of the nation. However, high-level discussions of a national institute surfaced in 1835, when American lawyer Richard Rush, son of Dr. Benjamin Rush, discovered that the United States was the ultimate beneficiary of British chemist James Smithson's will.[2] Smithson (CAT. 106) and his chemical experiments were known to the American scientific community, but this news was completely unexpected. What was even more unusual was the wording of Smithson's intent: that in the absence of any living heirs, his entire estate was to be bequeathed "to the United States of America, to found at Washington, under the name of the Smithsonian Institution, an Establishment for the increase & diffusion of knowledge among men."[3] Should the United States prevail in securing his bequest, and accept Smithson's terms, the nation would finally have a national institute.

The legal wrangling over whether and how those funds should be released to the Americans was considerable. The conceptual discussion over what exactly an institution "for the increase and diffusion of knowledge among men" would do had just begun. Although there was no consensus as to what Smithson's intentions had been, and no detailed instructions beyond the vague and open-ended wording of his bequest, a group of people enthusiastic in support of accepting the funds came with ideas at the ready. Spearheaded by John Quincy Adams, this group comprised many men who had met or corresponded with Humboldt. They were then charged with determining the best use of James Smithson's bequest. Joel Poinsett was the first to advocate that Smithson's bequest be used to found a national museum. Adams longed for a national observatory; John Vaughan, the librarian at the American Philosophical Society (APS), who had so admired Humboldt during his 1804 visit, now proposed a national library.[4] Others hoped for a botanical garden or suggested that George Washington's wish for a national university might now be realized.

Adams, the former president, was now a member of the U.S. House of Representatives. He argued passionately in Congress for the enormous potential of such a gift. In his diary he marveled,

A stranger to this country, knowing it only by its history, bearing in his person the blood of the Percys and the Seymours, brother to a nobleman of the highest rank in British heraldry, who fought against the revolution of our independence at Bunker's Hill—that he should be the man to found at the City of Washington for the United States of America, an establishment for the increase and diffusion of knowledge among men, is an event in which I see the finger of Providence compassing great results by incomprehensible means.[5]

HUMBOLDT AND SMITHSON

Woven into the rich tapestry of Humboldt's American networks is a thread that connects him with nearly all of the men associated with the founding of the Smithsonian Institution in Washington, D.C. In some respects this comes as no surprise; certainly the institution's credo, the increase and diffusion of knowledge, echoes Humboldt's own words and reflects his lifelong interest in developing networks of information linking scientific discoveries, exploration narratives, and data visualizations. However, Humboldt also had a direct connection to James Smithson (1765–1829) that drew the British chemist into his orbit. Their first meeting took place in London, almost ten years before Humboldt's five-year trip to the Americas. They met again during the turbulent months of 1814 to 1815 in Paris, when Smithson was considering his legacy and Humboldt was a vocal figure in liberal, pro-American circles. The nature of their meetings and the politics surrounding Smithson's estate planning conspired to make a gift to the United States an attractive prospect.

In the spring of 1790, Alexander von Humboldt and Georg Forster had made their way across Europe, including a tour of the British Isles. They arrived in London in May. For the next two months they circulated among the leading scientists and attended meetings of the Royal Society. They received a warm welcome from Sir Joseph Banks, president of the Royal Society, who had accompanied Captain James Cook on his first voyage around the world. Humboldt spent time at Banks's home and rapidly was pulled into the eminent scientist's inner circle.[6] As Banks and Forster compared stories, Humboldt's determination to travel widely deepened. There was another seminal moment for Humboldt on this trip. He made the acquaintance of Charles Blagden, the somewhat eccentric secretary of the Royal Society, who was able to arrange for the traveler to visit the renowned library of Sir Henry Cavendish, himself a legendary scientist who reputedly did not permit anyone to actually speak to him. Blagden urged his irrepressible Prussian visitor to examine the library but not to say a word to his host. As voluble as Humboldt could be, that must have seemed a Herculean task. Humboldt's propensity for engaging everyone in conversation was sorely tested, but he appears to have acceded to his host's request.

It is on this library visit that Humboldt likely met the young British scientist who was setting up his own laboratory inside Cavendish's spacious home and serving as his assistant. James Louis Macie was only four years older than Humboldt and a rising star in the new field of chemistry. In 1787 he had become the youngest member of the Royal Society after graduating from Oxford the year before.[7] William Thomson, his mineralogy professor, assessed him as "a valuable character in this country."[8] The illegitimate son of the Duke of Northumberland, Macie established his reputation as a talented chemist on the strength of his laboratory work. After his mother's death in 1800, and at her request, he asserted his parentage and assumed his biological father's name. In February 1801 the London newspapers made it official: James Louis Macie would be henceforth known as James Smithson.[9] Humboldt and Smithson had interests in common beyond Cavendish's library. Smithson had visited the island of Staffa, off the coast of Scotland, with its hexagonal basalt columns, in 1784.[10] Humboldt's first publication, "Mineralogical Observations on Some Basalts of the Rhine," had just been published in 1790.[11] The comparison of basalt formations from two distant sites provided the foundation for their professional relationship.

Humboldt also took notice of Smithson's current experiments on a whitish substance called tabasheer, which is found inside particular species of bamboo. Smithson published his first paper in 1791 on this substance.[12] Blagden disseminated copies of this paper to his colleagues in England and in Paris, notably Claude-Louis Berthollet, a leading French chemist who had been elected a fellow of the Royal Society in 1789. Smithson's paper was well received and was widely cited for the next decade in the scientific literature.[13] Humboldt received a copy as well. He encountered a similar substance during his travels in the Americas, and on at least two instances, while he was in Peru and Mexico in 1802 and 1803 respectively, he reported this discovery to a colleague in Europe. In both cases he cited Smithson's work.[14] Back in Europe, Humboldt kept up with Smithson and his British colleagues; in Paris in 1805, his close friend Marc-Auguste Pictet "continued to see Berthollet, to whom he passed on the discoveries of Smithson Tennant and Wollaston in England of osmium, iridium and palladium, and he dined with him at Arcueil."[15] Smithson Tennant was James Smithson's friend and distant relative; he later joined Smithson and Blagden in Paris in 1814, where Smithson once again was in Humboldt's company.

Blagden and Smithson arrived in Paris in May 1814.[16] Blagden kept a diary, his handwriting nearly as illegible as Humboldt's. His entries detail multiple visits to see Humboldt and suggest the two Englishmen quickly became absorbed into Humboldt's Parisian circle of pro-American intellectuals. Daytime meetings with Humboldt, François Arago, and Berthollet segued into evening soirées with Madame de Staël and Benjamin Constant. Parisian scientists thrived on these social gatherings. Cuvier hosted Saturday receptions; Antoine Lavoisier's widow, Marie-Anne Pierrette Paulze, who was a respected chemist in her own right, hosted concerts every two weeks and salons twice a week. Smithson hosted gatherings on Wednesdays, his archive of calling cards and signatures including those of scientists, liberal thinkers, philanthropists, social reformers, political radicals, and eccentrics.[17] Blagden had been a British army surgeon, stationed on a hospital ship in 1776 during the American Revolution, but he remained in the United States until 1780.[18] His time on shore was spent traveling between Rhode Island—where he is considered the state's first ornithologist—to South Carolina, collecting natural history specimens to ship back to Sir Joseph Banks.[19] In recognition of his accomplishments and his interest in the United States, Blagden was made a foreign honorary member of the American Academy of Arts and Sciences in 1789.

In Paris, Smithson traveled in the same social circle as did Humboldt and the Marquis de Lafayette. This pro-American enclave of French intellectuals must have embraced Smithson as well. From the men who described themselves as "practically American," Smithson would have understood Lafayette's enthusiasm for American democracy and Humboldt's optimism for the future prospects of American science and exploration. When news arrived in August that the British burned Washington, Blagden wrote in his diary that the assembled company had agreed that when "told of capture of Washington by English, blowing up the capitol, & proceeding to Baltimore We all agreed, that Baltimore's a proper subject of attack, but better have left Washington alone."[20] Smithson was traveling with a crowd sympathetic to the United States, often at Britain's expense.

By 1814 Smithson's London-based social circle revolved around Holland House, home to Lord Holland and his wife, whose opposition politics drew together a group of like-minded scientists, statesmen, and artists. Their salon represented a pocket of British intellectual society that quietly approved of the American colonies' break with their mother country. Lord and Lady Holland were particularly fond of Smithson, drawing him further into a network of pro-American advocates. Smithson also belonged to the liberal-leaning Coffee House Philosophical Society, several of whose members would end up immigrating to the United States.[21] Notable among them was fellow chemist Joseph Priestley. As he was being hounded out of his native England, Priestley was consoled by a colleague who reminded him, "But be cheerful, dear Sir, you are going to a happier world, the world of Washington and Franklin."[22] Priestley went on to become a revered member of the APS and to extol the American Revolution as having taken the country "from darkness to light, from superstition to sound knowledge, and from a most debasing servitude to a state of the most exalted freedom."[23]

In France in 1814, Charles Blagden introduced Smithson to a group that held similar views, the Société d'Arcueil. Blagden was staying with his friend, Berthollet, who hosted the October 2 dinner for the scientists who formed the Société d'Arcueil, with Blagden and Smithson in attendance. Alexander von Humboldt was there, along with three of his closest friends, Joseph Gay-Lussac, Pierre-Simon Laplace, and François Arago. The year before, British scientist Sir Humphry Davy had been impressed with Humboldt's warmth, writing to Blagden that he found him "one of the most agreeable men I have ever known; social, modest, full of intelligence, with facilities of every kind; almost *too fluent* in conversation."[24] Now Blagden and Smithson met the full force of Humboldt's enthusiasm. The conversation was lively, and Humboldt, as Blagden noted, "talked away as usual."[25] During their yearlong stay in Paris, Blagden and Smithson continued to spend time in Humboldt's company.[26]

Talking wasn't all that kept Smithson entertained; he and his cousin Tennant enjoyed the gaming tables. James Smithson was a compulsive gambler, although he wasn't very successful at it.[27] Letters from Smithson to Madame Récamier's husband enumerate amounts owed in round numbers, suggesting the payment of debts rather than the reimbursement of expenses.[28] Arago finally became so alarmed that he intervened, appealing to Smithson's reason over the emotional thrill of risking cash. Referring to the "inveterate mania for play," Arago recalled rationalizing with Smithson about breaking his habit, only to have his subject cheerfully agree that he was right, but not in the way Arago had intended. Arago explained,

[M]y calculations had completely convinced him; that he would no longer be the ignorant tributary of the gaming-

houses of Paris; that he would continue the same kind of life, but without the mad excitement of hope and fear that led him on before. "I shall no longer," he added, "be in ignorance of the fact that the 50,000 francs of my fortune consecrated to play will pass into other hands. I am resigned to it perfectly; but I shall no longer be, in the eyes of the world, the dupe of an absurd delusion. I shall continue to play, because the superfluous 50,000 francs expended in any other way would be unable to excite my feeble frame, undermined by suffering, the lively sensations alone aroused by the various combinations, fortunate and adverse, displayed every night on the green carpet!"[29]

That amount translated to just over $10,000 in 1814; Smithson's fortune must have been considerable.

SMITHSON'S FINAL GAMBLE

When Smithson returned to London, he continued to make a name for himself as an ambitious and highly accomplished scientist. However, Smithson's reckless streak resurfaced in 1825, when he changed his will and articulated a bequest that defied convention. Earlier that year he had a falling out with the Royal Society over the publication of some of his papers.[30] Louis Agassiz described these circumstances in 1855:

Smithson had already made his will and left his fortune to the Royal Society of London, when certain scientific papers were offered to that learned body for publication. Notwithstanding his efforts to have them published in their Transactions, they were refused, upon which he changed his will, and made his bequest to the United States. It would be easy to collect in London more minute information upon this occurrence, and should it appear desirable, I think I can put the committee in the way of learning all the circumstances. Nothing seems to me to indicate more plainly what were the testator's views respecting the best means of promoting science than this fact.[31]

Harvard botanist Asa Gray (fig. 97) reiterated Agassiz's story of Smithson's anger with the Royal Society over issues stemming from papers he wanted published. Smithson could be prickly, and this rupture apparently spurred him to change the conditions of his will. The language of his new will was unusual and carried a polemical message. If his surviving nephew, Henry James Hungerford, were to die without legitimate or illegitimate heirs, Smithson stated, the entire bequest would come to the United States to found the Smithsonian Institution. When Hungerford died unexpectedly young in 1835, and, in fact, appeared not to have left any heirs, Richard Rush began the legal process of securing Smithson's unorthodox bequest.[32]

James Smithson never seriously contemplated giving up the advantages of European society and moving to a country with far less polish or sophistication. However, America looked to be an excellent investment, especially for a man interested in the sciences. As one of Charles Blagden's friends opined,

Were I younger and in better health I would not have a shilling in this country [England] nor stay in it a fortnight. The best plan would be to have one's property in America which seems very flourishing, & to live, *not there, but in almost any part of Europe* except England.[33]

That attitude toward the United States aligned closely with Smithson's situation: invest in America and live in comfort in Europe.

When Smithson wrote his new will, he certainly presumed that his younger nephew would live a long life, reaping the benefits of Smithson's accrued fortune, and that he in turn would sire children, with or without the benefit of marriage. Smithson could thus afford to indulge himself and envision in his mind's eye an institution bearing his name that would rival the Royal Society. That he would choose the United States as the beneficiary appears to have been the effect of so many people he admired who were enthusiastic about its prospects, not the least of whom was Humboldt. Of all of Smithson's acquaintances, Humboldt likely cared the least about Smithson's parentage, being far more concerned with the caliber of a person's thinking. Humboldt had returned from South America with a reputation for distributing new specimens for others to research, allowing his friend, Jean-Antoine-Claude Chaptal, to assert that he walked with an entire scientific society in his shoes. His passion for justice and equality, reflected in the company that he kept, cut across the artificial strictures of class and rested solidly on merit.

As the unacknowledged, illegitimate son of the Duke of Northumberland, Smithson had made his way through the ranks on merit, denied the automatic memberships accorded to his half

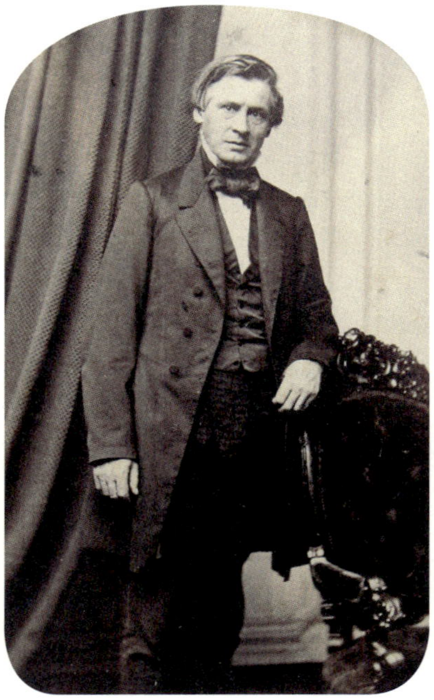

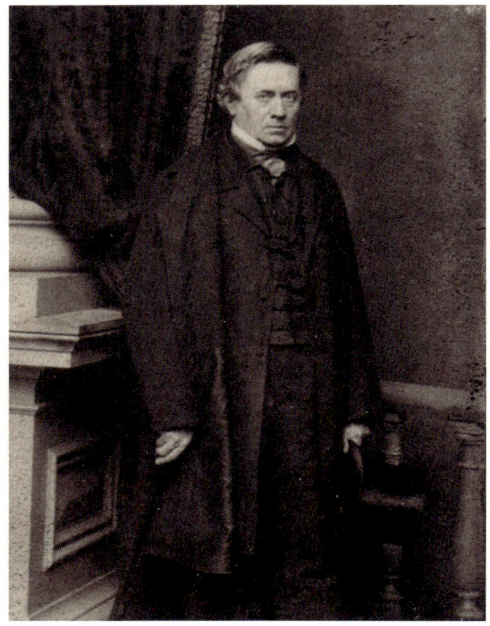

Fig. 97
John Adams Whipple, *Asa Gray* (1810–88), ca. 1865, carte de visite, 11 × 8 in., University of South Carolina, South Caroliniana Library, Ravenel Album of Distinguished Botanists

Fig. 98
Mathew Brady Studio, *Joseph Henry* (1797–1878) (detail), ca. 1860, modern albumen print from wet plate collodion negative, 3 7⁄16 × 2 5⁄16 in., National Portrait Gallery, Smithsonian Institution; Meserve Collection

brothers. He was good at masking his resentments in public, but the strict rules of British society regarding the circumstances of his birth accompanied him the rest of his life. In France or Italy he enjoyed a more relaxed view of his parentage, but he had to have known that in the United States there was more emphasis placed on accomplishments than on an accident of birth.[34] The APS had absorbed refugee scientists from Haiti during the revolution there and welcomed Joseph Priestley. There is little question that James Smithson would have been welcomed by Philadelphia's scientific community had he made the decision to immigrate there.

Smithson had watched talented colleagues forced out of England for their political beliefs who went on to thrive in the United States. Did this attitude color Smithson's decision to change his will? The United States was actively supporting scientific inquiry as a key component of its national identity, while Humboldt maintained close ties with and affection for many of its prominent statesmen and scientists. A bastard country born in revolution, the United States had sought its own legitimacy, which was but one element in the American courtship of Alexander von Humboldt beginning in 1804. As such, the United States was less likely to care about Smithson's parentage; yet with his bequest, the country could confer a kind of legitimacy missing in his own life. Smithson could now attach his name to every major scientific discovery emanating from the organization bearing his name. It was the United States, rather than his biological father, Percy Smithson, Duke of Northumberland, who conferred legitimacy upon James Smithson.

After the legal case was settled, Smithson's bequest arrived in the United States in 1836. The 500,000 gold sovereigns were worth roughly $15 million in today's dollars.[35] Congress would take another decade before it formally established the Smithsonian Institution in August of 1846. The following month, on September 7, the institution's founding board of regents hosted their first meeting. In December they resolved to hire a secretary to run the institution, offering the position to Joseph Henry, the eminent scientist and Princeton professor of electromagnetics, who was charged both with enacting Congressional plans and developing his own.[36]

Joseph Henry (fig. 98) had his own ties to Humboldt, stemming from their shared interest in terrestrial magnetism and galvanic theory. Both men knew Samuel F. B. Morse, as Henry had consulted with the prickly inventor on the development of the technology for the telegraph. During the first decade of the Smithsonian's existence, Henry further cemented the institution's ties with Humboldt. Among the men Henry appointed to serve on the new executive committee were Alexander Dallas Bache and Lorin Blodget. Bache was the lone trained scientist on the first board of regents and one of Henry's staunchest supporters.[37] Bache had first met Humboldt in 1837, when the two men were part of the "magnetic crusade."[38] His work in terrestrial magnetism and geophysics earned him Humboldt's admiration and a successful letter of support when funding for his U.S. Coastal Survey's future was in doubt. Honorary board members included Benjamin Silliman and Washington Irving, representing the broad spectrum of the Smithsonian's commitment to both science and the arts. Irving's presence in particular spoke to the long-standing friendship between the author and the Prussian naturalist, stemming back to their days together in Paris.[39]

George Perkins Marsh was another of Henry's founding trustees.[40] His own appreciation of Humboldt openly manifested in 1864, when he published his book *Man and Nature; or, Physical*

Geography as Modified by Human Action. In 1860 Marsh praised Humboldt as the last man whose "learning…embraced the whole past history and present phase of every branch of physical research… graced with the elegances of all literature and dignified with the comprehensive wisdom of all philosophy," his insights gleaned from his understanding of "the mutual interdependence of apparently unrelated knowledges."[41] Humboldt did not live to see the publication of the book that would, in effect, launch the American environmental movement, but Marsh ensured that Humboldt's concern for man's impact on the globe remained central to the debate. Spencer Baird, whom Henry appointed as assistant secretary in charge of museum collections, served with Balduin Möllhausen on the Amiel Whipple Expedition in 1853. Möllhausen's account of his time spent in the United States anchored his second career as a German novelist, which earned him the affectionate sobriquet of "the German Cooper." When Humboldt wrote the preface to Möllhausen's account of the Whipple Expedition, he noted with pride "the increasing favour with which my labours and exertions have been regarded in the United States," and was particularly pleased with the attention from "the admirable Smithsonian institution."[42] Möllhausen's friendship with Baird and his loyalty to Humboldt would play key roles in the Smithsonian's involvement with Humboldt's estate in 1859.

THE SMITHSONIAN AND HUMBOLDT

The founding of the Smithsonian ushered in what Benjamin Silliman termed the "Great Scientific Awakening….the beginning of a new era in the scientific life of America, which is still in active progress." He pointed to the centennial of the APS in 1843, Bache's U.S. Coastal Survey of 1845, Louis Agassiz's arrival in Boston in 1846, and the expansion of the American Association for the Advancement of Science in 1848, along with territorial expansion following the war with Mexico as indicative of a groundswell in American contributions to exploration and the sciences.[43] Humboldt was aware of all of these milestones. By 1847 he had been made a member of no fewer than seven American scientific and cultural organizations, each of which made a point of extolling his contributions to their fields of endeavor and to American cultural progress in the arts and sciences.

The founding of the Smithsonian in 1846 was yet another facet of this scientific era. One of Henry's earliest tasks was the design of a building to house whatever form the institution would take. Architect Robert Mills, who had designed the new Patent Office Building (now home to the Smithsonian American Art Museum) and the Washington Monument, contributed several early designs for a Smithsonian building, incorporating a "long room" reminiscent of Charles Willson Peale's original museum.[44] With a nod to European architectural messaging, Henry chose to add a castle to the Washington landscape, intending the architecture of the building to embody the intellectual spirit of Smithson's bequest.[45]

By 1851 Henry had engaged the esteemed horticulturist Andrew Jackson Downing to design the Smithsonian's grounds, corresponding with George Perkins Marsh regarding the environmental and educational purpose of the landscape.[46] Recalling the philosophy behind Samuel Vaughan's patriotic garden at the State House in Philadelphia and George Washington's western approach to Mount Vernon, Henry envisioned

> *a collection of all the trees that will grow in the climate of Washington, and, by having these trees plainly labeled with their popular and scientific names, to form a public museum of living trees and shrubs where every person visiting Washington could become familiar with the habits and growth of all the hardy trees.*[47]

It turned out to be easier to decide on the architecture and landscaping than on the Smithsonian's underlying mission. The endless debate over the parameters defining the Smithsonian's research, collections, and publications led Congress to appoint a select committee on the condition and management of the institution in 1855. These people evaluated and oversaw Henry's first decade of leadership. The committee chair was Rep. Charles Wentworth Upham of Massachusetts. Upham had advocated for a permanent home for the institution, but had been a vocal advocate for it to focus on building a national library instead of a research institute.[48] Other committee members disagreed. On his last day in office in 1855, Upham held a hearing during which he presented his committee's report. Included in that report was a letter sent by one of Humboldt's leading protégés in the United States, Professor Louis Agassiz from Harvard University. Agassiz articulated what he felt the Smithsonian should represent, reminding the committee of

the indisputable fact that the Smithsonian Institution is not an American institution. It was originated by the liberality of a high-minded English gentleman, intrusting his fortune to the United States to found in Washington an institution to increase and diffuse knowledge among men. America, in accepting the trust, has obtained the exclusive management of the most important and the most richly endowed scientific institution in the world: but it is at the same time responsible to the scientific world at large for the successful prosecution of the object of the trust, which is to increase and diffuse knowledge among men.[49]

Agassiz emphasized the global perspective that Humboldt had encouraged the United States to adopt fifty years earlier.

Agassiz felt very strongly about the need for the Smithsonian to adopt a Humboldtian commitment to encouraging and publishing research, rather than simply accumulating books or collections. In this regard, he and Henry were on the same wavelength. They both understood that Humboldt's legacy was on the pages of the books he had written, and by extension in the people who read them and whose lives were shaped by them. Humboldt's global reach was facilitated by those very publications; they formed his most durable and effective network of all. For the Smithsonian to achieve a similar stature, it needed to support and publish new research and disseminate that information. That was what they understood Smithson's "increase and diffusion of knowledge" to mean. It was, at its core, a Humboldtian precept.

Agassiz then brought Humboldt directly into his argument. Concerning how to define the parameters of an institution so founded, he continued,

The votaries of science may differ in their views about the best means of advancing science, according to the progress they have themselves made in its prosecution; but there is one standard of appreciation which cannot fail to guide rightly those who would form a candid opinion about it. I mean the lives of those who have most extensively contributed in enlarging the boundaries of knowledge. There are two individuals who may, without qualification, be as considered the most prominent scientific men of the nineteenth century, Cuvier and Humboldt. By what means have they given such powerful impulse to science? How have they succeeded, not only in increasing the amount of knowledge of their age, but also in founding new branches of science? It is by their own publications and by aiding in the publications of others; by making large collections of specimens and other scientific apparatus, and not by the accumulation of large libraries. Humboldt never owned a book, not even a copy of his own works, as I know from his own lips. 'He was too poor,' he once said to me, 'to secure a copy of them,' and all the works he receives constantly from his scientific friends are distributed by him to needy students.[50]

Here Agassiz deliberately blurred the truth, as Humboldt accumulated a vast library numbering more than 10,000 volumes at his death.[51] It is true, Humboldt noted wryly, that he could not have afforded to buy a complete set of his own books, based on how much they cost, and that publishing them had ruined him financially. However, he did own copies of his own publications. He was given many books inscribed to him by his scientific peers and grateful authors, and he sometimes paid out of his own pocket to provide books to students he admired.

Agassiz invoked Humboldt as an exemplar of what the Smithsonian should aspire to be.

Again, there is hardly a scientific man living on the continent of Europe who is not indebted to him for some recommendation in the proper quarters for assistance in the publication of their works. I mention, more particularly, these details about Humboldt, because he is happily still among the living, and his testimony may be asked in a matter of such deep importance to the real progress of science.[52]

Agassiz all but dared Congress to call Humboldt himself to testify in person as to the merits of the Smithsonian as a research institution and as an imprimatur for scientific publications. Not content to leave such matters in the hands of Congress, Agassiz simultaneously published his letter as an editorial in American and Canadian scientific journals to expand the reach of his argument.[53]

Joseph Henry had envisioned the Smithsonian as an institution that would make signature contributions to the advancement of science and publish its findings to address both directives to increase and diffuse knowledge. To that end he had titled the Smithsonian's compendia of publications "Smithsonian Contributions to Knowledge." Highly regarded European institutions hailed them for promoting "progress of science and the arts in the United States."[54] As Agassiz noted,

> *The distribution of the* Smithsonian Contributions to Knowledge *has already carried the name of the institution to all parts of the civilized world, and conveyed with them such evidence of the intellectual activity of America as challenges everywhere admiration; a result which could hardly be obtained by applying a large part of the resources of the institution to other purposes.*[55]

His point was clear: through its scholarly publications, the Smithsonian, in reach and esteem, was already America's answer to Humboldt. In 1810 Chaptal had defended Humboldt against Napoleon's jealousy, remarking that the entire French scientific academy marched in his shoes.[56] Forty-five years later, the analogy still held. Humboldt was, in essence, a learned society.

Fig. 99
Joseph Karl Stieler, *Alexander von Humboldt*, 1843, oil on canvas, 42 ⅛ × 34 ¼ in., Stiftung Preussische Schlösser & Gärten Berlin-Brandenburg

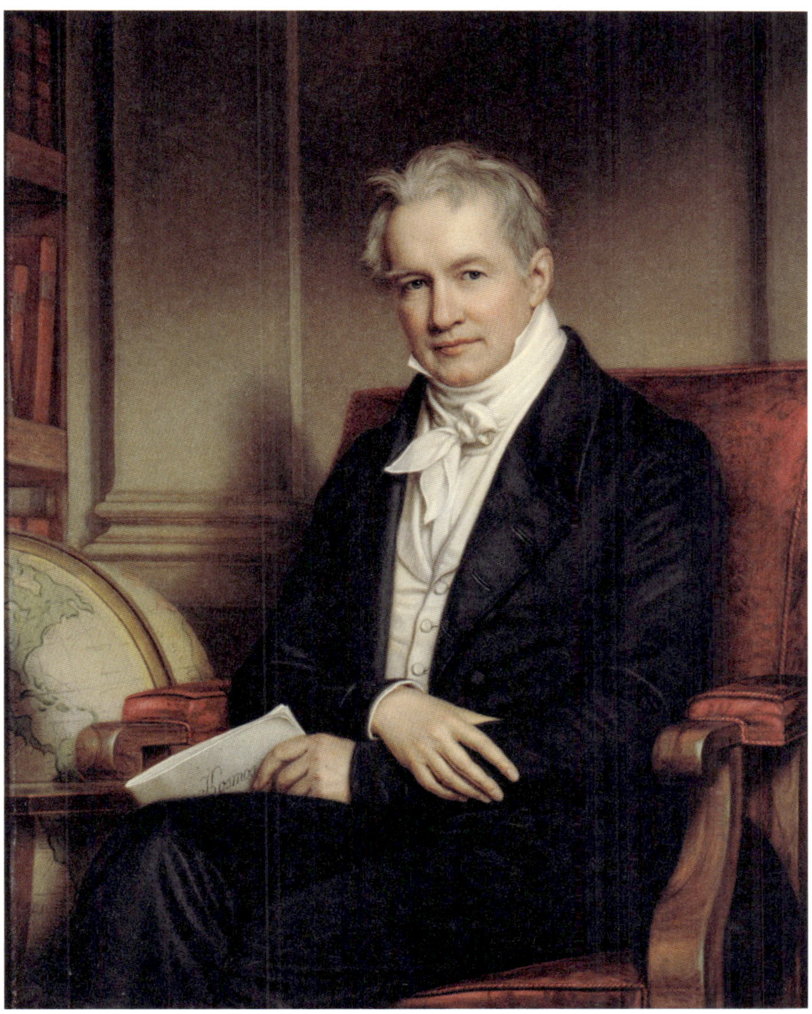

With the founding of the Smithsonian, the United States had established a direct channel to Alexander von Humboldt (fig. 99). Every volume of "Smithsonian Contributions to Knowledge" made its way into Humboldt's hands during his lifetime, and in every volume, Humboldt was cited as an authority on some matter or another.[57] Humboldt made note of the attention and in many cases provided a letter of praise commending the author for his contributions. The first publication of original research to carry the Smithsonian's name was Ephraim George Squier's *Ancient Monuments of the Mississippi Valley*, written with Edwin H. Davis, published in 1848. Squier and Davis had done much of their work with the support of the American Ethnological Society, which had been founded by Humboldt's friend Albert Gallatin. Gallatin now graciously agreed to allow this important volume to be published under the Smithsonian's auspices instead.[58] Humboldt praised Squier's work, proclaiming in a letter, "with Dr. Morton's *Crania Americana*, the work of Mr. Squier constitutes the most valuable contribution ever made to the archaeology and ethnology of America."[59]

Humboldt's imprimatur on American discoveries remained valuable for American scientists' careers. When the U.S. government considered cutting Alexander Dallas Bache's funding for the U.S. Coastal Survey, he appealed to Humboldt, who wrote a letter in 1851 arguing for the importance of his work. In 1856 the aging Humboldt sent Bache a chromolithograph showing the explorer in his study (CAT. 107), surrounded by his travel diaries and manuscripts, in a room decorated with some of the souvenirs of his travels. On the wall hangs a large map of the world, conveying the global reach of his ideas and his aspirations. Smithsonian climate scientist Lorin Blodget received a letter from Humboldt congratulating him for his work, but he was especially proud of the signed photograph he received from the august Prussian traveler (CAT. 108). Humboldt sat for a number of painted portraits over his lifetime, but few photographs of him have survived. Blodget's large-scale salted paper print is a blend of techniques, the image

CAT. 107 Eduard Hildebrandt, *Humboldt in His Study*, 1856, chromolithograph, image: 10 × 14 ¼ in., George Glazer Gallery, NY

CAT. 108 George Francis Schreiber, *Baron von Humboldt*, 1858, salted paper print, 20 × 16 ⅜ in., American Philosophical Society

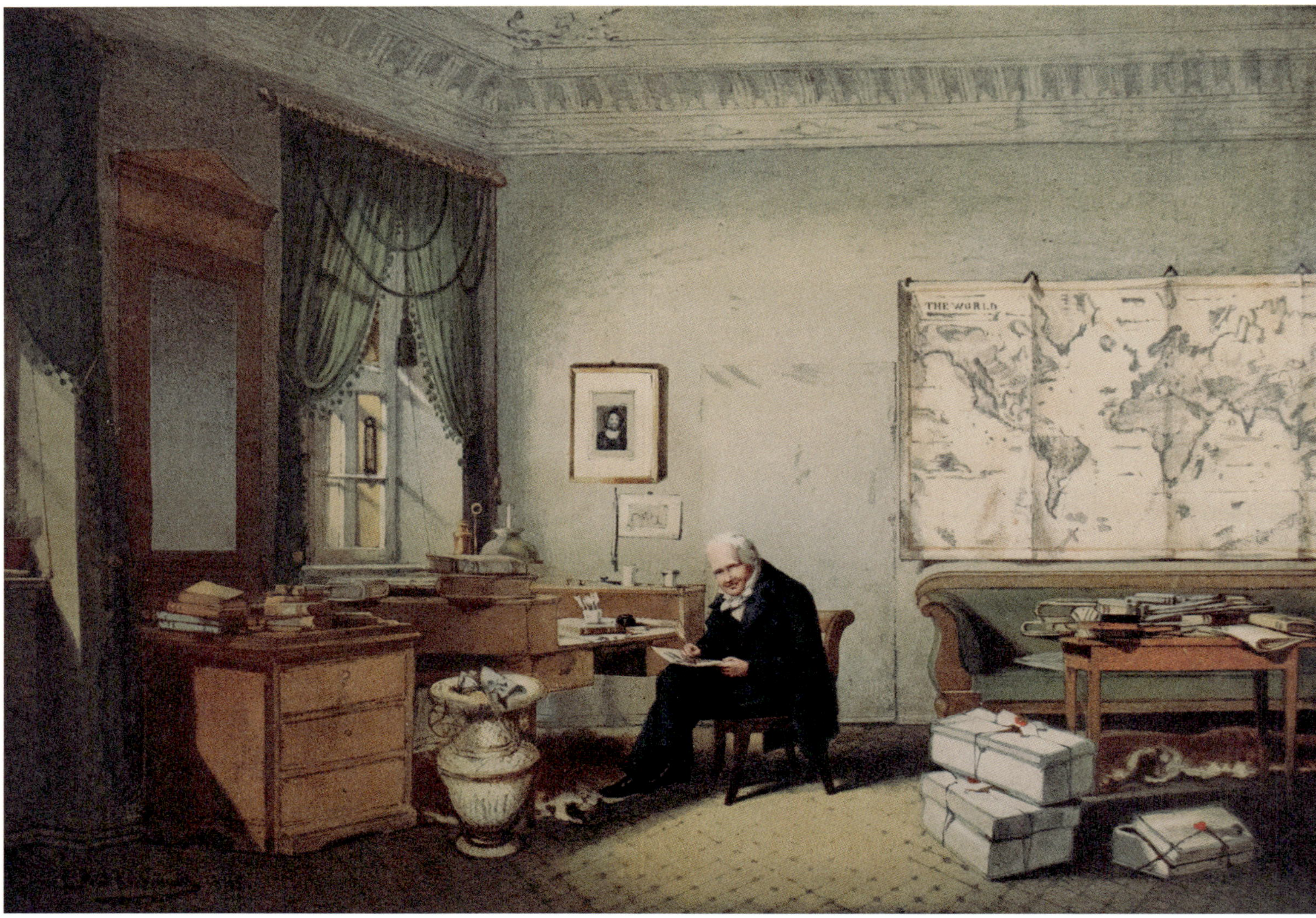

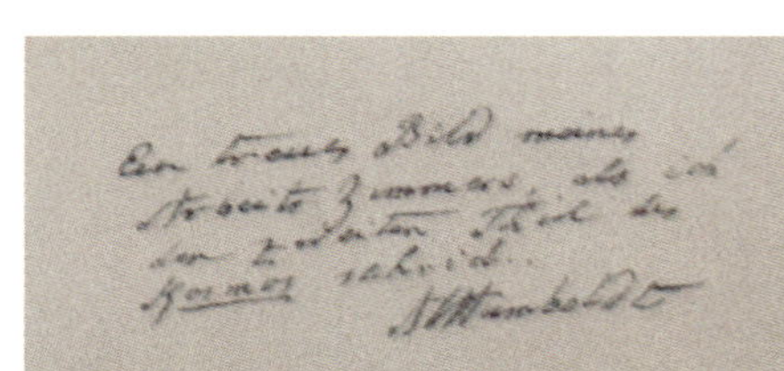

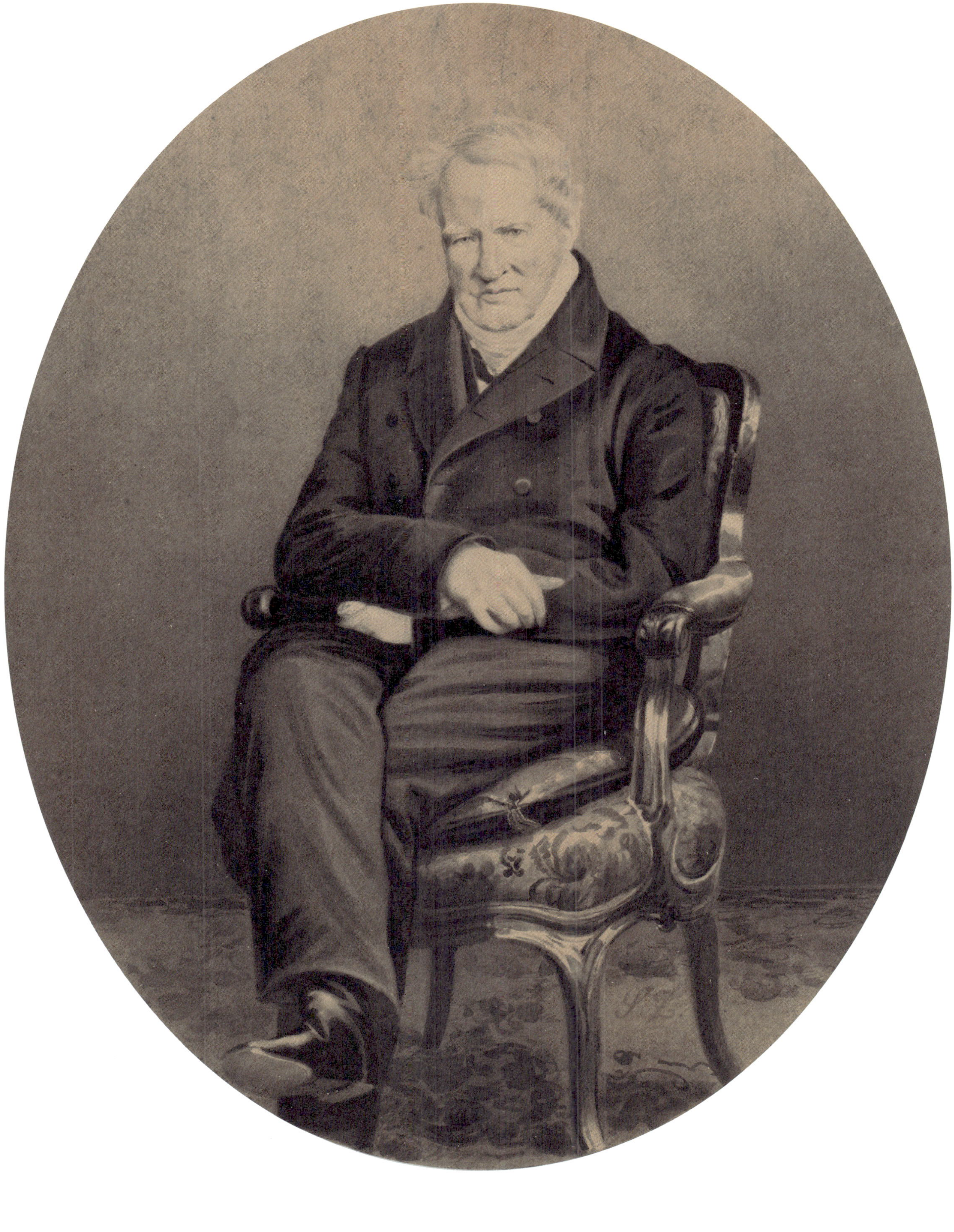

enhanced by painterly strokes most visible in Humboldt's clothing and the chair in which he sits.

Every early Smithsonian volume made its way to Humboldt via two men who became his cultural ambassadors to the United States. In 1839 German American lexicographer and diplomat Johann Gottfried Flügel was appointed the U.S. consul to Saxony in Leipzig. In 1847, after the Smithsonian was founded, Flügel became a representative for the institution to Europe. By 1849 he and Humboldt had begun a correspondence in which Humboldt queried Flügel about the development of scientific knowledge in the United States. Humboldt used Flügel as a conduit. Flügel arranged for the Smithsonian's publications to be sent to Humboldt; he also sent books by other Americans to his friend in Berlin. Humboldt often responded in letters to Flügel, asking him to convey to the author his critiques or praise. In this manner, Matthew Fontaine Maury, who had sent Humboldt his "Wind and Current Charts" in 1849, heard back via Flügel: "I beg you to express to Lieut. Maury, the author of the beautiful chart of the winds and currents, prepared with so much care and profound learning, my hearty gratitude and esteem." Humboldt deepened his praise, linking Maury's achievement favorably to his own work on isothermal lines.[60] Flügel also translated Humboldt's letter for Maury. That translation appeared in numerous American newspapers, and Maury included Humboldt's testimonial in his 1851 book publishing his charts.[61] When Maury and Joseph Henry became involved with the transatlantic cable project, Maury's participation came in part through Humboldt's estimation of his abilities.

Humboldt's old friend Friedrich von Gerolt continued to live in New York and serve as a conduit of information in his role as minister from Prussia. In 1855 Joseph Henry had his own chance to send an ambassador of sorts to Humboldt, when Smithsonian regent and Washington, D.C., art collector William Wilson Corcoran made plans to sail to Europe with his close friend, former president Millard Fillmore. It was in his official capacity that Gerolt had met Fillmore; it was through the Smithsonian that he came to know Corcoran. When Corcoran and Fillmore arrived in Berlin, Corcoran was carrying a letter on Smithsonian letterhead from Henry, announcing their arrival as emissaries of the institution. Humboldt was happy to greet them both. He arranged to have his protégé Balduin Möllhausen escort them around Potsdam and coordinated a dinner with the Prussian king. Two years later Humboldt called in the favor, asking Corcoran to do what he could to provide Möllhausen with his friendship and support as he sailed to the United States to join the Ives Expedition. For Humboldt, the Smithsonian became part of his network that served his personal, political, and scientific agendas.

HUMBOLDT'S LEGACY

Although he was now in his late eighties, it seemed as though Humboldt might live forever. He battled back from a minor stroke in February 1858, and he continued to see visitors, write letters, and work on the fifth volume of *Cosmos*. His death on May 6, 1859, was both inevitable and a shock. Newspapers across the United States ran lengthy, adulatory obituaries, often taking over the front page of the paper. In multiple cities mourners eulogized Humboldt in testimonials and at ceremonies, a measure of the extent of his influence. The *New York Times* and *Harper's Weekly* devoted their front pages for his eulogy—a show of respect more often reserved for American presidents. The writer for *Harper's Weekly* summed up the depth of Humboldt's influence in the United States, noting,

> *Between Americans and Humboldt there always existed a strong bond of sympathy. His early travels; the respect shown him on his visit to this country; his republican opinions, and his intimacy with some of our greatest men, always led him to feel a strong affection for the United States, and to show to American travellers more attention than to foreigners generally. It is not enough to say that his regard was reciprocated by the American people. No foreigner enjoyed such general veneration in this country as Baron Humboldt.*[62]

One of the immediate questions following Humboldt's death was the disposition of his estate. Less than a month after Humboldt's passing, Smithsonian Secretary Joseph Henry received a letter from Balduin Möllhausen offering the institution everything: books, specimens, maps, miscellaneous personal effects, and his massive stacks of travel journals, notes, and manuscript pages. It was an unparalleled trove of scientific data gathered over seven decades. What better place for Humboldt's legacy than an institution that Louis Agassiz had declared compatible with Humboldt's dedication to the increase and diffusion of knowledge? Möllhausen, working with Friedrich von Gerolt, wanted to place Humboldt's estate with the institution that most resembled him.

Gerolt had passed along similar letters from Möllhausen to Corcoran and James Lenox, whose book collection would become the foundation of the New York Public Library. In his letter to Corcoran (CAT. 109), Möllhausen pressed his case:

> *As it is our warmest wish, that the inheritance of Alexander von Humboldt shall not become scattered all over the world by auction; as we wish to see it placed in a way worthy of his memory, and that every member of human society may have a chance of looking with veneration upon things that have surrounded one of the greatest and best men of past and coming centuries, during lifetime and as further I hold proofs in my hands, that the illustrious deceased himself would consent with all his heart to: "His inheritance becoming the property of the United States of America," and as he has ordered "that nothing out of an erroneous delicacy shall be left unsold," I take the liberty of informing you: that Mr. Seifert is willing to dispose by way of sale, of the whole inheritance.*[63]

By purchasing his estate, the United States could claim Humboldt, the "half an American," more fully.

What transpired was a rapid exchange of letters over the summer and into the fall of 1859, as Corcoran and Lenox waited to hear if Henry intended to try to acquire the entire estate or only Humboldt's library for the Smithsonian.[64] The cost was staggering: $100,000, or roughly one-fifth of the entire Smithson bequest.[65] The timing could not have been worse; in October 1859 John Brown's raid on Harpers Ferry deflected attention to more pressing political matters in Congress.[66] Henry and Corcoran had no official authority to make such a deal without congressional approval. Gerolt recognized that and suggested that Corcoran might be interested in acquiring the works of art, and indicated a competing interest from William Backhouse Astor, of the Astor Library, for the books and manuscripts.[67]

Over the next year, several New York newspapers carried stories of the proposed sale of Humboldt's library to someone in the United States.[68] Public interest in the disposition of Humboldt's estate rippled across the country. By the summer of 1860 possession of the library had changed hands twice, ending up with Henry Stevens, an American bookseller based in London. The Humboldt library encompassed more than 10,000 volumes, among which were the larger part of his own production.[69] Ultimately Stevens arranged to sell the books over a monthlong public auction.

In advance of the auction, John C. Frémont contacted one of his lawyers, Simon Stevens, who happened to be Henry Stevens's brother. Writing a hasty memorandum, Frémont sent him an annotated list of the 153 specific volumes he hoped to acquire, focusing on his own publications (which he had inscribed to Humboldt), selected volumes of Humboldt's published work, and three custom-bound volumes containing all 181 of Humboldt's membership certificates, honorary degrees, and diplomas from professional societies and organizations worldwide.[70] Those volumes earned pride of place in the Frémonts' personal library, which became a kind of shrine to the explorer and his impact on Frémont's career. When they were forced to sell the contents of their brownstone to avert bankruptcy, the Frémonts' collection of Humboldt's books and certificates were in turn purchased by Ephraim George Squier, who cherished his acquisition. For Frémont and Squier, Humboldt had been a stalwart ally and a powerful influence. Owning his books, and in particular his albums of honors, was a capstone to each man's career. When Squier's library was sold in 1876, the three albums of diplomas (see cat. 14) ended up back in Berlin. Gerolt was still keeping an eye on Humboldt's legacy; one is left to wonder if he helped arrange for the repatriation of those volumes.

With Humboldt's death, it was as though America had indeed lost a member of the pantheon of great Americans that had graced Peale's Museum. In calling him "an ornament of our age," Jefferson had formally adopted Humboldt as an instrumental aspect of that emerging American identity. Humboldt's own pronouncements about being "half an American" were often made in the presence of visiting Americans eager to meet him. In

(copy)

Potsdam, June 1st 1859.

See page 231.

Sir,

By the last will of Alexander von Humbolt and by his legal donation, Mr Seifert, late master of the house of Alexander von Humbolt, now Royal Castellan in Berlin, has become the only inheritor of the whole property of his deceased master.

I am authorized by Mr Seifert, my father-in-law, to make this matter known throughout the United States.

As it is our warmest wish, that the inheritence of Alexander von Humbolt shall not become scattered all over the world by auction, as we wish to see it placed in a way worthy of his memory, and that every member of human society may have a chance of looking with veneration upon things that have surrounded one of the greatest and best men of past and coming centuries, during lifetime and as further I hold proofs in my hands, that the illustrious deceased himself would consent with all his heart to: "His inheritance becoming the property of the United States of America," and as he has

8311

CAT. 109 Letter from Balduin Möllhausen to William Wilson Corcoran, June 1, 1859, ink on paper, 10 ½ × 8 ⅝ in., Library of Congress, Manuscript Division, W. W. Corcoran Papers

short, for fifty years American statesmen, explorers, artists, authors, and scientists had made concerted and consistent efforts to cement the relationship between him and this country. Humboldt admired America's democratic ideals, and his impact on American cultural identity spanned the arts and sciences. What made Humboldt special were the twin impulses of nearly manic, nonstop curiosity tied to a need to know everything, and a collegial desire to encourage everyone else at the same time. Humboldt might talk all night, but he wrote thousands of letters supporting others in the pursuit of their ideas. A thread that runs through many of these remembrances was Humboldt's support for America's public institutions. The one that came closest to embodying the full range of Humboldt's interests was the Smithsonian Institution.

In October 1859 the American Geographical and Statistical Society held a memorial meeting in Humboldt's honor. The eulogies and reminiscences suggested the extent to which the Smithsonian and Humboldt seemed akin to each other. To Matthew Fontaine Maury, "Baron Humboldt was among scientists what our own Washington was among statesmen, upright and just, with attributes grand and lofty in their intellectual proportions."[70] Humboldt, he said, "used the power they gave him among men, not for self, but for the advancement of knowledge," and specifically, the encouragement of others.[71] Peter Dumont Vroom, who had served as America's minister to Prussia from 1853 to 1857, got to know Humboldt well. He noted,

> *It would seem indeed particularly fitting that this country should hold him in grateful remembrance; for although Baron von Humboldt was a Prussian by birth, and was devotedly attached to his king and his fatherland, he was a man of liberal sentiments, and took a lively interest in the progress and welfare of our Republic. Our literature; our advancement in science and the arts; our public institutions, and our growing power among the nations, were themes on which he often dwelt. He loved to speak of Henry, Bache, Maury, and Kane, with other distinguished scientific men of our land, whose works, he remarked, had done so much to elevate our national character.*[72]

Perhaps Edward Everett put it best when he sketched a picture of Humboldt that deepens the analogy of the Smithsonian as the embodiment of Humboldt's curiosity, his boundless enthusiasm, and his lifelong commitment to learning about the planet. In Everett's estimation,

> *The foundations of his fame were laid on this continent. . . . As there was no one speciality, to which he exclusively gave himself, so there is no disparagement in saying, that in almost every branch of science, there were individuals, who had pushed themselves beyond his. But it belonged to Humboldt to take an imperial survey of the whole field of Science, and to mould the mass of materials, derived from the individual researches of others, into one grand system;—himself an Intellectual Cosmos, combining the Geographer, the Antiquary, the Geologist, the Chemist, in short every separate title in his own person, akin to the scientific* Cosmos *of his own formation.*[73]

In 1869, to commemorate what would have been Humboldt's one-hundredth birthday, Louis Agassiz and other scientists and men of letters in Boston organized a daylong gathering with speeches delivered by Edward Everett and Ralph Waldo Emerson. Emerson described Humboldt as

> *the wonderful Humboldt, with his solid centre and expanded wings, marches like an army, gathering all things as he goes. How he reaches from science to science, from law to law, folding away moons and asteroids and solar systems in the clauses and parentheses of his encyclopedic paragraphs!*[74]

Agassiz's eulogy, delivered at the same celebration, concluded with the observation,

> *Every schoolboy is familiar with his methods now, but he does not know that Humboldt is his teacher. The fertilizing power of a great mind is wonderful; but as we travel farther from the sources, it is hidden from us by the very abundance and productiveness it has caused. How few of us remember that the tidal lines, the present mode of registering magnetic phenomena and oceanic currents, are but the application of Humboldt's researches, and of his graphic mode of recording them!*[75]

Agassiz emphatically made the point that Humboldt's ideas were ubiquitous to the point of obviousness. People in the United States, in fact those across the globe, had absorbed Humboldt so effectively that he himself had all but disappeared.

Commemoration of Humboldt did not stop with eulogies; public monuments soon followed. Gustav Blaeser cast an oversize bronze bust in Germany and shipped it to New York, where it was installed with great fanfare at East Drive and 59th Street; later it was moved to its current location at Central Park's "Naturalists' Gate," opposite the American Museum of Natural History. Smaller busts by one of Humboldt's favorite artists, Christian Daniel Rauch, joined the collections at Harvard and the German Cultural Garden of Cleveland. Full-length statues atop massive stone plinths appeared in Philadelphia and Pittsburgh (fig. 100), Chicago's newly established Humboldt Park, and the Missouri Botanical Garden, to name but a few locations across the United States.[76]

At the Smithsonian, Joseph Henry acquired a painted plaster bust sculpted by Henry Berger (CAT. 110), who patented his creation in 1860. This bust appears in one of Titian Ramsay Peale's photographs of the Smithsonian Castle, and with the expansion of the institution it moved to the new National Museum of Natural History. Today Humboldt presides over the botany department library, at home amid the published volumes that launched his reputation.

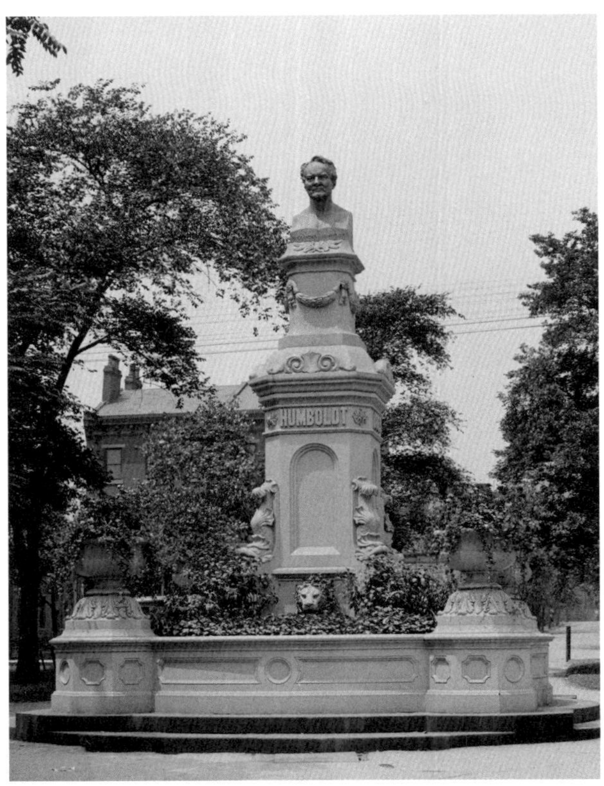

Fig. 100
Detroit Publishing Company, photo of Humboldt Statue, West Park, Pittsburgh, 1900–1920, dry-plate negative, 10 × 8 in., Library of Congress, Prints and Photographs Division

THE SMITHSONIAN AS COSMOS

The Smithsonian itself became a touchstone for many of the artists, inventors, and naturalists whose lives had been shaped by Humboldt's work. Titian Ramsay Peale had moved to Washington, D.C., in January 1843, "to review, arrange and label the collections" from the Wilkes Expedition, on which he had served as one of the naturalists.[77] The massive, uncatalogued collections had been deposited in the new Patent Office Building, where Emerson had seen the specimens on view and declared them "the best sight after the Capitol."[78] Peale applied for the inaugural position of curator for the newly founded Smithsonian, but in 1848 he became an assistant examiner with the U.S. Patent Office. Perhaps due to his interest in the Smithsonian and his family background with Peale's Museum, he and Joseph Henry developed a close friendship. In 1863 Peale climbed to the turret atop the Castle and took a series of photographs that formed a 360-degree panorama of the city of Washington.[79] Although he never joined the staff officially, Peale took numerous photographs of the offices and galleries in the Castle, most likely with Henry's encouragement.

George Catlin found his place in the Smithsonian after his 1872 return to the United States. Rebuffed so many times in his efforts to sell his Indian Gallery to the nation, the seventy-five-year-old artist and entrepreneur now had the offer of lodging and a studio within the Castle to display his paintings. Further, he now had support for a Congressional purchase of his gallery from Joseph Henry, who wrote,

> *I would respectfully urge the importance of purchasing these valuable records of the previous inhabitants of North America, which, if not secured at this time, will be dissipated and lost to the world. They will grow in importance with advancing years, and when the race of which they are the representation shall have entirely disappeared their value will be inestimable.*[80]

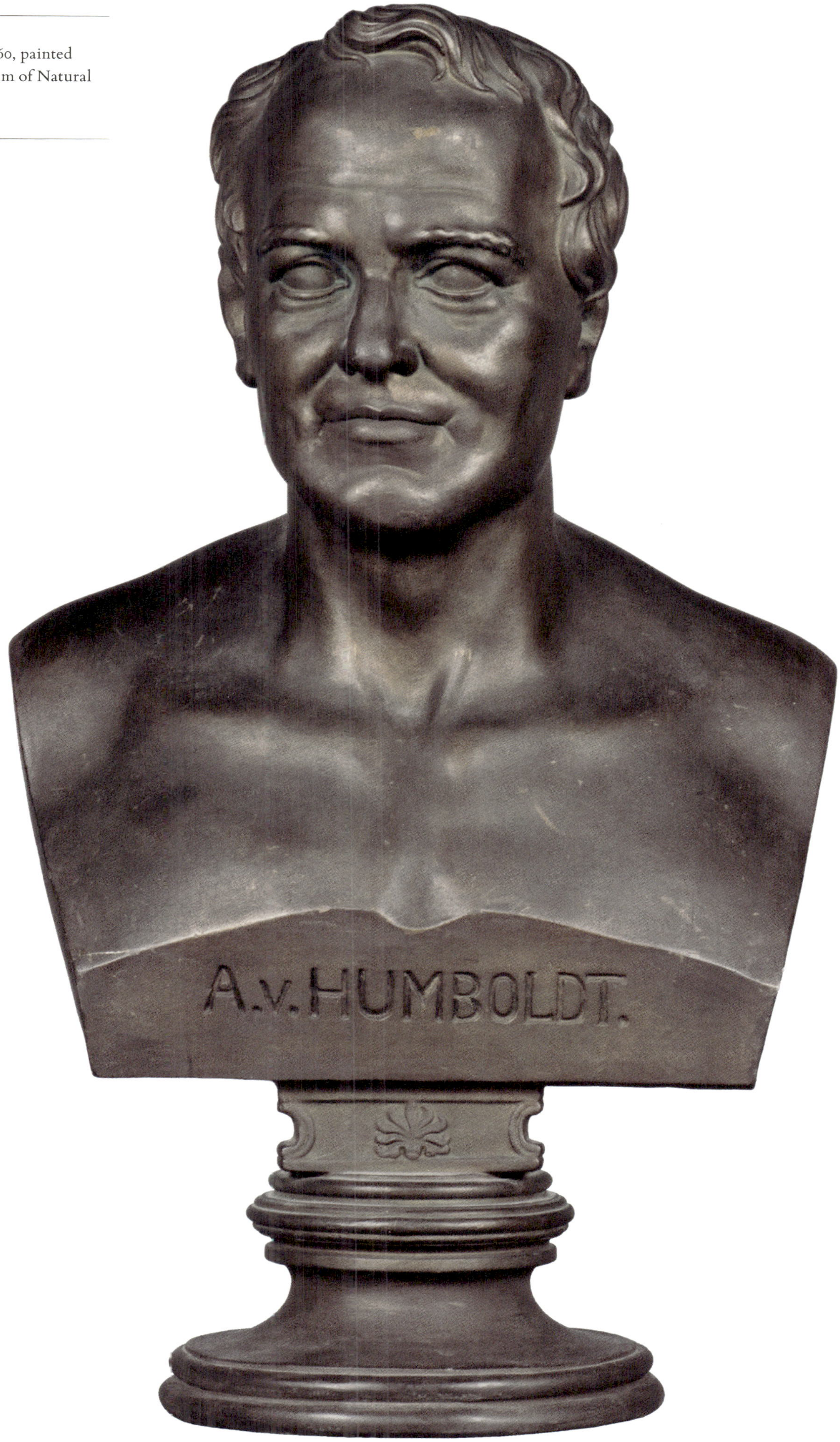

CAT. 110 Henry Berger, *Bust of Humboldt*, 1860, painted plaster, 27 × 16 × 12 ½ in., The National Museum of Natural History, Smithsonian Institution

Two years after Catlin's death, in 1879, Joseph Harrison's widow donated the more than five hundred paintings and associated artifacts to the Smithsonian. Catlin's Indian Gallery initially joined the Smithsonian's newly formed Bureau of Ethnology, headed by John Wesley Powell. Powell's regard for Humboldt inflected his work in ethnography as well as physical geology and exploration, tying the Smithsonian ever closer to Humboldt's multifaceted interests.

The impulse to tie national ambitions to natural landmarks gained renewed vigor under Powell's aegis. Powell had absorbed Humboldt's ideas about geological processes, as had the other men who led the post–Civil War surveys of the American West: Clarence King, George Wheeler, and Ferdinand Hayden. From 1867 through 1879, the four Great Surveys of the West originated in association with the Smithsonian and led to the creation of a separate federal agency, the U.S. Geological Survey. In true Humboldtian fashion, their teams comprised geologists, botanists, and cartographers. Taking a page from the Long Expedition and from Prince Maximilian's travels, each survey was accompanied by an artist: either a painter or a photographer, or both. The multipronged emphasis on the identification of mineral resources, accurate mapping and measurements, collection of ethnographic information, and the assessment of viable routes across the country placed each of them firmly in the camp of owing a legacy to Humboldt's enthusiasm for what might be found "west of these mountains." Additionally, Hayden's 1871 exploration of Yellowstone and Powell's two trips through the Grand Canyon in 1869 and 1874 set the stage for the protection of these landmark sites. Although the story of those surveys extends beyond the parameters of this study, it provides a rich field for understanding how deeply rooted Humboldt's ideas on physical geography and geological processes remained in the United States.[81]

Washington's leading scientific associations were also founded by Smithsonian scientists. In 1871 Joseph Henry founded the Philosophical Society of Washington, now the oldest scientific society in the city, with the aim of "the promotion of science, the advancement of learning, and the free exchange of views among its members on scientific subjects." Seven years later, in 1878, Powell founded the Cosmos Club, devoted to "the advancement of its members in science, literature, and art." Their founders were a roster of Humboldtian thinkers whose interests cut across disciplines and fostered the kind of intellectual curiosity that had fueled Humboldt's career and inspired so many of his protégés. Beginning in 1887 Henry's society held its meetings in the John Wesley Powell auditorium at the Cosmos Club. The organization named in Humboldt's honor also became the location of the founding of the National Geographic Society in 1888, "to increase and diffuse geographical knowledge," with many of its founding members overlapping with those of the Cosmos Club.

"HOW INTENSELY I DESIRE TO BE A HUMBOLDT!"

Humboldt's name appears on an impressive number and variety of features across the American West. John C. Frémont's campaign to name geographic landmarks for Humboldt resulted in permanent appellations that survive to this day. Humboldt's durable ethos lies behind Carleton Watkins's efforts to draw attention to Yosemite as a landscape worthy of protection and inspired John Muir to wage a lengthy campaign to preserve it, and nearby Hetch-Hetchy Valley, from development. In 1865 Muir announced, "How intensely I desire to be a Humboldt!"[82] Two years later he had begun a trek on foot that he ambitiously planned would take him from the East Coast of the United States all the way to South America. He founded the Sierra Club in 1892 as part of a larger effort to preserve the California wilderness landscape.

Perhaps the most influential Humboldt admirer at the turn of the century was President Theodore Roosevelt. Roosevelt got his start in politics after spending several years as a rancher in the Dakota Territory. Motivated by his concern about overgrazing and the disappearance of the bison and other big game due to overhunting, he gave his first major political speech in the Badlands on July 4, 1886. After invoking the Declaration of Independence, he emphasized the need for Americans to appreciate the wonders of the country. He said, "We have fallen heirs to the most glorious heritage a people ever received, and each one must do his part if we wish to show that the nation is worthy of its good fortune."[83] Roosevelt set the tone for his political career with this speech, framing his determination to preserve signature features in the American landscape in language Humboldt and Jefferson would have admired. During his 1901 to 1909 presidency, he continually urged Americans to protect the environment in order to preserve habitat for wildlife and foster an appreciation of nature.

In 1903 he and John Muir met and camped together in California's Eastern Sierra. The photograph of Roosevelt and Muir taken in Yosemite (fig. 101) captures the moment when two Humboldtian thinkers combined their interests to spearhead efforts that shaped the cultural landscape into the twentieth century in Humboldt's idiom. As president, Roosevelt established the U.S. Forest Service and created 150 national forests, fifty-one federal bird reserves, four national game preserves, five national parks, and eighteen national monuments that protected more than 230 million acres of public land. The subsequent establishment of the National Park Service in 1916 made permanent the patchwork of protections afforded national landmarks. To this day millions of Americans and foreign tourists alike make pilgrimages to stand face to face with natural icons—such as El Capitan at Yosemite, and the South Rim overlooks at the Grand Canyon—places that are concrete reminders of how inextricably intertwined America's cultural identity and the American landscape remain.

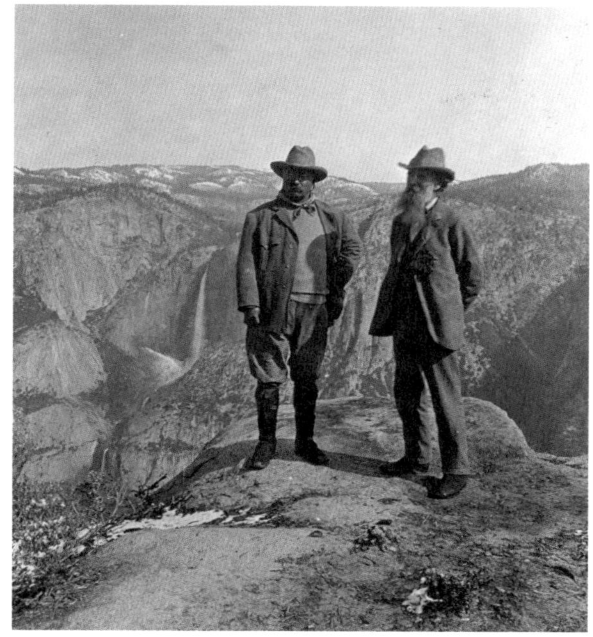

Fig. 101
Underwood & Underwood, *John Muir and Theodore Roosevelt on Glacier Point, Yosemite Valley, California* (detail), 1903, half of a sterograph, approx. 4 5/8 × 3 9/10 in., Library of Congress, Prints and Photographs Division

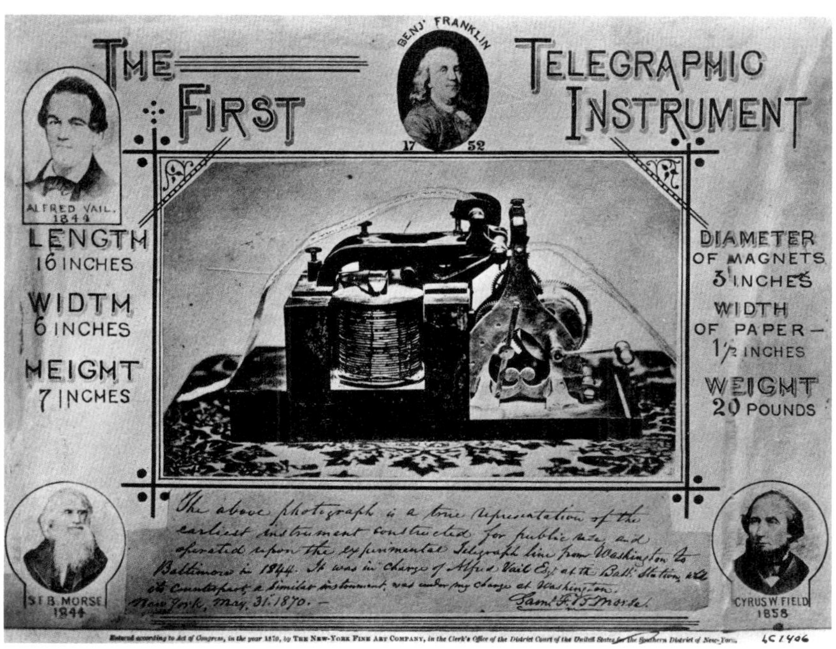

Fig. 102
The New-York Fine Art Company, *The First Telegraphic Instrument*, 1870, photomechanical print, Library of Congress, Prints and Photographs Division

Roosevelt followed through on another of Humboldt's cherished ideas, authorizing construction of the Panama Canal. Humboldt had broached the subject with Secretary of State James Madison in 1804 and often reminded his readers of his vision for this channel. He envisioned a network of waterways designed to bring commerce, people, and ideas together; however, this daunting construction project lagged behind the development of electronic communication relays developed by Morse and Field. In each of these instances it was Humboldt's American network of admirers who fulfilled his lifelong goals. In 1888 the Smithsonian began to acquire Samuel F. B. Morse's photographic equipment, multiple versions of his telegraphic equipment, and the paper tape from 1844, when Morse transcribed and returned his partner, Alfred Vail's, inaugural message, "What hath GOD wrought." An advertisement for the achievement (fig. 102), published after Morse went into business with Cyrus West Field, shows the three men and the telegraph register with its message. In 1890 Field's daughter, Isabella Judson, donated her father's memorabilia from the transatlantic cable project, including segments of the cable, the globe on which he had traced the projected route for it, and his related papers. The artifacts associated with the technologies that facilitated Humboldt's long-sought communications network thus became part of the Smithsonian's core collections.

In musing over Humboldt's disappearing legacy in the United States, Roosevelt made the trenchant comment that the educational system had shifted away from encouraging generalists in favor of specialists in each of the distinct branches of science. He said,

The average unfortunate student who has taken up scientific work in the colleges [has] been carefully trained not to do the field work which in the past has aided in producing men like Humboldt.[84]

The modern professionalization of the individual fields of scientific endeavor emanated from Humboldt's desire to become broadly knowledgeable about all of them in order for him to conceptualize *Cosmos*. Indeed, Humboldt may have lived at the last moment when it was possible to believe that a single man could know everything about the physical world. Despite Roosevelt's concern for the loss of the big picture, specialization has provided deep knowledge of our planet. Over time the expanding number of researchers and the increasing speed of communication have made it impossible for any one person to know and understand everything in the all-encompassing manner to which Humboldt aspired. Instead, in the era of ever more sophisticated and expensive technologies, institutions have undertaken his audacious goal. Today, the dozens of museums and research institutes and projects of the Smithsonian Institution have become, in effect, the physical manifestation of Humboldt's brain.

HUMBOLDT'S RELEVANCE

One of Alexander von Humboldt's animating desires was to understand the connections between our lives and the state of our planet. It is also his enduring legacy. He insisted that we consider the effect of human activity on the weather, terrain, and the viability of other species. Humboldt wrote of "mankind's mischief... which disturbs nature's order."[85] When Rachel Carson wrote *Silent Spring* in 1962, she did not mention Humboldt by name, but one cannot read her call to arms without recognizing Humboldt's ideas. Her eloquent plea for a deeper, more nuanced understanding of the downstream consequences of human action on the planet resulted in the campaign to ban the use of the pesticide DDT, and in so doing helped bring the endangered bald eagle—symbol of the nation—back from the brink of extinction. Humboldt's ideas and concerns—about plate tectonics, human-induced climate change, the possibilities of evolution, the essential equality of races, and the overarching need for us to appreciate, understand, and live in rhythm with our landscape—are as pressing and relevant now as they were when he first articulated them more than two hundred years ago.

The Smithsonian was founded to embrace the ideas Humboldt fostered, the hypotheses he promoted, and the scholarship he so admired. In 1853 Alexander Winchell, who later became one of the most renowned geologists of his day, exulted,

[T]o be once in the atmosphere of the Smithsonian Institution.... If I could once step *within its threshold, it seems to me my eager eyes and ears and understanding would drink in a world of wonders and knowledge.*[86]

Humboldt's observations, his data, and his books foster the sense of wonder Winchell expressed in the natural world. It has taken until our own lifetime to rediscover Humboldt's role in the founding of the Smithsonian and recognize his fingerprints on so many of the environmentally aware and ecocritical concerns of our time. We need only look back to Humboldt's *Naturgemälde*—his Picture of Nature—to see art and science, word and image, brought together as coeval aspects of nature enabling us to envision the planet as Humboldt described it: "one great whole, animated by the breath of life." He found excitement in his exploration of nature and believed aesthetic inspiration was as important as scientific evidence in understanding and celebrating all aspects of this planet. We would do well to rediscover Humboldt for ourselves, experiencing his joy in each discovery, finding renewed energy in each question or conundrum, and committing to understanding how that knowledge can inform our lives, inflect our culture, and enhance our stewardship of our world.

NOTES

1 S. M. Fuller, *Summer on the Lakes, in 1843* (Boston: Charles C. Little and James Brown, 1844), 233.

2 Richard Rush was a diplomat and lawyer who was appointed to represent the U.S. government in pursuing the United States' claim to Smithson's bequest. He had experience with the British legal system, having lived in London for eight years as U.S. ambassador to the Court of St James's. For details on Rush's life, see J. H. Powell, *Richard Rush: Republican Diplomat, 1780–1859* (Philadelphia: University of Pennsylvania Press, 1942). For the most detailed and thoughtful accounting of James Smithson's life and the context in which he lived, see Heather Ewing, *The Lost World of James Smithson: Science, Revolution, and the Birth of the Smithsonian* (New York: Bloomsbury, 2007).

3 Notice of Smithson's will is mentioned in the *New-York American* newspaper on January 26, 1830, six months after Smithson's death. Smithson's nephew, Hungerford, is the immediate beneficiary. News of Hungerford's death reached the United States in July 1835, and the legal proceedings to secure the bequest for the United States began immediately. Ewing, *Lost World of James Smithson,* 313, 316–17.

4 Edward C. Carter II, *"One Grand Pursuit": A Brief History of the American Philosophical Society's First 250 Years, 1743–1993* (Philadelphia: American Philosophical Society, 1993), 22.

5 John Quincy Adams diary entry, 10 January 1836, in *The Diaries of John Quincy Adams,* vol. 40, *1 June 1835–5 December 1836,* 159–60. See *The Diaries of John Quincy Adams: A Digital Collection* (Boston: Massachusetts Historical Society, 2004), http://masshist.org/jqadiaries/.

6 Sir Joseph Banks invited Humboldt to his home at 32 Soho Square, London, a distinct honor for the traveler, where Humboldt also met William Bligh. Maren Meinhardt, *A Longing for Wide and Unknown Things: The Life of Alexander von Humboldt* (London: Hurst, 2018), 44.

7 Smithson was elected to membership in the Royal Society on January 18, 1787; Blagden and Cavendish were two of Smithson's nominators. Ewing, *Lost World of James Smithson*, 20, 118.

8 William Thomson to Joseph Black, 15 November 1784, Edinburgh University Library MSS Collection, Gen 873/II/160–61; quoted in Ewing, *Lost World of James Smithson*, 69.

9 James Macie's mother died in May 1800; on 18 June 1800 Macie recorded his name as James Smithson at his bank, and the change of name was announced in the *London Gazette,* 16 February 1801, 202; cited in Ewing, *Lost World of James Smithson*, 357n3.

10 Joseph Banks had been the first to recognize the volcanic origins of the columns at Staffa, linking them to the features of the Giant's Causeway (in what is now Northern Ireland). Barthelemy Faujas de Saint-Fond, *A Journey through England and Scotland to the Hebrides in 1784*, ed. Sir Archibald Geikie (Glasgow: Hugh Hopkins, 1907), 1:123–24, 232. Heather Ewing notes that Faujas calls Smithson "Monsieur de Mecies." Ewing, *Lost World of James Smithson*, 76–77.

11 "Mineralogische Beobachtungen über einige Basalte am Rhein" (Brunswick, 1790); cited in Karl Bruhns, ed., *Life of Alexander von Humboldt. Compiled in Commemoration of the Centenary of His Birth, by J. Löwenberg, Robert Avé-Lallemant, and Alfred Dove*, trans. Jane Lassell and Caroline Lassell (London: Longmans, Green, 1873), 1:81.

12 James Louis Macie, "An Account of Some Chemical Experiments on Tabasheer," *Philosophical Transactions of the Royal Society of London* 81 (1791): 368–88.

13 Ewing, *Lost World of James Smithson*, 142, 144.

14 Humboldt to French chemist Jean-Baptiste-Joseph Delambre, Lima, 25 November 1802, in which Humboldt notes, "Nous rapportons aussi une substance siliceuse analogue au tabasheer des Indes Orientales, que M. Macié a analysé." "We also report a siliceous substance analogous to the tabasheer of the East Indies, which Mr. Macie analyzed." Humboldt to Antonio José Cavanilles, a leading Spanish botanist, Mexico, 22 April 1803, in which Humboldt refers him to "Vous vous rappellerez très bien sans doute cette substance siliceuse, ressemblant à l'opale, que M. Macié analysa en Angleterre." "You will no doubt recall that opaque-like siliceous substance that Mr. Macie analyzed in England." Both letters in Alexander von Humboldt, *Lettres americaines d'Alexandre de Humboldt*, with an introduction and notes by Dr. E. T. Hamy (Paris: E. Guilmoto, 1905), 147, 152. Translations by the author.

15 Pictet to A. Marcet, Geneva, 17 July 1806, Bibliothèque universitaire de Genève, MS fr. 666, f. 162. Pictet wrote of his admiration for Humboldt and their frequent visits to see each other. See letter from Pictet to Marcet, Paris, 19 January 1805, Bibliothèque universitaire de Genève, MS fr. 666, f. 151; both quoted in Maurice Crosland, *The Society of Arcueil: A View of French Science at the Time of Napoleon I* (Cambridge, MA: Harvard University Press, 1967), 139. Pictet was an associate member of the Société d'Arcueil and became friends with Berthollet and Blagden.

16 Ewing, *Lost World of James Smithson*, 271–72.

17 Ibid., 284.

18 Hannah Wills, "Charles Blagden's Diary: Information Management and British Science in the Eighteenth Century," *Notes and Records: The Royal Society Journal of the History of Science* 73, no. 1 (February 2019): 67, 68n44; accessed online at https://royalsocietypublishing.org/doi/10.1098/rsnr.2018.0016.

19 Reginald Heber Howe Jr., "Sir Charles Blagden, Earliest of Rhode Island Ornithologists," *American Naturalist* 39, no. 462 (June 1905): 397–404.

20 Charles Blagden diary, vol. 6 (1810–1814), entry for 29 September 1814, Charles Blagden Papers, Royal Society Archives, London.

21 See Ewing, *Lost World of James Smithson*, 114.

22 John Towill Rutt, ed., *The Life and Correspondence of Joseph Priestley* (London: R. Hunter, 1831), 2:218; quoted in Isaac Kramnick, "Eighteenth-Century Science and Radical Social Theory: The Case of Joseph Priestley's Scientific Liberalism," *Journal of British Studies* 25, no. 1 (January 1986): 1.

23 J. Priestley, "Letter xiii," in *Letters to the Right Honourable Edmund Burke, Occasioned by His Reflections on the Revolution in France, etc.* (Birmingham: Thomas Pearson, 1791), 141; quoted in Peter M. Jones, "Living the Enlightenment and the French Revolution: James Watt, Matthew Boulton, and Their Sons," *Historical Journal* 42, no. 1 (March 1999): 181.

24 Crosland, *Society of Arcueil*, 323. Humboldt met Sir Humphry Davy at Arcueil Sunday, November 7, 1813. On November 12, 1813, Davy wrote to Charles Blagden about this meeting. Royal Society Archives, Charles Blagden letters, H. 52. Davy's visit is recalled in John Davy, *Memoirs of the Life of Sir Humphry Davy* (London: Longman, Rees, Orme, Brown, Green & Longman, 1836), 1:469–70.

25 Charles Blagden diary, vol. 6 (1810–1814), entry for 2 October 1814, Charles Blagden Papers, Royal Society Archives, London.

26 Blagden's diary for 1814 mentions encounters with Humboldt on June 10, 11, 12, 14; July 29; September 26; October 2, 3, 4; November 6; and December 10, 17. His diary for 1815 notes encounters with Humboldt on January 2, 6, 9; February 7; and April 3. Charles Blagden diary, vol. 6 (1810–1814) and vol. 7 (1815–1818), Charles Blagden Papers, Royal Society Archives, London.

27 Blagden notes disapprovingly in his diary on February 5, 1815, "Tennant and Smithson at gaming house; former won latter lost." Charles Blagden diary, vol. 7 (1815–1818), Charles Blagden Papers, Royal Society Archives, London.

28 Heather Ewing has compiled financial information from Smithson's bank ledgers, with totals ranging from £100 to £1,300 each between December 1814 and March 1815. Ewing, *Lost World of James Smithson*, 401n22. According to one historical currency converter, the range in U.S. dollars would have been between $425 and $5,500. in today's currency.

29 François Arago, "Eulogy on Ampère," delivered in 1836 in French; translated for the Smithsonian and published in the *Annual Report of the Board of Regents of the Smithsonian Institution* (1872) (Washington, DC: Government Printing Office, 1873), 124–25. This story was recounted in "Anecdote of James Smithson," in *Putnam's Monthly Magazine of American Literature, Science and Art* 1, no. 1 (January 1853): 105–106. "Dr. B[enjamin]. A[pthorp]. Gould, Jr., in a lecture on 'The Theory of Probabilities', delivered, on the 24th of November, before the American Academy of Arts and Sciences, in Boston, illustrated a proposition by nattering an occurrence in the life of James Smithson." Gould was a friend of Arago's and knew Humboldt personally.

30 As Charles Wentworth Upham characterized this in his report, "Under the sting of resentment and wounded pride, he changed his will, and left his fortune to the United States of America. In this way a harvest of dissatisfaction and animosities is constantly maturing." Congressional Proceedings of the 33rd Congress, in William J. Rhees, comp. and ed., *The Smithsonian Institution: Documents Relative to Its Origin and History*, vol. 17 in *Smithsonian Miscellaneous Collections* (Washington, DC: Smithsonian Institution, 1880), 566.

31 Agassiz to Charles Upham, letter submitted to the U.S. Congress and read into the record. Agassiz's letter appeared as part of the report printed in *Smithsonian Miscellaneous Collections*. Ibid., 587.

32 Smithson's nephew, Hungerford, inherited the interest from Smithson's estate upon Smithson's death. The principal was held in trust for any children he might have. At Hungerford's death on June 5, 1835, in Italy (at age twenty-six or twenty-seven), he was unmarried and had no children. Hungerford's death led a group to come to the American chargé d'affaires in London to make him aware of the bequest. Letter from Clarke, Fynmore, and Fladgate to Aaron Vail, London, 21 July 1835, printed in Rhees, *Smithsonian Institution: Documents*, 3; cited in Ewing, *Lost World of James Smithson*, 315–16.

33 William Parson to Blagden, Paris, 20 February 1818, BLA p. 10, Royal Society Archives; quoted in Ewing, *Lost World of James Smithson*, 402n24.

34 Washington Irving echoed this when he wrote, "We have no hereditary distinctions, but we have classes of society arising from talents, manners, wealth, &c. This is a natural aristocracy." Washington Irving, *Journals*, ed. Sue Fields Ross (Boston: Twayne, 1986), 5:311; quoted in Peter Antelyes, *Tales of Adventurous Enterprise: Washington Irving and the Poetics of Western Expansion* (New York: Columbia University Press, 1990), 75–76.

35 I am grateful to Dr. Ellen Feingold, curator of numismatics, Smithsonian National Museum of American History, for her calculation of the modern monetary value of Smithson's bequest.

36 The unit of inductance, called "the henry," is named in Joseph Henry's honor.

37 Carter, *"One Grand Pursuit,"* 26.

38 For the significance of the "magnetic crusade," see chapter 4 in this volume.

39 For Washington Irving's friendship with Humboldt, see in particular chapters 2 and 4 in this volume.

40 George Perkins Marsh donated his extensive collection of prints to the Smithsonian; they are currently housed in the National Museum of American History. See Helena E. Wright, *The First Smithsonian Collection: The European Engravings of George Perkins Marsh and the Role of Prints in the U.S. National Museum* (Washington, DC: Smithsonian Institution Scholarly Press, 2015).

41 George Perkins Marsh, "The Study of Nature," *Christian Examiner* 68 (1860): 44; quoted in David Lowenthal, "Nature and Morality from George Perkins Marsh to the Millennium," *Journal of Historical Geography* 26, no. 1 (2000): 18.

42 Alexander von Humboldt, "Preface," in Balduin Möllhausen, *Diary of a Journey from the Mississippi to the Coasts of the Pacific with a United States Government Expedition*, trans. Mrs. Percy Sinnett (London: Longman, Brown, Green, Longmans & Roberts, 1858), 1: xxiii.

43 Benjamin Silliman, *American Contributions to Chemistry: An Address Delivered on the Occasion of the Celebration of the Centennial of Chemistry* (Philadelphia: Collins, 1874), 66–71; reprinted in "Centennial of Chemistry," *American Chemist* (August/September 1874): 93–96.

44 For a description of the architecture, see Richard E. Stamm, *The Castle: An Illustrated History of the Smithsonian Building*, 2nd ed. (Washington, DC: Smithsonian Books, 2012), 13.

45 Ibid., 18–21.

46 Henry to Marsh, 18 November 1851, Joseph Henry Papers, Smithsonian Institution Archives; quoted in Marc Rothenberg et al., eds., *The Papers of Joseph Henry*, vol. 8, *The Smithsonian Years: January 1850–December 1853* (Washington, DC: Smithsonian Institution, 1998), 8:260–62.

47 Records of the Commissioner of Public Buildings and Grounds, National Archives and Records Administration, letters received 3 March 1851, vol. 32; quoted in Stamm, *Castle*, 160.

48 Charles Wentworth Upham, U.S. Representative from Massachusetts, was born in Saint John, New Brunswick, Canada, and served in office from March 4, 1853 to March 3, 1855. A former Harvard classmate of Emerson's, he made enemies of the leading transcendentalists and had Nathaniel Hawthorne fired from his position with the Salem Custom House. Hawthorne exacted his literary revenge, patterning his evil Judge Pyncheon in *The House of Seven Gables* on Upham. Senator Charles Sumner once referred to Upham as "that smooth, smiling, oily man of God." Quoted in Carlos Baker, *Emerson among the Eccentrics: A Group Portrait* (New York: Viking Press, 1996), 123.

49 Louis Agassiz, "Letter on the Smithsonian Institution by Prof. Agassiz, Addressed to the Hon. Charles W. Upham," Miscellaneous Intelligence, *American Journal of Science and Arts*, 2nd series, 19, no. 56 (May 1855): 285.

50 Agassiz's letter appeared as part of the report printed as *Smithsonian Miscellaneous Collections,* vol. 17, in Rhees, *Smithsonian Institution: Documents*, 586–88. It is part of the Congressional Record.

51 See Henry Stevens, *Catalogue of the First Portion of the Humboldt Library* (London: J. Davy & Sons, 1862). The Stevens catalogue lists 11,164 separate items, more than 10,900 of which are books.

52 Louis Agassiz letter to Rep. Charles Upham of the U.S. Congress and read into the record as part of the report to the 33rd Congress, March 3, 1855—House of Representatives, in Rhees, *Smithsonian Institution: Documents*, 587.

53 Agassiz's letter in its entirety appeared in several scientific journals, among them the *American Journal of Science and Arts*, 2nd series, 19, no. 56 (May 1855): 284–85; and in *Canadian Journal* 3, no. 9 (1854 and 1855): 216–17, and in the April number for 1855.

54 C. C. Felton, professor of Greek at Harvard University, to Rep. Charles Upham of the U.S. Congress and read into the record. Rhees, *Smithsonian Institution: Documents*, 586.

55 Louis Agassiz to Rep. Charles Upham, submitted to the U.S. Congress and read into the record. Rhees, *Smithsonian Institution: Documents*, 588.

56 This story is recounted in chapter 3 in this volume.

57 *Smithsonian Contributions to Knowledge* volume 1, *Ancient Monuments of the Mississippi Valley by E. G. Squier, A.M. and E. H. Davis, M.D.* (New York: Bartlett & Welford, 1848) mentions Humboldt six times—(pp. 240, 246, 250, 253, 254, 285)—by last name only, to assert his importance and the ubiquity of his influence. No first name was necessary.

58 Extended discussion of this publication decision is in "Archaeology and the Smithsonian Institution: Editing and Publishing the Squier-Davis Manuscript," in Terry A. Barnhart, *Ephraim George Squier and the Development of American Archaeology* (Lincoln: University of Nebraska Press, 2005), 51–69.

59 Barnhart, *Ephraim George Squier and the Development of American Archaeology*, 100, 363n96; quoted in Don C. Seitz, *Letters from Francis Parkman to E. G. Squier, with biographical notes and a bibliography of E. G. Squier* (Cedar Rapids, IA: Torch Press, 1911), 12.

60 Humboldt to Flügel, Berlin, 19 June 1850; translated and quoted in Ingo Schwarz, "Alexander von Humboldt's Correspondence with Johann Gottfried Flügel," *Yearbook of German-American Studies* 46 (2011): 89. Most of what we know of the relationship between Humboldt and Flügel derives from this article.

61 Matthew Fontaine Maury, *Explanations and Sailing Directions to Accompany the Wind and Current Charts* (Washington, DC: C. Alexander, 1851), 3–4.

62 "The Late Baron Humboldt," *Harper's Weekly* 3, no. 126, 28 May 1859, 1–2.

63 Möllhausen to Corcoran, Potsdam, 1 June 1859, William Wilson Corcoran Papers, Library of Congress. In the letter, Möllhausen went on to enumerate exactly what was included: "In this is enclosed: The library of about 15,000 volumes, maps, charts, Chronometers of gold, clocks, astronomical instruments of all descriptions, meteorological instruments, copper and stone plates ready for use, works of art, among them a colossal marble bust of Alexander von Humboldt by David d'Anger, his portrait by Schrader, numerous busts, statuettes, engravings, drawings, oil paintings, portraits, precious stones, gold, silver and platina medals, collections of natural history, the whole furniture, clothes, in short everything that has been left by Alexander von Humboldt to his old, true servant."

64 Gerolt to Henry, Saratoga Springs, 2 August 1859; Lenox to Henry, Newport, RI, 14 August 1859; Möllhausen to Corcoran, Potsdam, 24 August 1859; Henry to Corcoran, Smithsonian Institution, 13 September 1859; Gerolt to Corcoran, Berlin, 13 October 1859. All letters are part of the William Wilson Corcoran Papers, Library of Congress. Möllhausen's letter to Henry was destroyed in the Smithsonian's Castle fire on January 26, 1865.

65 Möllhausen enumerated the terms of the sale: "If the citizens of the United States are inclined, to acquire for the benefit of themselves and of their descendants, the inheritance in a way and on conditions that are worthy of the memory of Alexander von Humboldt, they may appoint at once their agents and give them the necessary authority to arrange this matter speedily. To name the value of the whole undivided property is impossible, but it will be between eighty and hundred thousand dollars, insured it is with ninety thousand Prussian Thaler. As Mr. Seifert is obliged to dispose of his property by the fall of this year, I ask you most respectfully to communicate to me as soon as possible on this subject. Until the arrival of your answer, we will keep ourselves under the promise not to make any final arrangement neither, with Prussia or any other country, provided your decision will reach us before the first of September next." Möllhausen to Corcoran, Potsdam, 1 June 1859, William Wilson Corcoran Papers, Library of Congress.

66 Henry Stevens later recalled, "In 1860 I paid £4,000 for the Humboldt Library, but the war coming on after I had got the immense bulk in London, I was crushed by the weight of it. Dr. Cogswell had encouraged me to the venture, by giving me to understand that the trustees of the Astor Library would select at once between £2,000 and £3,000 worth, and Mr. Lenox had intimated that he should be glad to secure as usual certain of the rarities. On the first gun of Fort Sumter all these and many other clients shut up like clam shells, and began to practice those beautiful virtues of prudence and economy which protected themselves and at the same time ruined me." Unable to find a single buyer, Stevens arranged to hold a public auction in London in 1863. Henry Stevens, *Recollections of James Lenox and the Formation of His Library* (London, 1886; revised and elucidated by Victor H. Paltsis, New York: New York Public Library, 1951), 88–89.

67 William Backhouse Astor had been encouraged by his board member, Washington Irving, to pursue Humboldt's library. Gerolt noted this in a letter to Corcoran, in which he wrote, "having no doubt that you will feel some interest in the acquisition of A. von Humboldt's valuable library and relics of various kinds. The Astor Library would like to have the books and manuscripts of Humboldt but not the articles of art." Gerolt to Corcoran, 2 August 1859, William Wilson Corcoran Papers, Library of Congress.

68 "Literary," *New-York Daily Tribune,* 9 June 1860, 4–5. At the end of the lengthy column, this tidbit appears: "The library of Alexander von Humboldt, after all the bootless talk about securing it for America, has been definitely sold to Messrs. A. Asher & Co., the well-known booksellers of Berlin. It contains about 10,000 volumes, and many of the presentation copies of books are of the most valuable and expensive description. The sale, piecemeal, of the collection will be made toward the end of the present year."

69 The catalogue of Humboldt's library listed 158 separate entries for his own publications. Stevens, *Catalogue of the First Portion of the Humboldt Library*, 320–32, nos. 4563–721.

70 The Frémonts marked up a copy of the Henry Stevens catalogue and arranged to purchase 153 volumes prior to the public auction, including Frémont's autographed copies of his own publications, many of Humboldt's publications, and the three bound volumes of Humboldt's diplomas and certificates. That copy of the catalogue is now in the University of Illinois Library and has been digitized by HathiTrust. Simon Stevens's account ledgers for these purchases are in the University of Michigan Clements Library collection. Stevens, *Catalogue of the First Portion of the Humboldt Library.*

71 Joseph P. Thompson et al., "Proceedings: Alexander von Humboldt Commemoration," in "Humboldt Commemoration," special issue, *Journal of the American Geographical and Statistical Society* 1, no. 8 (October 1859): 226–27.

72 Letter from Vroom, Trenton, 31 May 1859. Peter Dumont Vroom was Governor of New Jersey. In 1853, Vroom accepted a federal appointment as minister to Prussia and lived in Berlin until 1857. He also served in Congress. See Thompson et al., "Proceedings: Alexander von Humboldt Commemoration," 229.

73 Edward Everett, "Number Thirty-One. The Illustrious Dead of 1859—Prescott, Bond, Hallam, von Humboldt," in *The Mount Vernon Papers* (New York: D. Appleton and Company, 1860), 285, 287–88.

74 Ralph Waldo Emerson, "Humboldt," in *The Works of Ralph Waldo Emerson: Miscellanies,* vol. 11 (Boston: Houghton, Mifflin, 1904), 457.

75 "Alexander von Humboldt: Celebration of the Centennial Anniversary of His Birth," *Littell's Living Age*, 2 October 1869, 48; reprinted from the *Boston Daily Advertiser*, 15 September 1859.

76 Three of the most notable surviving Humboldt monuments are as follows: In 1876 in West Fairmount Park, Philadelphia, along the Schuylkill River, completed as a gift from the city's German citizens and sculpted by German Friedrich Drake. Humboldt is portrayed wearing a long coat, holding a scroll inscribed COSMOS, his left hand on a globe with North America facing front. His index finger appears to land on Washington, D.C. In 1878 a statue was erected in Tower Grove Park, St. Louis, "in honor of the most accomplished traveller of this or any other age" [https://www.waymarking.com/waymarks/WMMM8K3_Alexander_Von_Humboldt_Tower_Grove_park_St_Louis_MO.]; it was a gift of Henry Shaw, St. Louis philanthropist and businessman who also founded the city's botanical garden. The sculpture is eleven feet tall. A plaque on the north side shows Chimborazo, and on the south side, the Amazon River. It was sculpted by Ferdinand von Miller II, director of the Academy of Fine Arts in Munich. Humboldt Park in Chicago had been established in 1869, and in 1892 the local German community erected a monument to him, located on North Humboldt Drive. This was sculpted by Felix Görling, cast in Europe, and shipped to the United States. Humboldt is portrayed with a globe here, too.

77 Letters from Charles Pickering, 8 January 1843, and Secretary of War A. P. Upshur, 11 January 1843, to Titian Ramsay Peale; quoted in Jessie Poesch, *Titian Ramsay Peale, 1799–1885, and His Journals of the Wilkes Expedition* (Philadelphia: American Philosophical Society, 1961), 96. It was a Herculean task made more difficult by the haphazard way they had been labeled and stored. The ensuing squabbles between the various members of the expedition and government officials damaged public interest in the findings. The fraught history of the Wilkes Expedition is a subject unto itself.

78 Ralph Waldo Emerson to Lidian Emerson, Washington, 14 January 1843, in *The Letters of Ralph Waldo Emerson,* ed. Ralph L. Rusk (New York: Columbia University Press, 1939), 3:123; quoted in Laura Dassow Walls, *The Passage to Cosmos: Alexander von Humboldt and the Shaping of America* (Chicago: University of Chicago Press, 2009), 141.

79 Titian Ramsay Peale, *The Amateur Photographic Print,* no. 7 (1 September 1863): 2; cited in Julie Link Haifley, "Capital Images: The Photography of Titian Ramsay Peale, 1855–1885," *Records of the Columbia Historical Society* 50 (1980): 240.

80 Henry to the Chairman of the Library Committee of Congress, 13 December 1873; quoted in Brian W. Dippie, *Catlin and His Contemporaries: The Politics of Patronage* (Lincoln: University of Nebraska Press, 1990), 383.

81 Aaron Sachs features Clarence King in *The Humboldt Current: Nineteenth-Century Exploration and the Roots of American Environmentalism* (New York: Viking, 2006).

82 Muir to Mrs. Carr, Trout's Mills, near Meaford, 13 September 1865, in William Frederic Badè, *The Life and Letters of John Muir* (New York: Houghton, Mifflin, 1924), 1:140.

83 National Park Service, *Theodore Roosevelt National Park, North Dakota,* "Theodore Roosevelt and Conservation," last modified 16 November 2017, https://www.nps.gov/thro/learn/historyculturetheodore-roosevelt-and-conservation.htm.

84 Roosevelt to Hugo Münsterberg, 7 May 1901; quoted in Sachs, *Humboldt Current*, 237; also cited in Keir B. Sterling, *Last of the Naturalists: The Career of C. Hart Merriam,* rev. ed. (New York: Arno Press, 1977), 392n38.

85 Alexander von Humboldt, travel diary entry for February 1800, in *Reise durch Venezuela: Auswahl aus den amerikanischen Reisetagebüchern*, ed. Margot Faak (Berlin: Akademie Verlag, 2000), 216; translated in Andrea Wulf, *The Invention of Nature: Alexander von Humboldt's New World* (New York: Knopf, 2015), 337.

86 Winchell to Spencer Baird, Eutaw, Alabama, 6 January 1853; quoted in William Healey Dall, *Spencer Fullerton Baird: A Biography* (Philadelphia: Lippincott, 1915), 290.

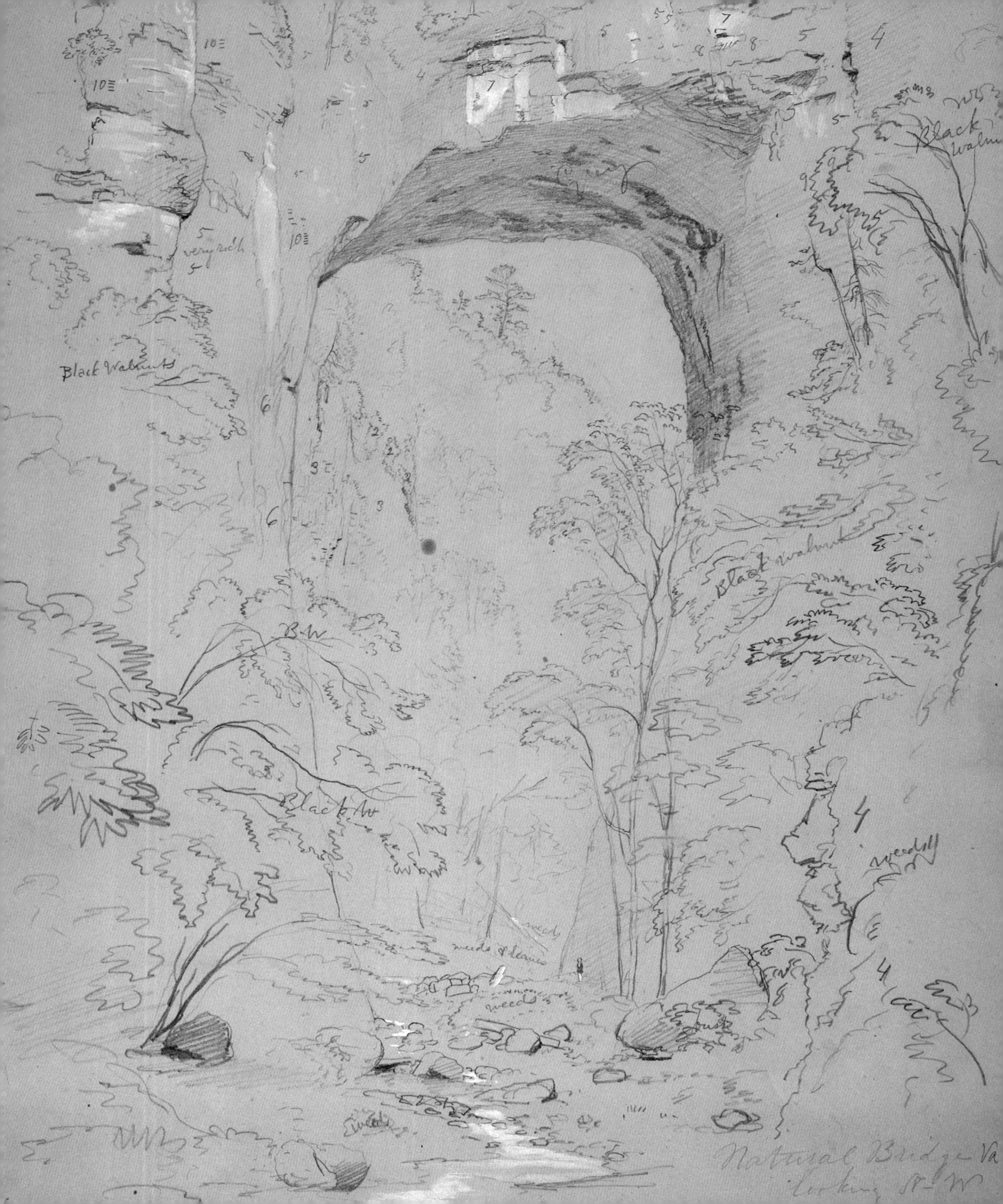
Black Walnuts
very rich
Black walnut
B-W
Black W
weeds & leaves
weeds
weedy
Natural Bridge Va

CATALOGUE OF THE EXHIBITION

CAT. 1

Friedrich Georg Weitsch (1758–1828)
Portrait of Alexander von Humboldt (1769–1859), 1806
oil on canvas, 49 ⅝ × 36 ⅜ in.
Staatliche Museen zu Berlin, Nationalgalerie

CAT. 2

Alexander von Humboldt
Baggage entry statement to the U.S. Customs, May 24, 1804
ink on paper, 9 ⅜ × 7 ⅞ in.
American Philosophical Society, Philadelphia

CAT. 3

Charles Willson Peale (1741–1827)
Self-Portrait with Mastodon Bone, 1824
oil on canvas, 26 ¼ × 22 in.
New-York Historical Society, Purchase, James B. Wilbur Fund

CAT. 4

Letter from Alexander von Humboldt to Thomas Jefferson, May 24, 1804
ink on paper, 9 ⅛ × 7 ¾ in.
Library of Congress, Manuscript Division, Thomas Jefferson Papers

CAT. 5

Rembrandt Peale (1778–1860)
Thomas Jefferson, 1805
oil on linen, 28 × 23 ½ in.
New-York Historical Society, Gift of Thomas Jefferson Bryan

CAT. 6

Copy after Alexander von Humboldt
General Chart of the Kingdom of New Spain between Parallels of 16 & 38° N., 1804
pencil and ink on tracing paper, 37 ¾ × 26 in.
Library of Congress, Geography and Map Division

CAT. 7

Eastman Johnson (1824–1906)
The Old Mount Vernon, 1857
oil on board, framed: 23 ⅜ × 34 ½ in.
Mount Vernon Ladies' Association, Purchased with funds courtesy of an anonymous donor and the Mount Vernon Licensing Fund, 2009

CAT. 8

Charles Willson Peale
Alexander von Humboldt, 1804
paper silhouette, 4 ⅝ × 4 in.
The Thomas Jefferson Foundation at Monticello, Charlottesville, VA

CAT. 9

Charles Willson Peale
Aimé Jacques Alexandre Bonpland, 1804
paper silhouette, 3 ¾ × 3 ¾ in.
The Thomas Jefferson Foundation at Monticello, Charlottesville, VA

CAT. 10

Charles Willson Peale
Anthony Fothergill, 1804
paper silhouette, 4 ⅝ × 3 ⅞ in.
The Thomas Jefferson Foundation at Monticello, Charlottesville, VA

CAT. 11

Charles Willson Peale
Nicholas Collin, 1804
paper silhouette, 4 ⅝ × 4 in.
The Thomas Jefferson Foundation at Monticello, Charlottesville, VA

CAT. 12

Raphaelle Peale (1774–1885)
Thomas Jefferson, 1804
paper silhouette, 4 ⅝ × 3 ⅞ in.
The Thomas Jefferson Foundation at Monticello, Charlottesville, VA

CAT. 13

Charles Willson Peale
Portrait of Baron von Humboldt, 1804
oil on canvas, 21 × 17 in.
The College of Physicians of Philadelphia

CAT. 14

American Philosophical Society
Alexander von Humboldt's membership certificate, 1804
ink on paper bound in an album,
closed: $24\frac{3}{16} \times 17\frac{5}{16} \times 1\frac{9}{16}$ in.
Berlin-Brandenburgischen Akademie der Wissenschaften, Archiv

CAT. 15

Alexander von Humboldt and Aimé Bonpland
Géographie des plantes Équinoxiales: Tableau physique des Andes et Pays voisins, from *Essai sur la géographie des plantes*, 1805
hand-colored print, 24 × 36 in.
Royal Botanic Gardens, Kew

CAT. 16

Alexander von Humboldt and Aimé Bonpland
Geographie der pflanzen in den Tropen-Ländern: ein Naturgemälde der Anden, 1811
black-and-white print, 24 × 36 in.
Royal Botanic Gardens, Kew

CAT. 17

Charles Willson Peale
Portrait of William Clark, 1807
oil on paper mounted on canvas, $25 \times 21\frac{1}{4}$ in.
Independence National Historic Park Collection, Philadelphia, PA

CAT. 18

Charles Willson Peale
Portrait of Meriwether Lewis, 1807
oil on wood panel, 25 × 20 in.
Independence National Historic Park Collection, Philadelphia, PA

CAT. 19

Alexander von Humboldt
Map of Mexico and the Surrounding Territories, plate 2, from *Atlas géographique et physique du royaume de la Nouvelle-Espagne*, Paris, 1811
engraving, open: 24 × 35 in., closed: 24 × 18 in.
Smithsonian Libraries

CAT. 20

Thomas Buchanan Read (1822–1872)
Portrait of John C. Frémont, 1856
oil on canvas, 40 × 30 in.
Sons of the Revolution in the State of California

CAT. 21

Thomas Buchanan Read
Portrait of Mrs. Jessie Benton Frémont, Wife of Gen. John Frémont, 1856
oil on canvas, $30 \times 24\frac{1}{2}$ in.
Braun Research Library Collection,
Autry Museum of the American West,
Gift of Elizabeth Benton Frémont, 18.G.2

CAT. 22

Benjamin O'Fallon (1793–1842)
Lewis and Clark's Special Map of the Missouri River, pages 20 and 27, 1804–32
watercolor and ink on paper, single page: $10\frac{3}{8} \times 15\frac{3}{4}$ in., overall: $10\frac{5}{8} \times 16\frac{1}{4}$ in.
Joslyn Art Museum, Omaha, Nebraska,
Gift of the Enron Art Foundation

CAT. 23

Karl Bodmer (1809–1893) and Prince Maximilian zu Wied-Neuwied (1782–1867)
Map to Illustrate the Route of Prince Maximilan of Wied in the Interior of North America, from *Reise in das innere Nord-America in den Jahren 1832 bis 1834*, Coblenz, 1839–41
hand-colored engraving on paper, 24 × 36 in.
Smithsonian Libraries, Gift of Charlotte and Lloyd Wineland

CAT. 24

Prince Maximilian zu Wied-Neuwied
American Fur Co., Manuscript Journal of Prince Maximilian's Travels in North America, Parts 7–14 (Vol. 2), 1833
ink and watercolor on paper, single page: $12\frac{3}{4} \times 7\frac{3}{4}$ in., overall: $13\frac{1}{8} \times 8\frac{1}{4} \times 1\frac{1}{4}$ in.
Joslyn Art Museum, Omaha, Nebraska,
Gift of the Enron Art Foundation, 511.NNG

CAT. 25

After Karl Bodmer, Charles Beyer, engraver, Friedrich Salathé, engraver
View of the Stone Walls on the Upper Missouri, 1840
hand-colored aquatint, image: $11\frac{13}{16} \times 17\frac{1}{4}$ in., plate mark: $15\frac{7}{8} \times 20\frac{3}{4}$ in.
Joslyn Art Museum, Omaha, Nebraska, Gift of the Enron Art Foundation, 1986.49.542.41

CAT. 26

Skeleton of the Mastodon,
excavated 1801–1802 by Charles Willson Peale
bone, wood, and papier mâché,
approx. 118 × 177 × 65 in.
Hessisches Landesmuseum,
Darmstadt, Germany

CAT. 27

Rembrandt Peale
Sketch of the Elephant (Mastodon), ca. 1801
pencil on paper, $15\frac{5}{16} \times 12\frac{3}{4}$ in.
American Philosophical Society, Philadelphia

CAT. 28

Charles Willson Peale
Exhumation of the Mastodon, ca. 1806–1808
oil on canvas, $49 \times 61\frac{1}{2}$ in.
Maryland Historical Society, Baltimore City Life Museum Collection, Gift of Bertha White in memory of her husband, Harry White, BCLM-MA.5911

CAT. 29

Charles Willson Peale
The Artist in His Museum, 1822
oil on canvas, $103\frac{3}{4} \times 79\frac{7}{8}$ in.
Pennsylvania Academy of Fine Arts, Philadelphia, Gift of Mrs. Sarah Harrison (The Joseph Harrison Jr. Collection)

CAT. 30

Titian Ramsay Peale (1799–1885)
The Gigantic Mastodon, 1821
watercolor wash on paper, $14\frac{1}{2} \times 18\frac{7}{8}$ in.
American Philosophical Society, Philadelphia

CAT. 31

Charles Willson Peale and Titian Ramsay Peale
The Long Room, Interior of Front Room in Peale's Museum, 1822
watercolor over pencil on paper, $14 \times 20\frac{3}{4}$ in.
Detroit Institute of Arts, Founders Society Purchase, Director's Discretionary Fund

CAT. 32

Alexander von Humboldt
Natural Bridge of the Icononzo, plate 4, from *Researches, Concerning the Institutions & Monuments of the Ancient Inhabitants of America*, London, 1814
engraving, open: 9 × 11 in., closed: 9 × 6 × 1 in.
Smithsonian Libraries, Gift of the Burndy Library

CAT. 33

Rembrandt Peale
Portrait of Alexander von Humboldt, 1809–12
oil and encaustic on canvas, 22 ½ × 27 in.
Collection of Robert W. Hoge and Immaculada Socias Hoge

CAT. 34

Samuel F. B. Morse's Passport, signed by Washington Irving, issued December 19, 1829
ink on paper, 16 ¼ in × 10 ½ in.
Smithsonian's National Museum of American History

CAT. 35

Samuel F. B. Morse (1791–1872)
Gallery of the Louvre, 1831–33
oil on canvas, 73 ¾ × 108 in.
Terra Foundation for American Art, Daniel J. Terra Collection, 1992.51

CAT. 36

Samuel F. B. Morse and Alfred Vail
Telegraph Key, ca. 1845
wood, brass, and ivory, overall: 3 ½ × 4 ½ × 8 in.
Smithsonian's National Museum of American History

CAT. 37

Samuel F. B. Morse and Alfred Vail
Telegraph Register, 1848
wood, brass, iron, and copper, overall: 7 ¼ × 4 × 11 ½ in.
Smithsonian's National Museum of American History

CAT. 38

Samuel F. B. Morse
Reproduction of *"What hath GOD wrought" Telegraph Message*, 1844
ink on paper, 1 ½ × 27 ¾ in.
Smithsonian's National Museum of American History

CAT. 39

Newton & Son
Newton's New and Improved Terrestrial Globe, 1852
wood, paper, brass, and glass, overall: 42 × 26 in.
Smithsonian's National Museum of American History

CAT. 40

Frederic Edwin Church (1826–1900)
Aurora Borealis, 1865
oil on canvas, 56 × 83 ½ in.
Smithsonian American Art Museum, Gift of Eleanor Blodgett, 1911.4.1

CAT. 41

Daniel Huntington (1816–1906)
The Atlantic Cable Projectors, 1895
oil on canvas, 87 × 108 ¼ in.
New York State Museum, Albany, NY

CAT. 42

Tiffany and Co.
Tiffany's Transatlantic Cable Souvenir, 1858
steel, brass, copper, and gutta-percha, 4 × ½ in. diam.
Smithsonian's National Museum of American History

CAT. 43

After Cyrus West Field
Facsimile letter accompanying the cable souvenir, 1858
ink on paper, 7 ½ × 4 ⅝ in.
Smithsonian's National Museum of American History

CAT. 44

After Eduard Hildebrandt (1818–1868)
Humboldt in His Library, 1856
chromolithograph on paper, 18 ⅝ × 26 ⅜ in.
Collection of Mr. and Mrs. Robert F. Norfleet Jr.

CAT. 45

Unidentified photographer
John C. Frémont with "Fremont and Freedom!" banner, 1856
ambrotype housed in half of original leatherette case, 3 ¼ × 3 ¾ in
Collection of Alan V. Weinberg

CAT. 46

Martin M. Lawrence (1808–1859)
John Brown, ca. 1858
salted paper print, 7 ⅜ × 5 ⁵⁄₁₆ in.
National Portrait Gallery, Smithsonian Institution, NPG.74.76

CAT. 47

John Rogers (1829–1904)
The Slave Auction, 1859
painted plaster, 13 ½ × 9 × 8 ½ in.
Albany Institute of History & Art, Gift of Mrs. Ledyard Cogswell Jr., from the Benjamin Walworth Arnold Collection

CAT. 48

John Rogers
The Wounded Scout, a Friend in the Swamp, 1864
painted plaster, 22 ⅛ × 11 ⅛ × 8 ¼ in.
Smithsonian American Art Museum, Gift of John Rogers and Son, 1882.1.5

CAT. 49

John Quincy Adams Ward (1830–1910)
The Freedman, 1863
bronze, 19 ½ × 14 ¹¹⁄₁₆ × 9 ⅝ in.
Boston Athenæum, gift of Elizabeth Frothingham (Mrs. William L.) Parker, 1922

CAT. 50

Carleton E. Watkins (1829–1916)
El Capitan, ca. 1860s
albumen silver print, 25 × 20 in.
Smithsonian American Art Museum, Gift of Isaac Lagnado, 1992.78

CAT. 51

Carleton E. Watkins
Cascade, Nevada Falls, Yosemite, California, ca. 1861
albumen silver print, 15 ⅝ × 20 ⅞ in.
Smithsonian American Art Museum, Museum purchase from the Charles Isaacs Collection made possible in part by the Luisita L. and Franz H. Denghausen Endowment, 1994.91.276

CAT. 52

Carleton E. Watkins
The Grizzly Giant, Mariposa Grove, Yosemite, California, 1861, printed after 1875
albumen silver print, 12 × 7 ⅞ in.
Smithsonian American Art Museum, Museum purchase from the Charles Isaacs Collection made possible in part by the Luisita L. and Franz H. Denghausen Endowment, 1994.91.278

CAT. 53

Carleton E. Watkins and Isaiah West Taber (1830–1912)
Yosemite Falls, ca. 1865–66, printed after 1875
albumen silver print, 12 × 8 in.
Smithsonian American Art Museum, Museum purchase from the Charles Isaacs Collection made possible in part by the Luisita L. and Franz H. Denghausen Endowment, 1994.91.282

CAT. 54

Albert Bierstadt (1830–1902)
Cho-Looke, The Yosemite Fall, 1864
oil on canvas, 34 ¼ in. × 27 ⅛ in.
Timken Museum of Art, Putnam Foundation

CAT. 55

Albert Bierstadt
Valley of the Yosemite, 1864
oil on paperboard, 11 ⅞ × 19 ¼ in.
Museum of Fine Arts, Boston, Gift of Martha C. Karolik for the M. and M. Karolik Collection of American Paintings, 1815–1865, 47.1236

CAT. 56

John Matthews
Alexander von Humboldt on the Fugitive Slave Law, Oct. 12, 1858, 1864
ink on paper, 4 ¾ × 2 ⅝ in.
Library of Congress, Manuscript Division, Abraham Lincoln Papers

CAT. 57

E. Haines
Alexander von Humboldt, 1858
albumen silver print, 7 3/16 × 5 ⅝ in.
National Portrait Gallery, Smithsonian Institution, S/NPG.77.28

CAT. 58

George Catlin (1796–1872)
Máh-to-tóh-pa, Four Bears, Second Chief, in Full Dress, 1832
oil on canvas, 29 × 24 in.
Smithsonian American Art Museum, Gift of Mrs. Joseph Harrison Jr., 1985.66.128

CAT. 59 (same journal as cat. 24)

Prince Maximilian zu Wied-Neuwied
Indian Ponies, Manuscript Journal of Prince Maximilian's Travels in North America, Parts 7–14 (Vol. 2), 1833
ink and watercolor on paper, single page: 12 ¾ × 7 ¾ in., overall: 13 ⅛ × 8 ¼ × 1 ¼ in.
Joslyn Art Museum, Omaha, Nebraska, Gift of the Enron Art Foundation, 511.NNG

CAT. 60

After Karl Bodmer, Alexandre Damien Manceau, engraver
Travellers Meeting Minatarre Indians near Fort Clark, 1842
hand-colored aquatint, image: 7 ¾ × 11 ⅝ in., plate mark: 11 15/16 × 13 ½ in.
Joslyn Art Museum, Omaha, Nebraska, Gift of the Enron Art Foundation, 1986.49.542.74

CAT. 61

After Karl Bodmer, Johann Hürlimann, engraver
Mató-Tópe, a Mandan Chief, 1839
hand-colored aquatint, image: 16 ¼ × 12 ¾ in., plate mark: 20 ¼ × 14 ⅝ in.
Joslyn Art Museum, Omaha, Nebraska, Gift of the Enron Art Foundation, 1986.49.517.13

CAT. 62

After Karl Bodmer, Alexandre Zschokke, engraver
Indian Utensils and Arms, 1840
hand-colored aquatint, image: 11 13/16 × 17 in., plate mark: 16 × 20 ¾ in.
Joslyn Art Museum, Omaha, Nebraska, Gift of the Enron Art Foundation, 1986.49.542.21

CAT. 63

Mató-Tópe (about 1784–1837)
Battle with a Cheyenne Chief, 1834
watercolor and pencil on paper, 12 ⅜ × 15 ⅜ in.
Joslyn Art Museum, Omaha, Nebraska, Gift of the Enron Art Foundation, 1986.49.384

CAT. 64

After Karl Bodmer, Louis René Lucien Rollet, engraver
Pehriska-Ruhpa, Moennitarri Warrior in the Costume of the Dog Dance, 1841
aquatint and engraving on paper, image: 18 5/16 × 13 ⅛ in., plate mark: 21 1/16 × 15 ¼ in.
Joslyn Art Museum, Omaha, Nebraska, Gift of the Enron Art Foundation, 1986.49.542.23

CAT. 65

Maximilian zu Wied-Neuwied
Pehriska-Ruhpa in Costume of Dog Dancer, 1834
watercolor and pencil on paper, 11 ⅝ × 7 ½ in.
Joslyn Art Museum, Omaha, Nebraska, Gift of the Enron Art Foundation, 1986.49.490.A

CAT. 66

After Karl Bodmer, Narciso Edmundo Jose Desmadryl, engraver
Interior of a Hut of a Mandan Chief, 1841
hand-colored aquatint, image: 11 1/16 × 16 ¼ in., plate mark: 16 ⅜ × 21 5/16 in.
Joslyn Art Museum, Omaha, Nebraska, Gift of the Enron Art Foundation, 1986.49.542.19

CAT. 67

George Catlin
Bull Dance, Mandan O-kee-pa Ceremony, 1832
oil on canvas, 23 ¼ × 28 in.
Smithsonian American Art Museum, Gift of Mrs. Joseph Harrison Jr., 1985.66.505

CAT. 68

After Karl Bodmer, Johann Hürlimann, engraver
Sih-Chida and Mahchsi-Karehde, Mandan Indians, 1841
hand-colored aquatint, image: 15 ⅝ × 14 ⅝ in., plate mark: 19 13/16 × 16 ½ in.
Joslyn Art Museum, Omaha, Nebraska, Gift of the Enron Art Foundation, 1986.49.517.20

CAT. 69

After Karl Bodmer, Alexandre Damien Manceau, engraver
Bison Dance of the Mandan Indians in front of Their Medicine Lodge in Mih-Tutta-Hankush, 1842
hand-colored aquatint, image: 12 1/16 × 17 ⅜ in., plate mark: 16 ½ × 21 5/16 in.
Joslyn Art Museum, Omaha, Nebraska, Gift of the Enron Art Foundation, 1986.49.542.18

CAT. 70

George Catlin
No-ho-mun-ya, One Who Gives No Attention, 1844
oil on canvas, 29 × 24 in.
Smithsonian American Art Museum, Gift of Mrs. Joseph Harrison Jr., 1985.66.522

CAT. 71

George Catlin
Neu-mon-ya, Walking Rain, War Chief, 1844–45
oil on canvas, 29 × 24 in.
Smithsonian American Art Museum, Gift of Mrs. Joseph Harrison Jr., 1985.66.518

CAT. 72

George Catlin
Wash-ka-mon-ya, Fast Dancer, a Warrior, 1844–45
oil on canvas, 29 × 24 in.
Smithsonian American Art Museum,
Gift of Mrs. Joseph Harrison Jr., 1985.66.520

CAT. 73

George Catlin
See-non-ty-a, an Iowa Medicine Man, 1844–45
oil on canvas, 27 15/16 × 22 13/16 in.
National Gallery of Art, Paul Mellon Collection

CAT. 74

George Catlin
Shon-ta-yi-ga, Little Wolf, a Famous Warrior, 1844–45
oil on canvas, 29 × 24 in.
Smithsonian American Art Museum,
Gift of Mrs. Joseph Harrison Jr., 1985.66.521

CAT. 75

George Catlin
Ru-ton-ye-wee-ma, Strutting Pigeon, Wife of White Cloud, 1844
oil on canvas, 29 × 24 in.
Smithsonian American Art Museum,
Gift of Mrs. Joseph Harrison Jr., 1985.66.525

CAT. 76

George Catlin
Ru-ton-wee-me, Pigeon on the Wing, 1844
oil on canvas, 29 × 24 in.
Smithsonian American Art Museum,
Gift of Mrs. Joseph Harrison Jr., 1985.66.526

CAT. 77

George Catlin
Koon-za-ya-me, Female War Eagle, 1844
oil on canvas, 29 × 24 in.
Smithsonian American Art Museum,
Gift of Mrs. Joseph Harrison Jr., 1985.66.528

CAT. 78

Karl Girardet (1813–1871)
Danse d'indiens Iowas devant le roi Louis-Philippe aux Tuileries (Dance of the Iowa Indians before the King Louis-Philippe at the Tuileries), 1845
oil on canvas, 15 9/16 × 21 1/16 in.
Établissement public du château, du musée et du domaine national de Versailles

CAT. 79

George Catlin
Catlin's Notes of Eight Years' Travels and Residence in Europe, with His North American Indian Collection (vols. 1 and 2), 1848
embossed and gilt cloth covers, engraving and letterpress printing
open: 9 × 11 1/4 in., closed: 9 × 6 × 1 in.
National Gallery of Art Library,
David K. E. Bruce Fund

CAT. 80

George Catlin
Buffalo Hunt under the Wolf-Skin Mask, 1832–33
oil on canvas, 24 × 29 in.
Smithsonian American Art Museum,
Gift of Mrs. Joseph Harrison Jr., 1985.66.414

CAT. 81

Alexander von Humboldt
Cosmos: A Sketch of a Physical Description of the Universe, 1849
ink on paper, open: 7 1/2 × 10 in.,
closed: 7 1/2 × 5 × 1 2/5 in.
Olana State Historic Site, New York State Office of Parks, Recreation and Historic Preservation, OL.1984.376.2

CAT. 82

Frederic Edwin Church
The Natural Bridge, Virginia, 1852
oil on canvas, 28 × 23 in.
The Fralin Museum of Art at the University of Virginia, Gift of Thomas Fortune Ryan

CAT. 83

Frederic Edwin Church
Tequendama Falls near Bogotá, Colombia, 1853
pencil and gouache on paper,
18 1/8 × 12 3/8 in.
Cooper Hewitt, Smithsonian Design Museum,
Gift of Louis P. Church, 1917-4-265

CAT. 84

Frederic Edwin Church
Tequendama Falls near Bogotá, Colombia, July 1853
pencil and gouache on paper, 18 1/8 × 12 1/2 in.
Cooper Hewitt, Smithsonian Design Museum,
Gift of Louis P. Church, 1917-4-260

CAT. 85

Frederic Edwin Church
The Falls of Tequendama, near Bogotá, New Grenada, 1854
oil on canvas, 60 7/16 × 48 1/16 in.
Cincinnati Art Museum,
The Edwin and Virginia Irwin Memorial

CAT. 86

Frederic Edwin Church
Studies of Trees and Plants seen from the Río Magdalena, Colombia, 1853
pencil on paper, 8 7/16 × 10 7/8 in.
Cooper Hewitt, Smithsonian Design Museum,
Gift of Louis P. Church, 1917-4-135

CAT. 87

Frederic Edwin Church
Botanical Sketches from Colombia: Bamboo Trees, Trunks and details, July 1853
pencil on paper, 4 7/8 × 7 13/16 in.
Cooper Hewitt, Smithsonian Design Museum,
Gift of Louis P. Church, 1917-4-85

CAT. 88

Frederic Edwin Church
A Group of Thirteen Botanical Studies from Colombia or Ecuador, 1853
pencil on paper, 10 13/16 × 18 1/16 in.
Cooper Hewitt, Smithsonian Design Museum,
Gift of Louis P. Church, 1917-4-879

CAT. 89

Frederic Edwin Church
Studies from Guacamaya, Bogotá, Colombia, June–July 1853
pencil and watercolor on paper, 10 3/4 × 14 5/8 in.
Cooper Hewitt, Smithsonian Design Museum,
Gift of Louis P. Church, 1917-4-877

CAT. 90

Alexander von Humboldt
Frontispiece to *Researches, Concerning the Institutions & Monuments of the Ancient Inhabitants of America,* London, 1814
engraving, open: 9 × 11 in., closed: 9 × 6 × 1 in.
Smithsonian Libraries

CAT. 91

Frederic Edwin Church
Cotopaxi, 1855
oil on canvas, 28 × 42 in.
Smithsonian American Art Museum,
Gift of Mrs. Frank R. McCoy, 1965.12

CAT. 92

Alexander Von Humboldt
Chimborazo from the Plains of Tapia, plate 25, from *Vues des Cordillères et monuments des peuples indigènes de l'Amérique,* Paris, 1810
color engraving, open: 23 × 34 in.,
closed: 23 × 18 in.
Smithsonian Libraries, Gift of the Burndy Library

CAT. 93
Frederic Edwin Church
Botanical sketches, May 1857
pencil on paper, 4 ¼ × 7 ⅛ in.
Cooper Hewitt, Smithsonian Design Museum, Gift of Louis P. Church, 1917-4-835-A

CAT. 94
Frederic Edwin Church
Niagara, 1857
oil on canvas, 40 × 90 ½ in.
National Gallery of Art, Corcoran Collection (Museum Purchase, Gallery Fund)

CAT. 95
Frederic Edwin Church
Horseshoe Falls, December 1856–January 1857
oil on paper mounted to canvas, 11 ½ × 35 ⅝ in.
Olana State Historic Site, New York State Office of Parks, Recreation and Historic Preservation, OL.1981.15.A

CAT. 96
Rafael Salas (1824–1906)
Portrait of Humboldt, 1857
oil on canvas, 27 ¾ × 22 in.
Private Collection

CAT. 97
Frederic Edwin Church
Mount Chimborazo, Ecuador, 1857
oil and pencil on paperboard, 13 7/16 × 20 9/16 in.
Cooper Hewitt, Smithsonian Design Museum, Gift of Louis P. Church, 1917-4-1296-b

CAT. 98
Frederic Edwin Church
Mount Chimborazo, Ecuador, 1857
oil and pencil on paperboard, 9 ¼ × 17 ½ in.
Cooper Hewitt, Smithsonian Design Museum, Gift of Louis P. Church, 1917-4-1296-a

CAT. 99
Frederic Edwin Church
Mount Chimborazo at Sunset, 1857
oil on academy board, 12 × 21 ⅖ in.
Olana State Historic Site, New York State Office of Parks, Recreation and Historic Preservation, OL.1980.1884

CAT. 100
Frederic Edwin Church
Study for "The Heart of the Andes," 1858
oil on canvas, 10 ¼ × 18 ¼ in.
Olana State Historic Site, New York State Office of Parks, Recreation and Historic Preservation, OL.1981.47.A.B

CAT. 101 (same book as cat. 92)
Alexander von Humboldt
View of Cayambe in the Region of Quito, Ecuador, plate 42, from *Vues des Cordillères et monuments des peuples indigènes de l'Amérique*, Paris, 1810
color engraving, open: 23 × 34 in., closed: 23 × 18 in.
Smithsonian Libraries, Gift of the Burndy Library

CAT. 102
Frederic Edwin Church
Cayambe, 1858
oil on canvas, 30 × 48 ⅛ in.
New-York Historical Society, The Robert L. Stuart Collection, the gift of his widow Mrs. Mary Stuart

CAT. 103
Frederic Edwin Church
Mount Chimborazo through Rising Mist and Clouds, 1857
oil and pencil on paperboard, 13 9/16 × 21 ⅛ in.
Cooper Hewitt, Smithsonian Design Museum, Gift of Louis P. Church, 1917-4-824

CAT. 104
William James Stillman (1828–1901)
The Philosophers' Camp in the Adirondacks, 1858
oil on canvas, 20 × 30 in.
Courtesy William Munroe Special Collections–Concord Free Public Library Corporation

CAT. 105
Frederic Edwin Church
Clouds Over Olana, August 1872
oil on paper, 8 ⅔ × 12 ⅛ in.
Olana State Historic Site, New York State Office of Parks, Recreation and Historic Preservation, OL.1976.1

CAT. 106
Henri-Joseph Johns (1761–1843)
James Smithson, May 11, 1816
gouache on ivory, 3 × 2 ¾ in.
National Portrait Gallery, Smithsonian Institution; transfer from the National Museum of American History, Conserved with funds from the Smithsonian Women's Committee, NPG.85.44

CAT. 107
Eduard Hildebrandt
Humboldt in His Study, 1856
chromolithograph, image: 10 × 14 ¼ in., sheet: 14 ½ × 17 ¾ in.
George Glazer Gallery, NY

CAT. 108
George Francis Schreiber (1803–1892)
Baron von Humboldt, 1858
salted paper print, 20 × 16 ⅜ in.
American Philosophical Society, Philadelphia

CAT. 109
Letter from Balduin Möllhausen to William Wilson Corcoran, June 1, 1859
ink on paper, 10 ½ × 8 ⅝ in.
Library of Congress, Manuscript Division, W. W. Corcoran Papers

CAT. 110
Henry Berger (1825–1925?)
Bust of Humboldt, 1860
painted plaster, 27 × 16 × 12 ½ in.
The National Museum of Natural History, Smithsonian Institution

SELECTED BIBLIOGRAPHY

ARCHIVAL SOURCES

American Philosophical Society, Philadelphia

Rare Books and Manuscripts Department, Boston Public Library

John Hay Library, Brown University, Providence, RI

Century Association Archives Foundation, New York City

Cornell Regional Archives, Ithaca, NY

Explorers Club Archives, New York

Historical Society of Pennsylvania, Philadelphia

Manuscripts Division, Huntington Library, Art Collections, and Botanical Gardens, San Marino, CA

Maximilian-Bodmer Collection, Margre H. Durham Center for Western Studies, Joslyn Art Museum, Omaha, NE

Library of Congress, Washington, DC

Maryland Historical Society, Baltimore

Massachusetts Historical Society, Boston

The Jefferson Library at Monticello, Charlottesville, VA

The Fred W. Smith National Library for the Study of George Washington, Mount Vernon, VA

National Archives and Records Administration, Washington, DC

New-York Historical Society

Olana State Historic Site, Hudson, NY

Peabody Museum of Archaeology and Ethnology, Harvard University, Cambridge, MA

The Royal Society, London

Staatsbibliothek zu Berlin, Preußischer Kulturbesitz

David Rumsey Map Center, Stanford University, Palo Alto, CA

Bancroft Library, University of California, Berkeley

Winterthur Museum, Garden, and Library, Winterthur, DE

Beinecke Rare Book and Manuscript Library, Yale University, New Haven, CT

Peabody Museum of Natural History, Yale University, New Haven, CT

DIGITAL ARCHIVES

Alexander von Humboldt-Chronologie

Archive.org

Biodiversity Heritage Library

Founders Online

Fultonhistory.com

Gale Historical Newspapers

Google Books

HathiTrust

JSTOR

Making of America, University of Michigan

The Montpelier Research Database

Project Gutenberg

ProQuest Historical Newspapers

Rotunda, University of Virginia Press

ALEXANDER VON HUMBOLDT

Ackerknecht, Erwin A. "George Forster, Alexander von Humboldt, and Ethnology." *Isis* 46, no. 2 (June 1955): 83–95.

Agassiz, Louis. *Address Delivered on the Centennial Anniversary of the Birth of Alexander von Humboldt.* Boston: Boston Society of Natural History, 1869.

Agassiz, Louis, and Thomas H. Huxley. *Darwin and Humboldt: Their Lives and Work.* Humboldt Library of Popular Science Literature 43. New York: J. Fitzgerald, 1883.

Alexander von Humboldt-Kommission der Deutschen Akademie der Wissenschaften zu Berlin. *Alexander von Humboldt, 14.9.1769–6.5.1859, Gedenkschrift zur 100: Weiderkehr seines Todestages.* Berlin: Akademie Verlag, 1959.

Andress, Reinhard. "Alexander von Humboldt und Carlos Montúfar als Reisegefährten: Ein Vergleich ihrer Tagebücher zum Chimborazo-Aufstieg." *Internationale Zeitschrift für Humboldt-Studien* 12, no. 22 (May 2011): 10–22.

Baron, Frank. "From Alexander von Humboldt to Frederic Edwin Church: Voyages of Scientific Exploration and Artistic Creativity." *Internationale Zeitschrift für Humboldt-Studien* 11, no. 10 (2005): 2–15.

"Baron Humboldt." *Weekly Visitor, or Ladies' Miscellany (1802–1806)* 2, no. 101 (8 September 1804):3.

"Baron Humboldt's Letters." *Eclectic Magazine of Foreign Literature* 50, no. 2 (October 1860): 230–37.

Beck, Hanno. *Gespräche Alexander von Humboldts.* Berlin: Akademie Verlag, 1959.

Belgum, Kirsten. "Reading Alexander von Humboldt: Cosmopolitan Naturalist with an American Spirit." In *German Culture in Nineteenth-Century America: Reception, Adaptation, Transformation*, edited by Lynne Tatlock and Matt Erlin, 107–27. Rochester, NY: Camden House, 2005.

Biermann, Kurt-Reinhard, Ilse Jahn, and Fritz G. Lange, eds. *Alexander von Humboldt: Chronologische Übersicht über wichtige Daten seines Lebens.* Vol. 1, Beiträge zur Alexander-von-Humboldt-Forschung. Berlin: Akademie Verlag, 1983.

Biermann, Kurt-Reinhard, and Ingo Schwarz. "Alexander von Humboldt—Half an American." *Alexander-von-Humboldt-Magazin* 67 (1996): 43–50.

Borch, Rudolf. *Alexander von Humboldt: Sein Leben in Selbstzeugnissen, Briefen, und Berichten.* Berlin: Druckhauses Tempelhof, 1948.

Botting, Douglas. *Humboldt and the Cosmos.* New York: Harper and Row, 1973.

Brotherston, Gordon. "Amazon and Cordillera: Humboldt's Tropical American Quest." In "Alexander von Humboldt and America," edited by Peter Hulme. Special issue, *Studies in Travel Writing* 15, no. 1 (February 2011): 53–60.

Bruhns, Karl, ed. *Life of Alexander von Humboldt. Compiled in Commemoration of the Centenary of His Birth, by J. Löwenberg, Robert Avé-Lallemant, and Alfred Dove.* Translated by Jane Lassell and Caroline Lassell, 2 vols. Boston: Lee and Shepard, 1873.

Bunkše, Edmunds V. "Humboldt and an Aesthetic Tradition in Geography." *Geographical Review* 71, no. 2 (April 1981): 127–46.

Buttimer, Anne. "Beyond Humboldtian Science and Goethe's Way of Science: Challenges of Alexander von Humboldt's Geography." *Erdkunde* 55, no. 2 (June 2001): 105–20.

Chateaubriand, François-René. "Voyages." *Le Conservateur: Le Roi, la Charte, et les Honnêtes Gens* 6 (1820): 31–36.

Clark, Rex, and Oliver Lubrich, eds. *Cosmos and Colonialism: Alexander von Humboldt in Cultural Criticism.* New York: Berghahn Books, 2012.

Clark, Rex, and Oliver Lubrich, eds. *Transatlantic Echoes: Alexander von Humboldt in World Literature.* New York: Berghahn Books, 2012.

Crampton, C. Gregory. "Humboldt's Utah, 1811." *Utah Historical Quarterly* 26, no. 3 (July 1958): 269–81.

De Havenon, Georgia Riley, Christina De Léon, Alicia Lubowski-Jahn, and Gabriela Rangel, eds. *Unity of Nature: Alexander von Humboldt and the Americas.* Bielefield: Kerber Verlag for the Americas Society, 2014.

Dettelbach, Michael. "The Face of Nature: Precise Measurement, Mapping, and Sensibility in the Work of Alexander von Humboldt." *Studies in the History and Philosophy of Biological and Biomedical Sciences* 30, no. 4 (1999): 473–504.

Dettelbach, Michael. "Global Physics and Aesthetic Empire: Humboldt's Physical Portrait of the Tropics." In *Visions of Empire: Voyages, Botany, and Representations of Nature*, edited by David Philip Miller and Peter Hanns Reill, 258–92. London: Cambridge University Press, 1996.

Dettelbach, Michael. "Humboldtian Science." In *Cultures of Natural History*, edited by Nicholas Jardine, James A. Secord, and Emma C. Spary, 287–304. New York: Cambridge University Press, 1996.

Du Bois-Reymond, Émil. *Alexander von Humboldt* (From a Memorial Address Delivered in the Hall of the University, August 3, 1883). Printed in *Popular Science Monthly* 10 (December 1883): 144–60.

Echenberg, Myron. *Humboldt's Mexico: In the Footsteps of the Illustrious German Scientific Traveller.* Montreal: McGill-Queens University Press, 2017.

Erickson, Raymond, Mauricio A. Font, and Brian Schwartz, eds. *Alexander von Humboldt: From the Americas to the Cosmos.* New York: Bildner Center for Western Hemisphere Studies, City University of New York, 2004.

Ette, Ottmar. "Alexander von Humboldt: The American Hemisphere and TransArea Studies." *Iberoamericana* 5, no. 20 (December 2005): 85–108.

Ette, Ottmar. "'Not just brought about by chance': Reflections on Globalization in Cornelius de Pauw and Alexander von Humboldt." In "Alexander von Humboldt and America," edited by Peter Hulme. Special issue, *Studies in Travel Writing* 15, no. 1 (February 2011): 3–25. Translated from German by Mark Minnes.

Ette, Ottmar. "Towards World Science? Humboldtian Science, World Concepts, and TransArea Studies." In *Dos culturas in diálogo. Historia cultural de la naturaleza, la técnica y las ciencias naturales en España y América Latina*, edited by Norbert Rehrman and Laura Ramírez Sáinz, 121–50. Frankfurt and Madrid: Iberoamericana Vervuert, 2007.

Ette, Ottmar. "TransTropics: Alexander von Humboldt and Hemispheric Constructions." In *Alexander von Humboldt and the Americas*, edited by Vera M. Kutzinski, Ottmar Ette, and Laura Dassow Walls, 209–36. Berlin: Verlag Walter Frey, 2012.

Ette, Ottmar. "Unterwegs zu einer Weltwißenschaft? Alexander von Humboldts Weltbegriffe und die transarealen Studien." Translated by Tobias Kraft. *Internationale Zeitschrift für Humboldt-Studien* 7, no. 13 (2006): 34–54.

Everett, Edward. "Alexander von Humboldt." In *Orations and Speeches on Various Occasions*, 4:170–77. Boston: Little, Brown, 1868.

Everett, Edward. "Art. 1.—"[Review of] *Voyage aux Regions equinoxiales du Nouveau Continent; fait en 1799, 1800, 1801, 1802, 1803, 1804, par Al. de Humboldt et A. Bonpland, redigé par Alexandre de Humboldt: Avec un Atlas Geographique et Physique*," Tomes VII & VIII. 8vo. Paris, 1822. *North American Review*, n.s., 13, no. 38 (January 1823): 1–30.

Everett, Edward. "The Illustrious Dead of 1859—Prescott, Bond, Hallam, von Humboldt." In *Mount Vernon Papers*, 284–92. New York: D. Appleton, 1860.

Ewan, Joseph. "Humboldt and American Botany." *Rhodora* 58, no. 691 (July 1956): 191–97.

Fernandes, Luiz Estevam de Oliveira. "Humboldt and the Writing of New Spain's History." In *Contributions to Theory and Comparative History of Historiography: German and Brazilian Perspectives*, edited by Luiz Estevam de Oliveira Fernandes, Luísa Rauter Periera, and Sérgio da Mata, 87–104. Frankfurt: Peter Lang, 2015.

Fiedler, Horst, and Ulrike Leitner. *Alexander von Humboldts Schriften: Bibliographie der selbständig erschienenen Werke*. In *Beiträge zur Alexander-von-Humboldt-Forschung*, vol. 20. Berlin: Akademie Verlag, 2000.

Foner, Philip S. "Alexander von Humboldt on Slavery in America." *Science and Society* 47, no. 3 (Fall 1983): 330–42.

Foner, Philip S., ed. *Alexander von Humboldt on Slavery in the United States*. Berlin: Humboldt-Universität, 1981.

Friis, Hermann R. "Alexander von Humboldts Besuch in den Vereinigten Staaten von Amerika vom 20 Mai bis zum 30 Juni 1804." In *Alexander von Humboldt: Studien zu Seiner universalen Geisteshaltung*, edited by Joachim H. Schultze, 142–95. Berlin: Walter de Gruyter, 1959.

Friis, Hermann R. "Baron Alexander von Humboldt's Visit to Washington, D.C., June 1 through June 13, 1804." *Records of the Columbia Historical Society of Washington, D.C.* 60/62 (1960–62): 1–35.

Galle, Helmut. "'Aesthetical Representation of Natural History Objects': Narrative Strategies in A. v. Humboldt's *Ansichten der Natur*." In *Contributions to Theory and Comparative History of Historiography: German and Brazilian Perspectives*, edited by Luiz Estevam de Oliviera Fernandes, Luísa Rauter Periera, and Sérgio da Mata, 41–58. Frankfurt: Peter Lang, 2015.

Gerassi-Navarro, Nina. "Picturing the Tropics from Humboldt to Darwin." In *Entangled Knowledge: Scientific Discourses and Cultural Difference*, edited by Klaus Hock and Gesa Mackenthun, 201–29. Vol. 4 in *Cultural Encounters and the Discourses of Scholarship*. Munich: Waxmann, 2012.

Gerhard, Caspar. "A Young Man from "Ultima Thule" Visits Jefferson: Alexander von Humboldt in Philadelphia and Washington." *Proceedings of the American Philosophical Society* 155, no. 3 (September 2011): 247–62.

Godlewska, Anne Marie Claire. "From Enlightenment Vision to Modern Science? Humboldt's Visual Thinking." In *Geography and Enlightenment*, edited by David N. Livingstone and Charles J. Withers, 236–79. Chicago: University of Chicago Press, 1999.

Görgens, Lutz H. "Humboldt's Example." *Amerikastudien/American Studies* 51, no. 1 (2006): 51–58.

Gould, Stephen Jay. "Humboldt and Darwin: The Tension and Harmony of Art and Science." In *Frederic Edwin Church*, edited by Franklin Kelly, 94–107. Washington, DC: National Gallery of Art, 1989.

Hagen, Victor Wolfgang von. "Was This the Fate of the Library of Alexander von Humboldt? An Inquiry." *Isis* 41, no. 2 (July 1950): 164–67.

Harvey, David. "The Humboldt Connection." *Annals of the Association of American Geographers* 88, no. 4 (December 1998): 723–30.

Heinzen, Karl. *The True Character of Humboldt: Oration, Delivered at the German Humboldt Festival in Boston*. Indianapolis: Association for the Propagation of Radical Principles, 1869.

Helferich, Gerard. *Humboldt's Cosmos: Alexander von Humboldt and the Latin American Journey That Changed the Way We See the World*. New York: Gotham Books, 2004.

Holl, Frank, ed. *Alexander von Humboldt. Netzwerke des Wißens*. Ostfildern: Hatje-Cantz, 1999.

Holl, Frank. "El científico independiente y su crítica al colonialismo." In *Alejandro de Humboldt y el mundo hispánico*, edited by Miguel Ángel Puig-Samper, 113–17. Madrid: Fundación Histórica Tavera, 2000.

Holl, Frank, and Cecelia Estrada, eds. *Alejandro de Humboldt: Una Nueva Visión del Mundo*. Madrid: Museo Nacional de Ciencias Naturales, 2005.

Howell, James F. "Recasting the Significant: The Transcultural Memory of Alexander von Humboldt's Visit to Philadelphia and Washington, D.C." *Humanities* 5, no. 3 (July 2016): 1–9.

Hulme, Peter, ed. "Alexander von Humboldt and America." Special issue, *Studies in Travel Writing* 15, no. 1 (February 2011).

Humboldt, Alexander von. *Alexander von Humboldt: Briefe aus Amerika 1799–1804*. Edited by Ulrike Moheit. Berlin: Akademie Verlag, 1993.

Humboldt, Alexander von, and Carl Ritter. *Alexander von Humboldt/Carl Ritter: Briefwechsel*. Edited by Ulrich Päßler. Berlin: Akademie Verlag, 2010.

Humboldt, Alexander von, and Johann Friedrich Cotta. *Alexander von Humboldt und Cotta: Briefwechsel*. Edited by Ulrike Leitner. Berlin: Akademie Verlag, 2009.

Humboldt, Alexander von. *Alexander von Humboldt und die Vereinigten Staaten von Amerika: Briefwechsel*. Edited by Ingo Schwarz. Berlin: Akademie Verlag, 2004.

Humboldt, Alexander von. *Aspects of Nature, in Different Lands and Different Climates, with Scientific Elucidations*. Translated by Elizabeth J. L. Sabine. 2 vols. London: Longman, Brown, Green, and John Murray, 1849.

Humboldt, Alexander von. *Atlas geográfico y físico del reino de la Nueva España*. Edited by Charles Minguet and Jaime Labastida with an introduction by Elías Trabulse. Mexico City: Siglo Veintiuno Editores, 2003.

Humboldt, Alexander von. "Baron Humboldt." Translated by John Vaughan. *Literary Magazine, and American Register* 2, no. 10 (July 1804): 321–27.

Humboldt, Alexander von, *Briefe von Alexander von Humboldt an Varnhagen von Ense, aus den Jahren 1827 bis 1858*. Edited by Ludmilla Assing. Second edition. Leipzig: F. A. Brockhaus, 1860.

Humboldt, Alexander von, and Heinrich Karl Wilhelm Berghaus. *Briefwechsel, Alexander von Humboldt's mit Heinrich Berghaus aus den Jahren 1825 bis 1858*. Leipzig: Hermann Costenoble, 1863.

Humboldt, Alexander von, and Emil du Bois-Reymond. *Briefwechsel zwischen Alexander von Humboldt und Emil du Bois-Reymond*. Edited by Ingo Schwarz and Klaus Wenig. Berlin: Akademie Verlag, 1997.

Humboldt, Alexander von, and Heinrich Christian Schumacher. *Briefwechsel zwischen Alexander von Humboldt und Heinrich Christian Schumacher*. Edited by Kurt-Reinhard Biermann. Berlin: Akademie Verlag, 1979.

Humboldt, Alexander von, and François Arago. *Correspondance d'Alexandre de Humboldt avec François Arago (1809–1853)*. With a preface and notes by Dr. E. T. Hamy. Paris: E. Guilmoto, 1907.

Humboldt, Alexander von. *Cosmos: A Sketch of a Physical Description of the Universe*. Translated by Elise C. Otté. 5 vols. London: Henry G. Bohn, 1848–59.

Humboldt, Alexander von. *Essai politique sur le royaume de la Nouvelle Espagne*. 5 vols. Paris: F. Schoell, 1811.

Humboldt, Alexander von. *Essai politique sur l'île de Cuba*. Paris: Gide, 1826.

Humboldt, Alexander von. *The Island of Cuba: A Political Essay*. Translated from Spanish with Notes and a Preliminary Essay by J. S. Thrasher. New York: Derby and Jackson, 1856.

Humboldt, Alexander von. *The Island of Cuba: A Political Essay. Translated from Spanish with Notes and a Preliminary Essay by J. S. Thrasher*. With an introduction by Luis Martínez-Fernández; "The Nature of Slavery" translated from German by Shelley L. Fritsch. Princeton, NJ: Markus Wiener Publishers, 2001.

Humboldt, Alexander von, "A Letter of Alexander von Humboldt, 1845." *American Historical Review* 7, no. 4 (July 1902): 704–6.

Humboldt, Alexander von, and Karl August Varnhagen von Ense. *Letters of Alexander von Humboldt, Written between the Years 1827 and 1858 to Varnhagen von Ense: Together with Extracts from Varnhagen's Diaries, and Letters from Varnhagen and Others to Humboldt*. Edited by Ludmilla Assing. London: Trübner, 1860.

Humboldt, Alexander von, and Karl August Varnhagen von Ense. *Letters of Alexander von Humboldt to Varnhagen von Ense, from 1827 to 1858, with Extracts from Varnhagen's Diaries, and Letters of Varnhagen and Others to Humboldt*. Translated by Friedrich Kapp. New York: Rudd and Carleton, 1860.

Humboldt, Alexander von. *Lettres américaines d'Alexandre de Humboldt (1798–1807)*. With a preface and notes by Dr. E. T. Hamy. Paris, E. Guilmoto, 1905.

Humboldt, Alexander von. *Political Essay on the Island of Cuba: A Critical Edition*. Edited by Vera M. Kutzinski and Ottmar Ette. Translated by J. Bradford Anderson, Vera M. Kutzinski, and Anja Becker. Chicago: University of Chicago Press, 2011.

Humboldt, Alexander von. *Political Essay on the Kingdom of New Spain... with Physical Sections and Maps, Founded on Astronomical Observations, and Trigonometrical and Barometrical Measurements*. Translated by John Black. 4 vols. London: Longman Hurst, Rees, Orne, and Brown, 1811.

Humboldt, Alexander von. *Reise auf dem Río Magdalena, durch die Anden und Mexiko*. Edited by Margot Faak. 2 vols. Berlin: Akademie Verlag, 1986–1990

Humboldt, Alexander von. *Researches, Concerning the Institutions and Monuments of the Ancient Inhabitants of America, with Descriptions and Views of Some of the Most Striking Scenes in the Cordilleras!* Translated by Helen Maria Williams. 2 vols. London: Longman, Hurst, Rees, Orme, and Brown, 1814.

Humboldt, Alexander von. *Views of the Cordilleras and Monuments of the Indigenous Peoples of the Americas: A Critical Edition*. Edited and translated by Vera M. Kutzinski and Ottmar Ette. Chicago: University of Chicago Press, 2013.

Humboldt, Alexander von. *Views of Nature*. Edited by Stephen T. Jackson and Laura Dassow Walls. Translated by Mark W. Person. Chicago: University of Chicago Press, 2014.

Humboldt, Alexander von. *Views of Nature, or Contemplations on the Sublime Phenomena of Creation, with Scientific Illustrations*. Translated by Elise C. Otté and Henry G. Bohn. 1850. Photostatic reprint. London: George Bell and Sons, 1902.

Humboldt, Alexander von, and Aimé Bonpland. *Correspondance, 1805–1858*. Edited by Nicolas Hossard. Paris: L'Harmattan, 2004.

Humboldt, Alexander von, and Aimé Bonpland. *Essay on the Geography of Plants*. Edited by Stephen T. Jackson. Translated by Sylvie Romanowski. Chicago: University of Chicago Press, 2009.

Humboldt, Alexander von, and Aimé Bonpland. *Personal Narrative of Travels to the Equinoctial Regions of America, during the Years 1799–1804*. Translated and edited by Thomasina Ross. 3 vols. London: Henry G. Bohn, 1852.

Humboldt, Alexander von, and Aimé Bonpland. *Personal Narrative of Travels to the Equinoctial Regions of the New Continent, during the Years 1799–1804*. Translated by Helen Maria Williams. 7 vols. London: Longman, Hurst, Rees, Orme, and Brown, 1814–29.

Humboldt, Alexander von, and Christian Carl Josias Freiherr von Bunsen. *Briefe von Alexander von Humboldt an Christian Carl Josias Freiherr von Bunsen*. Leipzig: F. A. Brockhaus, 1869.

Humboldt, Alexander von, and Wilhelm von Humboldt. *Briefe Alexander's von Humboldt an seinen Bruder Wilhelm*. Stuttgart: J. G. Cotta, 1880.

"Humboldt on Landscape Painting." *Crayon* 1, no. 13 (28 March 1855): 199–200.

Humboldt, Wilhelm von, and Caroline von Humboldt. *Wilhelm und Caroline von Humboldt in ihren Briefen*. Edited by Anna von Sydow. 7 vols. Berlin: E. S. Mittler, 1906–16.

Iannini, Christopher. "Humboldt's *Political Essay on the Island of Cuba* and the Natural History of Caribbean Slavery." In *Alexander von Humboldt and the Americas*, edited by Vera M. Kutzinski, Ottmar Ette, and Laura Dassow Walls, 161–82. Berlin: Verlag Walter Frey, 2012.

Keil, Hartmut. "Race and Ethnicity: Slavery and the German Radical Tradition." Talk sponsored by the Max Kade Institute for German-American Studies and the Center for the History of Print Culture, Madison WI, 3 February 1999. https://mki.wisc.edu/sites/mki.wisc.edu/files/graphics/Keil-RACE_AND_ETHNICITY.pdf.

Kortum, Gerhard. "M. F. Maury (1806–1873), A. v. Humboldt (1769–1859) und der Mythos des Telegraphen-Plateaus im Nordatlantischen Ozean." In *Geographie der Küsten und Meere. Beiträge zum Küstensymposium in Mainz, 14.–18. Oktober 1984*, edited by Rolf-Dieter Volmerig, Burkhard Hofmeister, Frithjof Voss, and Michael Brillat, 1–23. Berlin: Institut für Geographie der Technischen Universität Berlin, 1985.

Kortum, Gerhard, and Ingo Schwarz. "Alexander von Humboldt and Matthew Fontaine Maury—Two Pioneers of Marine Sciences." *Historisch-Meereskundliches Jahrbuch* 10 (2003–4): 157–85.

Kraft, Tobias. *Figuren des Wissens bei Alexander von Humboldt: Essai, Tableau, und Atlas im amerikanischen Reisewerk*. Berlin: Walter de Gruyter, 2014.

Kraft, Tobias. "From Total Impression to Fractal Representation: The Humboldtian *Naturbild*." In *Alexander von Humboldt and the Americas*, edited by Vera M. Kutzinski, Ottmar Ette, and Laura Dassow Walls, 144–60. Berlin: Verlag Walter Frey, 2012.

Kutzinski, Vera M. "Facts and Fictions: Alexander von Humboldt in Brazil." In *Contributions to Theory and Comparative History of Historiography: German and Brazilian Perspectives*, edited by Luiz Estevam de Oliveira Fernandes, Luísa Rauter Periera, and Sérgio da Mata, 59–86. Frankfurt: Peter Lang, 2015.

Kutzinski, Vera M. "Translations of Cuba: Fernando Ortiz, Alexander von Humboldt, and the Curious Case of John Sidney Thrasher." *Atlantic Studies* 6, no. 3 (December 2009): 303–26.

Kutzinski, Vera M., ed. *Alexander von Humboldt's Transatlantic Personae*. London: Routledge, 2012.

Kutzinski, Vera M., Ottmar Ette, and Laura Dassow Walls, eds. *Alexander von Humboldt and the Americas*. Berlin: Verlag Walter Frey, 2012.

Labastida, Jaime. "Humboldt, México y Estados Unidos: Historia de una intriga." In Alexander von Humboldt *Atlas geográfico y físico del reino de la Nueva España*, edited by Jaime Labastida, 131–47, with an introduction by Elías Trabulse. Mexico City: Siglo Veintiuno Editores, 2003.

Lack, H. Walter. *Alexander von Humboldt and the Botanical Exploration of the Americas*. New York: Prestel, 2009.

Lack, H. Walter. "The Botanical Field Notes Prepared by Humboldt and Bonpland in Tropical America." *Taxon* 53, no. 2 (May 2004): 501–10.

Lack, H. Walter. "The Plant Self-Impressions Prepared by Humboldt and Bonpland in Tropical America." *Curtis's Botanical Magazine* 18, no. 4 (November 2001): 218–29.

Large, Arlen J. "The Humboldt Connection." *We Proceeded On* 16, no. 4 (November 1990): 4–12.

Laszlo, Pierre. "A French Perspective." In *Alexander von Humboldt: From the Americas to the Cosmos*, edited by Raymond Erickson, Mauricio A. Font, and Brian Schwartz, 281–91. New York: Bildner Center for Western Hemisphere Studies, City University of New York, 2005.

Leske, Irene Prüfer. "Ancient and Modern Forms of Slavery." In *Alexander von Humboldt: From the Americas to the Cosmos*, edited by Raymond Erickson, Mauricio A. Font, and Brian Schwartz, 469–83. New York: Bildner Center for Western Hemisphere Studies, City University of New York, 2005.

"Lettres d'Alexandre de Humboldt à Marc-Auguste Pictet (1795–1824)." *Le Globe: Bulletin de la Société Géographie de Genève* 7 (1868): 129–204.

Lubowski-Jahn, Alicia. "A Comparative Analysis of the Landscape Aesthetics of Alexander von Humboldt and John Ruskin." *British Journal of Aesthetics* 51, no. 3 (July 2011): 321–33.

Lüdecke, Cornelia. "Beiträge zur maritimen Meteorologie der Südhemisphäre in der Tradition von Heinrich Berghaus (1797–1884) und Matthew Fontaine Maury (1806–1873)." *Historisch-Meereskundliches Jahrbuch* 10 (2003–4): 135–56.

Martin, Alison. *Nature Translated: Alexander von Humboldt's Works in Nineteenth-Century Britain*. In *Edinburgh Critical Studies in Critical Translation*, edited by Stuart Gillespie and Emily Wilson. Edinburgh: Edinburgh University Press, 2018.

Martin, Alison. "'These changes and accessions of knowledge': Translation, Scientific Travel Writing and Modernity—Alexander von Humboldt's *Personal Narrative*." In "Alexander von Humboldt and America," edited by Peter Hulme, special issue, *Studies in Travel Writing* 15, no. 1 (February 2011): 39–51.

Mathewson, Kent. "Alexander von Humboldt's Image and Influence in North American Geography, 1804–2004." *Geographical Review* 96, no. 3 (July 2006): 416–38.

Mattos, Claudia. "Landscape Painting between Art and Science." In *Alexander von Humboldt: From the Americas to the Cosmos*, edited by Raymond Erickson, Mauricio A. Font, and Brian Schwartz, 141–55. New York: Bildner Center for Western Hemisphere Studies, City University of New York, 2004.

McCrory, Donald. *Nature's Interpreter: The Life and Times of Alexander von Humboldt*. Cambridge: Lutterworth Press, 2010.

McCullough, David. *Brave Companions: Portraits in History*. New York: Prentice Hall Press, 1992.

Meinhardt, Maren. *A Longing for Wide and Unknown Things: The Life of Alexander von Humboldt*. London: Hurst, 2018.

Millán, Elizabeth. "Alexander von Humboldt's Poetry of Nature: Freedom and the Latin American Landscape." *Meridional: Revista Chilena de Estudios Latinoamericanos*, no. 3 (October 2014): 9–36.

Millán, Elizabeth. "The Quest for the Seeds of Eternal Growth: Goethe and Humboldt's Presentation of Nature." *Goethe Yearbook* 18, no. 1 (2011): 97–114.

Nelken, Halina. *Alexander von Humboldt: His Portraits and Their Artists, A Documentary Iconography*. Berlin: Dietrich Reimer Verlag, 1980.

Nelken, Halina. *Humboldtiana at Harvard*. Cambridge, MA: Widener Library, Harvard University, 1976.

Nichols, Sandra. "Why Was Humboldt Forgotten in the United States?" *Geographical Review* 96, no. 3 (July 2006): 399–415.

Nollendorfs, Cora Lee. "Alexander von Humboldt Centennial Celebrations in the United States: Controversies Concerning His Work." *Monatshefte* 80, no. 1 (Spring 1988): 59–66.

"[Notice of the Arrival of Baron von Humboldt]." *National Intelligencer and Washington Advertiser*, 8 June 1804.

Orovio, Consuelo Naranjo. "Humboldt en Cuba: Reformismo y abolición." In "Alejandro de Humboldt y el mundo hispánico," edited by Miguel Ángel Puig-Samper. Special issue, *Debate y Perspectivas*, no. 1 (2000): 183–201.

Ortiz, Fernando. "Humboldt's Translator in the Context of Cuban History." In *Transatlantic Personae,* edited and translated by Vera M. Kutzinski. London: Routledge, 2012.

Päßler, Ulrich. "Plantae des États-Unis: Rediscovered Pages from Alexander von Humboldt's United States Diary." *Internationale Zeitschrift für Humboldt-Studien* 18, no. 35 (2017): 6–12.

Puig-Samper, Miguel Ángel. "Humboldt, un prusiano en la corte del Rey Carlos IV." *Revista de Indias* 59, no. 216 (1999): 329–55.

Puig-Samper, Miguel Ángel, and Elisa Garrido. "The Presentation of the Results of Alexander von Humboldt's Voyage to Carlos IV." *Internationale Zeitschrift für Humboldt-Studien* 17, no. 32 (2016): 50–62.

Quiroz, Alfonso W. "The Scientist and the Patrician: Reformism in Cuba." In *Alexander von Humboldt: From the Americas to the Cosmos*, edited by Raymond Erickson, Mauricio A. Font, and Brian Schwartz, 111–27. New York: Bildner Center for Western Hemisphere Studies, City University of New York, 2004.

Rebok, Sandra. "Alexander von Humboldt, His Scientific Expedition through the Americas and the Impact of His Atlantic Experiences." In *Atlantic Biographies: Individuals and Peoples of the Atlantic World*, edited by Jeffrey A. Fortin and Mark Meuwese, vol. 27, in *The Atlantic World*, 69–97. Leiden: Brill, 2013.

Rebok, Sandra. "Between Nature and Culture: Thomas Jefferson's and Alexander von Humboldt's Interactions with the Natural World." *Bollettino Filosofico* 27 (2011–12): 275–88.

Rebok, Sandra. "Enlightened Correspondents: The Transatlantic Dialogue of Thomas Jefferson and Alexander von Humboldt." *Virginia Magazine of History and Biography* 116, no. 4 (2008): 328–69.

Rebok, Sandra. *Humboldt and Jefferson: A Transatlantic Friendship of the Enlightenment*. Charlottesville: University of Virginia Press, 2014.

Rebok, Sandra. "'In Whose Heads I Would Like My Ideas to Live': Humboldt's Interest in the United States." *Iberoamericana* 19, no. 70 (2019): 113–30.

Rebok, Sandra. "A New Approach: Alexander von Humboldt's Perception of Colonial Spanish America as Reflected in His Travel Diaries." *Itinerario* 31, no. 1 (2007): 61–88.

Rebok, Sandra. "Two Exceptional Witnesses of Latin American Independence: The Prussian Explorer Alexander Von Humboldt and the Virginian Politician Thomas Jefferson." *Revista historia de la educación latinoamericana* 16, no. 23 (2014): 205–16.

Rebok, Sandra. "Two Exponents of the Enlightenment: Transatlantic Communication by Thomas Jefferson and Alexander von Humboldt." *Southern Quarterly* 43, no. 4 (2006): 126–52.

Rupke, Nicholaas. *Alexander von Humboldt: A Metabiography*. Chicago: University of Chicago Press, 2008.

Rupke, Nicolaas. "Alexander von Humboldt and Revolution: A Geography of Reception of the Varnhagen von Ense Correspondence." In *Geography and Revolution*, edited by David N. Livingstone and Charles W. J. Withers, 336–50. Chicago: University of Chicago Press, 2005.

Rupke, Nicolaas. "A Geography of Enlightenment: The Critical Reception of Alexander von Humboldt's Mexico Work." In *Geography and Enlightenment*, edited by David N. Livingstone and Charles W. J. Withers, 319–39. Chicago: University of Chicago Press, 1999.

Sachs, Aaron. *The Humboldt Current: Nineteenth-Century Exploration and the Roots of American Environmentalism*. New York: Viking, 2006.

Sachs, Aaron. "The Ultimate 'Other': Post-Colonialism and Alexander von Humboldt's Ecological Relationship with Nature." *History and Theory* 42, no. 4 (December 2003): 111–35.

Schaumann, Caroline. "Who Measures the World? Alexander von Humboldt's Chimborazo Climb in the Literary Imagination." *German Quarterly* 82, no. 4 (Fall 2009): 447–68.

Schoenwaldt, Peter. "Alexander von Humboldt und die Vereinigten Staaten von Amerika." In *Alexander von Humboldt: Werk und Weltgeltung*, edited by Heinrich Peiffer, 431–82. Munich: Piper, 1969.

Schwarz, Ingo. "Alexander von Humboldts Bild von Latein- und Angloamerika im Vergleich." In *Nord und Süd in Amerika: Gegensätze, Gemeinsamkeiten, Europäischer Hintergrund*, edited by Wolfgang Reinhard and Peter Waldmann, vol. 2:1142–54. Freiburg: Rombach, 1992.

Schwarz, Ingo. "Alexander von Humboldt's Correspondence with Johann Gottfried Flügel." *Yearbook of German-American Studies* 46 (2011): 87–93.

Schwarz, Ingo. "Alexander von Humboldt—Socio-Political Views of the Americas." In *Ansichten Amerikas: Neuere Studien zu Alexander von Humboldt*, edited by Ottmar Ette and Walther L. Bernecker, 105–16. Frankfurt: Vervuert, 2001.

Schwarz, Ingo. "Alexander von Humboldt's Visit to Washington and Philadelphia: His Friendship with Jefferson, and His Fascination with the United States." In "Proceedings: Alexander von Humboldt's Natural History Legacy and Its Relevance for Today." Special issue, *Northeastern Naturalist* 1 (2001): 43–56.

Schwarz, Ingo. "Carl Gustav Carus und Alexander von Humboldt—Briefwechsel." *Internationale Zeitschrift für Humboldt-Studien* 10, no. 18 (2009): 5–29.

Schwarz, Ingo. "From Alexander von Humboldt's Correspondence with Thomas Jefferson and Albert Gallatin." *Berliner Manuskripte zur Alexander-von-Humboldt-Forschung* 2 (2000): 1–20.

Schwarz, Ingo. "The Second Discoverer of the New World and the First American Literary Ambassador to the Old World: Alexander von Humboldt and Washington Irving." *Acta Historica Leopoldina* 27 (1997): 89–97.

Schwarz, Ingo. "'Shelter for a Reasonable Freedom' or Cartesian Vortex: Aspects of Alexander von Humboldt's Relation to the United States of America." In "Alejandro de Humboldt y el mundo hispánico," edited by Miguel Ángel Puig-Samper. Special issue, *Debate y Perspectivas*, no. 1 (2000): 169–82.

Schwarz, Ingo. "Transatlantic Communication in the 19th Century: Aspects of the Correspondence between Alexander von Humboldt and George Ticknor." *Asclepio* 56, no. 2 (2004): 25–39.

Schwarzenberg, F. A. *Alexander von Humboldt, or, What May Be Accomplished in a Lifetime*. London: Robert Hardwicke, 1866.

Sherwood, Robert. *The Cartography of Alexander von Humboldt: Images of the Enlightenment in America*. Saarbrücken: Verlag Dr. Müller, 2008.

Sotheby, Wilkinson, & Hodge. *Catalogue of the Remains of the Humboldt Library: Which Will Be Sold by Auction on Tuesday, the 14th Day of March, 1871*. London: Messrs. Sotheby, Wilkinson, & Hodge, 1871.

Stephens, Mr. [John Lloyd], the Traveller. "An Hour with Alexander von Humboldt." *Littell's Living Age* 15, no. 180 (23 October 1847): 151–53.

Stevens, Henry. *Catalogue of the First Portion of the Humboldt Library*. London: J. Davy and Sons, 1862.

Stoddard, Richard Henry. *The Life, Travels, and Books of Alexander von Humboldt*. With an introduction by Bayard Taylor. New York: Rudd and Carleton, 1859.

Sugden, Edgar. "Simultaneity-across-Borders: Richard Henry Dana Jr., Alexander von Humboldt, Edgar Allan Poe." *J19: The Journal of Nineteenth-Century Americanists* 2, no. 1 (Spring 2014): 83–106.

Terra, Helmut de. "Alexander von Humboldt's Correspondence with Jefferson, Madison, and Gallatin." *Proceedings of the American Philosophical Society* 103, no. 6 (15 December 1959): 783–806.

Terra, Helmut de. *Humboldt: The Life and Times of Alexander von Humboldt, 1769–1859*. New York: Alfred A. Knopf, 1955.

Terra, Helmut de. "Motives and Consequences of Alexander von Humboldt's Visit to the United States (1804)." *Proceedings of the American Philosophical Society* 104, no. 3 (15 June 1960): 314–16.

Terra, Helmut de. "Studies of the Documentation of Alexander von Humboldt." *Proceedings of the American Philosophical Society* 102, no. 2 (30 April 1958): 136–41.

Terra, Helmut de. "Studies of the Documentation of Alexander von Humboldt: The Philadelphia Abstract of Humboldt's American Travels, Humboldt Portraits and Sculpture in the United States." *Proceedings of the American Philosophical Society* 102, no. 6 (15 December 1958): 560–89.

Théodoridès, Jean. "Humboldt and England." *British Journal for the History of Science* 3, no. 1 (June 1966): 39–55.

Théodoridès, Jean. "Les séjours aux États-Unis de deux savants européens du XIXe siècle: Alexander von Humboldt et Victor Jacquemont." *Archives internationales d'histoire des sciences* 64 (July–September 1963): 287–304.

Thompson, Joseph P., Francis Lieber, Charles P. Daly, A. D. Bache, Arnold Guyot, and George Bancroft. "Proceedings: Alexander von Humboldt Commemoration." *Journal of the American Geographical and Statistical Society* 1, no. 8 (October 1859): 225–46.

Tuckerman, Henry T. "Alexander von Humboldt." *Godey's Lady's Book* 40 (September 1850): 133–38.

Tuckerman, Henry T. *Characteristics of Literature, Illustrated by the Genius of Distinguished Writers*. Philadelphia Lindsay and Blakiston, 1851.

Villanova, Manuel. "Humboldt y Thrasher, I." *La Semana* (5 September 1887): 6–7.

Villanova, Manuel. "Humboldt y Thrasher, II." *La Semana* (12 September 1887): 8–9.

Villanova, Manuel. "Humboldt y Thrasher, III." *La Semana* (19 September 1887): 6.

Villanova, Manuel. "Humboldt y Thrasher, IV." *La Semana* (24 September 1887): 5–6.

Villanova, Manuel. "Humboldt y Thrasher, V." *La Semana* (3 October 1887): 6–7.

Walls, Laura Dassow. "The Birth of the Two Cultures." In *Alexander von Humboldt: From the Americas to the Cosmos*, edited by Raymond Erickson, Mauricio A. Font, and Brian Schwartz, 247–58. New York: Bildner Center for Western Hemisphere Studies, City University of New York, 2004.

Walls, Laura Dassow. "Introducing Humboldt's Cosmos." *Minding Nature* 2, no. 2 (August 2009): 3–15.

Walls, Laura Dassow. "'The Napoleon of Science': Alexander von Humboldt in Antebellum America." *Nineteenth Century Contexts* 14, no. 1 (1990): 71–98.

Walls, Laura Dassow. *The Passage to Cosmos: Alexander von Humboldt and the Shaping of America*. Chicago: University of Chicago Press, 2009.

Walls, Laura Dassow. "Rediscovering Humboldt's Environmental Revolution." *Environmental History* 10, no. 4 (October 2005): 758–60.

Wasserman, Felix M. "Six Unpublished Letters of Alexander von Humboldt to Thomas Jefferson." *Germanic Review: Literature, Culture, Theory* 29, no. 3 (October 1954): 191–200.

Werner, Anja. "Alexander von Humboldt's Footnotes: 'Networks of Knowledge' in the Sources of the 1826 *Essai politique sur l'île de Cuba*." *Internationale Zeitschrift für Humboldt-Studien* 15, no. 28 (2014): 59–67.

Werner, Petra. *Naturwahrheit und ästhetische Umsetzung: Alexander von Humboldt im Briefwechsel mit bildenden Künstlern*. Berlin: Akademie Verlag, 2013.

Werner, Ray Joseph. "In Connection with All Things: The Memory of Alexander von Humboldt and His Works in the Context of 19th Century Thought in Europe and America." Master's thesis, University of Northern Iowa, 2015.

Wilson, Jason. "Humboldt in the Cono Sur of South America: Writing, Genre, and the Temptations of Going Native." In "Alexander von Humboldt and America," edited by Peter Hulme, 27–38. Special issue, *Studies in Travel Writing* 15, no. 1 (February 2011), 27–38.

Wilson, Philip K. "Influences of 'Kosmos' in 'Earth and Man.'" In *Alexander von Humboldt: From the Americas to the Cosmos*, edited by Raymond Erickson, Mauricio A. Font, and Brian Schwartz, 371–84. New York: Bildner Center for Western Hemisphere Studies, City University of New York, 2005.

Wright, Joseph Albert. "Commemorative Remarks at the Geographical Society, Berlin, May 7, 1859." *Zeitschrift für Allgemeine Erdkunde*, n.s., 6 (1859): 414–15.

Wulf, Andrea. *The Invention of Nature: Alexander von Humboldt's New World*. New York: Alfred A. Knopf, 2015.

Zeuske, Michael. "¿Padre de la Independencia? Humboldt y la transformación a la Modernidad en la América española." In "Alejandro de Humboldt y el mundo hispánico," edited by Miguel Ángel Puig-Samper. Special issue, *Debate y Perspectivas*, no. 1 (2000): 67–100.

HISTORY AND CULTURE

Adams, William Howard, ed. *The Eye of Thomas Jefferson*. Washington, DC: National Gallery of Art, 1976.

Alderson, William T., ed. *Mermaids, Mummies, and Mastodons: The Emergence of the American Museum*. Washington, DC: American Association of Museums for the Baltimore City Life Museums, 1992.

Alexander, Edward P. *Museum Masters: Their Museums and Their Influence*. Nashville: American Association for State and Local History, 1983.

Allgor, Catherine. *Parlor Politics: In Which the Ladies of Washington Help Build a City and a Government*. Charlottesville: University of Virginia Press, 2000.

American Museum of Natural History. *First Annual Report of the American Museum of Natural History, January 1870*. New York: Major and Knapp for the American Museum of Natural History, 1870.

Anderson, Kirsten L. "German Americans, African Americans, and the Republican Party in St. Louis, 1865–1872." *Journal of American Ethnic History* 28, no. 1 (Fall 2008): 34–51.

Anti-Slavery Convention of American Women. *Turning the World Upside Down: The Anti-Slavery Convention of American Women, Held in New York City, May 9–12, 1837*. With an introduction by Dorothy Sterling. New York: Feminist Press at the City University of New York, 1987.

Appleton, Jay. *The Experience of Landscape*. Rev. ed. Chichester: John Wiley and Sons, 1996.

Arago, François. *Biographies of Distinguished Scientific Men*. Translated by Adm. W. H. Smyth, Rev. Baden Powell, and Robert Grant. 2 vols., first printed 1859. Freeport, NY: Books for Libraries Press, 1972.

Arnold, Bruce W. "French Donative Medals of King Louis Philippe I." In *Peace Medals: Negotiating Power in Early America*, edited by Robert B. Pickering, 63–72. Tulsa, OK: Gilcrease Museum, 2011.

Auchampaugh, Phillip G. "Political Techniques, 1856, or Why the *Herald* Went for Fremont." *Western Political Quarterly* 1, no. 3 (September 1948): 243–251.

Bachman, Catherine. *John Bachman, D.D., LL.D., Ph.D., the Pastor of St. John's Lutheran Church, Charleston*. Charleston, SC: Walker, Evans, and Cogswell, 1888.

Bachman, John. *An Examination of Professor Agassiz's Sketch of the Natural Provinces of the Animal World and Their Relation to the Different Types of Man*. Charleston, SC: James, Williams, and Gitsinger, 1855.

Bachman, John. *An Examination of the Characteristics of the Genera and Species as Applicable to the Doctrine of the Unity of the Human Race*. Charleston, SC: C. Canning, 1850.

Bachman, John. *A Notice of the "Types of Mankind" with an Examination of the Charges Contained in the Biography of Dr. Morton*. Charleston, SC: James, Williams, and Gitsinger, 1854.

Baltzell, Edward Digby. *Puritan Boston and Quaker Philadelphia: Two Protestant Ethics and the Spirit of Class Authority and Leadership*. New York: Free Press, 1979.

Bancroft, George. *Life and Letters of George Bancroft*. 2 vols. Edited by Mark Antony De Wolfe Howe. New York: Scribner's, 1908.

Bank, Rosemarie K. "Archiving Culture: Performance and American Museums in the Earlier Nineteenth Century." In *Performing America: Cultural Nationalism in American Theater*, edited by Jeffrey D. Mason and J. Ellen Gainor, 37–51. Ann Arbor: University of Michigan Press, 1999.

Barba, Preston A. *Balduin Möllhausen, the German Cooper*. Americana Germanica 17. Philadelphia: University of Pennsylvania, 1914.

Barnhart, Terry A. *Ephraim George Squier and the Development of American Anthropology*. Lincoln: University of Nebraska Press, 2005.

Barrow, Mark V., Jr. *Nature's Ghosts: Confronting Extinction from the Age of Jefferson to the Age of Ecology*. Chicago: University of Chicago Press, 2009.

Barton, Benjamin Smith. *Elements of Botany, or, Outlines of the Natural History of Vegetables*. Philadelphia: Robert DeSilver, 1827.

Basker, James G., ed. *Early American Abolitionists: A Collection of Anti-Slavery Writings, 1760–1820*. New York: Gilder Lehrman Institute of American History, 2005.

Baym, Nina. *American Women of Letters and the Nineteenth-Century Sciences: Styles of Affiliation*. New Brunswick, NJ: Rutgers University Press, 2002.

Bear, James A., ed. *Jefferson at Monticello: Recollections of a Monticello Slave and of a Monticello Overseer*. Charlottesville: University of Virginia Press, 1967.

Beckert, Sven, and Seth Rockman, eds. *Slavery's Capitalism: A New History of American Economic Development*. Philadelphia: University of Pennsylvania Press, 2016.

Bellion, Wendy. "City as Spectacle: William Birch's *Views* and the Chestnut Street Theatre." *Studies in the History of Gardens and Designed Landscapes* 32, no. 1 (2012): 15–34.

Bigelow, John. *Memoir of the Life and Public Services of John Charles Fremont, Including an Account of His Explorations, Discoveries and Adventures on Five Successive Expeditions across the North American Continent*. New York: Derby and Jackson, 1856.

Bigelow, Timothy. *Journal of a Tour to Niagara Falls in the Year 1805*. Boston: Wilson and Son, 1876.

Blackburn, Robin. *The American Crucible: Slavery, Emancipation, and Human Rights*. New York: Verso, 2011.

Bolzau, Emma Lydia. *Almira Hart Lincoln Phelps: Her Life and Work*. Philadelphia: University of Pennsylvania Press, 1936.

Boutwell, George S. "Report of the Council, Associate Fellows: John C. Fremont." *Proceedings of the American Academy of Arts and Sciences* 27 (May 1891–May 1892): 422–26.

Bowen, Margarita. *Empiricism and Geographical Thought: From Francis Bacon to Alexander von Humboldt*. Cambridge: Cambridge University Press, 1981.

Bridenbraugh, Carl, and Jessica Bridenbaugh. *Rebels and Gentlemen: Philadelphia in the Age of Franklin*. New York: Oxford University Press, 1962.

Brigham, David R. *Public Culture in the Early Republic: Peale's Museum and Its Audience*. Washington, DC: Smithsonian Institution Press, 1995.

Brodhead, Michael J. "The Work of Charles Lucien Bonaparte in America." *Proceedings of the American Philosophical Society* 122, no. 4 (August 1978): 198–203.

Brown, Gordon S. *Incidental Architect: William Thornton and the Cultural Life of Early Washington, D.C., 1794–1828*. Athens: Ohio University Press for the United States Capitol Historical Society, 2009.

Brown, Ralph H. "The American Geographies of Jedidiah Morse." *Annals of the Association of American Geographers* 31, no. 3 (September 1931): 145–217.

Brückner, Martin. *The Geographic Revolution in Early America: Maps, Literacy, and National Identity*. Chapel Hill: University of North Carolina Press for the Omohundro Institute of Early American History and Culture, 2006.

Burtt, Edward H., Jr., and William E. Davis. *Alexander Wilson: The Scot Who Founded American Ornithology*. Cambridge, MA: Belknap Press of Harvard University Press, 2013.

Busey, Samuel Clagett. *Pictures of the City of Washington in the Past*. Washington, DC: W. Ballantyne and Sons, 1898.

Calhoun, Daniel H. "Eyes for the Jacksonian World: William C. Woodbridge and Emma Willard." *Journal of the Early Republic* 4, no. 1 (Spring 1984): 1–26.

Callicott, J. Baird, and Michael P. Nelson, eds. *The Great New Wilderness Debate*. Athens: University of Georgia Press, 1998.

Cañizares-Esguerra, Jorge. *How to Write the History of the New World: Histories, Epistemologies, and Identities in the Eighteenth-Century Atlantic World*. Stanford, CA: Stanford University Press, 2002.

Carson, Hampton L., Esq. *The Centenary of the Wistar Party: An Historical Address delivered by appointment in the hall of the Historical Society of Pennsylvania, May 4th, 1918*. Philadelphia: Printed for the Wistar Association, 1918.

Carter, Edward C., II. *"One Grand Pursuit": A Brief History of the American Philosophical Society's First 250 Years, 1743–1993*. Philadelphia: American Philosophical Society, 1993.

Carter, Edward C., II, and Charles E. Brownell, eds. *Latrobe's Views of America, 1795–1820: Selections from the Watercolors and Sketches*. Second edition/printing. New Haven, CT: Yale University Press for the Maryland Historical Society, 1985.

Carter, Samuel. *Cyrus Field: Man of Two Worlds*. New York: Putnam, 1968.

Casper, Scott E. *Sarah Johnson's Mount Vernon: The Forgotten History of an American Shrine*. New York: Hill and Wang, 2008.

Chaffin, Tom. *Pathfinder: John Charles Frémont and the Course of American Empire*. New York: Hill and Wang, 2002.

Chaptal, Jean-Antoine-Claude, comte de Chanteloup. *Mes Souvenirs sur Napoléon*. Paris: E. Plon, Nourrit et Cie, Imprimeurs-Éditeurs, 1893.

Chardon. F. A. *Chardon's Journal at Fort Clark, 1834–1839*. Edited by Annie Heloise Abel. Lincoln: University of Nebraska Press, 1997.

Chinard, Gilbert. "The American Philosophical Society and the Early History of Forestry in America." *Proceedings of the American Philosophical Society* 89, no. 2, Symposium on Forestry and the Public Welfare (18 July 1945): 444–88.

Chinard, Gilbert. "La Correspondance de Madame de Staël avec Jefferson." *Revue de Littérature Comparée* 2, no. 4 (1922): 621–40.

Clements, William M. *Native American Verbal Art: Texts and Contexts*. Tucson: University of Arizona Press, 1996.

"Climatology of the United States." *United States Magazine and Democratic Review* 11, no. 53 (November 1842): 449–71.

Cohen, I. Bernard. *Science and the Founding Fathers: Science in the Political Thought of Thomas Jefferson, Benjamin Franklin, John Adams, and James Madison*. New York: W. W. Norton, 1995.

Corcoran, William Wilson. *A Grandfather's Legacy; Containing a Sketch of His Life and Obituary Notices of Some Members of His Family, Together with Letters from His Friends*. Washington, DC: Polkinhorn, 1879.

Cronon, William. "The Trouble with Wilderness; or Getting Back to the Wrong Nature." In *Uncommon Ground: Rethinking the Human Place in Nature*, edited by William Cronon, 69–90. New York: W. W. Norton, 1996.

Cunfer, Geoff, and Bill Waiser, eds. *Bison and People on the North American Great Plains: A Deep Environmental History*. College Station: Texas A&M University Press, 2016.

Cunningham, Andrew, and Nicholas Jardine, eds. *Romanticism and the Sciences*. Cambridge: Cambridge University Press, 1990.

Curtius [Henry Schoolcraft]. "Examination of an Article in the North American Review." *Literary World* 13 (20 August 1853): 51–54.

Cutler, William Parker, and Julia Perkins Cutler. *Life, Journal, and Correspondence of Rev. Manasseh Cutler, LL.D.* 2 vols. Cincinnati: Robert Clarke, 1888.

Daniels, Stephen. *Fields of Vision: Landscape Imagery and National Identity in England and the United States*. Princeton, NJ: Princeton University Press, 1993.

Davidson, Abraham A. "Catastrophism and Peale's 'Mammoth.'" *American Quarterly* 21, no. 3 (Autumn 1969): 620–29.

Davis, Richard Beale. "The Abbé Correa in America, 1812–1820: The Contributions of the Diplomat and Natural Philosopher to the Foundations of Our National Life." *Transactions of the American Philosophical Society*, n.s., 45, part 2 (1955): 87–197.

Deloria, Philip J. *Playing Indian*. New Haven, CT: Yale University Press, 1998.

Denslow, Ray V. *Out of the Past: The Story of George Henry Curzon Melody, Friend of General Lafayette, Governor Dewitt Clinton, Disraeli, Victor Hugo, and King Louis Philippe of France*. Saint Louis: Royal Arch Masons, 1942.

Denton, Sally. *Passion and Principle: John and Jessie Frémont, the Couple Whose Power, Politics, and Love Shaped Nineteenth-Century America*. New York: Bloomsbury, 2007.

Douglass, Frederick. *The Life and Times of Frederick Douglass*. Hartford, CT: Park Publishing, 1881.

Duncan, Nicholas. *Gallatin: America's Swiss Founding Father.* New York: New York University Press, 2010.

Dyson, Stephen L. *The Last Amateur: The Life of William J. Stillman.* Albany, NY: Excelsior Editions, 2014.

Ellis, Joseph J. *After the Revolution: Profiles of Early American Culture.* New York: W. W. Norton, 1979.

Ellis, Joseph J. *American Sphinx: The Character of Thomas Jefferson.* New York: Alfred A. Knopf, 1997.

Evans, Estwick. *A Pedestrious Tour of Four Thousand Miles through the Western States and Territories during the Winter and Spring of 1818.* Concord, NH: Joseph C. Spear, 1819.

Ewan, Joseph, and Nesta Dunn Ewan. *Benjamin Smith Barton: Naturalist and Physician in Jeffersonian America.* Saint Louis: Missouri Botanical Garden Press, 2007.

Faxon, Walter. "Relics of Peale's Museum." *Bulletin of the Museum of Comparative Zoology of Harvard College* 59, no. 3 (July 1915): 119–48.

Fiebig-von Hase, Ragnhild, and Ursula Lehmkuhl. *Enemy Images in American History.* Providence, RI: Berghahn Books, 1998.

Foster, Augustus. *Jeffersonian America: Notes on the United States of America Collected in the Years 1805–6–7 and 11–12 by Sir Augustus Foster.* San Marino, CA: Huntington Library, 1954.

Fox, Robert, ed. *The Culture of Science in France, 1700–1900.* Aldershot, Hampshire: Variorum, 1992.

Frankel, Oz. *States of Inquiry: Social Investigations and Print Culture in Nineteenth-Century Britain and the United States.* Baltimore: Johns Hopkins University Press, 2006.

Frémont, Jessie Benton. *The Letters of Jessie Benton Frémont.* Edited by Pamela Herr and Mary Lee Spence. Urbana: University of Illinois Press, 1993.

Furstenberg, François. *When the United States Spoke French: Five Refugees Who Shaped a Nation.* New York: Penguin Press, 2014.

Gallatin, Albert. *A Synopsis of the Indian Tribes.* Cambridge: Cambridge University Press, 1836.

Gallatin, Albert. *The Writings of Albert Gallatin.* Edited by Henry Adams. 3 vols. Philadelphia: J. B. Lippincott, 1879.

Gawalt, Gerard W. "'Strict Truth': The Narrative of William Armistead Burwell." *Virginia Magazine of History and Biography* 101, no. 1 (January 1993): 103–32.

Gelo, Daniel J., and Christopher J. Wickham with contributions by Heide Castañeda. *Comanches and Germans on the Texas Frontier: The Ethnology of Heinrich Berghaus.* College Station: Texas A&M University Press, 2018.

"Gen. Fremont's Collection.; Sale of His Library, Chinaware, and Paintings at Auction—Some of the Books Disposed Of." *New York Times,* 10 October 1877, 8.

"Gen. John C. Fremont: The Hero Who Should Be Far More Dear to the Negro than Abe Lincoln—He Issued the First Emancipation Proclamation—'Honor to Whom, etc.'" *Cleveland Gazette* 4, no. 36 (23 April 1887): 2.

Gentile, Nicholas W. "Philosophical Sympathy and 'Seeds of Extinction': Jefferson's French Correspondence and the Roots of His Presidential Indian Policy, 1785–1789." *We Proceeded On* 36, no. 2 (May 2010): 18–27.

Gilpin, William. *The Central Gold Region: The Grain, Pastoral, and Gold Regions of North America.* Philadelphia: Sower, Barnes, 1860.

Gilpin, William. *Mission of the North American People, Geographical, Social, and Political: Illustrated by Six Charts Delineating the Physical Architecture and Terminal Laws of All the Continents.* Philadelphia: J. B. Lippincott, 1873.

Gisel, Bonnie J. *Nature's Beloved Son: Rediscovering John Muir's Botanical Legacy.* Berkeley, CA: Heyday Books, 2008.

Gougeon, Len. *Virtue's Hero: Emerson, Antislavery, and Reform.* Athens: University of Georgia Press, 1990.

Greeley, Horace, ed. *Life of Col. Fremont.* New York: Greeley and M'Elrath, 1856.

Haber, Francis C. *David Bailie Warden: A Bibliographical Sketch of America's Cultural Ambassador in France, 1804–1845.* Washington, DC: Institut français de Washington, 1954.

Haines, Aubrey L. *The Yellowstone Story: A History of Our First National Park.* Yellowstone National Park, WY: Yellowstone Library and Museum Association, ca. 1977.

Harvey, Sean P. *Native Tongues: Colonialism and Race from Encounter to the Reservation.* Cambridge, MA: Harvard University Press, 2015.

Hatch, Peter. *A Rich Spot of Earth: Thomas Jefferson's Revolutionary Garden at Monticello.* New Haven, CT: Yale University Press, 2012.

Herr, Pamela. *Jessie Benton Frémont: A Biography.* Norman: University of Oklahoma Press, ca. 1987.

Hickey, Donald R. "America's Response to the Slave Revolt in Haiti, 1791–1806." *Journal of the Early Republic* 2, no. 4 (Winter 1982): 361–70

Hock, Klaus, and Gesa Mackenthun, eds. *Entangled Knowledge: Scientific Discourses and Cultural Difference.* Vol. 4 in *Cultural Encounters and Discourses of Scholarship.* Munich: Waxmann, 2012.

Hone, Philip. *Diary of Philip Hone, 1828–1851.* Edited by Allan Nevins. 2 vols. New York: Dodd, Mead, 1927.

Hunt, Alfred N. *Haiti's Influence on Antebellum America: Slumbering Volcano of the Caribbean.* Baton Rouge: Louisiana State University Press, 1988.

Huth, Hans. *Nature and the American: Three Centuries of Changing Attitudes.* Lincoln: University of Nebraska Press, 1957.

Iannini, Christopher P. *Fatal Revolutions: Natural History, West Indian Slavery, and the Routes of American Literature.* Chapel Hill: University of North Carolina Press for the Omohundro Institute of Early American History and Culture, 2012.

Iannini, Christopher P. "*Notes on the State of Virginia* and the Natural History of the Haitian Revolution," *Clio* 40, no. 1 (2010): 63–85.

Idzerda, Stanley J., Anne C. Loveland, and Marc H. Miller. *Lafayette, Hero of Two Worlds: The Art and Pageantry of His Farewell Tour of America, 1824–1825.* Hanover, NH: University of New England Press for the Queens Museum, 1989.

Jackson, Stuart W., ed. "Lafayette Letters and Documents in the Yale Cooper Collection." *Yale University Library Gazette* 8, no. 4 (April 1934): 113–46.

Jardine, Nicholas, James A. Secord, and Emma C. Spary, eds. *Cultures of Natural History*. Cambridge: Cambridge University Press, 1996.

Jefferson, Thomas. *The Family Letters of Thomas Jefferson*. Edited by Edwin M. Betts and James Adam Bear. Charlottesville: University of Virginia Press, 1986.

Jefferson, Thomas. *Message from the President of the United States Communicating Discoveries Made in Exploring the Missouri, Red River, and Washita by Captains Lewis and Clark, Doctor Sibley and Mr. Dunbar; with a Statistical Account of the Countries Adjacent, Feb. 19, 1806*. Washington, DC: A. and G. Way, 1806.

Jefferson, Thomas. *Notes on the State of Virginia*. Edited by William Peden. Chapel Hill: University of North Carolina Press for the Institute of Early American History and Culture, 1955. First published 1787.

Jefferson, Thomas. *The Papers of Thomas Jefferson*. Edited by Julian P. Boyd et al., vols. 1–20, 1943–80; Charles T. Cullen et al., vols. 21–23, 1980–86; John Catanzariti et al., vols. 24–28, 1986–98; Barbara B. Oberg et al., vols. 29–41, 1999–2014; James P. McClure et al., vols. 42–current, 2014–19. Princeton, NJ: Princeton University Press, 1943–2019.

Jefferson, Thomas. *The Papers of Thomas Jefferson: Retirement Series*. Edited by J. Jefferson Looney et al. Vols. 1–14 (continuing). Princeton, NJ: Princeton University Press, 2005–2019.

"John Charles Frémont." *Littell's Living Age* 26 (July 1850): 207–8.

Johnson, Amandus. *The Journal and Biography of Nicholas Collin, 1764–1831*. Philadelphia: New Jersey Society of Pennsylvania, 1936.

Johnston, Matthew N. *Narrating the Landscape: Print Culture and American Expansion in the Nineteenth Century*. Norman: University of Oklahoma Press, 2016.

Jones, Bessie Zaban, ed. *The Golden Age of Science: Thirty Portraits of the Giants of 19th-Century Science, by Their Scientific Contemporaries*. New York: Simon and Schuster, in association with the Smithsonian Institution, 1966.

Jones, Caroline A., and Peter Galison, eds. *Picturing Science, Producing Art*. New York: Routledge, 1998.

Joshi, S. T., ed. *Documents of American Prejudice: An Anthology of Writings on Race from Thomas Jefferson to David Duke*. New York: Basic Books, 1999.

Judson, Isabella Field, ed. *Cyrus Field, His Life and Work (1819–1892)*. New York: Harper and Brothers, 1896.

Jungnickel, Christa, and Russell McCormmach. *Cavendish*. Philadelphia: American Philosophical Society, 1996.

Kamphoefner, Walter D. "St. Louis Germans and the Republican Party, 1848–1860." *Mid-America: An Historical Review* 57, no. 2 (April 1975): 69–88.

Karnes, Thomas L. *William Gilpin: Western Nationalist*. Austin: University of Texas Press, 1969.

Kimball, Marie G. "Unpublished Correspondence of Mme. De Staël with Thomas Jefferson." *North American Review* 208, no. 752 (July 1918): 63–71.

Koelsch, William A. "Thomas Jefferson, American Geographers, and the Uses of Geography." *Geographical Review* 98, no. 2 (April 2008): 260–79.

Kovarsky, Joel. *The True Geography of Our Country: Thomas Jefferson's Cartographic Vision*. Charlottesville: University of Virginia Press, 2014.

Lafayette, Marie Joseph Paul Yves Roch Gilbert du Motier, marquis de. *Mémoires, correspondance et manuscrits du général Lafayette*. 6 vols. Paris: H. Fournier, 1837–38.

Landow, George P. "The Rainbow: A Problematic Image." In *Nature and the Victorian Imagination*, edited by U. C. Knoepflmacher and G. B. Tennyson, 341–69. Berkeley: University of California Press, 1977.

Levine, Robert S., and Samuel Otter, eds. *Frederick Douglass and Herman Melville. Essays in Relation*. Chapel Hill: University of North Carolina Press, 2008.

Lewis, G. Malcolm. "William Gilpin and the Concept of the Great Plains Region." *Annals of the Association of American Geographers* 56, no. 1 (1966): 33–51.

Lincoln [Phelps], Almira H. *Familiar Lectures on Botany*. Hartford, CT: H. and F. J. Huntington, 1829.

Livingstone, David N., and Charles W. J. Withers, eds. *Geography and Enlightenment*. Chicago: University of Chicago Press, 1999.

Lovejoy, Arthur O. *The Great Chain of Being: A Study of the History of an Idea*. Cambridge, MA: Harvard University Press, 1936.

Lowenthal, David. "George Perkins Marsh and the American Geographical Tradition." *Geographical Review* 43, no. 2 (April 1953): 207–13.

Lowenthal, David. *George Perkins Marsh: Prophet of Conservation*. Seattle: University of Washington Press, 2003.

Madison, James. *The Papers of James Madison, Secretary of State Series*. Vol. 7, *2 April–31 August 1804*. Edited by David B. Mattern, J. C. A. Stagg, Ellen J. Barber, Anne Mandeville Colony, Angela Kreider, and Jeanne Kerr Cross. Charlottesville: University of Virginia Press, 2005.

Marsh, George Perkins. *Man and Nature; or, Physical Geography as Modified by Human Action*. New York: Charles Scribner's Sons, 1864. Modern edition edited by David Lowenthal, Seattle: University of Washington Press, 2003.

Marsh, George Perkins. "The Study of Nature." *Christian Examiner* 68 (January 1860): 33–62.

Masur, Gerhard. *Simón Bolívar*. Albuquerque: University of New Mexico Press, 1948.

Mathews, Catharine Van Cortlandt. *Andrew Ellicott: His Life and Letters*. New York: Grafton Press, 1908.

McCandless, Peter. "The Political Evolution of John Bachman: From New York Yankee to South Carolina Secessionist." *South Carolina Historical Magazine* 108, no. 1 (January 2007): 6–31.

McCauley, Anne. "François Arago and the Invention of Photography." In *Multiple Views: Logan Grant Essays on Photography, 1983–89*, edited by Daniel P. Younger, 44–63. Albuquerque: University of New Mexico Press, 1991.

McClellan, James Edward, III. *Colonialism and Science: Saint Domingue and the Old Regime*. Chicago: University of Chicago Press, 1992.

McKinsey, Elizabeth. *Niagara Falls: Icon of the American Sublime*. Cambridge: Cambridge University Press, 1985.

Merrill, Elmer Drew. "Palisot de Beauvois as an Overlooked American Botanist." *Proceedings of the American Philosophical Society* 76, no. 6 (1936): 899–920.

Meyers, Amy R. W. "Imposing Order on the Wilderness: Natural History Illustration and Landscape Portrayal." In *Views and Visions: American Landscape before 1830*, edited by Edward J. Nygren with Bruce Robertson, 105–32. Washington, DC: Corcoran Gallery of Art, 1986.

Meyers, Amy R. W., ed. *Knowing Nature: Art and Science in Philadelphia, 1740–1840*. New Haven, CT: Yale University Press, 2012.

Meyers, Amy R. W., and Margaret Beck Pritchard, eds. *Empire's Nature: Mark Catesby's New World Vision*. Chapel Hill: University of North Carolina Press for the Omohundro Institute of Early American History and Culture, 1999.

Miller, Charles A. *Jefferson and Nature: An Interpretation*. Baltimore: Johns Hopkins University Press, 1988.

Miller, Lillian B. *Patrons and Patriotism: The Encouragement of the Fine Arts in the United States, 1790–1860*. Chicago: University of Chicago Press, 1966.

Miller, Sally M., and Daryl Morrison, eds. *John Muir: Family, Friends, and Adventures*. Albuquerque: University of New Mexico Press, 2005.

Miller, Sara Cedar. *Central Park: An American Masterpiece*. New York: Harry N. Abrams in association with the Central Park Conservancy, 2003.

Mitchell, W. J. T., ed. *Landscape and Power*. Chicago: University of Chicago Press, 1994.

Morgan, Jo-Ann. *Uncle Tom's Cabin as Visual Culture*. Columbia: University of Missouri Press, 2007.

Morgan, S. R. "Schelling and the Origins of His *Naturphilosophie*." In *Romanticism and the Sciences*, edited by Andrew Cunningham and Nicholas Jardine, 25–37. Cambridge: Cambridge University Press, 1990.

Morse, Jedidiah. *The American Gazetteer*. Boston: S. Hall, and Thomas and Andrews, 1797.

Morse, Jedidiah. *The American Geography; or, A View of the Present Situation of the United States of America*. Elizabethtown, NJ: Shepard Kollock, 1789.

Mott, Frank Luther. *American Journalism: A History of Newspapers in the United States through 250 Years, 1690–1940*. New York: Macmillan, 1941.

Muir, John. *John Muir: His Life and Letters and Other Writings*. Edited and with an introduction by Terry Gifford. London: Bâton Wicks, 1996.

Muir, John. *John Muir's Last Journey: South to the Amazon and East to Africa*. Edited by Michael P. Branch. Washington, DC: Island Press, 2001.

Muir, John. *My First Summer in the Sierra*. Boston and New York: Houghton Mifflin, 1911.

Muir, John. *Our National Parks*. Boston and New York: Houghton Mifflin, 1901.

Nash, Roderick F. "The American Invention of National Parks." *American Quarterly* 22 (Autumn 1970): 726–35.

Nash, Roderick F. *Wilderness and the American Mind*. Rev. ed. New Haven, CT: Yale University Press, 1973.

Nevins, Allan. *The Evening Post: A Century of Journalism*. New York: Boni and Liveright, 1922. Reprint, New York: Russell and Russell, 1968.

Oelschlaeger, Max. *The Idea of Wilderness from Prehistory to the Age of Ecology*. New Haven, CT: Yale University Press, 1991.

Olmsted, Frederick Law. *The Papers of Frederick Law Olmsted*. Vol. 5, *The California Frontier, 1863–1865*. Edited by Victoria Post Ranney. Baltimore: Johns Hopkins University Press, 1995.

Olson, Greg. "Slave, Trader, Interpreter, and World Traveler: The Remarkable Story of Jeffrey Deroine." *Missouri Historical Review* 107, no. 4 (July 2013): 222–30.

"Origin and Characteristics of the American Aborigines." *United States Magazine and Democratic Review* 11, no. 54 (December 1842): 603–21.

Ostrom, John H. "Mr. Peale's Missing Mastodon." *Discovery–Magazine of the Peabody Museum of Natural History* 17, no. 1 (1983–84): 3–9.

Paulson, George W. *William Thornton, M.D.: Gentleman of the Enlightenment*. Columbus: Ohio State University Press, 2007.

Penny, H. Glenn. *Kindred by Choice: Germans and American Indians since 1800*. Chapel Hill: University of North Carolina Press, 2013.

Pickering, Robert B., ed. *Peace Medals: Negotiating Power in Early America*. Tulsa, OK: Gilcrease Museum, 2011.

Pickett, Terry H. "The Friendship of Frederick Douglass with the German, Ottilie Assing." *Georgia Historical Quarterly* 73, no. 1 (Spring 1989): 88–105.

Pohlsander, Hans A. *German Monuments in the Americas: Bonds across the Atlantic*. New German-American Studies 33. Edited by Don Heinrich Tolzmann. New York: Peter Lang, 2010.

Pratt, Mary Louise. *Imperial Eyes: Travel Writing and Transculturation*. New York: Routledge, 1992.

Pritchard, Margaret, with the assistance of Lisa L. Ford. "A Protracted View: The Relationship between Mapmakers and Naturalists in Recording the Land." In *Knowing Nature: Art and Science in Philadelphia, 1740–1840*, edited by Amy R. W. Meyers, 8–35. New Haven, CT: Yale University Press, 2011.

Reddin, Paul. *Wild West Shows*. Urbana: University of Illinois Press, 1999.

Regis, Pamela. *Describing Early America: Bartram, Jefferson, Crèvecoeur, and the Rhetoric of Natural History*. DeKalb: Northern Illinois University Press, 1992.

Reilly, F. Kent, III. "Displaying the Source of the Sacred: Shell Gorgets, Peace Medals, and the Accessing of Supernatural Power." In *Peace Medals: Negotiating Power in Early America*, edited by Robert B. Pickering et al., 9–17. Tulsa, OK: Gilcrease Museum, 2011.

The Republican Scrap Book: Containing the Platforms, and a Choice Selection of Extracts, Setting Forth the Real Questions in Issue, the Opinions of the Candidates, the Nature and Designs of the Slave Oligarchy, as Shown by Their Own Writers, and the Opinions of Clay, Webster, Josiah Quincy, and Other Patriots, on Slavery and Its Extension. Boston: John P. Jewett, 1856.

Rice, Howard C., Jr. "Jefferson's Gift of Fossils to the Museum of Natural History in Paris." *Proceedings of the American Philosophical Society* 95, no. 6 (21 December 1951): 597–627.

Rigal, Laura. *The American Manufactory: Art, Labor, and the World of Things in the Early Republic*. Princeton, NJ: Princeton University Press, 1998.

Rigal, Laura. "Peale's Mammoth." In *American Iconology: New Approaches to Nineteenth-Century Art and Literature*, edited by David C. Miller, 18–38. New Haven: Yale University Press, 1993

Rolle, Andrew F. *John Charles Frémont: Character as Destiny*. Norman: University of Oklahoma Press, 1991.

Royce, Sarah. *A Frontier Lady*. Edited by Ralph H. Gabriel. New Haven, CT: Yale University Press, 1932.

Rudolph, Emanuel D. "Women in Nineteenth Century American Botany; A Generally Unrecognized Constituency." *American Journal of Botany* 69, no. 8 (September 1982): 1346–55.

Sand, George. *Correspondance*. Vol. 6, *1843–juin 1845*. Edited by Georges Lubin. Paris: Garnier Frères, 1969.

Sand, George. *Oeuvres Illustrées*. Paris: Hetzel-Blanchard Maresq, 1852–56.

Sandrock, Oliver, and Gabriele Gruber, eds. "Peale's Mastodon und die Enträtselung des *American incognitum*." *Kaupia: Darmstädter Beiträge zur Naturgeschichte* 21 (2016): 5–70.

Saunders, Barbara, and Lea Zuyderhoudt, eds. *The Challenges of Native American Studies: Essays in Celebration of the Twenty-Fifth American Indian Workshop*. Leuven: Leuven University Press, 2004.

Schlaer, Robert. *Sights Once Seen: Daguerreotyping Frémont's Last Expedition through the Rockies*. Santa Fe: Museum of New Mexico Press, 2000.

Schlett, James. *A Not Too Greatly Changed Eden: The Story of the Philosophers' Camp in the Adirondacks*. Ithaca, NY: Cornell University Press, 2015.

Schoelwer, Susan P., ed. *The General and the Garden: George Washington's Landscape at Mount Vernon*. Charlottesville: University of Virginia Press for the Mount Vernon Ladies' Association, 2015.

Schoelwer, Susan P., ed. *Lives Bound Together: Slavery at George Washington's Mount Vernon*. Mount Vernon, VA: Mount Vernon Ladies' Association, 2016.

Schönhofer, Matthias. *Letters from an American Botanist: The Correspondence of Gotthilf Heinrich Ernst Mühlenberg (1753–1815)*. Stuttgart: Franz Steiner Verlag, 2014.

Schönhofer, Matthias. "'Runs, Creeks, and Rivers Join': The Correspondence Network of Gotthilf Henry Ernst Mühlenberg." In *A Peculiar Mixture: German-Language Cultures and Identities in Eighteenth-Century America*, edited by Jan Stievermann and Oliver Scheiding, 150–80. University Park: Pennsylvania State University Press, 2013.

Schoolcraft, Henry Rowe. *Historical and Statistical Information Respecting the History, Condition, and Prospects of the Indian Tribes of the United States; Collected and Prepared under the Direction of the Bureau of Indian Affairs per Act of Congress of March 3rd, 1847*. Illustrated by S. Eastman. 5 vols. Philadelphia: Lippincott, Grambo, 1851–57.

Schulten, Susan. "The Civil War and the Origins of the Colorado Territory." *Western Historical Quarterly* 44, no. 1 (Spring 2013): 21–46.

Schulten, Susan. "Emma Willard and the Graphic Foundations of American History." *Journal of Historical Geography* 33 (2007): 542–64.

Schulten, Susan. *The Geographical Imagination in America, 1880–1950*. Chicago: University of Chicago Press, 2001.

Schulten, Susan. *Mapping the Nation: History and Cartography in Nineteenth-Century America*. Chicago: University of Chicago Press, 2012.

Schwartz, Philip J., ed. *Slavery at the Home of George Washington*. Charlottesville: University of Virginia Press, 2002.

Sears, John F. *Sacred Places: American Tourist Attractions in the Nineteenth Century*. New York: Oxford University Press, 1989.

Sellers, Charles Coleman. *Mr. Peale's Museum: Charles Willson Peale and the First Popular Museum of Natural Science and Art*. New York: W. W. Norton, 1980.

Semonin, Paul. *American Monster: How the Nation's First Prehistoric Creature Became a Symbol of National Identity*. New York: New York University Press, 2000.

Semonin, Paul. "Citizens and Strangers: The American Museum before Barnum." PhD diss., University of Oregon, 1992.

Shuler, Jay. *Had I the Wings: The Friendship of Bachman and Audubon*. Athens: University of Georgia Press, 1995.

Silliman, Benjamin. *A Visit to Europe in 1851*. 2 vols. New York: Putnam, 1853.

Silverberg, Robert. *Mammoths, Mastodons, and Man*. New York: McGraw-Hill, 1970.

Silverman, Kenneth. *Lightning Man: The Accursed Life of Samuel F. B. Morse*. New York: Alfred A. Knopf, 2003.

Simpson, George Gaylord. "The Beginnings of Vertebrate Paleontology in North America." *Proceedings of the American Philosophical Society* 86, no. 1 (September 1942): 130–88.

Simpson, George Gaylord, and H. Tobien. "Rediscovery of Peale's Mastodon." *Proceedings of the American Philosophical Society* 98, no. 4 (August 1954): 279–81.

Smith, Margaret Bayard. *The First Forty Years of Washington Society: Portrayed by the Family Letters of Mrs. Samuel Harrison Smith (Margaret Bayard) from the Collection of Her Grandson, J. Henley Smith*. Edited by Gaillard Hunt. New York: Charles Scribner's Sons, 1906.

Smith, Margaret Bayard. *A Winter in Washington; or, Memoirs of the Seymour Family*. 2 vols. New York: Bliss and White, 1824.

Smith, Paul Chaat. *Everything You Know about Indians Is Wrong*. Minneapolis: University of Minnesota Press, 2009.

Sprague, William Buell. *The Life of Jedidiah Morse*. New York: Anson D. F. Randolph, 1874.

Squier, E. G. *Catalogue of the Books, Manuscripts, Maps, Drawings, and Engravings Principally Relating to Central America, and Peru, American Antiquities, &c Belonging to Mr. E. G. Squier*. Edited by Joseph Sabin. New York: Charles C. Shelley, 1876.

Staël, Madame de. *Selected Correspondence*. Arranged by Georges Solovieff. Translated and edited by Kathleen Jameson-Cemper. Boston: Kluwer Academic Publishers, 2000.

Stafford, Barbara Maria. *Voyage into Substance: Art, Science, Nature, and the Illustrated Travel Account, 1760–1840*. Cambridge, MA: MIT Press, 1984.

Starr, Kevin. *Americans and the California Dream, 1850–1915*. New York: Oxford University Press, 1973.

Stauffer, John. "Frederick Douglass and the Aesthetics of Freedom." *Raritan* 25, no. 1 (Summer 2005): 114–36.

Stein, Susan R. *The Worlds of Thomas Jefferson at Monticello*. New York: Harry N. Abrams in association with the Thomas Jefferson Memorial Foundation, 1993.

Stephens, Lester D. *Science, Race, and Religion in the American South: John Bachman and the Charleston Circle of Naturalists, 1815–1895*. Chapel Hill: University of North Carolina Press, 2000.

Stetson, Sarah P. "William Hamilton and His 'Woodlands.'" *Pennsylvania Magazine of History and Biography* 73, no. 1 (January 1949): 26–33.

Stevens, Henry. *Recollections of James Lenox and the Formation of his Library*. London: Stevens and Son, 1886. Revised and elucidated by Victor H. Paltsits. New York: New York Public Library, 1951.

Stevenson, Winona. "Beggars, Chickabobbooags, and Prisons: Paxoche (Ioway) Views of English Society, 1844–45." *American Indian Culture and Research Journal* 17, no. 4 (1993): 1–23.

St. John de Crèvecoeur, J. Hector. *Letters from an American Farmer*. London: Thomas Davies, 1782. Reprint, Gloucester, MA: Smith, 1968.

Tatlock, Lynne, and Matt Erlin, eds. *German Culture in Nineteenth-Century America: Reception, Adaptation, Transformation*. Rochester, NY: Camden House, 2005.

Taylor, Bayard. *At Home and Abroad: A Sketch-Book of Life, Scenery, and Men*. New York: Putnam, 1860.

Thomson, Keith Stewart. *The Legacy of the Mastodon: The Golden Age of Fossils in America*. New Haven, CT: Yale University Press, 2008.

Thornton, Anna. "Diary of Mrs. William Thornton, 1800–1863." *Records of the Columbia Historical Society* 10 (1907): 88–226.

Thornton, William. *Papers of William Thornton*. Vol. 1, *1781–1802*. Edited by Charles M. Harris. Charlottesville: University of Virginia Press, 1995.

Ticknor, George. *Life, Letters, and Journals of George Ticknor*. 2 vols. Boston: James B. Osgood, 1876

Ticknor, George. *Life, Letters, and Journals of George Ticknor*. 2 vols, expanded edition. Boston: Houghton Mifflin, 1909.

Tompkins, Edmund, and J. Lee Davis. *The Natural Bridge and Its Historical Surroundings*. Natural Bridge: Natural Bridge of Virginia, 1939.

Torrence, Gaylord. *The Plains Indians: Artists of Earth and Sky*. Paris: Musée du Quai Branly, 2014.

Tower, Charlemagne. "Joseph Bonaparte in Philadelphia and Bordentown." *Pennsylvania Magazine of History and Biography* 42, no. 4 (1918): 289–309.

Upham, Charles Wentworth. *Life, Explorations, and Public Services of John Charles Fremont*. Boston: Ticknor and Fields, 1856.

Viereck, L. "Education Report, 1900–1901: German Instruction in American Schools." In *Congressional Series of United States Public Documents*, vol. 4299, chapter 14, 531–703. Washington, DC: Government Printing Office, 1902.

Wallace, Anthony F. C. *Jefferson and the Indians: The Tragic Fate of the First Americans*. Cambridge, MA: Belknap Press of Harvard University Press, 1999.

Wallace, Paul A. W. "Henry Ernest Muhlenberg." *Proceedings of the American Philosophical Society* 92, no. 2 (5 May 1948): 107–10.

Wallace, Paul A. W. *The Muhlenbergs of Pennsylvania*. Philadelphia: University of Pennsylvania Press, 1950.

Watts, Iain P. "Philosophical Intelligence: Letters, Print, and Experiment during Napoleon's Continental Blockade." *Isis* 106, no. 4 (2015): 749–70.

Wharton, Anne Hollingsworth. *Social Life in the Early Republic*. Philadelphia: J. B. Lippincott, 1902.

Willard, Emma. *Guide to the Temple of Time; and Universal History for Schools*. New York: A. S. Barnes, 1850.

Wilmerding, John. *Compass and Clock: Defining Moments in American Culture: 1800, 1850, 1900*. New York: Harry N. Abrams, 1999.

Woodbridge, William. *A System of Universal Geography on the Principles of Comparison and Classification Bound Together with Emma Willard, Ancient Geography, as Connected with Chronology, and Preparatory to the Study of Ancient History: Accompanied with an Atlas*. Hartford, CT: O. D. Cooke, 1829.

Wood, Marcus. *Blind Memory: Visual Representations of Slavery in England and America*. New York: Routledge, 2000.

Wulf, Andrea. *Founding Gardeners: The Revolutionary Generation, Nature, and the Shaping of the American Nation*. New York: Vintage Books, 2011.

Yokota, Kariann Akemi. *Unbecoming British: How Revolutionary America Became a Postcolonial Nation*. New York: Oxford University Press, 2014.

SCIENCE AND EXPLORATION

Allen, Deborah. "Acquiring 'Knowledge of Our Own Continent': Geopolitics, Science, and Jeffersonian Geography, 1783–1803." *Journal of American Studies* 40 (2006): 205–32.

American Museum of Natural History. *Pioneers of American Science. An Account of the Exercises Held and the Addresses Delivered at the American Museum of Natural History, December 29, 1906*. New York: American Museum of Natural History, 1906.

Auben, David, Charlotte Bigg, and H. Otto Siburn, eds. *The Heavens on Earth: Observatories and Astronomy in Nineteenth-Century Science and Culture*. Durham, NC: Duke University Press, 2010.

Avila-Pires, Fernando Dias de. "The Type Specimens of Brazilian Mammals Collected by Prince Maximilian zu Wied." *American Museum Novitates*, no. 2209 (17 February 1965): 1–21.

Bailey, Vernon. "Maximilian's Travels in the Interior of North America, 1832–1834." *Natural History* 13, no. 4 (July–August 1923): 337–43.

Balm, Roger. "Expeditionary Art: An Appraisal." *Geographical Review* 90, no. 4 (October 2000): 585–602.

Barber, Lynn. *The Heyday of Natural History, 1820–1870*. Garden City, NY: Doubleday, 1980.

Bartram, John. *Observations on the Inhabitants, Climate, Soil, Rivers, Productions, Animals, and Other Matters Worthy of Notice: Made by Mr. John Bartram, in His Travels from Pennsylvania to Onondago, Oswego and the Lake Ontario, in Canada. [To Which Is Annex'd, a Curious Account of the Cataracts of Niagara. By Mr. Peter Kalm, a Swedish Gentleman Who Travelled There.]* London: J. Whiston and B. White, 1751.

Bartram, William. *The Travels of William Bartram: Naturalists' Edition*. Edited by Francis Harper. New Haven, CT: Yale University Press, 1958.

Bartram, William. *Travels through North and South Carolina, Georgia, East and West Florida.* London: J. Johnson, 1792.

Bell, Whitfield J., Jr. "A Box of Old Bones: A Note on the Identification of the Mastodon, 1766–1806." *Proceedings of the American Philosophical Society* 93, no. 2 (16 May 1949): 169–77.

Bell, Whitfield J., Jr. *A Cabinet of Curiosities: Five Episodes in the Evolution of American Museums.* Charlottesville: University of Virginia Press, 1967.

Bell, Whitfield J., and Nicholas Collin. "Nicholas Collin's Appeal to American Scientists." *William and Mary Quarterly* 13, no. 4 (October 1956): 519–50.

"Benjamin Silliman." *Harper's New Monthly Magazine*, June–November 1862, 480–86.

Benson, Maxine, ed. *From Pittsburgh to the Rocky Mountains: Major Stephen Long's Expedition, 1819–1820.* Golden, CO: Fulcrum, 1988.

Bernhard, Herzog zu Sachsen-Weimar-Eisenach, and Heinrich Luden. *Reise Sr. Hoheit des Herzogs Bernhard zu Sachsen-Weimar-Eisenach nach Nord-Amerika in den Jahren 1825 und 1826.* 2 vols. Weimar: Wilhelm Hoffmann, 1828.

Bernhard, Karl, Duke of Saxe-Weimar-Eisenach. *Travels through North America, during the Years 1825 and 1826.* 2 vols. Philadelphia: Carey, Lea and Carey, 1828.

Biddle, Nicholas, ed. *History of the Expedition under the Command of Captains Lewis and Clark: To the Sources of the Missouri, Thence across the Rocky Mountains and Down the River Columbia to the Pacific Ocean. Performed during the Years 1804–5–6.* 2 vols. Philadelphia: Bradford and Inskeep, 1814.

Blodget, Lorin. *Climatology of the United States, and of the Temperate Latitudes of the North American Continent. Embracing a Full Comparison of These with the Climatology of the Temperate Latitudes of Europe and Asia....* 8 vols. Philadelphia: J. B. Lippincott, 1857.

Bray, Edmund, and Martha Coleman Bray, eds. *Joseph N. Nicollet on the Plains and Prairies: The Expeditions of 1838–39 with Journals, Letters, and Notes on the Dakota Indians, Translated from the French.* Saint Paul: Minnesota Historical Society Press, 1976.

Breidbach, Olaf, and André Karliczek. "Himmelblau—das Cyanometer des Horace-Bénédict de Saussure' (1740–1799)." *Sudhoffs Archiv* 95, no. 1 (2011): 3–29.

Brooks, Alden Finney. "Grand Trip across the Plains" diary, 19 May–23 August 1859. Newberry Library, Chicago.

Byrd, William, II. *Histories of the Dividing Line betwixt Virginia and North Carolina.* Edited by William K. Boyd. Raleigh: North Carolina Historical Commission, 1929.

Byrd, William, II. *The Writings of "Colonel" William Byrd of Westover in Virginia Esqr.* Edited by John Spencer Bassett. New York: Doubleday, 1901.

Cannon, Susan Faye. *Science in Culture: The Early Victorian Period.* New York: Science History Publications, 1978.

Carson, Rachel. *Silent Spring.* Boston: Houghton Mifflin, 1962.

Carter, Edward C., III, ed. *Surveying the Record: North American Scientific Exploration to 1930.* Philadelphia: American Philosophical Society, 1999.

Carvalho, Solomon Nunez. *Incidents of Travel and Adventure in the Far West; with Col. Frémont's Last Expedition across the Rocky Mountains: Including Three Months Residence in Utah, and a Perilous Trip across the Great American Desert, to the Pacific.* New York: Derby and Jackson, 1857. Reprint with an introduction by Ava Kahn. Lincoln: University of Nebraska Press, 2004.

Coues, Elliott, ed. *The Expeditions of Zebulon Montgomery Pike, to Headwaters of the Mississippi River, through Louisiana Territory, and in New Spain, during the Years 1805–6–7.* 3 vols. 1810. Reprint, New York: Francis P. Harper, 1895.

Crosland, Maurice P. *The Society of Arcueil: A View of French Science at the Time of Napoleon I.* Cambridge, MA: Harvard University Press, 1967.

Dougherty, Frank William Peter. *The Correspondence of Johann Friedrich Blumenbach.* Vol. 5, *1796–1800, Letters 966–1359.* Edited by Norbert Klatt. Göttingen: Norbert Klatt, 2013.

Dugatkin, Lee Alan. *Mr. Jefferson and the Giant Moose: Natural History in Early America.* Chicago: University of Chicago Press, 2009.

Egmond, Florike, and Peter Mason. "Skeletons on Show: Learned Entertainment and Popular Knowledge." *History Workshop Journal* 41, no. 1 (Spring 1996): 92–116.

Ehrenberg, Ralph E. "'Forming a General Geographical Idea of a Country': Mapping Louisiana from 1803 to 1820." In *Charting Louisiana: Five Hundred Years of Maps*, edited by Alfred E. Lemmon, John T. Magill, and Jason R. Wiese, 123–61. New Orleans: Historic New Orleans Collection, 2003.

Ehrenberg, Ralph E. "Nicholas King: First Surveyor of the City of Washington, 1803–1812." *Records of the Columbia Historical Society* 69/70 (1969–70): 31–65.

Ehrenberg, Ralph E., and Herman J. Viola. *Mapping the West with Lewis & Clark.* Delray Beach, FL: Levenger Press in association with the Library of Congress, 2015.

Ellicott, Andrew. *The Journal of Andrew Ellicott: Late Commissioner on Behalf of the United States... for Determining the Boundary between the United States and the Possessions of His Catholic Majesty in America.* Philadelphia: Budd and Bartram for Thomas Dodson, 1803. Reprint, Chicago: Quadrangle Books, 1962.

Everett, Edward. "Long's Expedition." *North American Review* 16 (April 1823): 30–45.

Faujas de Saint Fond, Barthélemy. *A Journey through England and Scotland to the Hebrides in 1784.* Edited by Sir Archibald Geikie. 2 vols. Glasgow: Hugh Hopkins, 1907.

Freeman, Thomas, and Peter Custis. *An Account of the Red River in Louisiana Drawn Up from the Returns of Messrs. Freeman and Custis to the War Office of the United States, Who Explored the Same in the Year 1806.* Washington, DC: n. p., 1807.

Frémont, John Charles. *Memoirs of My Life.* Vol. 1. New York and Chicago: Belford, Clarke, 1887.

Frémont, John Charles. *Report of the Exploring Expedition to the Rocky Mountains in the Year 1842, and to Oregon and North California in the Years 1843–44.* Washington: Blair and Rives, 1845.

Frémont, John Charles. *Report on an Exploration of the Country Lying between the Missouri River and the Rocky Mountains on the Line of the Kansas and Great Platte Rivers.* Buffalo, NY: Derby, 1851.

Gerassi-Navarro, Nina. *Women, Travel, and Science in Nineteenth-Century Americas: The Politics of Observation*. In Palgrave Studies in Literature, Science, and Medicine. London: Palgrave Macmillan, 2017.

Goetzmann, William H. *Army Exploration in the American West, 1803–1863*. New Haven, CT: Yale University Press, 1959.

Goetzmann, William. *Exploration and Empire: The Explorer and the Scientist in the Winning of the American West*. New York: Alfred A. Knopf, 1966.

Goetzmann, William. *New Lands, New Men: America and the Second Age of Discovery*. New York: Viking, 1986.

Gondorf, Bernhard. "Die Expedition von Maximilian Prinz zu Wied und Karl Bodmer in das innere Nordamerika." In *Prärie- und Plainsindianer: Die Reise in das innere Nord-America von Maximilian Prinz zu Wied und Karl Bodmer*, 37–60. Mainz: Verlag Hermann Schmidt, 1993.

Graf, Andreas. *Der Tod der Wölfe: Das abenteuerliche und das bürgerliche Leben des Romanschriftstellers und Amerikareisenden Balduin Möllhausen*. Berlin: Duncker and Humblot, 1991.

Hicks, Philip Marshall. *The Development of the Natural History Essay in American Literature*. Ph.D. diss., University of Pennsylvania, 1924.

Huseman, Ben W. *Wild River, Timeless Canyons: Balduin Möllhausen's Watercolors of the Colorado*. Fort Worth: Amon Carter Museum, 1995.

Jackson, Donald Dean, ed. *The Journals of Zebulon Montgomery Pike with Letters and Related Documents*. 2 vols. Norman: University of Oklahoma Press, 1966.

Jackson, Donald Dean, ed. *Letters of the Lewis and Clark Expedition with Related Documents, 1783–1854*. Urbana: University of Illinois Press, 1962.

James, Edwin. *Account of an Expedition from Pittsburgh to the Rocky Mountains, Performed in the Years 1819 and 1820; by Order of the Hon. J. C. Calhoun, Secretary of War, under the Command of Maj. S. H. Long, of the U.S. Top. Engineers, Compiled from the notes of Major Long, Mr. T. Say, and Other Gentlemen of the Party*. 3 vols. London: Longman, Hurst, Rees, Orme, and Brown, 1823.

James, Edwin. *James's Account of S. H. Long's Expedition to the Rocky Mountains, Performed in the Years 1819, 1820*. Vols. 14–17 in *Early Western Travels, 1748–1846*, edited by Reuben Gold Thwaites. Cleveland: Arthur H. Clark, 1905.

Jansen, Axel. *Alexander Dallas Bache: Building the American Nation through Science and Education in the Nineteenth Century*. New York: Campus Verlag, 2011.

Krumpelmann, John T. "Duke Bernhard of Saxe-Weimar: An Emissary of Goethe to the American South." *South Central Bulletin* 30, no. 4 (Winter 1970): 201–3.

LeConte, John. "Physical Studies of Lake Tahoe, I." *Overland Monthly* 2 (November 1883): 506–16.

Leopold, Aldo. *A Sand County Almanac: With Other Essays on Conservation from Round River*. New York: Oxford University Press, 1966.

Liebersohn, Harry. *Aristocratic Encounters: European Travelers and North American Indians*. Cambridge: Cambridge University Press, 1998.

Louis-Philippe, King of France. *Diary of My Travels in America*. Translated by Stephen Becker. New York: Delacorte Press, 1977.

Lyell, Sir Charles. *Life, Letters, and Journals of Sir Charles Lyell, Bart*. 2 vols. Edited by K. Lyell. London: J. Murray, 1881.

Marcou, Jules. *The Life, Letters, and Works of Louis Agassiz*. New York: Macmillan, 1896.

McLaughlin, Castle. *Arts of Diplomacy: Lewis & Clark's Indian Collection*. Seattle: University of Washington Press for the Peabody Museum of Archaeology and Ethnology at Harvard University, 2003.

Möllhausen, Balduin. *Diary of a Journey from the Mississippi to the Coasts of the Pacific, with a United States Government Expedition*. With an introduction by Alexander von Humboldt. Translated by Mrs. Percy Sinnett. 2 vols. London: Longman, Brown, Green, Longmans, and Roberts, 1858.

Muir, John. "Yosemite Glaciers." *New York Tribune*, 5 December 1871.

Nichols, Roger L., and Patrick L. Halley. *Stephen Long and American Frontier Exploration*. Newark: University of Delaware Press, 1980.

Nicollet, Joseph N. *Report Intended to Illustrate a Map of the Hydrographical Basin of the Upper Mississippi River*. Washington, DC: Blair and Rives, 1843.

Noll, Michael G. "Prince Maximilian's America: The Narrated Landscapes of a German Explorer and Naturalist." PhD diss., University of Kansas, 2000.

Noll, Michael G. "Prince Maximilian's Other Worlds." *Pennsylvania Geographer* 43 (2005): 65–83.

Peale, Rembrandt. *Account of the Skelton of the Mammoth, a Non-Descript Carnivorous Animal of Immense Size, Found in America*. London: E. Lawrence, 1802.

Peale, Rembrandt. "A Short Account of the Mammoth." *Philosophical Magazine* 14 (November 1802): 162–69.

Peale, Titian Ramsay. *The Butterflies of North America: Titian Peale's Lost Manuscript*. With text by David A. Grimaldi and Kenneth Haltman. New York: American Museum of Natural History in association with Abrams Books, 2015.

Peale, Titian Ramsay. *Diary of Titian Ramsay Peale: Oregon to California, Overland Journey, September and October, 1841*. Edited by Clifford Merrill Drury. Los Angeles: G. Dawson, 1957.

Pike, Zebulon M. *An Account of Expeditions to the Sources of the Mississippi, and through the Western Parts of Louisiana, to the Sources of the Arkansaw, Kans, La Platte, and Pierre Jaun, Rivers; Performed by Order of the Government of the United States during the Years 1805, 1806, and 1807. And a Tour through the Interior Parts of New Spain*. Philadelphia: C. and A. Conrad, 1810.

Poesch, Jessie. *Titian Ramsay Peale and His Journals of the Wilkes Expedition 1799–1885*. Memoirs of the American Philosophical Society 52. Philadelphia: American Philosophical Society, 1961.

Porter, Joseph C. "Marvelous Figures, Astonished Travelers: The Montana Expedition of Maximilian, Prince of Wied." *Montana: Magazine of Western History* 41, no. 4 (Autumn 1991): 36–53.

Porter, Joseph C. "Maximilian, Prince of Wied: A Biographical Sketch." In *Views of a Vanishing Frontier*, edited by John C. Ewers et al., 9–20. Omaha: Center for Western Studies and the Joslyn Art Museum, 1984.

Preuss, Charles. *Exploring with Frémont: The Private Diaries of Charles Preuss, Cartographer for John C. Frémont on His First, Second, and Fourth Expeditions to the Far West*. Translated and edited by Erwin G. Gudde and Elizabeth K. Gudde. Norman: University of Oklahoma Press, 1958.

Reingold, Nathan. "Alexander Dallas Bache: Science and Technology in the American Idiom." *Technology and Culture* 11 (1970): 163–77.

Review of Balduin Möllhausen's *Diary of a Journey from the Mississippi to the Coasts of the Pacific, with a United States Government Expedition*, in *Littell's Living Age* 60 (1859): 242–44.

Roth, Hermann Josef, ed. *Maximilian Prinz zu Wied: Jäger, Reisender, Naturforscher*. Vol. 17 in *Fauna und Flora in Rheinland-Pfalz*. Landau: Gesellschaft für Naturschutz und Ornithologie Rheinland-Pfalz, 1995.

Rowan, Steven. *The Baron in the Grand Canyon: Friedrich Wilhelm von Egloffstein in the West*. Columbia: University of Missouri Press, 2012.

Rupke, Nicolaas, and Gerhard Lauer, eds. *Johann Friedrich Blumenbach: Race and Natural History, 1750–1850*. London: Routledge, 2018.

Saindon, Robert A., ed. *Explorations into the World of Lewis and Clark: 194 Essays from the Pages of* We Proceeded On*, the Quarterly Journal of the Lewis and Clark Trail Heritage Foundation*. 3 vols. Great Falls, MT: Lewis and Clark Trail Heritage Foundation, 2003.

Schach, Paul. "Maximilan, Prince of Wied (1782–1867), Reconsidered." *Great Plains Quarterly* 14 (Winter 1994): 5–20.

Schach, Paul. "Maximilian, Prince of Wied, Encounters 'Landsleute' in the New World." In *Maximilian Prinz zu Wied: Jäger, Reisender, Naturforscher*. Vol. 17 in *Fauna und Flora in Rheinland-Pfalz*, edited by Hermann Joseph, 155–70. Landau: Gesellschaft für Naturschutz und Ornithologie Rheinland, 1995.

Shortland, Michael, ed. *Science and Nature: Essays in the History of the Environmental Sciences*. Stanford in the Vale: British Society for the History of Science, 1993.

Shortland, Michael, and Richard Yeo, eds. *Telling Lives in Science: Essays on Scientific Biography*. New York: Cambridge University Press, 1996.

Smith, Edgar F. *James Woodhouse: A Pioneer in Chemistry (1770–1809)*. Philadelphia: J. C. Winston, 1918. Reprint, New York: Arno Press, 1980.

Smith, Henry Nash. "Clarence King, John Wesley Powell, and the Establishment of the United States Geological Survey." *Mississippi Valley Historical Review* 34, no. 1 (June 1947): 37–58.

Stegner, Wallace. *Beyond the Hundredth Meridian: John Wesley Powell and the Second Opening of the West*. Boston: Houghton Mifflin, 1954.

Taft, Robert. "Heinrich Balduin Möllhausen." In *Artists and Illustrators of the Old West, 1850–1900*, 22–35. New York: Charles Scribner's Sons, 1953.

Taft, Robert. "The Pictorial Record of the American West: VI. Heinrich Balduin Möllhausen." *Kansas Historical Quarterly* 16, no. 3 (August 1948): 225–44.

Thomas, Sabina, and Joseph T. Hannibal. "Revisiting New Harmony in the Footsteps of Maximilian, the Prince of Wied; David Dale Owen; Charles Lesueur; and Other Early Naturalists." In *From the Cincinnati Arch to the Illinois Basin: Geological Field Excursions along the Ohio River Valley*, Field Guide 12, edited by Anton H. Maria and Ronald C. Counts, 25–45. Boulder, CO: Geological Society of America, 2008.

Tidball, Eugene C. *Soldier-Artist of the Great Reconnaissance: John C. Tidball and the 35th Parallel Pacific Railroad Survey*. Tucson: University of Arizona Press, 2004.

Vanzolini, Paolo E., and Charles W. Myers. "The Herpetological Collections of Maximilian, Prince of Wied (1782–1867), with Special Reference to Brazilian Materials." *Bulletin of the American Museum of Natural History*, no. 395 (26 June 2015): 1–155.

Vekerdy, Lilla, ed. *The Era of Experiments and the Age of Wonder: Scientific Expansion from the Seventeenth to the Nineteenth Centuries*. In Smithsonian Contributions to Knowledge. Washington, DC: Smithsonian Institution Scholarly Press, 2015.

Viola, Herman J., and Carolyn Margolis, eds. *Magnificent Voyagers: The U.S. Exploring Expedition, 1838–1842*. Washington, DC: Smithsonian Institution Press, 1985.

Wheat, Carl I. *Mapping the Transmississippi West, 1540–1861*. 4 vols. San Francisco: Institute of Historical Cartography, 1957.

Whipple, Amiel Weeks. *Report of Explorations for a Railway Route, Near the Thirty-Fifth Parallel of North Latitude, from the Mississippi River to the Pacific Ocean, 1853–1854*. With the assistance of Joseph Christmas Ives. Vol. 3. Washington, DC: O. P. Nicholson, 1856.

Wied, Maximilian, Prinz von. *The North American Journals of Prince Maximilian of Wied*. Edited by Stephen S. Witte and Marsha V. Gallagher. Translated by William J. Orr, Paul Schach, and Dieter Karch. 3 vols. Norman: University of Oklahoma Press, 2008–12.

Wied, Maximilian, Prinz von. *People of the First Man: Life among the Plains Indians in Their Final Days of Glory: The Firsthand Account of Price Maximilian's Expedition up the Missouri River, 1833–34*. Edited by David Thomas and Karin Ronnenfeldt. New York: Dutton, 1976.

Wied, Maximilian, Prinz von. *Reise in das innere Nord-America in den Jahren 1832 bis 1834*. 2 vols. Koblenz: Hoelscher, 1839–41.

Wied, Maximilian, Prinz von. *Travels in Brazil in the Years 1815, 1816, 1817*. London: Henry Colburn, 1820.

Wied, Maximilian, Prince of. *Travels in the Interior of North America, by Maximilian, Prince of Wied: With Numerous Engravings on Wood and a Large Map*. Translated by H. Evans Lloyd. London: Ackermann, 1843.

Wied, Maximilian, Prince of. *Travels in the Interior of North America, 1832–1834*. Translated from the German by Hannibal Evans Lloyd. Vols. 22–24 in *Early Western Travels, 1848–1846*, edited and with an introduction by Reuben Gold Thwaites. Cleveland: Arthur H. Clark, 1905.

Wied, Maximilian, Prinz zu. "Einige Bemerkungen über Geo. Catlin's Werk: 'Lettres and notes on the manners, customs, and condition of the North-American Indians.'" *Isis* 10 (1842): 726–41.

Wied, Maximilian, Prinz zu. *Unveröffentlichte Bilder und Handschriften zur Völkerkunde Brasiliens*. Edited by Joseph Röder and Hermann Trimborn with Josefine Huppertz, Udo Oberem, and Karl Viktor, Prinz zu Wied. Bonn: Dümmler, 1954.

Wilhelm, Paul, Duke of Württemberg. *Travels in North America, 1822–1824*. Edited by Savoie Lottinville, translated by W. Robert Nitske. 2 vols. Norman: University of Oklahoma Press, 1973.

Wilhelm, Paul, Herzog von Württemberg. *Erste Reise nach dem nördlichen Amerika, in den Jahren 1822 bis 1824*. Stuttgart: Verlag der J. G. Cotta, 1835.

Wills, Hannah. "Charles Blagden's Diary: Information Management and British Science in the Eighteenth Century." *Notes and Records: The Royal Society Journal of the History of Science* 73, no. 1 (February 2019): 61–81. https://royal-societypublishing.org/doi/10.1098/rsnr.2018.0016

Wills, Hannah. "Joseph Banks and Charles Blagden: Cultures of Advancement in the Scientific Worlds of Late Eighteenth-Century London and Paris." *Notes and Records: The Royal Society Journal of the History of Science* (2019). https://royalsocietypublishing.org/doi/10.1098/rsnr.2018.0060

Wislizenus, Friedrich A., M.D. *A Journey to the Rocky Mountains in the Year 1839, Translated…with a Sketch of the Author's Life by Frederick A. Wislizenus, Esq.* Saint Louis: Missouri Historical Society, 1912.

Woodman, Neal. "The Stephen H. Long Expedition (1819–1820), Titian R. Peale's Field Illustrations, and the Lost Holotypes of the North American Shrews *Sorex brevicaudus* Say and *Sorex parvus* Say (Mammalia: Soricidae) from the Philadelphia Museum." *Proceedings of the Biological Society of Washington* 122, no. 1 (2009): 117–29.

Wood, W. Raymond, and Gary E. Moulton. "Prince Maximilian and New Maps of the Missouri and Yellowstone Rivers by William Clark." *Western Historical Quarterly* 12, no. 4 (October 1981): 372–86.

SMITHSONIAN

Agassiz, Louis. "Letter on the Smithsonian Institution: The Late Professor Louis Agassiz." *Science*, n. s. 49, no. 1265 (28 March 1919): 300–301.

Arago, François. "Eulogy on Ampère." In *Annual Report of the Board of Regents of the Smithsonian Institution*, Showing the Operations, Expenditures, and Condition of the Institution for the Year 1872, 111–71. Washington, DC: Government Printing Office, 1873.

Bird, William L., Jr. "A Suggestion Concerning James Smithson's Concept of 'Increase and Diffusion.'" *Technology and Culture* 24, no. 2 (April 1983): 246–55.

Burleigh, Nina. *The Stranger and the Statesman: James Smithson, John Quincy Adams, and the Making of America's Greatest Museum: The Smithsonian*. New York: William Morrow, 2003.

Carmichael, Leonard. "Joseph Henry and the National Academy of Sciences." *Proceedings of the National Academy of Sciences* 58, no. 1 (15 July 1867): 1–10.

Donaldson, Thomas. *The George Catlin Indian Gallery in the U.S. National Museum (Smithsonian Institution), with Memoir and Statistics*. Washington, DC: Government Printing Office, 1887.

Ewing, Heather. *The Lost World of James Smithson: Science, Revolution, and the Birth of the Smithsonian*. New York: Bloomsbury, 2007.

Gray, Asa. "Joseph Henry." In *The Golden Age of Science: Thirty Portraits of the Giants of 19th-Century Science, by Their Scientific Contemporaries*, edited by Bessie Zaban Jones, 264–95. New York: Simon and Schuster, in association with the Smithsonian Institution, 1966.

Henry, Joseph. *The Papers of Joseph Henry*. Edited by Nathan Reingold. Washington, DC: Smithsonian Institution Press, 1972.

Howe, Reginald Heber, Jr. "Sir Charles Blagden, Earliest of Rhode Island Ornithologists." *American Naturalist* 39, no. 462 (June 1905): 397–404.

Irby, J. R. McD. "On the Works and Character of James Smithson." In *Smithsonian Miscellaneous Collections*, 21:143–55. Washington, DC: Smithsonian Institution, 1881.

Johnson, Walter R. "A Memoir of the Scientific Character and Researches of James Smithson, Esq., F.R.S." In *Smithsonian Miscellaneous Collections*, 21:123–41. Washington, DC: Smithsonian Institution, 1881.

Lindsay, Debra. *Science in the Subarctic: Trappers, Traders, and the Smithsonian Institution*. Washington, DC: Smithsonian Institution Press, 1993.

Macie, James. "An Account of Some Chemical Experiments on Tabasheer." *Philosophical Transactions of the Royal Society of London* 81, part 1 (1791): 368–88.

Matthews, Jean V. "Libraries, Books, and the Nature of America: The Creation of the Smithsonian Institution." *Journal of Library History* 16, no. 1 (Winter 1981): 152–65.

Portolano, Marlana. "Increase and Diffusion of Knowledge: Ethos of Science and Education in the Smithsonian's Inception." *Rhetoric Review* 18, no. 1 (Autumn 1999): 65–81.

Rhees, William J. "James Smithson and His Bequest." *Smithsonian Miscellaneous Collections*, 21:1–68. Washington, DC: Smithsonian Institution, 1881.

Rhees, William J., ed. *The Smithsonian Institution: Documents Relative to Its Origin and History, 1835–1899*. Vol. 1, *1835–1887, 24th Congress to 49th Congress*. Washington, DC: Government Printing Office, 1901.

Silliman, Benjamin. *American Contributions to Chemistry: An Address Delivered on the Occasion of the Celebration of the Centennial of Chemistry*. Philadelphia: Collins, 1874. Reprinted from *American Chemist* (August–September, December 1874).

Smithsonian Institution. *Annual Report of the Board of Regents of the Smithsonian Institution to July 1885*. Washington, DC: Government Printing Office, 1886.

Smithsonian Institution. *Memorial of Joseph Henry*. Washington, DC: Smithsonian Institution, 1881.

Smithsonian Institution. *Report from the Board of Regents, Submitted to Congress, March 3, 1847*. Washington, DC: Government Printing Office, 1847.

Smithson, James. *The Scientific Writings of James Smithson*. Edited by William J. Rhees. Washington, DC: Smithsonian Institution, 1879.

Stamm, Richard E. *The Castle: An Illustrated History of the Smithsonian Building*. Washington, DC: Smithsonian Institution, 2012.

Washburn, Wilcomb E., ed. *The Great Design: Two Lectures on the Smithson Bequest by John Quincy Adams, Delivered at Quincy and Boston in November 1839, Now First Published Together from Contemporary Printed and Manuscript Texts*. Washington, DC: Smithsonian Institution, 1965.

Wright, Helena E. *The First Smithsonian Collection: The European Engravings of George Perkins Marsh and the Role of Prints in the U.S. National Museum*. Washington, DC: Smithsonian Institution Scholarly Press, 2015.

ART AND LITERATURE

Adamson, Jeremy Elwell, ed. *Niagara: Two Centuries of Changing Attitudes, 1697–1901*. Washington, DC: Corcoran Gallery of Art, 1985.

Anderson, Nancy K., and Linda S. Ferber. *Albert Bierstadt: Art and Enterprise*. New York: Hudson Hills Press in association with the Brooklyn Museum, 1990.

"An Old Contributor" [Henry T. Tuckerman]. "New York Artists." *Knickerbocker* 48 (July 1865): 26–42.

Anonymous [Washington Irving]. "The Empire of the West." *Knickerbocker* 15 (March 1840): 260–61.

Antelyes, Peter. *Tales of Adventurous Enterprise: Washington Irving and the Poetics of Western Expansion*. New York: Columbia University Press, 1990.

Applebaum, Lauren. "Elusive Matter, Material Bodies: American Art in the Age of Electronic Mediation, 1865–1918." PhD diss., University of Illinois at Urbana-Champaign, 2019.

Avery, Kevin J. *Church's Great Picture:* The Heart of the Andes. New York: Metropolitan Museum of Art, 1993.

Avery, Kevin J. "'The Heart of the Andes' Exhibited: Frederic E. Church's Window on the Equatorial World." *American Art Journal* 8, no. 1 (Winter 1986): 52–72.

Avery, Kevin J. "Maya on the Hudson: Church's *Cayambe* and Cruger's 'Folly.'" *Hudson River Valley Review* 31, no. 1 (Autumn 2014): 50–61.

Avery, Kevin J. "Rally 'round the flag': Frederic Edwin Church and the Civil War." *Hudson River Valley Review* 27 (Spring 2011): 66–103.

Avery, Kevin J. "Uxmal-on-Hudson: Mayan Ruins on Cruger's Island in the 19th Century." Bard College Archives and Special Collections. http://www.bard.edu/archives/voices/uxmal/uxmal.php.

Badeau, Adam. *The Vagabond*. New York: Rudd and Carleton, 1859.

Bate, Jonathan. *Romantic Ecology: Wordsworth and the Environmental Tradition*. London: Routledge, 1991.

Baudelaire, Charles. *The Painter of Modern Life and Other Essays*. Translated by Jonathan Mayne. London: Phaidon Press, 1964.

Bedell, Rebecca Bailey. *The Anatomy of Nature: Geology and American Landscape Painting, 1825–1875*. Princeton, NJ: Princeton University Press, 2001.

Beetam, Robert. "George Catlin in France: His Relationship to Delacroix and Baudelaire." *Art Quarterly* 24 (1961): 129–45.

Bibbins, Arthur Barneveld "Charles Willson Peale's Painting, *The Exhumation of the First American Mastodon*." *Bulletin of the Geological Society of America* 18 (1906): 650–52.

Bleichmar, Daniela. *Visual Voyages: Images of Latin American Nature from Columbus to Darwin*. New Haven, CT: Yale University Press in association with the Huntington Library, Art Collections, and Botanical Gardens, 2017.

Bodmer, Karl. *Karl Bodmer's North American Prints*. Edited by Brandon K. Ruud. Lincoln: University of Nebraska Press, 2004.

Bodmer, Karl. *The Maximilian-Bodmer Collection: Travels in the Interior of North America in 1832–1834: Original Sketches by Carl Bodmer*. Evanston, IL: Northwestern University, 1966.

Bohan, Ruth. "'The Gathering of the Forces': Walt Whitman and the Visual Arts in Brooklyn in the 1850s." In *Walt Whitman and the Visual Arts*, edited by Geoffrey M. Sill and Roberta K. Tarbell, 1–27. New Brunswick, NJ: Rutgers University Press, 1992.

Bohan, Ruth. *Looking into Walt Whitman: American Art, 1850–1920*. University Park: Pennsylvania State University Press, 2006.

Bohan, Ruth. "Walt Whitman and the Sister Arts." *Walt Whitman Quarterly Review* 16, no. 3 (1999): 153–60

Boime, Albert. *The Magisterial Gaze: Manifest Destiny and American Landscape Painting, c. 1830–1865*. Washington, DC: Smithsonian Institution Press, 1991.

Bolz, Peter, and Hans-Ulrich Sanner. *Native American Art: The Collections of the Ethnological Museum Berlin*. Translated by Ann Leslie Davis. Berlin: Staatliche Museen zu Berlin, 1999.

Borst, Raymond R., ed. *The Thoreau Log: A Documentary Life of Henry David Thoreau, 1817–1862*. New York: G. K. Hall, 1992.

Boyle, Nicholas. *Goethe: The Poet and the Age*. Vol. 1, *The Poetry of Desire, 1749–1790*. Oxford: Clarendon Press, 1992.

Braddock, Alan C., and Christopher Irmscher, eds. *A Keener Perception: Ecocritical Studies in American Art History*. Tuscaloosa: University of Alabama Press, 2009.

Branch, Michael P. "Five Generations of Literary Coopers: Intergenerational Valuations of the American Frontier." In *Susan Fenimore Cooper: New Essays on Rural Hours and Other Works*, edited by Rochelle Johnson and Daniel Patterson, 61–80. Athens: University of Georgia Press, 2001.

Branch, Michael P., ed. *Reading the Roots: American Nature Writing before* Walden. Athens: University of Georgia Press, 2004.

Brickhouse, Anna. "'A Story of the Island of Cuba': William Cullen Bryant and the Hispanophone Americas." *Nineteenth-Century Literature* 56, no. 1 (June 2001): 1–22.

Brownlee, Peter John, ed. *Samuel F. B. Morse's* Gallery of the Louvre *and the Art of Invention*. New Haven, CT: Yale University Press for the Terra Foundation for American Art, 2014.

The Bryant Festival at "The Century," November 5, 1864. New York: D. Appleton, 1865.

Buell, Lawrence. *The Environmental Imagination: Thoreau, Nature Writing, and the Formation of American Culture*. Cambridge, MA: Belknap Press of Harvard University Press, 1995.

Callow, James T. *Kindred Spirits: Knickerbocker Writers and American Artists, 1807–1855*. Chapel Hill: University of North Carolina Press, 1967.

Carr, Gerald L. "American Art in Great Britain: The National Gallery Watercolor of *The Heart of the Andes*." *Studies in the History of Art* 12 (1982): 81–100.

Carr, Gerald L. *Frederic Edwin Church: Catalogue Raisonné of Works of Art at Olana State Historic Site*. 2 vols. Cambridge: Cambridge University Press, 1994.

Catlin, George. *Catlin's Notes of Eight Years' Travels and Residence in Europe, with His North American Indian Collection*. 2 vols. London: by the author, 1848.; New York: Burgess, Stringer, 1848.

Catlin, George. *A Descriptive Catalogue of Catlin's Indian Gallery: Containing Portraits, Landscapes, Costumes, &c., and Representations of the Manners and Customs of the North American Indians. Collected and Painted Entirely by Mr. Catlin, during Seven Years' Travel amongst 48 Tribes, Mostly Speaking Different Languages. Now Exhibiting in the Egyptian Hall, Piccadilly, London*. London: C. Adlard, 1840.

Catlin, George. *Last Rambles amongst the Indians of the Rocky Mountains and the Andes*. New York: D. Appleton, 1867.

Catlin, George. *Letters and Notes on the Manners, Customs, and Condition of the North American Indians*. 2 vols. London: by the author, 1841. Reprint, Minneapolis: Ross and Haines, 1965.

Catlin, George. *Life among the Indians*. London: Gall and Inglis, ca. 1870.

Catlin, George. *O-kee-pa: A Religious Ceremony; and Other Customs of the Mandans*. London: Trübner, 1867.

Catlin, George. *Photostatic Copies from the New Commercial Advertiser of Letters by George Catlin Describing the Manners, Customs, and Conditions of the North American Indians: July 24, 1832 to September 30, 1837*. Washington, DC: Smithsonian Institution, 1900.

Channing, William Ellery. *Thoreau: The Poet-Naturalist*. Boston: Roberts Bros., 1873.

Chateaubriand, François-René, vicomte de. *Recollections of Italy, England, and America, with Essays on Various Subjects, in Morals and Literature*. Philadelphia: M. Carey, 1816.

Clair, Jean, ed. *Cosmos: From Goya to De Chirico, from Friedrich to Kiefer: Art in Pursuit of the Infinite*. Miami: Bompiani, 2000.

Conwell, Russell H. *The Life, Travels, and Literary Career of Bayard Taylor*. Boston: B. B. Russell, 1879.

Cooper, James Fenimore. *Last of the Mohicans*. 2 vols. New York: Macmillan, 1898.

Cooper, James Fenimore. *Letters and Journals of James Fenimore Cooper*. Edited by James Franklin Beard. 6 vols. Cambridge, MA: Belknap Press of Harvard University Press, 1960–1968.

Cooper, James Fenimore. *The Pioneers, or Sources of the Susquehanna: A Descriptive Tale*. New York: D. Appleton and Company, 1872.

Cooper, James Fenimore. "The Prairie." From *The Leatherstocking Tales*. New York: Sully and Kleinteich, ca. 1898.

Cooper, Susan Fenimore. *Journal of a Naturalist in the United States, by Miss Cooper*. London: R. Bentley, 1855.

Cooper, Susan Fenimore. *Rural Hours*. Boston: Houghton Mifflin, 1887.

Dant, Elizabeth A. "Composing the World: Emerson and the Cabinet of Natural History." *Nineteenth-Century Literature* 44, no. 1 (June 1989): 18–44.

Davidson, Marshall B. "Carl Bodmer's Unspoiled West." In *The Maximilian-Bodmer Collection*, 1–23. Omaha, NE: Joslyn Art Museum, 1966. Reprinted from *American Heritage*, April 1963.

Dean, Bradley P. "Natural History, Romanticism, and Thoreau." In *American Wilderness: A New History*, edited by Michael Lewis, 73–90. Oxford: Oxford University Press, 2007.

Dippie, Brian W. *Catlin and His Contemporaries: The Politics of Patronage*. Lincoln: University of Nebraska Press, 1990.

Dobkin, John, Nicolai Cikovski Jr., and Paul J. Staiti. *Samuel F. B. Morse: Educator and Champion of the Arts in America*. New York: National Academy of Design, 1982.

Emerson, Ralph Waldo. *The Early Lectures of Ralph Waldo Emerson*. Edited by Stephen E. Whicher, Robert E. Spiller, and Wallace E. Williams. 3 vols. Cambridge, MA: Belknap Press of Harvard University, 1959.

Emerson, Ralph Waldo. *Emerson's Antislavery Writings*. Edited by Len Gougeon and Joel Myerson. New Haven, CT: Yale University Press, 1995.

Emerson, Ralph Waldo. *The Journals and Miscellaneous Notebooks of Ralph Waldo Emerson*. Edited by William H. Gilman, Alfred R. Ferguson, George P. Clark, Merrell R. Davis, et al. 16 vols. Cambridge, MA: Belknap Press of Harvard University Press, 1960–82.

Emerson, Ralph Waldo. *The Letters of Ralph Waldo Emerson*. Edited by Ralph L. Rusk, vols. 1–6; and Eleanor Tilton, vols. 7–10. 10 vols. New York: Columbia University Press, 1939–95.

Fabre, Daniel. "The Catlin Effect: Paris, 1845–1846." *Gradhiva*, no. 3 (Spring 2006): 55–75.

Fairchild, Sharon L. "George Sand and George Catlin—Masking Indian Realities." *Nineteenth-Century French Studies* 22, nos. 3–4 (Spring–Summer 1994): 439–49.

Feest, Christian. "James Reid Lambdin, George Catlin, and the 'Costume of a Sioux Warrior.'" *Native American Studies* 17, no. 2 (2003): 19–26.

Ferber, Linda S., ed. *Kindred Spirits: Asher B. Durand and the American Landscape*. New York: Brooklyn Museum, 2007.

"The Fine Arts; Mr. R. L. Stuart's House and Art Collection." *Home Journal*, 2 April 1859, 2.

Friedl, Peter. "The Impossible Museum." *e-flux* 23 (March 2011). https://www.e-flux.com/journal/23/67778/the-impossible-museum/.

Gado, Frank, ed. *William Cullen Bryant: An American Voice*. Hartford, CT: University Press of New England, 2006.

Gallagher, Marsha V., and John F. Sears. *Karl Bodmer's Eastern Views: A Journey in North America*. Omaha, NE: Joslyn Art Museum, 1996.

Gardner, Albert ten Eyck. "Scientific Sources of the Full-Length Landscape: 1850." *Metropolitan Museum of Art Bulletin*, n.s., 4, no. 2 (October 1945): 59–65.

Georgi, Karen L. "Summer Camp with William J. Stillman: Looking at Nature, between Ruskin and Emerson." *American Art* 32, no. 3 (Fall 2018): 22–41.

Green, Tyler. *Carleton Watkins: Making the West American*. Oakland: University of California Press, 2018.

Gurney, George, and Therese Thau Heyman, eds. *George Catlin and His Indian Gallery*. New York: W. W. Norton in association with the Smithsonian American Art Museum, 2002.

Haifley, Julie Link. *Titian Ramsay Peale, 1799–1885*. Washington, DC: George Washington University, 1981.

Haltman, Kenneth. "Figures in a Western Landscape: Reading the Art of Titian Ramsay Peale from the Long Expedition to the Rocky Mountains, 1819–1820." PhD diss., Yale University, 1992.

Haltman, Kenneth. *Looking Close and Seeing Far: Samuel Seymour, Titian Ramsay Peale, and the Art of the Long Expedition, 1818–1823*. University Park: Pennsylvania State University Press, 2008.

Harbert Petrulionis, Sandra, ed. *Thoreau in His Own Time: A Biographical Chronicle of His Life, Drawn from Recollections, Interviews, and Memoirs by Family, Friends, and Associates*. Iowa City: University of Iowa Press, 2012.

Harding, Walter. *Emerson's Library*. Charlottesville: University of Virginia Press, 1967.

Harding, Walter. *Thoreau as Seen by His Contemporaries*. New York: Dover Publications, 1989.

Harris, Neil. *The Artist in American Society: The Formative Years, 1790–1860*. Chicago: University of Chicago Press, 1966.

Hartmann, Horst. *George Catlin und Balduin Möllhausen: Zwei Interpreten der Indianer und des Alten Westens*. Berlin: Dietrich Reimer Verlag, 1963.

Harvey, Eleanor Jones. "Capturing the Cosmos." In *Frederic Church's Olana on the Hudson: Art, Landscape, Architecture*, edited by Karen Zukowski and Julia B. Rosenbaum, 91–101. New York: Rizzoli, 2018.

Harvey, Eleanor Jones. *The Civil War and American Art*. New Haven, CT: Yale University Press in association with the Smithsonian American Art Museum, 2012.

Harvey, Eleanor Jones. "Founding Landscape: Charles Willson Peale's *Exhumation of the Mastodon*." *American Art* 31, no. 2 (Summer 2017): 40–42.

Hausdoerffer, John. *Catlin's Lament: Indians, Manifest Destiny, and the Ethics of Nature*. Lawrence: University Press of Kansas, 2009.

Hawkins, Richmond Laurin. "Bernardin de Saint-Pierre and Peale's Philadelphia Museum: The History of a Portrait." *Romantic Review* 20, no. 1 (January–March 1929): 1–12.

Herring, Joseph B. "Selling the 'Noble Savage' Myth: George Catlin and the Iowa Indians in Europe, 1843–1845." *Kansas History: A Journal of the Central Plains* 29 (Winter 2006–7): 226–45.

Hevner, Carol Eaton, ed. *Rembrandt Peale, 1778–1860: A Life in the Arts*. Philadelphia: Historical Society of Pennsylvania, 1985.

Hight, Kathryn S. "'Doomed to Perish': George Catlin's Depictions of the Mandan." *Art Journal* 49, no. 2 (Summer 1990): 119–24.

Holland, Eugenia Calvert, ed. *Four Generations of Commissions: The Peale Collection of the Maryland Historical Society, March 3, 1975–June 29, 1975*. Baltimore: Maryland Historical Society, 1975.

Honour, Hugh. *The New Golden Land: European Images of American from the Discoveries to the Present Time*. New York: Pantheon Books, 1975.

Horton, Jessica L. "Ojibwa *Tableaux Vivants*: George Catlin, Robert Houle, and Transcultural Materialism." *Art History* 39, no. 1 (February 2016): 124–51.

Howat, John K. *Frederic Church*. New Haven, CT: Yale University Press, 2005.

Hunt, David C., and Marsha V. Gallagher, eds. *Karl Bodmer's America*. Lincoln: University of Nebraska Press, 1984.

Huntington, David C. "Frederic Church's *Niagara*: Nature and Nation's Type." *Texas Studies in Literature and Language* 25, no. 1 (Spring 1983): 100–138.

Huntington, David C. "Landscapes and Diaries: The South American Trips of F. E. Church." *Brooklyn Museum Annual* 5 (1963–64): 65–98.

Huntington, David C. *The Landscapes of Frederic Edwin Church: Vision of an American Era*. New York: George Brazilier, 1966.

Irving, Washington. *The Adventures of Captain Bonneville*. New York: Putnam, 1837.

Irving, Washington. *Astoria, or, Anecdotes of an Enterprise Beyond the Rocky Mountains*. London: R. Bentley, 1836.

Irving, Washington. "The Journal of the Letters of Washington Irving: Part 2—From Italy to Paris and London." Edited by Clara Kirk and Rudolf Kirk. *Rutgers University Library Journal* 9, no. 2 (1946): 36–58.

Irving, Washington. *A Tour on the Prairies*. London: John Murray, 1835.

Isernhagen, Karin, ed. *Karl Bodmer, 1809–1893: A Swiss Artist in America*. Translated by Nigel Stephenson. Zurich: Scheidegger and Spiess for the Nordamerika Native Museum, 2009.

Johnson, Linck C. "'Liberty Is Never Cheap': Emerson, 'The Fugitive Slave Law,' and the Antislavery Lecture Series at the Broadway Tabernacle." *New England Quarterly* 76, no. 4 (December 2003): 550–92.

Johnson, Rochelle, and Daniel Patterson, eds. *Susan Fenimore Cooper: Essays on Nature and Landscape*. Athens: University of Georgia Press, 2002.

Johnson, Rochelle, and Daniel Patterson, eds. *Susan Fenimore Cooper: New Essays on Rural Hours and Other Works*. Athens: University of Georgia Press, 2001.

Katz, Wendy. "Robert S. Duncanson, Race, and Auguste Comte's Positivism in Cincinnati." *American Studies* 53, no. 1 (2014): 79–115.

Kelly, Franklin. "Frederic Church in the Tropics." *Arts in Virginia* 26 (1986): 16–33.

Kelly, Franklin. *Frederic Edwin Church*. With contributions by Stephen Jay Gould, James Anthony Ryan, and Deborah Rindge. Washington, DC: National Gallery of Art, 1989.

Kelly, Franklin. *Frederic Edwin Church and the National Landscape*. Washington, DC: Smithsonian Institution Press, 1988.

Kelly, Franklin, and Gerald L. Carr. *The Early Landscapes of Frederic Edwin Church*. Fort Worth: Amon Carter Museum, 1987.

Kennedy, Elizabeth, and Olivier Meslay, eds. *American Artists and the Louvre*. Paris: Éditions Hazan for the Musée du Louvre and the Terra Foundation for American Art, 2006.

Kime, Wayne R. "Washington Irving and 'The Empire of the West.'" *Western American Literature* 5, no. 4 (Winter 1971): 277–85.

Kornhauser, Elizabeth Mankin, and Tim Barringer, eds. *Thomas Cole's Journey: Atlantic Crossings*. New Haven, CT: Yale University Press for the Metropolitan Museum of Art, 2018.

Kurth, Rosaly Torna. *Susan Fenimore Cooper: New Perspectives on Her Works*. Bloomington, IN: iUniverse, 2016.

Kusserow, Karl. "Memory, Metaphor, and Meaning in Daniel Huntington's *Atlantic Cable Projectors*." In *Picturing Power: Portraiture and Its Uses in the New York Chamber of Commerce*, edited by Karl Kusserow, 319–75. New York: Columbia University Press, 2013.

Kusserow, Karl. "Technology and Ideology in Daniel Huntington's *Atlantic Cable Projectors*." *American Art* 24, no. 1 (Spring 2010): 94–113.

Kusserow, Karl, and Alan C. Braddock, eds. *Nature's Nation: American Art and Environment*. New Haven, CT: Yale University Press for the Princeton Art Museum, 2018.

Läng, Hans. *Indianer waren meine Freunde: Leben und Werk Karl Bodmers, 1809–1893*. Bern and Stuttgart: Hallwag, 1976.

Leonard, David Charles. "The Cartesian Vortex in Moby-Dick." *American Literature* 51, no. 1 (March 1979): 105–9.

Lewis, Robert M. "Wild American Savages and the Civilized English: Catlin's Indian Gallery and the Shows of London." *European Journal of American Studies* 3, no. 1 (Spring 2008): 1–21.

Mabee, Carleton. *The American Leonardo: A Life of Samuel F. B. Morse*. New York: Alfred A. Knopf, 1943.

Mack, Anglea D., and Stephen G. Hoffius, eds. *Landscapes of Slavery: The Plantation in American Art*. Columbia: University of South Carolina Press, 2007.

Manthorne, Katherine Emma. *Creation and Renewal: Views of Cotopaxi by Frederic Edwin Church*. Washington, DC: Smithsonian Institution Press for the National Museum of American Art, 1985.

Manthorne, Katherine Emma. "Legible Landscapes: Text and Image in the Expeditionary Art of Frederic Church." In *Surveying the Record: North American Scientific Exploration to 1930*, edited by Edward C. Carter II, 133–45. Philadelphia: American Philosophical Society, 1999.

Manthorne, Katherine Emma. *Tropical Renaissance*. Washington, DC: Smithsonian Institution Press, 1989.

Marley, Anna O. "Landscapes of the New Republic at Thomas Jefferson's Monticello." In *Building the British Atlantic World: Spaces, Places, and Material Culture*, edited by Daniel Maudlin and Bernard L. Herman, 78–99. Chapel Hill: University of North Carolina Press, 2016.

McKinney, Katie. "William Roberts's 'Excursion over the Mountains': Backcountry Landscapes 'from the Pencil of a Virginian.'" *MESDA Journal of Early Southern Decorative Arts* 39 (January 2019): n.p. http://www.mesdajournal.org/2019/william-robertss-excursion-over-the-mountainsbackcountry-landscapes-from-the-pencil-of-a-virginian/.

McKinsey, Elizabeth. *Niagara Falls: Icon of the American Sublime*. Cambridge: Cambridge University Press, 1985.

Miller, Angela L. *The Empire of the Eye: Landscape Representation and American Cultural Politics, 1825–1875*. Ithaca, NY: Cornell University Press, 1993.

Miller, Angela L. "The Fate of Wilderness in American Landscape Art: The Dilemmas of 'Nature's Nation.'" In *A Keener Perception: Ecocritical Studies in American Art History*, edited by Alan C. Braddock and Christopher Irmscher, 91–112. Tuscaloosa: University of Alabama Press, 2009.

Miller, Daegen. "A Map of Radical Bewilderment: On the Liberation Cartography of Henry David Thoreau." *Places Journal* (March 2018). https://doi.org/10.22269/180306.

Miller, Lillian B. "Charles Willson Peale as History Painter: *The Exhumation of the Mastodon*." *American Art Journal* 13, no. 1 (Winter 1981): 47–68.

Miller, Lillian B., ed. *In Pursuit of Fame: Rembrandt Peale, 1778–1860*. Seattle: University of Washington Press for the National Portrait Gallery, 1992.

Miller, Lillian B., and David C. Ward, eds. *New Perspectives on Charles Willson Peale: A 250th Anniversary Celebration*. Pittsburgh: University of Pittsburgh Press for the Smithsonian Institution, 1991.

Miller, Perry. *Nature's Nation*. Cambridge, MA: Belknap Press of Harvard University Press, 1967.

Minor, Kimberly. "Stealing Horses and Hostile Conflict: 1833–1834: Drawings of Mato-Tope and Sih-Chida." Master's thesis, University of Nebraska, 2011.

Mitchell, Timothy. "Frederic Church's 'The Icebergs': Erratic Boulders and Time's Slow Changes." *Smithsonian Studies in American Art* 3, no. 4 (Autumn 1989): 2–23.

Morse, Samuel F. B. *Samuel F. B. Morse: His Letters and Journals*. Edited by Edward Lind Morse. 2 vols. Boston: Houghton Mifflin, 1914. Reprint, New York: Da Capo Press, 1973.

Mulvey, Christopher. "George Catlin in Europe." In *George Catlin and His Indian Gallery*, edited by George Gurney and Therese Thau Heyman, 63–91. New York: W. W. Norton in association with the Smithsonian American Art Museum, 2002.

Myers, Kenneth John. "Art and Commerce in Jacksonian America: The Steamboat Albany Collection." *Art Bulletin* 82, no. 3 (September 2000): 503–28.

Navas Sanz de Santamaría, Pablo. *The Journey of Frederic Edwin Church through Colombia and Ecuador, April–October 1853*. Bogotá: Universidad de los Andes, 2008.

Noble, Louis Legrand. *Church's Painting: The Heart of the Andes*. New York: D. Appleton, 1859.

Noble, Louis Legrand. *The Course of Empire, Voyage of Life, and Other Pictures of Thomas Cole*. New York: Cornish, Lamport, 1853.

Noble, Louis Legrand. *The Life and Works of Thomas Cole*. New York: Sheldon, Blakeman, 1856.

Novak, Barbara. *Nature and Culture: American Landscape Painting, 1825–1875*. London: Thames and Hudson, 1980.

Nygren, Edward J. "From View to Vision." In *Views and Visions: American Landscape before 1830*, edited by Edward J. Nygren with Bruce Robertson, 3–81. Washington, DC: Corcoran Gallery of Art, 1986.

Nygren, Edward J., with Bruce Robertson, eds. *Views and Visions: American Landscape before 1830*. Washington, DC: Corcoran Gallery of Art, 1986.

Oedel, William T. "After Paris: Rembrandt Peale's Apollodorian Gallery." *Winterthur Portfolio* 27, no. 1 (Spring 1992): 1–27.

O'Malley, Therese. "Landscape Gardening in the Early National Period." In *Views and Visions: American Landscape before 1830*, edited by Edward J. Nygren with Bruce Robertson, 133–59. Washington, DC: Corcoran Gallery of Art, 1986.

Orcutt, Kimberly, ed. *John Rogers: American Stories*. London: Philip Wilson Publishers, 2010.

Orr, William J. "Karl Bodmer: The Artist's Life." In *Karl Bodmer's America*, edited by David C. Hunt and Marsha V. Gallagher, 349–76. Lincoln: University of Nebraska Press, 1984.

Packer, Barbara L. *The Transcendentalists*. Athens: University of Georgia Press, 2007.

Parry, Ellwood C., III. *The Art of Thomas Cole: Ambition and Imagination*. Newark: University of Delaware Press, 1988.

Peale, Charles Willson. *The Selected Papers of Charles Willson Peale and His Family*. Vol. 2, *Charles Willson Peale: The Artist as Museum Keeper, 1791–1810*, parts 1–2. Edited by Lillian B. Miller, Sidney Hart, and David C. Ward. New Haven, CT: Yale University Press for the National Portrait Gallery, 1988.

Peale, Charles Willson. *The Selected Papers of Charles Willson Peale and His Family*. Vol. 1, *Charles Willson Peale: Artist in Revolutionary America, 1735–1791*. Edited by Lillian B. Miller, Sidney Hart, and Toby A. Appel. New Haven, CT: Yale University Press for the National Portrait Gallery, 1983.

Peck, H. Daniel. *Thoreau's Morning Work: Memory and Perception in "A Week on the Concord and Merrimack Rivers," the Journal, and "Walden."* New Haven, CT: Yale University Press, 1990.

Perry, Claire. *The Great American Hall of Wonders: Art, Science, and Invention in the Nineteenth Century*. London: D Giles for the Smithsonian American Art Museum, 2011.

Pinto, Holly Joan. *William Cullen Bryant and the Hudson River School*. Roslyn, NY: Nassau County Museum of Fine Art, 1981.

Pratt, Stephanie. "Objects, Performance and Ethnographic Spectacle: George Catlin in Europe." *Interventions: International Journal of Postcolonial Studies* 15, no. 2 (2013): 272–85.

Prime, Samuel Iranaeus. *Life of Samuel F. B. Morse, LL.D., Inventor of the Electro-Magnetic Recording Telegraph*. New York: D. Appleton, 1875.

Raab, Jennifer. "Details of Absence: Frederic Church and the Landscape of Post-Emancipation Jamaica." *Art History* 34, no. 4 (September 2011): 715–31.

Raab, Jennifer. *Frederic Church: The Art and Science of Detail*. New Haven, CT: Yale University Press, 2015.

Raab, Jennifer. "'Precisely These Objects': Frederic Church and the Culture of Detail." *Art Bulletin* 95, no. 4 (December 2013): 578–94.

Reynolds, Davis S. *Walt Whitman's America: A Cultural Biography*. New York: Alfred A. Knopf, 1995.

Robertson, Bruce. "The Picturesque Traveler in America." In *Views and Visions: American Landscape before 1830*, edited by Edward J. Nygren with Bruce Robertson, 187–211. Washington, DC: Corcoran Gallery of Art, 1986.

Robertson. David. *West of Eden: A History of the Art and Literature of Yosemite*. Berkeley, CA: Wilderness Press, 1984.

Roehm, Marjorie Catlin. *The Letters of George Catlin and His Family*. Berkeley: University of California Press, 1966.

Ronda, James P. "Washington Irving's West." *Historian* 66, no. 3 (Fall 2004): 546–51.

Ronda, James P., ed. *Washington Irving: Three Western Narratives*. New York: Library of America, 2004.

Rosengarten, Dale. "Portrait of Two Painters: The Work of Theodore Sidney Moïse and Solomon Nunes Carvalho." In *By Dawn's Early Light: Jewish Contributions to American Culture from the Nation's Founding to the Civil War*, edited by Adam D. Mendelsohn, 138–87. Princeton, NJ: Princeton University Press, 2016.

Ruskin, John. *Modern Painters*. 5 vols. London: Smith, Elder, 1846–70.

Schama, Simon. *Landscape and Memory*. New York: Vintage Books, 1995.

Schofield, Robert E. "The Science Education of an Enlightened Entrepreneur: Charles Willson Peale and His Philadelphia Museum, 1784–1827." *American Studies* 30, no. 2 (Fall 1989): 21–40.

Schuyler, David. *Sanctified Landscape: Writers, Artists, and the Hudson River Valley, 1820–1909*. Ithaca, NY: Cornell University Press, 2012.

Scott, Amy, ed. *Yosemite: Art of an American Icon*. Los Angeles: Autry Museum of the American West; Berkeley: University of California Press, 2006.

Sellers, Charles Coleman. *Artist of the Revolution: The Early Life of Charles Willson Peale*. Hebron, CT: Feather and Good, 1939.

Sellers, Charles Coleman. *Charles Willson Peale*. Vol. 2, *His Later Life*. Philadelphia: American Philosophical Society, 1947.

Sellers, Charles Coleman. "Charles Willson Peale and 'The Mammoth Picture.'" In *Exhuming the First American Mastodon, or, The Mammoth Picture*. Peale Museum Historical Series 7. Baltimore: Peale Museum, February 1951.

Sellers, Charles Coleman. "'Good Chiefs and Wise Men': Indians as Symbols of Peace in the Art of Charles Willson Peale." *American Art Journal* 7, no. 2 (November 1975): 10–18.

Sharp. Lewis I. *John Quincy Adams Ward: Dean of American Sculpture; With a Catalogue Raisonné*. Newark: University of Delaware Press, 1985.

Shlaer, Robert. *Sights Once Seen: Daguerreotyping Frémont's Last Expedition through the Rockies*. Santa Fe: Museum of New Mexico Press, 2000.

Siegel, Nancy. *Along the Juniata: Thomas Cole and the Dissemination of American Landscape Imagery*. Seattle: University of Washington Press for Juniata College Museum of Art, 2003.

Sill, Geoffrey M., and Roberta K. Tarbell, eds. *Walt Whitman and the Visual Arts*. New Brunswick, NJ: Rutgers University Press, 1992.

Staiti, Paul. "American Artists and the July Revolution." In *American Artists and the Louvre*, edited by Elizabeth Kennedy and Olivier Meslay, 54–71. Chicago: Terra Foundation for American Art, 2006.

Staiti, Paul. *Samuel F. B. Morse*. New York: Cambridge University Press, 1989.

Stein, Roger B. "Charles Willson Peale's Expressive Design: The Artist in His Museum." *Prospects* 6 (1981): 139–85.

Stein, Roger B. *John Ruskin and Aesthetic Thought in America, 1840–1900*. Cambridge, MA: Harvard University Press, 1967.

Stillman, William. *Autobiography of a Journalist*. 2 vols. New York: Houghton Mifflin, 1901.

Sturhahn, Joan. *Carvalho, Artist—Photographer—Adventurer—Patriot: Portrait of a Forgotten American*. Merrick, NY: Richwood Publishing, 1976.

Taylor, Colin F. *Catlin's O-kee-pa: Mandan Cultural and Ceremonial: The George Catlin O-kee-pa Manuscript in the British Museum*. Braunschweig: Verlag für Amerikanistik, 1996.

Thoreau, Henry David. *The Correspondence of Henry David Thoreau*. Edited by Walter Harding and Carl Bode. New York: New York University Press, 1958.

Thoreau, Henry David. *The Essays of Henry David Thoreau*. Edited by Lewis Hyde. New York: North Point Press, 2002.

Thoreau, Henry David. "Resistance to Civil Government." In *Æsthetic Papers*, edited by Elizabeth P. Peabody, 189–211. Boston: Editor; New York: Putnam, 1849.

Thoreau, Henry David. *Walden; or, Life in the Woods*. Boston: Ticknor and Fields, 1854.

Thoreau, Henry David. "Walking." *Atlantic Monthly* 9, no. 56 (June 1862): 657–74.

Thoreau, Henry David. "A Walk to Wachusett." *Boston Miscellany of Literature and Fashion* 1 (January to July, 1842): 31–36.

Todd, Edgeley W. "Indian Pictures and Two Whitman Poems." *Huntington Library Quarterly* 1 (November 1955): 5–11.

Truettner, William H. "The Genesis of Frederic Edwin Church's *Aurora Borealis*." *Art Quarterly* 31, no. 3 (1968): 267–83.

Truettner, William H. *The Natural Man Observed: A Study of Catlin's Indian Gallery*. Washington, DC: Smithsonian Institution Press in cooperation with the Amon Carter Museum of Western Art and the National Collection of Fine Arts, 1979.

Tyler, Ron. "Karl Bodmer and the American West." In *Karl Bodmer's North American Prints*, edited by Brandon K. Ruud, 1–45. Lincoln: University of Nebraska Press, 2004.

Tyler, Ron. "Karl Bodmer and the American West." *Studii și cercetări de istoria artei, Artă Plastică*, n.s., 5, no. 49 (2015): 9–43.

Wagner, Virginia L. "John Ruskin and Artistical Thought in America." *Winterthur Portfolio* 23 (1988): 151–67.

Wallace, D. H. *John Rogers: The People's Sculptor*. Middletown, CT: Wesleyan University Press, 1967.

Walls, Laura Dassow. *Emerson's Life in Science: The Culture of Truth*. Ithaca, NY: Cornell University Press, 2003.

Walls, Laura Dassow. "Greening Darwin's Century: Humboldt, Thoreau, and the Politics of Hope." *Victorian Review* 36, no. 2 (Fall 2010): 92–103.

Walls, Laura Dassow. *Henry David Thoreau: A Life*. Chicago: University of Chicago Press, 2017.

Walls, Laura Dassow. "Henry David Thoreau: Writing the Cosmos: Address for the Thoreau Society Annual Gathering." *Concord Saunterer*, n.s., 19/20 (2011–12): 1–21.

Walls, Laura Dassow. *Seeing New Worlds: Henry David Thoreau and Nineteenth-Century Natural Science*. Madison: University of Wisconsin Press, 1995.

Walsh, Rebecca Ann. "The 'Terraqueous' Globe: Walt Whitman and the Cosmological Geography of Humboldt and Somerville." In *The Geopoetics of Modernism*, 54–75, 166–167. Gainesville: University of Florida Press, 2015.

Ward, David C. *Charles Willson Peale: Art and Selfhood in the Early Republic*. Berkeley: University of California Press, 2004.

Werner, James V. "'Ground Moles' and Cosmic Flaneurs: Poe, Humboldt, and Nineteenth-Century Science." *Edgar Allan Poe Review* 3, no. 1 (Spring 2002): 45–69.

Whitman, Walt. *Leaves of Grass*. Brooklyn, NY: n.p., 1855.

Wied, Karl Viktor, Prinz zu. *Carl Bodmer Paints the Indian Frontier: A Traveling Exhibition of Watercolors and Drawings Circulated by the Smithsonian Institution, 1954–1955*. Washington, DC: H. K. Press for the Smithsonian Institution, 1954.

Willis, Nathaniel Parker, and William H. Bartlett. *American Scenery, or, Land, Lake, and River:Illustrations of Transatlantic Nature*. 2 vols. London: Virtue, 1840.

Winthrop, Theodore. *A Companion to* The Heart of the Andes. New York: D. Appleton, 1859.

Wolf, Bryan. *Romantic Re-Vision: Culture and Consciousness in Nineteenth-Century American Painting and Literature*. Chicago: University of Chicago Press, 1982.

Wood, W. Raymond, Joseph C. Porter, and David C. Hunt. *Karl Bodmer's Studio Art: The Newberry Library Bodmer Collection*. Urbana: University of Illinois Press, 2002.

Wynne, James. "Samuel F. B. Morse." *Harper's New Monthly Magazine* 24, no. 140 (January 1862): 224–32.

Zygmont, Brian J. "Charles Willson Peale's *The Exhumation of the Mastodon* and the Great Chain of Being: The Interaction of Religion, Science, and Art in Early Federal America." *Text Matters* 5, no. 5 (2015): 95–111.

INDEX

Page numbers in *italics* indicate illustrations. Names of artworks and publications other than those of Humboldt are listed under the name of the artist or author.

A

abolitionism. *See* slaves and slavery

Adalbert of Prussia, Prince, 126

Adams, John Quincy, 65, 117, 180, 374

Adams, Louisa, 65

Adirondack Philosophers' Club, 360, *361* (cat. 104), 362

Agassiz, Louis, 17–18, 32, 34, 183, 185–86, 192, *360*, 364, 377, 379–81, 385, 387–88

Alison, Archibald, 158, 167

American Fur Company, *84*, 116, *123*, 161, 260, 263

American Indians, 31–32, 255–305
- access granted to Catlin and Maximilian by, 260, *261*, 267–75, *268*, *270–74*
- belief in disappearance of, 257, 259, 280, 283
- Carvalho, *Cheyenne Village* (1853, daguerreotype), *115*
- James Fenimore Cooper and, 161, 259, 284
- cultural transgressions regarding, 275–80
- Humboldt meeting with, 298–300, *299* (cat. 79)
- Humboldt's views on, 12, 31–32, 64, 255–58
- Indian Removal Act (1830), 31, 257, 258, 259, 260, 280, 281, 374
- Jefferson and, 31, 256–57
- Long Expedition and, 105, 106
- *See also* Bodmer, Karl; Catlin, George; Maximilian zu Wied-Neuwied; *specific tribes*, *e.g.* Pawnee

American Philosophical Society (APS)
- centennial (1843), 379
- explorations of United States and, 95, 99, 105, 107, 119
- French refugees from Haitian revolution and, 219, 378
- Humboldt's membership certificate (cat. 14), 71, *74*, 75
- Humboldt's visit to U.S. and, 29, 45, 46–49, 52, 55, 61, 65, 69, 71, *74* (cat. 14)
- Lambdin portrait of Humboldt in, 207
- mammoth/mastodon bones and, 138–39
- Priestley's membership in, 376, 378

American Revolution, 49, 55, 62, 64, 117, 135, 136, 138, 144, 146, 157, 169, 178, 186, 187, 376

Arago, François, 3[illegible], 114, 166, 177, *192*, 193, 196, 367, 375, 376–77

Arikara (Sahnish) Indians, 260, 263, 302

art in America
- Frémonts' art collection, 231–40, *232*, *234*, *235*
- Hudson River school, 30, 135, 146, 163, 164, 183, 240, 314, 317
- Humboldt's influence on, 16, 25, 28, 30, 32
- at Metropolitan Fair of 1864, 240–48, *241*, *245–46*
- wilderness aesthetic and, 30, 35, 127, 135, 155–60, *156*, *157*, 162, 317
- *See also specific artists*

Aspects of Nature (Humboldt, 1808), 19, 27, 86, 94, 111, 114, 159, 177, 181.196–97, 204, 226, 358

Assing, Ottilie "Ottie," *228*

Astor, John Jacob, 116, 155, 161, 263

Astor, William Backhouse, 116, 263, 385

B

Bache, Alexander Dallas, 196, 197, 198, 202, 207, 378, 379, 381

Bachman, John, 18, 72

Baird, Spencer, 364, 379

Baltimore, 51, 193–96, 376. *See also* Peale museums in Philadelphia and Baltimore

Bancroft, George, 183, *222*

Banks, Sir Joseph, 375, 376

Banneker, Benjamin, 53, 99

Barnum, P. T., 284

Barton, Benjamin Smith, 46, 51, 149

Bartram, John, 45, 62, 157, 158

Bartram, William, 107, 157

Baudelaire, Charles, 298, 300–301

Beecher, Henry Ward, 233

Beethoven, Ludwig van, 177

Benton, Thomas Hart, 110, 111, 222

Berger, Henry, *Bust of Humboldt* (1860, cat. 110), 388, *389*

Berghaus, Heinrich, 220

Bernhard, Duke of Saxe-Weimar, 71, 117, 119, 262

Berthollet, Charles-Louis, 177, 375

Biddle, Nicholas, 103–104

Bierstadt, Albert, 236–40, *240*
- *Cho-Looke, the Yosemite Fall* (cat. 54), 240, *243*
- *The Rocky Mountains, Lander's Peak*, 240, 241
- *Valley of the Yosemite* (1864, cat. 55), 241, *245*

Bigelow, John, 226

Birch, Thomas, 153

Birch, William Russell, *Woodlands* (1809), *70*

Black, John, 33

Blackfeet Indians, 260, 267

Blaeser, Gustav, 388

Blagden, Charles, 375–76, 377

Blodget, Lorin, 204, 378, 381

Blodgett, William T., 316, 358

Blumenbach, Johann Friedrich, 27, 46, 61, 116, 117, 139, 149, 182, 216, 255, 262, 275–79

Bodmer, Karl, 31, 32, 86, 118–26, 257–58, 262–73, 275, 282–83, 300–302

Bison Dance of the Mandan Indians in front of Their Medicine Lodge in Mih-Tutta-Hankush (aquatint after Bodmer, 1842, cat. 69), 283, *285*

Indian Utensils and Arms (aquatint after Bodmer, 1840, cat. 62), 269, *270*

Interior of a Hut of a Mandan Chief (aquatint after Bodmer, 1841, cat.66), 273, *276*

Map to Illustrate the Route of Prince Maximilian of Wied in the Interior of North America (with Max zu Wied, 1839–41, cats. 23A & 23B), 121, *122*

Mató-Tópe, a Mandan Chief (aquatint after Bodmer, 1839, cat. 61), 267, *268*

Niagara Falls (aquatint after Bodmer, 1841), *121*

Pehriska-Ruhpa, Moennitarri Warrior in the Costume of the Dog Dance (1841, cat. 64), 269–73, *272*

Reise in das innere Nord-America (with Maximilian zu Wied, 1839–41), 124, 282–83, 300

Rock Formations on the Upper Missouri (1833), 121–24, *124*

Sih-Chida and Mahchsi-Karehde, Mandan Indians (aquatint after Bodmer, 1841, cat. 68), 275, *278*

Travellers Meeting Minatarre Indians near Fort Clark (aquatint after Bodmer, 1842, cat. 60), *254*, *266*, 267

View of the Stone Walls on the Upper Missouri (aquatint after Bodmer, 1840, cat. 25), 124, *125*

Bolívar, Simón, 20, 34, *178*, 248

Bonaparte, Charles Lucien Jules Laurent, 108, 119

Bonaparte, Joseph, 119

Bonaparte, Napoleon. *See* Napoleon

Bonpland, Aimé Jacques Alexandre

Essay on the Geography of Plants (Humboldt and Bonpland, 1805), 19, 27, 33, 86–90, *90–93*, 100, 165, 196, 362, 367

exploratory expeditions in U.S. and, 86, 87, 89, 111, 114, 116, 117, 126

natural features in U.S. named for, 111, 114

New Spain and North America, visit with Humboldt to, 15, 19, 42, 45, 46, 48, 49, 61, 65, *66*, 68, 72, 75, 326

Rhexia Speciosa engraving, (Humboldt and Bonpland, 1806), *107*

Botocudo (Aimoré) Indians, 31, 117, 118, 262

Brady, Mathew B., 115, *197*, *204*, *222*, *316*, *365*, *378*

Bread and Cheese Club, 164, 165, 187

Briggs, Isaac, 99

Brissot, Jacques Pierre, 64

Brown, John, 229, 230 (cat. 46), 231, 385

Bryant, William Cullen, 164–65, 187, 223, 224, 225, 226, 248

Buchanan, James, 198, 204, 223, 226, 228

Buffon, George-Louis Leclerc, comte de, 30, 52, 60, 136–37, 138, 142, 148, 255

Bull/Bison Dance (Mandan), 275, *277*, 283

Burke, Sir Edmund, 315

Burr, Aaron, 100

Burwell William, 55

C

Calhoun, John, 105, 106, 109

California

gold, discovery of, 197

statehood of, 31, 114, 222, 226

Watkins and Bierstadt, art of, 236–44, *237–39*, *242*, *243*

See also Yosemite

Carlos IV (king of Spain), 15, 42, 43, 217

Carson, Rachel, *Silent Spring* (1962), 34, 392

Cartesian vortex, 224

Carvalho, Solomon Nunes, *115*

Cary, Elizabeth Cabot, 17

Cass, Lewis, 260

Catherine the Great, 95

Catlin, George

American Indians and, 32, 207, 257–60, 263, 264, 267, 275, 279–305

Humboldt and, 298–300, *299* (cat. 79), 302–4

Indian Gallery, 32, 258, 259, 260, 280–82, 284, 287, 298–302, 373–74, 388–90

Lambdin's portrait of, in Indian dress (1833), *281*

Letters and Notes on the Manners, Customs, and Condition of the North American Indians (1841), 282

Notes of Eight Years' Travels and Residence in Europe (1848, cat. 79), *284*, 296, *299*, 300

O-Kee-Pa: A Religious Ceremony and Other Customs of the Mandans (1867), 304

Paris salons, exhibiting in, 300–301

performing Indians, touring with, 283–300

Smithsonian and, 388–90

Catlin, George, artworks

Buffalo Hunt under the Wolf-Skin Mask (1832–33, cat. 80), 302, *303*

Bull Dance, Mandan O-kee-pa Ceremony (1832, cat. 67), 275, *277*

Catlin and Iowa Indians with Jeffrey Deroine (lithograph, 1845), *284*

Koon-za-ya-me, Female War Eagle (1844, cat. 77), 289–96, *295*

Line drawing of Iowa Indians as a group (1848), *284*, 296

Mah-to-toh-pa, Four Bears, Second Chief, in Full Dress (1832, cat. 58), 260, *261*, 267

Mandan Indians (ca. 1854), 302, *304*

Neu-mon-ya, Walking Rain (1844–45, cat. 71), 287, *288*

No-ho-mun-ya, One Who Gives No Attention (1844, cat. 70), 284–87, *286*

Pipestone Quarry on the Coteau des Prairies (1836–73), *279*

Ru-ton-wee-me, Pigeon on the Wing (1844, cat. 76), 289, *294*

Ru-ton-ye-wee-ma, Strutting Pigeon, Wife of White Cloud (1844, cat. 75), 289, *293*

See-non-ty-a, an Iowa Medicine Man (1844–45, cat. 73), 289, *291*

Shon-ta-yi-ga, Little Wolf, a Famous Warrior (1844–45, cat. 74), 289, *292*, 300

Stu-mick-o-sucks, Buffalo Bull's Back Fat, Head Chief, Blood Tribe (1832), 300–301, *301*

Theodore Burr Catlin, in Indian Costume (1840–41), *282*

Wash-ka-mon-ya, Fast Dancer (1844–45, cat. 72), 289, *290*

Wa-ta-we-bu-ka-na, Commanding General, 296

The White Cloud, Head Chief of the Iowas (1844–45), *287*

Catlin, Theodore Burr, *282*

Cavendish, Sir Henry, 375

Century Club, 165, 248

Chaptal, Jean-Antoine-Claude, 177, 180, 377, 381

Charbonneau, Toussaint, 257, 263, 273

Charles (Carl) zu Wied, Prince, *116*, 117

Charpentier, Jean de, 17

Chase, Salmon P., 222

Chastellux, Marquis de, 256

Chateaubriand, François-René de, 137–38, 158, 181

Child, Lydia Maria, 233

Chimborazo (volcano), 19, 51, 89, 176, 328, 335, *337* (cat. 92), 345–54, *346–48* (cats. 97–99), *352* (cat. 103), 355

Chopin, Frédéric, 298

Chouteau, Pierre, Jr., 260, 263

Church, Frederic Edwin, 16, 28, 32, 313–67

Bierstadt compared, 240

botanical sketches and studies, 326, 328, *329–32* (cats.86–89), 335, *339* (cat. 93)

Brady photo of, *316*

Chimborazo and, 328, 335, *337* (cat. 92), 345–54, *346–48* (cats. 97–99), *352* (cat. 103), 355

Cosmos, influence of, 313, 314–17, *318* (cat. 81), 354, 356, 357, 367

Cotopaxi and, 333–38, *334* (cat. 90), *336* (cat. 91), 360

cultural identity of America in paintings of, 35

Frémont presidential campaign and, 228

Humboldtian persona adopted by, 28, 32, 314–16

in Labrador, 360

literature in America and, 360–67, *361–64*

never meeting or corresponding with Humboldt, 33, 317

Niagara Falls and, 338–45, *340–41* (cat. 94), *342–43* (cat. 95), 354, *355*

Olana and, 314, *366* (cat. 105), 367

Salas's portrait of Humboldt procured by (cat. 96), *344*, 345–53

in South America, 197, 314, 317–38, 345–54

Tequendama Falls and, *322*, 322–33, *323–25*, *327*, *329–32*

in Virginia, 319–22, *320*, *321*

Church, Frederic Edwin, artworks

The Andes of Ecuador (1855), *356*, 356–57

Aurora Borealis (1865, cat. 40), *6–7*, 198, *200*

Botanical Sketches (1857, cat. 93), 335, *339*

Botanical Sketches from Colombia: Bamboo Trees, Trunks and details (1853, cat. 87), 328, *330*

Cayambe (1858, cat. 102), *351*, 353

A Century Plant in Bloom (ca. 1864), *248*

Clouds Over Olana (1872, cat. 105), *366*, 367

Cotopaxi (1855), *37*, 335–38, *336*, 360

The Falls of Tequendama, near Bogotá, New Grenada (1854, cat. 85), 322–33, *327*, 338, 354, 355

A Group of Thirteen Botanical Studies from Colombia or Ecuador (1853, cat. 88), 328, *331*

Heart of the Andes (1859), 33, 240–41, *241*, 345, *354*, 354–60, *355*, 363, 364

Horseshoe Falls (1856–57, cat. 95), 338, *342–43*

The Icebergs, 360

Mount Chimborazo, Ecuador (1857, cat. 97), *312*, *346*, 353

Mount Chimborazo, Ecuador (1857, cat. 98), *347*, 353

Mount Chimborazo at Sunset (1857, cat. 99), *348*, 353

Mount Chimborazo through Rising Mist and Clouds (1857, cat. 103), *352*

The Natural Bridge, Virginia (painting, 1852, cat. 82), 320–22, *321*, 338

The Natural Bridge, Virginia (sketch, 1851), *320*, *396*

Niagara (1857, cat. 94), *2–3*, 338–45, *340–41*, 354, *355*

Rainy Season in the Tropics, 360

Studies from Guacamaya, Bogotá, Colombia (1853, cat. 89), 328, *332*

Studies of Trees and Plants seen from the Río Magdalena, Columbia (1853, cat. 86), 328, *329*

Study for "The Heart of the Andes" (1858, cat. 100), *349*, 353

Tequendama Falls near Bogotá, Colombia (sketches, 1853), *323–25*, 323–26

West Rock, New Haven (1849), *319*, 336, 338

cinchona bark and quinine, 48, 49, 65

Civil War, 198, 207, 220, 221–24, 229, 231–33, 240–49

Clark, William, 95–98, *96* (cat. 17), 100, 119–21, 126, 257, 259–60, 262, 263

Clay, Henry, 180

Clinton, DeWitt, 62, 158

Cole, Thomas, 146, 162–64, *163*, 166–69, 187, 316, 326, 358

The Course of Empire, 280

An Essay on American Scenery (1835), 166–67

Kaaterskill Falls (1826), *163*

Landscape View near the Falls of the Kaaterskill (1826), *164*

Landscape with Figures: A Scene from "Last of the Mohicans" (1826), *164*

Mount Aetna from Taormina (1843), *335*, 338

View from Mount Holyoke (The Oxbow) (1836), 167–69, *168*

Collin, Nicholas, 46–49, 61, 64, 65, *66* (cat. 11), 68, 105, 139

Collins, Zacchaeus, 45

Compromise of 1850, 222, 223–24

Constant, Benjamin, 186, 375

Constitutional Convention (1787), 62

Cook, Captain James, 27

Cooke, George, *The City of Washington from Beyond the Navy Yard* (1833), *56*

Cooper, James Fenimore, 31, 109, 160, *161*, 161–62, 163, 164, 165–66, 186–87, 190, 191, 259, 281, 284, 298, 360

The Last of the Mohicans (1826), 161, 164, 281, 282

The Pioneers (1823), 161, 162, 164, 281

The Prairie (1827), 161

Corcoran, William Wilson, *204*, 204–5, 384, 385, *386* (cat. 109)

Corsair (Iowa), 284, 289

Cortés de Alcocer, José, 345

Cosmos (Humboldt, 1845–59), 11, 25, 27, 28, 32, 33, 99, 126, 181, 182, 191, 196, 221–22, 305, 313, 314–17, *318* (cat. 81), 354, 356, 357, 362–65, 367, 384, 392

Cosmos Club, 390

Cotopaxi (volcano), 89, 248, 333–38, *334* (cat. 90), *336* (cat. 91), 360

Cotta, John Friedrich, 228

Cranch, Christopher Pearse, *362*

Critical examination of the history and geography of the new continent (Humboldt, 1836), 186

Crow Indians, 260, 282, 302

Cuba, annexation of, 220–21, 223–25, 249, 373

cultural identity of United States, 30

European degeneracy theory about New World life forms and civilizations, 26, 30, 52, 60, 135–38, 148, 156, 256–57

exploration contributing to, 95, 127

mammoth/mastodon bones and, 139–42, 142, 146, 149–52

natural icons of, 30, 60–61, 86, 135–38, 149–52, 155, 166, 169, 186, 317

Peale's museums and, 60–61

visit of Humboldt to U.S. and, 75

writings of Humboldt and, 313–14

Cuvier, Georges, 17, 46, 52, 60, 139, 144, 177, 375

D

Daguerre, Louis-Jacques-Mandé, 114–15, *187*

Dahkóta Indians, 275, 279

Darwin, Charles, 17, 20–21, 27, 88, 315

David, Jacques-Louis, *177*

Davis, Edwin H., 281

Davy, Sir Humphrey, 376

de Pauw, Cornelius, 136, 138, 255

Declaration of Independence, 52, 152, 216, 390

deforestation concerns/preservation of mature forests, 21, 26–27, 48–49, 119

Delacroix, Eugène, 298, 300

democracy in America, Humboldt's hopes for, 26, 31, 51, 175–76, 180, 207, 215, 220, 249, 387

DePeyster, John, 71

Deroine, Jeffrey, *284*, 296, 298, 300

Dessalines, Jean-Jacques, 43, *219*

Diderot, Denis, 136

Dipäuch (Mandan), 273, 283, 301

Dog Society and Dog Dance (Minnetaree), 269–73, *272*, *274*

Donné, Al, 300

Doughty, Thomas, 164

Douglass, Frederick, *228*, 229, 248

Downing, Andrew Jackson, 379

Dreidoppel, David, 118, 119, 262, 267

Dunbar, William, 99

Dunlap, William, 162, 280–81

Dupont, Pierre-Samuel, 136

Durand, Asher Brown, *163*, 164, 165

Duras, Clare de, Duchess, 186

E

The Earth and Man (Guyot, 1849), 18

Ebeling, Daniel Christoph, 27, 137

education in America

Catlin's Indian Gallery and, 258, 282, 300

Humboldt's role in, 183–86, *184*, *185*, 258

natural icons contributing to, 159, 360, 379

Peale museum and, 49, 60–61, 106–8, 118–19, 142, 373

specialization, shift toward, 392

State House Square specimen garden and, 46

Edward VII (king of England), 300

Eichholtz, Jacob, *69*

Elgin Gardens (now New York Botanical Garden), 70

Ellicott, Andrew, 53, *69*, 69–70, 99, 101

Elliott, Steven, 62

Emerson, Ralph Waldo, 15, 16, 28, 32, 34, 166–69, *167*, 183, 223, 240, 314, 315, 360–62, 367, 387

Nature (1836), 32, 166, 169, 314, 360, *362*

Encyclopédie (ed. Diderot, 1776), 136

Enlightenment, 15, 30, 33, 90, 241

environmental and conservation movements, 21, 34–35, 390–92

Essay on the Geography of Plants (Humboldt and Bonpland, 1805), 19, 27, 33, 86–90, *90–93*, 100, 165, 196, 362, 367

Europe, Humboldt in, 30–31, 175–207

approbation of Humboldt, persons seeking, 202–7

in Berlin, 116, 177, 191–93, 202, 204–7

education in America, role of Humboldt in, 183–86, *184*, *185*

in London, 180, 375

networking with scientific community, 175–82, 192–93

in Paris, 176–82, 191, 193

return of Humboldt from America to Europe, 87

telegraph and transatlantic cable, 31, 190–91, 193–202, *194–95* (cats. 36–38), *201–3*, 384, *391*

visited by traveling Americans, 175, 178, 180, 182–83, 186–91, 193, 196, 202–7

European degeneracy theory about New World life forms and civilizations, 26, 30, 52, 60, 135–38, 148, 156, 256–57

Evans, Estwick, 160

Everett, Edward, 109, 166, 182–83, *183*, 191, 204, 205, 207, 358, 387

expansionist goals of U.S., 86, 98, 161, 207, 217, 222, 223, 257, 280

exploration of America, Humboldt's influence on, 30, 85–128
- Adalbert of Prussia, Prince, *Travels in Brazil*, 126
- APS and, 95, 99, 105, 107
- *Aspects of Nature* (Humboldt, 1808), 86, 94, 111, 114
- attention paid by Humboldt to American expedition reports, 95, 99–100, 101–4, 114–15, 127
- cultural identity of United States and, 95, 127
- *Essay on the Geography of Plants* (Humboldt and Bonpland, 1805), 86–90, *90–93*, 94, 100
- expansionist goals of U.S. and, 86, 98
- intentions of Humboldt to return to explore America, 30, 85, 95–98, 116, 258, 302–4
- map of central North America by Humboldt and, *58* (cat. 6), 98, 100–104, *101*, *102* (cat. 19)
- *Naturgemälde* (Humboldt, 1805), 89–90, *91–93*
- *Personal Narrative* (Humboldt), 86, 94, 103, 104
- photography, invention and use of, 114–15, *115*, 127
- *Political Essay on the Kingdom of New Spain* (Humboldt), 86
- scientific explorer, Humboldt as, 15–16, 18–19
- scientific instrumentation, Humboldt's passion for, *99*, 110, 114
- scientific travel books, popularity of, 85–86, 94, 108
- writings and publications of Humboldt, 86–94, 105
- *See also specific expeditions*

extinction, 148–49

F

Fevret de Saint-Mémin, Charles Balthazar Julien, *106*

Fidler, Peter, 95

Field, Cyrus West, *197*, 197–98, *203*, 316, 319–38, 345, 358, 391

Field, Robert, 55

Fillmore, Millard, *204*, 204–5, 384

Flügel, Johann Gottfried, 197, 202–4, 384

Forster, Georg, 19, 27, 375

Fortieth Parallel Survey (1867), 127

Fortunat, Dantès, *219*

Fothergill, Anthony, 48, 49, 61, 64, 65, *66* (cat. 10), 68

Franklin, Benjamin, 45, 196, 376

Frémont, Jessie Benton, 110, *113* (cat. 21), 114, 115, 216, 225–26, 231, 233–36, 240, 241–44, 385

Frémont, John C. ("the Pathfinder")
- abolitionism of, 114, 204–5, 216, 225–28
- ambrotype of (1856, cat. 45), *227*
- expeditions of, 16, 29–30, 31, 86, 110–15, *111–13*, *115*, 225, 363
- Fagnani portrait (1867), 233, *236*
- Humboldt's library and, 385
- naming natural features for Humboldt, 110–11, 114, 390
- never meeting Humboldt in person, 33
- presidential campaign supported by Humboldt, 28, 31, 204–5, 216, 225–28, *227*, 358
- Read portrait (1856, cat. 20), *112*, 233–36

Frémonts' art collection, 231–40, *232*, *234*, *235*, 244

French Revolution, 43, 64, 65, 176, 178, 180

Friedrich Wilhelm III (king of Prussia), 177, 191, 300

Friedrich Wilhelm IV (king of Prussia), 193, 302

Fugitive Slave Law, 222–23, 224, 229, 244

Fuller, Margaret, 374

G

Gallatin, Albert, *57*, 59, 60, 65, 69, 72, 94, 100, 103, 176, 180, 182, 301, 381

Garrison, William Lloyd, 225, 229

Gauss, Carl Friedrich, 19

Gautier, Théophile, 298

Gay-Lussac, Joseph, 166, 177, 376

Geographical and Physical Atlas of the Kingdom of New Spain (1811), Humboldt's *Map of Mexico and the Surrounding Territories* (cat. 19) in, *102*, 103, 105, 322, 354

Gérard, Baron François Pascal Símon, 186, 187

German American community, 20, 228

Gerolt, Friedrich von, 118, 204, 225, 228, 384, 385

Ghent, Treaty of (1814), 180

Gilpin, William, 158

Girardet, Karl, *Dance of the Iowa Indians before the King Louis-Philippe at the Tuileries* (1845, cat. 78), 296–98, *297*

Godefroid, Marie-Eleonore, *181*

Goethe, Johann Wolfgang von, 27, 182, 192, 315
- *Sorrows of Young Werther*, 182

Graham, George R., and *Graham's Magazine*, 124–26

Grand Canyon, 127, 390, 391

Gray, Asa, 183, 377, *378*

Gray, Vincent, 40, 42–43

Great Chain of Being, 60, 88, 146, 149, 169

Greeley, Horace, 223, 226, 241–44

Greenough, Horace, 187

Guyot, Arnold, 17, 18, 21

H

Haines, E., *247*

Haitian revolution, 43, 215, 218–20, *219*, 229, 378

Hale, Sarah, 185

Hall, James, 281

Hamilton, Alexander, 55

Hamilton, William, 62, 70

Harlan, Richard, 262

Harrison, Joseph, 302, 390

Harrison, William Henry, 256

Harvey, George, 258

Hay, John, 244

Hayden, Ferdinand, 127, 390

Henry, Joseph, 198, 204, *378*, 378–80, 384, 388, 390

Hildebrandt, Eduard
- *Humboldt in His Library* (chromolithograph after Hildebrandt, 1856, cat. 44), 60, *174*, 204, *206*
- *Humboldt in His Study* (chromolithograph after Hildebrandt, 1856, cat. 107), 381, *382*

Hill, John, 162

Holland, Lord and Lady, 376

Holtei, Karl von, 192

Hone, Philip, 187, 281

hoodoos, 121–24, *124*

Hook, James, 259

Hooker, Joseph Dalton, 20

Hosack, David, 70

Hudson River school, 30, 135, 146, 163, 164, 183, 240, 314, 317. *See also* Church, Frederic Edwin; Cole, Thomas

Hugo, Victor, 298

Humboldt, Alexander von, 11–13, 25–35
- admiration and fondness of Americans for, 68–69, 75, 175–76
- American eulogies for, 387–88
- art in America, influence on, 16, 25, 28, 30, 32 (*See also* art in America; Church, Frederic Edwin; Cole, Thomas)
- biographical information, 26–28
- death of (1859), 19, 28, *29*, 33, 207, 359, 384
- diffusion and absorption of ideas of, 34–35
- as direct and indirect influence on American life and culture, 32–34
- estate, disposition of, 384–85, *386* (cat. 109)
- facets of Humboldt's influence in United States, 28–32
- as "half an American," 11, 16, 26, 31, 69, 175, 193, 207, 385
- handwriting of, 205, 244, *246*
- languages spoken by, 46–48, 193
- legacy of, 390–92
- library of, 204, *206*, 380
- membership certificate in APS (cat. 14), 71, *74*, 75
- naming of species, plants, and geographic features for, 25, 110–11, 114, 390
- networking with scientific community, 31, 34, 46, 59, 70, 75, 87, 175–82, 192–93, 305
- reputation, waxing and waning of, 11, 19–21, 25–26, 34–35, 392
- science in America and, 15–21
- travels in New Spain (1799–1804), 11, 16, 28, 40–43, 64
- *See also* American Indians; Europe, Humboldt in; exploration of America, Humboldt's influence on; natural icons in America; slaves and slavery; Smithson, James, and Smithsonian Institution; United States, Humboldt's visit to

Humboldt, Alexander von, artworks
- *Chimborazo from the Plains of Tapia* (1810, cat. 92), 335, *337*
- *Isothermal Chart* (1817), 183, *184*
- *Mexican Calendar in Relief on Basalt* (1814), *256*
- *Natural Bridge of the Icononzo* (1814, cat. 32), 153–55, *154*, 320
- *Rhexia Speciosa* (Humboldt and Bonpland, 1806), *107*
- *Tequendama Falls* (1810), *322*
- *View of Cayambe in the Region of Quito* (1806, cat. 101), *350*, 353
- *See also Naturgemälde*

Humboldt, Alexander von, images of
Berger, bust by (1860, cat. 110), 388, *389*
Blaeser, bust by, 388
on *Cosmos* frontispiece (cat. 81), *318*
frontispiece, Catlin's *Notes of Eight Years' Travels and Residence in Europe* (1848, cat. 79), *299*, 300
Haines's albumen silver print (1858, cat. 57), *247*, 248
Hildebrandt, *Humboldt in His Library* (chromolithograph after, 1856, cat. 44), 60, *174*, 204, *206*
Hildebrandt, *Humboldt in His Study* (chromolithograph after, 1856, cat. 107), 381, *382*
Lambdin's portrait (1856–57), *207*
Peale's Museum, portrait in, 119, 148
Charles Willson Peale's portrait of (cat. 13), 71–72, *73*, 144, 178
Rembrandt Peale's portrait of (cat. 33), 178–80, *179*
Peale's silhouette of (cat. 8), *66*, 68
public images throughout U.S., *388*
Rauch, bust by (1857), 204, *205*, 388
Salas's copy of Cortés de Alcocer portrait (1857, cat. 96), *24*, *344*, 345–53
Schreiber's salted paper print (1858, cat. 108), 381–84, *383*
Stieler's portrait of (1843), *381*
Weitsch's portrait of (cat. 1), 29, *41*
Humboldt, Alexander von, writings and publications, 27, 29–30
explorations of U.S. influenced by, 86–94, 105
illustrations from, 33–34
Map of Mexico and the Surrounding Territories, *102* (cat. 19)
on natural icons in America, 137–38
personal fortune spent on, 115–16, 191, 353, 380
research and writing methods, *90*, 191, 192
scientific travel books, popularity of, 85–86, 94, 108
summary of visit to U.S, 75, 100
in translation, 33, 94, 176, 181–82, 315
updates and changes to, 126–27
wilderness aesthetic in America and, 30
See also specific publications, under their English titles
Humboldt, Caroline von (sister-in-law), 186, 354
Humboldt, Wilhelm von (brother), 15, 34, 45, 160, 177, 186
Humboldt celebration (1869), 28, 387–88
Humboldt Current, 26
"Humboldtian Science," 106, 166
Hungerford, Henry James, 376
Hunter, William, 138
Huntington, Daniel, 163
The Atlantic Cable Projectors (1895, cat. 41), *14*, 198–202, *201*

I

Indian Removal Act (1830), 31, 257, 258, 259, 260, 280, 281, 374
Institut National, Paris, 45, 144, 176, 177, 183, 193
Iowa Indians (Báxoje), 32, *284*, 284–300, *286–88*, *290–95*, *297*
Irving, Washington, 31, 109, 160–61, *161*, 183, 186, 187, *188*, 281, 283, 358, 360, 378
Astoria (1836), 161, 186
Bonneville Adventures series, 161, 186
Life of Columbus, 186
The Sketchbook of Geoffrey Crayon, Gent., 160
A Tour of the Prairies (1832), 161, 186
isotherms, 26, 183, *184*
Ives, Joseph Christmas, *126*

J

Jackson, Andrew, 31, 257, 258, 259, 260, 280, 281, 283, 374
Jackson, William Henry, 127
James, Edwin, 106
Jardin des Plants, Paris, 46, 87, 136, 166
Jarvis, John Wesley, *161*
Jefferson, Thomas
acquisition of new American lands by, Humboldt's involvement in, 28, 29
American Indians and, 31, 256–57
APS, as president of, 61
arrival and reception of Humboldt in America and, 42–43
Bernhard, duke of Saxe-Weimar, and, 71, 117
central North America, Humboldt's map of (cat. 6), 28, 40, 43, 57–60, *58*, 69, 72
death of, 220
establishment of capitol in Washington, D.C., and, 55
European degeneracy theory about New World life forms and civilizations, refuting, 136–38, 156
farewell letter of Humboldt to, 72–75
Lafayette and, 178, 180
letter of introduction from Humboldt to (cat. 4), 43, 49, *50*, 51–53, 137–38
letter of introduction to Lafayette from, 75
Lewis and Clark Expedition and, 85, 95–100, *96*, *97*, 103–4, 149
Louisiana Purchase, 25, 29, 39, 40, 43, 52, 57, 59, 98, 99, 104, 152, 157, 217
mastodon/mammoth bones and, 60, 138, 139 144, 148, 149, 152, *153*
meetings with/friendship with Humboldt 11, 25, 28, 29, 39, 40, 55, 57–60, 61, 65, 69, 75, 87, 104, 182, 183, 385
Monticello, 68, 71, 95, 117, 149, 152–53
natural icons in America, interest in, 152–53
on Niagara Falls, 121
Peale's Museum, portrait in, 71
Rembrandt Peale's painting of (cat. 5), 38, *54*, 55, 180
Raphaelle Peale's silhouette of (cat. 12), *67*, 68
Pike's 1806 expedition and map, 30, 85–128, *101*, *102* (cat. 19)
polygraph machine, 51
salon society of Washington, D.C./Paris and, 65, 181
slavery and, 31, 62, 72, 215, 216–18, 224
Spain, negotiations with, 29, 40, 43, 57, 69, 99, 100
Madame de Staël and, 181
Washington, D.C., trees planted in, 53
wilderness aesthetic embraced by, 35
writings of Humboldt sent to, 94, 176, 177, 217–18
See also Notes on the State of Virginia
Johns, Henri-Joseph, *James Smithson* (1816, cat. 106), *372*
Johnson, Eastman
The Old Mount Vernon (1857, cat. 7), 61–65, *63*, *214*
Ralph Waldo Emerson (charcoal, 1846), *167*
Josephine (wife of Napoleon), 177

K

Kansas-Nebraska Act (1854), 223, 224
Kant, Immanuel, 315
Kemble, Fanny, 231
King, Charles Bird, *64*
King, Clarence, 127, 390
King, Thomas Starr, 226, 240, 244
Kipp, James, 273, 275, 304
Koon-za-ya-me (Female War Eagle, Iowa), 289–96, *295* (cat. 77)
Kunth, Karl Sigismund, 87
Kutzinski, Vera, 33

L

Lafayette, Marquis de, 31, 64, 75, 116, 117, 178–81, *180*, 182, 186–87, *187*, 191, 218, 376
Lambdin, James Reid, *207*, 262, 280, *281*
Lancaster, Pennsylvania, Humboldt in, 69–71
Lander, Frederick, 241
Laplace, Pierre-Simon, 376
Latrobe, Benjamin, *53*
Lavoisier, Antoine, 375
Law, Thomas, 55, 149
Lawrence, Martin M., *230*
Lee, William, 64
L'Enfant, Pierre, 53
Lenox, James, 316, 385
Lesueur, Charles, 106–7, 119, 262
Leutze, Emanuel, *Washington Crossing the Delaware*, 240–41
Lewis, Meriwether, 46, 55, 95, *97* (cat. 18), 98–99, 100, 103–4, *106*, 114, 257
Lewis and Clark Expedition, 85, 95–100, *96*, *97*, 103–4, 105, 106, 108–9, 110, 118, 124, 149, 257, 258
Lincoln, Abraham, 233, 240, 244, 248
Lincoln, Almira, *185*
literature in America
Europe, writers visiting Humboldt in, 186–87
exploration of America and popularity of scientific travel books, 85–86, 94, 108
influenced by Humboldt, 314
transcendentalists, 32, 166, 183, 360–65, *361–63*
wilderness aesthetic and, 159–62, 164–66
See also specific writers
Long, Stephen Harriman, and Long Expedition (1818), 29, 86, 104–10, *106–9*, 114, 118–19, 121, 124, 126, 148, 160, 161, 186, 257, 258, 259, 262, 263, 283, 390

Longfellow, Henry Wadsworth, 183

Louisiana Purchase, 25, 29, 39, 40, 43, 52, 57, 59, 98, 99, 104, 152, 157, 217

Louis-Philippe (king of France), 158, 287, 289, 296, 300

Louverture, Toussaint. *See* Toussaint Louverture, François-Dominique

Lovelock, James E., 16

Lyell, Charles, *Principles of Geology* (1830–33), 20, 27, 88

M

Macie, James Lewis. *See* Smithson, James

Madison, Dolley, *64*, 65, 68, 180

Madison, James
- explorations of U.S., Humboldt's influence on, 94, 95, 100, 116
- natural icons as American landmarks and, 155
- United States, Humboldt's visit to, 40, 42–43, 45, 51, 55, 57, 61, 65, 72, 216, 391

magnetic field/magnetic equator, detection of, 18–19, 27, 99, 177, 196

Magnus, Heinrich Gustav, 202

Ma-ka-tai-me-she-kia-kiak (Black Hawk, Sauk), 260, 263

mammoth. *See* mastodon/mammoth bones, discovery and display of

Mandan (Nųm´ąkh´ų·ki) Indians, 260, 263, 267–69, 273, 275, *276–78*, 283, *285*, 302

Manifest Destiny, 161, 280

map of central North America by Humboldt
- expansions goals of U.S. and, 222
- Pike expedition's plagiarism of, 98, 100–104, *101*, *102* (cat. 19)
- shared by Humboldt on visit to United States, 28, 40, 43, 57–60, *58* (cat. 6), 69, 72, 263

Marbois, François, marquis de Barbé-Marbois, 136, 137, 187

Margulis, Lynn, 16

Marsh, George Perkins, *Man and Nature* (1864), 16, 21, 35, 378–79

Mason, John Y., 223

Masten, John, 139, 149

mastodon/mammoth bones, discovery and display of, 138–52
- "American incognitum," known as, 138
- banquets held under skeleton, 72, 144
- cultural identity of United States and, 139–42, 144, 146, 149–52
- differentiation of mammoth and mastodon species, 144, 177
- European degeneracy theory, refuting, 138, 148
- exploration of America, Humboldt's influence on, 95, 100, 119, *134*, *145*
- extinction theory and, 148–49
- Humboldt's awareness of, 138–39, 142
- Jefferson and, 60, 138, 139, 144, 148, 149, 152, *153*
- in Monticello entry hall, 149, 152, *153*
- Peale, Charles Willson, *The Artist in His Museum* (1822, cat. 29), 146–48, *147*
- Peale, Charles Willson, drawings of bones by, 138–39
- Peale, Charles Willson, *Exhumation of the Mastodon* (ca. 1806–1808, cat. 28), 134, 144–46, 145, 149, 160, 162, 163, 166
- Peale, Charles Willson, *Self-Portrait with Mastodon Bone* (1824, cat. 3), 47
- Peale's Museum and, 60, 72, 139–43, 139–48, 145, 147, 150, 151, 236
- skeleton of mastodon excavated by Peale (1801–2, cat. 26), 139, 140–41
- tusks, placement and orientation of, 142, 142–44, *143*
- United States, Humboldt's visit to, 30, 47, 52, 60, 72

Mató-Tópe (Mandan), 260, *261* (cat. 58), 267–69, *268* (cat. 61), 273, 275, 302
- *Battle with a Cheyenne Chief* (1834, cat. 63), 269, *271*

Matthews, John, *Alexander von Humboldt on the Fugitive Slave Law* (1864, cat. 56), 244, *246*

Maury, Matthew Fontaine, 16–17, 197–98, 384, 387

Maximilian zu Wied-Neuwied, Prince, 31–32
- *American Fur Co., Manuscript Journal of Prince Maximilian's Travels in North America* (1833, cat. 24), *84*, *123*, 265, 269–73
- American Indians and, 117–18, 119–21, 262–80, *265*, *266*, *268*, *270–74*, *276*, *278*, 281, 282–83, 301, 302–5
- as explorer influenced by Humboldt, *84*, 86, 115–26, *116*, *120–25*, 206, 390
- *Indian Ponies* (1833, cat. 59), 263, *265*, 267
- *Line drawing of Pehriska-Ruhpa* (1834), 269–73, *273*
- *Map to Illustrate the Route of Prince Maximilian of Wied in the Interior of North America* (Karl Bodmer with Max zu Wied, 1839–41, cat. 23A & 23B), 121, *122*
- *Pehriska-Ruhpa in Costume of Dog Dancer* (1834), 273, *274*
- *Reise in das innere Nord-America* (with Bodmer, 1839–41), 124, 282–83, 300

McClure, John, 359

McClure, William, 107, 119

McKenzie, Alexander, 260

McKinney, Thomas, 281

Melody, George Curzon, 284, 296

Melville, Herman, *Moby-Dick* (1851), 224

Mercury, transit of (1799), 28, 70

Merry, Anthony, 55

Methua, John Guido, 248

Metropolitan Fair of 1864, 240–48, *241*, *245–46*

Metternich, Klemens von, 181, 193

Mew-hu-she-kaw (The White Cloud, Iowa), *287*

Mexican-American War (1846–48), 222, 373, 379

Michaelis, Christian Friedrich, 138–39

Michaux, François André, 46, 70, 157

Mifflin, John Francis, 71

Mignot, Louis Rémy, 345, 353

Mills, Robert, 379

Milton, John, 20

"Mineralogical Observations on Some Basalts of the Rhine" (Humboldt, 1790), 375

Minnetaree Indians, 269–73

Missouri Compromise (1820), 223–24, 249

Missouri Mounds, 279

Möllhausen, Balduin, *126*, 379, 384–85, *386* (cat. 109)

monogenism, 275–79

Monroe Doctrine, 59

Monticello, 68, 71, 95, 117, 149, 152–53

Montúfar y Larrea, Carlos, 42, 45, 46, 48, 61, 65, *68*, 75, 345

Moran, Thomas, 127

Morgan, John, 138–39

Morse, Jedidiah, 137, 187, 190

Morse, Samuel F. B., 31, 35, 164, *187*, 187–91, 193–98, *194*, *195*, 378, 391
- *Benjamin Silliman Sr.* (1825), *159*
- *Gallery of the Louvre* (1831–33, cat. 35), *10*, 187–90, *189*, 191, 202
- *Marquis de Lafayette* (1826), *187*
- passport (cat. 34), 187, *188*

Morton, Dr. Samuel George, *Crania Americana*, 17–18, 381

Mount Vernon, 53, 61–65, *62*, *63* (cat. 7), 149, 153, 158, 217, 320, 379

Mühlenberg, Gotthilf Heinrich Ernst, *69*, 70

Muir, John, 21, 390, *391*

Murray, Charles Augustus, 282

Muséum national d'Histoire naturelle, Paris, 87, 166

N

Napoleon, 31, 40, 42, 43, 117, 175–82, *177*, 186, 191, 381

National Academy of Design (NAD), 163, 166, 167, 190, 319, 338

National Intelligencer, 59, 60, 65, 98, 152, 155, 197, 216

Native Americans. *See* American Indians

Natural Bridge (Virginia), 29, 30, 94, 127, 135, 137, 152–55, *153*, *154* (cat. 32), *156*, *157*, 159, 163, 166, 236, *320*, 320–22, *321* (cat. 82)

natural icons in America, 30, 135–69
- Cole, Thomas, and Hudson River school, 162–64, *163*, *164*, 166–69, *168*
- cultural identity of United States and, 30, 60–61, 86, 135–38, 149–52, 155, 166, 169, 186, 317
- deforestation concerns/preservation of mature forests, 21, 26–27, 48–49, 119
- Emerson and *Nature*, 166–69
- environmental and conservation movements, 21, 34–35, 390–92
- European degeneracy theory about New World life forms and civilizations, 26, 30, 52, 60, 135–38, 148, 156, 256–57
- Jefferson's interest in, 152–53
- Louisiana Purchase and, 152
- Peale's museums and, 60–61
- "Serbonian bog" of early Washington, D.C., and, 56–57, 149
- travel narratives and American literature of Cooper and Irving, 159–62, 164–66
- wilderness aesthetic and, 30, 35, 127, 135, 155–60, *156*, *157*, 162, 317
- in writings and publications of Humboldt, 137–38

Naturgemälde (Humboldt, 1805), 33, 45, 89–90, 158, 315, 353, 354–55, 358, 363, 364, 392
  adapted for Lincoln's *Familiar Lectures on Botany* (1829), *185*
  black and white print (cat. 16), 89, *91*, 354
  colored print (cat. 15), 89, *91*, *92–93*, 354
Neu-mon-ya (Walking Rain, Iowa), 287, *288* (cat. 71), 296
New Harmony, Indiana, 119, 262–64, 279
New York [Daily] Tribune, 223, 226, 229, 233, 244
New York Evening Post, 164, 187, 223, 224, 225, 226, 357–58
New York Times, 28, 225, 359, 384
Newton's New and Improved Terrestrial Globe (1852) (cat. 39), 197, *199*
Niagara Falls, 29, 30, 94, *121*, 127, 135, 152, 155, *156*, *157*, 159, 163, 166, 236
Nicollet, Joseph Nicholas, 110
Noble, Louis Legrand, 358
No-ho-mun-ya (One Who Gives No Attention/Roman Nose, Iowa), 284–87, *286* (cat. 70)
North American Review, 109, 162, 182, 301–2
Notes on Saturn (Humboldt, ca. 1789–1830s), *90*
Notes on the State of Virginia (Jefferson, 1785/1788), 26, 30, 39, 40, 52, 60, 86, 94, 100, 136–39, 153, 155, 157, 169, 218, 256

O

O'Connor, William Douglas, 367
O'Fallon, Benjamin, 119, 259, 263
  Lewis and Clark's Special Map of the Missouri River (1804–32, cat. 22A & B), 119–21, *120*
Ojibwe (Anishinaabe) Indians, 260, 283–84, 300
O-Kee-Pa ceremony (Mandan), 275, 283, 302, 304
Oke-we-me (Female Bear That Walks on the Back of Another, Iowa), 284, 289
Olmsted, Frederick Law, 231, 240
Ord, George, 107
Ostend Manifesto, 223, 249
O'Sullivan, Timothy, 127
Otoe Indians, 105, 267
Otté, Elise, 33, 181–82
Owen, Robert, 119

P

Pallas, Peter Simon, 95
Panama Canal, 72, 196–97, 391
Parker, Theodore, 229, 231, 255
"the Pathfinder." *See* Frémont, John C.
Patterson, Robert, 46, 99, 148
Paul, duke of Württemberg, 126
Paulze, Marie-Anne Pierrette, 375–76
Pawnee Indians, 105, 257
Payne, John Howard, 280
Peale, Charles Willson
  cultural identity of America in paintings of, 35
  drawings of mastodon bones by, 138–39
  farewell party for Humboldt, 72
  greeting Humboldt on arrival in Philadelphia, 45
  on Humboldt's linguistic abilities, 48
  Lewis and Clark Expedition and, 106
  Long Expedition and, 106, 108
  polygraph machine, 51
  Saint-Hilaire, letter to, 75
  on salon society of Washington, D.C., 65
  silhouettes commemorating Humboldt's visit to U.S. (cats. 8–12), *66*, *67*, 68
  on slavery, 64
  on Gilbert Stuart, 55
  traveling to Washington, D.C., with Humboldt, 48, 51, 53, 55, 57, 61, 65
Peale, Charles Willson, artworks
  Aimé Jacques Alexandre Bonpland (1804, silhouette, cat. 9), *66*, 68
  Alexander von Humboldt (1804, silhouette, cat. 8), *66*, 68
  Anthony Fothergill (1804, silhouette, cat. 10), *66*, 68
  The Artist in His Museum (1822, cat. 29), 146–48, *147*
  Exhumation of the Mastodon (ca. 1806–1808, cat. 28), *134*, 144–46, *145*, 149, 160, 162, 163, 166
  The Long Room (with Titian Ramsay Peale, 1822, cat. 31), *8*, 148, *151*
  Nicholas Collin (1804, silhouette, cat. 11), *66*, 68
  Portrait of Baron von Humboldt (1804, cat. 13), 71–72, *73*, 144, 178
  Portrait of Gotthilf Heinrich Ernst Mühlenberg (1810), *69*
  Portrait of Meriwether Lewis (1807, cat. 18), *97*
  Portrait of William Clark (1807, cat. 17), *96*
  Self-Portrait with Mastodon Bone (1824, cat. 3), *47*
  Titian Ramsay Peale II (1819), *105*
Peale, Raphaelle, 71, 146
  Thomas Jefferson (1804, silhouette, cat. 12), *67*, 69
Peale, Rembrandt, 51, 60, 71, 142–44, 146, 178–80, 258
  Historical Disquisition on the Mammoth, 142, 144
  Mammoth, from the Philosophical Magazine, London (1802), *142*
  The Marquis de Lafayette (1825), *180*
  Portrait of Albert Gallatin (1805), *57*
  Portrait of Alexander von Humboldt (1800–12, cat. 33), 173–80, *179*
  Sketch of the Elephant (Mastodon) (ca. 1801, cat. 27), *143*
  Thomas Jefferson (1805, cat. 5), 38, *54*, 55, 180
Peale, Rubens, 51, 142
Peale, Titian Ramsay, *106*, 106–9, 118–19, 148, 258, 388
  Corvus pica (magpie) (1822), 107, *109*
  The Gigantic Mastodon (1821, cat. 30), 148, *150*
  The Long Room (with Charles Willson Peale, 1822, cat. 31), *8*, 148, *151*
  Monarch Butterfly (1817), 107, *109*
  Peale, Charles Willson, *Titian Ramsay Peale II* (1819), *106*
  Sunset on the Missouri River (1819), 107, *108*
  White Violet, Indiana (1819), *107*
Peale museums in Philadelphia and Baltimore
  American Indians and, 258
  Catlin and, 257, 258, 280, 281
  efforts of Peales to get U.S. government to purchase, 373
  explorations of America and, 106, 107–8, 118–19, 258
  Hildebrandt's chromolithograph of *Humboldt in His Library* (cat. 44) recalling, 204, *206*
  Humboldt's portrait in, 119, 148
  Lambdin's Pittsburgh museum inspired by, 262
  mastodon/mammoth bones, discovery and display of, 60, 72, *139–43*, 139–48, *145*, *147*, *150*, *151*, 236
  Smithsonian and, 379
  visit of Humboldt to U.S. and, 32, 49, 51, 60–61, 68, 71, 72, 142
Péhriska-Rúhpa (Two Ravens, Minnetaree), 269–73, *272–74*
Person, Mark W., 33
Personal Narrative of Travels to the Equinoctial Regions of the New Continent, during the Years 1799–1804 (Humboldt), 16, 20, 27, 86, 94, 103, 104, 105, 126, 159, 176, 181, 183, 196, 197, 248, 258–59, 322, 353, 362
Philadelphia
  ANS (Academy of Natural Sciences), 105, 107, 119
  Bartram's Gardens, 45, 62
  Chestnut Street Theatre tableaux, 153
  French refugees from Hispaniola in, 219
  Humboldt in, 45–48, 71–75
  magnetic data station at Girard College, 196
  State House Square specimen garden, 46, 62, 64, 379
  Washington, D.C., compared, 53
  See also American Philosophical Society; Peale museums in Philadelphia and Baltimore
Phillips, Wendell, 231
photography, invention and use of, 114–15, *115*, 127, 169
physionotrace, 68
Pichon, Louis-André, Baron, 55
Pictet, Marc-Auguste, 176, 362, 375
the picturesque, 158
Picturesque Atlas (Humboldt), 34
Pike, Zebulon, and Pike expedition, 100–104, *101*, 109
plate tectonics, 26, 88
Playfair, William, 89
Plumbe, John, Jr., *222*
Poe, Edgar Allan, 16, 315
Poinsett, Joel, 110, 374
Political Essay on the Island of Cuba (Humboldt, 1826), 27, 220–21, 224–25, 229
Political Essay on the Kingdom of New Spain (Humboldt, 1811), 27, 86, 103, 105, 183, 217
Polk, James K., 193–95, 222
polygenism, 18, 275–79
polygraph machine, 51
Port Folio, 152, 160, 162
Porter, J. M., 284

Powell, John Wesley, 390

Prescott, William, 183, 186, 202, 205, 224

Preuss, Charles, *111*

Priestley, Joseph, 55, 376, 378

R

racial equality. *See* American Indians; slaves and slavery

Ramsay, David, 104, 152, 216–17

Rankin, Arthur, 283–84

Rauch, Christian Daniel, 204, *205*, 388

Raymond, Rossiter, 207

Raynal, Guillaume-Thomas, 136

Read, Thomas Buchanan

Portrait of John C. Frémont (1856, cat. 20), *112*, 233–36

Portrait of Mrs. Jessie Benton Frémont (1856, cat. 21), *113*, 233–36

Récamier, Juliette, 181, 376

Reinhard of Solms, Prince, 117

Republican Party, 223, 225, 226, 228

Researches, Concerning the Institutions and Monuments of the Ancient Inhabitants of America (Humboldt, 1814), 155, *256*, *334* (cat. 90)

Revolutionary War. *See* American Revolution

revolutions of 1848, 18, 217, 240

Ritter, Carl, 19, 21, 207

Ritter, Charles, 114

Roberts, William, 153

Robinson, Andrew, 51

Robinson, Angelica Peale, 51

Rochambeau, comte de, 256

Rogers, John

The Slave Auction (sculpture, 1859, cat. 47), 231–33, *232*

The Wounded Scout (sculpture, 1864, cat. 48), 233, *235*

Romanowski, Sylvie, 33

Romanticism, 16, 33

Roosevelt, Theodore, 197, 390–92, *391*

Ross, Thomasina, 33

Royal Society, London, 45, 46, 48, 138, 375, 377

Rush, Benjamin, 46, 48, 71, 99, 374

Rush, Richard, 374, 377

Rush, William, 143

Ruskin, John, *Modern Painters*, 33, 316–17

Russia, Humboldt journeying across, 95–98, 196

Ru-ton-wee-me (Pigeon on the Wing, Iowa), 289, *294* (cat. 76)

Ru-ton-ye-wee-ma (Strutting Pigeon, Iowa), 289, *293* (cat. 75)

S

Sabine, Elizabeth Juliana Leeves, and Edward, 33, 181

Sacagawea (Bird Woman, Shoshone), 257

Sachs, Aaron, 20

Saint-Hilaire, Etienne Geoffroy, 75

Salas, Rafael, *178*, *344*

Salas Alzamora, Manuel, *68*

salon society in Paris and Washington, D.C., 65–68, 181, 182, 375–76

Sand, George, 298

Sartain, John, *185*

Saussure, Horace-Bénédict Benjamin de, *99*

Say, Thomas, 106, 107, 108, 109, 119, 262, 283

Schelling, Friedrich, 315

Schiller, Friedrich, 365

Schimper, Karl Friedrich, 17

Schinz, Rudolf, 116–17

Schoolcraft, Henry Rowe, 233, 301–2, 305

Schreiber, George Francis, *383*

science in America, contributions of Humboldt to, 15–21

scientific racism, 17–18

See-non-ty-a (Blister Feet, Iowa), 289, *291* (cat. 73)

"Serbonian bog," 56, 149

Seward, William H., 222

Seymour, Samuel, 106, 109, 118

Shaw, Joshua, 163

Shaw, Robert Gould, 231

Shon-ta-yi-ga (Little Wolf, Iowa), 289, *292* (cat. 74), 300

Sih-Chidä (Mandan), 275, *278*

silhouettes commemorating Humboldt's visit to U.S. (cats. 8–12), *66*, *67*, 68

Silliman, Benjamin, 158–59, *159*, 163, 167, 378

Sketch Club, 165

slaves and slavery, 12, 28, 31, 215–49

art as expression of freedom and, 231–48, *232*, *234*, *235*, *237–39*, *241–43*, *245–48*

Brown, John, and Harper's Ferry raid, 229, 230 (cat. 46), 231, 385

California entering union as free state, 31, 114, 222, 226

Civil War, Humboldt's concerns regarding possibility of, 207, 220, 221–24, 229

Cuba and, 220–21, 223–25, 249, 373

Emerson on, 167

expansionist goals of U.S. and, 217, 222, 223

Fillmore, remarks of Humboldt directed at, 204

Frémont's abolitionism and presidential campaign, 114, 204–5, 216, 225–28, *227*

Haitian revolution, 43, 215, 218–20, *219*, 229, 378

Humboldt's opposition to, 12, 28, 31, 215–18, 229, 248–49

Jefferson and, 31, 62, 72, 215, 216–19, 224

Lafayette on, 178, 218

Massachusetts Anti-Slavery Society, 229

Maximilian zu Wied on, 263

in *Political Essay on the Kingdom of New Spain* (Humboldt), 183, 217

scientific racism, 17–18

Madame de Staël on, 218

United States, Humboldt's visit to, 61–65, 72, 216–17

Smith, Margaret Bayard, *64*, 65–68

Smith, Samuel Harrison, 53, 59, 65–68, 217

Smithson, James, and Smithsonian Institution, 12, 32, 373–92

John Quincy Adams advocating for, 180, 374

Berger bust of Humboldt at (cat. 110), 388, *389*

estate of Smithson bequeathed to U.S., 374, 377–79

Flügel as Prussian/Smithsonian representative to, 197, 202–4, 384

founding and establishment of Smithsonian, 378–81

Humboldt and Smithson, 27, 374–77

Humboldt and Smithsonian, 381–84, 387, 388–90, 392

Humboldt's estate, disposition of, 384–85, *386* (cat. 109)

involvement of entire Smithsonian in Humboldt exhibition, 12–13, 27

Johns's portrait of Smithson (1816, cat. 106), *372*

national institute, decision to establish, 374

Peale's and Catlin's efforts to get U.S. government to take over their collections, 373–74, 388

Société d'Arcueil, 178, 376

Soulé, Pierre, 223

South America

Church in, 197, 314, 317–38, 345–54

Humboldt and Bonpland in, 11, 16, 28, 40–43, 64

Maximilian zu Wied in, 115–26

Spain, U.S. negotiations with, 29, 40, 43, 57, 69, 99, 100, 222, 223

Sparks, Jared, 155

Squier, Ephraim George, 381

Stadler, J. C., *Natural Bridge* (aquatint after William Roberts, 1808), *153*

Staël, Madame de, 180–81, *181*, 218, 375

Stebich, Stephanie, 13

Stegner, Walter, *The Sound of Mountain Water* (1947), 35

Steuben, Charles Auguste de, *192*

Stevens, Henry, 385

Stevens, Simon, 385

Stieler, Joseph Karl, *381*

Stillman, William James, *The Philosophers' Camp in the Adirondacks* (cat. 104), 360–62, *361*

Stoddard, Richard Henry, 207

Stone, William Oliver, *204*

Stowe, Harriet Beecher, *Uncle Tom's Cabin*, 231

Stuart, Gilbert, 55, *64*

Stuart, Robert L., 316, 353

Stu-mick-o-sucks (Buffalo Bull's Back Fat, Blood tribe), 300–301, *301*

Sues, Hans-Dieter, 21

Sully, Thomas, *183*

Sumner, George, 183, 222

T

Tanner, Henry S., *Map of North America* (engraving, 1825), 155, *156*, *157*

Ta-pa-ta-me (Wisdom, Iowa), 289, *293* (cat. 75)

Tappan, Lewis, 231

Taylor, Bayard, 359

telegraph and transatlantic cable, 31, 190–91, 193–202, *194–95* (cats. 36–38), *201–3*, 384, *391*

The Temple of Time (Emma Willard, 1846), 183, *184*

Tennant, Smithson, 375, 376

Thomson, William, 375

Thoreau, Henry David, 16, 32, 240, 314, 360, 362–65, *363*, 367

map of Walden pond, *363*, 363–64

Walden, 32, 314, 364

"A Walk to Wachusett" (1842), 362–63

Thornton, Anna Maria Brodeau, *64*, 65, 68, 72

Thornton, William, 26, 53, *55*, 65, 68–69, 72, 75, 85, 155, 176, 217, 220, 224

Thrasher, John Sidney, 224–25, 226, 229, 249

Ticknor, George, 152, 162, 182–83, *183*, 186, 192–93, 202, 205, 224

Tiffany's Transatlantic Cable Souvenir (1858) (cats. 42 and 43), 198–202, *201*, *203*

Toussaint Louverture, François-Dominique, 43, *219*

transatlantic cable, 196–202, *201–3*, 384

transcendentalists, 32, 166, 183, 360–65, *361–63*

transparent eyeball, *362*, 367

Traubel, Horace, 365

travel narratives

American literature and wilderness aesthetic, 159–62, 164–66

exploration of America and popularity of scientific travel books, 85–86, 94, 108

Tucker, St. George, 219

Tuckerman, Henry T., 317, 355

Twain, Mark, 358–59, 364

Tyler, John, 286

U

United States, Humboldt's visit to (1804), 25–26, 39–75

admiration and fondness of Americans for Humboldt, 68–69, 75, 175–76

arrival and reception in America, 42–45, *44*

baggage entry statement, U.S. Customs (cat. 2), *44*, 45

in Baltimore, 51

central North America, Humboldt's map of (cat. 6), 28, 40, 43, 57–60, *58*, 69, 72

farewell parties and departure, 71–75

intentions of Humboldt to return to America, 30, 85, 95–98, 116, 258, 302–4

in Lancaster, Pennsylvania, 69–71

letters of introduction to Jefferson (cat. 4) and Madison, 43, 45, 49, *50*, 51–53, 55, 57, 59, 137–38

Maximilian zu Wied and, 117

at Mount Vernon (cat. 7), 61–65, *62*, *63*, 217

natural icons as American landmarks and, 60–61

New Spain, Humboldt in (1799–1804), 11, 16, 28, 40–43, 64

Peale's museums and, 49, 51, 60–61

in Philadelphia, 45–48, 71–75

"philosophic party" traveling to Washington with Humboldt, 48–51

reasons for, 39–40

salon society of Washington, D.C., and, 65–68

significance of, 16, 29

silhouettes commemorating (cats. 8–12), *66*, *67*, 68

slavery, Humboldt's opposition to, 61–65, 72, 216–17

in Washington, D.C., *53*, 53–57, *56*, 65–68

unity of nature, Humboldt's concept of, 11, 13, 15–16, 20–21, 27, 35, 87–88, 126–27

Upham, Charles Wentworth, 379–80

U.S. Coastal Surveys, 197, 202, 378, 379, 381

U.S. Geological Survey, 390

U.S. Weather Bureau (now National Weather Service), 18

V

Vail, Alfred, 193, *194*, *195*, 391

Vanderlyn, John, 153

Varnhagen von Ense, Karl August, 27, 224, 228

Vattemare, Alexandre, 287, 298, 300

Vaughan, John, 46, 49, 68–69, 72, 75, 95, 100, 105, 119, 175, 176, 374

Vaughan, Samuel, 46, *62*, 64, 379

Vesey, Denmark, 219

Victoria (queen of England), 198, 282

View of the Stone Walls on the Upper Missouri (aquatint after Bodmer, 1840, cat. 25), 124, 125

Views of the Cordilleras (Humboldt, 1810), 34, 94, 153–55, 181, 320, 322, 333, 335, 354

Vroom, Peter Dumont, 387

W

Wadsworth, Daniel, 163, 167

Waldseemüller, Martin, 186

Walls, Laura Dassow, 20

War of 1812, 103, 104, 163, 180, 257, 376

Ward, John Quincy Adams, *The Freedman* (sculpture, 1865, cat. 49), 233, *234*

Washington, D.C.

Capitol Building, 55–56, 64, 149

Humboldt in, *53*, 53–57, *56*, 65–68, 149

Patent Office Building, 55, 379, 388

President's House (White House), *53*, 55, 56, 57, 60, 61, 65

salon society in, 65–68

Stelle's Hotel and City Tavern, 53

Washington, George, 53, 55, 61, 62, 64, 71, 138, 146, 148.153, 158, 177, 178, 180, 186, 207, 217, 218, 241, 256, 320, 374, 376, 379, 387

Wash-ka-mon-ya (Fast Dancer/Jim, Iowa), 289, *290* (cat. 72)

Wa-tan-ye (Foremost Man, Iowa), 296

Wa-ta-we-bu-ka-na (Commanding General, Iowa), 296

Watkins, Carleton, 169, 236–40, 390

Cascade, Nevada Falls, Yosemite, California (ca. 1861, cat. 51), *4*, 236, *238*

El Capitan (ca. 1860s, cat. 50), 236, *237*

The Grizzly Giant, Mariposa Grove, California (after 1875, cat. 52), 236, *239*

Yosemite Falls (ca. 1865–66, cat. 53), 240, *242*

Webb, James Watson, 226

Webster, Daniel, *222*, 222–23, 224, 225, 229, 244, 281

Weir, Robert Walter, 357–58

Weitsch, Friedrich Georg, *Portrait of Alexander von Humboldt* (1806, cat. 1), 29, *41*

West, Benjamin, 144

Wheeler, George, 390

Whipple, Amiel, and Whipple expedition, 126, 379

Whipple, John Adams, *378*

Whitman, Walt, 32, 314, *365*, 365–67

Leaves of Grass, 32, 314, 365–67

Whittier, John Greenleaf, 231, 316

Willdenow, Karl Ludwig, 19, 87

wilderness aesthetic, 30, 35, 127, 135, 155–60, *156*, *157*, 162, 317

Wilkes Expedition, 388

Wilkinson, James, 69, 100–101

Willard, Emma, 183–85, *184*

Williams, Helen Maria, 33, 181

Williams, Moses, 143

Willis, Nathaniel Parker, 320

Wilmot Proviso, 223

Wilson, Alexander, 108, 146

Winchell, Alexander, 392

Winthrop, Theodore, 228, 358

Wistar, Caspar, Jr., 46, 49, 59, 71, 99, 107

women

in American Association of Arts and Sciences, 185

attending Humboldt's lectures, 19, 182

salon society of Paris/Washington, D.C., and, *64*, 65–68, 181

translating Humboldt's works, 181–82

Woodbridge, William, 183, *184*

Woodhouse, James, 46, 48, 61, 64, 65, 146

Woodlands (estate of William Hamilton), 62, *70*

Wordsworth, William, 20

Wright, Joseph A., 207

Wulf, Andrea, 20

Y

Yellowstone, Hayden Survey (1871) of, 127, 390

Yosemite, 31, 127, 169, 225–26, 236–40, *237–39*, *242*, *243*, 244, 390–91, *391*

ACKNOWLEDGMENTS

Alexander von Humboldt left his mark on nearly every aspect of American culture during the early nineteenth century. Much like Humboldt himself, over the years of my research I have benefited from the writings, advice, and friendship of many, accruing a substantial number of debts, and it is a pleasure to acknowledge them here.

Humboldt's multifaceted interests have led me down pathways familiar and unfamiliar, which began with the encouragement of my college geology professors, the late Charles Almy and Cyril Harvey at Guilford College. I began retracing Humboldt's steps in the summer of 1979 when the late Daniel Appleman of the Department of Mineral Sciences at the National Museum of Natural History assigned me to update the locations for the basalt collections from Prussia and Saxony. Professor Marjorie Balge at the University of Virginia turned my attention to American landscape painting, where I discovered Humboldt's influence in the United States. At Yale University, Jules Prown encouraged my interest in art and science, as did Bryan Wolf. Jules and Bryan taught me to develop and articulate complex ideas, skills essential to this project.

Many fine scholars have devoted their careers to enhancing our understanding of Humboldt's historical influence and his contemporary relevance. Humboldt would be justifiably proud that so many of them share his collaborative spirit. I have been fortunate to get to know Frank Baron, Andreas Daum, Georgia de Havenon, Frank Holl, Alicia Lubowski-Jahn, Ulrich Päßler, Sandra Rebok, Ingo Schwarz, and Andrea Wulf. In particular, Andrea's biography of Humboldt, *The Invention of Nature: Alexander von Humboldt's New World* (2016), reminded me of the importance of bringing the people at the core of my project to life. I have benefited from her friendship and her encouragement over the years. Ingo Schwarz's scholarship has been an inspiration, as was his advice to me when we met that it was time for me to go home and write. That was exactly what I needed to hear.

Although I have not had the pleasure of meeting them, I owe a debt to the scholarship of Ottmar Ette, Tobias Kraft, Vera M. Kutzinski, Nicolaas Rupke, Aaron Sachs, and Laura Dassow Walls. Their books inspired me to believe that pursuing my questions about Humboldt's influence on the United States was a worthwhile endeavor. In particular Aaron Sachs's *The Humboldt Current: Nineteenth-Century Exploration and the Roots of American Environmentalism* (2007) set up the search for recognizing how much Humboldt was hiding in plain sight in the American landscape and in our environmental ideas. Laura Dassow Walls's *The Passage to Cosmos: Alexander von Humboldt and the Shaping of America* (2009) is a compelling deep dive into Humboldt's influence on American literature, and by extension American culture, that alternately made me wonder if I needed to write this book and inspired me to do so. For those of us reading Humboldt in English, Vera Kutzinski's leadership in the Alexander von Humboldt in English (HiE) project is an admirable and welcome initiative to publish updated translations of all of Humboldt's writings. I hope each of these scholars finds this book a worthy companion to their accomplishments.

A great editor makes an author sound better than she deserves. If ever there were a collaborator who should share the credit as a coauthor, it is my editor, Tiffany Farrell. She took a dense manuscript packed with information and provided the direction and inspiration to turn each chapter into a compelling set of stories. She dove into every facet of this project and made editing this book an invaluable opportunity to improve my ability to convey my ideas. Together we shaped this book, and I am deeply grateful for her dedication and resilience.

If writing is often a solitary experience, research provides an opportunity for collaborative creative thinking. I am grateful to Bethany Palkovitz, who spent three years conducting the research with me at the core of this exhibition. She became the third hemisphere of my brain as this project expanded to encompass Humboldt's broad reach across early American culture. During the six years I spent researching this book and exhibition I also worked with a number of impressive interns and assistants who brought significant insights and research skills to this project. I am grateful to Jennifer Bauman, Amanda Beck, Becca Bollman, Jenny Dally, Emily Francisco, Patrick Jackson, Jasmine Jennings, Ellie McIntosh, Rebecca Singerman, and Shannon Sullivan. When it came time to fact check every source in this book, the museum's curatorial assistants Anne Hyland and Stacy Mince provided critical assistance. I am grateful to all of them for ensuring that any errors here belong to me alone.

Archivists and librarians are indispensable partners for projects of this scope and magnitude. Anne C. Evenhaugen, Alexandra Reigle, and Patricia Reid at the Smithsonian American Art Museum and Portrait Gallery Library found arcane research materials, and Wanda West at the National Museum of Natural History fulfilled voluminous interlibrary loan requests. All of them provided me with the resources I needed at critical moments. I am also deeply appreciative of the staffs at the many libraries I visited. My thanks go to Megan Hosie, Melanie Rinehart, and Kyle Roberts, American Philosophical Society Library, Philadelphia; Damon Talbot, the H. Furlong Baldwin Library, Maryland Historical Society; Melanie Lower, Jack Robertson, and Endrina Tay, the Jefferson Library, Monticello; Mary V. Thompson, the Fred Smith Library, Mount Vernon; Rupert Baker, the Royal Society Archives, London; and Everardus Overgaauw, Dominik Erdmann, and Felicias Rink at the Staatsbibliothek, Berlin. I enjoyed my days researching in each of your institutions and thank you for making me feel so welcome.

Many of my colleagues have heard more about Humboldt than they bargained for. Several of them read early versions of this manuscript, and their comments and suggestions have made my thinking measurably better and my prose significantly stronger. I am especially grateful to Randall Griffin, University Distinguished Professor of Art History at Southern Methodist University in Dallas, who has always been a treasured friend and a rigorous critic. When I mentioned I felt I was channeling Humboldt's *Cosmos* when I should have been aiming for *Aspects of Nature*, he responded with an inspirational quote that captured exactly what I was thinking. Nancy Siegel, professor of art history at Towson University, coached me though the nuances of Jeffersonian-era America, a mammoth task into which she dove with enthusiasm. Peter Betjemann, director of the School of Writing, Literature, and Film at Oregon State University, guided me through the literary landscape—from James Fenimore Cooper and Washington Irving to the transcendentalists—as I wrote this ninja training manual. Dan Peck, professor emeritus of English at Vassar College, introduced me to Thoreau's "Kalendar" and encouraged me to dive deeper into the literary giants influenced by Humboldt. Hans-Dieter Sues, chair of paleobiology, senior research geologist, and curator of vertebrate paleontology at the National Museum of Natural History and author of the preface to this book, was a kindred spirit in his dual appreciation of art and science in Humboldt's work, and I am delighted to have him involved in this project. Annika Johnson, associate curator of Native American art at the Joslyn Art Museum, provided a valued perspective on the native communities Humboldt and his protégés encountered in the Americas. I also want to thank my anonymous peer reviewers, as I benefitted tremendously from their insights, criticisms, and encouragement.

I am deeply indebted to my colleagues at lending institutions who provided critical help in securing the loans for this exhibition and providing information on their works. I would like to thank Vera Enke, Wolf-Hagen Krauth, Ulrich Päßler, and Ingo Schwarz, Berlin-Brandenburgische Akademie der Wissenschaften; Ralph Gleis and Francisca Cruz, Alte Nationalgalerie, Berlin; Tammis Groft, Albany Institute of History and Art; Patrick Spero, David J. Gary, and Valerie-Ann Lutz, American Philosophical Society, Philadelphia; W. Richard West Jr., Autry Museum of the American West, Los Angeles; Elizabeth E. Barker, John Buchtel, Charlotte Emans Moore, and Casey Riley, Boston Athenæum; Matthew Teitelbaum, Layla Bermeo, and Erica E. Hirshler, Museum of Fine Arts, Boston; Cameron Kitchin and Julie Aronson, Cincinnati Art Museum; Kerry Cronin and O. Mario Favorito, Concord Free Public Library, Massachusetts; Caroline Baumann, Perry Choe, Sarah Coffin, Caitlin Condell, Mir Finkelman, Laura Fravel, and Caroline O'Connell, Cooper Hewitt, Smithsonian Design Museum; Salvador Salort-Pons, Nancy Barr, Douglas Bulka, Clare Rogan, and Kenneth Myers, Detroit Institute of Arts; Matthew McLendon, Leah Hildreth Stearns, and Sandra Cintrón Goitía, the Fralin Museum of Art at the University of Virginia; George Glazer, George Glazer Gallery, New York; Martin Faass and Oliver Sandrock, Hessisches Landesmuseum, Darmstadt, Germany; Karie Diethorn and Doris Fanelli, Independence National Historical Park, Philadelphia; Jack Becker, Toby Jurovics, Marsha Gallagher, Erin Bradford, and Sarah Haines, Joslyn Art Museum, Omaha, Nebraska; Fiona Ainsworth, Lynn Parker, and Jonathan Farley, Royal Botanic Gardens, Kew; Carla Hayden, Paulette Haiser, Ryan Moore, and Patrick Shepler, Library of Congress; Mark Letzer and Alexandra Deutsch, Maryland Historical Society, Baltimore; Leslie Greene Bowman and Susan Stein, the Thomas Jefferson Foundation, Monticello; Doug Bradburn, Kathryn Priebe, and Susan P. Schoelwer, George Washington's Mount Vernon; Anna N. Dhody, Lowell Flanders, Robert Hicks, and George M. Wohlreich, the Mütter Museum of the College Physicians of Philadelphia; Kaywin Feldman, Nancy Kay Anderson, Barbara Berrie, Sarah Cash, Franklin Kelly, Suzanne Lomax, and Yuri Long, National Gallery of Art, Washington; Anthea Hartig, Nancy Bercaw, Bonnie Campbell Lilienfeld, Kristie DaFoe, Ellen Feingold, Bernard Finn, Susan Fruchter, Mazie Harris, Steve Hemlin, Bob Horton, James Oakley, Sarah Oakman, Kay Peterson, Steve Velasquez, Hal Wallace, and Helena Wright, Smithsonian's National Museum of American History; Kirk R. Johnson, Michael Braun, Robin Everly, Candace Greene, Catharine Hawks, Adrienne Kaeppler, and Hans-Dieter Sues, Smithsonian National Museum of Natural History; Kim Sajet, Robyn Asleson, Brandon Brame Fortune, Beth Isaacson, Ellen Miles, Ann Shumard, and Leslie J. Ureña, National Portrait Gallery, Smithsonian Institution; Louise Mirrer, Wendy Ikemoto, Roberta Olson, and Kimberly Orcutt, New-York Historical Society; Mark Schaming and Karen Quinn, the New York State Museum, Albany; Sean E. Sawyer, Valerie Balint, Ida Brier,

William L. Coleman, Steve Paneccasio, Mark Prezorski, Julia B. Rosenbaum, Evelyn Trebilcock, and Karen Zukowski, Olana State Historic Site, Hudson, New York; David R. Brigham, Brooke Davis Anderson, Anna O. Marley, and Mary McGinn, Pennsylvania Academy of the Fine Arts, Philadelphia; Nancy E. Gwinn, Mary Augusta Thomas, Morgan Aronson, Richard Greene, Allie Newman, Leslie Overstreet, Kirsten van der Veen, and Lilla Vekerdy, Smithsonian Institution Libraries; Richard H. Breithaupt Jr., President Emeritus, Sons of the Revolution in the State of California American Heritage Library and Museum; Elizabeth Glassman, Katherine M. Bourguignon, Peter John Brownlee, and Cathy Ricciardelli, Terra Foundation for American Art, Chicago; Megan Pogue and Derrick Cartwright, Timken Museum of Art, San Diego; Laurent Salomé, Château de Versailles, Paris; and Alex Nyerges and Stephen Bonadies, Virginia Museum of Fine Arts.

Over the years, many individuals have provided enthusiastic support for this project, among them my colleagues at museums and in the academy, art dealers, private collectors, and independent scholars. I am grateful to Lauren Applebaum, Toledo Museum of Art; Dr. Anita Auerbach; Kevin J. Avery; Tom Barwick; Eric and Katherine Baumgartner; Marisa Bourgoin and Emily Dana Shapiro, Archives of American Art; Kim Bjorgo-Thorne, West Virginia Wesleyan College; Stephanie Buffum and Cyrus West Field IV; Bill Burns; Gerald L. Carr; Emily Casey, St. Mary's College of Maryland; Christian Cloke, University of Maryland; Matthew Cushman, Winterthur Museum, Garden & Library; Scott V. Edwards and Jeremiah Trimble, Museum of Comparative Zoology, Harvard University; Christian Feest; James Glisson, the Huntington Library, Art Museum, and Botanical Gardens; Cassandra Good, Marymount College; Adam Goodheart, Michael Buckley, Patrick Nugent, and Jean Wortman, the C. V. Starr Center for the Study of the American Experience, Washington College, Maryland; George and Diana Gravlee; Tyler Green; the late Jonathan Harding and Jay Cantor, the Century Association, New York; Judith Hernstadt; Lisa Hodermarsky, Yale University Art Gallery; Patricia Hoge; Robert and Immaculada Socias Hoge; Wendy Bellion and Jennifer Van Horn, University of Delaware; the late Tony Horwitz; Wendy Katz, University of Nebraska; Elizabeth M. Kornhauser and Sylvia Yount, the Metropolitan Museum of Art, New York; Sara Hutchinson; Marci Kwon, Stanford University; Thomas Lovejoy, George Mason University; Mary Lublin; Katie McKinney, Colonial Williamsburg Foundation; Castle McLaughlin, Diana D. Loren, Constance Rinaldo, Kara Schneiderman, and Meredith Vasta, Peabody Museum of Archaeology and Ethnology, Harvard University; Barbara E. Mundy, Fordham University; Janie and Robert F. Norfleet Jr.; Bill O'Connor; Greg Olson, the State Historical Society of Missouri; Fran and Doug Parker; David W. Penney, Smithsonian's National Museum of the American Indian; James Prosek; Gregory Raml, American Museum of Natural History, New York; Dario Robleto; Alexis Rockman; Timothy Rooks; Dale R. Rosengarten, College of Charleston, South Carolina; Jill Vaum Rothschild; Kirk Savage, Lily Brewer, Alison Langmead, and Meredith North, University of Pittsburgh; Kimia Shahi; Fronia W. Simpson; Tom Slater, Heritage Auctions, Dallas; Ed Smith, National Zoological Park; Carol Soltis, Philadelphia Museum of Art; Allison Stagg; Rick Stamm, Smithsonian Castle; Timothy J. Standring, Lori Iliff, and Emily Willkom, Denver Art Museum; George Steinmann; Chantal and Paul Stern; Lisa Strong, Georgetown University; Edward J. Sullivan, New York University; Phillip Troutman, George Washington University; Robin Veder, *American Art* journal; Emily Voelker; David C. Ward, U.S. Capitol Visitor Center; Alan V. Weinberg; Jochen Wierich, Mississippi Museum of Art; and Hannah Wills.

Support across the Smithsonian American Art Museum has been essential to every facet of this endeavor. It takes a small army of people to make projects of this magnitude a reality. I am grateful for director Stephanie Stebich's staunch and enthusiastic support for this project. Deputy director David Voyles has supervised the budget along with Doug Wilde and Krista Aniel, while exhibitions coordinator Erin J. Bryan has managed the myriad details. My fellow curators, especially chief curator Virginia Mecklenburg, the Lucy S. Rhame Curator of Sculpture Karen Lemmey, the McEvoy Family Curator for Photography John Jacob, former film and media arts curator Michael Mansfield, deputy chief curator and curator of Latinx art E. Carmen Ramos, and curator emeritus Bill Truettner have provided encouragement and advice. Jean Lavery provided essential managerial support and arranged my research travel. Development officers Elizabeth Daoust, Kate Earnest, and Donna Rim have helped raise the substantial funds required to make this book and exhibition possible, and I am deeply grateful for their ability to make this project a reality.

The logistics needed to produce this book required its own small army. Chief of publications Theresa Slowik orchestrated the schedule and negotiated the arrangements with Princeton University Press. Aubrey Vinson handled photographic rights with enthusiasm, tracking down and securing the images. Karen Siatras created an elegant design for the book, ably assisted by Denise Arnot, and Helen McNiell of Studio A in Alexandria, Virginia. Theresa Dowell Blackinton, Kelly Jerkins, and Julianna White provided essential support preparing the manuscript for publication. Magda Nakassis proofread every chapter and the bibliography, no small feat for a project this complex. I recognize the amount of effort each dedicated to preparing this manuscript.

The book and the exhibition are like fraternal twins: born from the same gene pool but developing with distinct personalities. Chief registrar Melissa Kroning and exhibition registrars Jenni Lee and Heather Delemarre handled the loans, indemnification, and the negotiations to secure Peale's mastodon. Conservators Amber Kerr, Martin Kotler, Gwen Manthey, Catherine Maynor, and Ariel O'Connor, along with Denise Wamaling and Thomas Irion, ensured that every work in this exhibition received appropriate care. David DeAnna, Matthew Bacon, and Craig Pittman made sure that everything arrived safely and managed the installation of each object. I have benefitted greatly from working with exhibition designer Stefan Gibson, who oversaw every facet of construction in the galleries and created a compelling and elegant installation. Grace Lopez made sure the words on the walls enhanced the visitors' appreciation of the experience, while lighting designer Scott Rosenfeld created the ambiance that allowed every work to be displayed to advantage.

Three signature interpretive elements anchor the exhibition, and my thanks go to Joanna Marsh, deputy education chair and head of interpretation and audience research, for leading the team that developed these components of the exhibition. They provided visitors with an enhanced experience introducing Humboldt and connecting him to the Smithsonian and the ecological issues of our day. At SAAM I am grateful to Anne Showalter, Melissa Hendrickson, and Bethany Palkovitz, along with Jason Orfanon at Smithsonian Digital Studios for developing the amazing interactive presentation of Frederic Church's *Heart of the Andes*. The Smithsonian Channel embraced Humboldt as a mission-critical project, and I am grateful to Linda Goldman at the channel and director Tilman Remme at GTV in Germany for developing *Humboldt: Epic Explorer*, a docudrama recounting Humboldt's ideas and accomplishments.

Exhibitions benefit from a robust series of programs, and I thank Kayleigh Bryant-Greenwell and Gloria Kenyon for coordinating the symposium that launched this exhibition. The speakers spanned Humboldt's influence on the arts and the sciences, making the point that Humboldt continues to inspire transdisciplinary responses to contemporary environmental issues. I am grateful to Randall Griffin, Thomas Lovejoy, Dario Robleto, George Steinmann, and Andrea Wulf for making the case for Humboldt's continued relevance in our lives. The public awareness of exhibitions flourish thanks to effective publicity, and I am grateful to Laura Baptiste, chief of public affairs and communications, and Amy Hutchins for their dedication to the increase and diffusion of this project.

Humboldt believed that he did his best work in the company of others. This project has been a pan-institutional endeavor, with colleagues across the Smithsonian playing key roles in the success of this exhibition and book. My former director Betsy Broun provided early encouragement and suggestions for keeping Humboldt visible in each aspect of this project. Kirk Johnson, the Sant Director at the National Museum of Natural History, was an early and enthusiastic supporter of all things Humboldt, as were former National Museum of American History director John Gray and former National Museum of African American History and Culture director and current Smithsonian secretary Lonnie G. Bunch III. A number of the units at the Smithsonian have supported generous loans from their museums, including the Cooper Hewitt, Smithsonian Design Museum; National Museum of American History; National Museum of Natural History; National Portrait Gallery; and Smithsonian Libraries. Provost John Davis helped secure institutional support for the research, as did Joanne Flores and Michelle Delaney. The archivists at the Smithsonian, notably Kathleen Dorman, Heather Ewing, and Pamela Henson steered me to resources on Humboldt and founding donor of the Smithsonian Institution, James Smithson. Craig Blackwell in the Office of the General Counsel provided helpful legal advice for international loans.

The most essential and complex loan for this exhibition is the mastodon skeleton from Charles Willson Peale's Museum in Philadelphia. Since 1854 this behemoth has been a cherished part of the permanent collection at the Hessisches Landesmuseum in Darmstadt, Germany. I am profoundly grateful to director Martin Faass and chief curator and paleontologist Oliver Sandrock for making it possible to bring Peale's mastodon back to the United States for the first time since 1847. Your confidence and enthusiasm have been inspiring.

Finally, I could not have written this book or organized this exhibition without the support of my family. Thank you to my mother, Peggy Jones, and my children, Caroline and Duncan, for tolerating long stretches of time when I was absorbed in research or writing to the exclusion of normal interactions. Most of all, I owe a tremendous debt of gratitude to my husband, Steve, for enduring my obsession with Alexander von Humboldt. Of all the projects I have undertaken, this one has been the most demanding of my time, my attention, and my stamina. I dedicate this volume to him, in heartfelt appreciation for his unwavering support and love.

EJH

IMAGE CREDITS

FRONT MATTER

(through Preface)

10 Photography ©Terra Foundation for American Art, Chicago

24 Artwork Documentation: Steven Paneccasio

37 Photo by Gene Young

CHAPTER 1

38 Photography ©New-York Historical Society

41 bpk Bildagentur / Nationalgalerie, Staatliche Museen, Berlin, Germany / Klaus Goeken / Art Resource, NY

47 Photography ©New-York Historical Society, negative #8736c

50 Courtesy of the Library of Congress, Manuscript Division, Thomas Jefferson Papers, Series 1: General Correspondence. 1651–1827

53 Courtesy of the Library of Congress, Prints and Photographs Division, Frances Benjamin Johnston Collection, LC-USZ62-63254

54 Photography ©New-York Historical Society

56 ©White House Historical Association

58 Courtesy of the Library of Congress, Geography and Map Division, G4410 1804 .H8

62–63 Courtesy of the Mount Vernon Ladies' Association

64 (center): ©White House Historical Association

66–67 ©Thomas Jefferson Foundation at Monticello

68 The Picture Art Collection / Alamy Stock Photo

69 (bottom): Photography ©New-York Historical Society, ID 1916.3, negative #6843c

73 The image of the Baron von Humboldt is used by kind permission of The College of Physicians of Philadelphia. Photograph by Constance Mensh, ©2019 by The College of Physicians of Philadelphia. Used by permission.

CHAPTER 2

84 Photograph ©Bruce M. White, 2019

90 bpk Bildagentur / Staatsbibliothek zu Berlin, Stiftung Preussischer Kulturbesitz, Berlin, Germany / Art Resource, NY

91–93 ©The Board of Trustees of the Royal Botanic Gardens, Kew

99 Courtesy of the Bibliothèque de Genève, Arch. Saussure 66_07 piece 08

101 Courtesy, The David Rumsey Map Collection, David Rumsey Map Center, Stanford Libraries: http://purl.stanford.edu/qg223yz5508

106 (top): Photography ©New-York Historical Society, ID 1971.125

111 Courtesy of the Library of Congress, Geography and Map Division, G4210 1848 .F74

113 Courtesy of the Braun Research Library Collection, Autry Museum, Los Angeles; 81.G.2

115 (left to right): Courtesy of the Library of Congress, Prints and Photographs Division, LC-DIG-ppmsca-39308; Courtesy of the Library of Congress, Prints and Photographs Division, LC-USZ62-110045

116–25 Photographs ©Bruce M. White, 2019

CHAPTER 3

140–41 ©Hessisches Landesmuseum Darmstadt, Photo: Wolfgang Fuhrmannek

156–57 Courtesy, The David Rumsey Map Collection, David Rumsey Map Center, Stanford Libraries: https://purl.stanford.edu/wx433gv2995

161 (bottom): Photograph by Richard Walker

164 Photography ©Terra Foundation for American Art, Chicago

168 Image ©The Metropolitan Museum of Art. Image source: Art Resource, NY

CHAPTER 4

174 Photo: Travis Fullerton, Courtesy Virginia Museum of Fine Arts

177 Photo: Hervé Lewandowski ©RMN-Grand Palais / Art Resource, NY

178 De Agostini Picture Library / M. Seemuller / Bridgeman Images

180 Image ©The Metropolitan Museum of Art. Image source: Art Resource, NY

181 ©RMN-Grand Palais / Art Resource, NY

184 (top, bottom right): From The New York Public Library, image ID ps_grd_350; Courtesy of the Library of Congress, Geography and Map Division, G1019 .W713 1831 Copy 1

187 (bottom right): Photograph by Glenn Castellano

189 Photography ©Terra Foundation for American Art, Chicago

192 Photo by Sylvain Pelly ©Observatoire de Paris

200 Photo by Gene Young

202 Courtesy of the Library of Congress, Prints and Photographs Division, LC-DIG-pga-00117

205 Photo ©Carola Radke, Museum für Naturkunde Berlin

206 Photo: Travis Fullerton, Courtesy Virginia Museum of Fine Arts

CHAPTER 5

219 (top to bottom): From The New York Public Library, image ID 1267380; From The New York Public Library, image ID 1267381

222 (right): Courtesy of the Library of Congress, Prints and Photographs Division, Brady-Handy photograph collection, LC-BH82- 4506 B

227 Photo courtesy Heritage Auctions

228 (left): Courtesy of the National Park Service, Frederick Douglass National Historic Site, FRDO_4804

234 Photo by Mildred Baldwin

235 Photograph by Jerry L. Thompson for the Boston Athenæum

237–39 Photos by Mindy Barrett

240 Carleton E. Watkins photograph collection, PC-RM-Watkins; Box 34, Folder 2; California Historical Society

241 Photography ©New-York Historical Society, ID 61263

242 Photo by Mindy Barrett

245 Photograph ©2020 Museum of Fine Arts, Boston

246 Courtesy of the Library of Congress, Manuscript Division, Abraham Lincoln Papers: Series 2. General Correspondence. 1858–1864

CHAPTER 6

254 Photograph ©Bruce M. White, 2019

261 Photo by Gene Young

265–76 Photographs ©Bruce M. White, 2019

277 Photo by Gene Young

278 Photograph ©Bruce M. White, 2019

279 Photo by Mildred Baldwin

281 Photograph courtesy Sumpter Priddy III, Inc.

282 Photo by Gene Young

284 (top): ©The Trustees of the British Museum

285 Photograph ©Bruce M. White, 2019

286 Photo by Mildred Baldwin

288, 290, 292–93 Photos by Gene Young

294 Photo by Mildred Baldwin

295 Photo by Gene Young

297 Photo: Christophe Fouin ©RMN-Grand Palais / Art Resource, NY

301–303 Photos by Gene Young

304 bpk Bildagentur / Ethnologisches Museum, Staatliche Museen, Berlin, Germany / Dietrich Graf / Art Resource, NY

CHAPTER 7

312 Photo: Matt Flynn ©Cooper Hewitt, Smithsonian Design Museum

316 Courtesy of the Library of Congress, Prints and Photographs Division, Brady-Handy photograph collection, LC-BH82-5209 B

323–24 Photos: Matt Flynn ©Cooper Hewitt, Smithsonian Design Museum

325, 329–32 Photos ©Cooper Hewitt, Smithsonian Design Museum

335 Allen Phillips/Wadsworth Atheneum

336 Photo by Gene Young

339 Photo ©Cooper Hewitt, Smithsonian Design Museum

344 Artwork Documentation: Steven Paneccasio

346–47 Photos: Matt Flynn ©Cooper Hewitt, Smithsonian Design Museum

351 Digital image created by Oppenheimer Editions

354 Image ©The Metropolitan Museum of Art. Image source: Art Resource, NY

356 Courtesy of Reynolda House Museum of American Art, Affiliated with Wake Forest University

360 Courtesy Harvard University Archives, HUP Agassiz, Louis (16a), olvwork360592

362 Image ©The Metropolitan Museum of Art. Image source: Art Resource, NY

363 (right): From The New York Public Library, image ID 5189870

365 Courtesy National Archives, photo no. NWDNS-111-B-1672

CHAPTER 8

381 bpk / Stiftung Preussische Schlösser und Gärten Berlin-Brandenburg / Gerhard Murza / Art Resource, NY

382 Courtesy George Glazer Gallery, www.georgeglazer.com

386 Courtesy of the Library of Congress, Manuscript Division, W.W. Corcoran Papers, Incoming Correspondence, Box 9, 8311-8312

388 Courtesy of the Library of Congress, Prints and Photographs Division, LC-DIG-det-4a23352

391 (left to right): Courtesy of the Library of Congress, Prints and Photographs Division, LC-USZ62-1406; Courtesy of the Library of Congress, Prints and Photographs Division, LC-USZ62-1406

ALEXANDER VON

HUMBOLDT

and the United States

ART, NATURE, AND CULTURE

by Eleanor Jones Harvey
with a preface by Hans-Dieter Sues

Published in conjunction with the exhibition of the same name, on view at the Smithsonian American Art Museum, Washington, D.C., March 20 to August 16, 2020.

Produced by the Publications Office, Smithsonian American Art Museum, Washington, D.C., AmericanArt.si.edu

Theresa J. Slowik, Chief of Publications
Tiffany D. Farrell, Senior Editor
Julianna C. White, Editor
Theresa Dowell Blackinton, Editor
Karen Siatras, Senior Designer
Denise Arnot, Designer
Helen McNiell of Studio A, Designer
Aubrey Vinson, Permissions Coordinator
Magda Nakassis, Proofreader

The Smithsonian American Art Museum is home to one of the largest collections of American art in the world. Its holdings—more than 43,000 works—tell the story of America through the visual arts and represent the most inclusive collection of American art of any museum today.

It is the nation's first federal art collection, predating the 1846 founding of the Smithsonian Institution. The Museum celebrates the exceptional creativity of the nation's artists, whose insights into history, society, and the individual reveal the essence of the American experience.

For more information, write to:
Office of Publications
Smithsonian American Art Museum
MRC 970, PO Box 37012
Washington DC 20013–7012

SAAM Smithsonian American Art Museum

Cover: Frederic Edwin Church, *The Natural Bridge, Virginia*, 1852. See cat. 82, p. 321.

Endpapers: Alexander von Humboldt, *Map of Mexico and the Surrounding Territories*, from *Atlas géographique et physique du royaume de la Nouvelle-Espagne*, plate 2, 1811. See cat. 19, p. 102

Frontispiece: Frederic Edwin Church, *Niagara* (detail), 1857. See cat. 94, p. 340–41.

Page 4: Carleton E. Watkins, *Cascade, Nevada Falls, Yosemite, California* (detail), ca. 1861. See cat. 51, p. 38.

Pages 6–7: Frederic Edwin Church, *Aurora Borealis* (detail), 1865. See cat. 40, p. 200.

Page 8: Charles Willson Peale and Titian Ramsay Peale, *The Long Room, Interior of Front Room in Peale's Museum* (detail), 1822. See cat. 31, p. 151.

Page 10: Samuel F. B. Morse, *Gallery of the Louvre* (detail), 1831–33. See cat. 35, p. 189.

Page 14: Daniel Huntington, *The Atlantic Cable Projectors* (detail), 1895. See cat. 41, p. 201.

Page 24: Rafael Salas, *Portrait of Humboldt* (detail), 1857. See cat. 96, p. 344.

Page 37: Frederic Edwin Church, *Cotopaxi* (detail), 1855. See cat. 91, p. 336.

Page 396: Frederic Edwin Church, *Natural Bridge, Virginia*, June 1851. See fig. 86, p. 320.

Back cover: Friedrich Georg Weitsch, *Portrait of Alexander von Humboldt (1769–1859)*, (detail), 1806. See cat. 1, p. 41.

Printed in Italy by Verona Libri

Typeset in Bembo Book MT Pro, Remsen Script, Engravers Roman, and Copperplate and printed on Munken Polar 130 gsm

Published by the Smithsonian American Art Museum in association with Princeton University Press, Princeton and Oxford, press.princeton.edu.

Library of Congress
Cataloging-in-Publication Data

Names:
Harvey, Eleanor Jones, author. | Sues, Hans-Dieter, 1956– writer of preface.

Title:
Alexander von Humboldt and the United States: art, nature, and culture / Eleanor Jones Harvey; with a preface by Hans-Dieter Sues.

Description:
Washington DC: Smithsonian American Art Museum; Princeton: in association with Princeton University Press, 2020. | "Published in conjunction with the exhibition of the same name, on view at the Smithsonian American Art Museum, Washington, D.C., March 20 to August 16, 2020." | Includes bibliographical references and index.

Identifiers:
LCCN 2019051714 | ISBN 9780937311875 (cloth over board) | ISBN 9780691200804 (paper over board)

Subjects:
LCSH: Arts, American—19th century—Exhibitions. | Arts, American—German influences—Exhibitions. | Nature and civilization—United States—Exhibitions. | Humboldt, Alexander von, 1769–1859—Influence—Exhibitions.

Classification:
LCC NX503.7 .H37 2020 | DDC 700.973—dc23
LC record available at https://lccn.loc.gov/2019051714

Furthermore: a program of the J. M. Kaplan Fund

This catalogue is supported by Furthermore: a program of the J. M. Kaplan Fund

The decorative letters at the start of each chapter were specially designed for this book. They represent eleven trees and one shrub native to the eastern United States and present at the time of Humboldt's visit. In order of appearance, they are: American beech, balsam poplar, sugar maple, American elm, black locust, black walnut, American linden, sweetgum, willow oak, mountain laurel, scarlet oak, and magnolia.

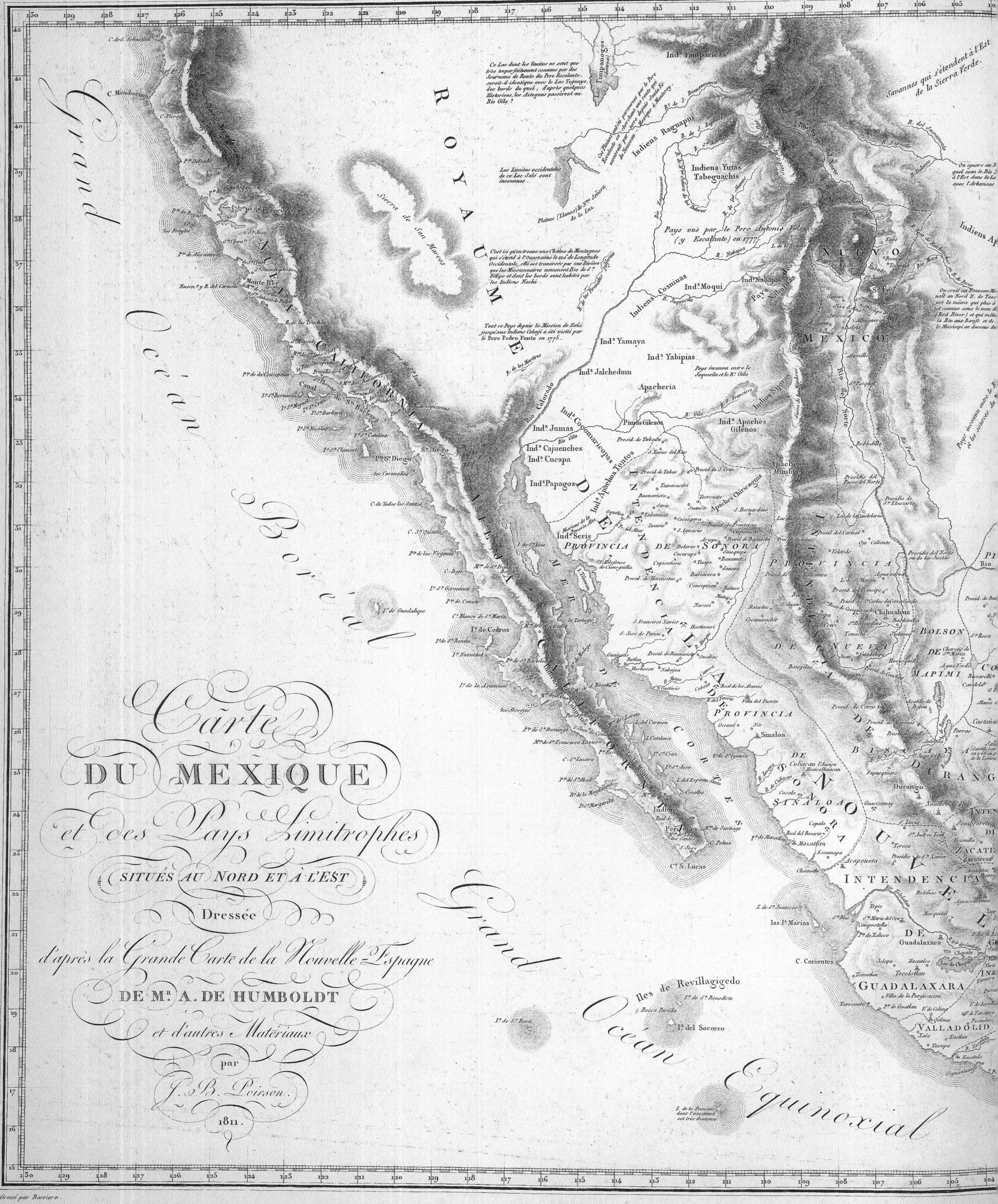

Gravé par Barriere.